GRAVITATION
FOUNDATIONS AND FRONTIERS

Covering all aspects of gravitation in a contemporary style, this advanced textbook is ideal for graduate students and researchers in all areas of theoretical physics.

The 'Foundations' section develops the formalism in six chapters, and uses it in the next four chapters to discuss four key applications – spherical space-times, black holes, gravitational waves and cosmology. The six chapters in the 'Frontiers' section describe cosmological perturbation theory, quantum fields in curved spacetime, and the Hamiltonian structure of general relativity, among several other advanced topics, some of which are covered in-depth for the first time in a textbook.

The modular structure of the book allows different sections to be combined to suit a variety of courses. More than 225 exercises are included to test and develop the readers' understanding. There are also over 30 projects to help readers make the transition from the book to their own original research.

T. PADMANABHAN is a Distinguished Professor and Dean of Core Academic Programmes at the Inter-University Centre for Astronomy and Astrophysics (IUCAA), Pune. He is a renowned theoretical physicist and cosmologist with nearly 30 years of research and teaching experience both in India and abroad. Professor Padmanabhan has published over 200 research papers and nine books, including six graduate-level textbooks. These include the *Structure Formation in the Universe* and *Theoretical Astrophysics*, a comprehensive three-volume course. His research work has won prizes from the Gravity Research Foundation (USA) five times, including the First Prize in 2008. In 2007 he received the Padma Shri, the medal of honour from the President of India in recognition of his achievements.

GRAVITATION

Foundations and Frontiers

T. PADMANABHAN

IUCAA, Pune,
India

CAMBRIDGE
UNIVERSITY PRESS

CAMBRIDGE
UNIVERSITY PRESS

University Printing House, Cambridge CB2 8BS, United Kingdom

One Liberty Plaza, 20th Floor, New York, NY 10006, USA

477 Williamstown Road, Port Melbourne, VIC 3207, Australia

314-321, 3rd Floor, Plot 3, Splendor Forum, Jasola District Centre, New Delhi - 110025, India

79 Anson Road, #06-04/06, Singapore 079906

Cambridge University Press is part of the University of Cambridge.

It furthers the University's mission by disseminating knowledge in the pursuit of education, learning and research at the highest international levels of excellence.

www.cambridge.org
Information on this title: www.cambridge.org/9780521882231

© T. Padmanabhan 2010

First published 2010

A catalogue record for this publication is available from the British Library

Library of Congress Cataloging in Publication data
Padmanabhan, T. (Thanu), 1957-
Gravitation : foundations and frontiers / T. Padmanabhan.
p. cm.
ISBN 978-0-521-88223-1 (Hardback)
1. Gravitation–Textbooks. 2. Relativity (Physics)–Textbooks. 3. Gravitation–Problems, exercises, etc. I. Title.
QC178.P226 2010
531´.14–dc22

2009038009

ISBN 978-0-521-88223-1 Hardback

Dedicated to the fellow citizens of India

Contents

List of exercises

List of projects

Preface

There is a need for a *comprehensive, advanced level,* textbook dealing with all aspects of gravity, written *for the physicist* in a *contemporary* style. The italicized adjectives in the above sentence are the key: most of the existing books on the market are either outdated in emphasis, too mathematical for a physicist, not comprehensive or written at an elementary level. (For example, the two unique books – L. D. Landau and E. M. Lifshitz, *The Classical Theory of Fields*, and C. W. Misner, K. S. Thorne and J. A. Wheeler (MTW), *Gravitation* – which I consider to be masterpieces in this subject are more than three decades old and are out of date in their emphasis.) The current book is expected to fill this niche and I hope it becomes a standard reference in this field. Some of the features of this book, including the summary of chapters, are given below.

As the title implies, this book covers both Foundations (Chapters 1–10) and Frontiers (Chapters 11–16) of general relativity so as to cater for the needs of different segments of readership. The Foundations acquaint the readers with the basics of general relativity while the topics in Frontiers allow one to 'mix-and-match', depending on interest and inclination. This modular structure of the book will allow it to be adapted for different types of course work.

For a specialist researcher or a student of gravity, this book provides a comprehensive coverage of all the contemporary topics, some of which are discussed in a textbook for the first time, as far as I know. The cognoscenti will find that there is a fair amount of originality in the discussion (and in the Exercises) of even the conventional topics.

While the book is quite comprehensive, it also has a structure which will make it accessible to a wide target audience. Researchers and teachers interested in theoretical physics, general relativity, relativistic astrophysics and cosmology will find it useful for their research and adaptable for their course requirements. (The section *How to use this book*, just after this Preface, gives more details of this aspect.) The discussion is presented in a style suitable for physicists, ensuring that it caters for the current interest in gravity among physicists working in related areas. The large number (more than 225) of reasonably nontrivial Exercises makes it ideal for self-study.

Another unique feature of this book is a set of Projects at the end of selected chapters. The Projects are advanced level exercises presented with helpful hints to show the reader a direction of attack. Several of them are based on research literature dealing with key open issues in different areas. These will act as a bridge for students to cross over from textbook material to real life research. Graduate students and grad school teachers will find the Exercises and Projects extremely useful. Advanced undergraduate students with a flair for theoretical physics will also be able to use parts of this book, especially in combination with more elementary books.

Here is a brief description of the chapters of the book and their inter-relationship.

Chapters 1 and 2 of this book are somewhat unique and serve an important purpose, which I would like to explain. A student learning general relativity often finds that she simultaneously has to cope with (i) conceptual and mathematical issues which arise from the spacetime being curved and (ii) technical issues and concepts which are essentially special relativistic but were never emphasized adequately in a special relativity course! For example, manipulation of four-dimensional integrals or the concept and properties of the energy-momentum tensor have nothing to do with general relativity *a priori* – but are usually not covered in depth in conventional special relativity courses. The first two chapters give the student a rigorous training in four-dimensional techniques in flat spacetime so that she can concentrate on issues which are genuinely general relativistic later on. These chapters can also usefully serve as modular course material for a short course on advanced special relativity or classical field theory.

Chapter 1 introduces special relativity using four-vectors and the action principle right from the outset. Chapter 2 introduces the electromagnetic field through the four-vector formalism. I expect the student to have done a standard course in classical mechanics and electromagnetic theory but I do *not* assume familiarity with the relativistic (four-vector) notation. Several topics that are needed later in general relativity are introduced in these two chapters in order to familiarize the reader early on. Examples include the use of the relativistic Hamilton–Jacobi equation, precession of Coulomb orbits, dynamics of the electromagnetic field obtained from an action principle, derivation of the field of an arbitrarily moving charged particle, radiation reaction, etc. Chapter 2 also serves as a launch pad for discussing spin-0 and spin-2 interactions, using electromagnetism as a familiar example.

Chapter 3 attempts to put together special relativity and gravity and explains, in clear and precise terms, why it does not lead to a consistent picture. Most textbooks I know (except MTW) do not explain the issues involved clearly and with adequate detail. For example, this chapter contains a detailed discussion of the spin-2 tensor field which is not available in textbooks. It is important for a student to realize that the description of gravity in terms of curvature of spacetime is inevitable and natural. This chapter will also lay the foundation for the description of the spin-2 tensor field h_{ab}, which will play an important role in the study of gravitational waves and cosmological perturbation theory later on.

Having convinced the reader that gravity is related to spacetime geometry, Chapter 4 begins with the description of general relativity by introducing the metric tensor and extending the ideas of four-vectors, tensors, etc., to a nontrivial background. There are two points that I would like to highlight about this chapter. First, I have introduced every concept with a physical principle rather than in the abstract language of differential geometry. For example, direct variation of the line interval leads to the geodesic equation *through* which one can motivate the notion of Christoffel symbols, covariant derivative, etc., in a simple manner. During the courses I have taught over years, students have found this approach attractive and simpler to grasp. Second, I find that students sometimes confuse issues which arise when curvilinear coordinates are used in flat spacetime with those related to genuine curvature. This chapter clarifies such issues.

Chapter 5 introduces the concept of the curvature tensor from three different perspectives and describes its properties. It then moves on to provide a complete description of electrodynamics, statistical mechanics, thermodynamics and wave propagation in curved spacetime, including the Raychaudhuri equation and the focusing theorem.

Chapter 6 starts with a clear and coherent derivation of Einstein's field equations from an action principle. I have provided a careful discussion of the surface term in the Einstein–Hilbert action (again not usually found in textbooks) in a manner which is quite general and turns out to be useful in the discussion of Lanczos–Lovelock models in Chapter 15. I then proceed to discuss the general structure of the field equations, the energy-momentum pseudo-tensor for gravity and the weak field limit of gravity.

After developing the formalism in the first six chapters, I apply it to discuss four key applications of general relativity – spherically symmetric spacetimes, black hole physics, gravitational waves and cosmology – in the next four chapters. (The only other key topic I have omitted, due to lack of space, is the physics of compact stellar remnants.)

Chapter 7 deals with the simplest class of exact solutions to Einstein's equations, which are those with spherical symmetry. The chapter also discusses the orbits of particles and photons in these spacetimes and the tests of general relativity. These are used in Chapter 8, which covers several aspects of black hole physics, concentrating mostly on the Schwarzschild and Kerr black holes. It also introduces important concepts like the maximal extension of a manifold, Penrose–Carter diagrams and the geometrical description of horizons as null surfaces. A derivation of the zeroth law of black hole mechanics and illustrations of the first and second laws are also provided. The material developed here forms the backdrop for the discussions in Chapters 13, 15 and 16.

Chapter 9 takes up one of the key new phenomena that arise in general relativity, viz. the existence of solutions to Einstein's equations which represent disturbances in the spacetime that propagate at the speed of light. A careful discussion of gauge invariance and coordinate conditions in the description of gravitational waves is provided. I also make explicit contact with similar phenomena in the case of electromagnetic radiation in order to help the reader to understand the concepts better. A detailed discussion of the binary pulsar is included and a Project at the end of the chapter explores the nuances of the post-Newtonian approximation.

Chapter 10 applies general relativity to study cosmology and the evolution of the universe. Given the prominence cosmology enjoys in current research and the fact that this interest will persist in future, it is important that all general relativists are acquainted with cosmology at the same level of detail as, for example, with the Schwarzschild metric. This is the motivation for Chapter 10 as well as Chapter 13 (which deals with general relativistic perturbation theory). The emphasis here will be mostly on the geometrical aspects of the universe rather than on physical cosmology, for which several other excellent textbooks (e.g. mine!) exist. However, in order to provide a complete picture and to appreciate the interplay between theory and observation, it is necessary to discuss certain aspects of the evolutionary history of the universe – which is done to the extent needed.

The second part of the book (Frontiers, Chapters 11–16) discusses six separate topics which are reasonably independent of each other (though not completely). While a student or researcher specializing in gravitation should study all of them, others could choose the topics based on their interest after covering the first part of the book.

Chapter 11 introduces the language of differential forms and exterior calculus and translates many of the results of the previous chapters into the language of forms. It also describes briefly the structure of gauge theories to illustrate the generality of the formalism. The emphasis is in developing the key concepts rapidly and connecting them up with the more familiar language used in the earlier chapters, rather than in maintaining mathematical rigour.

Chapter 12 describes the $(1 + 3)$-decomposition of general relativity and its Hamiltonian structure. I provide a derivation of Gauss–Codazzi equations and Einstein's equations in the $(1 + 3)$-form. The connection between the surface term in the Einstein–Hilbert action and the extrinsic curvature of the boundary is also spelt out in detail. Other topics include the derivation of junction conditions which are used later in Chapter 15 while discussing the brane world cosmologies.

Chapter 13 describes general relativistic linear perturbation theory in the context of cosmology. This subject has acquired major significance, thanks to the observational connection it makes with cosmic microwave background radiation. In view of this, I have also included a brief discussion of the application of perturbation theory in deriving the temperature anisotropies of the background radiation.

Chapter 14 describes some interesting results which arise when one studies standard quantum field theory in a background spacetime with a nontrivial metric. Though the discussion is reasonably self-contained, some familiarity with simple ideas of quantum theory of free fields will be helpful. The key result which I focus on is the intriguing connection between thermodynamics and horizons. This connection can be viewed from very different perspectives not all of which can rigorously be proved to be equivalent to one another. In view of the importance of this result, most of this chapter concentrates on obtaining this result using different techniques and interpreting it physically. In the latter part of the chapter, I have added a discussion of quantum field theory in the Friedmann universe and the generation of perturbations during the inflationary phase of the universe.

Chapter 15 discusses a few *selected* topics in the study of gravity in dimensions other than $D = 4$. I have kept the discussion of models in $D < 4$ quite brief and have spent more time on the $D > 4$ case. In this context – after providing a brief, but adequate, discussion of brane world models which are enjoying some popularity currently – I describe the structure of Lanczos–Lovelock models in detail. These models share several intriguing features with Einstein's theory and constitute a natural generalization of Einstein's theory to higher dimensions. I hope this chapter will fill the need, often felt by students working in this area, for a textbook discussion of Lanczos–Lovelock models.

The final chapter provides a perspective on gravity as an emergent phenomenon. (Obviously, this chapter shows my personal bias but I am sure that is acceptable in the *last* chapter!) I have tried to put together several peculiar features in the standard description of gravity and emphasize certain ideas which the reader might find fascinating and intriguing.

Because of the highly pedagogical nature of the material covered in this textbook, I have not given detailed references to original literature except on rare occasions when a particular derivation is not available in the standard textbooks. The annotated list of Notes given at the end of the book cites several other text books which I found useful. Some of these books contain extensive bibliographies and references to original literature. The selection of books and references cited here clearly reflects the bias of the author and I apologize to anyone who feels their work or contribution has been overlooked.

Discussions with several people, far too numerous to name individually, have helped me in writing this book. Here I shall confine myself to those who provided detailed comments on earlier drafts of the manuscript. Donald Lynden-Bell and Aseem Paranjape provided extensive and very detailed comments on most of the chapters and I am very thankful to them. I also thank A. Das, S. Dhurandar, P. P. Divakaran, J. Ehlers, G. F. R. Ellis, Rajesh Gopakumar, N. Kumar, N. Mukunda, J. V. Narlikar, Maulik Parikh, T. R. Seshadri and L. Sriramkumar for detailed comments on selected chapters.

Vince Higgs (CUP) took up my proposal to write this book with enthusiasm. The processing of this volume was handled by Laura Clark (CUP) and I thank her for the effort she has put in.

This project would not have been possible without the dedicated support from Vasanthi Padmanabhan, who not only did the entire TEXing and formatting but also produced most of the figures. I thank her for her help. It is a pleasure to acknowledge the library and other research facilities available at IUCAA, which were useful in this task.

How to use this book

This book can be adapted by readers with varying backgrounds and requirements as well as by teachers handling different courses. The material is presented in a fairly modular fashion and I describe below different sub-units that can be combined for possible courses or for self-study.

1 **Advanced special relativity**

   Chapter 1 along with parts of Chapter 2 (especially Sections 2.2, 2.5, 2.6, 2.10) can form a course in advanced special relativity. No previous familiarity with four-vector notation (in the description of relativistic mechanics or electrodynamics) is required.

2 **Classical field theory**

   Parts of Chapter 1 along with Chapter 2 and Sections 3.2, 3.3 will give a comprehensive exposure to classical field theory. This will require familiarity with special relativity using four-vector notation which can be acquired from specific sections of Chapter 1.

3 **Introductory general relativity**

   Assuming familiarity with special relativity, a basic course in general relativity (GR) can be structured using the following material: Sections 3.5, Chapter 4 (except Sections 4.8, 4.9), Chapter 5 (except Sections 5.2.3, 5.3.3, 5.4.4, 5.5, 5.6), Sections 6.2.5, 6.4.1, 7.2.1, 7.4.1, 7.4.2, 7.5. This can be supplemented with selected topics in Chapters 8 and 9.

4 **Relativistic cosmology**

   Chapter 10 (except Sections 10.6, 10.7) along with Chapter 13 and parts of Sections 14.7 and 14.8 will constitute a course in relativistic cosmology and perturbation theory from a contemporary point of view.

5 **Quantum field theory in curved spacetime**

   Parts of Chapter 8 (especially Sections 8.2, 8.3, 8.7) and Chapter 14 will constitute a first course in this subject. It will assume familiarity with GR but not with detailed properties of black holes or quantum field theory. Parts of Chapter 2 can supplement this course.

6 Applied general relativity

For students who have already done a first course in GR, Chapters 6, 8, 9 and 12 (with parts of Chapter 7 not covered in the first course) will provide a description of *advanced* topics in GR.

Exercises and Projects

None of the Exercises in this book is trivial or of simple 'plug-in' type. Some of them involve extending the concepts developed in the text or understanding them from a different perspective; others require detailed application of the material introduced in the chapter. There are more than 225 exercises and it is strongly recommended that the reader attempts as many as possible. Some of the nontrivial exercises contain hints and short answers.

The Projects are more advanced exercises linking to original literature. It will often be necessary to study additional references in order to comprehensively grasp or answer the questions raised in the projects. Many of them are open-ended (and could even lead to publishable results) but all of them are presented in a graded manner so that a serious student will be able to complete most parts of any project. They are included so as to provide a bridge for students to cross over from the textbook material to original research and should be approached in this light.

Notation and conventions

Throughout the book, the Latin indices $a, b, \ldots i, j \ldots$, etc., run over $0, 1, 2, 3$ with the 0-index denoting the time dimension and $(1, 2, 3)$ denoting the standard space dimensions. The Greek indices, $\alpha, \beta, \ldots$, etc., will run over 1, 2, 3. Except when indicated otherwise, the units are chosen with $c = 1$.

We will use the vector notation for both three-vectors and four-vectors by using different fonts. The four-momentum, for example, will be denoted by $\boldsymbol{p}$ while the three-momentum will be denoted by p.

The signature is $(-, +, +, +)$ and curvature tensor is defined by the convention $R^a_{\ bcd} \equiv \partial_c \Gamma^a_{bd} - \cdots$ with $R_{bd} = R^a_{\ bad}$.

The symbol $\equiv$ is used to indicate that the equation defines a new variable or notation.

1

Special relativity

1.1 Introduction

This chapter introduces the special theory of relativity from a perspective that is appropriate for proceeding to the general theory of relativity later on, from Chapter 4 onwards.[1] Several topics such as the manipulation of tensorial quantities, description of physical systems using action principles, the use of distribution function to describe a collection of particles, etc., are introduced in this chapter in order to develop familiarity with these concepts within the context of special relativity itself. Virtually all the topics developed in this chapter will be generalized to curved spacetime in Chapter 4. The discussion of Lorentz group in Section 1.3.3 and in Section 1.10 is somewhat outside the main theme; the rest of the topics will be used extensively in the future chapters.[2]

1.2 The principles of special relativity

To describe any physical process we need to specify the spatial and temporal coordinates of the relevant *event*. It is convenient to combine these four real numbers – one to denote the time of occurrence of the event and the other three to denote the location in space – into a single entity denoted by the four-component object $x^i = (x^0, x^1, x^2, x^3) \equiv (t, \boldsymbol{x}) \equiv (t, x^\alpha)$. More usefully, we can think of an event $\mathcal{P}$ as a point in a four-dimensional space with coordinates x^i. We will call the collection of all events as *spacetime*.

Though the actual numerical values of x^i, attributed to any given event, will depend on the specific coordinate system which is used, the event $\mathcal{P}$ itself is a geometrical quantity that is independent of the coordinates used for its description. This is clear even from the consideration of the spatial coordinates of an event. A spatial location can be specified, for example, in the Cartesian coordinates giving the coordinates (x, y, z) or, say, in terms of the spherical polar coordinates by providing (r, θ, ϕ). While the numerical values (and even the dimensions) of these coordinates are different, they both signify the same geometrical point in three-dimensional space. Similarly, one can describe an event in terms of any suitable set of four independent numbers and one can transform from any system of coordinates to another by well-defined coordinate transformations.

Among all possible coordinate systems which can be used to describe an event, a subset of coordinate systems, called the *inertial coordinate systems* (or inertial frames), deserve special attention. Such coordinate systems are defined by the property that a material particle, far removed from all external influences, will move with uniform velocity in such frames of reference. This definition is convenient and practical but is inherently flawed, since one can never operationally verify the criterion that no external influence is present. In fact, there is no fundamental reason why any one class of coordinate system should be preferred over others, except for mathematical convenience. Later on, in the development of general relativity in Chapter 4, we shall drop this restrictive assumption and develop the physical principles treating all coordinate systems as physically equivalent. For the purpose of this chapter and the next, however, we shall *postulate* the existence of inertial coordinate systems which enjoy a special status. (Even in the context of general relativity, it will turn out that one can introduce inertial frames in a sufficiently small region around any event. Therefore, the description we develop in the first two chapters will be of importance even in a more general context.) It is obvious from the definition that any coordinate frame moving with uniform velocity with respect to an inertial frame will also constitute an inertial frame.

To proceed further, we shall introduce two empirical facts which are demonstrated by experiments. (i) It turns out that all laws of nature remain identical in all inertial frames of reference; that is, the equations expressing the laws of nature retain the same form under the coordinate transformation connecting any two inertial frames. (ii) The interactions between material particles do not take place instantaneously and there exists a maximum possible speed of propagation for interactions. We will denote this speed by the letter c. Later on, we will show in Chapter 2 that ordinary light waves, described by Maxwell's equations, propagate at this speed. Anticipating this result we may talk of light rays propagating in straight lines with the speed c. From (i) above, it follows that the maximum velocity of propagation c should be the same in all inertial frames.

Of these two empirically determined facts, the first one is valid even in non-relativistic physics. So the key new results of special relativity actually originate from the second fact. Further, the existence of a uniquely defined speed c allows one to express time in units of length by working with ct rather than t. We shall accordingly specify an event by giving the coordinates $x^i = (ct, x^\alpha)$ rather than in terms of t and x^α. This has the advantage that all components of x^i have the same dimension when we use Cartesian spatial coordinates.

The two facts, (i) and (ii), when combined together, lead to a profound consequence: they rule out the absolute nature of the notion of simultaneity; two events which appear to occur at the same time in one inertial frame will not, in general, appear to occur at the same time in another inertial frame. For example, consider two inertial frames K and K' with K' moving relative to K along the x-axis with the speed V. Let B, A and C (in that order) be three points along the common x-axis with $AB = AC$ in the *primed* frame, K'. Two light signals that start from a point A and go towards B and C will reach B and C at the same instant of time as observed in K'. But the two events, namely arrival of signals at B and C, cannot be simultaneous for an observer in K. This is because, in the frame K, point B moves towards the signal while C moves away from the signal; but the speed of the signal is postulated to be the same in both frames. Obviously, when viewed in the frame K, the signal will reach B before it reaches C.

In non-relativistic physics, one would have expected the two light beams to inherit the velocity of the source at the time of emission so that the two light signals travel with *different* speeds $(c \pm V)$ towards C and B and hence will reach them simultaneously in both frames. It is the constancy of the speed of light, independent of the speed of the source, which makes the notion of simultaneity frame dependent.

The concept of associating a time coordinate to an event is based entirely on the notion of simultaneity. In the simplest sense, we will attribute a time coordinate t to an event – say, the collision of two particles – if the reading of a clock indicating the time t is simultaneous with the occurrence of the collision. Since the notion of simultaneity depends on the frame of reference, it follows that two different observers will, in general, assign different time coordinates to the same event. This is an important conceptual departure from non-relativistic physics in which simultaneity is an absolute concept and all observers use the same clock time.

The second consequence of the constancy of speed of light is the following. Consider two infinitesimally separated events $\mathcal{P}$ and $\mathcal{Q}$ with coordinates x^i and $(x^i + dx^i)$. We define a quantity ds – called the *spacetime interval* – between these two events by the relation

$$ds^2 = -c^2 dt^2 + dx^2 + dy^2 + dz^2. \tag{1.1}$$

If $ds = 0$ in one frame, it follows that these two infinitesimally separated events $\mathcal{P}$ and $\mathcal{Q}$ can be connected by a light signal. Since light travels with the same speed c in all inertial frames, $ds' = 0$ in any other inertial frame. In fact, one can prove the stronger result that $ds' = ds$ for any two infinitesimally separated events, not just those connected by a light signal. To do this, let us treat ds^2 as a function of ds'^2 we can expand ds^2 in a Taylor series in ds'^2, as $ds^2 = k + a ds'^2 + \cdots$. The fact that $ds = 0$ when $ds' = 0$ implies $k = 0$; the coefficient a can only be a function of the relative velocity V between the frames. Further, homogeneity and isotropy of space requires that only the magnitude $|V| = V$ enters into this function. Thus we conclude that $ds^2 = a(V)ds'^2$, where the coefficient $a(V)$ can depend only on the absolute value of the relative velocity between the inertial frames. Now consider three inertial frames K, K_1, K_2, where K_1 and K_2 have relative velocities V_1 and V_2 with respect to K. From $ds_1^2 = a(V_1)ds^2, ds_2^2 = a(V_2)ds^2$ and $ds_2^2 = a(V_{12})ds_1^2$, where V_{12} is the relative velocity of K_1 with respect to K_2, we see that $a(V_2)/a(V_1) = a(V_{12})$. But the magnitude of the relative velocity V_{12} must depend not only on the magnitudes of V_1 and V_2 but also on the angle between the velocity vectors. So, it is impossible to satisfy this relation unless the function $a(V)$ is a constant; further, this constant should be equal to unity to satisfy this relation. It follows that the quantity ds has the same value in all inertial frames; $ds^2 = ds'^2$, i.e. the infinitesimal spacetime interval is an invariant quantity. Events for which ds^2 is less than, equal to or greater than zero are said to be separated by *timelike, null* or *spacelike* intervals, respectively.

With future applications in mind, we shall write the line interval in Eq. (1.1) using the notation

$$ds^2 = \eta_{ab}dx^a dx^b; \quad \eta_{ab} = \text{diag}\,(-1, +1, +1, +1) \tag{1.2}$$

in which we have introduced the *summation convention*, which states that any index which is repeated in an expression – like a, b here – is summed over the range of values taken by the index. (It can be directly verified that this convention is a consistent one and leads to expressions which are unambiguous.) In defining ds^2 in Eq. (1.1) and Eq. (1.2) we have used a negative sign for $c^2 dt^2$ and a positive sign for the spatial terms dx^2, etc. The sequence of signs in η_{ab} is called *signature* and it is usual to say that the signature of spacetime is $(-+++)$. One can, equivalently, use the signature $(+---)$ which will require a change of sign in several expressions. This point should be kept in mind while comparing formulas in different textbooks.

A continuous sequence of events in the spacetime can be specified by giving the coordinates $x^a(\lambda)$ of the events along a parametrized curve defined in terms of a suitable parameter λ. Using the fact that ds defined in Eq. (1.1) is invariant, we can define the analogue of an (invariant) arc length along the curve, connecting two events $\mathcal{P}$ and $\mathcal{Q}$, by:

$$s(\mathcal{P}, \mathcal{Q}) = \int_{\mathcal{P}}^{\mathcal{Q}} |ds| = \int_{\lambda_1}^{\lambda_2} \frac{|ds|}{d\lambda} d\lambda \equiv \int_{\lambda_1}^{\lambda_2} d\lambda \left| \left(\frac{d\boldsymbol{x}}{d\lambda}\right)^2 - c^2 \left(\frac{dt}{d\lambda}\right)^2 \right|^{1/2}. \tag{1.3}$$

The modulus sign is introduced here because the sign of the squared arc length ds^2 is indefinite in the spacetime. For curves which have a definite sign for the arclength – i.e. for curves which are everywhere spacelike or everywhere timelike – one can define the arclength with appropriate sign. That, is, for a curve with $ds^2 < 0$ everywhere, we will define the arc length with a flip of sign, as $(-ds^2)^{1/2}$. (For curves along the path of a light ray the arc length will be zero.) This arc length will have the same numerical value in all inertial frames and will be independent of the parametrization used to describe the curve; a transformation $\lambda \rightarrow \lambda' = f(\lambda)$ leaves the value of the arc length unchanged.

Of special significance, among all possible curves in the spacetime, is the one that describes the trajectory of a material particle moving along some specified path, called the *worldline*. In three-dimensional space, we can describe such a trajectory by giving the position as a function of time, $x(t)$, with the corresponding velocity $v(t) = (dx/dt)$. We can consider this as a curve in *spacetime* with $\lambda = ct$ acting as the parameter so that $x^i = x^i(t) = (ct, x(t))$. Further, given the existence of a maximum velocity, we must have $|v| < c$ everywhere making the curve everywhere timelike with $ds^2 < 0$. In this case, one can provide a direct physical interpretation for the arc length along the curve. Let us consider a clock (attached to the particle) which is moving relative to an inertial frame K on an *arbitrary* trajectory. During a time interval between t and $(t + dt)$, the clock moves through a distance $|dx|$ as measured in K. Consider now another inertial coordinate system K', which – at that instant of time t – is moving with respect to K with the same velocity as the clock. In this frame, the clock is momentarily at rest, giving $dx' = 0$. If the clock indicates a lapse of time $dt' \equiv d\tau$, when the time interval measured in K is dt, the invariance of spacetime intervals implies that

$$ds^2 = -c^2 dt^2 + dx^2 + dy^2 + dz^2 = ds'^2 = -c^2 d\tau^2. \tag{1.4}$$

Or,

$$d\tau = \frac{[-ds^2]^{1/2}}{c} = dt\sqrt{1 - \frac{v^2}{c^2}}. \tag{1.5}$$

Hence $(1/c)(-ds^2)^{1/2} \equiv |(ds/c)|$, defined with a flip of sign in ds^2, is the lapse of time in a moving clock; this is called the *proper time* along the trajectory of the clock. The arclength in Eq. (1.3), divided by c, viz.

$$\tau = \int d\tau = \int_{t_1}^{t_2} dt\sqrt{1 - \frac{v^2(t)}{c^2}} \tag{1.6}$$

now denotes the total time that has elapsed in a moving clock between two events. It is obvious that this time lapse is smaller than the corresponding coordinate time interval $(t_2 - t_1)$ showing that moving clocks slow down. We stress that these results hold for a particle moving in an *arbitrary* trajectory and not merely for one moving with uniform velocity. (Special relativity is adequate to describe the physics involving accelerated motion and one does not require general relativity for that purpose.)

1.3 Transformation of coordinates and velocities

The line interval in Eq. (1.1) is written in terms of a special set of coordinates which are natural to some inertial frame. An observer who is moving with respect to an inertial frame will use a different set of coordinates. Since the concept of simultaneity has no invariant significance, the coordinates of any two frames will be related by a transformation in which space *and* time coordinates will, in general, be different.

It turns out that the invariant speed of light signals allows us to set up a possible set of coordinates for any observer, moving along an arbitrary trajectory. In particular, if the observer is moving with a uniform velocity with respect to the original inertial frame, then the coordinates that we obtain by this procedure satisfy the condition $ds = ds'$ derived earlier. With future applications in mind, we will study the *general* question of determining the coordinates appropriate for an *arbitrary* observer moving along the x-axis and then specialize to the case of a uniformly moving observer.

Before discussing the procedure, we emphasize the following aspect of the derivation given below. In the specific case of an observer moving with a uniform velocity, the resulting transformation is called the *Lorentz transformation*. It is possible to obtain the Lorentz transformation by other procedures, such as, for example, demanding the invariance of the line interval. But once a transformation from a set of coordinates x^a to another set of coordinates x'^a is obtained, we would also like to understand the operational procedure by which a particular observer can set up the corresponding coordinate grid in the spacetime. Given the constancy of the speed of light, the most natural procedure will be to use light signals to set up the coordinates. Since special relativity is perfectly capable of handling accelerated observers we should be able to provide an operational procedure by which *any* observer – moving along an arbitrary trajectory – can set up a suitable coordinate system. To stress this fact – and with future applications in mind – we will first obtain the coordinate transformations for the general observer and then specialize to an observer moving with a uniform velocity.

Let (ct, x, y, z) be an inertial coordinate system. Consider an observer travelling along the x-axis in a trajectory $x = f_1(\tau), t = f_0(\tau)$, where f_1 and f_0 are specified functions and τ is the proper time in the clock carried by the observer. We can assign a suitable coordinate system to this observer along the following lines. Let $\mathcal{P}$ be some event with inertial coordinates (ct, x) to which this observer wants to assign the coordinates (ct', x'), say. The observer sends a light signal from the

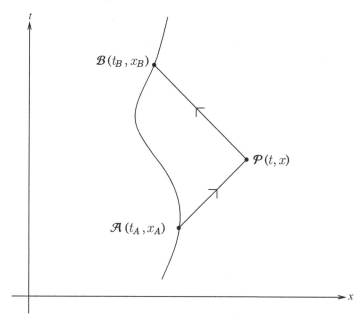

Fig. 1.1. The procedure to set up a natural coordinate system using light signals by an observer moving along an arbitrary trajectory.

event $\mathcal{A}$ (at $\tau = t_A$) to the event $\mathcal{P}$. The signal is reflected back at $\mathcal{P}$ and reaches the observer at event $\mathcal{B}$ (at $\tau = t_B$). Since the light has travelled for a time interval $(t_B - t_A)$, it is reasonable to attribute the coordinates

$$t' = \frac{1}{2}(t_B + t_A); \quad x' = \frac{1}{2}(t_B - t_A)c \tag{1.7}$$

to the event $\mathcal{P}$. To relate (t', x') to (t, x) we proceed as follows. Since the events $\mathcal{P}(t, x)$, $\mathcal{A}(t_A, x_A)$ and $\mathcal{B}(t_B, x_B)$ are connected by light signals travelling in forward and backward directions, it follows that (see Fig. 1.1)

$$x - x_A = c(t - t_A); \quad x - x_B = -c(t - t_B). \tag{1.8}$$

Or, equivalently,

$$x - ct = x_A - ct_A = f_1(t_A) - cf_0(t_A) = f_1\left[t' - (x'/c)\right] - cf_0\left[t' - (x'/c)\right], \tag{1.9}$$

$$x + ct = x_B + ct_B = f_1(t_B) + cf_0(t_B) = f_1\left[t' + (x'/c)\right] + cf_0\left[t' + (x'/c)\right]. \tag{1.10}$$

Given f_1 and f_0, these equations can be solved to find (x, t) in terms of (x', t'). This procedure is applicable to *any* observer and provides the necessary coordinate transformation between (t, x) and (t', x').

1.3.1 Lorentz transformation

We shall now specialize to an observer moving with uniform velocity V, which will provide the coordinate transformation between two inertial frames. The trajectory is now $x = Vt$ with the proper time given by $\tau = t[1 - (V^2/c^2)]^{1/2}$ (see Eq. (1.6) which can be trivially integrated for constant V). So the trajectory, parameterized in terms of the proper time, can be written as:

$$f_1(\tau) = \frac{V\tau}{\sqrt{1 - (V^2/c^2)}} \equiv \gamma V\tau; \qquad f_0(\tau) = \frac{\tau}{\sqrt{1 - (V^2/c^2)}} \equiv \gamma\tau, \quad (1.11)$$

where $\gamma \equiv [1 - (V^2/c^2)]^{-1/2}$. On substituting these expressions in Eqs. (1.9) and (1.10), we get

$$\begin{aligned} x \pm ct &= f_1\left[t' \pm (x'/c)\right] \pm cf_0\left[t' \pm (x'/c)\right] \\ &= \gamma\left[\left[Vt' \pm (V/c)x'\right] \pm \left[ct' \pm x'\right]\right] \\ &= \sqrt{\frac{1 \pm (V/c)}{1 \mp (V/c)}}(x' \pm ct'). \end{aligned} \qquad (1.12)$$

On solving these two equations, we obtain

$$t = \gamma\left(t' + \frac{V}{c^2}x'\right); \qquad x = \gamma\left(x' + Vt'\right). \qquad (1.13)$$

Using Eq. (1.13), we can now express (t', x') in terms of (t, x). Consistency requires that it should have the same form as Eq. (1.13) with V replaced by $(-V)$. It can be easily verified that this is indeed the case. For two inertial frames K and K' with a relative velocity V, we can always align the coordinates in such a way that the relative velocity vector is along the common (x, x') axis. Then, from symmetry, it follows that the transverse directions are not affected and hence $y' = y, z' = z$. These relations, along with Eq. (1.13), give the coordinate transformation between the two inertial frames, usually called the *Lorentz transformation*.

Since Eq. (1.13) is a linear transformation between the coordinates, the coordinate differentials (dt, dx^μ) transform in the same way as the coordinates themselves. Therefore, the invariance of the line interval in Eq. (1.1) translates to finite values of the coordinate separations. That is, the Lorentz transformation leaves the quantity

$$s^2(1, 2) = |\boldsymbol{x}_1 - \boldsymbol{x}_2|^2 - c^2(t_1 - t_2)^2 \qquad (1.14)$$

invariant. (This result, of course, can be verified directly from Eq. (1.12).) In particular, the Lorentz transformation leaves the quantity $s^2 \equiv (-c^2t^2 + |\boldsymbol{x}|^2)$ invariant since this is the spacetime interval between the origin and any event $(t, \boldsymbol{x})$. A quadratic expression of this form is similar to the length of a vector in three dimensions which – as is well known – is invariant under rotation of the coordinate axes. This suggests that the transformation between the inertial frames can be thought of as a 'rotation' in four-dimensional space. The 'rotation' must be in the tx-plane characterized by a parameter, say, χ. Indeed, the Lorentz transformation in Eq. (1.13) can be written as

$$x = x' \cosh \chi + ct' \sinh \chi, \qquad ct = x' \sinh \chi + ct' \cosh \chi, \qquad (1.15)$$

with $\tanh \chi = (V/c)$, which determines the parameter χ in terms of the relative velocity between the two frames. These equations can be thought of as rotation by a complex angle. The quantity χ is called the *rapidity* corresponding to the speed V and will play a useful role in future discussions. Note that, in terms of rapidity, $\gamma = \cosh \chi$ and $(V/c)\gamma = \sinh \chi$. Equation (1.13) can be written as

$$(x' \pm ct') = e^{\mp \chi}(x \pm ct), \qquad (1.16)$$

showing the Lorentz transformation compresses $(x + ct)$ by $e^{-\chi}$ and stretches $(x - ct)$ by e^{χ} leaving $(x^2 - c^2 t^2)$ invariant.

Very often one uses the coordinates $u = ct - x$, $v = ct + x$, $u' = ct' - x'$, $v' = ct' + x'$, instead of the coordinates (ct, x), etc., because it simplifies the algebra. Note that, even in the general case of an observer moving along an arbitrary trajectory, the transformations given by Eq. (1.9) and Eq. (1.10) are simpler to state in terms of the (u, v) coordinates:

$$-u = f_1(u'/c) - cf_0(u'/c); \qquad v = f_1(v'/c) + cf_0(v'/c). \qquad (1.17)$$

Thus, even in the general case, the coordinate transformations do not mix u and v though, of course, they will not keep the form of $(c^2 t^2 - |x|^2)$ invariant. We will have occasion to use this result in later chapters.

The non-relativistic limit of Lorentz transformation is obtained by taking the limit of $c \to \infty$ when we get

$$t' = t, \quad x' = x - Vt, \quad y' = y, \quad z' = z. \qquad (1.18)$$

This is called the *Galilean transformation* which uses the same absolute time coordinate in all inertial frames. When we take the same limit $(c \to \infty)$ in different laws of physics, they should remain covariant in form under the Galilean transformation. This is why we mentioned earlier that the statement (i) on page 2 is not specific to relativity and holds even in the non-relativistic limit.

Exercise 1.1

Light clocks A simple model for a 'light clock' is made of two mirrors facing each other and separated by a distance L in the rest frame. A light pulse bouncing between them will provide a measure of time. Show that such a clock will slow down exactly as predicted by special relativity when it moves (a) in a direction transverse to the separation between the mirrors or (b) along the direction of the separation between the mirrors. For a more challenging task, work out the case in which the motion is in an arbitrary direction with constant velocity.

1.3.2 Transformation of velocities

Given the Lorentz transformation, we can compute the transformation law for any other physical quantity which depends on the coordinates. As an example, consider the transformation of the velocity of a particle, as measured in two inertial frames. Taking the differential form of the Lorentz transformation in Eq. (1.13), we obtain

$$dx = \gamma \left(dx' + V\,dt' \right), \quad dy = dy', \quad dz = dz', \quad dt = \gamma \left(dt' + \frac{V}{c^2}dx' \right),$$

(1.19)

and forming the ratios $v = dx/dt, v' = dx'/dt'$, we find the transformation law for the velocity to be

$$v_x = \frac{v_x' + V}{1 + (v_x'V/c^2)}, \quad v_y = \gamma^{-1}\frac{v_y'}{1 + (v_x'V/c^2)}, \quad v_z = \gamma^{-1}\frac{v_z'}{1 + (v_x'V/c^2)}.$$

(1.20)

The transformation of velocity of a particle moving along the x-axis is easy to understand in terms of the analogy with the rotation introduced earlier. Since this will involve two successive rotations in the t–x plane it follows that we must have additivity in the rapidity parameter $\chi = \tanh^{-1}(V/c)$ of the particle and the coordinate frame; that is we expect $\chi_{12} = \chi_1(v_x') + \chi_2(V)$, which correctly reproduces the first equation in Eq. (1.20). It is also obvious that the transformation law in Eq. (1.20) reduces to the familiar addition of velocities in the limit of $c \to \infty$. But in the relativistic case, none of the velocity components exceeds c, thereby respecting the existence of a maximum speed.

The transverse velocities transform in a non-trivial manner under Lorentz transformation – unlike the transverse coordinates, which remain unchanged under the Lorentz transformation. This is, of course, a direct consequence of the transformation of the time coordinate. An interesting consequence of this fact is that the *direction* of motion of a particle will appear to be different in different inertial frames. If $v_x = v\cos\theta$ and $v_y = v\sin\theta$ are the components of the velocity in the coordinate frame K (with primes denoting corresponding quantities in the frame

K'), then it is easy to see from Eq. (1.20) that

$$\tan\theta = \gamma^{-1}\frac{v'\sin\theta'}{v'\cos\theta' + V}. \tag{1.21}$$

For a particle moving with relativistic velocities ($v' \approx c$) and for a ray of light, this formula reduces to

$$\tan\theta = \gamma^{-1}\frac{\sin\theta'}{(V/c) + \cos\theta'}, \tag{1.22}$$

which shows that the direction of a source of light (say, a distant star) will appear to be different in two different reference frames. Using the trigonometric identities, Eq. (1.22) can be rewritten as $\tan(\theta'/2) = e^{-\chi}\tan(\theta/2)$, where χ is the rapidity corresponding to the Lorentz transformation showing $\theta' \ll \theta$ when $\chi \gg 1$. This result has important applications in the study of radiative processes (see Exercise 1.4).

Exercise 1.2
Superluminal motion   A far away astronomical source of light is moving with a speed v along a direction which makes an angle θ with respect to our line of sight. Show that the *apparent* transverse speed v_{app} of this source, projected on the sky, will be related to the actual speed v by

$$v_{\mathrm{app}} = \frac{v\sin\theta}{1 - (v/c)\cos\theta}. \tag{1.23}$$

From this, conclude that the apparent speed can exceed the speed of light. How does v_{app} vary with θ for a constant value of v?

1.3.3 Lorentz boost in an arbitrary direction

For some calculations, we will need the form of the Lorentz transformation along an arbitrary direction n with speed $V \equiv \beta c$. The time coordinates are related by the obvious formula

$$x'^0 = \gamma(x^0 - \boldsymbol{\beta}\cdot\boldsymbol{x}) \tag{1.24}$$

with $\boldsymbol{\beta} = \beta\boldsymbol{n}$. To obtain the transformation of the spatial coordinate, we first write the spatial vector $\boldsymbol{x}$ as a sum of two vectors: $\boldsymbol{x}_{\parallel} = \mathbf{V}(\mathbf{V}\cdot\boldsymbol{x})/V^2$, which is parallel to the velocity vector, and $\boldsymbol{x}_{\perp} = \boldsymbol{x} - \boldsymbol{x}_{\parallel}$, which is perpendicular to the velocity vector. We know that, under the Lorentz transformation, we have $\boldsymbol{x}'_{\perp} = \boldsymbol{x}_{\perp}$ while $\boldsymbol{x}'_{\parallel} = \gamma(\boldsymbol{x}_{\parallel} - \mathbf{V}t)$. Expressing everything again in terms of $\boldsymbol{x}$ and $\boldsymbol{x}'$, it is easy to show that the final result can be written in the vectorial form as

$$\boldsymbol{x}' = \boldsymbol{x} + \frac{(\gamma - 1)}{\beta^2}(\boldsymbol{\beta}\cdot\boldsymbol{x})\boldsymbol{\beta} - \gamma\boldsymbol{\beta}x^0. \tag{1.25}$$

Equations (1.24) and (1.25) give the Lorentz transformation between two frames moving along an arbitrary direction.

Since this is a linear transformation between x'^i and x^j, we can express the result in the matrix form $x'^i = L^{i'}_j x^j$, with the inverse transformation being $x^i = M^i_j x'^j$ where the matices $L^{i'}_j$ and M^i_j are inverses of each other. Their components can be read off from Eq. (1.24) and Eq. (1.25) and expressed in terms of the magnitude of the velocity $V \equiv \beta c$ and the direction specified by the unit vector n^α:

$$L^{0'}_0 = \gamma = \left(1 - \beta^2\right)^{-1/2}; \qquad L^{0'}_\alpha = -L^{\alpha'}_0 = -\gamma\beta n^\alpha;$$

$$L^{\alpha'}_\beta = L^{\beta'}_\alpha = (\gamma - 1)\, n^\alpha n^\beta + \delta^{\alpha\beta}; \qquad M^a_{b'}(\beta) = L^{b'}_a(-\beta). \qquad (1.26)$$

The matrix elements have one primed index and one unprimed index to emphasize the fact that we are transforming from an unprimed frame to a primed frame or vice versa. Note that the inverse of the matrix L is obtained by changing β to $-\beta$, as to be expected. We shall omit the primes on the matrix indices when no confusion is likely to arise. From the form of the matrix it is easy to verify that

$$L^{a'}_i M^j_{a'} = \delta^i_j; \quad L^{a'}_i M^i_{b'} = \delta^{a'}_{b'}; \quad L^{a'}_i L^{b'}_j \eta'_{ab} = \eta_{ij}; \quad \det |L^{a'}_b| = 1. \qquad (1.27)$$

An important application of this result is in determining the effect of two consecutive Lorentz transformations. We saw earlier that the Lorentz transformation along a given axis, say x^1, can be thought of as a rotation with an imaginary angle χ in the t–x^1 plane. The angle of rotation is related to the velocity between the inertial frames by $v = c\tanh\chi$. Two successive Lorentz transformations with velocities v_1 and v_2, *along the same direction* x^1, will correspond to two successive rotations in the t–x^1 plane by angles, say, χ_1 and χ_2. Since two rotations in the same plane about the same origin commute, it is obvious that these two Lorentz transformations commute and this is equivalent to a rotation by an angle $\chi_1 + \chi_2$ in the t–x^1 plane. This results in a single Lorentz transformation with a velocity parameter given by the relativistic sum of the two velocities v_1 and v_2.

The situation, however, is quite different in the case of Lorentz transformations along two different directions. These will correspond to rotations in two different planes and it is well known that such rotations will not commute. The order in which the Lorentz transformations are carried out is important if they are along different directions. Let $L(\boldsymbol{v})$ denote the matrix in Eq. (1.26) corresponding to a Lorentz transformation with velocity $\boldsymbol{v}$. The product $L(\boldsymbol{v_1})L(\boldsymbol{v_2})$ indicates the operation of two Lorentz transformations on an inertial frame with velocity $\boldsymbol{v_2}$ followed by velocity $\boldsymbol{v_1}$. Two different Lorentz transformations will commute only if $L(\boldsymbol{v_1})L(\boldsymbol{v_2}) = L(\boldsymbol{v_2})L(\boldsymbol{v_1})$; in general, this is not the case.

We will demonstrate this for the case in which both $\boldsymbol{v}_1 = v_1 \boldsymbol{n}_1$ and $\boldsymbol{v}_2 = v_2 \boldsymbol{n}_2$ are infinitesimal in the sense that $v_1 \ll c$, $v_2 \ll c$. Let the first Lorentz transformation take x^b to $L_b^j(\boldsymbol{v}_1)x^b$ and the second Lorentz transformation take this further to $x_{21}^a \equiv L_j^a(\boldsymbol{v}_2)L_b^j(\boldsymbol{v}_1)x^b$. Performing the same two Lorentz transformations in reverse order leads to $x_{12}^a \equiv L_j^a(\boldsymbol{v}_1)L_b^j(\boldsymbol{v}_2)x^b$. We are interested in the difference $\delta x^a \equiv x_{21}^a - x_{12}^a$ to the lowest nontrivial order in (v/c). Since this involves the product of two Lorentz transformations, we need to compute $L_j^a(\boldsymbol{v}_1)L_b^j(\boldsymbol{v}_2)$ keeping all terms up to *quadratic* order in v_1 and v_2. Explicit computation, using, for example, Eq. (1.25) gives

$$x_{21}^0 \approx [1 + \frac{1}{2}(\boldsymbol{\beta}_2 + \boldsymbol{\beta}_1)^2]x^0 - (\boldsymbol{\beta}_2 + \boldsymbol{\beta}_1) \cdot \boldsymbol{x}$$

$$\boldsymbol{x}_{21} \approx \boldsymbol{x} - (\boldsymbol{\beta}_2 + \boldsymbol{\beta}_1)x^0 + [\boldsymbol{\beta}_2(\boldsymbol{\beta}_2 \cdot \boldsymbol{x}) + \boldsymbol{\beta}_1(\boldsymbol{\beta}_1 \cdot \boldsymbol{x})] + \boldsymbol{\beta}_2(\boldsymbol{\beta}_1 \cdot \boldsymbol{x}) \quad (1.28)$$

accurate to $\mathcal{O}(\beta^2)$. It is obvious that terms which are symmetric under the exchange of 1 and 2 in the above expression will cancel out when we compute $\delta x^a \equiv x_{21}^a - x_{12}^a$. Hence, we immediately get $\delta x^0 = 0$ to this order of accuracy. In the spatial components the only term that survives is the one arising from the last term which gives

$$\delta\boldsymbol{x} = [\boldsymbol{\beta}_2(\boldsymbol{\beta}_1 \cdot \boldsymbol{x}) - \boldsymbol{\beta}_1(\boldsymbol{\beta}_2 \cdot \boldsymbol{x})] = \frac{1}{c^2}(\boldsymbol{v}_1 \times \boldsymbol{v}_2) \times \boldsymbol{x}. \quad (1.29)$$

Comparing this with the standard result for infinitesimal rotation of coordinates $\delta\boldsymbol{x} = \boldsymbol{\Omega} \times \boldsymbol{x}$, we find that the net effect of two Lorentz transformations leaves a residual *spatial rotation* about the direction $\boldsymbol{v}_1 \times \boldsymbol{v}_2$. This result has implications for the structure of Lorentz group which we will discuss in Section 1.10.

1.4 Four-vectors

Equations like $\boldsymbol{F} = m\boldsymbol{a}$, which are written in vector notation, remain valid in any three-dimensional coordinate system without change of form. For example, consider two Cartesian coordinate systems with the same origin and the axes rotated with respect to each other. The *components* of the vectors $\boldsymbol{F}$ and $\boldsymbol{a}$ will be different in these two coordinate systems but the equality between the two sides of the equation will continue to hold.

If the laws of physics are to be expressed in a form which remains covariant under Lorentz transformation, we should similarly use vectorial quantities with four components and treat Lorentz transformations as rotations in a four-dimensional space. Such vectors are called *four-vectors* and will have one time component and three spatial components. The spatial components, of course, will form an ordinary three-vector and transform as such under spatial rotations with the time component remaining unchanged.

Let us denote a generic four-vector as A^i with components $(A^0, \boldsymbol{A})$ in some inertial frame K. The simplest example of such a four-vector is the four-velocity u^i of a particle, defined as $u^i = dx^i/d\tau$, where $x^i(\tau)$ is the trajectory of a particle parametrized by the the proper time τ shown by a clock carried by the particle. Since $d\tau$ is Lorentz invariant and dx^i transforms as $dx'^i = L_j^{i'} dx^j$ under the Lorentz transformation, it follows that u^i transforms as $u'^i = L_j^{i'} u^j$. We shall use this transformation law to *define* an arbitrary four-vector $A^i = (A^0, \boldsymbol{A})$ as a set of four quantities, which, under the Lorentz transformation, changes as $A'^i = L_j^{i'} A^j$. Explicitly, for Lorentz transformation along the x-axis, the components transform as

$$A'^0 = \gamma \left(A^0 - \frac{V}{c} A^1 \right), \quad A'^1 = \gamma \left(A^1 - \frac{V}{c} A^0 \right), \quad A'^2 = A^2, \quad A'^3 = A^3.$$

$$(1.30)$$

It is obvious from our construction that, under these transformations, the square of the 'length' of the vector defined by $-(A^0)^2 + |\boldsymbol{A}|^2$ remains invariant.

It is also convenient to introduce at this stage two different types of components of any four-vector denoted by A^i and A_i with $A^i \equiv (A^0, \boldsymbol{A})$ and $A_i \equiv (-A^0, \boldsymbol{A})$. In other words, 'lowering of index' changes the sign of the *time* component. More formally, this relation can be written as

$$A_i = \eta_{ij} A^j, \tag{1.31}$$

where $\eta_{ij} = \text{diag}\,(-1, 1, 1, 1)$ was introduced earlier in Eq. (1.2) and we have used the summation convention for the repeated index j. (In what follows, this convention will be implicitly assumed.) To obtain A^i from A_i we can use the inverse matrix $\eta^{ij} = dia(-1, 1, 1, 1)$ which is the inverse of the matrix η_{ij} – though it has the same entries – and write $A^j = \eta^{jk} A_k$. It can be trivially verified that the components η^{ik} and η_{ik} have the same numerical value in all inertial frames. We shall call A_i the *covariant* components of a vector and A^i the *contravariant* components. Given this definition, we can write the squared 'length' of the vector as

$$A^i A_i = \eta_{ij} A^i A^j = \eta^{ij} A_i A_j. \tag{1.32}$$

Explicitly, $A^i A_i$ stands for the expression

$$A^i A_i \equiv \sum_{i=0}^{3} A^i A_i = A^0 A_0 + A^1 A_1 + A^2 A_2 + A^3 A_3 = -(A^0)^2 + |\boldsymbol{A}|^2. \tag{1.33}$$

Unlike the squared norm of a three-vector, this quantity need *not* be positive definite. A four-vector is called *timelike, null* or *spacelike* depending on whether this quantity is negative, zero or positive, respectively.

More generally, given two four-vectors $A^i = (A^0, \boldsymbol{A})$ and $B^i = (B^0, \boldsymbol{B})$, we can define a 'dot product' between them by a similar rule as $A^i B_i$, with

$$A^i B_i = A^0 B_0 + A^1 B_1 + A^2 B_2 + A^3 B_3 = -A^0 B^0 + \boldsymbol{A} \cdot \boldsymbol{B}. \qquad (1.34)$$

Under a Lorentz transformation, we have $A'^i B'_i = L^{i'}_j M^k_{i'} A^j B_k = A^j B_j$, since $L^{i'}_j M^k_{i'} = \delta^k_j$. Thus the dot product is invariant under Lorentz transformations. The squared length of the vector, of course, is just the dot product of the vector with itself. Note that a null vector has a vanishing norm (i.e. vanishing dot product with itself) even though the vector itself is nonzero.

Given the four-vector with superscript index A^i, we introduced another set of components A_i by the definition in Eq. (1.31). There is an important physical context in which a covariant vector arises naturally, which we shall now describe. Just as the four-vector A^i is defined in analogy with the transformation of dx^i, we can define the quantities with subscripts such as A_i, in terms of the derivative operator $\partial/\partial x^i$ which is 'dual' to dx^i. (In Chapter 11 we will see that covariant vectors arise naturally in terms of certain quantities called *1-forms*, which will make this notion more formal.) To define the corresponding four-dimensional object, we only have to note that the differential of a scalar quantity $d\phi = (\partial\phi/\partial x^i)dx^i$ is also a scalar. Since this expression is a dot product between dx^i and $(\partial\phi/\partial x^i)$, it follows that the latter quantity transforms like a covariant four-vector under Lorentz transformations. Explicitly, the components of the four-gradient of a scalar are given by the covariant components of a four-vector

$$v_i \equiv \frac{\partial\phi}{\partial x^i} = \left(\frac{1}{c}\frac{\partial\phi}{\partial t}, \nabla\phi\right) \equiv \partial_i\phi. \qquad (1.35)$$

[We will often use the notation ∂_i to denote $(\partial/\partial x^i)$.] This is a direct generalization of the ordinary three-dimensional gradient $\nabla = [(\partial/\partial x), (\partial/\partial y), (\partial/\partial z)]$, which transforms as a vector under spatial rotations. Note that $(\partial\phi/\partial x_i) = \partial^i\phi = \eta^{ij}\partial_j\phi$ are the contravariant components of the gradient.

As an example, consider the notion of a *normal* to the surface. A three-dimensional surface in four-dimensional space is given by an equation of the form $f(x^i) = $ constant. The normal vector $n_i(x^a)$ at any event x^a on this surface is given by $n_i = (\partial f/\partial x^i)$ which is a natural example of a four-gradient. [For any displacement dx^i confined to the surface $f = $ constant we have $df = 0 = dx^i(\partial f/\partial x^i) = dx^i n_i$. This shows that n_i is orthogonal to any displacement on the surface and hence is indeed a normal but, in general, it is not a unit normal.] It is conventional to call a surface itself spacelike, null or timelike at x^a, depending on whether n_i is timelike, null or spacelike at x^a, respectively. Spacelike surfaces have timelike normals and vice versa, while null surfaces have null normals. When n_i is not null we can construct another vector $\hat{n}_i \equiv n_i(\pm n_j n^j)^{-1/2}$ such that $\hat{n}_i \hat{n}^i = \pm 1$ depending on whether n_i is spacelike or timelike. The $\hat{n}_i$ has unit normal.

In the above discussion, we defined the four-vector using the transformation law for $u^i = dx^i/d\tau$, which is the same as the transformation law for the infinitesimal coordinate differentials dx^i. For any parametrized curve $x^i(\lambda)$, the quantity $dx^i = (dx^i/d\lambda)d\lambda$ is an infinitesimal vector in the direction of the tangent to the curve and hence this definition is purely local. On the other hand, if we work with the standard inertial *Cartesian* coordinates (t, x, y, x) in the spacetime, then x^i also transforms as a four-vector, just like dx^i. We have refrained from using the latter – or even calling x^i a four-vector – since the definition based on dx^i continues to hold in more general contexts we will encounter in later chapters.

Exercise 1.3
The strange world of four-vectors We will call two vectors a^i and b^i orthogonal if $a^i b_i = 0$. Show that (a) the sum of two vectors can be spacelike, null or timelike independent of the nature of the two vectors; (b) only non-spacelike vectors, which are orthogonal to a given nonzero null vector, must be multiples of the null vector. (c) Find four linearly independent null vectors in the Minkowski space.

Exercise 1.4
Focused to the front An interesting application of the transformation law for the four-vectors is in the study of radiation from a moving source. This exercise explores several feature of it. Consider a null four-vector $k^a = (\omega, \omega \boldsymbol{n})$ which represents a photon (or light ray) of frequency ω propagating in the direction $\boldsymbol{n}$. Since the components of this four-vector determine both the frequency and direction of propagation of the wave, it follows that different observers will see the wave as having different frequencies and directions of propagation. As a specific example, consider two Lorentz frames S ('lab frame') and S' ('rest frame'), with S' moving along the positive x-axis of S with velocity v. A plane wave with frequency ω_L is travelling along the direction (θ_L, ϕ_L) in the lab frame. The corresponding quantities in the rest frame will be denoted with a subscript R.

(a) Show that

$$\omega_R = \gamma \omega_L [1 - (v/c) \cos \theta_L]; \qquad \mu_R = \frac{\mu_L - (v/c)}{1 - (v\mu_L/c)}, \qquad (1.36)$$

where $\mu_R \equiv \cos \theta_R$, $\mu_L \equiv \cos \theta_L$. Show that the second equation is equivalent to Eq. (1.22).

(b) Plot μ_L against μ_R for an ultra-relativistic speed $v \to c$ and show that the motion of a source 'drags' the wave forward. A corollary to this result is that a charged particle, moving relativistically, will 'beam' most of the radiation it emits in the forward direction.

(c) For several applications in radiative processes, etc., one will be interested in computing the transformation of an element of solid angle around the direction of propagation of a light ray. For example, consider a source which is emitting radiation with some angular distribution in its own rest frame so that $(d\mathcal{E}'/dt'd\Omega') \equiv f(\theta', \phi')$ represents the energy emitted per unit time into a given solid angle. If this source is moving with a velocity v

in the lab frame, we will be interested in computing the corresponding quantity in the lab frame $(d\mathcal{E}/dtd\Omega)$. Show that

$$d\Omega' == \frac{1}{\gamma^2} \frac{d\Omega}{(1 - v\cos\theta)^2}; \qquad d\mathcal{E}' = \gamma d\mathcal{E}(1 - v\cos\theta). \tag{1.37}$$

Combining them, show that

$$\left(\frac{d\mathcal{E}}{dtd\Omega}\right)_{\text{lab}} = \frac{(1 - v^2)^2}{(1 - v\cos\theta)^3} \left(\frac{d\mathcal{E}'}{dt'd\Omega'}\right)_{\text{rest}}. \tag{1.38}$$

As a check, consider the case in which $(d\mathcal{E}'/dt'd\Omega')$ is independent of (θ', ϕ') – i.e. the emission is isotropic in the rest frame. In this case show that

$$\left(\frac{d\mathcal{E}}{dt}\right)_{\text{lab}} = 4\pi \left(\frac{d\mathcal{E}'}{dt'd\Omega'}\right)_{\text{rest}} = \left(\frac{d\mathcal{E}'}{dt'}\right)_{\text{rest}}. \tag{1.39}$$

Interpret this result.

1.4.1 Four-velocity and acceleration

To illustrate the above ideas, we shall consider some examples of four-vectors starting with a closer study of the four-velocity itself. From the definition $u^i = dx^i/d\tau$, it follows that u^i has the components:

$$u^i = (\gamma, \gamma v); \qquad \gamma \equiv (1 - v^2)^{-1/2}. \tag{1.40}$$

(Here, and in most of what follows, we shall choose units with $c = 1$.) Further, $u^i u_i = dx^i dx_i/d\tau^2 = -1$, which can also be verified directly using the components in Eq. (1.40), so that the four-velocity has only three independent components determined by the three-velocity v. For practical computations, one often needs to convert between the three-velocity and the four-velocity. Such conversions are facilitated by the following formulas which are fairly easy to prove:

$$u^0 \equiv (1 + u^\beta u_\beta)^{1/2}; \quad v^\beta = u^\beta/u^0 = u^\beta(1 + u^\alpha u_\alpha)^{-1/2}; \quad |v| = [1 - (u^0)^{-2}]^{1/2}. \tag{1.41}$$

We will also often use the relation

$$\frac{d}{d\tau} = \left(\frac{dt}{d\tau}\right)\frac{d}{dt} = (1 - v^2)^{-1/2}\frac{d}{dt}. \tag{1.42}$$

As an application of the definition of four-velocity let us determine the relative velocity v between two frames of reference. To obtain this, let us consider two frames S_1 and S_2 which move with velocities v_1 and v_2 with respect to a third inertial frame S_0. We can associate with these three-velocities, the corresponding four-velocities, given by $u_1^i = (\gamma_1, \gamma_1 v_1)$ and $u_2^i = (\gamma_2, \gamma_2 v_2)$ with all the components being measured in S_0. On the other hand, with respect to S_1, these four-vector will have the components $u_1^i = (1, 0)$ and $u_2^j = (\gamma, \gamma v)$, where v (by definition) is the relative velocity between the frames. To determine the magnitude of this quantity, we note that in this frame S_1 we can write $\gamma = -u_{1i} u_2^i$. But since this expression is Lorentz invariant, we can evaluate it in any inertial frame. In S_0, with $u_1^i = (\gamma_1, \gamma_1 v_1)$, $u_2^i = (\gamma_2, \gamma_2 v_2)$, this has the value

$$\gamma = (1 - v^2)^{-1/2} = \gamma_1 \gamma_2 - \gamma_1 \gamma_2 v_1 \cdot v_2. \tag{1.43}$$

Simplifying this expression we get

$$v^2 = \frac{(1 - v_1 \cdot v_2)^2 - (1 - v_1^2)(1 - v_2^2)}{(1 - v_1 \cdot v_2)^2} = \frac{(v_1 - v_2)^2 - (v_1 \times v_2)^2}{(1 - v_1 \cdot v_2)^2}. \tag{1.44}$$

The concept of relative velocity can be used further to introduce an interesting structure in the velocity space. Let us consider a three-dimensional abstract space in which each point represents a velocity. Two nearby points correspond to velocities v and $v + dv$ which differ by an infinitesimal quantity. If the space is considered to be similar to the usual three-dimensional flat space, one would have assumed that the 'distance' between these two points is just $|dv|^2 = dv_x^2 + dv_y^2 + dv_z^2$. In non-relativistic physics, this distance corresponds to the magnitude of the relative velocity between the two frames. However, we have just seen that the relative velocity between two frames in relativistic mechanics is different and given by Eq. (1.44). It is more natural to define the distance between two points in the velocity space to be the relative velocity between the respective frames. In that case, the infinitesimal 'distance' between the two points in the velocity space will be given by

$$dl_v^2 = \frac{(dv)^2 - (v \times dv)^2}{(1 - v^2)^2}. \tag{1.45}$$

Using the relations

$$(v \times dv)^2 = v^2 (dv)^2 - (v \cdot dv)^2; \quad (v \cdot dv)^2 = v^2 (dv)^2 \tag{1.46}$$

and writing $(dv)^2 = dv^2 + v^2 (d\theta^2 + \sin^2 \theta \, d\phi^2)$, where θ, ϕ are the polar and azimuthal angles of the direction of v, we get

$$dl_v^2 = \frac{dv^2}{(1 - v^2)^2} + \frac{v^2}{1 - v^2} (d\theta^2 + \sin^2 \theta \, d\phi^2). \tag{1.47}$$

If we use the rapidity χ in place of v through the equation $v = \tanh \chi$, the line element in Eq. (1.47) becomes

$$dl_v^2 = d\chi^2 + \sinh^2 \chi (d\theta^2 + \sin^2 \theta \, d\phi^2). \tag{1.48}$$

This is an example of a *curved* space within the context of special relativity. This particular space is called (three-dimensional) Lobachevsky space.

In a manner similar to the definition of four-velocity, we can define the four-acceleration to be $a^i = d^2x^i/d\tau^2 = du^i/d\tau$. Differentiating the relation $u^iu_i = -1$ with respect to τ, we see that $a^iu_i = 0$. It follows that in the instantaneous rest frame of the particle, in which the four-velocity has the components $u^a = (1,0)$, the four-acceleration is purely spatial with $a^i = (0, \mathbf{a})$. The $\mathbf{a}$ is in fact the usual Newtonian acceleration in the comoving frame. For example, if an observer moving along some arbitrary trajectory releases a particle in the comoving frame (so that the particle remains stationary with respect to the comoving Lorentz frame), then the observer will pick up a velocity $d\mathbf{v} = \mathbf{a}d\tau$ with respect to the particle in a small time interval $d\tau$. For one-dimensional motion along the x-axis, $u^i = (\cosh\chi, \sinh\chi)$, where $\chi(\tau)$ is the rapidity. The corresponding acceleration, $a^i = (d\chi/d\tau)(\sinh\chi, \cosh\chi)$, has the magnitude $a^ia_i = |d\chi/d\tau|$.

One elementary, but important, application of the condition $a^iu_i = 0$ is the following. In non-relativistic physics, we are accustomed to equations of motion of the form $ma^\alpha = -\partial^\alpha\Phi$ (with $\alpha = 1, 2, 3$) for a particle of mass m moving in a potential Φ. One might think that a special relativistic generalization of this equation could be $ma^j = -\partial^j\Phi$ (with $j = 0, 1, 2, 3$). But the condition $a^iu_i = 0$ now implies that $u^i\partial_i\Phi = d\Phi/d\tau = 0$, implying that the potential should stay constant along the trajectory of the particle. Since this cannot be satisfied, the simplest generalization of Newton's second law of motion, in a conservative force field, will fail in relativity. Hence, the forces which act on particles in any relativistic theory will necessarily have to be velocity dependent.

1.5 Tensors

At the next level of structure, one can define 'four-tensors' as quantities which transform like the product of four-vectors. For example, consider a two-index object C_{ik} defined to be $C_{ik} = A_iB_k$, where A_i and B_k are four-vectors. Knowing the transformation law for the four-vectors, one can easily determine how the components of C_{ik} get mixed under a Lorentz transformation. We find that $C'_{ij} = M^a_{i'}M^b_{j'}C_{ab}$. A second rank tensor T_{ik} is *defined* to be a set of $4 \times 4 = 16$ quantities which transform like the product A_iB_k of two four-vectors under a Lorentz transformation. That is, T_{ij} transforms as $T'_{ij} = M^a_{i'}M^b_{j'}T_{ab}$ under a Lorentz transformation. (Of course, a general second rank tensor T_{ik} cannot be expressed as a product of two four-vectors.) Since we have defined two types of components for the four-vectors, we can have second rank tensors with different placement of indices like T^{ik}, $S^i_{\;k}$ or J_{ik}, etc. An index occurring as a superscript will be called a *contravariant index* and an index occurring as a subscript will be called a *covariant index*. These ideas generalize to tensors with an arbitrary number of contravariant and covariant indices in a natural and obvious manner. The η_{ab} used so far is indeed a second rank tensor and transforms as $\eta'_{ab} = M^j_{a'} M^k_{b'} \eta_{jk}$; but Eq. (1.27) ensures that $\eta'_{ab} = \eta_{ab}$, as expected.

In the case of vectors, we can raise and lower the index so that the same physical quantity can be expressed as A^i or $A_j = \eta_{ij} A^i$. The same is true for tensors of arbitrary rank. The raising and lowering of tensor indices follow obvious generalization of the above rule; e.g. $T^i_{\ k} = \eta_{ak} T^{ia}$, etc. Whenever an index corresponding to a time coordinate is raised or lowered, the sign of the component changes while lowering or raising of a spatial index leaves the value of the component unchanged.

Multiplying two tensors produces another tensor with a different rank depending on the nature of the two tensors; for example, $S^a_{\ bcd} = K^a_{\ b} U_{cd}$ indicates a product of a mixed tensor $K^a_{\ b}$ and a covariant tensor U_{cd} leading to another mixed tensor with four indices. We can also sum over one (or more) pair(s) of the indices while performing the multiplication; for example, $S^a_{\ bad} = K^a_{\ b} U_{ad} \equiv N_{bd}$ in which we have summed over the index a. In this process, one pair of indices vanishes and the resulting tensor has a lower rank. This operation is called *contraction* and we say that we have contracted on the index a in the above example. The contracted index is called a dummy index since it is summed over and can be replaced by any other index; e.g. $A_{ij} S^{ipq} = A_{kj} S^{kpq}$, etc. In fact, the elementary operation of lowering (or raising) of an index, $V_i = \eta_{ij} V^j$, is a simple example of multiplication of η_{ij} and V^k followed by a contraction over a pair of indices by setting $k = j$. Another example – which we will encounter frequently – is the contraction of a pair of indices of a second rank tensor, say, $T^a_{\ b}$, leading to a scalar: $T = T^a_{\ a} = \eta_{ij} T^{ij}$. If $T^a_{\ b}$ is thought of as a matrix, then T is the trace of the matrix.

A tensor is symmetric (antisymmetric) on two indices (a, b), if interchanging the indices leaves the value of the component the same (changes the sign of the component). Any tensor can be written as the sum of a symmetric and an antisymmetric part with respect to any pair of indices which are (both) covariant or contravariant. For example, $T^{ik\cdots} = A^{ik\cdots} + S^{ik\cdots}$, where $A^{ik\cdots} \equiv (T^{ik\cdots} - T^{ki\cdots})/2$ is antisymmetric, and $S^{ik\cdots} = (T^{ik\cdots} + T^{ki\cdots})/2$ is symmetric in (i, k), with ... denoting more indices, if any.

One of the key results regarding the contraction which will be used extensively in the latter chapters is the following. If A^{ik} is antisymmetric in i and k and J_{ik} is a general tensor, then the contraction $A^{ik} J_{ik}$ can be expressed in the form

$$A^{ik} J_{ik} = \frac{1}{2} A^{ik} J_{ik} + \frac{1}{2} A^{ki} J_{ki} = \frac{1}{2} A^{ik} (J_{ik} - J_{ki}). \qquad (1.49)$$

The first equality follows from interchanging i and k which are dummy indices and the second equality follows from $A^{ik} = -A^{ki}$. This shows that, in a contraction of an antisymmetric and general tensor, only the antisymmetric part of the latter contributes. Similarly, if S^{ik} is a symmetric tensor, we have

$$S^{ik} J_{ik} = \frac{1}{2} S^{ik} J_{ik} + \frac{1}{2} S^{ki} J_{ki} = \frac{1}{2} S^{ik} (J_{ik} + J_{ki}), \qquad (1.50)$$

showing that only the symmetric part contributes. An immediate corollary is that the contraction of a symmetric and antisymmetric tensor vanishes; i.e. $A^{ik} S_{ik} = 0$. The same result holds even if there are more indices in these tensors as long as the indices on which the contraction takes place have the specific symmetry.

In several applications we will need to determine the number of independent components of a tensor (with a certain number of indices) when some additional symmetry restrictions are imposed. With future applications in mind, we will consider the general case of a tensor with r indices in an N-dimensional space.

If the tensor has no symmetries, then each of the indices can take N different values. Therefore, it has N^r independent components. Suppose that the tensor is completely symmetric in s of these indices, with no further restrictions on $(r - s)$ indices. We can completely specify the symmetric part by just giving how many of the s indices are zeros, how many of them are ones, etc. So the problem reduces to partitioning s objects into N sets ignoring the relative positioning. Such a partitioning requires introducing $(N - 1)$ 'boundaries' amongst the s objects. These boundaries and the objects together can be arranged in $(N + s - 1)!$ possible different ways. Of these, the boundaries can be permuted amongst themselves in $(N - 1)!$ ways and the objects can be permuted amongst themselves in $s!$ ways. Therefore the number of inequivalent ways of choosing the values of these s indices from N possibilities is given by $(N + s - 1)!/(N - 1)!s!$. The remaining indices can be chosen in N^{r-s} ways. So the total number of independent components in this case is given by $N^{r-s}(N + s - 1)!/(N - 1)!s!$.

Consider now the case in which the tensor is completely antisymmetric in a indices with no restrictions on the remaining $(r - a)$ indices. Again, the number of ways of choosing the a indices is equal to the number of ways of choosing a objects from N objects but without repetition, because the tensor will be zero when any two antisymmetric indices take the same value. This is given by $^N C_a$ so that the total number of independent components is given by $N^{r-a} N!/(N - a)!a!$.

In particular, a completely antisymmetric tensor with r indices in an N-dimensional space has $^N C_r$ independent components. So, in an N-dimensional space, a completely antisymmetric tensor with r indices or with $N - r$ indices has the same number of independent components. (This fact will play a crucial role in the development of differential forms and exterior derivatives to be discussed in Chapter 11.) It particular, if $N = r$ there is only one independent component (and when $r > N$ all the components must vanish). This unique, completely antisymmetric, tensor is usually denoted by the symbol ϵ with $\epsilon_{\mu\nu}, \epsilon_{\alpha\beta\mu}, \epsilon_{ijkl} \cdots$ indicating the completely antisymmetric tensors in two, three and four dimensions. We shall now discuss several useful properties of these tensors.

In three dimensions the completely antisymmetric three-tensor $\epsilon_{\alpha\beta\gamma}$ is defined by the relation $\epsilon_{123} = 1$ in Cartesian coordinates. All other components can, of course, be obtained from this one since we know that there is only one independent component for any such tensor. The most familiar use of such a tensor is in defining the cross product between two vectors A^α and B^μ. It can be easily shown that, if we define a three-vector C_α by

$$C_\alpha = \epsilon_{\alpha\beta\gamma}(A^\beta B^\gamma) = \frac{1}{2}\epsilon_{\alpha\beta\gamma}(A^\beta B^\gamma - A^\gamma B^\beta) \equiv \frac{1}{2}\epsilon_{\alpha\beta\gamma}C^{\beta\gamma}, \qquad (1.51)$$

this result is equivalent to the relation $\boldsymbol{C} = \boldsymbol{A} \times \boldsymbol{B}$. (The raising and lowering of the indices in three dimensions is done with the Kronecker delta $\delta_{\alpha\beta}$ and hence $A^\alpha = A_\alpha$, etc. The second equality follows from the antisymmetry in $\beta\gamma$.) We see that, in $N = 3$ dimensions, the antisymmetric tensor of second rank, $C^{\beta\gamma} \equiv A^\beta B^\gamma - A^\gamma B^\beta$ has the same number of components as a tensor of rank 1, which is just a vector. It is this fact that allows us to define a cross product of two vectors as another vector in three dimensions but not in higher dimensions. More generally, the quantity $\epsilon_{\alpha\beta\gamma}K^{\beta\gamma}$ is called the *dual* of an antisymmetric tensor $K^{\beta\gamma}$.

The products of two ϵ-tensors are of use in several computations and we will give the results related to them for the purpose of reference. The product $\epsilon_{\alpha\beta\gamma}\, \epsilon^{\lambda\mu\nu}$ can be expressed as a determinant of a 3×3 matrix whose elements are Kronecker deltas. The first column of the matrix has $(\delta_\alpha^\lambda, \delta_\beta^\lambda, \delta_\gamma^\lambda)$. The next two columns have the same structure with λ replaced by μ and ν, respectively. The product of two ϵ-tensors with one index contracted can be expressed as $\epsilon_{\alpha\beta\gamma}\, \epsilon^{\lambda\mu\gamma} = \delta_\alpha^\lambda\delta_\beta^\mu - \delta_\alpha^\mu\delta_\beta^\lambda$ (which can be expressed as the determinant of a 2×2 matrix made of Kronecker deltas). Contracting this relation further we find that $\epsilon_{\alpha\beta\gamma}\, \epsilon^{\lambda\beta\gamma} = 2\delta_\alpha^\lambda$ and $\epsilon_{\alpha\beta\gamma}\, \epsilon^{\alpha\beta\gamma} = 6$.

The same ideas generalize to four dimensions and we shall define ϵ_{ijkl} by fixing $\epsilon_{0123} = +1$. This tensor can be used to obtain a third rank tensor as the dual of a four-vector (and vice versa) by the relation $B_{ijk} = \epsilon_{ijkl}\, A^l$. For a second rank antisymmetric tensor A^{kl} we obtain a dual that is another second rank antisymmetric tensor given by $B_{ij} = \epsilon_{ijkl}\, A^{kl}$, which is often denoted by a 'star': $B_{ij} \equiv (*A)_{ij}$. The product of ϵ tensors in four dimensions can again be expressed in terms of Kronecker deltas as in the case of three dimensions. The product $\epsilon^{iklm}\, \epsilon_{prst}$ will be the determinant of a matrix in which each entry is a Kronecker delta. The first column will be made of $(\delta_p^i, \delta_p^k, \delta_p^l, \delta_p^m)$; the second column has a similar structure with p replaced by r, etc. When the product is contracted on one index, we get $\epsilon^{iklm}\, \epsilon_{prsm}$ which can be similarly expressed as a determinant of a 3×3 matrix built out of Kronecker deltas. Finally, $\epsilon^{iklm}\epsilon_{prlm} = -2(\delta_p^i\,\delta_r^k - \delta_r^i\,\delta_p^k)$ and $\epsilon^{iklm}\epsilon_{pklm} = -6\delta_p^i$. This tensor is also useful in defining the determinant of a matrix in terms of its components. It is easy to prove that

$$\epsilon^{prst}A_{ip}A_{kr}A_{ls}A_{mt} = -A\epsilon_{iklm}; \qquad \epsilon^{iklm}\epsilon^{prst}A_{ip}A_{kr}A_{ls}A_{mt} = 24A, \quad (1.52)$$

where $A = \det|A_{ij}|$.

Exercise 1.5
Transformation of antisymmetric tensors Write down the transformation law for a two index object A^{ik} under the Lorentz transformations; i.e. express the components A'^{ik} explicitly in terms of A^{ik}. How does it simplify if $A^{ik} = -A^{ki}$?

Exercise 1.6
Practice with completely antisymmetric tensors (a) Prove the relations regarding the ϵ tensors stated in the text.

(b) Show that $\epsilon_{abcd} = -\epsilon^{abcd}$.

(c) Show that $V_i V^i = -(1/3!)(*V)_{abc}(*V)^{abc}$ where $(*V)_{abc} = \epsilon_{abcd}V^d$ is the dual of the vector V^d. Also show that taking the dual twice leads to the same tensor except for a sign.

(d) The tensor $\delta_{a\ldots j}^{m\ldots n}$ is defined as the determinant of a matrix made of Kronecker deltas where the first row is made of $(\delta_a^m, \ldots, \delta_a^n)$ and so on with the last row being $(\delta_j^m, \ldots, \delta_j^n)$. Show that if there are more than four upper indices, the tensor vanishes identically. Further show that

$$\delta_m^a = -\frac{1}{6}\epsilon^{ablp}\epsilon_{mblp};$$

$$\delta_{lk}^{mn} = -\frac{1}{2}\epsilon^{mnps}\epsilon_{lkps} = \delta_l^m\delta_k^n - \delta_k^m\delta_l^n;$$

$$\delta_{mnl}^{abc} = -\epsilon^{abcp}\epsilon_{mnlp}. \qquad (1.53)$$

Show that $\delta_{mnl}^{abc} = 1$ if abc is an even permutation of mnl, -1 if abc is an odd permutation of mnl and 0 otherwise.

1.6 Tensors as geometrical objects

In the above discussion, we have worked with vectors, tensors, etc., using their components. This approach is quite adequate for all calculational purposes and – in fact – concrete calculations often require expressing all quantities in terms of components in some coordinate systems. But we know from elementary vector analysis in three dimensions that one can also think of vectors as abstract geometrical entities like $\boldsymbol{v}, \boldsymbol{u}, \ldots$, etc. If one has a Cartesian coordinate system with three basis vectors $\boldsymbol{e}_x, \boldsymbol{e}_y, \boldsymbol{e}_z$ then one can resolve a vector $\boldsymbol{v}$ into components by writing $\boldsymbol{v} = v^\alpha \boldsymbol{e}_\alpha$, etc. In these expressions, the superscript α in v^α denotes a *component* of a vector, while the subscript α in e_α denotes *which* vector; nevertheless, we shall assume summation convention over α in this and similar expressions. For many purposes, it is convenient to think of a vector as an abstract geometrical object v without introducing its components.

All these ideas generalize to four dimensions and four-vectors. We shall now describe this formalism; even though we will not need it in the first two chapters, it will be useful in the study of general relativity.

At any given event $\mathcal{P}$ in the spacetime, we can introduce a linear vector space $T(\mathcal{P})$ spanned by a set of four orthonormal basis vectors $\boldsymbol{e}_i$ usually called a *tetrad*. (We will employ the vector notation for both three-vectors and four-vectors by using different fonts. The four-momentum, for example, will be denoted by $\boldsymbol{p}$ while the three-momentum will be denoted by p. In most cases, the context will also make clear whether it is a three-vector or a four-vector.) The jth component of $\boldsymbol{e}_i$ is taken to be δ_i^j which ensures the orthonormality. Any other four-vector can now be expanded in terms of this basis by $\boldsymbol{v} = v^i \boldsymbol{e}_i$ thereby defining the contravariant components. All contravariant tensors of higher rank can also be defined in a similar manner using the direct product of the basis vectors to expand them. For example, a third rank contravariant tensor can be thought of as a geometric object $\boldsymbol{T}$ with components $\boldsymbol{T} = T^{ijk} \boldsymbol{e}_i \otimes \boldsymbol{e}_j \otimes \boldsymbol{e}_k$, etc.

The idea of covariant components arises in a somewhat different manner. They are related to certain geometrical objects called *1-forms* in a general manifold. However, in the presence of a metric tensor, one can introduce covariant components of the vector in a simpler way. To do this, let us consider another linear vector space $T^*(\mathcal{P})$ at any given event $\mathcal{P}$ with a new set of orthonormal basis vectors denoted by $\boldsymbol{w}^i$. A vector in this linear vector space is expanded as $\boldsymbol{u} = u_i \boldsymbol{w}^i$. Given an element $\boldsymbol{u}$ of $T^*(\mathcal{P})$, and an element $\boldsymbol{v}$ of $T(\mathcal{P})$, we can construct a real number by the rule

$$\langle \boldsymbol{u} | \boldsymbol{v} \rangle \equiv u_i v^i. \tag{1.54}$$

This operation is bi-linear on the two elements and hence allows us to determine the result once we know $\langle \boldsymbol{w}^i | \boldsymbol{e}_j \rangle$. Since the components of basis vectors are Kronecker deltas, it follows that

$$\langle \boldsymbol{w}^i | \boldsymbol{e}_j \rangle = w_k^i e_j^k = \delta_k^i \delta_j^k = \delta_j^i. \tag{1.55}$$

This operation does not require the metric tensor η_{ab}. But we know that, given two vectors $\boldsymbol{p}$ and $\boldsymbol{v}$ which are elements of $T(\mathcal{P})$, one can construct a real number by taking the dot product $\eta_{ij} p^i v^j$. On the other hand, there will be an element $\boldsymbol{u}$ of $T^*(\mathcal{P})$ such that $\langle \boldsymbol{u} | \boldsymbol{v} \rangle = \eta_{ij} p^i v^j$. That is, both these operations lead to the same real number. Then we can associate with every vector $\boldsymbol{p}$ (which is an element of $T(\mathcal{P})$) another vector $\boldsymbol{u}$ (which is an element of $T^*(\mathcal{P})$) such that the above result holds. Taking components, we immediately see that $u_j = \eta_{ij}p^i$; i.e. u_j is obtained by the standard procedure of lowering the index of p^i. Since there is a one-to-one correspondence, it makes sense to use the same symbol for both these and simply write $p_j = \eta_{ij}p^i$, etc. This is the origin of covariant components of a vector.

It follows that the covariant component of a tensor can be similarly defined by using the direct product basis of $\boldsymbol{w}^j$s; for example, we have $\boldsymbol{T} = T_{ijk}\boldsymbol{w}^i \otimes \boldsymbol{w}^j \otimes \boldsymbol{w}^k$. A mixed tensor has a similar expansion: $\boldsymbol{S} = S^k_{ij}\boldsymbol{w}^i \otimes \boldsymbol{w}^j \otimes \boldsymbol{e}_k$. In this language, Lorentz transformation corresponds to the rotation of the tetrad basis by $\boldsymbol{e}'_i = M^j_{i'}\boldsymbol{e}_j$. The vector $\boldsymbol{v}$ or any other tensorial object does not change under such rotation of coordinates because they are geometrical constructs with an intrinsic meaning; but when the basis vectors are rotated, the *components* of a vector (or a tensorial object) will change:

$$\boldsymbol{v} = v^i\boldsymbol{e}_i = v'^j\boldsymbol{e}'_j = v'^j M^i_{j'}\boldsymbol{e}_i, \tag{1.56}$$

showing that the components change as $v^i = v'^j M^i_{j'}$ which was our original definition of a four-vector. This is, of course, precisely what happens in ordinary three-dimensional vector analysis under the rotation of axes.

It is also possible to make another interesting and useful association between vectors and directional derivative operators. Consider a parameterized curve $\mathcal{C}$ $[x^i(\lambda)]$ in the spacetime with the tangent vector $u^i = dx^i/d\lambda$ near some event $\mathcal{P}$. Let $f(x)$ be a function defined in the spacetime in the neighbourhood of $\mathcal{C}$. The variation of f along $\mathcal{C}$ can be written in the form

$$\frac{df}{d\lambda} = \frac{dx^i}{d\lambda}\,\partial_i f = u^i\,\partial_i f. \tag{1.57}$$

This shows that we can build from the vector components u^i an invariant scalar operator $u^i\partial_i$ which allows us to determine how scalar functions vary along the direction of the vector. This operator is clearly Lorentz invariant, linear in the components of the vector and contains the same information as the vector $\boldsymbol{u}$. Hence, mathematically, one can identify a vector $\boldsymbol{u}$ with a directional derivative operator $u^i\partial_i$. But since we already have the result $\boldsymbol{u} = u^i\boldsymbol{e}_i$, it follows that we can identify the basis vectors $\boldsymbol{e}_i$ with the derivative operators $\partial/\partial x^i$. In such an approach, all vector relations are interpreted as equalities between operators acting on scalar functions and the basis which we have introduced is called a coordinate basis. Similarly, the basis vectors $\boldsymbol{w}^j$ of the dual vector space $T^*(\mathcal{P})$ can be identified with the coordinate differentials $\boldsymbol{w}^j = \boldsymbol{d}x^j$. The scalar product between the basis vectors now corresponds to the relation $df = (dx^i)(\partial f/\partial x^i)$. These notions will be useful in the context of general relativity.

1.7 Volume and surface integrals in four dimensions

The infinitesimal volume element in four dimensions is $dV = d^4x = cdt\,dx\,dy\,dz$, which is a direct generalization from three dimensions. Under a Lorentz transformation, we have $dx'^i = L_j^{i'} dx^j$. Since $\det|L_j^{i'}| = 1$ the Jacobian of this transformation is unity and $d^4x' = d^4x$. This will be the Lorentz invariant measure for integration over a volume in four dimensions.

In several calculations, we will need to perform integrations over a given three-dimensional surface in the four-dimensional spacetime and we shall now introduce the formal machinery needed to do this. A three-dimensional 'surface' (which is actually a volume element, in the conventional three-dimensional terminology) in a four-dimensional space can be described in parametric form by the four functions $x^i = x^i(a, b, c)$ of three parameters a, b and c. [This is equivalent to specifying the surface by an equation $f(x^i) = 0$; for example, a curve in a two-dimensional space can be specified either by an equation $f(x, y) = 0$ or in parametrized form as $x(s), y(s)$.] An infinitesimal volume element of this three-dimensional subspace is given by

$$d^3\sigma_i = \frac{1}{3!}\epsilon_{ijkl}\left[\frac{\partial(x^j, x^k, x^l)}{\partial(a, b, c)}\right] da\, db\, dc. \tag{1.58}$$

In particular, consider the spacelike hypersurface $x^0 = $ constant which represents the ordinary 3-space at a given time, with $x^1 = a, x^2 = b, x^3 = c$. In this case, the only surviving term in Eq. (1.58) will be $d^3\sigma_0 = da\, db\, dc = d^3x$. To see this, note that for each value of i there are 3! arrangements of j, k and l which are not equal to i that will keep ϵ_{ijkl} nonzero. This fact allows us to ignore the 3! in the denominator and consider just one representative sample of each permutation in studying $d^3\sigma_i$. In evaluating $d^3\sigma_\alpha$ for the spatial indices, one of the indices in the set j, k, l will take the value zero and hence the Jacobian will vanish. The only surviving term will be $d^3\sigma_0$, which will give d^3x.

For an observer moving with four-velocity u^i, the proper three-volume element is given by $d^3V = u^0 d^3x$ which is a scalar invariant. To prove this, note that the quantity $d^4V = dx\, dy\, dz\, dt$ is a scalar. Multiplying this by $1 = u^0(d\tau/dt)$ and noting that $d\tau$ is invariant, we conclude that $d^3V = u^0 d^3x$ is an invariant. In the rest frame (with $u^0 = 1$), this obviously represents the spatial volume element and being a Lorentz invariant quantity, the result holds in any other frame.

Integrals over lower dimensional surfaces (two dimensions and one dimension) can also be defined in an analogous manner in four-dimensional space. The integration along a parameterized curve $x^i(\lambda)$ uses the measure $dx^i = (dx^i/d\lambda)d\lambda$. To define a two-dimensional surface integral in four-dimensional space, we use the infinitesimal element of 2-surface $x^i = x^i(a, b)$ parameterized in terms of two

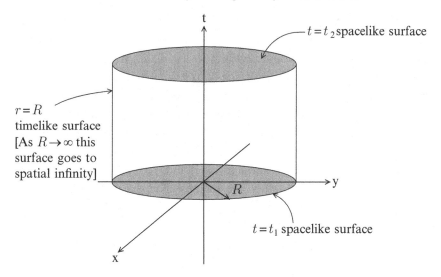

Fig. 1.2. The Gauss theorem in a spacetime volume is illustrated. The vertical axis is time and the two horizontal axes denote space coordinates with one dimension suppressed. In the most common application of the Gauss theorem, we will (a) use a four-dimensional region $\mathcal{V}$ made of two spacelike surfaces $t = (t_1, t_2)$, which are shaded in the figure, and a timelike surface at a large radii $r = R \to \infty$, and (b) assume that the contributions from the surface at spatial infinity vanishes.

parameters a and b and define

$$d^2\sigma_{ij} = \frac{1}{2!}\epsilon_{ijkl}\left[\frac{\partial(x^k, x^l)}{\partial(a,b)}\right] da\ db \tag{1.59}$$

in a manner completely analogous to the three-dimensional surface.

These results play a crucial role in the generalization of the Gauss theorem to four dimensions, which we will now discuss. Given the gradient operator and a vector, one can define the four-dimensional divergence as $(\partial A^i/\partial x^i) \equiv \partial_i A^i$ with summation over index i. (This is an obvious generalization of the ordinary three-dimensional divergence, $\nabla \cdot A = \partial_\alpha A^\alpha$.) Then the Gauss theorem in four dimensions can be expressed as

$$\int_{\mathcal{V}} d^4x\ \partial_i A^i = \int_{\partial\mathcal{V}} d^3\sigma_i\ A^i, \tag{1.60}$$

where $\mathcal{V}$ is a region of four-dimensional space bounded by a 3-surface $\partial\mathcal{V}$ and $d^3\sigma_i$ is an element of 3-surface defined earlier in Eq. (1.58). The left hand side is a four-dimensional volume integral and the right hand side is a three-dimensional surface integral. The proof uses exactly the same approach as in three-dimensional vector calculus.

While this result holds for an arbitrary bounded region of the four-dimensional spacetime, it is often used in the following context (see Fig. 1.2). Let us take the boundaries of a four-dimensional region $\mathcal{V}$ to be made of the following components: (i) two three-dimensional surfaces at $t = t_1$ and $t = t_2$ both of which are spacelike; the coordinates on these surfaces are the regular spatial coordinates (x, y, z) or (r, θ, ϕ); (ii) one timelike surface at a large spatial distance $(r = R \rightarrow \infty)$ at all time t is in the interval $t_1 < t < t_2$; the coordinates on this three-dimensional surface could be (t, θ, ϕ). In the right hand side of Eq. (1.60) the integral has to be taken over the surfaces in (i) and (ii). If the vector field A^j vanishes at large spatial distances, then the integral over the surface in (ii) vanishes for $R \rightarrow \infty$. For the integral over the surfaces in (i), the volume element can be parametrized as $d\sigma_0 = d^3x$. It follows that

$$\int_{\mathcal{V}} d^4x \, \partial_i A^i = \int_{t=t_2} d^3\boldsymbol{x} \, A^0 - \int_{t=t_1} d^3\boldsymbol{x} \, A^0 \qquad (1.61)$$

with the minus sign arising from the fact that the normal has to be always treated as outwardly directed. It follows that if $\partial_i A^i = 0$ then the integral of A^0 over all space is conserved in time. The same result, of course, can be obtained by writing the equation $\partial_i A^i = 0$ in the form $(\partial A^0/\partial(ct)) + \nabla \cdot \boldsymbol{A} = 0$ – which is the familiar form of continuity equation in 3-dimensional vector analysis – and integrating the terms over all space.

While Eq. (1.60) is usually used in the context of four-vectors (and it is usually used in three-dimensions in the context of three-vectors) the result actually has nothing to do with the transformation properties of A^i and holds for any set of four functions (A^0, A^1, A^2, A^3) when calculated in a specific coordinate system. Of course, in such a generalized context, the results will depend on the coordinate system in which it is evaluated.

One can provide similar results for the integrals over two-dimensional surfaces. These can be transformed into integrals over three-dimensional surfaces that span the two-dimensional surface by the replacement $d\sigma_{ik} \rightarrow d\sigma_i \partial_k - d\sigma_k \partial_i$. In particular, for any antisymmetric tensor A^{ik} we have the result

$$\frac{1}{2} \int A^{ik} d\sigma_{ik} = \frac{1}{2} \int \left(d\sigma_i \partial_k A^{ik} - d\sigma_k \partial_i A^{ik} \right) = \int d\sigma_i \left(\partial_k A^{ik} \right). \qquad (1.62)$$

If a vector J^i is conserved ($\partial_i J^i = 0$), we can always find an antisymmetric tensor A^{ik} such that $J^i = \partial_k A^{ik}$. Then Eq. (1.62) shows that the conserved charge can be expressed in the form

$$\int d\sigma_i J^i = \int d\sigma_i \left(\partial_k A^{ik} \right) = \frac{1}{2} \int A^{ik} d\sigma_{ik}. \qquad (1.63)$$

This relation will be useful in our later work.

Exercise 1.7

A null curve in flat spacetime Let (r, θ, ϕ) be arbitrary functions of a parameter λ. Consider the parameterized curve $x^i(\lambda)$ where

$$x = \int r \cos\theta \cos\phi \, d\lambda; \quad y = \int r \cos\theta \sin\phi \, d\lambda; \quad z = \int r \sin\theta \, d\lambda; \quad t = \int r \, d\lambda.$$

$$(1.64)$$

Show that $x^i(\lambda)$ is a null curve.

Exercise 1.8

Shadows are Lorentz invariant Show that the cross-sectional area of a parallel beam of light is invariant under a Lorentz transformation.

[Hint. Argue as follows: if k^i is the null four-vector along which the light beam is travelling, the cross-sectional area is defined by two other purely spacelike vectors a^i and b^i such that $k^i a_i = k^i b_i = 0$. Take the area to be a small square so that $a^i b_i = 0$. A different observer will have the corresponding vectors a'^i and b'^i. Argue that one must have $a'^i = a^i + \alpha k^i, b'^i = b^i + \beta k^i$. Determine α and β by the condition that the primed vectors must be orthogonal to u^i which is the four-velocity of the observer. Compute the area determined by a'^i, b'^i and show that it is the same as the one determined by a^i, b^i.]

1.8 Particle dynamics

So far we have been concerned with the kinematical aspects of relativity which arise from the nature of the transformation between the inertial frames. We shall now turn to the question of determining the laws governing the dynamics of the particle in accordance with Lorentz invariance.[3]

We shall do this using the principle of least action which should be familiar from the study of classical mechanics. Since these ideas will be used extensively in the following sections, we shall briefly recall and summarize the key results in classical mechanics before proceeding further. The starting point is an action functional defined as an integral (over time) of a Lagrangian:

$$\mathcal{A} = \int_{t_1, q_1}^{t_2, q_2} dt \, L(\dot{q}, q).$$

$$(1.65)$$

The Lagrangian depends on the function $q(t)$ and its time derivative $\dot{q}(t)$ and the action is defined for all functions $q(t)$ which satisfy the boundary conditions $q(t_1) = q_1, q(t_2) = q_2$. For each of these functions, the action $\mathcal{A}$ will be a pure number; thus the action can be thought of as a function in the space of functions and one usually says that the action is a *functional* of $q(t)$. Very often, the limits of integration on the integral will not be explicitly indicated or will be reduced to just t_1 and t_2 for notational convenience. Let us now consider the change in the action when the form of the function $q(t)$ is changed from $q(t)$ to $q(t) + \delta q(t)$. The variation gives

$$
\begin{aligned}
\delta \mathcal{A} &= \int_{t_1}^{t_2} dt \left[\frac{\partial L}{\partial q} \delta q + \frac{\partial L}{\partial \dot{q}} \delta \dot{q} \right] \\
&= \int_{t_1}^{t_2} dt \left[\frac{\partial L}{\partial q} - \frac{d}{dt} \left(\frac{\partial L}{\partial \dot{q}} \right) \right] \delta q + \int_{t_1}^{t_2} dt \frac{d}{dt} \left(\frac{\partial L}{\partial \dot{q}} \delta q \right) \\
&= \int_{t_1}^{t_2} dt \left[\frac{\partial L}{\partial q} - \frac{dp}{dt} \right] \delta q + p \delta q \Big|_{t_1}^{t_2} .
\end{aligned}
\tag{1.66}
$$

In arriving at the second equality we have used $\delta \dot{q} = (d/dt)\delta q$ and have carried out an integration by parts. In the third equality we have defined the *canonical momentum* by $p \equiv (\partial L/\partial \dot{q})$. Let us first consider variations δq which preserve the boundary conditions so that $\delta q = 0$ at $t = t_1$ and $t = t_2$. In that case, the $p\delta q$ term vanishes at the end points. If we now demand that $\delta \mathcal{A} = 0$ for arbitrary choices of δq in the range $t_1 < t < t_2$, we arrive at the equation of motion

$$
\left[\frac{\partial L}{\partial q} - \frac{dp}{dt} \right] = 0.
\tag{1.67}
$$

It is obvious that two Lagrangians L_1 and $L_1 + (df(q,t)/dt)$, where $f(q,t)$ is an arbitrary function, will lead to the same equations of motion.

The Hamiltonian for the system is defined by $H \equiv p\dot{q} - L$ with the understanding that H is treated as a function of p and q (rather than a function of $\dot{q}$ and q). By differentiating H with respect to time and using Eq. (1.67) we see that $(dH/dt) = 0$.

We will also introduce another type of variation which is useful in several contexts and allows us to determine the canonical momentum in terms of the action itself. To do this, we shall treat the action as a function of the upper limits of integration (which we denote simply as q and t rather than as q_2, t_2) but evaluated for a particular solution $q_c(t)$ which satisfies the equation of motion in Eq. (1.67). This makes the action a function of upper limits of integration; i.e. $\mathcal{A}(q,t) = \mathcal{A}[q, t; q_c(t)]$. We can then consider the variation in the action when the value of q at the upper limit of integration is changed by δq. In this case, the first term in the third line of Eq. (1.66) vanishes and we get $\delta \mathcal{A} = p\delta q$ so that

$$
p = \frac{\partial \mathcal{A}}{\partial q}.
\tag{1.68}
$$

This description forms the basis for the Hamilton–Jacobi equation in classical mechanics. From the relations

$$
\frac{d\mathcal{A}}{dt} = L = \frac{\partial \mathcal{A}}{\partial t} + \frac{\partial \mathcal{A}}{\partial q} \dot{q} = \frac{\partial \mathcal{A}}{\partial t} + p\dot{q}
\tag{1.69}
$$

we find that

$$\frac{\partial \mathcal{A}}{\partial t} + p\dot{q} - L = \frac{\partial \mathcal{A}}{\partial t} + H = 0. \tag{1.70}$$

In this equation we can express $H(p, q)$ in terms of the action by substituting for p by $\partial \mathcal{A}/\partial q$ thereby obtaining a partial differential equation for $\mathcal{A}(q, t)$ called the *Hamilton–Jacobi* equation:

$$\frac{\partial \mathcal{A}}{\partial t} + H\left(\frac{\partial \mathcal{A}}{\partial q}, q\right) = 0. \tag{1.71}$$

This equation has the same physical content as the equations of motion for the system. Integrating this equation will lead to the function $\mathcal{A}(q, t; k)$, where k is an integration constant. It is known from the theory of canonical transformations in classical mechanics that equating $(\partial \mathcal{A}/\partial k)$ to another constant will lead to an equation determining the trajectory of the particle. Very often this approach provides the quickest route to obtaining and solving the equations of motion.

After this preamble, we shall now return to the question of determining the dynamics of a free particle in special relativity. To determine the laws governing the motion of a *free* particle, we need an expression for the action which can be varied. This action should be constructed from the trajectory $x^i(\tau)$ of the particle and should be invariant under Lorentz transformations. The only possibility is some quantity proportional to the integral of $d\tau$; so the action must be

$$\mathcal{A} = -\alpha \int_a^b d\tau = -\int_{t_1}^{t_2} \alpha\sqrt{1 - \frac{v^2}{c^2}}\,dt, \tag{1.72}$$

where α is a constant. In arriving at the second equality, we have expressed $d\tau$ in terms of dt using Eq. (1.5), which shows that the Lagrangian is given by $L \equiv d\mathcal{A}/dt = -\alpha\sqrt{1 - v^2/c^2}$. When $c \to \infty$, this Lagrangian reduces to $L = \alpha v^2/2c^2 +$ constant. Comparing this with the Lagrangian $(1/2)mv^2$ for a free particle in non-relativistic mechanics, we find that $\alpha = mc^2$ where m is the mass of the particle. Substituting back in Eq. (1.72), the action for a relativistic particle becomes

$$\mathcal{A} = -mc^2 \int d\tau = -\int_{t_1}^{t_2} mc^2\sqrt{1 - \frac{v^2}{c^2}}\,dt, \tag{1.73}$$

where the second equation identifies the Lagrangian to be $L = -mc^2\sqrt{1 - v^2/c^2}$. This action in relativistic mechanics corresponds to the arc length of the curve connecting the two points and thus has a clear geometric meaning, unlike its non-relativistic counterpart.

It is worthwhile comparing this derivation with the corresponding one for a free particle in Newtonian mechanics, assuming that the laws should be covariant under

Galilean transformation. For a free particle, homogeneity and isotropy of the space and time translational invariance forces the Lagrangian $L(\boldsymbol{x}, \boldsymbol{v}, t) = f(\boldsymbol{v}^2)$ to be a function of just v^2. Up to this point the argument holds for both relativistic and Newtonian mechanics. In the latter case, we demand that the action should be invariant under Galilean transformations in Eq. (1.18), which leaves t invariant but changes $\boldsymbol{v}$ to $(\boldsymbol{v} + \boldsymbol{V})$ and hence v^2 to $v^2 + V^2 + 2\boldsymbol{v} \cdot \boldsymbol{V}$. If the Lagrangian is linear in v^2, such a transformation merely adds a total time derivative (of a function of t and $\boldsymbol{x}$) to the Lagrangian and hence will leave the equations of motion invariant. If the Lagrangian is not linear in v^2, it is easy to see that the invariance under the Galilean transformation cannot be maintained. Hence we must have $L \propto v^2$; the coefficient of proportionality is conventionally taken as $(m/2)$, where m is called the mass of the particle. A comparison of this argument with the one that led to Eq. (1.73) clearly shows how the symmetry considerations determine the form of action (and thus the dynamics) in both Newtonian and relativistic mechanics. In relativistic mechanics, the Lagrangian is *invariant* under the Lorentz transformation while in Newtonian mechanics the Lagrangian picks up a total time derivative under the Galilean transformation but leaves the equations of motion invariant.

To determine the equations of motion, we vary the action in Eq. (1.73) with respect to the trajectory $x^i(\tau)$ and get

$$\delta \mathcal{A} = -mc^2 \int_a^b \delta(d\tau) = -mc \int_a^b \delta(\sqrt{-dx_i dx^i}) = m \int_a^b \frac{dx_i \delta dx^i}{d\tau}. \quad (1.74)$$

Using $\delta dx^i = d\delta x^i$, writing $(dx_i/d\tau)$ as u_i and doing an integration by parts we get

$$\delta \mathcal{A} = m u_i \delta x^i \big|_a^b - m \int_a^b \delta x^i \frac{du_i}{d\tau} d\tau. \quad (1.75)$$

If we now assume that δx^i vanishes at the end points, we obtain the equations of motion $du^i/d\tau = 0$, which is a generalization of the force-free equation of motion to relativistic mechanics. Further, if we treat the action as a function of the end points of a trajectory which satisfies the equation of motion, then the second term in Eq. (1.75) vanishes and we get $\delta \mathcal{A} = m u_i \delta x^i$ so that $(\partial \mathcal{A}/\partial x^i) = m u_i$. Since the derivative of the action with respect to the end point coordinate defines the momentum, the *four*-momentum vector is given by

$$p_i = \frac{\partial \mathcal{A}}{\partial x^i} = m u_i = (-\gamma mc, \gamma m\boldsymbol{v}) \equiv (-E/c, \boldsymbol{p}), \quad (1.76)$$

and the corresponding contravariant components are:

$$p^i = m u^i = (\gamma mc, \gamma m\boldsymbol{v}) = (E/c, \boldsymbol{p}). \quad (1.77)$$

To obtain the physical significance of the 'time-component', $E = \gamma mc^2$, we note that, in the non-relativistic limit, this expression reduces to $E \approx mc^2 + mv^2/2$.

This suggests that E corresponds to the relativistic energy of the particle. Such an identification is further justified by the fact that, for a Lagrangian $L = -mc^2/\gamma$ in Eq. (1.73), we get $(\partial L/\partial v) = p = \gamma m v$ and

$$H = p \cdot v - L = \gamma m v^2 + \gamma^{-1} mc^2 = mc^2 \gamma, \tag{1.78}$$

which should be numerically the same as E. We thus conclude that the three-momentum $p = \gamma m v$ and energy E (divided by c) form the components of a four-vector. The relation in Eq. (1.76) now reads

$$E = -\frac{\partial \mathcal{A}}{\partial t}; \qquad p = \nabla \mathcal{A}, \tag{1.79}$$

which – remarkably – has the *same form* as the ones used in the Hamilton–Jacobi theory of Newtonian physics but now reveals their four-dimensional basis.

The four-momentum of a particle is proportional to its four-velocity and hence many of the results we obtained for four-velocity in Section 1.4 lead to similar results for four-momentum. Since $u^i u_i = -c^2$, it follows that $p^i p_i = -m^2 c^2$, giving the following relations connecting momentum, energy and velocity:

$$E = \sqrt{p^2 c^2 + m^2 c^4}; \qquad p = E\left(\frac{v}{c^2}\right). \tag{1.80}$$

In particular, the first relation allows for the existence of massless particles like photons with $m = 0$, $E = pc$ and travelling with the speed of light $v = (pc^2/E) = c$.

One is often interested in expressing the energy and other variables of a particle as measured by different observers in a covariant manner. Consider, for example, a particle of mass m and four-momentum p^i observed by someone moving with a four-velocity u^i. By working out the components in the rest frame of the observer, it is easy to verify the following relations. (i) The energy measured by the observer will be $E = -p^i u_i$. (ii) The magnitude of the three-momentum measured by the observer will be $|p| = [(p^i u_i)^2 + p^i p_i]^{1/2}$. (iii) The three-velocity will have a magnitude $|v| = [1 + (p^i p_i/(p^i u_i)^2)]^{1/2} c$. (iv) Using these results, one can construct a four-vector v^i such that in the observer's rest frame $v^0 = 0$ and the spatial components agree with the ordinary three-velocity of the particle; that is, $v^\alpha = dx^\alpha/dt$. This four-vector is given by $v^i = -u^i - c^2 p^i (p^j u_j)^{-1}$.

Finally, the Hamilton–Jacobi equation for the relativistic particle can be obtained from the definition $p_i = (\partial \mathcal{A}/\partial x_i)$ and the condition $p^i p_i = -m^2 c^2$; we get

$$\left(\frac{\partial \mathcal{A}}{\partial x^i}\right)\left(\frac{\partial \mathcal{A}}{\partial x_i}\right) = -\frac{1}{c^2}\left(\frac{\partial \mathcal{A}}{\partial t}\right)^2 + \left(\frac{\partial \mathcal{A}}{\partial x}\right)^2 + \left(\frac{\partial \mathcal{A}}{\partial y}\right)^2 + \left(\frac{\partial \mathcal{A}}{\partial z}\right)^2 = -m^2 c^2. \tag{1.81}$$

One can verify that this reduces to the correct non-relativistic Hamilton–Jacobi equation in the appropriate limit. Since a relativistic particle has an extra term to the energy $E_0 = mc^2$, the relativistic and non-relativistic actions will differ by a term $-E_0 t = -mc^2 t$. So, to obtain the non-relativistic limit of this equation, we substitute $\mathcal{A} = -mc^2 t + S(x^i)$ into Eq. (1.81). Simplification then gives

$$\frac{1}{2m} (\nabla S)^2 + \frac{\partial S}{\partial t} = \frac{1}{2mc^2} \left(\frac{\partial S}{\partial t} \right)^2 \cong 0, \tag{1.82}$$

where the last equality arises in the limit of $c \to \infty$. This is exactly the Hamilton–Jacobi equation for the free particle in the non-relativistic theory.

Exercise 1.9

Hamiltonian form of action – Newtonian mechanics An alternative action principle in non-relativistic mechanics uses the action expressed in the form

$$\mathcal{A} = \int_{t_1}^{t_2} dt \, [p\dot{q} - H(p, q)], \tag{1.83}$$

in which the functions $p(t)$ and $q(t)$ are considered as independent and $H(p, q)$ is a given Hamiltonian. Vary $p(t)$ and $q(t)$ independently in this action and show that the demand $\delta\mathcal{A} = 0$ will lead to the following equations of motion

$$\dot{p} = -\frac{\partial H}{\partial q} ; \qquad \dot{q} = \frac{\partial H}{\partial p}, \tag{1.84}$$

provided $\delta q = 0$ at the end points but δp is arbitrary. Convince yourself that this equation is equivalent to the standard equations of motion in classical mechanics. (Note that the action principle itself has the ability to tell us which quantities are to be kept fixed at the end points for leading to sensible equations of motion.)

Exercise 1.10

Hamiltonian form of action – special relativity In the case of a special relativistic particle, show that the corresponding Hamiltonian form of the action (in units with $c = 1$) is given by

$$\mathcal{A} = \int_{\lambda_1}^{\lambda_2} d\lambda \, \left[p_a \dot{x}^a - \frac{1}{2} C \left(\frac{H}{m} + m \right) \right], \tag{1.85}$$

where $H = \eta_{ab} p^a p^b$ and C is an auxiliary variable. The parameter λ is treated as arbitrary in the action. Show that varying x^a, p^a, C independently leads to the correct equations of motion for a free particle if we make the choice $C = 1$ in the end. Explain why this is allowed.

Exercise 1.11

Hitting a mirror A mirror moves in a direction perpendicular to its plane with a three-velocity v. A ray of light of frequency ν_1 is incident on the mirror at an angle of incidence θ and is reflected at an angle of reflection ϕ with frequency ν_2. Show that

$$\frac{\tan(\theta/2)}{\tan(\phi/2)} = \frac{c+v}{c-v}; \qquad \frac{\nu_2}{\nu_1} = \frac{(c + v\cos\theta)}{(c - v\cos\phi)}. \tag{1.86}$$

What happens if the mirror was moving in a direction parallel to its plane?

Exercise 1.12

Photon–electron scattering (a) Use four-vector techniques to show that when a photon of wavelength λ scatters off a stationary electron of mass m_e, its wavelength will change to λ' such that $\lambda' - \lambda = (h/m_e c)(1 - \cos\theta)$ where θ is the scattering angle.

(b) A related process, called inverse Compton scattering, occurs when a charged particle of mass m and energy E (in the lab frame) collides head-on with a photon of frequency ν. Show that when $E \gg mc^2$, the maximum energy that is transfered to the photon is given by $E[1 + (m^2 c^4/4h\nu E)]^{-1}$.

Exercise 1.13

More practice with collisions Prove the following results.

(a) The threshold of energy for the production of an $e^+ e^-$ pair in a collision between a photon and an electron at rest is $4m_e c^2$.

(b) A high energy electron strikes an electron at rest in an elastic encounter and the two electrons share the energy equally. Then the angle between their directions of travel will be $\pi/2$ in non-relativistic scattering but will be less than $\pi/2$ in relativistic mechanics.

(c) If a particle of mass M hits a stationary target of mass m, the γ factor of the incident particle after the collision cannot exceed $(m^2 + M^2)/2mM$. Compare this with the corresponding situation in the non-relativistic situation.

Exercise 1.14

Relativistic rocket A relativistic rocket has a variable rest mass $m(\tau)$ and obeys the equation of motion $d(mu^i)/d\tau = J^i$ where J^i is the rate of emission of four-momentum through the burning of the fuel.

(a) Show that this requires the condition

$$m_{\text{final}} < m_{\text{initial}} \exp\left(\int g(\tau)\, d\tau\right), \tag{1.87}$$

where g is the magnitude of the acceleration.

(b) Consider a motion in (1+1) dimension with $g(\tau) = d\chi/d\tau$ where χ is the rapidity. If the rocket starts from rest and reaches a final velocity v_{final} show that

$$m_{\text{final}} < m_{\text{initial}} \sqrt{\frac{1 - v_{\text{final}}}{1 + v_{\text{final}}}}. \tag{1.88}$$

1.9 The distribution function and its moments

So far, we have discussed the dynamics of a single, free particle. Often in physics, one has to deal with a large collection of particles undergoing nearly identical physical processes. In non-relativistic mechanics, we deal with this situation using a distribution function. It is necessary to generalize this concept in a Lorentz invariant manner to take into account a system of relativistic particles.

In order to do that, we shall first obtain several Lorentz invariant quantities which will serve as basic building blocks. Let us consider a set of N particles, each of mass m, described by a distribution function $f(p^i)$ at any given location in space. The total number of particles can be written in terms of the distribution function as

$$N = \int d^4p \, \theta(p^0) \delta_D \left(p^a p_a + m^2 c^2 \right) f(p^i), \tag{1.89}$$

where $d^4p = dp^0 d^3\boldsymbol{p}$; the Dirac delta function $\delta_D(p^a p_a + m^2 c^2)$ ensures that all the particles have mass m and the theta function $\theta(p^0)$ (which is unity for $p^0 > 0$ and vanishes for $p^0 < 0$) ensures that $p^0 > 0$ so that the energy is positive. The quantities N, d^4p, θ and $\delta_D(p^a p_a + m^2 c^2)$ are all individually Lorentz invariant, implying f is Lorentz invariant. (It is obvious from their definitions that $N, d^4p, \theta(p^0)$ are Lorentz invariant. To prove that the Dirac delta function is invariant we only need to use the fact that Lorentz transformation has unit Jacobian.) Introducing the energy $E_{\boldsymbol{p}} \equiv (m^2 c^4 + p^2 c^2)^{1/2}$ corresponding to momentum $\boldsymbol{p}$, we write the Dirac delta function as

$$\delta_D \left(p_i p^i + m^2 c^2 \right) \equiv \delta_D \left(p_0^2 - \frac{E_{\boldsymbol{p}}^2}{c^2} \right) = \frac{c}{2E_{\boldsymbol{p}}} \left[\delta_D \left(p^0 - \frac{E_{\boldsymbol{p}}}{c} \right) + \delta_D \left(p^0 + \frac{E_{\boldsymbol{p}}}{c} \right) \right].$$
$$\tag{1.90}$$

Noting that integration over dp^0 in Eq. (1.89) will merely replace p^0 by $(E_{\boldsymbol{p}}/c)$ due to the condition $p^0 > 0$, we get

$$N = \int d^3p \, dp^0 \theta(p^0) \frac{c}{2E_{\boldsymbol{p}}} \left[\delta_D \left(p^0 - \frac{E_{\boldsymbol{p}}}{c} \right) + \delta_D \left(p^0 + \frac{E_{\boldsymbol{p}}}{c} \right) \right] f \left(p^0, \boldsymbol{p} \right)$$
$$= \frac{c}{2} \int \frac{d^3\boldsymbol{p}}{E_{\boldsymbol{p}}} f(p^0 = E_{\boldsymbol{p}}/c, \boldsymbol{p}). \tag{1.91}$$

Since N and f are invariant, the combination $(d^3p/E_{\boldsymbol{p}})$ must be invariant under Lorentz transformations.

We noted earlier (see page 26) that $u^0 d^3\boldsymbol{x} = d^4x/d\tau$ is Lorentz invariant. Since $E = mcu^0$, it follows that the combination $E_{\boldsymbol{p}} d^3\boldsymbol{x}$ is also an invariant. Combined with the result that $d^3\boldsymbol{p}/E_{\boldsymbol{p}}$ is Lorentz invariant, we conclude that the product $(E_{\boldsymbol{p}} d^3\boldsymbol{x})(d^3\boldsymbol{p}/E_{\boldsymbol{p}}) = d^3\boldsymbol{x} d^3\boldsymbol{p}$ is Lorentz invariant. In other words, an element of phase volume is Lorentz invariant even though neither the spatial volume nor the volume in momentum space is individually invariant.

This result allows us to introduce distribution functions in relativistic theory in exact analogy with non-relativistic mechanics. We define the distribution function f such that

$$dN = f(x^i, \boldsymbol{p}) d^3\boldsymbol{x} d^3\boldsymbol{p} \tag{1.92}$$

represents the number of particles in a small phase volume $d^3x d^3p$. The x^i here has the components $(ct, \boldsymbol{x})$ while $\boldsymbol{p}$ is the *three*-momentum vector; the fourth component of the momentum vector $(E_{\boldsymbol{p}}/c)$ does not appear since it is completely determined by $\boldsymbol{p}$ and mass m of the particle. Each of the quantities dN, f and $d^3x d^3p$ are individually Lorentz invariant.

Given the Lorentz invariant distribution function f, one can construct several other invariant quantities by taking moments of this function. Of particular importance are the moments constructed by integrating the distribution function over various powers of the four-momentum. We shall now construct a few such examples.

The simplest Lorentz invariant quantity which can be obtained from the distribution function by integrating out the momentum, is the harmonic mean $\bar{E}_{\text{har}}$ of the energy of the particles at an event x^i. This is defined by the relation

$$\frac{1}{\bar{E}_{\text{har}}(x^i)} \equiv \int \frac{d^3\boldsymbol{p}}{E_{\boldsymbol{p}}} f(x^i, \boldsymbol{p}), \tag{1.93}$$

which is clearly Lorentz invariant because of our earlier results. Unfortunately, this quantity does not seem to play any important role in physics.

Taking the first power of the four-momentum, we can define the four-vector

$$S^a(x^i) \equiv c \int \frac{d^3\boldsymbol{p}}{E_{\boldsymbol{p}}} p^a f(x^i, \boldsymbol{p}). \tag{1.94}$$

The components of this vector are $(S^0, \boldsymbol{S})$ where

$$S^0(x^i) = \int d^3\boldsymbol{p} f(x^i, \boldsymbol{p}) \equiv n(x^i);$$

$$\boldsymbol{S}(x^i) = \frac{1}{c} \int d^3\boldsymbol{p} f(x^i, \boldsymbol{p}) \boldsymbol{v} \equiv c^{-1} n(x^i) \langle \boldsymbol{v} \rangle, \tag{1.95}$$

where we have used the relation $(p^\alpha/E) = (v^\alpha/c^2)$. The time component of this vector, S^0, gives the particle number density n in a given frame; the spatial components give the flux of the particles in each direction. The factor c was introduced in the definition Eq. (1.94) to facilitate such an interpretation.

Taking quadratic moments allows us to define the quantity

$$T^{ab}(x^i) \equiv c^2 \int \frac{d^3\boldsymbol{p}}{E_{\boldsymbol{p}}} p^a p^b f(x^i, \boldsymbol{p}), \tag{1.96}$$

called the *energy-momentum tensor* of the system. This tensor is clearly symmetric. When one of the indices is zero, we get,

$$T^{b0}(x^i) = T^{0b}(x^i) = c \int \frac{d^3\boldsymbol{p}}{E_{\boldsymbol{p}}} (E_{\boldsymbol{p}} p^b) f(x^i, \boldsymbol{p}) = c \int d^3\boldsymbol{p} \, p^b f(x^i, \boldsymbol{p}), \tag{1.97}$$

which is (c times) the sum of the four-momentum of all the particles per unit volume. The time–time component, $T^{00}(x^i)$, gives the energy density and the time–space component, $T^{0\alpha}(x^i)$, gives the density of the α-component of the three-momentum. The total four-momentum of the system is defined as the integral over all space:

$$P^i = \int d^3x \, T^{0i}. \tag{1.98}$$

The space–space components of the energy-momentum tensor represent the stresses within the medium. The component $T^{\alpha\beta}$ is

$$T^{\alpha\beta}(x^i) \equiv c^2 \int \frac{d^3p}{E_p} p^\alpha p^\beta f(x^i, p) = \int d^3p \, v^\alpha p^\beta f(x^i, p) = \int d^3p \, v^\beta p^\alpha f(x^i, p).$$

$$\tag{1.99}$$

Since f denotes the phase space density of particles, $p^\alpha f$ represents the density of the α-component of the momentum and $v^\beta p^\alpha f$ denotes the flux of this momentum. Equation (1.99) gives the α-component of the momentum that crosses a unit area orthogonal to the β direction per unit time. Therefore, $T^{\alpha\beta}$ represents the α-component of the net force acting across a unit area of a surface, the normal to which is in the direction denoted by β. The symmetry of $T^{\alpha\beta}$ implies that this is also equal to the β-component of the net force acting across a unit area of a surface the normal to which is in the direction denoted by α.

The symmetry of the energy-momentum tensor is necessary – in general – for the angular momentum of the system to be conserved. In three dimensions, angular momentum is usually defined through the cross product $(\boldsymbol{x} \times \boldsymbol{p})$. But as we saw in Section 1.5 the cross product of two vectors is a special construction which works only in three dimensions. It is therefore better to think of the components of the angular momentum J^μ in three dimensions as the dual (see Eq. (1.51)) of the tensor product $J_{\alpha\beta} \equiv (x_\alpha p_\beta - x_\beta p_\alpha)$ defined by:

$$J^\mu = \frac{1}{2} \epsilon^{\mu\alpha\beta}(x_\alpha p_\beta - x_\beta p_\alpha) = \frac{1}{2} \epsilon^{\mu\alpha\beta} J_{\alpha\beta} = (\boldsymbol{x} \times \boldsymbol{p})^\mu. \tag{1.100}$$

In four dimensions, the tensor product generalizes to an antisymmetric tensor $J^{ik} = x^i p^k - x^k p^i$. (But, of course, we cannot take its dual to get another vector which only works in three dimensions.) When we proceed from a single particle to a continuous medium, we need to work with an integral over $dp^a = d^3x \, T^{0a}$ etc. So the angular momentum tensor is now defined as:

$$J^{ik} \equiv \int d^3\sigma_l \, (x^i T^{kl} - x^k T^{il}) = \int d^3x \, (x^i T^{k0} - x^k T^{i0}) \equiv \int d\sigma_l \, M^{ikl}. \tag{1.101}$$

The second equality shows that J^{ik} is indeed the moment of the momentum density integrated over all space and hence represents the total angular momentum.

The conservation of this quantity requires $\partial_l M^{ikl} = 0$. A simple computation now shows that this requires $T^{ab} = T^{ba}$ and – in particular – we need $T^{\alpha\beta} = T^{\beta\alpha}$. This symmetry ensures that the angular momentum of an isolated system is conserved and the internal stresses cannot spontaneously rotate a body.

The angular momentum tensor J^{ik} is clearly antisymmetric and hence has six independent components. Its spatial components have clear meaning as the angular momentum of the system since they essentially generalize the expression $\boldsymbol{x} \times \boldsymbol{p}$. The other three components

$$J^{0\alpha} = tP^{\alpha} - \int d^3x \, x^{\alpha} T^{00}, \tag{1.102}$$

where P^{α} is the total three-momentum of the system, however, do not play an important role. They give the location of the centre of mass at $t = 0$. It is possible to choose the coordinate system such that at $t = 0$ the integral in the above expression vanishes.

While the angular momentum tensor is Lorentz covariant, it changes under the translation of coordinates $x^i \to x'^i = x^i + \epsilon^i$. It is easy to see that

$$J^{ik} \to J'^{ik} = J^{ik} + \epsilon^i P^k - \epsilon^k P^i. \tag{1.103}$$

This result arises because J^{ik} includes the orbital angular momentum of the system as well as any intrinsic angular momentum and the former depends on the choice of origin of coordinates. It is, however, straightforward to obtain the intrinsic angular momentum of the system by defining a spin four-vector as

$$\Sigma_a \equiv \frac{1}{2}\epsilon_{abcd} J^{bc} \left(\frac{P^d}{(-P_j P^j)^{1/2}} \right) \equiv \frac{1}{2}\epsilon_{abcd} J^{bc} U^d. \tag{1.104}$$

This quantity is expressed in terms of the (dimensionless) four-velocity U^i of the system which, in turn, is defined in terms of the total four-momentum. Under the translation of the coordinates, when J^{bc} changes as in Eq. (1.103), Σ_k does not change because of the antisymmetry of the ϵ-tensor. In the centre of mass frame of the system in which $U^i = (1, \boldsymbol{0})$, each spatial component of the spin vector Σ_α are related to the spatial components of the angular momentum tensor by $\Sigma_\alpha = (1/2)\epsilon_{\alpha\beta\gamma} J^{\beta\gamma}$; the time component vanishes, $\Sigma_0 = 0$. In any frame, the definition in Eq. (1.104) ensures that $U^i \Sigma_i = 0$ so that the spin vector has only three independent components.

Given a distribution function, we can construct the current four-vector $S^a(x^i)$ at any given event, through Eq. (1.94). It is also always possible to choose a Lorentz frame such that the spatial components of this vector vanish at that event (i.e. $\langle \boldsymbol{v} \rangle = 0$) so that an observer at rest in that Lorentz frame does not see any mean flux of particles around a given event. If the gradient of the mean velocity $\langle \boldsymbol{v} \rangle$ is sufficiently small, then such a Lorentz frame can be defined even globally for the whole system. (Such a definition is approximate; it is valid and useful when physical processes which depend on the gradients of mean velocity, mean kinetic energy, etc., are ignored; also see Project 1.1.) Let us suppose that we are working in such a Lorentz frame and also that the distribution function is isotropic in momentum in this frame; that is, it depends only on the magnitude, p, of the momentum $\boldsymbol{p}$. In such a frame,

$$S^0 = \int d^3\boldsymbol{p}\, f(x^i, \boldsymbol{p}) = 4\pi \int_0^\infty p^2 dp\, f(x^i, p); \qquad S^\alpha = 0, \qquad (1.105)$$

and

$$T^{00} = \int d^3\boldsymbol{p}\, E_{\boldsymbol{p}} f(x^i, \boldsymbol{p}) = 4\pi \int_0^\infty p^2 E(p)\, f(x^i, p) dp; \qquad T^{0\alpha} = 0. \quad (1.106)$$

As regards the space–space part of the energy-momentum tensor, it has to be an isotropic, symmetric, three-dimensional tensor. Hence, T_β^α must have the form $T_\beta^\alpha = P(x^i)\delta_\beta^\alpha$, since δ_β^α is the only tensor available satisfying these conditions. (The symbol P should not be confused with the total four-momentum P^i used earlier.) To find an expression for $P(x^i)$, note that

$$T_\alpha^\alpha = P(x^i)\delta_\alpha^\alpha = 3P(x^i) = c^2 \int \frac{d^3\boldsymbol{p}}{E_{\boldsymbol{p}}} p^2 f(x^i, \boldsymbol{p}) = 4\pi c^2 \int_0^\infty dp\, \frac{p^4}{E(p)} f(x^i, p).$$
$$(1.107)$$

Hence,

$$P(x^i) = \frac{4\pi c^2}{3} \int_0^\infty dp\, \frac{p^4}{E(p)} f(x^i, p). \qquad (1.108)$$

This quantity represents the pressure of the fluid and has simple limits in two extreme cases. In the non-relativistic limit, the energy of the particle is $E(p) \cong mc^2 + (p^2/2m)$. Substituting in the expression for T^{00}, we find that the energy density can be written $T^{00} \cong mc^2 n + \epsilon_{\mathrm{nr}}$ where the non-relativistic contribution ϵ_{nr} to the kinetic energy is

$$\epsilon_{\mathrm{nr}} \equiv 4\pi \int_0^\infty p^2 \frac{p^2}{2m} f(p) dp = \frac{2\pi}{m} \int_0^\infty p^4 f(p) dp. \qquad (1.109)$$

In the same limit, the expression Eq. (1.108) for pressure reduces to

$$P_{\mathrm{nr}} \cong \frac{4\pi c^2}{3} \int_0^\infty dp\, \frac{p^4}{mc^2} f(p) = \frac{4\pi}{3m} \int_0^\infty p^4 f(p) dp. \qquad (1.110)$$

Comparing the two expressions, Eq. (1.109) and Eq. (1.110), we see that $P_{\mathrm{nr}} = (2/3)\epsilon_{\mathrm{nr}}$ which is the relation between energy density and pressure in non-relativistic theory. (Note that pressure has nothing to do, *a priori*, with inter-particle collisions but is defined in terms of the momentum transfer across a surface.) In the other extreme limit of highly relativistic particles we have

$$E(p) \cong pc. \tag{1.111}$$

Then

$$\rho \equiv T_{\text{rel}}^{00} = 4\pi c \int_0^\infty p^3 f(x^i, p) dp; \qquad P = \frac{4\pi c}{3} \int_0^\infty p^3 f(x^i, p) dp, \tag{1.112}$$

which shows that, for extreme relativistic particles, the pressure and energy density are related by

$$P = \frac{1}{3}\rho. \tag{1.113}$$

In particular, this equation is exact for particles with zero mass (e.g. a gas of photons) for which $E(p) = pc$ is an exact relation.

Given the components of the energy-momentum tensor in the special frame in which bulk flow vanishes, it is easy to obtain the results in any other frame in which the observer has a four-velocity u^a. The result, obtained by a Lorentz transformation (with $c = 1$ for simplicity), is

$$T_b^a = (P + \rho)u^a u_b + P\delta_b^a; \qquad S^a = n_{\text{prop}} u^a. \tag{1.114}$$

Here n_{prop} is the proper number density – i.e. the number density in the frame comoving with the particles – and is a scalar; it is related to n in Eq. (1.95) by $n = \gamma n_{\text{prop}}$. This energy-momentum tensor is usually called the energy-momentum tensor of an *ideal* fluid. The trace of this energy-momentum tensor $T \equiv T_a^a = 3P - \rho$ and vanishes for a fluid of ultra-relativistic particles or radiation with the equation of state $P = (1/3)\rho$.

This energy-momentum tensor in Eq. (1.114) can be expressed in a different form which brings out its physical meaning more clearly. We can write

$$T_b^a = \rho u^a u_b + P(\delta_b^a + u^a u_b) = \rho u^a u_b + P\mathcal{P}_b^a, \tag{1.115}$$

where the symmetric tensor $\mathcal{P}_b^a = \delta_b^a + u^a u_b$ is called the *projection tensor*. When any other other vector v^a is contracted on one of the indices of this tensor, the resultant vector $\mathcal{P}_j^a v^j$ will be the part of v^a which is orthogonal to u^i. Mathematically, for any four-vector v^j, we have

$$v_\perp^a \equiv \mathcal{P}_j^a v^j = v^a + u^a(v^j u_j). \tag{1.116}$$

Since $v^j u_j$ is the component of vector v^a along the vector u^a (note that the latter has a norm $u^i u_i = -1$), this expression is clearly the part of the vector v^a which is orthogonal to u^a and we do get $v_\perp^a u_a = 0$ from the above equation as expected. The projection tensor $\mathcal{P}_j^a$ itself is orthogonal to the four-velocity u^i in the sense that $P^a_j u^j = 0$. Therefore, in the instantaneous rest frame of the particle in which $u^i = (1, 0)$, the tensor P^a_j has only (nonzero) *spatial* components. In this frame Eq. (1.115) shows a clear separation of the two contributions to the energy-momentum tensor: the time–time component arises from the first term and is equal to ρ. The second term involving the projection tensor has only spatial contribution and along each of the three axes it contributes a pressure P.

In the absence of collisions or external forces, the distribution function $f(x^a, \boldsymbol{p})$ satisfies the equation $(df/d\tau) = 0$ (called the Vlasov equation) which can be written in four-dimensional notation as

$$\frac{df}{d\tau} = \frac{dx^i}{d\tau} \partial_i f = u^i \partial_i f = -\frac{E}{m} \left[\frac{\partial f}{\partial t} - \frac{\boldsymbol{p}}{E} \cdot \nabla f \right]$$

$$= -\frac{E}{m} \left[\frac{\partial f}{\partial t} - \boldsymbol{v} \cdot \nabla f \right] = 0, \tag{1.117}$$

where we have used $\boldsymbol{v} = (\boldsymbol{p}/E)$. Since the proper time derivative along a streamline of a fluid is $(d/d\tau) = (u^i \partial_i)$, this shows that f is conserved along the streamlines.

It is also easy to show that the current vector S^a as well as the energy-momentum tensor T^{ab} are conserved; that is, $\partial_a S^a = 0, \partial_a T^{ab} = 0$. More generally, these equations will lead to the standard equations governing the dynamics of the fluid. To see this, we substitute the explicit form of T^{ab} in Eq. (1.115) into $\partial_a T^{ab} = 0$ and simplify the terms to obtain

$$u^m u^n \partial_m (\rho + P) + (\rho + P) \left[u^n (\partial_m u^m) + u^m (\partial_m u^n) \right] = -\eta^{mn} \partial_m P. \tag{1.118}$$

On the other hand, differentiating the relation $u^j u_j = -1$ we get $u_n \partial_m u^n = 0$. (This condition is equivalent to $a^j u_j = 0$.) This suggests projecting Eq. (1.118) along u_n and perpendicular to it. Taking the dot product of Eq. (1.118) with u_n and collecting terms, we get

$$\partial_m (\rho u^m) + P \partial_m u^m = 0. \tag{1.119}$$

This is the relativistic generalization of the continuity equation in fluid mechanics. Using this in Eq. (1.118) we get

$$(\rho + P) u^m \partial_m u^n = (\eta^{mn} + u^m u^n) \partial_m P = P^{mn} \partial_m P. \tag{1.120}$$

This is the relativistic Euler equation giving the acceleration of the fluid element in terms of the pressure gradient along the spatial directions. The occurrence of the projection tensor makes this clear. In normal units, ρ has the same dimensions as P/c^2 and the combination $(\rho + P/c^2)$ becomes just ρ in the $c \to \infty$ limit. In this case, the equations reduce to $\partial_m (\rho u^m) \approx 0$ and $\rho u^m \partial_m u^n \approx P^{mn} \partial_m P$, which can be easily shown to be equivalent to the standard continuity equation and Euler equation of non-relativistic fluid mechanics.

In the study of radiative processes, one often has to deal with a photon gas using our formalism. Considering its practical utility, we shall briefly describe this special case. If the number of photons in a phase space volume $d^3x d^3p$ is dN, then we have

$$
\begin{aligned}
dN &= f(x^i, \boldsymbol{p})\, d^3x d^3\boldsymbol{p} = f[x^i, (h\nu/c)\hat{\boldsymbol{k}}]\, d^3x d^3\boldsymbol{p} \\
&= n\,(x^i, \boldsymbol{p})\,\frac{d^3x d^3\boldsymbol{p}}{(2\pi\hbar)^3} = n[x^i, (h\nu/c)\hat{\boldsymbol{k}}]\,\frac{d^3x d^3\boldsymbol{p}}{(2\pi\hbar)^3},
\end{aligned} \tag{1.121}
$$

where n is the number of photons in a particular quantum state labelled by the wave vector $\boldsymbol{k}$ and momentum $(h\nu/c)\hat{\boldsymbol{k}}$, where $\hat{\boldsymbol{k}}$ is the unit vector in the direction of propagation. In conformity with the usual practice, we are now using the frequency $\nu = (\omega/2\pi)$ instead of energy. The energy-momentum tensor corresponding to this distribution function is

$$
T^{ab}(x^i) = \int \frac{d^3\boldsymbol{p}}{E(p)} c^2 p^a p^b f(x^i, \boldsymbol{p}). \tag{1.122}
$$

The integration over p in $d^3p = p^2 dp d\Omega$ can be converted into an integration over the frequency ν by using $p = (h\nu/c)$. Defining the symbol $\hat{k}^a = k^a/k^0$, where k^a is the wave vector of the photons, T^{ab} becomes

$$
T^{ab}(x^i) = \int \frac{h^4 \nu^3}{c^3} \hat{k}^a \hat{k}^b f(x^i, \nu, \hat{\boldsymbol{k}})\, d\nu d\Omega. \tag{1.123}
$$

This expression suggests defining a quantity (called the *specific intensity* of radiation) by

$$
I_\nu(x^i, \hat{\boldsymbol{k}}) = (h^4 \nu^3/c^2) f = (h\nu^3/c^2) n, \tag{1.124}
$$

so that the energy-momentum tensor becomes

$$
T^{ab}(x^i) = \frac{1}{c} \int d\nu d\Omega\, \hat{k}^a \hat{k}^b I_\nu(x^i, \hat{\boldsymbol{k}}). \tag{1.125}
$$

Note that $\hat{k}^a$ (which is *not* a four-vector) has the four components $(1, \hat{\boldsymbol{k}})$. Since $T^{00} = (dE/dV)$ is the energy per unit volume, it is clear that $I_\nu = (cdE/dV d\nu d\Omega) = (dE/dtd A d\nu d\Omega)$ is the energy flowing per unit area per second per unit frequency range into a solid angle $d\Omega$. The units for I_ν will be erg cm^{-2} s^{-1} Hz^{-1} steradian^{-1}, and is extensively used in astrophysics when dealing with radiative processes. From the definition of intensity I_ν in terms of the photon occupation number, we also find that $I_\nu \propto \nu^3 n$. Since n is Lorentz invariant it follows that (I_ν/ν^3) is invariant.

Exercise 1.15
Practice with equilibrium distribution functions Consider a distribution function, describing particles in thermal equilibrium, given by

$$f(x^i, \boldsymbol{p}) = \frac{dN}{d^3x\, d^3p} = \frac{2j+1}{h^3} \left[\exp(-\theta - \beta p^i u_i) - \epsilon\right]^{-1}, \qquad (1.126)$$

where h is the Planck constant, j is the spin of the particle, u^i is the mean four-velocity of the gas, $\epsilon = 1, 0, -1$ for the Bose–Einstein, Maxwell–Boltzmann or Fermi–Dirac statistics, $\beta = (1/k_B T)$ and θ is a parameter independent of p^i.

(a) Obtain integral expressions for S^a and T^{ab}. Using these express n, ρ and P as one-dimensional integrals.

(b) Manipulate the expressions to show that $dP = [(\rho + P)/T]dT + nk_B T d\theta$.

(c) Show that $\theta k_B T$ is actually the chemical potential $\mu = (\rho + P)/n - Ts$, where s is the entropy density.

(d) For an MB gas, show that $P = nk_B T$. Also find an exact expression for ρ/n.

[Hint. The required expressions can be obtained by using appropriate dot products like $n = -u_i S^i$, $P = (1/3)\mathcal{P}_{ab} T^{ab}$ and $\rho - 3P = -\eta_{ab} T^{ab}$. Using the variable $\chi = \sinh^{-1}(p/m)$, one gets the integral expressions

$$n = \frac{4\pi g m^3}{h^3} \int_0^\infty \frac{\sinh^2 \chi \cosh \chi d\chi}{\exp(\beta \cosh \chi - \theta) - \epsilon}$$

$$P = \frac{4\pi g m^4}{3h^3} \int_0^\infty \frac{\sinh^4 \chi\, d\chi}{\exp(\beta \cosh \chi - \theta) - \epsilon}$$

$$\rho - 3P = \frac{4\pi g m^4}{h^3} \int_0^\infty \frac{\sinh^2 \chi\, d\chi}{\exp(\beta \cosh \chi - \theta) - \epsilon}. \qquad (1.127)$$

Part (b) can be proved directly from these expressions. For part (c) evaluate $d\mu$ from the definition of μ and use the result of part (b). Part (d) can be obtained directly by putting $\epsilon = 0$. The exact expression for ρ/n when $\epsilon = 0$ is given by

$$\frac{\rho}{n} = m \left[\frac{K_1(\beta)}{K_2(\beta)} + \frac{3}{\beta}\right], \qquad (1.128)$$

where $K_n(z)$ is the modified Bessel function.]

Exercise 1.16
Projection effects Let S be a surface with normal n_i. Show that $\mathcal{P}_b^a = \delta_b^a + n^a n_b$ is the projection tensor when S is a spacelike surface, while $\mathcal{P}_b^a = \delta_b^a - n^a n_b$ is the projection tensor when S is a timelike surface. Is there a unique projection tensor associated with a null surface?

Exercise 1.17
Relativistic virial theorem Using the conservation law $\partial_i T^{ij} = 0$, show that for any system which exists in a finite region of space (i.e. $T^{ij} = 0$ outside a compact region in space) we have:

$$\frac{d^2}{dt^2} \int d^3x\, T^{00} x^\alpha x^\beta = 2 \int d^3x\, T^{\alpha\beta}. \qquad (1.129)$$

Interpret this result. [Answer. Using the conservation law $\partial_0 T^{00} = -\partial_\mu T^{0\mu}$, we can write:

$$\left(\partial_0 T^{00}\right)\left(x^\alpha x^\beta\right) = -\left(\partial_\mu T^{0\mu}\right)\left(x^\alpha x^\beta\right) = -\partial_\mu\left[T^{0\mu}x^\alpha x^\beta\right] + \left(T^{0\alpha}x^\beta + T^{0\beta}x^\alpha\right).$$
(1.130)

Taking one more time derivative, and using the same trick, we get:

$$\left(\partial_0^2 T^{00}\right)\left(x^\alpha x^\beta\right) = -\partial_\mu\left[\left(\partial_0 T^{0\mu}\right)x^\alpha x^\beta\right] - \left(\partial_\nu T^{\nu\alpha}\right)x^\beta - \left(\partial_\nu T^{\nu\beta}\right)x^\alpha$$
$$= -\partial_\mu\left[\left(\partial_0 T^{0\mu}\right)x^\alpha x^\beta + T^{\mu\alpha}x^\beta + T^{\mu\beta}x^\alpha\right] + 2T^{\alpha\beta}.$$
(1.131)

Integrating over the source and noting that the divergence term vanishes on the surface, we get Eq. (1.129). This result will be needed in Chapter 9.]

1.10 The Lorentz group and Pauli matrices

We shall now take a closer look at the notion of Lorentz transformations, along with spatial rotations, forming a group. In addition to the intrinsic importance, this analysis will also allow us to introduce a simple 2×2 matrix notation for Lorentz transformation (and rotations) and demonstrate a curious effect known as *Thomas precession*. We will use units with $c = 1$ in this section.[4]

It is obvious from our result in Section 1.3.3 that the set of all Lorentz transformations do *not* constitute a group (since the combination of two infinitesimal Lorentz transformations, in general, involves a spatial rotation) while the set of all Lorentz transformation *and* rotations will form a group, called the *Lorentz group*. In abstract terms, each element of a Lorentz group corresponds to either a Lorentz boost or a spatial rotation. The group structure crucially depends on the fact that the resultant of two operations corresponding to any two elements g_1, g_2 of the group will lead to another unique element of the group usually denoted by the composition law $g_1 \circ g_2$.

We can provide a matrix representation to any group by associating with each element of the group a matrix such that combining two operations corresponding to two elements of the group is mapped to the operation of multiplying the two matrices. That is, the group element $g_1 \circ g_2$ will be associated with a matrix that is obtained by multiplying the matrices for g_1 and g_2. In the case of a Lorentz group, a set of $k \times k$ matrices $D(L)$ will provide a k-dimensional representation of the Lorentz group if $D(L_1)D(L_2) = D(L_1 \circ L_2)$ for any two elements of the Lorentz group L_1, L_2, where L_1, L_2, etc., could correspond to either Lorentz boosts or rotations. We can also introduce a set of k quantities ψ_A with $A = 1, 2, ...k$, forming a row vector, on which these matrices act such that, under the action of an element g_1 of Lorentz group (which could be either a Lorentz transformation or a rotation), the ψ_As undergo a linear transformation of the form $\psi'_A = D_A^B(g_1)\psi_B$, where $D_A^B(g_1)$ is a $k \times k$ matrix representing the element g_1. By doing this, we have also generalized the idea of Lorentz transformation from 4-component objects to k-component objects.

An *infinitesimal* element of a Lorentz group will correspond to the transformation of spacetime coordinates by $x'^a = (\delta^a_b + \omega^a_b)x^b$, where ω^a_b are treated as first order infinitesimal quantities. The first condition in Eq. (1.27), which is required to preserve the form of η_{ab}, now requires $\omega_{ab} = -\omega_{ba}$. This condition implies that ω_{ab} has six independent parameters; three of which ($\omega_{0\mu}$) correspond to Lorentz boosts and the other three ($\omega_{\mu\nu}$) representing spatial rotations. Using this concept we can associate with this infinitesimal element of the Lorentz group the operator

$$D = 1 + (1/2)\omega^{ab}\sigma_{ab}, \tag{1.132}$$

where σ_{ab} are a set of operators that generate infinitesimal Lorentz transformations. As described above, we can also think of these operators as represented by $k \times k$ matrices acting on k-component objects in a specific representation. (This is analogous to the situation in quantum mechanics in which we work with abstract operators as well as their matrix representations, depending on the context.) Since ω^{ab} is antisymmetric, we can take σ_{ab} also as antisymmetric without any loss of generality.

Equation (1.132) can be expressed more transparently by separating out σ_{ab} into two sets: $\sigma_{0\alpha}$ corresponding to Lorentz boosts and $\sigma_{\beta\alpha}$ corresponding to spatial rotations. We associate two vector operators with each of these sets: $\sigma_{0\alpha} \propto K_\alpha$, which will be the three operators generating the boosts, and $\sigma_{\mu\nu} \propto \epsilon_{\mu\nu\rho}J^\rho$, where J^ρ will be the three operators generating the rotations. Normalizing them suitably for future convenience, we can write the operator corresponding to the infinitesimal element of the Lorentz group as

$$D = 1 + \frac{i}{2}K_\alpha v^\alpha + \frac{i}{2}J_\alpha \theta^\alpha. \tag{1.133}$$

The second term on the right hand side generates Lorentz boosts with an infinitesimal three-velocity v while the third term generates infinitesimal spatial rotations. This is, of course, identical to Eq. (1.132) expressed in terms of a different set of parameters which are more convenient.

The structure of the Lorentz group is determined by the commutation rules for these six operators K_α and J_α. These commutation rules can be found most conveniently by calculating the effect of rotations and Lorentz boosts on functions and obtaining an operator representation for J_α and K_α in the space of functions. For example, ordinary infinitesimal rotations by an angle θ in the x–y plane changes the coordinates according to $x' \approx x - y\theta$, $y' \approx y + x\theta$. For any function $f(x, y, z)$ simple Taylor expansion shows that

$$f(\boldsymbol{x}') - f(\boldsymbol{x}) \approx \theta\,[x\partial_y - y\partial_x]f. \tag{1.134}$$

We can write this relation as $f(\boldsymbol{x}') = [1 + (iJ_z\theta/2)]f(\boldsymbol{x})$ with the operator identification

$$J_z = -2i(x\partial_y - y\partial_x). \tag{1.135}$$

Similar results will hold for the other two components showing that J_α are just the angular momentum operators familiar from quantum mechanics.

As regards the Lorentz boost along the x-axis, say, the infinitesimal coordinate transformations are $x' \approx x - vt$, $t' \approx t - vx$. Carrying out a similar analysis we can identify the operator for the boost to be

$$K_x = 2i(t\partial_x + x\partial_t). \tag{1.136}$$

It is now trivial to work out the commutation rules between all these generators of the Lorentz group. We get

$$[J_\alpha, J_\beta] = i\epsilon_{\alpha\beta\gamma} J_\gamma; \quad [J_\alpha, K_\beta] = i\epsilon_{\alpha\beta\gamma} K_\gamma; \quad [K_\alpha, K_\beta] = -i\epsilon_{\alpha\beta\gamma} J_\gamma. \tag{1.137}$$

These relations have a simple interpretation. The first one is the standard commutation rule for angular momentum operators. The second one is equivalent to saying that K_α behaves like a three-vector under rotations. The crucial relation is the third, which shows that the commutator of two boosts is a rotation (with an important minus sign) which we have already discussed in Section 1.3.3.

We now consider the issue of providing explicit matrix representations for the Lorentz group. Since any finite group element can be obtained from the ones which are close to identity by repeated action, we only have to provide a matrix representation for the infinitesimal generators of the Lorentz group in Eq. (1.137). That is, we have to find all matrices which satisfy these commutation relations.

To do this, we introduce the linear combination $a_\alpha = (1/2)(J_\alpha + iK_\alpha)$ and $b_\alpha = (1/2)(J_\alpha - iK_\alpha)$. This allows the commutation relation in Eq. (1.137) to be separated into two sets

$$[a_\alpha, a_\beta] = i\epsilon_{\alpha\beta\mu} a^\mu; \quad [b_\alpha, b_\beta] = i\epsilon_{\alpha\beta\mu} b^\mu; \quad [a_\mu, b_\nu] = 0. \tag{1.138}$$

These are the familiar commutation rules for a *pair* of independent angular momentum matrices in quantum mechanics. We therefore conclude that each irreducible representation of the Lorentz group is characterized by two numbers n, m each of which can be an integer or half-integer with the dimensionality $(2n + 1)$ and $(2m + 1)$. So the representations can be characterized in increasing dimensionality as $(0, 0)$, $(1/2, 0)$, $(0, 1/2)$, $(1, 0)$, $(0, 1)$, ..., etc.

The smallest nontrivial representation corresponding to $(1/2, 0)$ or $(0, 1/2)$ will be in terms of 2×2 matrices, which we will now discuss in detail. Since the mathematical structure is very similar in Lorentz transformations and ordinary rotations, we shall begin by briefly reviewing the case of ordinary rotations in three-dimensional Euclidean space before describing the corresponding results for Lorentz transformations.

A rotation in three-dimensional space can be defined by specifying the unit vector n in the direction of the axis of rotation and the angle θ through which the axes are rotated. (We shall use the standard right hand rule to define the orientation of n.) We shall associate with this rotation a 2×2 matrix

$$R(\theta) = \cos\left(\frac{\theta}{2}\right) - i(\boldsymbol{\sigma} \cdot \boldsymbol{n})\sin\left(\frac{\theta}{2}\right) = \exp -\frac{i\theta}{2}(\boldsymbol{\sigma} \cdot \boldsymbol{n}), \qquad (1.139)$$

where σ_α are the standard Pauli matrices and the first term, $\cos(\theta/2)$, is multiplied by the unit matrix though it is not explicitly indicated. The second equality can be demonstrated by expanding the exponential in a power series and using the easily proved relation $(\boldsymbol{\sigma} \cdot \boldsymbol{n})^2 = 1$. (Incidentally, the occurrence of the angle $\theta/2$ has a simple geometrical origin: a rotation through an angle θ about a given axis may be visualized as the consequence of successive reflections in two planes which meet along the axis at an angle $\theta/2$.) Using the properties of the Pauli matrices, it is easy also to show that $\text{Tr}(\boldsymbol{\sigma} \cdot \boldsymbol{n}) = 0$, $dR/d\theta = -(i/2)(\boldsymbol{\sigma} \cdot \boldsymbol{n})R$ and that R commutes with $(\boldsymbol{\sigma} \cdot \boldsymbol{n})$. We can also associate with a three-vector $\boldsymbol{x}$ the 2×2 matrix $X = \boldsymbol{x} \cdot \boldsymbol{\sigma}$. The effect of any rotation can be concisely described by the matrix relation $X' = RXR^*$.

Using the explicit form of $R(\theta)$, one can characterize the matrix corresponding to an *infinitesimal* rotation by an angle $d\theta$ as

$$R = 1 - (id\theta/2)(\boldsymbol{\sigma} \cdot \boldsymbol{n}). \qquad (1.140)$$

From this form, it is clear that the Pauli matrices can be thought of as providing the 2×2 matrix representation of the generators of infinitesimal rotations. These generators satisfy the standard commutation relations $[\sigma_\alpha, \sigma_\beta] = \epsilon_{\alpha\beta\gamma}\sigma_\gamma$. It can be easily verified that the relation $X' = RXR^*$ with R in Eq. (1.140) reproduces the standard result $\boldsymbol{x}' = [\boldsymbol{x} + (d\theta)\boldsymbol{n} \times \boldsymbol{x}]$ for an infinitesimal rotation.

All these results generalize, in a natural fashion, to Lorentz transformations. We shall associate with a Lorentz transformation in the direction $\boldsymbol{n}$, with the speed $v = c\tanh\chi$, the 2×2 matrix

$$L = \cosh(\chi/2) + (\boldsymbol{n} \cdot \boldsymbol{\sigma})\sinh(\chi/2) = \exp\frac{1}{2}(\boldsymbol{\chi} \cdot \boldsymbol{\sigma}). \qquad (1.141)$$

The change from trigonometric functions to hyperbolic functions is in accordance with the fact that Lorentz transformations correspond to rotation by an *imaginary* angle. Just as in the case of rotations, we can associate to any event $x^i = (x^0, \boldsymbol{x})$ a (2×2) matrix $P \equiv x^i\sigma_i$ where σ_0 is the identity matrix and σ_α are the Pauli matrices. Under a Lorentz transformation along the direction $\hat{n}$ with speed V, the event x^i goes to $x^{i'}$ and P goes P'. (By convention σ_is do not change.) They are related by

$$P' = LPL^*, \tag{1.142}$$

where L is given by Eq. (1.141).

With this formalism, it is straightforward, though algebraically a bit tedious, to determine the effect of consecutive Lorentz transformations along different directions. From the discussion in Section 1.3.3, we know that the combined effect of two Lorentz transformations is equivalent to a spatial rotation plus a Lorentz transformation. This allows us to write

$$L(\boldsymbol{v}_1)L(\boldsymbol{v}_2) = R(\theta\hat{\boldsymbol{n}})L(\boldsymbol{v}_3). \tag{1.143}$$

Expanding out the form of the matrices and using the relation

$$(\boldsymbol{A} \cdot \boldsymbol{\sigma})(\boldsymbol{B} \cdot \boldsymbol{\sigma}) = \boldsymbol{A} \cdot \boldsymbol{B} + i(\boldsymbol{A} \times \boldsymbol{B}) \cdot \boldsymbol{\sigma}, \tag{1.144}$$

which is valid for any two vectors $\boldsymbol{A}$ and $\boldsymbol{B}$, one can determine the angle of rotation $\theta\hat{\boldsymbol{n}}$ as well as the velocity $\boldsymbol{v}_3$. In particular, one finds that the angle θ is given by the relation

$$\tan(\theta/2) = \frac{-\sinh(\chi_2/2)\sinh(\chi_1/2)\sin\gamma}{\cosh(\chi_1/2)\cosh(\chi_2/2) + \cos\gamma\sinh(\chi_2/2)\sinh(\chi_1/2)}, \tag{1.145}$$

where γ is the angle between the two velocity vectors $\boldsymbol{v}_1$ and $\boldsymbol{v}_2$. While this expression is not illuminating, it leads to an interesting physical phenomena called *Thomas precession*, which we shall now discuss.[5]

Thomas precession arises in the context of an object with an intrinsic spin which moves in an orbit with variable velocity – an example being an electron orbiting the nucleus in an atom treated along classical lines. The effective energy of coupling between spin and orbital angular momentum of an atomic electron picks up an extra factor of (1/2) due to this effect and, of course, has experimentally verifiable consequences. One might have thought that any special relativistic effect should lead to a correction which is of the order of $(v/c)^2$ for an electron in hydrogen atom. This is indeed true. But experimentally observable effects of the spin–orbit interaction are also relativistic effects arising from the Coulomb field (Ze^2/r) transforming to a $(v/c)(Ze^2/r)$ magnetic field in the rest frame of the electron and coupling to the magnetic moment $(e\hbar/2m_e c)$ of the electron. So any other effect at $\mathcal{O}(v^2/c^2)$ will change the observable consequences by order unity factors.

Consider a frame S_0 which is an inertial laboratory frame and let $S(t)$ be a Lorentz frame comoving with a particle (which has spin) at time t. These two frames are related to each other by a Lorentz transformation with a velocity $\boldsymbol{v}$. Consider a pure Lorentz boost in the *comoving* frame which changes its velocity relative to the lab frame from v to $v + dv$. We know that the resulting final config-uration cannot be reached from S_0 by a pure boost and we require a rotation by an angle $\delta\theta = \omega dt$ followed by a simple boost. This leads to the relation in terms of the 2×2 matrices corresponding to the rotation and Lorentz transformations

$$L(v + dv)R(\omega dt) = L_{\text{comov}}(dv)L(v). \tag{1.146}$$

On the right hand side of Eq. (1.146), $L_{\text{comov}}(dv)$ has a subscript 'comov' to stress the fact that this corresponds to a pure boost *only* in the comoving frame but *not* in the lab frame. To determine its form, we can proceed as follows. We first bring the particle to rest by applying the inverse Lorentz transformation operator $L^{-1}(v) = L(-v)$. Then we apply a boost $L(a_{\text{comov}}d\tau)$, where a_{comov} is the acceleration of the system in the comoving frame. Since the object was at rest initially, this can be characterized by a pure boost. Finally, we transform back from the lab to the moving frame by applying $L(v)$. Therefore we have the relation

$$L_{\text{comov}}(dv) = L(v)L(a_{\text{comov}}d\tau)L(-v). \tag{1.147}$$

Using this in Eq. (1.146), we get $L(v + dv)R(\omega dt) = L(v)L(a_{\text{comov}}d\tau)$. In this equation, the unknowns are ω and a_{comov}. Moving the unknown terms to the left hand side, we have the equation,

$$R(\omega dt)L(-a_{\text{comov}}d\tau) = L(-[v + dv])L(v), \tag{1.148}$$

which can be solved for ω and a_{comov}. If we denote the rapidity parameters for the two infinitesimally separated Lorentz boosts by χ and $\chi' \equiv \chi + d\chi$ and the corresponding directions by n and $n' \equiv n + dn$ then this matrix equation can be expanded to first order quantities to give

$$1 - (idt\omega + d\tau a) \cdot \frac{\sigma}{2} =$$
$$[\cosh(\chi'/2) - (n' \cdot \sigma)\sinh(\chi'/2)][\cosh(\chi/2) - (n \cdot \sigma)\sinh(\chi/2)]. \tag{1.149}$$

Performing the necessary Taylor series expansion on the right hand side and identifying the corresponding terms on both sides, we find that:

$$a_{\text{comov}} = \hat{n}_\alpha \frac{d\chi}{d\tau} + (\sinh\chi)\frac{d\hat{n}_\alpha}{d\tau}; \qquad \omega = \left(2\sinh^2\frac{\chi}{2}\right)\left(\frac{d\hat{n}_\alpha}{dt} \times \hat{n}_\alpha\right), \tag{1.150}$$

with $\tanh\chi = v$. Expressing everything in terms of the velocity, it is easy to show that the expression for ω is equivalent to

$$\omega = \frac{\gamma^2}{\gamma + 1}\frac{a \times v}{c^2} = (\gamma - 1)\frac{(v \times a)}{v^2}, \tag{1.151}$$

where we have temporarily re-introduced the c-factor. At the lowest order, this gives a precession angular velocity $\omega \cong (1/2c^2)(a \times v)$ which the spin will undergo because of the non-commutativity of Lorentz transformations in different directions.

The purpose of the above derivation was to indicate the purely kinematic origin of the Thomas precession. It is possible to work out the same effect more formally by writing down an equation of motion for the spin of a particle moving in a given trajectory. To do this, we shall introduce the concept of a spin four-vector S^j such that, in the rest frame of the particle, it has purely spatial components which coincide with the standard three-dimensional spin vector S; that is, in the rest frame of the particle, $S^j = (0, S)$. Since the four-velocity in the rest frame is $u^i = (1, 0)$, this condition can be stated in an invariant manner as $S^j u_j = 0$. Further, in the rest frame of the particle, if there are no torques, we will have $dS^j/d\tau = 0$. This fact can be expressed as a covariant equation of motion for the spin in the form $dS^j/d\tau = ku^j$, where k is some quantity which needs to be determined. Differentiating the condition $S^i u_i = 0$, we get $0 = ku^i u_i + S^i a_i$, where $a_i = du_i/d\tau$ is the acceleration. This determines $k = S^i a_i$ so that the equation of motion for the spin can be expressed in the form

$$\frac{dS^j}{d\tau} = u^j (S^k a_k). \tag{1.152}$$

Let us apply this to a particle moving in a trajectory $x^i(\tau)$. The instantaneous rest frame of the particle can be obtained from the lab frame by a Lorentz transformation having the velocity $v(\tau) = c\beta(\tau)$. Since the spin vector has the form $S^a = (0, S(\tau))$ in the rest frame, its components in the lab frame are:

$$S^k = \left(\gamma\beta \cdot S, \; S + \beta\frac{\gamma^2}{\gamma+1}\beta \cdot S \right), \tag{1.153}$$

where we have used Eq. (1.25). Further, from $u^i = (\gamma, \gamma\beta)$, we find that $a^i = (\dot{\gamma}, \dot{\gamma}\beta + \gamma\dot{\beta})$. This gives

$$S^k a_k = \gamma \left(\dot{\beta} \cdot S + \frac{\gamma^2}{\gamma+1}(\dot{\beta} \cdot \beta)(\beta \cdot S) \right). \tag{1.154}$$

Substituting this into Eq. (1.152) and separating the space and time components leads to

$$\frac{d}{d\tau}(\gamma\beta \cdot S) = \gamma^2 \left(\dot{\beta} \cdot S + \frac{\gamma^2}{\gamma+1}(\dot{\beta} \cdot \beta)(\beta \cdot S) \right)$$

$$\frac{d}{d\tau}\left(S + \beta\frac{\gamma^2}{\gamma+1}\beta \cdot S \right) = \beta\gamma^2 \left(\dot{\beta} \cdot S + \frac{\gamma^2}{\gamma+1}(\dot{\beta} \cdot \beta)(\beta \cdot S) \right). \tag{1.155}$$

Somewhat lengthy but straightforward algebra will now allow these equations to be transformed into the form

$$\frac{d\boldsymbol{S}}{d\tau} = \boldsymbol{S} \times \boldsymbol{\omega}, \tag{1.156}$$

with $\boldsymbol{\omega}$ given by Eq. (1.151). Equation (1.156) shows that the spin precesses with the angular velocity $\boldsymbol{\omega}$.

Exercise 1.18
Explicit computation of spin precession Consider an electron (with spin) moving in a circular orbit in the x–y plane with $x = r\cos\omega t$, $y = r\sin\omega t$. Determine the four-velocity as well as the four-acceleration from this trajectory. Solve Eq. (1.152) with the initial condition $S^x = \hbar/\sqrt{2}$, $S^y = 0$, $S^z = (1/2)\hbar$ (so that $S^2 = (3/4)\hbar^2$) and show that

$$S^x + iS^y = \frac{\hbar}{\sqrt{2}} \left[e^{-i(\gamma-1)\omega t} + i(1-\gamma)\sin(\omega\gamma t)e^{i\omega t} \right]. \tag{1.157}$$

The first term leads to a Thomas precession around the z-axis with the angular velocity $(\gamma - 1)\omega$ while the second term is negligibly small for the electron in an atom.

Exercise 1.19
Little group of the Lorentz group In some inertial frame, a photon has the four-momentum $p^i = (\omega, \omega, 0, 0)$. The *little group* G of p^i is a special class of Lorentz transformation which leaves these components unchanged. A pure rotation in the y–z plane is, of course, an element of G. Find a sequence of pure boost and pure rotation which is not a pure rotation in the y–z plane but is still an element of G. [Hint. Think of a boost in the y–z plane followed by (i) a pure rotation to realign the spatial momentum along the x-axis again and (ii) a final boost to get the magnitude back to original value.]

PROJECT

Project 1.1

Energy-momentum tensor of non-ideal fluids

The T^{ik} and J^i for an ideal fluid was obtained in Section 1.9, ignoring the gradients of temperature (T), number density (n) and bulk velocity (u^i) of the fluid. At the next order of approximation, in which these gradients are taken into account, we expect T_{ik} and J_i to contain terms which are proportional to the gradients $(\partial T/\partial x^i)$ and $(\partial u_k/\partial x^i)$. The density gradient in space will lead to diffusion, the temperature gradient will lead to thermal conduction and the velocity gradient will lead to viscous effects. The aim of this project is to generalize the form of the energy momentum tensor by including terms containing these gradients. We will write these expressions, correct to linear order in the gradients, as

$$T_{ik} = wu_iu_k + P\eta_{ik} + \tau_{ik}; \quad J_i = nu_i + h_i, \tag{1.158}$$

where $w = (P + \rho)$ is usually called the enthalpy. In a relativistic theory, since all energy fluxes involve equivalent mass fluxes, it is necessary to define h_i, etc., more precisely.

(a) Argue that the following procedure will lead to a consistent description. In the proper rest frame of the fluid element demand that: (i) the momentum of the fluid element should be zero and (ii) the energy should be expressible in terms of other thermodynamic variables in the same functional form as in the absence of dissipative processes. This requires that, in the proper frame, $\tau^{0i} = 0$, which can be written in a Lorentz invariant form as $\tau_{ik}u^i = 0$. Similarly, demand $h_i u^i = 0$ so that, in the rest frame, n^0 is same as the proper number density n.

(b) Using these conditions and the form of the expressions in Eq. (1.158) show that

$$\frac{\partial}{\partial x^i}\left(su^i - \frac{\mu}{T}h^i\right) = -h^i\frac{\partial}{\partial x^i}\left(\frac{\mu}{T}\right) + \frac{\tau_i^k}{T}\frac{\partial u^i}{\partial x^k}, \tag{1.159}$$

where s is the entropy density and μ is the chemical potential. The left hand side is the divergence of the entropy current $[su^i - (\mu/T)h^i]$, which was zero in the absence of dissipative terms.

(c) In the presence of dissipation, the entropy is expected to increase and the right hand side of Eq. (1.159) should be positive. Further, since we are computing the first order corrections, the quantities τ_{ik} and h_i must be linear in the gradients $(\partial u^i/\partial x^k)$ and $(\partial(\mu/T)/\partial x^k)$. Taking, $\tau_{ab} = M_{abik}(\partial u^i/\partial x_k), h^i = N^{ik}(\partial(\mu/T)/\partial x^k)$, substituting into Eq. (1.159) and using the conditions $\tau_{ik}u^i = 0, h_i u^i = 0$ along with the positivity of right hand side, determine the forms of τ_{ik} and h_i to be:

$$\tau_{ik} = -\eta\left(\frac{\partial u_i}{\partial x^k} + \frac{\partial u_k}{\partial x^i} + u_k u^l\frac{\partial u_i}{\partial x^l} + u_i u^l\frac{\partial u_k}{\partial x^l}\right) - \left(\zeta - \frac{2}{3}\eta\right)\frac{\partial u^l}{\partial x^l}\left(\eta_{ik} + u_i u_k\right), \tag{1.160}$$

$$h_i = -\kappa\left(\frac{nT}{w}\right)^2\left[\frac{\partial}{\partial x^i}\left(\frac{\mu}{T}\right) + u_i u^k\frac{\partial}{\partial x^k}\left(\frac{\mu}{T}\right)\right], \tag{1.161}$$

where the coefficients η and ζ describe viscosity – arising from velocity gradients – and κ describes thermal conduction – arising from temperature gradient.

(d) What is the non-relativistic limit of this expression?

2

Scalar and electromagnetic fields in special relativity

2.1 Introduction

This chapter develops the ideas of classical field theory in the context of special relativity. We use a scalar field and the electromagnetic field as examples of classical fields. The discussion of scalar field theory will allow us to understand concepts that are unique to field theory in a somewhat simpler context than electromagnetism; it will also be useful later on in the study of topics such as inflation, quantum field theory in curved spacetime, etc. As regards electromagnetism, we concentrate on those topics that will have direct relevance in the development of similar ideas in gravity (gauge invariance, Hamilton–Jacobi theory for particle motion, radiation and radiation reaction, etc.).

The ideas developed here will be used in the next chapter to understand why a field theory of gravity – developed along similar lines – runs into difficulties. The concept of an action principle for a field will be extensively used in Chapter 6 in the context of gravity. Other topics will prove to be valuable in studying the effect of gravity on different physical systems.[1]

2.2 External fields of force

In non-relativistic mechanics, the effect of an external force field on a particle can be incorporated by adding to the Lagrangian the term $-V(t, \boldsymbol{x})$, thereby adding to the action the integral of $-V\,dt$. Such a modification is, however, not Lorentz invariant and hence cannot be used in a relativistic theory. Our first task is to determine the form of interactions which are permitted by the Lorentz invariance.

The action for the free particle was the integral of $d\tau$ (see Eq. (1.72)), which is Lorentz invariant. We can modify this expression to the form

$$\mathcal{A} = -\int \mathcal{L}(x^a, u^a)\,d\tau, \qquad (2.1)$$

where $\mathcal{L}(x^a, u^a)$ is a Lorentz invariant scalar dependent on the position and velocity of the particle, and still maintain Lorentz invariance. A possible choice for $\mathcal{L}(x^a, u^a)$ is obtained by taking the polynomial in u^a, as

$$\mathcal{L} = mc^2 + \lambda\phi(x) - \frac{q}{c}A_i(x)u^i + \mu g_{ab}(x)u^a u^b + \cdots, \qquad (2.2)$$

where $\phi(x)$ is a scalar, $A_i(x)$ is a four-vector, $g_{ab}(x)$ is a second rank tensor, etc., $\lambda, q, \mu...$, etc., are constants that have been introduced, with some choices for signs, for later convenience. Quantities like ϕ depend on the four-vector x^i but for simplicity of notation we shall write $\phi(x)$ instead of $\phi(x^i)$. In this expansion, ϕ, A_i, g_{ab}, etc., are externally specified fields which influence the trajectory of the particle. Of the three terms, the scalar field ϕ can be included in the term with g_{ab} by adding a part $\phi(x)\eta_{ab}$. Nevertheless, we keep it separate for future convenience.

If we assume that the Lagrangian should have only terms up to the quadratic order in the four-velocity, we cannot have more terms in Eq. (2.2) and this indeed turns out to be a valid assumption. In nature, we only come across a vector field A_i describing electromagnetism and a second rank tensor field $g_{ab}(x)$ describing gravity; that is, the Taylor series expansion of $\mathcal{L}$ in the variable u^a terminates after the quadratic term and no higher degree terms arise. We shall postpone the study of $g_{ab}(x)$ (which could describe the gravitational field) to later chapters and will discuss the other two – scalar field ϕ and vector field $A_i(x)$ – in this chapter. Of these two, the really important case corresponds to A_i, which describes electromagnetism, but we shall start with the scalar field ϕ since it is mathematically simpler and will have applications in Chapters 13 and 14.

2.3 Classical scalar field

2.3.1 Dynamics of a particle interacting with a scalar field

The action for a particle influenced by a scalar field is described by the first two terms in Eq. (2.2) of which the first term is the free particle Lagrangian used in the last chapter and the second term describes the influence of the scalar field. Using $d\tau = \gamma^{-1}dt$, we can identify the corresponding Lagrangian as

$$L = -\sqrt{1 - (v^2/c^2)}(mc^2 + \lambda\phi) \approx -mc^2 + \frac{1}{2}mv^2 - \lambda\phi + \mathcal{O}(1/c^2), \quad (2.3)$$

where the second expression is obtained by taking a Taylor series expansion in $1/c$. Except for an irrelevant constant term $(-mc^2)$, this is identical to the standard Lagrangian in classical mechanics for a particle moving in a potential $V(t, x) \equiv \lambda\phi$. Thus a scalar field can indeed describe a particle moving in some potential in the non-relativistic limit. However, in the fully relativistic situation, the equation of motion resulting from the exact Lagrangian is quite different. To obtain it we need to vary $x^i(\tau)$ in the action

$$\mathcal{A} = -\int_{\tau_1}^{\tau_2} d\tau (m + \lambda\phi) \tag{2.4}$$

obtained from Eq. (2.2) by retaining only the first two terms (and using the units with $c = 1$ for convenience). On using $\delta\phi = [\partial_a\phi]\delta x^a$ and recalling the derivation of Eq. (1.76), we get

$$\delta\mathcal{A} = \int_{\tau_1}^{\tau_2} (m + \lambda\phi)u_i \, d\delta x^i - \lambda \int_{\tau_1}^{\tau_2} d\tau (\partial_i\phi\delta x^i)$$

$$= -\int_{\tau_1}^{\tau_2} d\tau \left(\lambda\partial_i\phi + \frac{d[u^i(m + \lambda\phi)]}{d\tau} \right) \delta x^i + (m + \lambda\phi)u_i\delta x^i \Big|_{\tau_1}^{\tau_2}. \tag{2.5}$$

If we make the usual assumption that the variation δx^i vanishes at the end points, the second term goes to zero and we get the equations of motion

$$\frac{du^i}{d\tau} = -\lambda\frac{\partial^i\phi}{(m + \lambda\phi)} - \lambda u^i u^j \frac{\partial_j\phi}{(m + \lambda\phi)}, \tag{2.6}$$

where we have used $(d\phi/d\tau) = u^i\partial_i\phi$. We see that, in the fully relativistic case, the equations of motion are fairly complicated and satisfy $u_i a^i = 0$ identically. (Of course, in the $c \to \infty$ limit, the spatial part of Eq. (2.6) reduces to $m(d\boldsymbol{v}/dt) = -\lambda\nabla\phi$, which is the equation for a non-relativistic particle moving in the potential $\lambda\phi$.)

Since such a scalar field does not seem to exist in nature, we shall not pursue this analysis further, except to make a couple of comments which will be of relevance in the case of the electromagnetic field as well. First, we see that there are velocity dependent forces in Eq. (2.6), which is a generic feature of relativistic Lagrangians. In the case of electromagnetism, we will see that a similar analysis leads to the velocity dependent Lorentz force. Second, the expression for the canonical momentum now picks up a field dependent term. We saw earlier that the canonical momentum of the particle can be obtained by treating the action as a function of the end points for a trajectory which satisfies the equation of motion and computing $p_i = (\partial\mathcal{A}/\partial x^i)$. In this case, the first term in Eq. (2.5) vanishes and we get

$$p_i = \frac{\partial\mathcal{A}}{\partial x^i} = \left(m + \frac{\lambda\phi}{c^2} \right) u_i, \tag{2.7}$$

where we have temporarily restored the c-factor. This result shows that the canonical momentum picks up a field dependent term in the fully relativistic case. (In the non-relativistic limit, the ϕ dependent term vanish because of the $(1/c^2)$ factor.) We will see that such an effect also arises in the case of the electromagnetic field.

The Hamilton–Jacobi equation for the particle can be obtained from the above expression for the canonical momentum by using $p_i = \partial_i \mathcal{A}$ and $u_i u^i = -1$. We get

$$\eta^{ij}\partial_i \mathcal{A}\partial_j \mathcal{A} = -\left(mc + \frac{\lambda\phi}{c}\right)^2 \tag{2.8}$$

for a massive particle. We will have occasion to comment on this in Chapter 3.

We shall now take up the more important issue related to this model which has to do with the dynamics of the scalar field itself.

2.3.2 Action and dynamics of the scalar field

The action principle developed above couples the particle to the scalar field but treats the field ϕ as an externally specified entity. Such an external field can act on the particle and change its energy, momentum, angular momentum, etc. But since the conservation of these quantities is assured for a closed system from general symmetry considerations, it is clear that the scalar field must possess energy, momentum and angular momentum which should also get changed during the interaction with the particle. In other words, the field must be a dynamic entity that changes in response to the interaction, obeying certain equations of motion.

The action in Eq. (2.4) does not allow one to determine the evolution of the field. To do this, we should treat the field as a dynamical variable and add a term to the action in Eq. (2.4) which will produce the equations of motion determining the field, when the action is varied with respect to the field variables. The total action will now be of the form:

$$\mathcal{A} = -\int_{\tau_1}^{\tau_2} d\tau\, m - \lambda \int_{\tau_1}^{\tau_2} d\tau\, \phi - \int d^4x\, L_{field}, \tag{2.9}$$

where the last term depends only on the scalar field. Varying the field ϕ in this action will lead to the dynamics of the field. Our first task will be to write down a suitable Lagrangian L_{field} for such a scalar field.

The procedure we will adopt is a direct generalization from classical mechanics with some significant new features. (See Table 2.1 for a comparison.) In classical mechanics, the action is expressed as an integral of the Lagrangian over a time coordinate with the measure dt. In relativity, while dealing with a field, we will generalize this to an integral over the spacetime coordinates with a measure d^4x in any inertial Cartesian system. Further, in classical mechanics, the Lagrangian for a closed system depends on the dynamical variable $q(t)$ and its first time derivative $\dot{q}(t) \equiv \partial_0 q$. In relativity, one cannot treat the time coordinate preferentially in

Table 2.1. *Comparison of action principles in classical mechanics and field theory*

Property	Mechanics	Field theory
Independent variable	t	$(t, \boldsymbol{x})$
Dependent variable	$q(t)$	$\phi(t, \boldsymbol{x})$
Definition of action	$\mathcal{A} = \int dt L$	$\mathcal{A} = \int d^4 x L$
Form of Lagrangian	$L = L(\partial_0 q, q)$	$L = L(\partial_i \phi, \phi)$
Domain of integration	$t \in (t_1, t_2)$ one-dimensional interval	$x^i \in \mathcal{V}$ four-dimensional region
Boundary of integration	two points; $t = t_1, t_2$	three-dimensional surface $\partial\mathcal{V}$
Canonical momentum	$p = \dfrac{\partial L}{\partial(\partial_0 q)}$	$\pi^j = \dfrac{\partial L}{\partial(\partial_j \phi)}$
General form of the variation	$\delta\mathcal{A} = \int_{t_1}^{t_2} dt \mathcal{E}[q]\delta q$ $+ \int_{t_1}^{t_2} dt \partial_0(p\delta q)$	$\delta\mathcal{A} = \int_{\mathcal{V}} d^4 x \mathcal{E}[\phi]\delta\phi$ $+ \int_{\mathcal{V}} d^4 x \partial_j(\pi^j \delta\phi)$
Form of $\mathcal{E}$	$\mathcal{E}[q] = \dfrac{\partial L}{\partial q} - \partial_0 p$	$\mathcal{E}[\phi] = \dfrac{\partial L}{\partial\phi} - \partial_j \pi^j$
Boundary condition to get equations of motion	$\delta q = 0$ at the boundary	$\delta\phi = 0$ at the boundary
Equations of motion	$\mathcal{E}[q] = \dfrac{\partial L}{\partial q} - \partial_0 p = 0$	$\mathcal{E}[\phi] = \dfrac{\partial L}{\partial\phi} - \partial_j \pi^j = 0$
Form of $\delta\mathcal{A}$ when $\mathcal{E} = 0$ gives momentum	$\delta\mathcal{A} = (p\delta q)\big|_{t_1}^{t_2}$	$\delta\mathcal{A} = \int_{\partial\mathcal{V}} d^3\sigma_j(\pi^j \delta\phi)$
Energy	$E = p\partial_0 q - L$	$T_b^a = -[\pi^a \partial_b \phi - \delta_b^a L]$

a Lorentz invariant manner; the dynamical variable describing a field $\phi(x^i)$ will depend on both time and space and the Lagrangian will depend on the derivatives of the dynamical variable with respect to both time and space, $\partial_i \phi$. Hence the action for the field has the generic form

$$\mathcal{A}_{field} = \int_{\mathcal{V}} d^4 x \, L_{field}(\partial_a \phi, \phi). \tag{2.10}$$

The integration is over a four-dimensional region $\mathcal{V}$ in spacetime, the boundary of which will be a three-dimensional surface, denoted by $\partial\mathcal{V}$. (This generalizes the notion in classical mechanics in which integration over time is in some interval $t_1 \leq t \leq t_2$.) We stress that, in this action, the dynamical variable is the field $\phi(t, \boldsymbol{x})$ and $x^i = (t, \boldsymbol{x})$ are just parameters.

During the variation of ϕ in Eq. (2.9), the first term does not change; but the second term is expressed as an integral over $d\tau$ while the third term is expressed as an integral over d^4x. For evaluating the variation it will be convenient if the second term can also be expressed as an integral over the spacetime volume d^4x. This can be done by using the fact that, if a point particle follows a trajectory $z^i(\tau)$, then the 'particle density' contributed by this particle is given by

$$n(x^i) = \int_{-\infty}^{+\infty} d\tau\, \delta_D[x^i - z^i(\tau)], \tag{2.11}$$

where the four-dimensional Dirac delta function $\delta_D[x^i] \equiv \delta(x^0)\delta(x)\delta(y)\delta(z)$ is a product of four Dirac delta functions on each Cartesian component of the coordinate x^i. This incorporates the fact that the density is zero everywhere except on the world line of the particle. It is now possible to express the second term in the action in Eq. (2.9) as

$$-\lambda \int d\tau\, \phi = -\lambda \int d^4x\, n(x)\phi(x). \tag{2.12}$$

Substituting Eq. (2.11) into Eq. (2.12), it is easy to verify this equality.

So, as far as the dynamics of the field is concerned, we need to vary the dynamical variable ϕ in the last two terms in Eq. (2.9) with $n(x)$ treated as an externally specified quantity. Writing these two terms together as

$$\mathcal{A} = -\lambda \int_{\tau_1}^{\tau_2} d\tau\, \phi + \int_{\mathcal{V}} d^4x\, L_{field}(\partial_a\phi, \phi)$$

$$= \int_{\mathcal{V}} d^4x\, [L_{field}(\partial_a\phi, \phi) - \lambda n\phi] \equiv \int_{\mathcal{V}} d^4x\, L(\partial_a\phi, \phi), \tag{2.13}$$

where $L \equiv L_{\text{field}} + \lambda n\phi$, and performing the variation, we get (in a manner very similar to the corresponding calculation in classical mechanics):

$$\delta\mathcal{A} = \int_{\mathcal{V}} d^4x \left(\frac{\partial L}{\partial\phi}\delta\phi + \frac{\partial L}{\partial(\partial_a\phi)}\delta(\partial_a\phi) \right)$$

$$= \int_{\mathcal{V}} d^4x \left[\frac{\partial L}{\partial\phi} - \partial_a\left(\frac{\partial L}{\partial(\partial_a\phi)} \right) \right]\delta\phi + \int_{\mathcal{V}} d^4x\, \partial_a\left[\frac{\partial L}{\partial(\partial_a\phi)}\delta\phi \right]. \tag{2.14}$$

In obtaining the second equality, we have used the fact that $\delta(\partial_a\phi) = \partial_a(\delta\phi)$ and have performed an integration by parts. The last term in the second line is an integral over a four-divergence, $\partial_a[\pi^a\delta\phi]$ where $\pi^a \equiv [\partial L/\partial(\partial_a\phi)]$. This quantity π^a generalizes the expression $[\partial L/\partial(\partial_0 q)] = [\partial L/\partial \dot{q}]$ from classical mechanics and can be thought of as the analogue of canonical momentum. In fact, the 0th component of this quantity is indeed $\pi^0 = [\partial L/\partial \dot{\phi}]$, as in classical mechanics. Using the four-dimensional divergence theorem (see Eq. (1.60)), we can convert this into a surface term

$$\delta\mathcal{A}_{\text{sur}} \equiv \int_{\mathcal{V}} d^4x\,\partial_a(\pi^a\delta\phi) = \int_{\partial\mathcal{V}} d\sigma_a\,\pi^a\delta\phi \rightarrow \int_{t=cons} d^3x\,\pi^0\delta\phi, \qquad (2.15)$$

where the last expression is valid if we take the boundary to be the spacelike surfaces defined by $t =$ constant and assume that the surface at spatial infinity does not contribute. (In classical mechanics, the corresponding analysis leads to $p\delta q$ at the end points $t = t_1$ and $t = t_2$. Since the integration is over one dimension, the 'boundary' in classical mechanics is just two points. In the relativistic case, the integration is over four dimensions leading to a boundary term which is a three-dimensional integral.) We see that we can obtain sensible dynamical equations for the field ϕ by demanding $\delta\mathcal{A} = 0$ if we consider variations $\delta\phi$ which vanish everywhere on the boundary $\partial\mathcal{V}$. (This is similar to demanding $\delta q = 0$ at $t = t_1$ and $t = t_2$ in classical mechanics.) For such variations, the demand $\delta\mathcal{A} = 0$ leads to the field equations

$$\partial_a\left(\frac{\partial L}{\partial(\partial_a\phi)}\right) = \partial_a\pi^a = \frac{\partial L}{\partial\phi}. \qquad (2.16)$$

Given the form of the Lagrangian, this equation determines the dynamics of the field.

We can also consider the change in the action when the field configuration is changed on the boundary $\partial\mathcal{V}$ assuming that the equations of motion are satisfied. In classical mechanics this leads to the relation $p = (\partial\mathcal{A}/\partial q)$, where the action is treated as a function of its end points. In our case, Eq. (2.15) can be used to determine different components of π^a by choosing different surfaces. In particular, if we take the boundary to be $t =$ constant, we get $\pi^0 = (\delta\mathcal{A}/\delta\phi)$ on the boundary, where the symbol $(\delta\mathcal{A}/\delta\phi)$ is called the *functional derivative* and is defined through the second equality in Eq. (2.15). This provides an alternative justification for interpreting π^a as the canonical momentum.

2.3.3 Energy-momentum tensor for the scalar field

In classical mechanics, if the Lagrangian has no explicit dependence on time t, then one can prove that the energy defined by $E = (p\dot{q}) - L$ is conserved. By analogy, when the relativistic Lagrangian has no explicit dependence on the spacetime coordinate x^i, we will expect to obtain a suitable conservation law. In this case, we will expect $\dot{q}$ to be replaced by $\partial_i\phi$ and p to be replaced by π^a. This suggests, considering a generalization of $E = (p\dot{q}) - L$ to the second rank tensor,

$$T^a{}_i \equiv -[\pi^a(\partial_i\phi) - \delta^a_i L]. \qquad (2.17)$$

Again, we see that the component $-T^0{}_0 = \pi^0\dot{\phi} - L$ is identical in structure to E in classical mechanics making $T_{00}(= -T^0_0)$ the positive definite energy density. The overall sign in Eq. (2.17) has been chosen to facilitate this. To check the conservation law, we calculate $\partial_a T^a{}_i$ treating L as an implicit function of x^a through ϕ and $\partial_i\phi$. Explicit computation gives

$$-\partial_a T^a{}_i = (\partial_i\phi)(\partial_a\pi^a) + \pi^a\partial_a\partial_i\phi - \frac{\partial L}{\partial \phi}\partial_i\phi - \pi^a\partial_i\partial_a\phi$$

$$= (\partial_i\phi)\left[\partial_a\pi^a - \frac{\partial L}{\partial \phi}\right] = 0. \qquad (2.18)$$

In arriving at the second equality, we have used $\partial_i\partial_a\phi = \partial_a\partial_i\phi$ to cancel out a couple of terms and the last equality follows from the equations of motion, Eq. (2.16). It is obvious that the quantity $T^a{}_i$ is conserved when the equations of motion are satisfied. It is also obvious that if we accept the expression for $T^a{}_i$ given in Eq. (2.17), then demanding $\partial_a T^a{}_i = 0$ will lead to the equations of motion for the scalar field. Integrating the conservation law $\partial_a T^a{}_i = 0$ over a four-volume and using the Gauss theorem we find that the quantity

$$P^i = \int d\sigma_k T^{ki} = \int d^3x\, T^{0i} \qquad (2.19)$$

is a constant which does not vary with time (see Eq. (1.61)). We will identify P^i with the total four-momentum of the field.

Since the absence of x^i in the Lagrangian is equivalent to the four-dimensional translational invariance of the Lagrangian, we see that the symmetry of four-dimensional translational invariance leads to the conservation of both energy and momentum at one go. In classical mechanics, time translation invariance leads to energy conservation and spatial translation invariance leads to momentum conservation, separately. But since the Lorentz transformation mixes space and time coordinates, the conservation law in relativity is for the four-momentum.

There is, however, one difficulty with this procedure. For a general Lagrangian, the quantity $T^{ab} = \eta^{bj}T^a{}_j$ obtained from Eq. (2.17) will not be symmetric in a and b. However, we have seen in Section 1.9 that the angular momentum, defined by Eq. (1.101), will be conserved only if the energy-momentum tensor is symmetric. To remedy this difficulty, we first note that, if T^{ik} is conserved, then any other tensor of the form

$$T^{ik}_{\text{new}} = T^{ik}_{\text{old}} + \partial_l S^{ikl}, \qquad S^{ikl} = -S^{ilk} \qquad (2.20)$$

is also conserved since $\partial_k \partial_l S^{ikl} = 0$ due to the antisymmetry of S^{ikl} in k and l. Further, this modification does not change the definition of total momentum. To see this, note that the modification in Eq. (2.20) changes the definition of total momentum by the term

$$\int d\sigma_k \partial_l S^{ikl} = \frac{1}{2} \int \left(d\sigma_k \partial_l S^{ikl} - d\sigma_l \partial_k S^{ikl} \right) = \frac{1}{2} \int S^{ikl} d\sigma_{ik}, \qquad (2.21)$$

where the final integration is over a two-dimensional surface which bounds the three-dimensional volume of space and thus is located at spatial infinity (see Eq. (1.62)). With the usual assumption that all fields vanish sufficiently fast at spatial infinity, this integral may be taken to be zero. It is then possible to choose a suitable form of S^{ikl} in order to make the tensor T_{new}^{ik} symmetric.

The results obtained for a scalar field Lagrangian can be directly generalized to any multi-component field. If the dynamical variable is made of an N-component object ϕ_A with $A = 1, 2, ...N$, then all the expressions obtained earlier hold for each component independently. For example, the energy-momentum tensor for a multi-component field will be a sum over the corresponding tensors for each of the components treated separately. This allows the results to be used for any other field such as, for example, a vector field with $N = 4$ independent components.

2.3.4 Free field and the wave solutions

Having described the general formalism, let us consider the explicit form for the Lagrangian for a scalar field. If the Lagrangian is Lorentz invariant and depends on the first derivatives at most quadratically, then its most general form is given by

$$L_{field} = -\frac{1}{2}\partial_a\phi \, \partial^a\phi - U(\phi), \qquad (2.22)$$

where $U(\phi)$ is an arbitrary scalar function of ϕ. (A seemingly more general Lagrangian with a kinetic term $(1/2)M(\phi)\partial_i\phi\partial^i\phi$ can be converted into this form by a field redefinition $\phi \to \psi$ with $\sqrt{|M|}\,d\phi = d\psi$.) So

$$L = L_{field} - \lambda n\phi \equiv -\frac{1}{2}\partial_a\phi \, \partial^a\phi - V(\phi) \qquad (2.23)$$

with $V = U + \lambda n\phi$. This form is analogous to $L = (1/2)\dot{q}^2 - V$ in classical mechanics.

The sign of the 'kinetic energy' term $\partial_a\phi\partial^a\phi$ term is chosen so as to ensure that the square of the time derivative appears with a positive sign. The factor of a half is introduced for future convenience and any other constant can be eliminated by rescaling the field ϕ. However, this choice fixes the dimension of ϕ and we will now introduce the standard convention regarding the dimensions of fields. The action has dimensions of angular momentum making $\mathcal{A}/\hbar$ dimensionless where $\hbar$ is the Planck constant. It is convenient therefore to choose units such that $c = \hbar = 1$ making action dimensionless. With this choice, all physical quantities can be expressed in length units; for example, mass $m(c/\hbar)$ and energy $E(1/c\hbar)$ have the dimensions of inverse length. For any field, the kinetic energy term in the action will be an integral over four-volume (with dimension L^4) of a quantity quadratic in the first derivatives of the field. It follows that *all* fields must have the dimension $1/L$ in units with $c = \hbar = 1$ if there are no other dimensional factor multiplying the action. Thus, our choice in Eq. (2.22) gives ϕ the natural dimension of $1/L$. Since $(\lambda\phi)\, d\tau$ is dimensionless, it follows that λ itself is dimensionless.

Let us next consider the field equations for the Lagrangian in Eq. (2.23). In this case, the field equations in Eq. (2.16) reduce to

$$\partial_a\partial^a\phi \equiv \Box\phi = \nabla^2\phi - \frac{\partial^2\phi}{\partial t^2} = \frac{\partial V}{\partial\phi} \tag{2.24}$$

while the canonical momentum has the components

$$\pi^a = -\partial^a\phi = -\eta^{ab}\partial_b\phi = (\dot{\phi}, -\nabla\phi). \tag{2.25}$$

The field equations can – in principle – be solved if $V(\phi)$ (that is, $U(\phi)$ and the external source n) is specified and we will come across specific cases in later chapters. Here we shall discuss the two simple cases.

The first one corresponds to $\lambda = 0$ and $U(\phi) = V(\phi) = m^2\phi^2$, which is quadratic in ϕ. When $\lambda = 0$ there is no coupling to the particle and we are studying a 'free' field. In this case, the field equations become $(\Box - m^2)\phi(x) = 0$. This equation is easily solved by introducing the four-dimensional Fourier transform of $\phi(x)$ by

$$\phi(x) = \int \frac{d^4k}{(2\pi)^4}\, \phi(k)\, e^{ikx} \tag{2.26}$$

in which the condensed notation kx stands for $k_i x^i$. We will use the same symbol $\phi(k), n(k), ...$, etc., to denote the Fourier transform of $\phi(x), n(x), ...$, etc., when no confusion is likely to arise. Substituting into the field equation we find that nontrivial solutions exist only for $k^i k_i = -m^2$ thereby determining $k^0 = \pm\omega_{\boldsymbol{k}} \equiv \pm\sqrt{\boldsymbol{k}^2 + m^2}$. Hence the general solution is given by

$$\phi(t, \boldsymbol{x}) = \int \frac{d^3\boldsymbol{k}}{(2\pi)^3}\, e^{i\boldsymbol{k}\cdot\boldsymbol{x}}\left(A(\boldsymbol{k})e^{-i\omega_{\boldsymbol{k}}t} + B(\boldsymbol{k})e^{i\omega_{\boldsymbol{k}}t}\right), \tag{2.27}$$

where $A(\boldsymbol{k})$ and $B(\boldsymbol{k})$ are arbitrary functions satisfying $A^*(\boldsymbol{k}) = B(-\boldsymbol{k})$ to ensure that ϕ is real. It is clear that the solution represents a superposition of waves with wave vector $\boldsymbol{k}$ and the frequency of the wave is given by the dispersion relation $\omega_{\boldsymbol{k}}^2 = \boldsymbol{k}^2 + m^2$ corresponding to a four-vector k^i with $k^i k_i = -m^2$. This allows the interpretation of k^i as the momentum four-vector of particles with mass m. This forms the basis of quantum field theory, which describes particles as excitations of an underlying field. In particular, when $m = 0$, the field ϕ describes massless particles. In normal units, the dispersion relation is $\omega^2 = k^2 c^2 + (m^2 c^4/\hbar^2)$; the parameter m has dimensions of inverse length in natural units.

The energy-momentum tensor in Eq. (2.17) is now given by the expression

$$T^a_b = [\partial^a \phi \, \partial_b \phi + \delta^a_b L] , \qquad (2.28)$$

which is manifestly symmetric; hence we do not have to resort to the procedure described in Eq. (2.20). Working out the components, we find that the energy density is given by

$$T_{00} = T^{00} = -T^0_0 = \frac{1}{2}\dot{\phi}^2 + \frac{1}{2}|\nabla\phi|^2 + V, \qquad (2.29)$$

which will be of use in several later chapters.

2.3.5 Why does the scalar field lead to an attractive force?

The second case we want to discuss corresponds to one with $\lambda \neq 0$, thereby coupling the source to the field. In this case $n(x)$ will generate a scalar field (just as a charged particle will generate an electromagnetic field). If we further take $U = 0$, then $V = \lambda n \phi$ and the field equation reduces to $\Box \phi = \lambda n$. Given the trajectory of the particle $z^i(\tau)$ in Eq. (2.11), we can solve this equation and determine the field produced by an arbitrarily moving particle. We shall not discuss it since it seems to have no practical relevance. However, there is an interesting question one can ask regarding the nature of the interaction between any two particles mediated through the ϕ field (which is analogous to the electromagnetic interaction between two charged particles): we want to determine whether the 'like charges' in such a theory attract each other or repel each other:

To analyse the issue, let us consider a Lagrangian with a slightly modified form:

$$L = -\frac{\nu}{2}\partial_a\phi\partial^a\phi - \lambda n\phi, \qquad (2.30)$$

in which we have added a parameter ν in the first term to analyse the effect of various possible choices for the relative signs in the Lagrangian. (Our original Lagrangian had $\nu = +1$.) The field equations now have the form $\nu \Box \phi = \lambda n$. Given a particular source distribution n, we can solve this and obtain the field ϕ. The simplest case corresponds to a static source distribution for which the equation reduces to $\nu \nabla^2 \phi = \lambda n$ with the solution

$$\phi(\boldsymbol{x}) = -\frac{\lambda}{4\pi\nu} \int d^3y \, \frac{n(\boldsymbol{y})}{|\boldsymbol{x} - \boldsymbol{y}|}. \qquad (2.31)$$

We want to compute the energy of this static configuration. For a Lagrangian in Eq. (2.30) the energy density for a static configuration is $T_{00} = (1/2)[\nu(\nabla\phi)^2 + 2\lambda n\phi]$. Therefore, the total energy is given by the integral

$$E = \int d^3x\, T_{00} = \int d^3x\, \frac{1}{2}[\nu(\nabla\phi)^2 + 2\lambda n\phi]. \tag{2.32}$$

We will now write $(\nabla\phi)^2 = \nabla \cdot (\phi\nabla\phi) - \phi\nabla^2\phi$ and use the Gauss theorem to convert the first term to a surface term at infinity. This will vanish if the field vanishes at infinity. In the remaining term we again use the field equation $\nu\nabla^2\phi = \lambda n$ to obtain

$$E = \frac{1}{2}\int d^3x\,[-\nu\phi\nabla^2\phi + 2\lambda n\phi] = \frac{\lambda}{2}\int d^3x\, n\phi. \tag{2.33}$$

Using the solution in Eq. (2.31), we finally get:

$$E = -\frac{\lambda^2}{8\pi\nu}\int d^3y\, d^3x\, \frac{n(\boldsymbol{y})n(\boldsymbol{x})}{|\boldsymbol{x} - \boldsymbol{y}|}. \tag{2.34}$$

There are two interesting features to note about this expression. First, our result is independent of the sign of λ and we could have taken either $\lambda = +1$ or $\lambda = -1$. (This is easy to see from the following argument as well. In Eq. (2.23), if we rescale ϕ to another field $\psi \equiv \lambda\phi$, the coupling constant λ disappears in the interaction term but $(1/\lambda^2)$ appears in the kinetic energy term for ψ. So, obviously the theory depends only on λ^2 and not on the sign of λ.) Second, if $\nu = +1$ the potential energy in Eq. (2.34) is negative and hence 'like charges' attract. On the other hand, if $\nu = -1$, the like charges repel. We saw that the field equation for the static source has the form $\nu\nabla^2\phi = \lambda n$. This equation has the form of the Poisson equation for the gravitational field, say, for both $\nu = +1, \lambda = 1$ or for $\nu = -1$ and $\lambda = -1$. Hence, just using the criterion that the equation should reduce to $\nabla^2\phi = n$ when $c \to \infty$ limit we cannot decide between the two possibilities. Thus, in non-relativistic field theory, one can have a scalar field which can produce either attraction or repulsion between like charges.

The situation, however, is different in the fully relativistic theory. Here the Lagrangian also must have a $\dot{\phi}^2$ term and it is necessary to have this term appearing with positive sign in the Lagrangian if the energy has to be bounded from below. (For example, we would like to have the plane wave solutions of the free field to have positive energy.) This requires $\nu = +1$ and we must have $\lambda = +1$ to get $\nabla^2\phi = n$. Such a theory has $E < 0$ for like charges showing that the like charges attract. This happens to be a special case of a general procedure that can be used to analyse any field. It will turn out that a vector field theory – like electromagnetism – will lead to repulsion, a second rank tensor field theory will lead to attraction, etc. We will say more about this in later chapters.

With future applications in mind we shall introduce an alternative way of writing the action functional when the Lagrangian has only up to quadratic terms of the field. Consider, for example, the action

$$A = -\frac{1}{2} \int d^4x [\partial_a \phi \, \partial^a \phi + m^2 \phi^2 + 2\lambda n \phi] \rightarrow -\frac{1}{2} \int d^4x [\phi(-\Box + m^2)\phi + 2\lambda n \phi],$$
$$(2.35)$$

where we have integrated the kinetic energy term by parts and neglected the surface term to arrive at the second expression. We now write $\phi(x)$ in terms of its Fourier transform $\phi(k)$ (see Eq. (2.26)) and also introduce the corresponding Fourier transform $n(k)$ for $n(x)$. This gives the action in the momentum space to be

$$A = -\frac{1}{2} \int \frac{d^4k}{(2\pi)^4} [(k^2 + m^2)|\phi(k)|^2 + 2\lambda n(-k)\phi(k)], \qquad (2.36)$$

where we have used the result $n^*(k) = n(-k)$ for real $n(x)$. One could have worked with this action and varied the $\phi(k)$ instead of working with our original action and varying $\phi(x)$. This action, of course, has no derivatives and is a quadratic polynomial. The variation will give $(k^2 + m^2)\phi(k) = -\lambda n(k)$ (which is just the Fourier space version of the real space equation $(\Box - m^2)\phi = \lambda n(x)$) with the particular solution

$$\phi(k) = -\lambda \frac{n(k)}{(k^2 + m^2)} \qquad (2.37)$$

relating the source to the field. Formally, we can think of this solution as $\phi(x) = (\Box - m^2)^{-1}[\lambda n(x)]$ where the inverse of the operator in the square bracket is defined in Fourier space. The existence of this inverse assures us that a unique solution can be obtained. These ideas will be useful in later discussion.

2.4 Electromagnetic field

We shall now take up the third term in Eq. (2.2) which couples a particle to a vector field A_i through a coupling constant q which we will call the *electric charge* of the particle. This theory can be developed exactly in analogy with the scalar field theory and will lead to electromagnetism. As in the case of the scalar field, we will first describe the dynamics of a (charged) particle coupled to a given external field. Then we shall introduce the action for the field A_i itself and study the nature of the electromagnetic field produced by a charged particle in different states of motion. Unless mentioned otherwise, we shall use units with $c = 1$.

2.4.1 Charged particle in an electromagnetic field

The action for a particle influenced by a vector field can be obtained from Eq. (2.2) by retaining the first and the third term. That is, we concentrate on

$$\mathcal{A} = \int_a^b \left(-md\tau + qA_i dx^i \right), \tag{2.38}$$

where we have used the fact that $u^i = (dx^i/d\tau)$. Introducing the components of A^i as $A^i = (\phi, \mathbf{A})$ with $A_i = (-\phi, \mathbf{A})$ we can read off the corresponding Lagrangian:

$$L = -m\sqrt{1 - v^2} - q\phi + q\mathbf{A} \cdot \mathbf{v}. \tag{2.39}$$

As in the case of a scalar field, we can find the equation of motion for the particle, by varying the action Eq. (2.38) with respect to the trajectory $x^i(\tau)$. This gives:

$$\delta\mathcal{A} = \int_a^b \left(m\frac{dx_i d\delta x^i}{d\tau} + qA_i d\delta x^i + q\delta A_i dx^i \right). \tag{2.40}$$

Integrating the first two terms by parts and using the relations

$$\delta A_i = \frac{\partial A_i}{\partial x^k}\delta x^k, \qquad dA_i = \frac{\partial A_i}{\partial x^k}dx^k, \tag{2.41}$$

we find that

$$\delta\mathcal{A} = \int_a^b \left(-m\frac{du_i}{d\tau}\delta x^i - q(\partial_k A_i)u^k \delta x^i + q(\partial_k A_i)u^i \delta x^k \right) d\tau$$

$$+ (mu_i + qA_i)\,\delta x^i \Big|_a^b. \tag{2.42}$$

In the third term, we interchange indices i and k (which changes nothing, since they are summed over), to obtain

$$\delta\mathcal{A} = \int_a^b \left[-m\frac{du_i}{d\tau} + qF_{ik}u^k \right]\delta x^i d\tau + (mu_i + qA_i)\,\delta x^i \Big|_a^b, \tag{2.43}$$

where we have defined the second rank, antisymmetric, tensor:

$$F_{ik} \equiv \partial_i A_k - \partial_k A_i. \tag{2.44}$$

As usual we will first consider variations in which $\delta x^i = 0$ at the end points. Demanding $\delta\mathcal{A} = 0$ for such variations leads to the equation of motion for the charged particle in a given external vector field:

$$m\frac{du^i}{d\tau} = qF^{ik}u_k. \tag{2.45}$$

Before discussing the implications of this result, let us also consider the second type of variation we are familiar with. If we treat the action as a function of the end points for a trajectory that satisfies the equation of motion, then we get the canonical momentum to be:

$$P_i = \frac{\partial \mathcal{A}}{\partial x^i} = mu_i + qA_i = p_i + qA_i. \tag{2.46}$$

We shall now describe several features of these results.

To understand these equations in more familiar terms, we shall substitute the components of the four-vector $A^i = (\phi, \boldsymbol{A})$ into the definition of F_{ik} in Eq. (2.44). Since it is an antisymmetric tensor it has only six independent components which can be separated into spacetime components $F^{0\alpha} \equiv E^\alpha$ and the space–space components $F^{\mu\nu}$. In three dimensions, the antisymmetric components can be expressed in terms of another three-vector B_α by $F^{\mu\nu} = \epsilon^{\mu\nu\alpha} B_\alpha$. Thus we can interpret everything in terms of the three components of $\boldsymbol{E}$ (electric field) and three components of $\boldsymbol{B}$ (magnetic field). It is easy to verify from the definition in Eq. (2.44) that the electric field, $\boldsymbol{E}$, is given in terms of the components of the vector field by:

$$\boldsymbol{E} = -\frac{\partial \boldsymbol{A}}{\partial t} - \operatorname{grad} \phi. \tag{2.47}$$

Similarly, the magnetic field is given by:

$$\boldsymbol{B} = \operatorname{curl} \boldsymbol{A}. \tag{2.48}$$

In terms of the components of $\boldsymbol{E}$ and $\boldsymbol{B}$, the matrix structure of F_{ik} is

$$F_{ik} = \begin{pmatrix} 0 & -E_x & -E_y & -E_z \\ E_x & 0 & B_z & -B_y \\ E_y & -B_z & 0 & B_x \\ E_z & B_y & -B_x & 0 \end{pmatrix}, \quad F^{ik} = \begin{pmatrix} 0 & E_x & E_y & E_z \\ -E_x & 0 & B_z & -B_y \\ -E_y & -B_z & 0 & B_x \\ -E_z & B_y & -B_x & 0. \end{pmatrix}. \tag{2.49}$$

The definition of $\boldsymbol{E}$ and $\boldsymbol{B}$ in terms of ϕ and $\boldsymbol{A}$, implies that

$$\nabla \times \boldsymbol{E} = -\frac{\partial \boldsymbol{B}}{\partial t}; \qquad \operatorname{div} \boldsymbol{B} = 0, \tag{2.50}$$

which can be directly verified from Eq. (2.47) and Eq. (2.48). This is equivalent to the Lorentz covariant equation satisfied by F_{ik}:

$$\partial_a F_{bc} + \partial_b F_{ca} + \partial_c F_{ab} = 0, \tag{2.51}$$

which can be verified from the definition in Eq. (2.44).

This equation can be written in a more convenient form by introducing the dual $(*F)^{cd}$ of the tensor F_{ab} by the standard definition

$$(*F)^{cd} = \epsilon^{cdab} F_{ab} = 2\epsilon^{cdab} \partial_a A_b, \tag{2.52}$$

where we have used the antisymmetry of ϵ_{cdab} in a, b. Again, from the antisymmetry of ϵ_{cdab} in c, a leads to the identity:

$$\partial_c(*F)^{cd} = 2\epsilon^{cdab}\partial_c\partial_a A_b = 0. \tag{2.53}$$

This is equivalent to Eq. (2.50). We will see later in Chapter 11 that this equation has a geometric interpretation in terms of certain structures called exterior derivatives.

Expressing the components of F^{ik} in terms of $\boldsymbol{E}$ and $\boldsymbol{B}$ in the equations of motion, the spatial part of the equation Eq. (2.45) can be written in three-dimensional form as

$$\frac{d\boldsymbol{p}}{dt} = q\boldsymbol{E} + q\boldsymbol{v} \times B. \tag{2.54}$$

(Note that the left hand side is $(d\boldsymbol{p}/dt)$ and not $(d\boldsymbol{p}/d\tau)$.) In the non-relativistic limit, $\boldsymbol{p} \approx m\boldsymbol{v}$ and this reduces to the familiar Lorentz force equation for a charged particle in an electromagnetic field. Since this equation completely determines the motion of the charged particle in a given electromagnetic field, it follows that the zeroth component of Eq. (2.45) should not give any new information. This is indeed true. Using the expression for the energy $\mathcal{E}$ of the particle (we use the symbol $\mathcal{E}$ to avoid confusion with the electric field E) in terms of the momentum, $\mathcal{E}^2 = p^2 + m^2$, we have

$$\frac{d\mathcal{E}}{dt} = \frac{1}{\mathcal{E}}\,\boldsymbol{p} \cdot \frac{d\boldsymbol{p}}{dt} = \boldsymbol{v} \cdot \frac{d\boldsymbol{p}}{dt}. \tag{2.55}$$

Hence the time component of the Eq. (2.45) always gives the rate of change of the energy of the particle as equal to the work done by the external field. In our case

$$\frac{d\mathcal{E}}{dt} = q\boldsymbol{E} \cdot \boldsymbol{v}, \tag{2.56}$$

which, incidentally, shows that $\mathcal{E}$ is a constant for a particle moving in a purely magnetic field and the work is done only by the electric field.

It is also easy to write down the three-dimensional expression for any other physical quantity. For example, the spatial component $\boldsymbol{P}$ of the canonical momentum, defined by Eq. (2.46), is given by

$$\boldsymbol{P} = \gamma m\boldsymbol{v} + q\boldsymbol{A} = \boldsymbol{p} + q\boldsymbol{A}. \tag{2.57}$$

The corresponding Hamiltonian, $\mathcal{H} \equiv \boldsymbol{P} \cdot \boldsymbol{v} - L$, expressed in terms of the canonical momenta (with the c-factor reintroduced) will be

$$\mathcal{H} = \sqrt{m^2c^4 + c^2\left(\boldsymbol{P} - \frac{q}{c}\boldsymbol{A}\right)^2} + q\phi. \tag{2.58}$$

The non-relativistic limit of $\mathcal{H}$ is obtained by taking the $c \to \infty$ limit and subtracting the rest energy mc^2. This gives

$$\mathcal{H}_{\mathrm{NR}} = \frac{1}{2m}\left(\boldsymbol{P} - \frac{q}{c}\boldsymbol{A}\right)^2 + q\phi, \tag{2.59}$$

which is the Hamiltonian that governs the interaction of non-relativistic particles with the electromagnetic field.

The Hamilton–Jacobi equation for a charged particle in an electromagnetic field can be obtained by replacing P_i by $(\partial \mathcal{A}/\partial x^i)$ in the relation $p_i p^i = -m^2$. This equation is equivalent to writing Eq. (2.58) in the form

$$[\mathcal{H} - q\phi]^2 - (\boldsymbol{P} - q\boldsymbol{A})^2 = m^2 \tag{2.60}$$

and substituting $\mathcal{H} = -\partial \mathcal{A}/\partial t$ and $\boldsymbol{P} = \nabla \mathcal{A}$. This leads to the relativistic Hamilton–Jacobi equation for a charged particle in an electromagnetic field:

$$(\nabla \mathcal{A} - q\boldsymbol{A})^2 - \left(\frac{\partial \mathcal{A}}{\partial t} + q\phi\right)^2 + m^2 = 0. \tag{2.61}$$

The equations of motion for charged particles depend only on F_{ik} and not directly on the vector field A_j. This implies that by measuring the trajectories of charged particles we can only determine F_{ik} and not the vector field A_j. In fact, several different A_i can lead to the same F_{ab} and this fact leads to an important concept of *gauge invariance*. It is obvious from the definition in Eq. (2.44) that two different vector fields related by $A'_j \equiv A_j + \partial_j f$ for some function $f(x)$ lead to the same F_{ab}. This is called a gauge transformation and will play a key role in our later discussions.

Exercise 2.1
Measuring the $F^a{}_b$ At a given instant, we want to measure the components of $F^a{}_b$ by measuring the coordinate acceleration of a set of nearby test particles moving in the field. Show that three particles are required to do this reliably.

Exercise 2.2
Schrödinger equation and gauge transformation Consider the Schrödinger equation for a particle in an electromagnetic field expressed in the form $i\hbar(\partial \psi/\partial t) = \mathcal{H}_{\mathrm{NR}}\psi$, where $\mathcal{H}_{\mathrm{NR}}$ is given by the operator corresponding to the expression in Eq. (2.59). Show that, under the gauge transformation $A_i \to A_i + \partial_i f$, the wave function transforms to $\psi \to \psi \exp[iqf]$.

Exercise 2.3
Four-vectors leading to electric and magnetic fields The electric and magnetic field *cannot* be thought of as the spatial components of any intrinsic four-vector. But it is possible to do this if we also use the four-velocity u^a of an observer. We can define two four-vectors

$$E^a = F^{ab}u_b, \qquad B^a = \frac{1}{2}\epsilon^{abcd}\, u_b\, F_{cd} \tag{2.62}$$

such that the spatial components gives the electric and magnetic fields as measured by the observer with four-velocity u^i. (a) Show that both these four-vectors are orthogonal to the world line of the observer; i.e. $E^i u_i = B^i u_i = 0$. (b) More importantly, show that F^{ab} can be constructed from E^i and B^i by

$$F^{ab} = u^a E^b - E^a u^b - \epsilon^{ab}{}_{cd} u^c B^d. \tag{2.63}$$

Exercise 2.4
Hamiltonian form of action – charged particle Show that the Hamiltonian form of the action for a charged particle is given by

$$\mathcal{A} = \int_{\lambda_1}^{\lambda_2} d\lambda \left[P_a \dot{x}^a - \frac{1}{2} C \left(\frac{H}{m} + m \right) \right], \tag{2.64}$$

where $H = \eta_{ij}(P^i - qA^i)(P^j - qA^j)$ and C is an auxiliary variable. The parameter λ is treated as arbitrary in the action. Prove that varying x^a, P^a, C independently (with C set to unity at the end) leads to the correct equations of motion for the charged particle and fixes the parameter λ to be the proper time.

Exercise 2.5
Three-dimensional form of the Lorentz force Show that the acceleration of a charged particle moving in an electromagnetic field, in the three-dimensional notation, is:

$$\frac{d\boldsymbol{v}}{dt} = \frac{q}{m} \sqrt{1 - \frac{v^2}{c^2}} \left[\boldsymbol{E} + \boldsymbol{v} \times \boldsymbol{B} - \frac{1}{c^2}(\boldsymbol{v} \cdot \boldsymbol{E})\boldsymbol{v} \right]. \tag{2.65}$$

Exercise 2.6
Pure gauge imposters A vector potential of the form $A_j = \partial_j f$ should represent a pure gauge with $F_{ik} = 0$. This is true as long as f is a sensible, single valued, function but every once in a while one needs to be careful about imposters which look like pure gauge. Consider for example, a four-vector potential with $A^0 = 0$ and $A_\alpha = \partial_\alpha f(x, y)$ with $f(x, y) = \tan^{-1}(y/x)$. Evaluate the line integral of $\boldsymbol{A}$ around a circle of unit radius in the x–y plane and show that it is nonzero, implying there is a nonzero magnetic field. Explain why this $\boldsymbol{A}$ is not a pure gauge mode, in spite of the appearance.

2.4.2 Lorentz transformation of electric and magnetic fields

Since a second rank tensor like F^{ik} transforms like the product of two four-vectors, we can easily find how the electric and magnetic fields change under the Lorentz transformation. A simple calculation shows that the components of the field parallel to the velocity $\boldsymbol{V}$ are unchanged while the components perpendicular to the velocity gets modified. This result can be expressed in a concise manner as

$$E' = \gamma E + \gamma V \times B - \frac{\gamma^2}{\gamma + 1} V(V \cdot E)$$

$$B' = \gamma B - \gamma V \times E - \frac{\gamma^2}{\gamma + 1} V(V \cdot B). \tag{2.66}$$

Clearly, the electric and magnetic fields are not Lorentz invariant quantities; the vanishing of the electric or magnetic field in one frame does not necessarily imply its vanishing in other inertial frames.

An interesting application of this result is to determine the electromagnetic field of a uniformly moving charged particle. This can be most easily obtained by transforming the Coulomb field in the rest frame to a moving frame. A charged particle at rest, at the origin of an inertial frame, produces the field $E = qR/R^3$, $B = 0$, where R is a vector from the position of the charge to the field point. Let us next consider the field produced by the same charge moving with a velocity V in the laboratory frame K. Taking the x-axis to be the direction of the velocity, we introduce another frame K' in which the charge is at rest. In K' we have the Coulomb field; transforming from K' to K using Eq. (2.66), and expressing the coordinates of K' in terms of that in K by a Lorentz transformation, we can easily compute the electromagnetic fields in K:

$$E = \frac{qR}{R^3} \frac{(1 - V^2/c^2)}{\left(1 - (V^2/c^2) \sin^2 \theta\right)^{3/2}}; \quad B = \frac{1}{c} V \times E, \tag{2.67}$$

where θ is the angle between the direction of motion and the radius vector R. The vector R now has the components $(x - Vt, y, z)$. It is worth noting that, in this particular case, the electric field is radially directed from the *instantaneous* position of the charged particle.

Even though electric and magnetic fields are not Lorentz invariant, there are two combinations we can form out of the electric and magnetic fields which remain invariant under Lorentz transformations. This is obvious from the fact that we can construct from the second rank antisymmetric tensor F_{ik} the invariant quantities $F_{ik}F^{ik}$ and $\epsilon_{iklm}F^{ik}F^{lm}$. Working out these expressions in terms of electric and magnetic fields, we find that

$$F_{ik}F^{ik} = F_{0\alpha}F^{0\alpha} + F_{\mu 0}F^{\mu 0} + F_{\mu\nu}F^{\mu\nu}$$
$$= 2(-E^2) + \epsilon_{\mu\nu\rho}\epsilon^{\mu\nu\sigma}B^\rho B_\sigma = 2(B^2 - E^2), \tag{2.68}$$

where we have used $F^{\mu\nu} = \epsilon^{\mu\nu\rho}B_\rho$ and $\epsilon^{\mu\nu\rho}\epsilon_{\mu\nu\sigma} = 2\delta^\rho_\sigma$. Similarly

$$\epsilon_{iklm}F^{ik}F^{lm} = 4\epsilon_{0\alpha\mu\nu}F^{0\alpha}F^{\mu\nu} = 4\epsilon_{0\alpha\mu\nu}E^\alpha(\epsilon^{\mu\nu\rho}B_\rho) = -8(E \cdot B). \tag{2.69}$$

In the first equality we have used the fact that any one of the four indices can be 0 leading to the factor 4; in arriving at the last equality we have used $\epsilon_{0\alpha\mu\nu}\epsilon^{\mu\nu\rho} = -2\delta_\alpha^\rho$. Both these combinations are invariant under Lorentz transformations. There is no other independent quadratic invariant. This is because any invariant quantity should also be invariant under spatial rotations and hence has to be made from E^2, B^2 and $\boldsymbol{E}\cdot\boldsymbol{B}$. If there is another independent invariant, then we can express B^2, say, in terms of these invariants. But since Eq. (2.66) shows that B^2 changes under Lorentz transformations, this cannot be true. Therefore, we cannot have another independent quadratic invariant.

2.4.3 Current vector

So far we have considered a single charge in an external electromagnetic field. If there are several charges, we have to add up the terms for each of the particles which will give, in place of Eq. (2.38),

$$\mathcal{A} = -\sum_A \int m_A d\tau_A + \sum_A \int q_A A_k u^k d\tau_A, \tag{2.70}$$

where m_A, q_A and τ_A corresponds to the mass, charge and proper time of the Ath particle. When a large number of charges are present, it is convenient to introduce a charge density ρ such that $dq = \rho dV$ gives the amount of charge in an infinitesimal region of volume dV. Multiplying both sides of this relation by dx^i we can write

$$dq dx^i = \rho dV dx^i = \rho dV dt \frac{dx^i}{dt}. \tag{2.71}$$

The left hand side is a four-vector (since dq is Lorentz invariant) and on the right hand side $d^4x = dt dV$ is a scalar; so the combination,

$$J^i = \rho \frac{dx^i}{dt}, \tag{2.72}$$

must be a four-vector and is called the *current vector*. In the action, the summation over the charges q_A involving $q_A u^k d\tau_A$ can be replaced by an integration over $u^k d\tau_A \rho dV = dx^k (\rho dV dt)/dt = J^k d^4x$. Hence we can write the action Eq. (2.70) as

$$\mathcal{A} = -\sum_A \int m_A d\tau_A + \int A_i J^i d^4x. \tag{2.73}$$

This form – which is analogous to Eq. (2.12) in the case of the scalar field – is useful for further generalizations. In fact, one can express the current vector for a single charge with a trajectory $z^a(\tau)$ as an integral over proper time as

$$J^i(x^a) = q \int_{-\infty}^{\infty} d\tau \, \delta_D[x^a - z^a(\tau)] \left(\frac{dz^i}{d\tau} \right). \tag{2.74}$$

(This is analogous to Eq. (2.11) for the scalar field; recall that $\delta_D[x^a - z^a(\tau)]$ is a condensed notation for the product of four Dirac delta functions for each of the components of $x^i = (t, x, y, z)$ in Cartesian coordinates.) For any function $F(\tau)$, we have the result

$$\int F(\tau) \, \delta[t - z^0(\tau)] d\tau = \int F(\tau) \, \delta[t - z^0(\tau)] \frac{d\tau}{dt} \, dt = \frac{F(\tau[t])}{u^0}, \tag{2.75}$$

which allows us to rewrite Eq. (2.74) as

$$J^i = q \left(\frac{u^i}{u^0} \right) \delta_D [\boldsymbol{x} - \boldsymbol{z}(t)] = \rho \frac{dx^i}{dt}, \tag{2.76}$$

in agreement with Eq. (2.72).

We saw earlier that the equations of motion for a charged particle depend only on F_{ab} and hence are invariant under the gauge transformation $A_j \to A_j + \partial_j f$. In the action in Eq. (2.73), the gauge transformation leads to the change

$$\mathcal{A} \to \mathcal{A} + \int J^i \partial_i f d^4 x = \mathcal{A} + \int \partial_i (J^i f) d^4 x - \int f (\partial_i J^i) d^4 x. \tag{2.77}$$

The term with the total divergence $\partial_i(f J^i)$ can be transformed to a surface integral in infinity. Let us choose f such that it vanishes sufficiently fast at large distances so that this term vanishes. Then, the invariance of the action under gauge transformation requires that the third term must vanish. Since f is arbitrary, this requires $\partial_i J^i = 0$ which is the same as the requirement of charge conservation. This shows that the vector field can be coupled only to a conserved current if gauge invariance is to be respected.

Exercise 2.7

Pure electric or magnetic fields Let the electric and magnetic fields at a given event be $\boldsymbol{E}$ and $\boldsymbol{B}$. We attempt to make a Lorentz transformation to a different frame such that, in the neighbourhood of this event, the electromagnetic field is either purely electric or purely magnetic in the transformed frame.

(a) When is this impossible?

(b) When it is possible, obtain a Lorentz invariant condition on $\boldsymbol{E}$ and $\boldsymbol{B}$ which decides whether the field will be purely electric or purely magnetic in the new frame.

(c) Express the velocity of the new Lorentz frame (in which the field is purely electric or magnetic) in terms of $\boldsymbol{E}$ and $\boldsymbol{B}$.

2.5 Motion in the Coulomb field

As a first application of the formalism we have developed, let us consider the motion of a charged particle (with charge q, mass m) in a *Coulomb field* given by $\phi = e/r$, $A = 0$. In the non-relativistic limit, this would correspond to the (electrostatic) Kepler problem and the trajectory will be a conic section. Bound orbits will be closed ellipses and the unbound orbits will correspond to the Rutherford scattering in the Coulomb potential. The situation is quite different in the ultra-relativistic case and the techniques we develop here to analyse this problem will be of use later on in the study of particle motion in general relativity in Chapter 7.

The most convenient procedure to study this problem is to use the Hamilton–Jacobi equation. In the exact, relativistic, case the Hamilton–Jacobi equation (see Eq. (2.61)) is

$$- \left(\frac{\partial \mathcal{A}}{\partial t} + \frac{\alpha}{r} \right)^2 + \left(\frac{\partial \mathcal{A}}{\partial r} \right)^2 + \frac{1}{r^2} \left(\frac{\partial \mathcal{A}}{\partial \theta} \right)^2 + m^2 = 0, \qquad (2.78)$$

where (r, θ) are the polar coordinates on the plane of the motion and $\alpha \equiv qe$. When $\alpha > 0$ both charges have the same sign and the force is repulsive; when $\alpha < 0$ the charges have opposite signs and the force is attractive. Since the energy ($\mathcal{E}$) and the angular momentum (J) are conserved for this motion, we can separate the action as $\mathcal{A} = -\mathcal{E}t + J\theta + f(r)$. (This is, of course, obvious from the structure of Eq. (2.78) as well.) Substituting this into Eq. (2.78) and solving for $f(r)$, we find that the action is given by

$$\mathcal{A} = -\mathcal{E}t + J\theta + \int dr \sqrt{\frac{1}{c^2} \left(\mathcal{E} - \frac{\alpha}{r} \right)^2 - \frac{J^2}{r^2} - m^2 c^2}, \qquad (2.79)$$

where we have reintroduced the c-factor. The trajectory $r(\theta)$ can be determined by differentiating this expression with respect to J and equating it to another constant, say, $-\theta_0$. The time dependence of $r(t)$ can be determined, similarly, by differentiating with respect to $\mathcal{E}$ and equating it to another constant, say, t_0. Simplifying these two resulting equations we find that the orbit is determined by

$$\left(\frac{dr}{d\theta} \right)^2 = \frac{r^4}{J^2} \left[\frac{1}{c^2} \left(\mathcal{E} - \frac{\alpha}{r} \right)^2 - \frac{J^2}{r^2} - m^2 c^2 \right], \qquad (2.80)$$

while the time dependence is decided by

$$\left[\mathcal{E} - \frac{\alpha}{r} \right]^2 \left(\frac{dr}{cdt} \right)^2 = \left[\mathcal{E}^2 - m^2 c^4 - \frac{J^2 c^2}{r^2} \left(1 - \frac{\alpha^2}{J^2 c^2} \right) - \frac{2\alpha \mathcal{E}}{r} \right]. \qquad (2.81)$$

It is now obvious that, in general, the behaviour is quite different from the non-relativistic Kepler problem. The net sign of the $(1/r^2)$ term in Eq. (2.80) or Eq. (2.81) will depend on whether Jc is greater than $|\alpha|$ or not. When $Jc < |\alpha|$, the third term in the square brackets in Eq. (2.81) has a $(-1/r^2)$ behaviour near the origin. Hence the trajectory will spiral to the origin; this is quite unlike the non-relativistic case in which a nonzero angular momentum, however small, will prevent the orbit from reaching the origin. This arises because the sign of the $(1/r^2)$ term is always positive in the non-relativistic case but can be negative in the relativistic case. This possible change of sign makes the angular momentum inadequate – in general – for preventing the collapse to the origin.

Equation (2.80) can be integrated in closed form for all the range of parameters and we will briefly describe one case and mention the results for the others. If we introduce the variable $u = 1/r$ into Eq. (2.80) and differentiate the resulting equation again with respect to θ, we get

$$\frac{d^2 u}{d\theta^2} + \omega^2 u = -\frac{\alpha \mathcal{E}}{c^2 J^2}; \qquad \omega^2 \equiv 1 - \frac{\alpha^2}{J^2 c^2}. \tag{2.82}$$

This is a harmonic oscillator equation with a constant forcing term and can be solved easily. Let us concentrate on the case with $Jc > |\alpha|$; in this case, if $\alpha = -|\alpha|$ is negative, the field is attractive and we have bound motion for $\mathcal{E} < mc^2$. The trajectory obtained by solving Eq. (2.82) can be expressed in the form

$$\frac{1}{r} = \frac{1}{R} \cos(\omega \theta) - \frac{\mathcal{E}\alpha}{c^2 J^2 \omega^2}, \tag{2.83}$$

where

$$R \equiv \frac{J\omega^2}{mc} \left[\left(\frac{\mathcal{E}}{mc^2} \right)^2 - 1 + \frac{\alpha^2}{c^2 J^2} \right]^{-1/2} \tag{2.84}$$

is a constant. In a more familiar form, the trajectory is $l/r = (1 + e \cos \omega \theta)$ with

$$l = \frac{c^2 J^2 \omega^2}{\mathcal{E}|\alpha|}; \qquad e^2 = \frac{J^2 c^2}{\alpha^2} \left[1 - \frac{m^2 c^4 \omega^2}{\mathcal{E}^2} \right]. \tag{2.85}$$

It is easy to verify that, when $c \to \infty$, this reduces to the standard equation for an ellipse in the Kepler problem. In terms of the non-relativistic energy $E = \mathcal{E} - mc^2$, we get, to leading order, $\omega \approx 1, l \approx J^2/m|\alpha|$ and $e^2 \approx 1 + (2EJ^2/m\alpha^2)$, which are the standard results. In the fully relativistic case all these expressions change but the key new effect arises from the fact that $\omega \neq 1$. Because of this reason, the trajectory is not closed and the ellipse precesses. When $\omega \neq 1$ the r in Eq. (2.83) does not return to the value at $\theta = 0$ when $\theta = 2\pi$; instead, we need a further turn by $\Delta\phi$ (the 'angle of precession') for r to return to the original value. This is determined by the condition $(2\pi + \Delta\phi)\omega = 2\pi$. From Eq. (2.83) we find that the orbit precesses by the angle

$$\Delta\phi = 2\pi \left[\left(1 - \frac{\alpha^2}{c^2 J^2} \right)^{-1/2} - 1 \right] \simeq \frac{\pi\alpha^2}{c^2 J^2} \tag{2.86}$$

per orbit, where the second expression is valid for $\alpha^2 \ll c^2 J^2$. This is a purely relativistic effect and vanishes when $c \to \infty$.

The orbit equation can be integrated for other cases as well and – for the sake of completeness – we quote the results. For the other two cases $Jc < |\alpha|$ and $Jc = |\alpha|$ the trajectories are:

$$(\alpha^2 - c^2 J^2) \frac{1}{r} = \pm c\sqrt{(J\mathcal{E})^2 + m^2 c^2 (\alpha^2 - J^2 c^2)} \cosh\left(\theta\sqrt{\frac{\alpha^2}{c^2 J^2} - 1} \right) + \mathcal{E}\alpha, \tag{2.87}$$

$$\frac{2\mathcal{E}\alpha}{r} = \mathcal{E}^2 - m^2 c^4 - \theta^2 \left(\frac{\mathcal{E}\alpha}{cJ} \right)^2, \tag{2.88}$$

respectively. We have also chosen the initial conditions such that $\theta_0 = 0$ for simplicity. For $Jc < |\alpha|$, we take the positive root for $\alpha < 0$ and negative root for $\alpha > 0$.

Exercise 2.8

Elegant solution to non-relativistic Coulomb motion Consider a particle moving under a central force $f(r)\hat{r}$ in non-relativistic mechanics. It follows from the symmetries that the total energy E and the angular momentum J are conserved. Since the particle has a six-dimensional phase space, the conservation of these four quantities should confine the motion to a $(6 - 4) = 2$ dimensional space. The projection of the phase space trajectory on to the real space, say the x–y plane, will also be two-dimensional and the orbit will fill a finite two-dimensional region in this space. This is what happens in general.

(a) Show that the time derivative of the quantity $p \times J$ is given by

$$\frac{d}{dt}(p \times J) = -mf(r)r^2 \frac{d\hat{r}}{dt}, \tag{2.89}$$

where $\hat{r} = r/r$. Hence show that, for the non-relativistic, attractive, Coulomb motion problem with $f(r)r^2 = -|\alpha|$ there is an extra conserved quantity given by

$$e \equiv \frac{A}{m|\alpha|} \equiv \frac{1}{m|\alpha|} p \times J - \hat{r}. \tag{2.90}$$

(b) Prove that this *eccentricity vector* e points along the major axis of the ellipse and has a magnitude equal to the eccentricity of the orbit – that is, $|e|^2 = 1 + (2EJ^2/m\alpha^2)$, where $E = p^2/2m - |\alpha|/r$ is the conserved energy for the motion. (This vector was first introduced by Hamilton and is known by many names including the Runge–Lenz vector.) Hence argue that the motion in this case will be one-dimensional and not two-dimensional.

(c) Show that $e \cdot r$ has the value $er\cos\theta = (J^2/m|\alpha|) - r$. Hence determine the trajectory $r(\theta)$ without solving any differential equation.

(d) Assume now that the potential has a small $(1/r^2)$ perturbation and is given by $V = -(|\alpha|/r) - (\beta/r^2)$. Argue that the change in the eccentricity vector per orbit is given by

$$\Delta \boldsymbol{A}\big|_{\text{orbit}} = 2\beta m \int_0^{2\pi} \frac{\boldsymbol{r}}{r^3} \frac{dr}{d\theta} d\theta. \tag{2.91}$$

Using this, show that the precession of the perihelion of the ellipse per orbit is given by

$$\Delta\theta = \frac{\Delta A_y}{A} = -\frac{2\beta m}{A} \int_0^{2\pi} \sin\theta \frac{du}{d\theta} d\theta = \frac{2\pi\beta m}{J^2}. \tag{2.92}$$

Find the exact solution to the orbit equation for the potential $V = -(|\alpha|/r + \beta/r^2)$ and compare the results.

There is another situation that can be discussed using our result in Eq. (2.83). Consider the unbound motion of a particle which scatters off the Coulomb potential. In this case, we are interested in the deviation of the trajectory from the straight line path as the particle comes in from a large distance and returns to a large distance, and we can simplify the expressions by assuming that $\alpha^2 \ll c^2 J^2$. Then the trajectory in Eq. (2.83) simplifies to

$$\frac{1}{r} \cong \frac{1}{R} \cos\theta - \frac{\alpha\mathcal{E}}{c^2 J^2}; \qquad R \cong \frac{J}{p_\infty}, \tag{2.93}$$

where p_∞ is the momentum of the particle, corresponding to energy $\mathcal{E}$, when it is far away from the origin. In the absence of the field ($\alpha = 0$), the trajectory is a straight line with $r\cos\theta = x = R$ so that the quantity R is the impact parameter for the scattering of the two charged particles. The second term $(\alpha\mathcal{E}/c^2 J^2)$ on the right hand side of Eq. (2.93) introduces a small correction to this straight line trajectory when $\alpha^2 \ll c^2 J^2$. In this limit, the above trajectory describes the (small) deflection of a charged particle in the external field. The asymptotic directions, $\pm\theta_c$, of the charged particle can be obtained by setting $r^{-1} = 0$ in the trajectory, which gives $\cos\theta_c \approx (\alpha R\mathcal{E}/c^2 J^2)$. If we write $\theta_c = (\pi/2) - \psi$, so that $\cos\theta_c = \sin\psi \cong \psi = (\alpha R\epsilon/c^2 J^2)$, the total deflection χ is

$$\chi = 2\psi \cong \frac{2\alpha}{Rc^2} \frac{\mathcal{E}}{p_\infty^2} = \frac{2\alpha}{mRv_\infty^2} \left(1 - \frac{v_\infty^2}{c^2}\right)^{1/2}, \tag{2.94}$$

where v_∞ denotes the speed corresponding to the momentum p_∞. The deflection is $\chi = [(\alpha/R)/(mv^2/2)]$ in the non-relativistic case (which is the ratio between the electrostatic potential energy at the impact parameter and the kinetic energy at infinity) and *goes to zero* for an ultra-relativistic particle as $v_\infty \to c$. We shall see later (see Exercise 6.14 and Eq. (7.142)) that the corresponding result in the case of gravity has a factor $[1 + (v_\infty^2/c^2)]$ so that, in the $v_\infty \to c$ limit, the result is twice the $(v_\infty/c) \ll 1$ limit.

2.6 Motion in a constant electric field

As a second example, we shall consider the motion of a charged particle in a constant electric field. This corresponds to a motion with constant acceleration in special relativity (in the sense that $a^i a_i$ = constant) and will play a crucial role in our future discussions. Taking the direction of the electric field to be along the x-axis and the plane of motion as the x–y plane, the force equation Eq. (2.54) becomes $\dot{p}_x = qE$, $\dot{p}_y = 0$, which may be integrated to give $p_x = qEt$, $p_y = p_0$ with a suitable choice of integration constants. The energy of the particle is given by

$$\mathcal{E} = \sqrt{m^2 + p_0^2 + (qEt)^2} \equiv \sqrt{\mathcal{E}_0^2 + (qEt)^2}. \tag{2.95}$$

Using Eq. (1.80), we find that v_x is given by

$$v_x = \frac{dx}{dt} = \frac{p_x}{\mathcal{E}} = \frac{qEt}{\sqrt{\mathcal{E}_0^2 + (qEt)^2}}. \tag{2.96}$$

Integrating again, we obtain the trajectory which is a hyperbola

$$x^2 - c^2 t^2 = \left(\frac{\mathcal{E}_0}{qE}\right)^2, \tag{2.97}$$

with a suitable choice for the integration constant. The motion along the y-axis can also be determined by the same method to give

$$y = \frac{p_0}{qE} \sinh^{-1}\left(\frac{qEt}{\mathcal{E}_0}\right). \tag{2.98}$$

Eliminating t between these equations, we find that the trajectory in the x–y plane is given by $x = (\mathcal{E}_0/qE)\cosh(qEy/p_0 c)$, which is a catenary. This is a simple example of motion with constant acceleration in special relativity. When $c \to \infty$, the trajectory will reduce to a parabola, which is a well known non-relativistic result.

Of particular importance is the one-dimensional motion of a particle along the x-axis with a uniform acceleration g. This can be obtained from the above results by setting $p_0 = 0$, $\mathcal{E}_0 = m$ and $(qE/m) = g$. Using the units with $c = 1$ for convenience, the trajectory now becomes $x^2 - t^2 = g^{-2}$, which is a hyperbola in the x–t plane asymptotically approaching the light cone $x = \pm t$ at very early and very late times. The proper time τ shown by a clock in such a uniformly accelerated motion can be related to the coordinate time by the standard result

$$\tau = \int_0^t \sqrt{1 - v^2}\, dt = \frac{1}{g}\sinh^{-1}(gt). \tag{2.99}$$

Using this result, we can express the trajectory in a parameterized form in terms of the proper time as

$$gx = \cosh(g\tau); \quad gt = \sinh(g\tau). \tag{2.100}$$

Further, this leads to the magnitude of the velocity being $v = \tanh(g\tau)$ with the rapidity $\chi = g\tau$ increasing linearly with proper time.

Exercise 2.9

More on uniformly accelerated motion   Derive the following results for a uniformly accelerated motion along the x-axis. (We have reintroduced the c-factor to indicate appropriate limits.) These results will be useful in several later sections.

(a) Show that the trajectory can be taken to be

$$t(\tau) = \frac{c}{g} \sinh \frac{g\tau}{c}, \quad x(\tau) = \frac{c^2}{g}\left(\cosh \frac{g\tau}{c} - 1\right), \quad y(\tau) = 0, \quad z(\tau) = 0 \tag{2.101}$$

to ensure the correct $c \to \infty$ limit. (This differs from the one given in the text by an integration constant.) Obtain the non-relativistic limit of this trajectory.

(b) Show that the four-velocity is given by

$$u^i = \left(c \cosh \frac{g\tau}{c}, \ c \sinh \frac{g\tau}{c}, 0, 0\right). \tag{2.102}$$

(c) A photon with frequency ν_0 is emitted from the origin of the inertial frame. If the trajectory of the photon intersects that of the uniformly accelerated observer at proper time τ, show that (s)he will attribute to it a frequency

$$\nu(\tau) = \nu_0 \exp\left(-\frac{g\tau}{c}\right). \tag{2.103}$$

Exercise 2.10

Motion of a charge in an electromagnetic plane wave   Consider a plane electromagnetic wave pulse, travelling along the x-axis and described by a vector potential of the form $A^j = \delta^j_x F(x - t)$, where F is some arbitrary function sharply peaked around the origin. (We will discuss these solutions in the next section.) A charged particle is moving under the action of this wave pulse. Use the Hamilton–Jacobi method to study the motion of this charged particle and show that its trajectory can be represented as

$$x(\tau) = \frac{q^2}{2m^2} \int_0^\tau F^2(-\tau') \, d\tau'; \quad y(\tau) = -\frac{q}{m} \int_0^\tau F(-\tau') \, d\tau';$$
$$t(\tau) = \tau + x(\tau); \quad z(\tau) = 0. \tag{2.104}$$

Compute the amount of energy, momentum and angular momentum conveyed to the charged particle if F is a sharply peaked function describing a wave packet.

2.7 Action principle for the vector field

The action in Eq. (2.38) describes the effect of an external electromagnetic field on a charged particle and is similar to that in Eq. (2.4) for the scalar field. This action, however, is incomplete for the reasons described in the beginning of Section 2.3.2 and we need to add a term to the action describing the dynamics of the field. (This is analogous to the addition of $\mathcal{A}_{\text{field}}$ in Eq. (2.10) in the case of a scalar field.)

The action representing the field will be expressible as an integral over the four-volume d^4x of some scalar Lagrangian (density) $L = L(A^i, \partial_j A^i)$, which could be a function of the potential A^i and its first derivative. We have seen earlier that the equations of motion for the charged particle and the coupling of a conserved current J^i to the electromagnetic field respect the gauge transformation: $A'_i = A_i + \partial_i f$. Since the potential is not directly observable we shall demand that the action for the field should also be invariant under the gauge transformation.

Further, experiments show that electromagnetic fields obey the *principle of superposition*, viz. that the field due to two independently specified charge distributions is the sum of the fields produced by each of them in the absence of the other. For this to be true the differential equations governing the dynamics have to be linear in the field; alternatively, the Lagrangian can be at most quadratic in the field variable. We will now determine the form of this action. In the case of electromagnetism, one can obtain the correct result more simply by other routes. We will, however, provide a formal analysis since it will be useful in the study of a spin-2 tensor field in Chapter 3.

Clearly, the Lagrangian must have the structure $M^{ijkl}\partial_i A_j \partial_k A_l$, where M^{ijkl} is a suitable fourth rank tensor which has to be built out of the only two covariant tensors η^{ij} and ϵ^{ijkl}. One obvious choice, of course, is $M^{ijkl} = \epsilon^{ijkl}$. To determine the terms that can be constructed from η^{ij}, we note that there are essentially three different ways of pairing the indices in $\partial_i A_j \partial_k A_l$ with a product of two ηs. One can contract: (1) i with j and k with l; (2) i with k and j with l; (3) i with l and j with k. Adding up all these possible terms with arbitrary coefficients leads to a Lagrangian of the form

$$L = c_1 \epsilon^{ijkl}\partial_i A_j \partial_k A_l + c_2 (\partial_i A^i)^2 + c_3 (\partial_i A_j \partial^i A^j) + c_4 (\partial_i A_j \partial^j A^i). \quad (2.105)$$

It is convenient at this stage to express $\partial_i A_j$ as $(1/2)[F_{ij} + S_{ij}]$, where S_{ij} is the symmetric tensor $S_{ij} \equiv \partial_i A_j + \partial_j A_i$ which complements the information contained in the antisymmetric part F_{ij}. The last three terms in Eq. (2.105) can be expressed in terms of these two tensors and, using the fact that $F_{ij}S^{ij} = 0$, one can easily work out the Lagrangian to be of the form

$$4L = 4c_1\epsilon^{ijkl}\partial_i A_j \partial_k A_l + c_2(S_k^k)^2$$
$$+ c_3(F_{ik}F^{ik} + S_{ik}S^{ik}) + c_4(S_{ik}S^{ik} - F_{ik}F^{ik})$$
$$= 4c_1\epsilon^{ijkl}\partial_i A_j \partial_k A_l + c_2(S_k^k)^2 + (c_3 - c_4)F_{ik}F^{ik} + (c_3 + c_4)S_{ik}S^{ik}.$$

$$(2.106)$$

In the first term with c_1, we can replace $\partial_i A_j \partial_k A_l$ by $F_{ij}F_{kl}$ since the ϵ_{ijkl} assures that only the antisymmetric part contributes. Therefore, this term is clearly gauge invariant. But we can also write this term as

$$\epsilon^{ijkl}\partial_i A_j \partial_k A_l = \partial_i[\epsilon^{ijkl}A_j\partial_k A_l] + \epsilon^{ijkl}A_j\partial_i\partial_k A_l = \partial_i[\epsilon^{ijkl}A_j\partial_k A_l], \quad (2.107)$$

where the second equality arises from the fact that $\partial_i\partial_k A_l$ is symmetric in i and k but ϵ_{ijkl} is completely antisymmetric, making the contraction vanish. The surviving term in Eq. (2.107) is a four-divergence which – when integrated over all space, with the usual assumption that all fields vanish at spatial infinity – will contribute only at the two boundaries at $t = (t_1, t_2)$. So we only need to deal with the surface terms of the kind:

$$\int d^4x \partial_i[\epsilon^{ijkl}A_j\partial_k A_l] = \int_{t=con} d^3x[\epsilon^{0jkl}A_j\partial_k A_l] = \int_{t=con} d^3x[\epsilon^{0\alpha\beta\mu}A_\alpha\partial_\beta A_\mu].$$

$$(2.108)$$

(As an aside, note that the integrand in the above expression is essentially $\boldsymbol{A} \cdot \boldsymbol{B}$; it is usually called the *Chern–Simons term*.) In the variational principle we use to get the equations of motion, we will consider variations with A_μ fixed at the $t = t_1, t_2$ surfaces. If A_μ is fixed everywhere on this surface, its *spatial* derivatives are also fixed and hence the term in Eq. (2.108) will not vary. Hence this term does not make a contribution. (It is important to realize that *only the spatial derivatives* of the field variables remain fixed at the $t = $ constant boundary surface and *not* the time derivatives. In this particular case, we do not have surviving time derivative terms on the boundary and hence ignoring this term is acceptable.) Among the remaining three terms in Eq. (2.106), the term with $(c_3 - c_4)$ also remains invariant under a gauge transformation but the other two terms do not. Hence, if we want the Lagrangian to be gauge invariant we must have $c_3 = -c_4 = a_1$, say, and $c_2 = 0$. Thus, only the scalar $F_{ik}F^{ik}$ survives as a possible choice for the Lagrangian of the electromagnetic field and the action will be proportional to the integral of this term over d^4x. It is conventional to write this part of the action as

$$\mathcal{A}_f = -\frac{1}{16\pi}\int F_{ik}F^{ik}d^4x = -\frac{1}{2}\int M^{lmik}\partial_l A_m \partial_i A_k d^4x$$

$$= \frac{1}{8\pi}\int (\boldsymbol{E}^2 - \boldsymbol{B}^2)\,d^4x, \quad (2.109)$$

where $4\pi M^{lmik} = (\eta^{li}\eta^{mk} - \eta^{lk}\eta^{mi})$. The magnitude of the constant in front is arbitrary and merely decides the units used for measuring the electromagnetic field. As explained earlier in the case of a scalar field, we have taken this to be a dimensionless numerical factor $(1/16\pi)$, thereby making the field A_i have the dimensions of inverse length in natural units and making q dimensionless. (In normal units, q stands for $q/(c\hbar)^{1/2}$.) The sign is chosen so that the term $(\partial \boldsymbol{A}/\partial t)^2$ has a positive coefficient. This is needed to ensure that the energy of the plane wave solutions should be positive. The second equality in Eq. (2.109) allows us to identify the Lagrangian for the field as the integral over $d^3\boldsymbol{x}$ of the quantity $\mathcal{L} \equiv (8\pi)^{-1}(\boldsymbol{E}^2 - \boldsymbol{B}^2)$.

The above analysis simplifies significantly, if we work with the action expressed in the Fourier space in terms of the Fourier transform of $A_j(x)$ which we will denote by $A_j(k)$. The partial differentiation with respect to coordinates becomes multiplication by k_i in Fourier space. The most general quadratic action in the Fourier space must have the form $M^{ij}(k,\eta)A_iA_j$, where M^{ij} is built from k^m and η^{ab}. Since the Lagrangian is quadratic in first derivatives of $A_i(x)$ the expression $M^{ij}(k,\eta)A_iA_j$ must be quadratic in k_i. Hence $M_{ij}A^iA^j$ must have the form $[\alpha k_i k_j + (\beta k^2 + \gamma)\eta_{ij}]A^iA^j$, where α, β, and γ are constants. The gauge transformation has the form $A_j(k) \rightarrow A_j(k) + k_j f(k)$ in the Fourier space. Demanding that $M_{ij}A^iA^j$ should be invariant under such a transformation – except for the addition of a term that is independent of $A_i(k)$ – we find that $\alpha = -\beta, \gamma = 0$. Therefore, the action in the Fourier space will be

$$\mathcal{A}_f \propto \int \frac{d^4k}{(2\pi)^4} A_i[k^ik^j - k^2\eta^{ij}]A_j \propto \int \frac{d^4k}{(2\pi)^4}[k_jA_i - k_iA_j]^2, \qquad (2.110)$$

which reduces to Eq. (2.109) in the real space.

Exercise 2.11
Something to think about: swindle in Fourier space? We saw earlier that one possible gauge invariant term in the action is given by Eq. (2.107). It may be a total divergence but it is not zero. But this term never came up in the Fourier space discussion. What is happening here? Similarly both the Lagrangians $L = \phi\Box\phi$ and $L = \partial_a\phi\partial^a\phi$ seem to give $-k^2|\phi(k)|^2$ in Fourier space but they are not equal in real space. Figure these out.

2.8 Maxwell's equations

The full action for a system of charged particles interacting with the electromagnetic field now reads as

$$\mathcal{A} = -\sum \int m\, d\tau + \int A_i J^i d^4x - \frac{1}{16\pi} \int F_{ik}F^{ik}d^4x, \qquad (2.111)$$

where the first two terms are the same as in Eq. (2.73). To find the equations of motion for the field, we have to vary the potentials A_i in this action. The first term is independent of A_i and hence does not contribute; the next two terms give

$$\delta \mathcal{A} = - \int \left(-J^i \delta A_i + \frac{1}{8\pi} F^{ik} \partial_i \delta A_k - \frac{1}{8\pi} F^{ik} \partial_k \delta A_i \right) d^4 x, \qquad (2.112)$$

where we have used the definition in Eq. (2.44) of F_{ik} and the relation $F^{ik} \delta F_{ik} = F_{ik} \delta F^{ik}$. In the second term in Eq. (2.112) we interchange i and k and replace F_{ki} by $-F_{ik}$ to obtain

$$\delta \mathcal{A} = \int \left(J^i \delta A_i + \frac{1}{4\pi} F^{ik} \partial_k \delta A_i \right) d^4 x. \qquad (2.113)$$

Finally, we integrate the second term by parts and convert the resulting four-divergence into a surface integral:

$$\delta \mathcal{A} = - \int \left(-J^i + \frac{1}{4\pi} \partial_k F^{ik} \right) \delta A_i d^4 x + \frac{1}{4\pi} \int F^{ik} \delta A_i d\sigma_k. \qquad (2.114)$$

Using the – by now familiar – procedure, we shall first consider variations such that $\delta A_i = 0$ on the surface. This leads to the field equations

$$\partial_k F^{ik} = 4\pi J^i. \qquad (2.115)$$

Taking the components, with $J^i = (\rho, \boldsymbol{J})$, these equations can be written in three-dimensional notation as

$$\nabla \times \boldsymbol{B} - \frac{\partial \boldsymbol{E}}{\partial t} = 4\pi \boldsymbol{J}; \qquad \text{div } \boldsymbol{E} = 4\pi \rho. \qquad (2.116)$$

These are the equations connecting the field to its source. These equations, along with Eq. (2.50), determine the evolution of electromagnetic field.

If we differentiate Eq. (2.115) with respect to x^i, the left hand side $\partial_i \partial_k F^{ik}$ vanishes identically due to the antisymmetry F^{ik}. Hence we get the constraint, $\partial_i J^i = 0$ on the current vector showing that the integral of ρ over all space is conserved. In three-dimensional form, this becomes the familiar continuity equation:

$$\frac{\partial \rho}{\partial t} + \text{div } \boldsymbol{J} = 0. \qquad (2.117)$$

Integrating this equation over a large three-volume $d^3 x$ and using Gauss's theorem, we can see that the total charge does not change with time. Thus, Maxwell's equations *imply* the conservation of electric charge.

We next consider the variation $\delta \mathcal{A}$ in Eq. (2.114) when the field equations are satisfied but $\delta A_i \neq 0$ on the boundary. Then the first term vanishes and we get

$$\delta \mathcal{A} = \frac{1}{4\pi} \int F^{ik} \delta A_i d\sigma_k, \qquad (2.118)$$

so that the four-momentum density conjugate to the dynamical variable A_i is

$$P^{(i)k} = (F^{ik}/4\pi) = -(F^{ki}/4\pi). \tag{2.119}$$

In particular, the zeroth component of the four-momentum that is conjugate to A_i is $P^{(i)0} = (\partial L/\partial \dot{A}_i) = -(F^{0i}/4\pi)$. This result has some peculiar features. First, we see that $P^{(\alpha)0}$ corresponding to $(A)^\alpha$ is $-(E)^\alpha/4\pi$ so that one can think of $-(E/4\pi)$ as analogous to the momentum $p = (\partial L/\partial \dot{q})$ in classical mechanics. Second, $P^{(0)0}$ corresponding to $\phi = A^0$ *vanishes identically* because $F^{00} = 0$. To understand this result note that, in the electromagnetic Lagrangian $F_{ab}F^{ab}$, the time derivatives arise only from $F_{0\alpha}$ which has only the time derivatives of A_α (again, because $F_{00} = 0$). So the electromagnetic Lagrangian does not contain time derivatives of ϕ at all; therefore the canonical momentum $(\partial L/\partial \dot{\phi})$ corresponding to ϕ vanishes. A closely related feature is that, in the field equations involving $\partial_a F^{ab}$, one can have the second time derivatives only through the term $\partial_0 F^{0b}$. So, again, the field equations do not contain second time derivatives of ϕ but have the second time derivatives of A.

If we consider the surface term in Eq. (2.118) at the surfaces $t =$ constant, we can determine which quantities need to be fixed for a valid variational principle. On this surface, we have

$$\delta \mathcal{A} = \frac{1}{4\pi} \int F^{i0} \delta A_i d\sigma_0 = \frac{1}{4\pi} \int F^{\alpha 0} \delta A_\alpha d^3 x = -\frac{1}{4\pi} \int \mathbf{E} \cdot \delta \mathbf{A} d^3 x. \tag{2.120}$$

So it might seem that we need to fix $\mathbf{A}$ on the $t =$ constant surface. However, one can have variations of $\mathbf{A}$ arising from a gauge transformation $\delta \mathbf{A} = \nabla f$ without affecting the equations of motion. Such changes will only add a surface term to Eq. (2.120) when $\nabla \cdot \mathbf{E} = 0$. Therefore we need to fix $\mathbf{A}$ at the boundary, only up to a gradient term; this is equivalent to fixing just $\nabla \times \mathbf{A} = \mathbf{B}$. In this sense, the magnetic field is the 'coordinate' in the electromagnetic theory and the electric field is the 'momentum'.

Using the relation between F_{ik} and A_j, Eq. (2.115) can be written in terms of the potential A_i as

$$\frac{\partial}{\partial x^k}\left(\frac{\partial A^k}{\partial x_i} - \frac{\partial A^i}{\partial x_k}\right) = \frac{\partial}{\partial x_i}\left(\frac{\partial A^k}{\partial x^k}\right) - \frac{\partial^2}{\partial x^k \partial x_k}A^i = 4\pi J^i. \tag{2.121}$$

This equation determines the field produced by some given source J^i. As in the case of the scalar field, we can attempt to solve this equation for a general source by Fourier transforming both sides. (See the discussion around Eq. (2.37).) This will give, in the Fourier space, the equation

$$(-k^m k_n + \delta^m_n k^2)A^n(k) = 4\pi J^m(k). \tag{2.122}$$

Solving this equation will require finding the inverse of the matrix $G_n^m = (-k^m k_n + \delta_n^m k^2)$ so that one can write $A = G^{-1} J$. However, it is easy to see that $G_n^m k^n = 0$; therefore the matrix G_n^m has a zero eigenvalue and is not invertible. In other words, there is no unique solution to Eq. (2.122); if $A^j(k)$ is a solution in the Fourier space, so is $A^j + k^j f(k)$ with any arbitrary function $f(k)$. This is, of course, to be expected since our formalism and the field equation Eq. (2.121) are gauge invariant and if A^j is a solution to Eq. (2.121), so is $A^j + \partial^j f$ for function f. So we can determine A^j only up to a gauge transformation.

So to solve Eq. (2.121) or Eq. (2.122), we need to impose a gauge condition, the choice of which is essentially determined by the context. One convenient, Lorentz invariant, condition is $\partial_k A^k = 0$ in real space (called the *Lorenz gauge*).[2] One can easily show that we can always impose this condition. If $A'_j = A_j + \partial_j f$, then $\partial^j A'_j = 0$ requires us to choose f such that $\Box f = -\partial^j A_j$; such an equation always has local solutions. So one can always impose the Lorenz gauge condition. In momentum space, this corresponds to $k_n A^n = 0$ so that the first term in Eq. (2.122) vanishes, leading to the solution

$$A^m(k) = 4\pi \frac{J^m(k)}{k^2}. \tag{2.123}$$

Our condition $k_m A^m = 0$ requires $k_m J^m = 0$, which is just the Fourier space version of $\partial_m J^m = 0$, once again showing the close connection between the conservation of charge and gauge invariance. In real space the Lorenz gauge reduces the Maxwell equations to

$$-\frac{\partial^2}{\partial x^k \partial x_k} A^i \equiv \left(\frac{\partial^2}{\partial t^2} - \nabla^2 \right) A^i \equiv -\Box A^i = \frac{4\pi}{c} J^i, \tag{2.124}$$

which has the standard form of a wave equation with a source for each of the components. It may be noted that the condition $\partial_i A^i = 0$ does not completely fix the gauge. We can still add to A_i a gradient $\partial_i \lambda$ where λ is a solution of $\Box \lambda = 0$.

One application of Eq. (2.124) is to determine the nature of the interaction between the like charges, just as we did in the case of the scalar field in Section 2.3.2. The crucial difference is that, for the static source, we now have $-\nabla^2 \phi = 4\pi \rho$, while in the case of the scalar field we had $\nabla^2 \phi = 4\pi \rho$. The coupling term in the Lagrangian has the same form in both $(-\rho\phi)$. It is obvious that the total energy for a static configuration will now be positive, showing that like charges repel. One simple way of understanding this result is as follows. The interaction term for the scalar field had the form $\rho \Box^{-1} \rho$, while it is $J^a \Box^{-1} J_a$ in the case of the electromagnetic field. The static source (involving J^0) will require the lowering of one 0-component in the case of a vector source, flipping the sign with respect to the scalar field. This interpretation also works for higher spin fields.

In the absence of charged particles, $J^i = 0$ and F^{ik} satisfies the equation $\partial_k F^{ik} = 0$. Expressing F^{ik} in terms of the potential A^i and imposing Lorenz gauge, we get

$$-\Box\, A^i = -\frac{\partial^2}{\partial x^k \partial x_k} A^i = \left(\frac{\partial^2}{\partial t^2} - \nabla^2\right) A^i = 0. \tag{2.125}$$

Thus each component A^i satisfies a wave equation with the speed of propagation equal to c, which is unity in our notation.

To solve Eq. (2.125), we use the ansatz $A^j = a^j \exp i(k_b x^b)$ (as in Eq. (2.26)) where k_b and a^j are constant four-vectors. Substituting this ansatz in the wave equation, we find that $k^b k_b = 0$. The Lorenz gauge condition $\partial_j A^j = 0$ implies that $k_b a^b = 0$. Thus the solution is parametrized by two four-vectors subject to two constraints. Denoting the components of k^b by $(\omega, \boldsymbol{k})$ the constraint $k^a k_a = 0$ becomes $|\boldsymbol{k}|^2 = \omega^2$. When $J^i = 0$, we can use the residual gauge freedom to make $A^0 = 0$. To do this, we have to choose a solution to $\Box \lambda = 0$ such that $\partial_0 \lambda = -A_0$. This is easy to do by choosing in Fourier space, $\lambda(k) = A_0(k)/i\omega = A_0(k)/i|\boldsymbol{k}|$. Note that since $\Box A_0 = 0$, we also have $\Box \lambda = 0$.

Choosing the amplitude to be $a^b = (0, \boldsymbol{a})$ the constraint $k_b a^b = 0$ becomes $\boldsymbol{k} \cdot \boldsymbol{a} = 0$, and the solution to the wave equation becomes

$$\boldsymbol{A_k}(t, \boldsymbol{x}) = \boldsymbol{a_k} \exp i(\pm \omega_k t - \boldsymbol{k} \cdot \boldsymbol{x}); \quad \omega_k = |\boldsymbol{k}|, \tag{2.126}$$

with $\boldsymbol{a}$ confined to a plane perpendicular to the direction of $\boldsymbol{k}$; thus the vector potential has only two independent components which are transverse to the direction of propagation. The *general* solution to Eq. (2.125) can be found by superposing the plane wave solutions. The general solution can be expressed as

$$\boldsymbol{A}(x) = \int \frac{d^3 k}{(2\pi)^3}\, \boldsymbol{A}(k) e^{ikx}; \quad \omega_k = |\boldsymbol{k}|; \quad \boldsymbol{k} \cdot \boldsymbol{A}(k) = 0. \tag{2.127}$$

By its very construction, the wave equation in Eq. (2.125) is invariant under the Lorentz transformations with the same velocity of propagation in all inertial frames. For comparison, it is instructive to consider the transformation of the wave operator $\Box \equiv -(\partial^2/\partial t^2) + \nabla^2$ under the *Galilean* transformation in Eq. (1.18). A simple calculation shows that the operator changes to

$$\Box' \equiv \left(1 - \frac{V^2}{c^2}\right) \frac{\partial^2}{\partial x^2} - \frac{2V}{c^2}\frac{\partial^2}{\partial x \partial t} - \frac{1}{c^2}\frac{\partial^2}{\partial t^2} + \nabla_\perp^2, \tag{2.128}$$

where $\nabla_\perp^2$ is the operator in the transverse direction. It is clear that the behaviour of light waves under the Galilean transformation will be quite different from that under the Lorentz transformations. Maxwell's equations describing electromagnetic phenomena, from which the wave equation was derived, are clearly not invariant under the Galilean transformations. This was, in fact, the original historical motivation to introduce the Lorentz transformations.

Exercise 2.12

Hamiltonian form of action – electromagnetism   In the case of classical mechanics, we found that (see Ex. 1.9) one can obtain the equations of motion from a Hamiltonian form of action as well, in which the momenta and the coordinates are varied independently. It is possible to do this in the case of the electromagnetic field as well. Consider the action principle given by

$$4\pi \mathcal{A} = \int d^4x \left[-\mathbf{E} \cdot \dot{\mathbf{A}} + \phi \nabla \cdot \mathbf{E} - \frac{1}{2} \left(E^2 + (\nabla \times \mathbf{A})^2 \right) \right] \qquad (2.129)$$

in which $\mathbf{E}(t, \mathbf{x}), \mathbf{A}(t, \mathbf{x})$ and $\phi(t, \mathbf{x})$ are considered to be three independent functions that are to be varied in the action.

(a) Convince yourself that the Lagrangian in this action principle has a $(p\dot{q} - H)$ form as in the case of classical mechanics.

(b) Vary $\mathbf{E}(t, \mathbf{x}), \mathbf{A}(t, \mathbf{x})$ and $\phi(t, \mathbf{x})$ and show that, with suitable boundary conditions, one can obtain the following equations:

$$\mathbf{E} = -\dot{\mathbf{A}} - \nabla \phi \, ; \qquad \dot{\mathbf{E}} = \nabla \times (\nabla \times \mathbf{A}) \equiv \nabla \times \mathbf{B} \, ; \qquad \nabla \cdot \mathbf{E} = 0. \qquad (2.130)$$

Prove that these are equivalent to standard Maxwell equations.

(c) To obtain the above equations, it is necessary to fix the following quantity on the $t = $ constant boundaries

$$S = \int_t d^3\mathbf{x} \, \mathbf{E} \cdot \delta \mathbf{A}. \qquad (2.131)$$

This might suggest that $\mathbf{A}(t, \mathbf{x})$ need to be fixed at the boundary. Show that this is not the case and only $\mathbf{B}(t, \mathbf{x}) = \nabla \times \mathbf{A}(t, \mathbf{x})$ needs to be fixed at the boundary. [Hint. Consider what happens to $\mathbf{A}(t, \mathbf{x})$ under a gauge transformation.]

Exercise 2.13

Eikonal approximation   To obtain the propagation of light rays in straight lines from the propagation of an electromagnetic wave described by Maxwell's equations, we need to take the high frequency limit of the wave solution (called the *Eikonal approximation*). Assume that the wave is described by a vector potential of the form $A^i = a^i(x) \exp[i\omega S(x)]$, where $a^i(x)$ is a slowly varying amplitude and $\omega S(x)$ is a rapidly varying phase. Substitute this ansatz into the wave equation $\Box A^i = 0$ and equate the coefficient of ω^2 and ω to zero separately. Show that the resulting equations can be interpreted as the propagation of a light ray with the wave vector $k_i = \partial_i S$ and an intensity proportional to $|a_i a^i|$.

Exercise 2.14

General solution to Maxwell's equations   Since Maxwell's equations are linear in the electromagnetic field, it is possible to provide the most general solution to these equations in the Fourier space. To do this, it is convenient to introduce the variables $\phi_{\mathbf{k}}(t), \mathbf{A}_{\mathbf{k}}(t), \mathbf{E}_{\mathbf{k}}(t), \mathbf{B}_{\mathbf{k}}(t), \rho_{\mathbf{k}}(t)$ and $\mathbf{J}_{\mathbf{k}}(t)$ which are the *spatial* Fourier transforms of the scalar potential, vector potential, electric field, magnetic field, charge density and current density, respectively.

(a) Write down the gauge condition $\nabla \cdot \mathbf{A} = 0$, law of conservation of charge and the Maxwell's equations in Fourier space.

(b) Manipulate these to show the following. (i) The scalar potential is related to the charge by

$$\phi_{\mathbf{k}}(t) = \frac{4\pi}{k^2} \rho_{\mathbf{k}}(t). \tag{2.132}$$

How do you reconcile Eq. (2.132) with the fact that signals cannot travel with infinite speed? (ii) The vector potential is related to the transverse part of the current by

$$\ddot{\mathbf{A}}_{\mathbf{k}}^{\perp} + c^2 k^2 \mathbf{A}_{\mathbf{k}}^{\perp} = 4\pi c \mathbf{J}_{\mathbf{k}}^{\perp}; \qquad \mathbf{A}_{\mathbf{k}}^{\|} = 0, \tag{2.133}$$

where the symbol $\perp$ indicates the component perpendicular to $\mathbf{k}$, etc.

[Note. This analysis shows that the propagating degree of freedom of an electromagnetic field is the transverse component $\mathbf{A}_{\mathbf{k}}^{\perp}$ of the vector potential. The scalar potential $\phi_{\mathbf{k}}$ is completely determined by the charge density instantaneously through Eq. (2.132) and the longitudinal component $\mathbf{A}_{\mathbf{k}}^{\|}$ of the vector potential can be made to vanish by a gauge condition. Thus the electromagnetic field has only two independent degrees of freedom per space point. Since the magnetic field has a one-to-one correspondence with the transverse part of the vector potential, the true dynamical degrees of freedom of the electromagnetic field are contained in the (divergence-free) magnetic field.]

Exercise 2.15

Gauge covariant derivative In the text we discussed the coupling between electromagnetic field and charged particles. It is also possible to couple the electromagnetic field to other *fields* like, for example, a complex scalar field. This exercise explores some aspects of such a coupling and introduces the important concept of the covariant derivative in the context of electromagnetism.

(a) Consider a complex scalar field ϕ described by an action

$$\mathcal{A} = -\frac{1}{2} \int d^4 x \left[\partial_a \phi \, \partial^a \phi^* + m^2 |\phi|^2 \right]. \tag{2.134}$$

Vary ϕ and ϕ^* independently to obtain the equations of motion. Rewrite the action in terms of the real and imaginary parts, vary them independently and obtain the resulting equations of motion.

(b) The above action is invariant under the transformation $\phi \to e^{-iq\alpha}\phi$, where q and α are real constants. On the other hand, if $\alpha = \alpha(x)$ is a function of spacetime coordinates (with q constant), the action is not invariant because the derivatives $\partial_i \phi$ are not invariant. We want to modify the action by replacing $\partial_i \phi$ by another quantity $D_i \phi$ (called the *gauge covariant derivative*) such that the action is invariant. Assume that $D_i = \partial_i + iqA_i(x)$, where A_i is a four-vector field. Demand that, when $\phi \to \phi' = e^{-iq\alpha(x)}\phi$, the A_i transforms to A_i' such that $D_i'\phi' = e^{-iq\alpha(x)} D_i \phi$. Use this criterion to determine how A_i transforms and show that

$$A_i' = A_i + \partial_i \alpha. \tag{2.135}$$

(c) Equation (2.135) shows that A_i undergoes standard gauge transformation when we demand that the action should be invariant under spacetime dependent changes in phase. Write down the action

$$\mathcal{A} = -\frac{1}{2} \int d^4 x \left[D_i \phi \, D^i \phi^* + m^2 |\phi|^2 \right] \tag{2.136}$$

by expanding out the gauge covariant derivative and explicitly displaying the coupling between A_i and the scalar field. Use this to identify the current four-vector of the complex scalar field. Is this current conserved ?

Exercise 2.16
Massive vector field Consider an action of the form

$$\mathcal{A} = -\frac{1}{4} \int d^4x \, F_{ik} F^{ik} - \frac{1}{2} \int d^4x \, m^2 A_i A^i, \qquad (2.137)$$

with $F_{ik} = \partial_i A_k - \partial_k A_i$.

(a) Show that when $m \neq 0$ the theory is not gauge invariant. Vary the action to obtain the equations of motion.

(b) Show that the equation of motion now *implies* the condition $\partial_i A^i = 0$. Compare this with the case of electromagnetism. Can you think of a physical origin for this condition when $m \neq 0$?

(c) Using the result of (b), show that the equation of motion becomes $(\Box - m^2)A^i = 0$. Solve this equation in the Fourier space and explain why m can be thought of as the mass of the field excitations.

(d) Add a term to the action coupling A^i to a current J^i. Is it necessary that this current is conserved? Solve the field equations in the presence of J^i in the Fourier space and show that the matrix corresponding to the generalization of Eq. (2.122) now has a well-defined inverse. Can you take the limit $m \to 0$ smoothly?

Exercise 2.17
What is c if there are no massless particles? Show that, if the photon has a tiny mass m, then we will have an extra term $m^2 A_j A^j$ in the electromagnetic Lagrangian and the velocity of the electromagnetic waves will not be universal but will depend on the frequency. Suppose there are no massless particles at all in nature (and the photon, for example, has a very tiny mass). How will one then formulate special relativity and interpret the universal speed c ?

2.9 Energy and momentum of the electromagnetic field

By including the electromagnetic fields in the action we are treating them as dynamical entities with degrees of freedom of their own. When a charge moves under the influence of an electromagnetic field, its momentum and energy will change. Since the total momentum (or energy) of a closed system is a constant, it follows that the change of energy and momentum of the charged particle must be compensated for by a change of the energy and momentum of the field. To show this explicitly, we need to obtain the expressions for the energy and momentum of the electromagnetic field.

Let us consider the energy-momentum tensor corresponding to the last term in Eq. (2.111) describing the action for the electromagnetic field. We have already seen that the canonical momentum corresponding to the vector field A_l is given by $P^{(l)k} = -(1/4\pi)F^{kl}$. One should interpret this expression as giving the four-momentum (indexed by k) for each component of the field (indexed by l). The energy-momentum tensor in Eq. (2.17) is now defined through

$$T_i^k = -[\partial_i A_l P^{(l)k} - \delta_i^k L]. \tag{2.138}$$

Note that in the first term we are summing over l, which indicates the different components of the vector field. This is an illustration of the comment we made earlier (at the end of Section 2.3.1) that, in the case of a multi-component field, we merely sum up the expression for the energy-momentum tensor for each component. Substituting $P^{(l)k} = -(1/4\pi)F^{kl}$ into this expression we get

$$T^{ik} = \frac{1}{4\pi} F^k_{\ l} \partial^i A^l - \frac{1}{16\pi} \eta^{ik} F_{lm} F^{lm}. \tag{2.139}$$

This is an example of an energy-momentum tensor which is *not* symmetric. However, we can make it symmetric by adding the quantity

$$-\frac{F^k_{\ l}}{4\pi} \partial^l A^i = -\frac{1}{4\pi} \partial_l (A^i F^{kl}), \tag{2.140}$$

which is of the form given in Eq. (2.20). (We have used the fact that, in the absence of charges, $\partial_l F^{kl} = 0$.) This leads to the symmetrized energy-momentum tensor of the form:

$$T^{ik} = \frac{1}{4\pi} \left(F^{il} F^k_{\ l} - \frac{1}{4} \eta^{ik} F_{lm} F^{lm} \right). \tag{2.141}$$

While we have obtained a useful expression, the procedure that is adopted is inherently ambiguous and unsatisfactory. For example, there is no assurance that some other form of the energy-momentum tensor – obtained by adding some other tensor – will not be more appropriate physically. The fundamental reason for this ambiguity is that the energy-momentum tensor is not directly observable *in the absence of gravitational interactions*. What is usually relevant is the integral of this expression over a volume, which is quite unambiguous. However, since energy density is equivalent to mass density, we would expect different energy-momentum tensors to lead to different gravitational fields. In any theory which includes gravity, we need an unambiguous procedure to obtain a symmetric energy-momentum tensor. We will see that general relativity indeed provides such a prescription and leads (unambiguously) to the same expression as that in Eq. (2.141). This is the real justification for our choice.

Let us now consider the properties of T^{ik}. This tensor is obviously symmetric and is traceless: i.e. $T_i^i = 0$. Further, it has the components

$$T^{00} \equiv W = \frac{E^2 + B^2}{8\pi}, \quad T^{0\alpha} \equiv S^\alpha = \frac{1}{4\pi} (\mathbf{E} \times \mathbf{B})^\alpha, \tag{2.142}$$

and

$$T_{\alpha\beta} = \frac{1}{4\pi}\left[-E_\alpha E_\beta - B_\alpha B_\beta + \frac{1}{2}\delta_{\alpha\beta}\left(E^2 + B^2\right)\right]. \qquad (2.143)$$

It is possible to relate the divergence of the energy-momentum tensor to the source J^i when the latter is nonzero. To do this, we differentiate the expression in Eq. (2.141), getting

$$\partial_k T_i^k = -\frac{1}{4\pi}\left(\frac{1}{2}F^{lm}\partial_i F_{lm} - F^{kl}\partial_k F_{il} - F_{il}\partial_k F^{kl}\right). \qquad (2.144)$$

On using Eq. (2.115) as well as Eq. (2.51) we can rewrite this in the form

$$\partial_k T_i^k = \frac{1}{4\pi}\left(\frac{1}{2}F^{lm}\partial_l F_{mi} + \frac{1}{2}F^{lm}\partial_m F_{il} + F^{kl}\partial_k F_{il} + \frac{4\pi}{c}F_{il}J^l\right). \qquad (2.145)$$

By permuting indices, we can easily show that the first three terms on the right cancel one another, thereby leading to

$$\partial_k T_i^k = -F_{ik}J^k. \qquad (2.146)$$

This equation relates the change in electromagnetic energy-momentum to the work done by the field on the charged particles. In particular, the $i = 0$ component of this equation gives:

$$\frac{\partial}{\partial t}\left(\frac{E^2 + B^2}{8\pi}\right) + \nabla\cdot\boldsymbol{S} = -\boldsymbol{J}\cdot\boldsymbol{E}. \qquad (2.147)$$

Integrating this expression over a three-dimensional volume and applying the Gauss theorem to the second term on the left, we get

$$\frac{\partial}{\partial t}\int\frac{E^2 + B^2}{8\pi}dV + \oint\boldsymbol{S}\cdot d\boldsymbol{a} = -\int\boldsymbol{J}\cdot\boldsymbol{E}\,dV. \qquad (2.148)$$

The term on the right hand side can be written as $\sum q_a\boldsymbol{v}_a\cdot\boldsymbol{E}$, where the sum is over all charges. Since this represents the amount of work done on the charged particles by the electromagnetic field, it is equal to the rate of change of the kinetic energy $\mathcal{E}$ of the charges so that Eq. (2.148) becomes

$$\frac{\partial}{\partial t}\left(\int\frac{E^2 + B^2}{8\pi}dV + \mathcal{E}\right) = -\oint\boldsymbol{S}\cdot d\boldsymbol{a}. \qquad (2.149)$$

The left hand side can be interpreted as the rate of change of the total energy of the system (made of the electromagnetic field and the charged particles) contained in a volume and the right hand side gives the flux of this energy through the surface bounding the given volume, demonstrating the conservation of energy. (Also see Exercise 2.18.)

Exercise 2.18

Conserving the total energy (a) Consider a system of particles with charges q_A and masses m_A (with $A = 1, 2, \dots$ labelling the particle) which are interacting through electromagnetic forces. Argue that the energy-momentum tensor for the particles can be taken to be

$$T_{\text{part}}^{ab} = \sum_A m_A \int d\tau_A \, u_A^a(\tau_A) u_A^b(\tau_A) \, \delta_D[x - z_A(\tau_A)]. \tag{2.150}$$

(b) Compute $\partial_a T_{\text{part}}^{ab}$. Hence show that

$$\partial_a [T_{\text{part}}^{ab} + T_{\text{EM}}^{ab}] = 0, \tag{2.151}$$

where T_{EM}^{ab} is the energy-momentum tensor of the electromagnetic field.

Exercise 2.19

Stresses and strains Consider an electric field E which is constant along the x-axis. Show that the energy-momentum tensor has the components $T^{00} = (E^2/8\pi)$, $T^{xx} = -(E^2/8\pi)$, $T^{yy} = T^{zz} = (E^2/8\pi)$. Consider now another frame moving with speed $v = \beta c$ along the y-axis. Show that the components in this frame are given by

$$T'^{00} = (E^2/8\pi)\gamma^2(1 + \beta^2), \qquad T'^{0y} = -(E^2/4\pi)\gamma^2\beta \tag{2.152}$$

$$T'^{xx} = -(E^2/8\pi) \quad T'^{yy} = (E^2/8\pi)\gamma^2(1 + \beta^2), \quad T'^{zz} = (E^2/8\pi). \tag{2.153}$$

This shows that, in the ultra-relativistic limit, the pressure in the y-direction dominates over the tension in the x- or z-directions. Explain why.

Exercise 2.20

Everything obeys Einstein Consider a parallel plate capacitor with plates of area A located normal to the x-direction with a small separation d. Assume that the capacitor is charged to give a uniform electric field E between the plates and ignore all the edge effects. The electromagnetic mass of the system is $E^2 A d/8\pi$. If this capacitor moves along the x-direction, the electric field does not change but the separation undergoes Lorentz contraction so that the electrostatic energy decreases to $E^2 A d/8\pi\gamma$. Consider next the force required to hold the plates apart. One possibility will be to fill the region between the plates by an ideal gas of proper mass density ρ_0 such that its pressure provides the necessary force. The total rest mass of the system is now $M = E^2 A d/8\pi + \rho_0 A d$. Show that under the Lorentz transformation this quantity goes over to γM as it should.

Exercise 2.21

Practice with the energy-momentum tensor (a) Prove that $W^2 - |S|^2$ is Lorentz invariant. Note that W and S are *not* components of a four-vector.

(b) You saw in Exercise 2.7 that, except when $(E \cdot B)^2 + (E^2 - B^2)^2 = 0$, it is possible to make a Lorentz transformation to a frame which will make E and B parallel. (That is, if you did that exercise.) Show that, in this frame, the energy-momentum tensor has the form $T_b^a = W \operatorname{dia}(-1, -1, +1, +1)$.

(c) When $E \cdot B = E^2 - B^2 = 0$ show that one can always find a Lorentz frame in which the field has the form $E = (0, f, 0)$ and $B = (0, 0, f)$. (Such a field is called null.) What is the form of energy-momentum tensor in this case?

(d) Show that the square of the electromagnetic energy-momentum tensor T_b^a, treated as a matrix, is proportional to the unit matrix. More explicitly, show that:

$$T_a^m T_n^a = \frac{\delta_n^m}{64\pi^2}[(E^2 - B^2)^2 + 4(E \cdot B)^2]. \qquad (2.154)$$

(e) A symmetric 4×4 tensor T_b^a, when represented as a matrix, has two properties: (i) $T = 0$ and (ii) $T_b^a T_c^b = (M^4/64\pi^2)\delta_c^a$, where M is a constant. Show that by a suitable choice of an inertial frame, such a tensor can be reduced to the form in which the matrix elements of T_b^a are as given in parts (c) or (b) above.

(f) Show that the determinant of $F^a{}_b$ treated as a matrix is det $F^a{}_b = E \cdot B$. Use this as well as the results of (b) above to determine the eigenvalues of $F^a{}_b$. [Hint. Because of the antisymmetry of F_{ab}, we know that if λ is an eigenvalue so is $-\lambda$. So the equations determining the eigenvalues must have the form $\lambda^4 + a\lambda^2 + b = 0$. Show that $a = (1/2)F_{ab}F^{ab}$ and $b = -(1/16)(\epsilon_{abcd}F^{ab}F^{cd})^2$.]

(g) If an electromagnetic wave is propagating along a definite direction with unit vector n (so that $k = \omega n$), it is called a monochromatic plane wave. (i) Show that the energy-momentum tensor for the plane wave which is given by $T^{ab} = (W/\omega^2)k^a k^b$, where $W = (E^2 + B^2)/8\pi = E^2/4\pi$. Hence show that the combination (E/ω) is Lorentz invariant. (This is a nontrivial result since it is valid for any Lorentz transformation – not merely for the boosts along the direction of propagation.) (ii) Find the momentum flux S of a plane wave and show that momentum density and energy density are related by $S = W$. Can you interpret this result?

(h) The standard duality operation, indicated by a $*$, is defined as $(*F)_{ab} = \epsilon_{abcd}F^{cd}$. The operation – called *duality rotation* – is defined through $\exp(*\alpha) = \cos\alpha + *(\sin\alpha)$. How do the electric and magnetic fields change under a duality rotation? How does the electromagnetic energy-momentum tensor change under duality rotation?

(i) Define a complex antisymmetric tensor by $W_{ab} = F_{ab} + i(*F)_{ab}$. How does W_{ab} transform under duality rotation? Show that $8\pi T_{ab} = -W_{ac}\overline{W}_b{}^c$ where the overbar denotes complex conjugation.

(j) Express the energy-momentum tensor in terms of the electric and magnetic fields defined in Exercise 2.3. How do you interpret this result?

Exercise 2.22

Poynting–Robertson effect A spherical particle of mass m and effective cross-sectional area A scatters all the electromagnetic radiation incident on it isotropically in its rest frame. Determine the equation of motion for this particle when it is hit by a constant radiation field of intensity S (erg s^{-1} cm^{-2}) incident from a given direction. What is the solution to this equation if the particle was initially at rest?

[Hint. Argue that if the four-velocity of the particle is u^i and the four-vector along the direction of propagation of radiation is k^i, then the particle will absorb a four-momentum flux $-AT^{ab}u_a = -ASk^b(k^a u_a)$. The energy absorbed by the particle in the rest frame will be the zeroth component of this expression: $SA(k^a u_a)^2$. This energy will be re-radiated away isotropically in the rest frame of the particle. Therefore, the equation of motion for the particle will be

$$\frac{dp^i}{ds} = -SA[(k^a u_a)k^i + (k^a u_a)^2 u^i]. \qquad (2.155)$$

The equation can be integrated by changing variables to $k^a u_a$ and the final solution is

$$t = \frac{1}{6C}[1 + 2Cs]^{3/2} + \frac{1}{2C}[1 + 2Cs]^{1/2} - \frac{2}{3C}$$
$$x = \frac{1}{6C}[1 + 2Cs]^{3/2} - \frac{1}{2C}[1 + 2Cs]^{1/2} + \frac{2}{3C}, \qquad (2.156)$$

where $C = SA/m$.]

Exercise 2.23
Moving thermometer (a) Consider an observer moving with a velocity v through a radiation bath of temperature T_0. Show that the observer will see an anisotropy in the radiation field with the effective temperature in a direction making an angle θ with the direction of motion being $T(\theta) = T_0[\gamma(1 + v\cos\theta)]^{-1}$.

(b) A thermally conducting black sphere is moving through an isotropic blackbody radiation field of temperature T_0 with velocity v. There is a thermometer attached to this sphere. Show that, in equilibrium, the thermometer reading will be $T = [\gamma^2(1+v^2/3)]^{1/4}T_0$. [Hint. Argue that the equilibrium temperature will be $\langle T(\theta)^4 \rangle^{1/4}$ where the averaging is over all angles.]

2.10 Radiation from an accelerated charge

The electric field of a stationary point charge, as well as that of a charge moving with uniform velocity, falls as r^{-2}. (See Eq. (2.67).) The situation, however, changes drastically when a charged particle moves with an acceleration. The field of an accelerated charge picks up a part which falls only as $(1/r)$, usually called the *radiation field*. A field with $E \propto r^{-1}$ has an energy flux $S \propto E^2 \propto r^{-2}$; since the surface area of a sphere increases as r^2, the same amount of energy will flow through spheres of different radii. This fact allows the accelerating charge to transfer energy to large distances and thus provides the radiation field with an independent dynamical existence. This issue of radiation arises both in electromagnetism and in gravity and we shall provide a fairly detailed discussion here, partly as a warm up for discussing gravitational waves in Chapter 9.

To find the field due to a given source $J^a(x)$, we have to solve the equation $\Box A^a = -4\pi J^a$. To solve an equation of the type $\Box Q = P(x)$, one usually uses the method of Green's function. (We shall denote four-vectors $x^a, y^a, \ldots$ by $x, y, \ldots$ and $x^i x_i$ by x^2, etc., when no confusion is likely to arise.) We define a 'retarded' Green function D_R to be the solution to the equation $\Box D_R = \delta(x)$. The subscript R implies that we choose the boundary conditions so as to ensure $D_R(t, \boldsymbol{x}) = 0$ for $t < 0$. Given $D_R(x)$, we can relate Q to P by

$$Q(x) = \int d^4y\, D_R(x - y)P(y). \qquad (2.157)$$

Thus we only need to find the retarded Green function $D_R(x)$.

There is a simple way of determing $D_R(x)$. Assume for a moment we are working in four-dimensional *Euclidean* space (rather than *Minkowskian* space) so that the distance from the origin to x^i is $s^2 = (\tau^2 + |x|^2)$ rather than $(-t^2 + x^2)$ with $\tau = it$. In such Euclidean space, it is trivial to verify that the spherically symmetric solution to $\Box D_R = 0$ is proportional to s^{-2} except at the origin (just as in three dimensions the Green function is proportional to $|x|^{-1}$). Consider now the four-volume integral of $\Box D_R$ over a region bounded by a sphere of radius R. We have

$$\int d^4x \,\Box\, D_R = \int d^3x \,\hat{n}\cdot\nabla D_R = \left(2\pi^2 R^3\right)\left(-\frac{2}{R^3}\right) = -4\pi^2. \quad (2.158)$$

In arriving at the last result, we have used the fact that the 'surface' area of a 3-sphere of radius R is $2\pi^2 R^3$ and $\nabla D_R = \left(-2/R^3\right)\hat{n}$. It follows that

$$\Box\left(\frac{-1}{4\pi^2 s^2}\right) = \delta\left(x\right), \quad (2.159)$$

giving $D_R = \left(-4\pi^2 s^2\right)^{-1}$. Using Eq. (2.157), the solution to the equation of the form $\Box Q = P$ can be written as

$$Q\left(x\right) = -\int \frac{d^4y}{4\pi^2} \frac{P(y)}{(x-y)^2}, \quad (2.160)$$

where x stands for x^i, etc. This is in exact analogy with the solution to the Poisson equation in three dimensions and should be intuitively obvious. If we now analytically continue from the Euclidean to the Minkowski space using $\left(d^4y\right)_E = i\left(dt d^3y\right)_M$ we get

$$Q(x) = -i\int \frac{d^4y}{4\pi^2} \frac{P(y)}{(x-y)^2}, \quad (2.161)$$

with the understanding that we evaluate the contribution from the pole of the integrand at the retarded time $x^0 - y^0 = +|x - y|$. Integrating over y^0 and using residue theorem, this gives

$$Q(t,x) = (-2\pi i)(-i)\int \frac{d^3y}{4\pi^2} \frac{P(t-|x-y|,y)}{2|x-y|} = -\frac{1}{4\pi}\int d^3y \frac{P(t-|x-y|,y)}{|x-y|} \quad (2.162)$$

which is, of course, the standard result.

There is an alternative route to Eq. (2.162) which uses a powerful physical argument. We first note that the equation $\Box Q = P$ is linear and hence, if we know the solution for an infinitesimal amount of source located around the origin, we can find the result for any finite source by straightforward integration. In the case of an infinitesimal source confined close to the origin, the right hand side vanishes everywhere except at the origin and the solution in empty space has the form $Q(t, \boldsymbol{R}) = F(t - R)/R$, where $\boldsymbol{R}$ is the position vector to the field point from the source point with the latter taken at the origin. This is just the solution to the wave equation corresponding to radially outgoing waves. We now need to choose F in such a way that the solution satisfies $\Box Q = P$ at the origin when $Pd^3\boldsymbol{y} = dq(t)$ is an infinitesimal quantity that is nonzero only at the origin but could be varying with time. (We have denoted by $d^3\boldsymbol{y}$ an infinitesimal volume element around the origin.) When we take the limit of $R \to 0$, the field increases rapidly and hence the $\nabla^2 Q$ term will dominate over the $\partial^2 Q/\partial t^2$ term. This means that near the origin we essentially have to solve a Poisson equation and the solution must go over to Coulomb's law. To satisfy this condition, the solution for an infinitesimal source must be of the form

$$Q(t, \boldsymbol{R}) = -\frac{1}{4\pi R}dq\Big|_{t-R} \equiv -\frac{P(t - R)d^3\boldsymbol{y}}{4\pi R}. \tag{2.163}$$

This solution shows that the effect at time t depends on the behaviour of the source at a retarded time $t_r \equiv t - R$ with a radial fall-off $1/R$ which is characteristic of Coulomb potential. It is now easy to obtain the full solution by integrating over a finite source and we will get the result in Eq. (2.162).

We shall now apply this result to obtain the electric and magnetic fields produced by a charged particle moving in an arbitrary trajectory by directly integrating Maxwell's equations, in a four-dimensional notation.[3] Consider a charge q moving along a trajectory $z^a(\tau)$ where τ is the proper time. It contributes a current

$$J^a(x) = q \int_{-\infty}^{\infty} d\tau \delta_D \left[x - z(\tau)\right] u^a(\tau), \tag{2.164}$$

where $u^a(\tau)$ is the four-velocity of the particle. In the case of a point charge, it is somewhat more convenient to retain the four-dimensional form in Eq. (2.161) which becomes, in the case of Maxwell's equations:

$$A^a(x) = \frac{i}{\pi} \int d^4y \frac{J^a(y)}{(x - y)^2} = \frac{iq}{\pi} \int_{-\infty}^{\infty} d\tau \frac{u^a(\tau)}{s^2}, \tag{2.165}$$

where we have used Eq. (2.164), integrated over y eliminating the delta function and defined

$$s^2 = (x - z(\tau))^2 \equiv R^a R_a; \qquad R^a \equiv x^a - z^a(\tau). \tag{2.166}$$

As before, the integral in Eq. (2.165) should be interpreted as providing the complex residue at the first order pole which satisfies the retarded condition $s^2 = 0$ and $z^0 < x^0$.

We now convert the integration over τ to integration over s^2 by using the Jacobian $(ds^2/d\tau) = 2R^a(-u_a) \equiv -2l$. This gives

$$A^a(x) = \frac{iq}{\pi} \int ds^2 \frac{u^a(\tau)}{(ds^2/d\tau)s^2} = -\frac{iq}{2\pi} \int ds^2 \frac{(u^a/l)}{s^2}. \qquad (2.167)$$

Taking the contribution from the first order pole at $s^2 = 0$ as $(-2\pi i)$ times the residue at the retarded time, the integral gives

$$A^a(x) = \left(-\frac{qu^a}{l}\right)_{\text{ret}} = \left(-\frac{qu^a}{R^b u_b}\right)_{\text{ret}}; \quad s^2[x, z(\tau)] = 0; \quad x^0 > z^0. \quad (2.168)$$

This four-potential, produced by an arbitrarily moving charged particle, is called the *Liénard–Wiechert* potential. Its three-dimensional form is

$$\phi = \frac{q}{(R - \boldsymbol{v} \cdot \boldsymbol{R})}, \qquad \boldsymbol{A} = \frac{q\boldsymbol{v}}{(R - \boldsymbol{v} \cdot \boldsymbol{R})}, \qquad (2.169)$$

where all the quantities on the right side are evaluated at the retarded time.

To find the electromagnetic field, $F^{ba} = \partial^b A^a - \partial^a A^b$ we have to evaluate $\partial^b A^a$. From Eq. (2.165), this is given by

$$\partial^b A^a = -\frac{iq}{\pi} \int_{-\infty}^{\infty} d\tau \frac{u^a(\tau)}{s^4} \frac{\partial s^2}{\partial x_b} = -\frac{2iq}{\pi} \int_{-\infty}^{\infty} d\tau \frac{R^b u^a}{s^4}. \qquad (2.170)$$

We now use the result

$$\frac{d}{d\tau}\left(\frac{1}{s^2}\right) = -\frac{1}{s^4}\frac{ds^2}{d\tau} = \frac{2l}{s^4} \qquad (2.171)$$

to substitute for $(1/s^4)$ in the integrand, and do an integration by parts to obtain

$$\partial^b A^a = -\frac{2iq}{\pi} \int_{-\infty}^{\infty} d\tau \left(\frac{R^b u^a}{2l}\right) \frac{d}{d\tau}\left(\frac{1}{s^2}\right) = \frac{iq}{\pi} \int_{-\infty}^{\infty} \frac{d\tau}{s^2} \frac{d}{d\tau}\left(\frac{R^b u^a}{l}\right).$$
$$(2.172)$$

The integral is exactly in the same form as that in Eq. (2.165) with u^a replaced by $d(R^b u^a/l)/d\tau$. Therefore, in analogy with the result obtained in Eq. (2.168), we get

$$\partial^b A^a = \left[-\frac{q}{l}\frac{d}{d\tau}\left(\frac{R^b u^a}{l}\right)\right]_{\text{ret}}. \qquad (2.173)$$

Antisymmetrizing the right hand side, we find that the electromagnetic field of an arbitrarily moving charged particle is given by the manifestly four-dimensional expression

$$F^{ba} = \left[-\frac{q}{l}\frac{d}{d\tau}\left(\frac{R^b u^a - R^a u^b}{l}\right)\right]_{\text{ret}}. \qquad (2.174)$$

This completely solves the problem. To obtain explicit expressions, we have to carry out the differentiation in the above equation which can be done using the result

$$
\frac{d}{d\tau}\left(\frac{R^{[k}u^{i]}}{l}\right) = \frac{R^{[k}a^{i]}}{l} - \frac{R^{[k}u^{i]}}{l^2}\frac{dl}{d\tau} = \frac{R^{[k}a^{i]}}{l} - \frac{R^{[k}u^{i]}}{l^2}\left(R_b a^b + 1\right),\quad (2.175)
$$

where the square brackets denote antisymmetrization. Using this expression, we can separate the electromagnetic field in Eq. (2.174) into two parts – one which falls as $(1/l^2)$ and the other which falls as $(1/l)$. For this, it is also convenient to write

$$
R^i = -l(n^i + u^i); \qquad u^i n_i = 0; \qquad n^i n_i = 1. \quad (2.176)
$$

Then we get $F^{ab} = F^{ab}_{\mathrm{coul}} + F^{ab}_{\mathrm{rad}}$ with

$$
F^{ab}_{\mathrm{coul}} = \frac{q}{l^2}\,u^{[a}n^{b]}; \qquad F^{ab}_{\mathrm{rad}} = -\frac{q}{l}\left[a^{[a}u^{b]} - n^{[a}a^{b]} - (n^i a_i)\,n^{[a}u^{b]}\right]. \quad (2.177)
$$

The part in F^{ab}_{coul} falls as the square of the distance and is independent of the acceleration of the charge. On the other hand, F^{ab}_{rad} falls linearly with distance and is also linear in acceleration of the charge. This allows us to interpret the former as the Coulomb field and the latter as the radiation field. In the radiation field, one can further introduce covariant notions of electric and magnetic fields, measured at the instantaneous rest frame of the charged particle, by the definitions

$$
E^i_{\mathrm{rad}} = F^{ij}u_j = \frac{q}{l}\left[a^i - n^i(n^j a_j)\right]
$$
$$
B^i_{\mathrm{rad}} = \frac{1}{2}\,\epsilon^{ijkl}u_j F_{kl} = \frac{q}{l}\,\epsilon^{ijkl}n_k a_l\, u_j. \quad (2.178)
$$

Note that the electric field depends on the component of the acceleration which is *transverse* to n^i as expected for a radiation field. Since $u_i E^i = u_i B^i = 0$, it is clear that both of them are three-vectors in the rest frame of the charged particles.

Given the electric and magnetic fields of the radiation field, one can also compute the Poynting vector. Once again, it can be expressed in four-dimensional form as

$$
S^b = -\epsilon^{bijk} E_i B_j u_k = \frac{q^2}{4\pi l^2}\left[a^i a_i - (n^i a_i)^2\right] n^b. \quad (2.179)
$$

It is obvious that E^i and B^i are orthogonal to n^i and that S^i, E^i and B^i form pair-wise orthogonal vectors.

The expression in Eq. (2.174) was obtained for the case of a point charge moving in an arbitrary manner while the first equality in Eq. (2.165) gives the four-vector potential generated by an arbitrary current distribution J^a. It is possible to use this expression to find F^{ab} for an arbitrary source in an exactly similar manner as we have given the derivation for a point charge. The result in this case is given by

$$F^{mn} = \frac{4}{i\pi} \int d^4y \, \frac{\partial^n J^m(y) - \partial^m J^n(y)}{(x-y)^2}. \tag{2.180}$$

In the case of a point charge, it can be verified that this result leads to the expression in Eq. (2.174).

Exercise 2.24

Standard results about radiation (a) Obtain the three-dimensional expressions corresponding to Eq. (2.177), given in standard text books, by converting from proper time to coordinate time. This can be done by using the result of differentiating the relation $x^0 = z^0 + |R|$, which gives

$$l_{\text{ret}} = R^0 u^0 - R \cdot u = |R|u^0 (1 - v_R); \quad dt = dx^0 = dz^0(1 - v_R), \tag{2.181}$$

where $v_R = (v \cdot R/R)$. Combining these, you get:

$$l_{\text{ret}} d\tau = dz^0(1 - v_R) = R dt, \tag{2.182}$$

which is useful in carrying out the differentiation. Performing the differentiation, show that

$$E = \left[\frac{qn}{R^2} + q\frac{R}{c}\frac{d}{dt}\left(\frac{n}{R^2}\right) + \frac{q}{c^2}\frac{d^2n}{dt^2} \right]_{\text{ret}}. \tag{2.183}$$

[Note. This formula for the electric field has a curious interpretation.[4] The first term represents the Coulomb field of the particle evaluated at the retarded time. The second term can be thought of as a 'first-order-correction' to the retarded Coulomb field: this term is obtained by multiplying the time taken for the light signal to travel the distance R and the time rate of change of the Coulomb term (n/R^2). These two terms fall as R^{-2}. The third term depends on the acceleration of the charge and represents the radiation field of the charge.]

(b) Show that this result can also be written in the form

$$E = \frac{q(1 - v^2/c^2)}{R^2(1 - v_R)^3}\left(n - \frac{v}{c}\right) + \frac{q}{R(1 - v_R)^3}n \times \left[\left(n - \frac{v}{c}\right) \times \frac{dv}{c^2 dt}\right], \tag{2.184}$$

where all quantities on the right hand side refer to the retarded time. The first term in the electric field is independent of the acceleration and depends only on the velocity. Compare this result with the ones in Eq. (2.177) and Eq. (2.178). Show that this is the field of a charged particle moving with uniform velocity.

2.11 Larmor formula and radiation reaction

Since the fields at large distance fall as $1/r$, the Poynting vector $S \propto E \times B$ falls as $1/r^2$. The amount of energy flowing into a solid angle $d\Omega$ in unit time, $|S|r^2d\Omega$, is independent of r and represents an irreversible transfer of energy from the near field to the far field. Converting the expression in Eq. (2.179) into the three-dimensional notation, the term $[a^i a_i - (n^i a_i)^2]$ will give $|a|^2 - (n \cdot a)^2 = a^2 \sin^2 \theta$, where θ is the angle between a and n, and we get the energy flux to be:

$$\frac{d\mathcal{E}}{dt d\Omega} = |S| r^2 = \frac{q^2 |a|^2}{4\pi c^3} \sin^2 \theta . \tag{2.185}$$

In Eq. (2.185) the right hand side should be evaluated at the retarded time. The total energy radiated, found by integrating over the solid angle, will be

$$\frac{d\mathcal{E}}{dt} = \frac{q^2 |a|^2}{4\pi c^3} \int_0^\pi 2\pi \left(\sin^2 \theta \right) \left(\sin \theta d\theta \right) = \frac{2q^2}{3c^3} |a|^2 . \tag{2.186}$$

This is called the *Larmor formula*.

Using this, one can also obtain the expression for the four-momentum radiated by a charged particle. The Larmor formula (Eq. (2.186)) for radiation can be written in the form

$$d\mathcal{E} = \frac{2}{3} \frac{q^2}{c^3} a^2(t') dt, \tag{2.187}$$

where $t' = (t - r/c)$ is the retarded time. Since $d\mathcal{E}$ and dt are fourth components of four-vectors, this form suggests a simple relativistic generalization along the following lines. Let us choose an instantaneous rest frame for the charge in which this non-relativistic formula is valid. Because of symmetry, the net momentum radiated, dP, will vanish in this instantaneous rest frame. Clearly this result should be valid even for relativistic motion, if it is expressed in a Lorentz covariant manner. If a^i is the four-acceleration so that $a^2 = a^i a_i$ in the instantaneous rest frame of the charge, then we can express Eq. (2.187), as well as the condition $dP = 0$, by the relation

$$dP^k = \frac{2}{3} \frac{q^2}{c} \left(a^i a_i \right) dx^k = \frac{2}{3} \frac{q^2}{c} \left(a^i a_i \right) u^k d\tau, \tag{2.188}$$

where dP^k is the four-momentum radiated by the particle during the proper time interval $d\tau$. Being relativistically covariant, this result is true for arbitrary velocities.

The radiation of electromagnetic waves transports energy to large distances. This energy has to be eventually supplied by the agency which is accelerating the charged particle. Hence, there has to be an extra drag force f acting on the charged particle due to the fact that it is radiating energy. The rate of work done against this drag, $f \cdot v$, should account for the energy radiated.

We will obtain the radiation reaction force by a fairly simple-minded approach. In particular, the formulas derived below can only be used when the motion of the particle is bounded; that is, the particle should be confined to a finite region of space at all times. We shall first derive the radiation damping force for non-relativistic motion and then generalize the result for the relativistic case.

If the damping force is f, then the work done by the damping force is expected to be equal to the mean power radiated except for a sign, when averaged over a period of time, i.e.

$$\langle f \cdot v \rangle = \left\langle \left(\frac{\Delta \mathcal{E}}{\Delta t} \right) \right\rangle = - \left\langle \frac{2q^2 a^2}{3c^3} \right\rangle . \tag{2.189}$$

Averaging a^2 over a time interval T, we get:

$$\langle a^2 \rangle = \frac{1}{T} \int_0^T dt a^2 = \frac{1}{T} \int_0^T dt \, (\dot{v} \cdot \dot{v})$$
$$= \frac{1}{T} \int_0^T dt \left[\frac{d}{dt} (v \cdot \dot{v}) - v \cdot \ddot{v} \right] = \frac{1}{T} \left[v \cdot \dot{v} \right]_0^T - \langle v \cdot \ddot{v} \rangle . \tag{2.190}$$

The first term vanishes as $T \to \infty$ for any bounded motion, giving $\langle a^2 \rangle = -\langle v \cdot \ddot{v} \rangle$. It follows that, *if* the damping force is

$$f_{\text{damp}} = \frac{2}{3} \frac{q^2 \ddot{v}}{c^3} = \frac{2}{3} \frac{q^2}{c^3} \dddot{x}, \tag{2.191}$$

then the average work done by the damping force exactly accounts for the mean energy radiated.

Note that this expression was obtained after averaging the energy radiated over a period of time. In other words, this expression is capable of maintaining equality between the total energy radiated during a finite interval of time and the total amount of work done during the same period of time. Careless use of this formula can easily lead to wrong results. For example, consider a charged particle that is moving with uniform acceleration a. Since the radiation reaction force depends on the second derivative of v, it vanishes in the case of uniform acceleration. But the Larmor formula implies that a charge undergoing uniform acceleration will be radiating energy at a steady rate. This contradiction arises because a charge moving with uniform acceleration *at all times* will not be bounded and we cannot apply the results obtained above. If the uniform acceleration occurs only for a finite duration of time, then a more careful consideration of the conditions at the beginning and ending of the acceleration will reveal that there is no inherent conceptual difficulty (see Project 2.3).

Let us next consider the case of radiation reaction for a particle moving with relativistic velocities. In this case, the radiation reaction force f can be expressed in terms of a four-force g^i with components $(\gamma f \cdot v, \gamma f)$. Therefore, we need to find a four-vector g^i which reduces to $\left(0, (2/3) q^2 \ddot{v} \right)$ in the rest frame of the charge.

This condition is satisfied by any vector of the form

$$g^i = \left(\frac{2q^2}{3}\right)\left[\left(\frac{d^2u^i}{d\tau^2}\right) + Au^i\right] \tag{2.192}$$

where A is to be determined. To find A, we use the condition that $g^i u_i = 0$, which should be valid for any particle trajectory. This gives $A = u^k\left(d^2u_k/d\tau^2\right)$, leading to

$$g^i = \left(\frac{2q^2}{3}\right)\left[\frac{d^2u^i}{d\tau^2} + u^i u^k\frac{d^2u_k}{d\tau^2}\right]. \tag{2.193}$$

The second term can be rewritten using the result

$$u^k\frac{d^2u_k}{d\tau^2} = u^k\frac{da_k}{d\tau} = \frac{d}{d\tau}\left(u^k a_k\right) - a^k a_k = -a^k a_k \tag{2.194}$$

since $u^k a_k = 0$. This gives another, equivalent, form for g^i:

$$g^i = \frac{2}{3}q^2\left[\frac{d^2u^i}{d\tau^2} - u^i\left(a^k a_k\right)\right] = \frac{2}{3}q^2\left[\frac{da^i}{d\tau} - u^i\left(a^k a_k\right)\right]. \tag{2.195}$$

This is the final form of the radiation reaction force in special relativity.

Exercise 2.25
Radiation drag A more complicated situation exhibiting radiation drag arises when a charged particle with velocity v is moving through a region of space containing an isotropic bath of radiation with energy density $U_{\rm rad}$. Since the charge has a nonzero velocity v, the scattering of the radiation by the charge will be anisotropic and the net momentum transfer to the charged particle will be in the direction opposite to the velocity.

(a) Show that when the acceleration of a charge is caused by the action of an electromagnetic field, the radiation reaction force can be written in terms of the electromagnetic energy-momentum tensor T^{ab} as:

$$g^i = \left(\frac{\sigma_T}{c}\right)\left[T^{ij}u_j - \left(T^{ab}u_a u_b\right)u^i\right], \tag{2.196}$$

where $\sigma_T \equiv (8\pi/3)\left(q^2/mc^2\right)^2$ is called the *Thomson scattering cross-section* and we have reintroduced the c-factor.

(b) Show that the energy-momentum tensor for a isotropic bath of radiation with energy density $U_{\rm rad}$ is given by $T^{ab} = U_{\rm rad}$ dia $(1, 1/3, 1/3, 1/3)$. For a particle moving through this bath with a four-velocity $u^i = (\gamma, \gamma v)$ show that

$$T^{ab}u_a u_b = U_{\rm rad}\gamma^2\left(1 + \frac{1}{3}v^2\right); \qquad T^{ab}u_b = \left(U_{\rm rad}\gamma, -\frac{1}{3}U_{\rm rad}\gamma v\right). \tag{2.197}$$

(c) Using this, show that the drag force has the components $g^i = (\gamma\boldsymbol{f}\cdot\boldsymbol{v}, \gamma\boldsymbol{f})$, where

$$\boldsymbol{f} = -\frac{4}{3}\sigma_T U_{\rm rad}\gamma^2\left(\frac{\boldsymbol{v}}{c}\right); \qquad -\boldsymbol{f}\cdot\boldsymbol{v} = \frac{4}{3}\sigma_T U_{\rm rad}\gamma^2\left(\frac{v^2}{c^2}\right)c. \tag{2.198}$$

This result is valid for any isotropic radiation field with energy density U_{rad}. [The work done by this drag force, $\boldsymbol{f}_{drag} \cdot \boldsymbol{v} = -(4/3)\,\sigma_T U_{\text{rad}}\gamma^2\,(v/c)^2$, will reduce the kinetic energy and hence the velocity of the charged particle. This loss of energy by the particle will appear as net gain of energy of radiation given by

$$\frac{dE}{dt} = \frac{4}{3}\sigma_T U_{\text{rad}}\gamma^2 \left(\frac{v}{c}\right)^2 c. \tag{2.199}$$

Thus a charged particle, moving relativistically through a radiation bath, can transfer its kinetic energy to the radiation. This process is called *inverse Compton scattering*.]

(d) The radiation drag on a particle, Eq. (2.199), can also be obtained as follows. Treat a radiation bath as equivalent to randomly fluctuating $\boldsymbol{E}$ and $\boldsymbol{B}$ fields with $\langle\boldsymbol{E}\rangle = \langle\boldsymbol{B}\rangle = 0$; $\langle\boldsymbol{E}^2\rangle = \langle\boldsymbol{B}^2\rangle = 4\pi U_{\text{rad}}$. Evaluate the mean Lorentz force on the charge and show that the mean power radiated by the charge is

$$\left(\frac{dE}{dt}\right)_{\text{rad}} = \sigma_T c\gamma^2 \left(1 + \frac{v^2}{c^2}\right) U_{\text{rad}}. \tag{2.200}$$

Argue that the radiation absorbed by the charge is $(dE/dt)_{\text{abs}} = \sigma_T c U_{\text{rad}}$. Hence calculate the net energy transfer from the charge to the radiation bath and obtain Eq. (2.199).

PROJECTS

Project 2.1

Third rank tensor field

A possible generalization of electromagnetism will be to a potential B_{ij} (in place of A_i) and a field $H_{mnp} = \partial_{[m}B_{np]}$ (in place of F_{ij}) where $[\,]$ indicates a completely antisymmetric tensor. The source for this field could be an antisymmetric current $J_{ab} = -J_{ba}$. The theory is expected to be invariant under the generalized gauge transformation $B_{mn} \to B_{mn} + \partial_{[m}f_{n]}$.

(a) Show that an appropriate gauge invariant action for this field could be

$$\mathcal{A} = \int d^4x \left[-H^{mnp}H_{mnp} + \frac{1}{2}\epsilon^{mnps}B_{mn}J_{ps}\right]. \tag{2.201}$$

What is the condition on J_{ab} for the action to be gauge invariant?

(b) Derive the equations of motion for the field and its energy-momentum tensor.

(c) Develop this theory along the lines of electromagnetism as far as you can.

Project 2.2

Hamilton–Jacobi structure of electrodynamics

In the case of classical mechanics, the Hamilton–Jacobi equation gives the quickest route to solve equations of motion and also provides a window into quantum theory. The purpose of this project is to review this and generalize it for electrodynamics.

(a) In classical mechanics, we can consider the action $\mathcal{A}$ as a function of the upper limit of integration (t, x) when evaluated on the extremum trajectory. Suppose the end points are varied by $x \rightarrow x + \delta x$, $t \rightarrow t + \delta t$. Show that this leads to a variation in the action given by

$$\delta \mathcal{A} = L\,\delta t + p(\delta x - \dot{x}\,\delta t) = p\delta x - (p\dot{x} - L)\delta t = p\delta x - H\delta t. \qquad (2.202)$$

Make sure you understand the origin of $\dot{x}\delta t$ term in the first equality. This shows that $p = (\partial \mathcal{A}/\partial x)$, $H = -(\partial \mathcal{A}/\partial t)$. This will directly lead to the Hamilton–Jacobi equation.

(b) In the case of a field theory, we would like to generalize the notion of the time derivative to an infinitesimal variation of a spacelike surface. Suppose an event on a spacelike surface has the coordinates $[x, y, z, t(x, y, z)]$. If we displace this surface slightly, this event will have the coordinates $[x, y, z, t + \delta t(x, y, z)]$. (i) Show that the four-volume δV included between the two surfaces can be expressed in terms of a suitable four-vector δn^i as $\delta V = \delta n^i\,d^3\sigma_i$. (ii) Consider the change in the electromagnetic action under the combined influence of changing the fields on a spacelike hypersurface as well as a shifting of the spacelike hypersurface. Show that the net change can be expressed in the form

$$\delta \mathcal{A} = \int d^3 x\,\frac{F^{\alpha 0}}{4\pi}\,\delta A_\alpha - \int \frac{1}{8\pi}(E^2 + B^2)\delta V. \qquad (2.203)$$

From this we can identify the canonical momentum $\pi^\alpha = \delta \mathcal{A}/\delta A_\alpha = -E^\alpha/4\pi$ and the Hamiltonian $H = -\delta \mathcal{A}/\delta V = (1/8\pi)(E^2 + B^2)$. Use these results to write down the Hamilton–Jacobi equation for the electromagnetic action as

$$-\frac{\delta \mathcal{A}}{\delta V} = 2\pi \left(\frac{\delta \mathcal{A}}{\delta \boldsymbol{A}}\right)^2 + \frac{1}{8\pi}(\nabla \times \boldsymbol{A})^2. \qquad (2.204)$$

Explain how – in principle – this equation can be used to find solutions to electromagnetic field equations (though this may not be the best procedure in field theory).

(c) The extremal value of the action should not change under a gauge transformation. Show that this leads to the constraint equation $\nabla \cdot \boldsymbol{E} = 0$.

(d) From the Hamilton–Jacobi equation of classical mechanics, one can make a transition to quantum mechanics by obtaining the Schrödinger equation for the wave function $\psi(t, x)$. Investigate a similar procedure to obtain a functional Schrödinger equation in the case of quantum electrodynamics.

Project 2.3
Does a uniformly accelerated charge radiate?

The question in the title of the project is one of the 'eternal issues' in special relativistic electrodynamics with a large amount of literature, running over decades.[5] The purpose of this project is to acquaint you with this problem in the special relativistic context so that more complex questions can be handled in the later chapters. This project is closely related to Project 5.2 and Project 7.3.

(a) Consider a charged particle which is moving in a uniformly accelerated trajectory along the X-axis from $T = -\infty$ to $T = +\infty$. While this motion is clearly unphysical, one can treat this problem purely as an exercise in the study of partial differential equations and ask: what is the A_i and F^{ik} produced by a charge moving in such a trajectory? Determine these fields paying special attention to the surface $X = -T$. Do you think such a charged particle radiates? If so, where does the energy come from, since the radiation reaction force in Eq. (2.193) vanishes for this trajectory?

(b) Consider next a charged particle which was at rest till $T = 0$, moves with uniform acceleration g during the time $0 \leq T \leq T_0$ and moves with uniform velocity for $T > T_0$.

(i) Compute the total amount of energy radiated by the particle to infinity. (ii) Determine when the particle experiences the radiation reaction force? (iii) Compute the work done by the radiation reaction force on the particle. Is this equal to the energy radiated by the particle? (iv) Sketch roughly the behaviour of electromagnetic fields at all distances from the particle at different stages of motion. How exactly is the energy transfered from finite distance to infinity?

(c) Is it possible to obtain the results in part (a) above as a limiting case of part (b) above?

3

Gravity and spacetime geometry: the inescapable connection

3.1 Introduction

In the previous chapter we described how the dynamics of the scalar and vector fields can be described in a manner consistent with the special theory of relativity. Given the fact that Newtonian gravity is described by a gravitational potential $\phi_N(t, \boldsymbol{x})$, which satisfies the Poisson equation $\nabla^2 \phi_N = 4\pi G \rho_m$ (where ρ_m is the mass density), it might seem that one could construct a theory for gravity consistent with special relativity by suitably generalizing the Poisson equation for the gravitational potential. It turns out, however, that this is not so straightforward. The natural description of the gravitational field happens to be completely different and is intimately linked with the geometrical properties of the spacetime. We will be concentrating on such a description from Chapter 4 onwards in this book.

The key purpose of the present chapter is to explain in physical terms why such a geometrical description for gravity is almost inevitable.[1] We shall first describe several difficulties that arise in any attempt to provide a purely field theoretic description of gravity in flat spacetime. We will then give a series of simple thought experiments that illustrate an intimate connection between gravitational fields and spacetime geometry. None of this can be thought of as a mathematically rigorous proof that gravity *must* be described as spacetime geometry; however, it goes a long way in showing that such a description is *most natural* and, of course, consistent with all known facts about gravity.

3.2 Field theoretic approaches to gravity

To understand the issues involved, we begin by investigating the possibility of providing a description of gravity which is consistent with special relativity. This is done in a spirit similar to the description of scalar and vector fields in the four-dimensional language given in the previous chapter. Such a description should reduce to ordinary Newtonian gravity described by a gravitational potential $\phi_N(t, \boldsymbol{x})$, satisfying the Poisson equation

$$\nabla^2 \phi_N = 4\pi G \rho_m \qquad (3.1)$$

in the limit $c \to \infty$. In addition, the theory should satisfy three more requirements. (1) The theory should be Lorentz covariant. (2) The gravitational force between the sources must always be attractive. (3) The description must obey the *principle of equivalence* which, for the purpose of this chapter, can be stated as follows: the trajectories of particles (having the same initial conditions) in a given gravitational field must be independent of the properties of the particle. We shall now attempt to determine the ingredients of such a theory.

The Lorentz invariance of the theory along with the appropriate Newtonian approximation can be ensured in several ways. In the simplest approach, one can think of the gravitational potential as a Lorentz invariant scalar ϕ and modify the left hand side of Eq. (3.1) to $\Box\phi$. In the limit of slowly moving sources, we will obtain $\nabla^2\phi$ from $\Box\phi$ when we take the limit $c \to \infty$. While this is the simplest description, it is not unique. In principle, one could have also thought of the Newtonian potential as the zeroth component of a four-vector $V^i = (\phi, \boldsymbol{V})$ and attempted to construct a suitable theory for the Lorentz covariant vector field V^i. The Newtonian potential ϕ_N can also arise from the time–time component of a second rank tensor field $H_{ab} = (H_{00}, H_{0\alpha}, H_{\alpha\beta}) \equiv (\phi, H_{0\alpha}, H_{\alpha\beta})$. In the cases of vector and tensor fields, one can arrange factors of c in the scaling of the components such that the time components dominate over spatial components in the limit $c \to \infty$. Obviously, this idea can be generalized to higher rank tensor fields. Therefore the existence of the Newtonian approximation and Lorentz invariance alone do not determine whether the theory of gravity should be based on, for example, a scalar or a vector or a second rank tensor field. We need to make further assumptions or examine the consequences of each possibility and compare the results with experiments to make this choice. We shall now consider each of these cases one by one.

3.3 Gravity as a scalar field

The simplest Lorentz invariant description of gravity would be based on treating gravity as a scalar field along the lines described in Section 2.3.1 and Section 2.3.2. For such a theory, we found that the equation of motion for a particle of mass m is given by Eq. (2.6). With the c-factors reintroduced, using the fact that the dimensions of mc^2 and $\lambda\phi$ are the same, it reads:

$$\frac{du^i}{d\tau} = -\lambda \frac{\partial^i \phi}{(mc^2 + \lambda\phi)} - \lambda u^i u^j \frac{\partial_j \phi}{(mc^2 + \lambda\phi)}. \qquad (3.2)$$

The trajectory of the particle in a given gravitational field (described by ϕ) depends explicitly on the mass of the particle m and the coupling constant λ. In order to be consistent with the principle of equivalence, the right hand side of Eq. (3.2) has to be independent of the mass of the particle. This demands that the coupling constant λ must be proportional to the mass m. In natural units ($\hbar = c = 1$), the coupling constant is dimensionless and $m \, (= mc/\hbar)$ has the dimension of inverse length. So we can take $\lambda = lm$ where l is a fundamental constant with dimensions of length. (With this choice, we treat l as a universal constant while the coupling parameter λ will vary with the mass of the particle.) It is also convenient at this stage to define a new field $\Phi \equiv l\phi$ which is dimensionless in natural units and has the dimensions of c^2 in normal units (which will match with the dimension of Newtonian gravitational potential, ϕ_N). Then, Eq. (3.2) becomes

$$\frac{du^i}{d\tau} = -\frac{\partial^i(\Phi/c^2)}{1 + (\Phi/c^2)} - u^i u^j \frac{\partial_j(\Phi/c^2)}{1 + (\Phi/c^2)} \tag{3.3}$$

which shows that all particles follow the same trajectory in a given gravitational potential Φ if they start with the same initial condition. With this choice of coupling, the action for a particle moving in a gravitational field becomes (see Eq. (2.4)):

$$A = -mc^2 \int_{\tau_1}^{\tau_2} d\tau \left(1 + \frac{\Phi}{c^2}\right). \tag{3.4}$$

Since the mass of the particle appears only as an overall scaling factor, it is obvious that, as Eq. (3.3) shows, the equations of motion derived from this action will be independent of m.

To determine the field equations for gravity (which should reduce to the Poisson equation at the appropriate limit) we need to introduce a Lagrangian for the field with a kinetic energy term. The simplest choice for this is a Lagrangian which is proportional to $L_{\text{field}} = -(1/2)\partial_a \phi \partial^a \phi$. In terms of Φ this becomes

$$L_{\text{field}} = -\frac{1}{2l^2} \partial_a \phi \partial^a \phi \equiv -\frac{1}{8\pi Gc} \partial_a \Phi \partial^a \Phi, \tag{3.5}$$

where we have rewritten the constant l as $l = \sqrt{4\pi Gc}$ so that in the appropriate limit G can be identified with the Newtonian gravitational constant. (In natural units $G(\hbar/c^3)$ has the dimensions of square of the length; the c-factor is introduced to cancel the c-factor in $d^4x = cdtd^3x$ when we define the action as an integral of the Lagrangian over d^4x.) We can now obtain the field equations for the system along the lines described in Section 2.3.2. Using Eq. (2.24) with $V = \lambda n\phi/c = mn\Phi/c$ we will get

$$\Box\Phi = 4\pi Gmn = 4\pi G\rho_m. \tag{3.6}$$

In the notation of Chapter 2, the n in Eq. (2.11) is the *number* density rather than mass density so that $\rho_m = mn$. It is obvious that, in the $c \to \infty$ limit, Eq. (3.6) reduces to Eq. (3.1).

The above description works fine when the source of the gravitational field is made up of a system of particles. However, such a description needs to be generalized in the context of special relativity because of the equivalence between mass and energy. For example, we know that a system made of equal numbers of electrons and positrons can annihilate to produce pure radiation. If the mass density of electrons and positrons, treated as point particles with world lines, produces a gravitational field, then it seems natural that the energy density of radiation resulting from their annihilation should also contribute to the gravitational field. This requires us to generalize the source on the right hand side of Eq. (3.6) in a suitable manner to take care of any physical system with energy and momentum. We have seen in Chapter 1 that each physical system possesses its own energy-momentum tensor T_{ab}. The only Lorentz invariant scalar that can be constructed from T_{ab} and η_{ab} is the trace $T \equiv T_{ab}\eta^{ab} = T^a_a$. In the case of pressureless point particles, Eq. (1.114) shows that the trace of the energy-momentum tensor is given by $T = -\rho_m c^2$ where ρ_m is the mass density of the particles. For such a system of pressureless point particles, Eq. (3.6) can equivalently be written as

$$\Box \Phi = -\frac{4\pi G}{c^2} T. \tag{3.7}$$

This equation again has the correct Newtonian approximation and can now be taken to describe the coupling of *any* physical system with an energy-momentum tensor T_{ab} to gravity. It can be obtained by varying Φ in the action functional given by

$$A = \int d^4x \left[-\frac{1}{2}\partial_a\phi\partial^a\phi + l\frac{\phi T}{c^3} + L_{\text{source}} \right]$$
$$= \int d^4x \left[-\frac{1}{8\pi Gc}\partial_a\Phi\partial^a\Phi + \frac{\Phi T}{c^3} + L_{\text{source}} \right] \tag{3.8}$$

Here, the first term describes the kinetic energy of the gravitational field Φ, the second term describes the interaction between the gravitational field and some source for gravity which has an energy-momentum tensor T_{ab} with $T = T^a_a$, and L_{source} describes the Lagrangian for the source which is independent of Φ. This defines a scalar theory of gravity which is Lorentz invariant. Unfortunately, there are some key difficulties with such a theory which we shall now describe.

The first problem arises from the fact that the natural source for gravity in a scalar theory of gravity is the trace of the energy-momentum tensor T. As we noticed in Chapter 2, the electromagnetic field has an energy-momentum tensor with a vanishing trace. This implies that the electromagnetic field will not couple to gravity in this theory. Therefore, the electromagnetic field will neither produce a gravitational field around it nor will it be affected by the gravitational field. The Maxwell equations in the presence of a nonzero gravitational field Φ will have the same form as the Maxwell equations in the case of $\Phi = 0$. In particular, light rays (which are particular solutions to Maxwell equations) will be unaffected by the gravitational field. This contradicts the observational evidence that trajectories of light rays bend in the presence of gravitationally attracting masses like, for example, the Sun. The above fact is sufficient reason to abandon pursuing a scalar theory of gravity based on Eq. (3.8). (Also see Project 3.2.)

There is also a conceptual issue associated with this theory which is worth mentioning. The Lagrangian in Eq. (3.8) couples the gravitational field Φ to the trace of the energy-momentum tensor of all sources *other than gravity*. But the energy-momentum tensor of the gravitational field Φ itself has a nonzero trace which does not act as its own source in the above description. Though one could postulate the existence of such a field, it is conceptually unnatural because it requires separating the trace of the total energy-momentum tensor of the system into two parts, the one due to the gravitational field and the second due to all fields other than gravity, and treating them differently. A better description will be to couple the gravitational field to the trace of its own energy-momentum tensor as well in a self-consistent manner. This will make the theory nonlinear (see Project 3.1) but will not solve the problem that the electromagnetic field is unaffected by gravity.

While the scalar field theory of gravity is observationally untenable, it illustrates a very important concept which we will come back to in Section 3.5. Let us take a closer look at the action in Eq. (3.4) for a point particle in this model. This action shows that the equation of motion can be obtained by replacing the line interval $ds = cd\tau$ by

$$ds_\Phi \equiv cd\tau \left(1 + \frac{\Phi}{c^2} \right) = ds \left(1 + \frac{\Phi}{c^2} \right). \tag{3.9}$$

Since $ds^2 = \eta_{ik} dx^i dx^k$, this replacement is equivalent to:

$$\eta_{ik} \rightarrow g_{ik}(t, \boldsymbol{x}) \equiv \eta_{ik} \left[1 + \frac{\Phi(t, \boldsymbol{x})}{c^2} \right]^2. \tag{3.10}$$

In other words, our theory (at least as far as the interaction with massive point particles are concerned) is equivalent to working with a modified line interval between the events given by $ds^2 = g_{ik}(t, \boldsymbol{x}) dx^i dx^k$ which depends on the gravitational field. Such a description, of course, is purely geometrical and g_{ik} is called the metric tensor. This alternative, geometrical, interpretation is possible only because the mass of the particle scales out as an overall multiplication constant in the action in Eq. (3.4) which, in turn, is a direct consequence of the principle of equivalence.

The same result can also be obtained from the Hamilton–Jacobi equation for the particle moving in the scalar gravitational field given by Eq. (2.8). With $\lambda = m$, this equation can be written in the form

$$g^{ik}\partial_i\mathcal{A}\partial_k\mathcal{A} = -m^2c^2; \quad g^{ik} \equiv \eta^{ik}\left[1 + \frac{\Phi(t,\boldsymbol{x})}{c^2}\right]^{-2}, \tag{3.11}$$

where g^{ik} is the inverse of the matrix g_{ik}.

These are the first indications that the gravitational field might have a purely geometric description. In fact, this feature provides a stronger conceptual reason for not pursuing a scalar version of gravitation theory; we will see in later chapters that a relativistic theory of gravity can be given a purely geometric description in terms of a metric tensor. From the above analysis we see that a description in terms of a scalar field is just a special case of a more general description when the metric tensor has the specific form in Eq. (3.10).

Exercise 3.1

Motion of a particle in the scalar theory of gravity For a massive particle located at the origin, one can take the solution of Eq. (3.7) to be spherically symmetric and static. It follows from standard Newtonian gravity result that, in this case, $\Phi = -GM/r$. We will now investigate the motion of a relativistic particle in this gravitational field. This can be done by solving Eq. (3.2) but it is simpler and quicker to use the Hamilton–Jacobi equation, given by Eq. (3.11).

(a) Show, from symmetry considerations, that the motion of the particle can be assumed to be in a plane, which we will take to be $\theta = \pi/2$. Expand the Hamilton–Jacobi equation to obtain

$$(1 + \Phi)^{-2}\left[\left(\frac{\partial\mathcal{A}}{\partial t}\right)^2 - \left(\frac{\partial\mathcal{A}}{\partial r}\right)^2 - \frac{1}{r^2}\left(\frac{\partial\mathcal{A}}{\partial\psi}\right)^2\right] - m^2 = 0, \tag{3.12}$$

where we have temporarily set $c = 1$ and indicated the polar angle in the plane by ψ to avoid notational conflict with the scalar field ϕ. Show that the relevant solution to this equation is given by $\mathcal{A} = -Et + L\psi + \mathcal{A}_r$, where

$$\mathcal{A}_r = \int\left[E^2 - \frac{L^2}{r^2} - m^2(1 + \Phi)^2\right]^{1/2}dr$$

$$= \int\left[E^2 - m^2 + \frac{2GMm^2}{r} - \frac{1}{r^2}(L^2 + G^2M^2m^2)\right]^{1/2}dr. \tag{3.13}$$

Interpret E and L as the conserved energy and angular momentum of the particle.

(b) The key new effect arises from the change in the coefficient of a $(1/r^2)$ term which is just L^2 in standard Newtonian theory. Argue that this will lead to a systematic precession of planetary orbits. Perform an analysis similar to that done in Section 2.5 and determine the rate of precession of the perihelion of a particle, $\delta\psi$, and show that

$$\delta\psi = -\frac{G^2M^2m^2}{L^2}\pi \tag{3.14}$$

per orbit. [Note. This is $(-1/6)$ times the prediction in Einstein's general theory of relativity; see Eq. (7.122).]

Let us next consider the possibility of describing gravity by a vector field. Given the fact that the scalar potential in electrodynamics obeys a Poisson equation with charge density as the source, one could also consider the possibility of a theory of gravity in which the gravitational field arises as the time component of a four-vector. It is, however, easy to see that this idea does not work. The key problem is that, in a relativistically invariant theory of a vector field, like charges will repel each other (see the discussion on page 86) while in gravity like 'charges' attract each other.

3.4 Second rank tensor theory of gravity

We shall next consider the possibility of providing a Lorentz covariant theory of gravity based on a second rank, symmetric, tensor field H_{ab}. (As we shall see, the symmetric second rank tensor has an adequate number of components to be considered a viable candidate for describing gravity; this is one of the reasons we do not consider a more general, asymmetric, second rank tensor.) Such a theory is closest to the exact description of gravity developed in the later chapters and the reason for its non-viability is fairly subtle. In view of this, we shall provide a fairly detailed and careful description of the situation and will also use this context to illustrate several other important aspects of field theory which are of interest in their own right. We shall first briefly re-examine the vector field theory and the issue of gauge invariance from a specific perspective and then generalize these concepts to a second rank symmetric tensor field.

Let us recall that, in the case of a scalar field, the field equation $\Box \phi = J$ can be obtained from the Lagrangian density

$$L = -(1/2)\partial_a \phi \partial^a \phi - J\phi. \tag{3.15}$$

The minus sign in the kinetic energy term of L is crucial because we require $\dot{\phi}^2$ to appear with positive sign in the Lagrangian. If it does not, the plane wave solutions to the wave equation $\Box \phi = 0$ in the absence of source ($J = 0$) will carry negative energy, which is unphysical.

Next, we want to construct a theory for a vector field $V^a(x)$ which obeys the field equation $\Box V^a = J^a$, where J^a is a source. That is, we want each of the components of the four-vector to satisfy the same equation as the scalar field. In Chapter 2, we obtained the field equation $\Box A^i = -4\pi J^i$ (see Eq. (2.124)) by first obtaining the gauge invariant equation $\partial_k F^{ki} = -4\pi J^i$ (see Eq. (2.115)) and then imposing the gauge condition $\partial_i A^i = 0$. It might seem that one can obtain the field equation $\Box V^i = J^i$ more simply by taking the Lagrangian to be

$$L = -\frac{1}{2}\partial_a V_b \partial^a V^b - J_a V^a = +\frac{1}{2}\partial_a V^0 \partial^a V^0 - \frac{1}{2}\partial_a \mathbf{V} \cdot \partial^a \mathbf{V} - J_a V^a. \quad (3.16)$$

There is, however, a serious problem with this approach. We see from Eq. (3.16) that, in the kinetic energy term of the Lagrangian, while the spatial components have the correct overall sign, the V^0 component comes with the wrong sign. So in the absence of the source ($J_a = 0$), the V^0 degree of freedom will propagate as a wave carrying negative energy. (If the overall sign of the Lagrangian is changed, then the V^0 component can have the correct sign but the spatial components $\mathbf{V}$ will have a wrong sign and will carry negative energy.) Hence the indefinite signature of the Lorentz metric and the requirement of Lorentz invariance show that the Lagrangian in Eq. (3.16) is untenable.

So, to obtain a viable physical theory, in which all the propagating physical degrees of freedom carry positive energy, we should be able to eliminate the V^0 component in the absence of the source. (In the presence of the source, we make the time component a non-propagating mode; we will not discuss this complication since we do not need it for our argument.) Thus the theory *must* have an extra invariance which depends on an arbitrary, single, function degree of freedom. One can then use this symmetry to eliminate the unwanted modes from the theory. This is most easily achieved if we make the theory invariant under the transformation

$$V_a \to V_a + \partial_a F, \quad (3.17)$$

where $F(x)$ is an arbitrary function. This allows us to eliminate one degree of freedom from the theory which carries negative energy. This is the key physical reason for the existence of an extra symmetry in the form of gauge invariance in the case of a massless vector field. The kinetic energy term in Eq. (3.16) should now be modified to preserve the gauge invariance. In Chapter 2, we showed that such a gauge invariant Lagrangian for the vector field has a kinetic energy term proportional to $F_{ab}F^{ab}$ and obtained Maxwell's equations. We also noted that the interaction term $J_a V^a$ can maintain gauge invariance only if the source is conserved: $\partial_a J^a = 0$.

The situation is very similar in the case of a symmetric tensor field H_{ab} which we shall now discuss. We begin by attempting to construct a theory in which each component satisfies the equation $\Box H_{ab} = lS_{ab}$, where S_{ab} is some suitable source and l is a coupling constant. If we succeed in that, then we can arrange matters such that in the $c \to \infty$ limit, only the 00 component dominates and the relevant equation reduces to $\nabla^2 H_{00} \approx lS_{00}$. This will reproduce Eq. (3.1) if we take $H_{00} \propto \phi_N$ and $S_{00} \propto \rho_m$ and we will have a relativistically covariant description of gravity. In natural units, we expect S_{ab} to have dimensions of L^{-4} corresponding to mass density and H_{ab} to have dimensions L^{-1} like any other field; then the coupling constant l will have dimensions of length.

The simplest Lagrangian which can lead to the field equation $\Box H_{ab} = lS_{ab}$ is again given by

$$L = -\frac{1}{4}\partial_i H_{ab}\partial^i H^{ab} - \frac{l}{2}S_{ab}H^{ab} + L_{\text{source}}(q_A). \qquad (3.18)$$

Here the first term is the kinetic energy term for the gravitational field, the last term is the Lagrangian for the source which is *independent of H_{ab}* but depends on some source variables (symbolically denoted as q_A) and the second term is the coupling between the source and gravity. The factors $1/4$ and $1/2$ in the two terms, of course, can be rescaled; they are chosen with the aim of simplifying future computations. With this choice, H_{ab} has the dimensions of inverse length, just like any other field, and S_{ab} (built from the source variables) has L^{-4} and l is a length. Varying the H_{ab} in the first two terms will lead to the field equation.

This Lagrangian, however, has the same sign problem as the Lagrangian in Eq. (3.16). Once again, we need to eliminate the unphysical degrees of freedom (which carry negative energies due to the wrong sign in the Lagrangian) by using some extra symmetry analogous to the one in Eq. (3.17) for the vector field. Here the three components $H_{0\alpha}$ are unphysical and we need to have a symmetry involving at least three function degrees of freedom to eliminate these. The minimal relativistically covariant object which has at least three function degrees of freedom is a four-vector, which has four function degrees of freedom. Therefore, the symmetry could be based on an arbitrary four-vector ξ_a and a unique choice for this symmetry is

$$H_{ab}(x) \rightarrow H_{ab}(x) + \partial_a\xi_b(x) + \partial_b\xi_a(x). \qquad (3.19)$$

This transformation in Eq. (3.19) for the tensor field is a natural generalization of the gauge transformation in Eq. (3.17) for a vector.

The kinetic energy term in the Lagrangian in Eq. (3.18) has now to be modified to maintain invariance under transformation in Eq. (3.19), just as we needed to modify the kinetic energy term in the Lagrangian in Eq. (3.16) to $F_{ab}F^{ab}$ for maintaining invariance under the transformation in Eq. (3.17). Constructing such a suitable kinetic energy term is completely analogous to the manner in which we constructed the Maxwell Lagrangian in Chapter 2, except that it is more complicated algebraically. We shall now describe how this can be done.[2]

For convenience, we will rescale the variables and introduce a dimensionless field $h_{ab} \equiv lH_{ab}$ so that, in the Newtonian approximation, we can possibly identify h_{00} with ϕ_N/c^2 up to a numerical constant. The kinetic energy term in the action for the symmetric tensor field h_{ab} is built out of scalars which are quadratic in the derivatives $\partial_a h_{bc}$. The most general expression will be the sum of different scalars obtained by contracting pairs of indices in $\partial_a h_{bc} \partial_i h_{jk}$ in different manners. Since this product is symmetric in (b,c) and (j,k) and also under the interchange $(a,b,c) \rightarrow (i,j,k)$, it is easy to figure out that, *a priori*, seven different contractions are possible. For example, if a is contracted with i, then there are two possibilities for contracting b (with either c or with j; contracting b with k is the same as contracting b with j). These contractions will lead to the terms

$$c_1 \partial_a h_{bc} \partial_i h_{jk} \eta^{ai} \eta^{bc} \eta^{jk} + c_2 \partial_a h_{bc} \partial_i h_{jk} \eta^{ai} \eta^{bj} \eta^{ck} = c_1 \partial_a h_b^b \partial^a h_j^j + c_2 \partial_a h_{bc} \partial^a h^{bc}$$
(3.20)

in the Lagrangian with as yet undetermined constants (c_1, c_2). For brevity, we will denote these two terms symbolically as $(ai, bc, jk), (ai, bj, ck)$. Next, if a is contracted with b, there are again two inequivalent possibilities for contracting c leading to $(ab, ci, jk), (ab, ck, ij)$. Finally if a is contracted with k, there are three possible ways of contracting b giving $(ak, bj, ci), (ak, bc, ij), (ak, bi, cj)$.

Of these, the contraction (ak, bc, ij) is the same as (ab, ci, jk) since $(ak, bc, ij) = (ic, jk, ab)$ under $(a,b,c) \leftrightarrow (i,j,k)$ and, of course, $(ic, jk, ab) = (ab, ci, jk)$. Similarly, $(ak, bj, ci) = (ic, jb, ka) = (ib, jc, ka)$; the first equality comes from $(a,b,c) \leftrightarrow (i,j,k)$ symmetry while the second arises from $b \leftrightarrow c$ symmetry. Since $(ib, jc, ka) = (ak, bi, cj)$ trivially, we need to retain only the first two out of the three possibilities in the last set. Thus dropping the two contractions (ab, ci, jk) and (ak, bi, cj) out of the seven possibilities, we are left with five different contractions: $(ai, bc, jk), (ai, bj, ck), (ab, ck, ij), (ak, bj, ci), (ak, bc, ij)$. This will correspond to a Lagrangian for the second rank tensor field of the form

$$L = \partial_a h_{bc} \partial_i h_{jk} \left[c_1 \eta^{ai} \eta^{bc} \eta^{jk} + c_2 \eta^{ai} \eta^{bj} \eta^{ck} \right.$$
$$\left. + c_3 \eta^{ab} \eta^{ck} \eta^{ij} + c_4 \eta^{ak} \eta^{bj} \eta^{ci} + c_5 \eta^{ak} \eta^{bc} \eta^{ij} \right],$$
(3.21)

which can be expressed as:

$$L = \left[c_1 \partial_a h_b^b \partial^a h_j^j + c_2 \partial_a h_{bc} \partial^a h^{bc} + c_3 \partial_a h^{ab} \partial_i h_b^i \right.$$
$$\left. + c_4 \partial_a h_{bc} \partial^c h^{ba} + c_5 \partial_a h_b^b \partial_i h^{ia} \right].$$
(3.22)

Each term in the Lagrangian in Eq.(3.22) is of the kind $J^{abcijk}(\eta)\, \partial_a h_{bc} \partial_i h_{jk}$, where J^{abcijk} is a cubic in η^{lm} and hence is constant (i.e. all the components are 0 or ± 1). This allows one to 'swap' the derivatives ∂_i and ∂_a by adding a total divergence, using the identity:

$$J^{abcijk} [\partial_a h_{bc} \partial_i h_{jk} - \partial_i h_{bc} \partial_a h_{jk}] = \partial_a [(J^{abcijk} - J^{ibcajk}) h_{bc} \partial_i h_{jk}].$$
(3.23)

Using this result, one can convert the c_3 term to the c_4 term and rewrite the Lagrangian as

$$L = \left[c_1 \partial_a h_b^b \partial^a h_j^j + c_2 \partial_a h_{bc} \partial^a h^{bc} + (c_3 + c_4) \partial_a h_{bc} \partial^c h^{ba} + c_5 \partial_a h_b^b \partial_i h^{ia} \right] + L_{\text{div}}$$

$$\equiv L_h + L_{\text{div}},$$
(3.24)

where

$$L_{\text{div}} = c_3 \left[\partial_a h^{ab} \partial_i h_b^i - \partial_a h_{bc} \partial^c h^{ba} \right] = c_3 \partial_a [h^{ab} \partial_i h_b^i - h^{ib} \partial_i h_b^a] \qquad (3.25)$$

which, being a total divergence, does not contribute to the equations of motion if suitable boundary conditions are imposed. Notice that there are *no* further ambiguities related to 'swapping' of derivatives in the action in Eq. (3.22); this is clearly not possible in the c_1, c_2 or c_5 terms, since the swapping leads to identical terms. Hence the only ambiguity is in the choice between the c_3 term or c_4 term. The final action, obtained by integrating the Lagrangian over d^4x and ignoring the divergence term in L_{div} now has four constants: $c_1, c_2, c_5, (c_3 + c_4)$. From this stage, it is more convenient to focus on the action functional rather than on the Lagrangian.

Interestingly enough, these constants in Eq. (3.24) can all be determined except for an overall scaling by the requirement that the field equations should be invariant under the gauge transformation in Eq. (3.19). Carrying out this transformation in the action and demanding invariance fixes the constants to be $c_1 = -c_2 = 1$; $c_3 + c_4 = -c_5 = 2$ except for an overall scaling. We will write the overall scaling as $(1/4l^2)$, where l is a coupling constant with the dimensions of length so that, in natural units, the action is dimensionless. (The algebra is straightforward but a bit lengthy.) The resulting expression for the quadratic part of the action can be written in different forms:

$$
\begin{aligned}
A_h &= \frac{1}{4l^2} \int d^4x \, \partial_a h_{bc} \partial_i h_{jk} \left[\eta^{ai} \eta^{bc} \eta^{jk} - \eta^{ai} \eta^{bj} \eta^{ck} + 2\eta^{ak} \eta^{bj} \eta^{ci} - 2\eta^{ak} \eta^{bc} \eta^{ij} \right] \\
&= \frac{1}{4l^2} \int d^4x \left[\partial_i h_a^a \partial^i h_j^j - \partial_a h_{bc} \partial^a h^{bc} + 2\partial_a h_{bc} \partial^c h^{ba} - 2\partial_a h_b^b \partial_i h^{ia} \right] \\
&= \frac{1}{4l^2} \int d^4x \left[\frac{1}{2} \partial_i \bar{h}_a^a \partial^i \bar{h}_j^j - \partial_a \bar{h}_{bc} \partial^a \bar{h}^{bc} + 2\partial_a \bar{h}_{bc} \partial^c \bar{h}^{ba} \right].
\end{aligned}
\qquad (3.26)
$$

In the last line, we have introduced the *trace-reversed* tensor:

$$\bar{h}_{ab} \equiv h_{ab} - \frac{1}{2} \eta_{ab} h_i^i. \qquad (3.27)$$

(The fact $\bar{h} \equiv \bar{h}_a^a = -h$ explains the terminology, 'trace-reversed'.) We see that the expressions are simpler in terms of $\bar{h}_{ab}$. We shall hereafter use the shorter notation:

$$A_h = \frac{1}{4} \int d^4x \, M^{abcijk}(\eta^{mn}) \, \partial_a H_{bc} \partial_i H_{jk} = \frac{1}{4l^2} \int d^4x \, M^{abcijk}(\eta^{mn}) \, \partial_a h_{bc} \partial_i h_{jk}, \qquad (3.28)$$

where $h_{ab} = l H_{ab}$ and the tensor $M^{abcijk}(\eta^{mn})$ is symmetric in bc, jk and under the triple exchange $(a, b, c) \leftrightarrow (i, j, k)$ and is given by:

$$M^{abcijk}(\eta^{mn}) = \left[\eta^{ai} \eta^{bc} \eta^{jk} - \eta^{ai} \eta^{bj} \eta^{ck} + 2\eta^{ak} \eta^{bj} \eta^{ci} - 2\eta^{ak} \eta^{bc} \eta^{ij} \right]_{\text{symm}}, \qquad (3.29)$$

where the subscript 'symm' indicates that the expression inside the square bracket should be suitably symmetrized in bc, jk and under the exchange $(a, b, c) \leftrightarrow (i, j, k)$. In the expression for the action, since M^{abcijk} is multiplied by $\partial_a h_{bc} \partial_i h_{jk}$, we need not worry about symmetrization and use the expression given inside the square bracket in Eq. (3.29) as it is. This expression is analogous to the one in Eq. (2.109) for the case of electromagnetism. Just as the symmetries of M^{abcd} in Eq. (2.109) prevent the occurrence of time derivatives of A^0 in that Lagrangian, the symmetries of M^{abcijk} prevent the occurrence of time derivatives of either H_{00} or $H_{0\alpha}$ in the Lagrangian for the spin-2 field in Eq. (3.29). Thus the A^0 in the case of spin-1 field and the H_{0k} in the case of spin-2 field are not propagating dynamical variables.

The symmetries of the theory are somewhat easier to see in the momentum space, which can be done by introducing the Fourier components $f_{ab}(p)$ of $h_{ab}(x)$ defined as usual by:

$$h_{ab}(x) \equiv \int \frac{d^4 p}{(2\pi)^4} f_{ab}(p) e^{ipx}. \tag{3.30}$$

The action becomes

$$A_h = \frac{1}{4l^2} \int \frac{d^4 p}{(2\pi)^4} f_{bc} f_{jk}^* \left[p_a p_i M^{abcijk}(\eta^{mn}) \right] \equiv \frac{1}{4l^2} \int \frac{d^4 p}{(2\pi)^4} f_{bc} f_{jk}^* N^{bcjk} \tag{3.31}$$

with

$$\begin{aligned}
N^{bcjk} &= \left[p^2 \left(\eta^{bc} \eta^{jk} - \eta^{bj} \eta^{ck} \right) + 2 p^k \left(p^c \eta^{bj} - p^j \eta^{bc} \right) \right]_{\text{symm}} \\
&= \left(p^2 \eta^{bc} \eta^{jk} - p^k p^j \eta^{bc} - p^c p^b \eta^{jk} \right) - \frac{p^2}{2} \left(\eta^{bj} \eta^{ck} + \eta^{cj} \eta^{bk} \right) \\
&\quad + \frac{1}{2} \left(p^k p^c \eta^{bj} + p^j p^c \eta^{bk} + p^k p^b \eta^{cj} + p^j p^b \eta^{ck} \right),
\end{aligned} \tag{3.32}$$

where we have exhibited the symmetrized expression in full. (N^{bcjk} is symmetric in bc, jk and under the pair exchange $(b, c) \leftrightarrow (j, k)$.) The gauge transformation of the spin-2 field, given by Eq. (3.19) is equivalent to $f_{ab} \to f_{ab} + p_a \xi_b + p_b \xi_a$ in the Fourier space. Using this in Eq. (3.31) we find that A_h changes to

$$A_h \to \frac{1}{4l^2} \int \frac{d^4 p}{(2\pi)^4} (f_{bc} + 2 p_b \xi_c)(f_{jk}^* + 2 p_j \xi_k^*) N^{bcjk}. \tag{3.33}$$

Straightforward computation now shows that N^{abcd} satisfies the identities:

$$p_b N^{bcjk} = 0 = p_j N^{bcjk} \tag{3.34}$$

making A_h invariant under the gauge transformations. This is, of course, built-in by the choice of the coefficients c_i in the original action. More importantly, the condition in Eq. (3.34) translates, in coordinate space, to the *identity*

$$M^{abcijk}\partial_b\partial_a\partial_i h_{jk} = 0 = \partial_b[M^{abcijk}\partial_a\partial_i h_{jk}], \tag{3.35}$$

which will play a crucial role in our discussion below.

To obtain the full action with the source, we have to add to A_h the action for the sources, $A_{\text{source}}(q_A, \eta^{ab})$, which is independent of h_{ab} and the interaction term in Eq. (3.18):

$$A_{\text{int}} = \frac{l}{2}\int d^4x\, S_{ab}H^{ab} = \frac{1}{2}\int d^4x\, S_{ab}h^{ab}. \tag{3.36}$$

The S_{ab} is a second rank symmetric tensor built out of the source variables. In the limit of $c \to \infty$ we expect the h_{00} component to dominate over others when the interaction Lagrangian should have the form proportional to $\rho_m\phi_N$. Therefore, we need $h_{00} \propto \phi_N$ and $S_{00} \propto \rho_m$. The most natural second rank symmetric tensor built out of matter variables which satisfies this condition is just the energy-momentum tensor; therefore, we write $S_{ab} = T_{ab}/c$ where the c-factor takes care of the dimensions, keeping h_{ab} dimensionless. Then the total action is given by the sum

$$A_{\text{tot}} = A_h + A_{\text{int}} + A_{\text{source}} \tag{3.37}$$

$$= \frac{1}{4l^2}\int d^4x\, M^{abcijk}(\eta^{mn})\,\partial_a h_{bc}\partial_i h_{jk} + \frac{1}{2c}\int d^4x\, T_{ab}h^{ab} + A_{\text{source}}. \tag{3.38}$$

Varying this action with respect to h_{ab} will now lead to the field equations in the theory to be:

$$M^{abcijk}\partial_a\partial_i h_{jk} = (l^2/c)T^{bc}. \tag{3.39}$$

More explicitly, written in terms of $\bar{h}_{ab}$, this equation reads:

$$-\partial^a\partial_a\bar{h}^{mn} - \eta^{mn}\partial_b\partial_a\bar{h}^{ab} + \partial_a\partial^n\bar{h}^{ma} + \partial_a\partial^m\bar{h}^{na} = (l^2/c)T^{mn}. \tag{3.40}$$

In the $c \to \infty$ limit only the 00 component of this equation will dominate and by taking $h_{00} \propto \phi_N$ we will be able to get the correct Newtonian approximation. The left hand side of Eq. (3.40) is invariant under the transformation in Eq. (3.19). This gauge invariance ensures that degrees of freedom which carry negative energy can be eliminated by suitable gauge conditions. Several aspects of this theory, including its Newtonian approximation, are explored in Exercise 3.2 and Exercise 3.3. (In particular, it can be shown that $l^2 = 16\pi G/c^3$, where G is the Newtonian gravitational constant.) We seem to have achieved our goal of providing a Lorentz covariant theory of gravity based on a second rank tensor field.

Unfortunately, there is a serious problem with Eq. (3.39) which makes this theory unviable. The problem arises because the gauge invariance, which is needed

to eliminate the unphysical degrees of freedom, implies the identity in Eq. (3.35).
This, in turn, requires the consistency condition

$$\partial_b T^{bc} = 0 \tag{3.41}$$

on the source. (It can be easily verified from Eq. (3.40) that the left hand side has
identically vanishing divergence.) This equation, however, cannot hold – except
as an approximation – since it implies that the energy-momentum tensor of the
source is conserved even when the source is coupled to the gravitational field.
Alternatively, since the conservation of T_{ab} is equivalent to the validity of equations of motion for the source, this implies that the equations of motion for the
source are unaffected by the presence of gravity, which is inconsistent with the
source producing the gravitational field. Thus our theory can be valid only as a first
order approximation and indeed we will see in Chapter 6 that Einstein's theory of
gravity reduces to this theory in the appropriate limit.

We stress that the gauge invariance and conservation of the source are intimately
linked (see Table 3.1). The same phenomenon arises in electromagnetism, as discussed
in the last chapter as well. The left hand side of the Maxwell equation, $\partial_a F^{ab}$, satisfies
the identity $\partial_b \partial_a F^{ab} = 0$ which is analogous to the identity in Eq. (3.35) for the tensor
field. Both arise owing to the gauge invariance of the action. When coupled to the
source, the Maxwell equations become $\partial_a F^{ab} = J^b$ and the identity $\partial_b \partial_a F^{ab} = 0$
requires the source to be conserved: $\partial_b J^b = 0$. This is again analogous to the condition
$\partial_b T^{bc} = 0$ we obtained above. The crucial difference between the theories arises
from the following fact: conservation of charge does not lead to any inconsistencies
because electromagnetic field does not carry a charge and thus will not affect charge
conservation. But since the gravitational field can do work on the source and carries
energy, it affects the conservation of energy for the source. We cannot maintain
$\partial_a T^{ab} = 0$ for the sources alone in the exact theory.

It might seem that this problem could possibly be tackled as follows. In the above
discussion, we have separated the total energy-momentum tensor of the system into
that of gravitational field and that due to other sources, and have coupled gravity to
the latter. This is similar to what we did in the case of the scalar field by coupling
the scalar field to the trace of the energy-momentum tensor of *other* sources. As
we mentioned in that context, this is conceptually problematic and it would be
more natural to couple the scalar field to the trace of its own energy-momentum
tensor as well. The resulting theory (described in Project 3.1) is nonlinear due to
the self-coupling. In the similar spirit, one can try to couple the field H_{ab} to its own
energy-momentum tensor in a self-consistent manner. The idea is to compute the
energy-momentum tensor $t_{ab}^{(0)}$ of the spin-2 field due to the lowest order Lagrangian
$L^{(0)}$ and then add a coupling term $t_{ab}^{(0)} H^{ab}$ to modify the Lagrangian to obtain the

Table 3.1. *Comparison of spin-1 and spin-2 field theories*

	Spin-1 field (A^i)	Spin-2 field (H_{ab})
Possible Lagrangian	$L = -\dfrac{1}{2}\partial_a A_b \partial^a A^b$	$L = -\dfrac{1}{2}\partial_a H_{bc}\partial^a H^{bc}$
Trouble with this Lagrangian	A^0 carries negative energy	$H^{0\alpha}$ carries negative energy
Possible solution	Need a symmetry with at least one degree of freedom to eliminate A^0	Need a symmetry with at least 3 degrees of freedom to eliminate $H^{0\alpha}$
Symmetry needed	$A_j \to A_j + \partial_j f$	$H_{ab} \to H_{ab} + \partial_a \xi_b + \partial_b \xi_a$
Lagrangian with this symmetry	$L = -\dfrac{1}{2}M^{ijkl}\partial_i A_j \partial_k A_l$	$L = \dfrac{1}{4}M^{abcijk}\partial_a H_{bc}\partial_i H_{jk}$
	(see Eq. (2.109) for M^{ijkl})	(see Eq. (3.29) for M^{abcijk})
Field equation with source	$M^{ijkl}\partial_i \partial_k A_l = J^j$	$M^{abcijk}\partial_a \partial_i H_{jk} = T^{bc}$
Gauge condition to eliminate unphysical modes	$\partial_i A^i = 0$	$\partial_a H^{ab} = 0$
Identity obeyed by the LHS of the field equation	$M^{ijkl}\partial_j \partial_i \partial_k A_l = 0$	$M^{abcijk}\partial_b \partial_a \partial_i H_{jk} = 0$
The resulting constraint on the source	$\partial_s J^s = 0$	$\partial_a T^{ab} = 0$
Is the constraint physically reasonable?	Yes	No

next order one $L^{(1)}$. The t_{ab} will now get modified and we need to repeat the procedure again. Once such an iteration is completed we will obtain a nonlinear theory which couples to its own energy-momentum tensor and one hopes that the problem described above disappears.

This procedure does *not* work without additional inputs because of several difficulties. (We will comment more fully about this approach in Chapter 6.) To begin with, it turns out that one cannot obtain an expression for the energy-momentum tensor for the spin-2 field that is unique and gauge invariant. So one can obtain a large class of theories by this procedure without any uniqueness. If we know beforehand that the final field equations have to match with those in Einstein's theory, it is possible to introduce extra assumptions to obtain it; but these extra assumptions are essentially equivalent to the result we are attempting to derive. It turns out that the gauge invariance of the spin-2 field is closely related to the invariance of the full theory under general coordinate transformations. If one uses this knowledge, it is possible to modify the form of the gauge transformation order-by-order and thus arrange the self-coupling in such a way as to reproduce Einstein's theory; however, such an approach relies heavily on our knowledge of the Einstein's theory to guide us and does not provide any fresh insight. Second, even if one pursues such an approach, the resulting final expression for the *action* remains non-gauge invariant though the field equations maintain gauge invariance. Finally, even after doing all these, one essentially obtains a theory for which the most natural interpretation is geometrical and not field theoretical. In view of these, we shall not pursue this further except for making some more comments about this approach in Chapter 6.

Interestingly enough, the geometrical interpretation of the scalar field theory, discussed on the basis of Eq. (3.10), can also be presented in terms of a second rank tensor field because the results are indistinguishable in the $c \to \infty$ limit. To see this, we note that the line interval using the g_{ik} in Eq. (3.10) has the form:

$$ds^2 = \left[1 + \frac{\phi(t, \boldsymbol{x})}{c^2}\right]^2 \eta_{ik} dx^i dx^k = \left[1 + \frac{\phi(t, \boldsymbol{x})}{c^2}\right]^2 (-c^2 dt^2 + |d\boldsymbol{x}|^2). \quad (3.42)$$

In the weak field limit ($c \to \infty; \phi/c^2 \ll 1$), we can replace $[1 + (\phi/c^2)]^2$ by $[1 + (2\phi/c^2)]$. Then the coefficient of dt^2 will become $-(c^2 + 2\phi)$ while the coefficient of $|d\boldsymbol{x}|^2$ will be $[1 + (2\phi/c^2)]$. Retaining terms to $\mathcal{O}(\phi/c^2)$, we can now approximate the line interval by

$$ds^2 \approx -\left[1 + \frac{2\phi(t, \boldsymbol{x})}{c^2}\right] c^2 dt^2 + |d\boldsymbol{x}|^2. \quad (3.43)$$

Hence, in this limit, one could have thought of the line interval as given by $ds^2 = g_{ik} dx^i dx^k$ with $g_{ik} = \eta_{ik} + h_{ik}$ where h_{ik} is second rank tensor field with only the non-vanishing component being $h_{00} = -2\phi/c^2$. This shows again that, in the Newtonian approximation, one can think of a second rank tensor field having a geometric description with the metric being described by $g_{ik} = \eta_{ik} + h_{ik}$. We shall pursue this idea further in Section 3.5.

Exercise 3.2
Field equations of the tensor theory of gravity (a) Work out the left hand side of Eq. (3.39) explicitly and rewrite the resulting equation in terms of $\bar{h}_{ab}$. Show that the resulting equation has the form in Eq. (3.40).

(b) Verify explicitly that the divergence of the left hand side of Eq. (3.40) vanishes identically.

(c) Show that, under the gauge transformations in Eq. (3.19), the $\bar{h}_{mn}$ changes to

$$\bar{h}_{mn} \to \bar{h}_{mn}^{(\text{new})} = h_{mn} + \partial_n \xi_m + \partial_m \xi_n - \eta_{mn} \partial_a \xi^a. \quad (3.44)$$

Use this to show that one can always choose ξ_n such that the gauge transformed tensor satisfies the condition $\partial_a \bar{h}^{ab} = 0$. Assume that such a gauge has been chosen and show that, in this gauge, the field equation simplifies drastically and is given by

$$\Box \bar{h}^{mn} \equiv \partial_a \partial^a \bar{h}^{mn} = -(l^2/c)T^{mn}. \tag{3.45}$$

(d) Solve the field equation for a static point mass located at the origin for which the only nonzero component of T^{mn} is $T^{00} = Mc^2\delta^{(3)}[\boldsymbol{x}]$ and obtain $\bar{h}_{mn}$. From this calculate h_{mn} and show that

$$h_{00} = \frac{2GM}{c^2 r}; \qquad h_{0\gamma} = 0; \qquad h_{\alpha\gamma} = \delta_{\alpha\gamma}\frac{2GM}{c^2 r} \tag{3.46}$$

if we set $l^2 = 16\pi G/c^3$. Using the form of h_{ab} in Eq. (3.43) in the Newtonian approximation conclude that G is indeed the Newtonian gravitational constant.

Exercise 3.3

Motion of a particle in tensor theory of gravity The energy-momentum tensor for a particle of mass m moving on a trajectory $z^a(\tau)$ is given by

$$T^{mn}(\boldsymbol{x}) = \int m \frac{dz^m}{d\tau} \frac{dz^n}{d\tau} \delta^4 [\boldsymbol{x} - \boldsymbol{z}(\tau)] \, d\tau. \tag{3.47}$$

The trajectory of a particle in a given gravitational field can be obtained by varying the trajectory $z^a(\tau)$ in the full action in Eq. (3.38). The kinetic energy term for the gravitational field does not depend on $z^a(\tau)$ and hence we only need to work with

$$A = -m \int ds + \frac{1}{2c} \int d^4x \, h_{ab} T^{ab}, \tag{3.48}$$

where the first term is the A_{source} for a particle. Using the form T^{ab} for a point particle and the fact that $\eta_{ab}\dot{z}^a \dot{z}^b = -1$ (where the dot denotes a derivative with respect to τ), we can write this action in the form

$$A = \frac{1}{2}m \int (\eta_{mn} + h_{mn}) \, \dot{z}^m \dot{z}^n \, d\tau. \tag{3.49}$$

(a) Consider this action for the motion of a particle in the field of a mass M located at the origin which was obtained in Exercise 3.2. Show that this is equivalent to a Lagrangian given by

$$L = \frac{1}{2}mc^2 \left[-\left(1 - \frac{2GM}{c^2 r}\right)\dot{t}^2 + \frac{1}{c^2}\left(1 + \frac{2GM}{c^2 r}\right)(\dot{r}^2 + r^2\dot{\phi}^2) \right] \tag{3.50}$$

if we take the motion of a particle to be in the equatorial plane. Obtain two conserved quantities α and γ for this motion where $mc\alpha = J$ is the angular momentum of the particle and $-mc^2\gamma = E$ is the energy of the particle. Hence reduce the problem to quadrature and show that the orbit is determined by the equation

$$\left(\frac{du}{d\phi}\right)^2 + u^2 = (\gamma^2 - 1 + 2u)\left(\frac{GM}{c^2\alpha}\right)^2 \left[\frac{1 + 2u}{1 - 2u}\right], \tag{3.51}$$

where $u = (GM/c^2 r)$. Ignoring the cubic and higher powers of u in this equation, obtain the perihelion precession of a nearly elliptic orbit. Show that this shift is given by

$$\delta\phi = \frac{8G^2 M^2 m^2}{J^2}\pi \tag{3.52}$$

per orbit. [Note. This is 4/3 times the result obtained in general relativity; see Eq. (7.122).]

Exercise 3.4
Velocity dependence of effective charge for different spins We have seen that the source for a scalar field, vector field and tensor field can be written in the forms

$$\rho(x) = q_0 \int \delta^4[x^m - z^m(\tau)] \, d\tau \tag{3.53}$$

$$J^m(x) = q_1 \int u^m \delta^4[x^p - z^p(\tau)] \, d\tau \tag{3.54}$$

$$T^{mn}(x) = q_2 \int u^m u^n \delta^4[x^p - z^p(\tau)] \, d\tau. \tag{3.55}$$

The q_1 is the electric charge and in a tensor theory of gravity $q_2 = m$ is the mass but in general, they are just parameters. This suggests that a field of spin-s can be coupled to a point particle through the source given by

$$T^{mn...r}(x) = q_s \int u^m u^n ....u^r \delta^4[x^p - z^p(\tau)] \, d\tau. \tag{3.56}$$

We are interested in determining the Lorentz transformation properties of the 'effective charge' associated with the sources for different spins.

(a) The 'effective charge' can be defined in the rest frame of the particle (in which $u^i = \delta_0^i$) by

$$q_s = \int T^{00...0} d^3x. \tag{3.57}$$

Verify that this definition makes sense for $s = 1$ and $s = 2$.

(b) We next want to consider the effective charge attributed to the source by an arbitrary observer moving with a four-velocity v_i. Show that this is given by:

$$q = (-1)^s \int T^{ab...c} v_a v_b...v_c \, d^3x. \tag{3.58}$$

(b) Compute this quantity for the source for a spin-s field and show that:

$$q = q_s \int d^3x \, dt \left(\frac{d\tau}{dt}\right) (-\mathbf{u} \cdot \mathbf{v})^s \, \delta^4[x^p - z^p(\tau)]. \tag{3.59}$$

Hence prove that

$$q = q_s \gamma^{s-1} \approx q_s \left[1 + \frac{s-1}{2} \frac{v^2}{c^2}\right], \tag{3.60}$$

where the second result holds for $v \ll c$.

(c) This result shows that the effective charge is Lorentz invariant for spin-1 field and picks up no contribution from velocity. On the other hand, the effective charge *increases* with velocity for a spin-2 field while it *decreases* with velocity for a spin-0 field. Explain this result.

3.5 The principle of equivalence and the geometrical description of gravity

The discussion in the previous section shows that it is not straightforward to obtain a consistent field theoretic description of gravity. What is more, we have seen that both the scalar and tensor theories of gravity lend themselves to a geometrical description in terms of a suitably defined metric tensor g_{ik}. In the Newtonian approximation, when $\phi \ll c^2$, the modified metric has the form in Eq. (3.43). This geometrical interpretation is a peculiar feature not shared by any other field theory and we shall now take a closer look at this feature. It turns out that there is a direct connection between the principle of equivalence and the possibility of describing gravity as a geometrical phenomena. Our aim in this section will be to elucidate this connection and provide several simple thought experiments to clarify its origin.[3]

In Newtonian gravity, the kinetic and potential energy terms of the Lagrangian $L = (1/2)mv^2 - m\phi$ are proportional to the mass of the particle m and hence m has no influence on the equations of motion for a particle. Therefore, the trajectories $q(t)$ of material particles (with the same initial conditions) will be independent of the properties of the particle and will depend only on the gravitational potential $\phi(x^i)$.

We expect this feature to hold in any relativistic generalization of the description of gravity. In the limit $c \to \infty$, all particles at an event $\mathcal{P}$ will experience the *same* acceleration $g^\alpha \equiv -(\partial\phi/\partial x^\alpha)|_{\mathcal{P}}$, where $\alpha = 1, 2, 3$. If we now choose a new set of spatial coordinates

$$\xi^\alpha = x^\alpha - q^\alpha(t) \cong x^\alpha - \frac{1}{2}g^\alpha t^2 \tag{3.61}$$

near $\mathcal{P}$, then in the new frame the particles will experience no acceleration. It follows that the trajectories of particles in a given gravitational field are indistinguishable *locally* from the trajectories of free particles viewed from an accelerated frame. Hence a gravitational field is locally indistinguishable from a suitably chosen non-inertial frame (at least) as far as the laws of mechanics are concerned. Notice that we could arrive at this result only because the trajectories $q(t)$ are independent of any property of the particle. This is *not* the case, for example, in electromagnetism in which the trajectories depend on the (e/m) ratio for the particles.

Since accelerated frames seem to mimic the effects of gravitational fields locally, understanding the features of accelerated frames will be of some value in arriving at the correct description of gravity. We begin by constructing coordinate systems which are appropriate for observers who are not inertial. In particular, we want to generalize our concept of Lorentz transformations so as to include observers who are moving with a constant acceleration with respect to an inertial frame and study the consequences.

3.5.1 Uniformly accelerated observer

To do this, we shall adopt the procedure developed in Section 1.3. Let (T, X, Y, Z) be an inertial coordinate system in flat spacetime. Consider an observer travelling along the X-axis in a trajectory $X = f(\tau), T = h(\tau)$, where f and h are specified functions and τ is the proper time in the clock carried by the observer. Let $\mathcal{P}$ be some event with Minkowski coordinates (T, X) to which the observer assigns the coordinates (t, x). Then, it was shown in Section 1.3 that the two coordinates are related by

$$X - cT = f(t - x/c) - c\,h(t - x/c) \tag{3.62}$$

$$X + cT = f(t + x/c) + c\,h(t + x/c). \tag{3.63}$$

Given f and h, these equations can be solved to find (X, T) in terms of (x, t).

Let us now apply this to an observer travelling along the X-axis with a *uniform* acceleration g. The equation of motion

$$\frac{d}{dT}\left(\frac{v}{\sqrt{1 - v^2/c^2}}\right) = g \tag{3.64}$$

is identical to one in a constant electric field, discussed in Section 2.6. We found that, with suitable initial conditions, the trajectory can be expressed as:

$$gT = \sinh g\tau \equiv g\,h(\tau); \quad gX = \cosh g\tau \equiv g\,f(\tau) \tag{3.65}$$

in units with $c = 1$ (see Eq. (2.100)). Equations (3.62) and (3.63) now become

$$X - T = g^{-1}\exp\left[-g(t - x)\right]; \quad X + T = g^{-1}\exp\left[g(t + x)\right], \tag{3.66}$$

giving

$$X = g^{-1}e^{gx}\cosh gt; \quad T = g^{-1}e^{gx}\sinh gt. \tag{3.67}$$

This provides the transformation between the inertial coordinate system and that of a uniformly accelerated observer. The coordinate frame based on (t, x) is called the *Rindler frame*.

The crucial point is that the transformations in Eq. (3.67) are nonlinear and hence do not preserve the form of the line element ds^2. Using

$$dT^2 - dX^2 = d(T - X)\,d(T + X) = e^{2gx}\left(dt^2 - dx^2\right) \tag{3.68}$$

we get the line element in the accelerated frame:

$$ds^2 = -dT^2 + dX^2 + dY^2 + dZ^2 = e^{2gx}\left(-dt^2 + dx^2\right) + dy^2 + dz^2. \tag{3.69}$$

If we change to a new space coordinate $\bar{x}$ with $(1 + g\bar{x}) = e^{g x}$ and $e^{g x} dx = d\bar{x}$ we can write this metric in the form

$$ds^2 = -\left(1 + \frac{g\bar{x}}{c^2}\right)^2 c^2 dt^2 + d\bar{x}^2 + dy^2 + dz^2. \tag{3.70}$$

This result is quite general; transformations to non-inertial frames always lead to line intervals of the form $ds^2 = g_{ik}(x) dx^i dx^k$ where $g_{ik}(x)$ will, in general, depend on t and x. Since $g\bar{x} = \phi$ is the Newtonian potential corresponding to an acceleration $(-g)$ experienced by the observer in the accelerated frame, it follows that the metric in Eq. (3.70) does have the form in Eq. (3.43).

The two results we have obtained – namely, that (i) gravitational fields are locally indistinguishable from accelerated frames and that (ii) accelerated frames are described by a line element of the form in Eq. (3.70) – combine to provide an important conclusion: *the gravitational field affects the rate of clocks in such a way that the clocks slow down in strong gravitational fields.* To the lowest order in (ϕ/c^2), Eq. (3.70) shows that

$$\Delta T = \Delta t \left(1 + \frac{\phi}{c^2}\right), \tag{3.71}$$

where Δt is the time interval measured by a clock in the absence of the gravitational field and ΔT is the corresponding interval measured by a clock located in the gravitational potential ϕ. This conclusion has far reaching implications. Consider the line interval ds between two infinitesimally separated events in spacetime. When the spatial interval dx between the two points vanishes, $d\tau = (-ds^2)^{1/2}$ measures the lapse of proper time at a given spatial location. If a global Lorentz time exists, then $d\tau = dt$ which is the same for all observers at rest. If the gravitational field affects the rate of flow of clocks, then ds should vary from point to point even when $dx = 0$. This is possible only if $ds^2 = g_{00}(x)dt^2$, where g_{00} is a nontrivial function of the spatial coordinates. From our previous arguments, we can even fix the form of g_{00}; to arrive at the correct flow of time, we need $g_{00}(x) = -(1 + 2\phi/c^2)$. It follows that the line interval between two events in the spacetime cannot have the form $ds^2 = -dt^2 + dx^2$ but should *at least* be modified to a form

$$ds^2 = -\left(1 + \frac{2\phi}{c^2}\right) c^2 dt^2 + dx^2 \tag{3.72}$$

in the presence of a gravitational field. This is precisely the conclusion we reached earlier in the case of scalar and tensor field theories of gravity. The direct connection between the principle of equivalence and a geometrical description of gravity (at least in the case of dynamics of point particles) is now apparent.

Exercise 3.5
Another form of the Rindler metric Show that the metric in Eq. (3.70) can be transformed
to the form:

$$ds^2 = -\left(\frac{2gl}{c^2}\right)c^2dt^2 + \frac{dl^2}{(2gl/c^2)} + dy^2 + dz^2. \tag{3.73}$$

This form of the metric will play a crucial role in our discussion of black holes in Chapter
8. [Answer. Use the coordinate transformation $1 + (g\bar{x}/c^2) = (2gl/c^2)^{1/2}$.]

Exercise 3.6
Alternative derivation of the Rindler metric There is an alternative way of obtaining the
metric in Eq. (3.70) which is more direct. Consider an accelerated observer with the trajectory $h(\tau), f(\tau)$ and a coordinate velocity $u(\tau) \equiv df/dh$. At any given instant, there
exists a Lorentz frame (t, x) with: (a) the three coordinates axes coinciding with the axes
of the accelerating observer and (b) the origin coinciding with the location of the observer.
Show that the Lorentz transformations (with suitable translation of origin) from the global
inertial frame coordinates (T, X) to this instantaneously comoving frame is given by (with
$c = 1$)

$$X = f(\tau) + \gamma(u)(x + ut); \qquad T = h(\tau) + \gamma(u)(t + ux). \tag{3.74}$$

We now define the coordinates for the accelerated observer such that at $t = 0$ the coordinate
labels in the accelerated frame coincide with those in the comoving Lorentz frame. This
gives

$$X = f(\tau) + \gamma(u)x; \qquad T = h(\tau) + \gamma(u)ux. \tag{3.75}$$

Show that, if we take x and τ as the coordinates in the accelerated frame, then the resulting
metric has the form in Eq. (3.70).

3.5.2 Gravity and the flow of time

Considering the importance of this result, we shall provide a more physical argument leading to the same conclusion. We know that the annihilation of electrons
and positrons can lead to γ-rays and that under suitable conditions one can produce
e^+e^- pairs from the radiation. This fact can be used to devise a suitable thought
experiment which will prove that gravitational field must necessarily affect the rate
of flow of clocks. We will first prove that conservation of energy demands a redshift of the frequency of photons propagating in a static gravitational field. Using
this result and the relation between frequency and clock time, we will demonstrate
that the rate of flow of clocks must be affected by the gravitational field.

For the purpose of the argument given below, we may assume that the gravitational potential near the surface of Earth varies linearly with height; then the
potential difference between two points A and B, separated by height L, will be

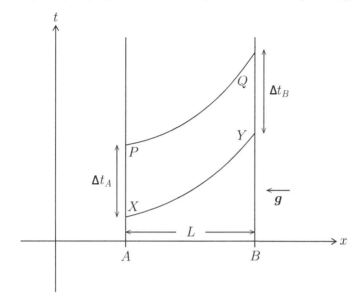

Fig. 3.1. The vertical lines at A and B are the world lines of two spatial locations near Earth separated by height L with A being closer to the Earth's surface. (In this spacetime diagram, the height from Earth's surface is measured horizontally.) Some simple thought experiments using this arrangement show that gravitational field must affect clock rates.

gL, where g is the acceleration due to gravity. (Figure 3.1 shows the situation in a spacetime diagram. The two vertical lines at the points A and B denote the world lines of two spatial locations separated by a height L; the gravitational field is along the negative x-axis in the diagram.) Let us assume that several pairs of sufficiently high energy photons, each of energy $\hbar\omega$, were converted into e^+e^- pairs at the point A – which is closer to the surface of the Earth than B. These particles move from point A to point B losing the gravitational potential energy mgL per particle. At B we annihilate the particles and produce photons which are sent back to A. The sequence of events described above has restored the original configuration. (To be rigorous, one also has to arrange matters such that the momentum balance of the particles is also taken care of. This can be easily done with a suitable mechanism and hence we will not bother about this.[4]) Since the amount of energy available in the form of particles was lower at B than at A (because of the change in the gravitational potential), it follows that the energy of the photons must change in going from B to A. If this is not the case, we could use the above sequence of events to increase the energy of the system repeatedly. Since the energy of a photon is proportional to its frequency, the frequencies of photons at A and B must be related (for $gL/c^2 \ll 1$) by

$$\omega_A \cong \omega_B \left(1 + \frac{gL}{c^2} \right). \tag{3.76}$$

Because the frequency shift derived above is independent of the Planck constant $\hbar$ (even though the original argument involving pair creation and annihilation was quantum mechanical), it follows that the result should be true even in the classical limit. But in the classical limit, one can consider photons as making up an electromagnetic wave with some frequency. This frequency can be determined by counting the number of crests N of a wave train which crosses an observer in a time interval Δt. Let us suppose that such a wave train, made of N crests (and troughs), was emitted from point A towards point B. The head of the wave train travels along some trajectory XY in Fig. 3.1. In the absence of a gravitational field, the radiation will travel along null lines (at $45°$ to the axes) but we do not want to make any assumption regarding how gravity affects the trajectory of the rays. Therefore, the curve XY at this stage is just some monotonic continuous curve connecting the points X and Y. Let the tail of the wave train leave the point A after a time interval Δt_A as measured by a local clock located at A. This will travel along the curve PQ. Since the gravitational field is static, nothing changes with time and the curves XY and PQ are locally parallel to each other. It follows from elementary geometry that the time interval between Y and Q measured by a local clock at B is $\Delta t_B = \Delta t_A$. The number of crests in a wave cannot change during the propagation either. Since the observers at A and B will attribute to the wave frequencies $\omega_A = N/\Delta t_A$ and $\omega_B = N/\Delta t_B$ the equality $\Delta t_A = \Delta t_B$ leads to $\omega_A = \omega_B$ which contradicts our previous result in Eq. (3.76).

This paradox arises because we had tacitly assumed that the clocks at A and B will run at the same rate as indicated by the coordinate interval Δt at the two locations. But if the line interval in the presence of gravity gets modified to the form in Eq. (3.72), then the clocks at rest ($|d\boldsymbol{x}|^2 = 0$) in the gravitational field will run at a rate determined by $d\tau = \sqrt{-g_{00}(\boldsymbol{x})}\,dt$ which will differ from location to location. Thus even though the coordinate time differences between the arrival of the head and tail of the wave at A and B are the same, the *proper* time interval shown by clocks corresponding to Δt_A and Δt_B are different. Hence the frequencies will differ. It is easy to see that if the line interval has the form in Eq. (3.72), then the corresponding frequencies do obey the redshift relation in Eq. (3.76) to the lowest order in (gL/c^2).

All the above features show that the spacetime interval cannot retain its usual special relativistic form in the presence of a gravitational field. A modification of the metric tensor is absolutely essential if we have to obtain physically reasonable conclusions. For this reason, it is natural to treat the gravitational field as geometrical in origin.

PROJECTS

Project 3.1
Self-coupled scalar field theory of gravity

In the text, we discussed a scalar theory of gravity in which the scalar field ϕ was coupled to the trace, $T_{(bg)}$, of the energy-momentum tensor of other sources which we will call the background fields. As we mentioned on page 111, such a coupling requires us to make an artificial distinction between the trace of the energy-momentum tensor of the background fields and that of the scalar field. The purpose of this project is to explore ways of removing this distinction and couple the scalar field to itself in a consistent manner. This can be done iteratively order by order in the coupling constant starting from the lowest order action given in Eq. (3.8). However, it is possible to do this at one go along the following lines.

(a) Assume that the Lagrangian for the scalar field is of the form

$$L = -\frac{1}{2}A(\phi)\partial_i\phi\partial^i\phi + \lambda T_{(bg)}B(\phi), \tag{3.77}$$

where $A(\phi)$ and $B(\phi)$ are functions that need to be determined. The choice $A = 1, B = \phi$ leads to the lowest order Lagrangian in Eq. (3.8). Compute the energy-momentum tensor for the Lagrangian in Eq. (3.77) and show that its trace is given by $T = -A\partial_i\phi\partial^i\phi + 4\lambda BT_{(bg)}$.

(b) Demand that the Lagrangian in Eq. (3.77) should be expressible in the alternative form as

$$L = -\frac{1}{2}\partial_i\phi\partial^i\phi + \lambda(T_{(bg)} + T)\phi, \tag{3.78}$$

which can be interpreted as the Lagrangian in Eq. (3.8) with $T_{(bg)}$ replaced by $(T_{(bg)} + T)$. This is *one way* of demanding self-consistency. Show that this requires

$$A = \frac{1}{1 - 2\lambda\phi}; \qquad B = \frac{\phi}{1 - 4\lambda\phi}. \tag{3.79}$$

(c) Investigate the resulting theory fully. Work out the field equations, motion of a point particle and precession of perihelion in a centrally symmetric problem.

(d) There is a different way of demanding self-consistent coupling for a scalar field. To see this, work out the field equations corresponding to the Lagrangian in Eq. (3.77) and show that they reduce to

$$\Box\phi = -\frac{1}{2}\frac{A'}{A}\partial_i\phi\partial^i\phi - \lambda\frac{B'}{A}T_{(bg)}. \tag{3.80}$$

We can now insist that this field equation should be the same as

$$\Box\phi = -\lambda(T_{(bg)} + T), \tag{3.81}$$

where T is the trace of the energy-momentum tensor of the full Lagrangian. Show that these lead to the equations $A'/A = -2\lambda A$ and $B'/A = (1 + 4\lambda B)$. Solve these to obtain

$$A = \frac{1}{1 + 2\lambda\phi}; \qquad B = \phi + \lambda\phi^2. \tag{3.82}$$

(e) Again work out all the physical consequences of the resulting theory. Explain why the Lagrangians arrived at in parts (b) and (d) above are different. Which of the two models is physically more reasonable?

Project 3.2

Is there hope for scalar theories of gravity?

One of the key reasons we abandoned the scalar theory of gravity is the fact that it does not couple to the electromagnetic field. This, in turn, was because we coupled the field to the trace T of the source. What happens if we try more complicated forms of coupling? Do you think a coupling like $F(\phi)T_{ab}T^{ab}$ could lead to a viable theory? (While you're at it, show that a coupling like $T^{ab}\partial_a\phi\partial_b\phi$ will not lead to a viable theory.) In a broader context, this project invites you to construct your own (viable) scalar field theory of gravity and investigate its consequences. (If successful, publish!)

Project 3.3

Attraction of light

The action in Eq. (3.26) can be simplified further if we impose the gauge condition $\partial_a \bar{h}^{ab} = 0$ (called the harmonic gauge condition, which is analogous to $\partial_i A^i = 0$ in electromagnetism) and integrate the third term by parts discarding the total divergence. This will lead to

$$A_{\rm hg} = \frac{1}{4l^2} \int d^4x \left[\frac{1}{2}\partial_i \bar{h}_a^a \partial^i \bar{h}_j^j - \partial_a \bar{h}_{bc}\partial^a \bar{h}^{bc} \right]. \tag{3.83}$$

(a) Show that this theory leads to a Green function in the Fourier space, given by

$$D_{mnls}(k) = \frac{1}{2k^2}(\eta_{ml}\eta_{ns} + \eta_{ms}\eta_{nl} - \eta_{mn}\eta_{ls}). \tag{3.84}$$

(b) Argue that the interaction term between two sources described by energy-momentum tensors $T_{(1)}^{mn}$ and $T_{(2)}^{ls}$ is proportional to $T_{(1)}^{mn}D_{mnls}T_{(2)}^{ls}$ in Fourier space. Expand this out and show that

$$T_{(1)}^{mn} D_{mnls}T_{(2)}^{ls} \propto \frac{1}{k^2}\left(2T_{(1)}^{mn}T_{(2)mn} - T_{(1)}T_{(2)}\right). \tag{3.85}$$

(c) For non-relativistic matter, T_{00} dominates over all other components. Show that, in this limit, the interaction term becomes $(1/k^2)\left(T_{(1)}^{00}T_{(2)}^{00}\right)$ and we recover Newton's law. But, for relativistic matter (or radiation) with a vanishing trace for the energy-momentum tensor $(T_{(1)} = 0)$ interacting with normal matter (for which T_{00} is the only non-vanishing component), the corresponding result will be $(2/k^2)\left(T_{(1)}^{00}T_{(2)}^{00}\right)$, which is twice as strong. Explain what this implies for the gravitational effect of the Sun on a light ray.

(d) Another curious result which can be obtained using Eq. (3.84) is about the grav-itational interaction between two beams of light. Consider the scattering of two photons $(k_1 + k_2 \to p_1 + p_2)$, where the symbols denote the four-momenta of the photons in the initial and final state. Compute the interaction strength $T_{(1)}^{mn} D_{mnls} T_{(2)}^{ls}$ in the limit of small momentum transfer; that is, when $\boldsymbol{q} \equiv \boldsymbol{p}_1 - \boldsymbol{k}_1$ is small. Explain why this will correspond to interaction between the photons at large distances. Show that, in this limit, the interaction term is proportional to $(\boldsymbol{k}_1 \cdot \boldsymbol{k}_2)(\boldsymbol{p}_1 \cdot \boldsymbol{p}_2)/q^2$. Hence show that two parallel beams of light do not couple to each other gravitationally! (We will obtain this result using the weak field limit of Einstein's theory in Chapter 6.)

Project 3.4
Metric corresponding to an observer with variable acceleration

In Section 3.5.1 we considered an observer moving with a uniform acceleration at all times and obtained the corresponding coordinate system. Such an observer, however, is not physically realizable since the observer's speed asymptotically approaches the speed of light as $|t| \to \infty$. In this project, we will consider an observer who is moving with a time dependent acceleration $g(t)$ along the x-axis, which is a physically reasonable situation.

(a) Obtain the trajectory of such an observer in terms of integrals over a given function $g(t)$.

(b) Repeat the analysis in Exercise 3.6 and show that in this case the coordinate transformations between the inertial frame (X^a) and the non-inertial frame (x^a) given by Eq. (3.75) can be rewritten as:

$$X = \int' \sinh \chi(t) dt + x \cosh \chi(t) = \int dt \, [1 + g(t)x] \sinh \chi(t)$$

$$T = \int' \cosh \chi(t) dt + x \sinh \chi(t) = \int dt \, [1 + g(t)x] \cosh \chi(t), \tag{3.86}$$

where the function $\chi(t)$ is related to the time dependent acceleration $g(t)$ by $g(t) = (d\chi/dt)$. Show that the line element in these coordinates is remarkably simple and is given by

$$ds^2 = -(1 + g(t)x)^2 dt^2 + dx^2 + dy^2 + dz^2. \tag{3.87}$$

(c) Repeat the analysis of Section 3.5.1 for this case. Show that, now, the metric in the T–X plane becomes

$$ds^2 = -dT^2 + dX^2 = e^{[\chi(t+x) - \chi(t-x)]} (-dt^2 + dx^2). \tag{3.88}$$

In terms of $U = (T - X)$, $V = (T + X)$, etc., the metric is given by the still simpler form: $dU dV = e^{-\chi(u)} e^{\chi(v)} du dv$. Also show that the relevant coordinate transformations are:

$$U = \int du \, e^{-\chi(u)}; \qquad V = \int dv \, e^{\chi(v)}. \tag{3.89}$$

(d) Describe the trajectories $X = $ constant in terms of the coordinates in the accelerated frame and the trajectories $x = $ constant in terms of the coordinates in the inertial frame. For $g(t)$ which vanishes outside a finite interval, are these coordinates well-defined for the entire range of values?

Project 3.5
Schwinger's magic

We showed in the text that the spin-2 field approach does *not* produce the correct precession for the perihelion of an elliptic orbit. It is, however, possible to obtain the correct result by an interesting reasoning due to Julian Schwinger.[5] Even though this does not constitute a rigorous derivation, it provides an interpretation of the corrections to Newtonian theory which is of some value. This project examines several features of this approach.

(a) Converting the result obtained in Project 3.3 into real space, show that the interaction energy between a massive body (like the Sun) with mass M and another test body with energy-momentum tensor T_{ab} can be expressed in the form

$$E_{\text{int}}(t) = -GM \int d^3x \, \frac{1}{|\boldsymbol{x}|} \left[2T^{00}(\boldsymbol{x},t) + T(\boldsymbol{x},t) \right], \tag{3.90}$$

where $T_{ab} = \rho u_a u_b$ and T is the trace of the energy-momentum tensor. Note that $2T^{00} + T = T^{00} + T_\alpha^\alpha$.

(b) Schwinger's strategy was to use this expression to obtain corrections to effective Newtonian potential starting from the lowest order term $V = -GMm/r$, where m is the mass of the test body ('planet'). Argue that, for small speeds of the test body with $p^2/2m \ll mc^2$, we can approximate $T_\alpha^\alpha \approx (p/m)^2 T^{00} \approx (2K/m)T^{00}$ where K is the kinetic energy. Further show that the integral over T^{00} will now give, to the lowest order of accuracy, $m + K$. Hence argue that these effects modify the Newtonian potential to the form

$$V \left(1 + \frac{2K}{m} \right) \left(1 + \frac{K}{m} \right) \simeq V \left(1 + \frac{3K}{m} \right). \tag{3.91}$$

(c) Compute the correction to the kinetic energy from special relativity to the quadratic order of accuracy and show that $(p^2 + m^2)^{1/2} - m \approx K - K^2/2m$.

(d) The really crucial correction that Schwinger brings in is from the energy density T^{00} associated with the gravitational interaction between the planet and the Sun that is obtained from the gradient of the gravitational field. Argue that a possible form for this energy density is given by

$$T_{\text{int}}^{00}(\boldsymbol{x}) = -\frac{G}{4\pi} \nabla \left(\frac{M}{|\boldsymbol{x}|} \right) \cdot \nabla \left(\frac{m}{|\boldsymbol{x} - \boldsymbol{R}|} \right) \tag{3.92}$$

by showing that the integral of T_{int}^{00} over all space gives $(-GMm/R)$. Now compute the energy of interaction between the mass m and this distributed energy density and show that it gives

$$-GM \int d^3x \, \frac{1}{|\boldsymbol{x}|} \left(-\frac{GMm}{4\pi} \right) \nabla \frac{1}{|\boldsymbol{x}|} \cdot \nabla \left(\frac{1}{|\boldsymbol{x} - \boldsymbol{R}|} \right)$$

$$= \frac{G^2 M^2 m}{4\pi} \int d^3x \, \frac{1}{2} \nabla \frac{1}{|\boldsymbol{x}|^2} \cdot \nabla \left(\frac{1}{|\boldsymbol{x} - \boldsymbol{R}|} \right)$$

$$= \frac{1}{2} \frac{G^2 M^2 m}{R^2} = \frac{V^2}{2m}. \tag{3.93}$$

(e) Combining all these, the correction to the Newtonian potential becomes $V(3K/m) - K^2/2m + V^2/2m$. Assuming that $E = K + V$ is conserved, eliminate K in favour of V. In the resulting potential argue that overall rescaling does not lead to any interesting effect (except a rescaling of the orbit) and that the precession is caused by an effective potential of the form $V_{\text{eff}} = V - 3V^2/m$. Show that, for orbits in this effective potential, we get *exactly* the same precession as in Einstein's theory (see Eq. (7.122)).

(f) Critically examine each link in the above argument paying special attention to the following questions. (i) Have we taken into account all the corrections up to the required order? (ii) What is the effect of each of the terms individually? (iii) Is the distribution of field energy in part (d) unique and can it be obtained from any fundamental reasoning? (iv) Can this approach be used to obtain any other result in full general relativity in a simplified manner?

4

Metric tensor, geodesics and covariant derivative

4.1 Introduction

We begin our study of general relativity and curved spacetime in this chapter. Chapters 4 and 5 will develop the necessary mathematical apparatus to deal with curved spacetime. As in the case of electromagnetism, the study of gravity can be divided into two separate – but interconnected – aspects. In this chapter and the next, we will study the influence of gravity on other physical systems (like particles, photons, ideal fluids, fields, etc.) without worrying about how a given gravitational field is generated – which will be discussed in Chapter 6.

All the topics introduced in this chapter will be required in the subsequent chapters and form core material for general relativity. In particular, we will start introducing index-free vector notation more liberally in the coming chapters and familiarity with the ideas and notation developed in Section 4.6.1 will be crucial. We will use units with $c = 1$ unless otherwise indicated.[1]

4.2 Metric tensor and gravity

The arguments presented in the previous chapter suggest that a weak gravitational field cannot be distinguished from a modified spacetime interval as far as mechanical phenomena are concerned. We shall now generalize this result by postulating that *all aspects of gravitational fields allow a geometrical description*. We thus extend the tentative conclusion of the previous chapter to include arbitrarily strong gravitational fields and *all* physical phenomena. This leads to Einstein's theory of gravitation, which is the most beautiful of all existing physical theories. In this chapter, we shall start by exploring the new features that arise in such a geometric description of gravitational phenomena.

To begin with, this postulate implies that, in the presence of a gravitational field, the spacetime interval will be modified to the form

$$ds^2 = g_{ik}(x)\, dx^i dx^k, \tag{4.1}$$

where the metric tensor $g_{ik}(x)$ characterizes the gravitational field. It can be taken to be symmetric without loss of generality and hence contains ten functions of spacetime. (As before we shall use the notation $g_{ab}(x)$ to indicate functional dependence on the four coordinates without specifically adding a superscript like $g_{ab}(x^i)$, etc.) We have to first understand the geometrical features of a spacetime described by such a metric.

A metric tensor like $g_{ab}(x)$ – which depends on the coordinates – can arise in two separate contexts. If the flat spacetime is described in non-inertial coordinates, then $g_{ab}(x)$ will not be constant. (We saw this result in the previous chapter for a uniformly accelerated frame; see Eq. (3.69).) On the other hand, a genuinely curved spacetime will *necessarily* have a non-constant $g_{ab}(x)$. To illustrate these two different situations in a familiar context, let us consider a two-dimensional space with coordinates (α, β) and the line interval

$$ds^2 = (d\alpha)^2 + (\alpha)^2 (d\beta)^2, \tag{4.2}$$

with $0 \le \alpha < \infty,\ 0 \le \beta < 2\pi$. This metric clearly depends on one of the coordinates α. However, this space is really the flat two-dimensional space expressed in polar coordinates and, if we transform to two other coordinates (X, Y) with $X = \alpha \cos(\beta),\ Y = \alpha \sin(\beta)$, then the metric will take the familiar form: $ds^2 = dX^2 + dY^2$. That is, we can transform the metric in Eq. (4.1) to the standard form in Cartesian coordinates globally. The coordinate dependence of the individual metric components in Eq. (4.2) arises only because we were using non-Cartesian (usually called *curvilinear*) coordinates. The underlying two-dimensional space is flat which is obvious when we use the Cartesian coordinates.

On the other hand, let us consider a line element of the form

$$ds^2 = (d\alpha)^2 + \sin^2(\alpha)(d\beta)^2 \tag{4.3}$$

with $0 \le \alpha \le \pi$ and $0 \le \beta \le 2\pi$, which represents the line interval on the surface of a sphere of unit radius. This is a genuinely curved surface and this line element cannot be reduced to the Cartesian form by any coordinate transformation. The dependence of metric coefficients on the coordinates in this case is due to the genuine curvature of the two-dimensional space which it represents.

The same situation arises in four dimensions as well. The Cartesian form of the spacetime interval $ds^2 = \eta_{ab} dx^a dx^b$ is maintained only under Lorentz transformations. But a transformation to curvilinear coordinates – like those which are appropriate for a uniformly accelerated observer, discussed in the previous chapter – will lead to the line interval of the form in Eq. (4.1) with the metric components depending on the coordinates. But, *in general*, a line element of the form in Eq. (4.1) describes a curved spacetime and not a flat spacetime expressed in curvilinear coordinates. This is most easily seen by counting the number of independent functional degrees of freedom in the metric. Suppose we start with flat spacetime described in inertial Cartesian coordinates X^i and metric $\eta_{ik} = $ dia$(1, -1, -1, -1)$, and make a coordinate transformation from X^i to a new set of coordinates x^k which are arbitrary functions of the original coordinates. The line interval in the new coordinates now becomes

$$ds^2 = \eta_{ik} dX^i dX^k = \eta_{ik} \frac{\partial X^i}{\partial x^a} \frac{\partial X^k}{\partial x^b} dx^a dx^b \equiv g_{ab}(x) dx^a dx^b. \qquad (4.4)$$

However, the transformation from X^i to x^k involves only four arbitrary functions. Therefore the g_{ab} in the line interval can contain only four independent functions. In general, the symmetric second rank tensor $g_{ab}(x)$ has ten independent functions. It is, therefore, impossible to obtain an arbitrary set of $g_{ab}(x)$ by transforming the coordinates from the inertial frame to the non-inertial frame. Conversely, it is not possible to reduce an arbitrary metric $g_{ab}(x)$ to η_{ab} by a coordinate transformation. More generally, if we are working in an N-dimensional spacetime, the metric tensor g_{ab} will have $(1/2)N(N+1)$ components while the coordinate transformation will only involve N functions. For all $N \geq 2$, a general metric contains more information than can be obtained from a coordinate transformation.

In flat spacetime, the Cartesian coordinates (and inertial frames associated with them) enjoy a special status. Any modification of the metric tensor, arising from the use of curvilinear coordinates, can be given an unambiguous meaning in the flat spacetime because we can always compare it with the preferred form of the line element in the Cartesian coordinates. But, in the presence of a gravitational field – described by a spacetime metric of the form in Eq. (4.1) – we have to deal with a curved spacetime. In a genuinely curved spacetime one cannot reduce the line intervals to the Cartesian form globally; therefore, no system of coordinates can be given a preferred status at a fundamental level in a curved spacetime. It follows that we need to formulate our theories in a manner such that the equations are covariant under *arbitrary* coordinate transformations. This generalizes similar ideas introduced in Chapter 1 in which we formulated the laws of physics in a form that remained covariant under Lorentz transformations. As a bonus, we will arrive at a formulation of physics valid in curvilinear coordinates in flat spacetime when it is mathematically convenient to do so.

One of the key issues in the study of gravitational effects is to distinguish those features which arise from the genuine curvature of the spacetime from those features which primarily have their origin in the use of curvilinear coordinates.

In this chapter we shall concentrate on features which are common to curvilinear coordinates *as well as* curved spacetime. The concept of curvature itself will be introduced in the next chapter.

Though the metric in Eq. (4.1) cannot be reduced to the Cartesian form in general, it is possible to reduce it to a rather special form *around any event* in spacetime by a suitable choice of coordinates. This reduction also allows us to understand – in an indirect fashion – some features arising from the curvature. We shall now describe how this can be done.

Let us consider a coordinate transformation from the coordinates x^i to x'^i in the neighbourhood of an event $\mathcal{P}$, which we will take to be the origin in the old coordinates. The coordinate transformation gives the new coordinates x'^i as functions of the old coordinates x^i (and vice versa) in the neighbourhood of the origin. We will choose the transformation such that the event $\mathcal{P}$ remains at the origin of new coordinates and expand the functions x'^i in terms of x^i in a Taylor series as

$$x'^i = B_k^i x^k + C_{jk}^i x^j x^k + D_{jkl}^i x^j x^k x^l + \cdots, \tag{4.5}$$

where $C_{jk}^i, D_{jkl}^i...$, etc., are completely symmetric in the lower indices. We shall now attempt to choose the coordinate transformation in order to make the metric as close to the Cartesian form as possible near the origin. To do this, we must make $g_{ab} = \eta_{ab}$ at the origin and make as many derivatives of the metric tensor ($\partial_i g_{ab}, \partial_i \partial_j g_{ab}...$, etc.) vanish at the origin as possible. That is, we will try to: (i) choose the values of B_k^i such that the transformed metric tensor g'_{ik} at the event $\mathcal{P}$ is the same as η_{ik}; (ii) choose C_{jk}^i such that the first derivatives of g'_{ik} at $\mathcal{P}$ vanishes and (iii) choose D_{jkl}^i such that the second derivatives of g'_{ik} at $\mathcal{P}$ vanishes, etc. Let us see whether this can be done. With future applications in mind, we shall work this out in N dimensions.

The first requirement ($g_{ab} = \eta_{ab}$) amounts to $(1/2)N(N+1)$ conditions while we have N^2 independent coefficients in B_j^i. The B_j^i can be chosen so as to impose the condition ($g_{ab} = \eta_{ab}$) with

$$N^2 - \frac{1}{2}N(N+1) = \frac{1}{2}N(N-1) = {}^N C_2 \tag{4.6}$$

parameters to spare. This is precisely the number of free parameters specifying Lorentz transformations and spatial rotations in N dimensions. Taking Lorentz boosts as complex rotations in the relevant planes, we have the freedom to rotate the Cartesian coordinates in ${}^N C_2$ planes which matches with the above result. This is expected because, after reducing the metric to Cartesian form, we can *still* introduce Lorentz boosts and rotations while retaining its form.

The second requirement – viz., that the first derivatives of metric ($\partial g'_{ik}/\partial x'^a$) should vanish at the origin – imposes $N \times (1/2)N(N+1) = (1/2)N^2(N+1)$

conditions. In C^i_{jk}, we have precisely $N \times (1/2)N(N+1) = (1/2)N^2(N+1)$ independent parameters, since $C^i_{jk} = C^i_{kj}$ by definition. These parameters C^i_{jk} can be used to impose the conditions $(\partial g'_{ik}/\partial x'^a) = 0$. Hence one can always choose the coordinates in such a way that the metric reduces to the inertial form in an infinitesimal *region* around an event $\mathcal{P}$. Later on, we shall provide an explicit construction which achieves this.

The next step will require choosing the coefficients D^i_{jkl} (which are completely symmetric in j, k and l) in order to make all the second derivatives of the metric vanish around the chosen event. This, however, turns out to be impossible. To see this, let us first count the number of independent components in D^i_{jkl}. For a given value of i, the number of possible independent combinations of jkl is $^{N+2}C_3$. (This was proved in Chapter 1, Section 1.5.) Multiplying by the N values that the index i can take, we have $(1/6)N^2(N+2)(N+1)$ independent coefficients in D^i_{jkl}. On the other hand, the number of independent components in the second derivatives $(\partial^2 g'_{ik}/\partial x^{a'}\partial x'^b)$ is the product of the independent combinations available for the symmetric pair of indices i, k multiplied by the same for the symmetric pair a, b. This is given by $(1/2)N(N+1) \times (1/2)N(N+1) = (1/4)N^2(N+1)^2$. Clearly this is larger than the number of independent coefficients in D^i_{jkl} we have at our disposal. After making as many second derivatives of the metric vanish as possible, we will be left with

$$\frac{1}{4}N^2(N+1)^2 - \frac{1}{6}N^2(N+2)(N+1) = \frac{1}{12}N^2(N^2-1) \qquad (4.7)$$

independent non-vanishing second derivatives of the metric at any given event. In the case of $N = 4$, we will have 20 independent components which cannot be reduced to zero. These surviving second derivatives of the metric characterize the genuine curvature of the spacetime and can be identified with the independent components of a particular tensor (called *curvature tensor*) which we will describe in the next chapter.

To summarize, we can always choose coordinates in such a way that the metric has the Cartesian form at a given event and its first derivatives vanish at the event. Such a coordinate system around an event $\mathcal{P}$ is called a *local inertial frame* at $\mathcal{P}$. It is obvious that all effects of gravity, arising due to the non-trivial nature of the metric tensor, will vanish to first order in the local inertial frame. More precisely, let us assume that a given metric tensor varies over a region of size L in the sense that, for a typical component of the metric tensor, $\partial g/g \approx L^{-1}$. Then we can identify a region of size $\mathcal{O}(L)$ around any event $\mathcal{P}$ in which gravitational effects will vanish and the standard laws of special relativity will hold to the accuracy of $\mathcal{O}(L)$. Physically, one can think of the local inertial frame as cubical boxes of size

$\mathcal{O}(L)$ freely falling in a gravitational field. The principle of equivalence assures us that, within such a box, gravitational effects will vanish to $\mathcal{O}(L)$ accuracy.

Exercise 4.1

Practice with metrics Preparing a planar map of the world requires projecting from the spherical surface of the Earth to a flat two-dimensional surface. It is obvious that we cannot do this without some distortion and different kinds of maps use projections with different properties. This exercise explores some aspects of this.

(a) Let λ, ϕ be the latitude and longitude on the surface of the sphere. Show that the *stereographic projection* to a plane with x, y coordinates involves mapping $x = 2\tan(\theta/2)\cos\phi$, $y = 2\tan(\theta/2)\sin\phi$, where $\theta = \pi/2 - \lambda$ and the radius of the globe is taken to be unity. Show that the resulting metric in (x, y) coordinates is

$$ds^2 = \cos^4\left(\frac{\theta}{2}\right)(dx^2 + dy^2). \tag{4.8}$$

Can you think of any use for the map made using this projection?

(b) A more familiar one is the Mercator projection. In this projection, a straight line on the map will be a line of constant compass bearing on the globe which makes such maps valuable for navigation. Show that this requires the transformation $x = \phi, y = \ln\cot\theta/2$ leading to the metric

$$ds^2 = \text{sech}^2 y\,(dx^2 + dy^2). \tag{4.9}$$

Show that the great circles on the sphere, which are paths of least distance between any two points on the globe, are mapped to curves given by the equation $\sinh y = \alpha\sin(x + \beta)$, where α and β are constants, except for the two special cases $y = 0$ or $x = $ constant. Draw this curve on the plane, connecting two points on the globe at the same latitude in the northern hemisphere.

4.3 Tensor algebra in curved spacetime

Since g_{ik} cannot be reduced globally to any preassigned form, no coordinate system has a preferred status in the presence of a gravitational field. Hence all laws of physics must be expressed in terms of physical quantities which are coordinate independent. When we developed special relativity in Chapter 1, we saw that four-vectors, tensors, etc., satisfy this requirement as far as Lorentz transformations are concerned. We have to now generalize the definitions of these quantities to incorporate *arbitrary* coordinate transformations. It turns out that extending the ideas of tensor *algebra* to curved spacetime or to arbitrary coordinate systems is fairly straightforward, while the concepts in tensor *calculus* (involving differentiation and integration) require more careful handling. We shall begin with tensor algebra.

Let us start with the definition of a four-vector which was introduced in Chapter 1 using the four-velocity $u^i = (dx^i/d\tau)$ of a material particle as the prototype. This definition carries over directly to curved spacetime. (Throughout this chapter, we shall use the term curved spacetime to indicate a spacetime with a nontrivial metric of the form in Eq. (4.1) even though it could also represent flat spacetime in curvilinear coordinates. So, more precisely, we should say 'curved spacetime or curvilinear coordinates' but we shall not bother to do so when no confusion is likely to arise.) One can define a parameterized curve $x^i(\lambda)$ in any coordinate system in the spacetime by just providing four functions. Given this curve, one can also define a tangent vector v^i to this curve by $v^i = (dx^i/d\lambda)$. In the special case of the curve being traced out by the motion of a particle and the parameter λ being chosen to be the proper time τ, shown by a clock attached to the particle, the tangent vector becomes the four-velocity of the particle. When the coordinate system is changed from x^i to x'^i, the components of the tangent vector v^i to a curve $x^i(\lambda)$ changes to

$$v'^i = \frac{dx'^i}{d\lambda} = \frac{\partial x'^i}{\partial x^a}\frac{dx^a}{d\lambda} = \frac{\partial x'^i}{\partial x^a}v^a. \tag{4.10}$$

We can now define a general four-vector as a set of four quantities which transform in the same manner as the tangent vector v^i. That is, the four-vector $q^a(x)$ is defined to be a set of four functions of space and time which change according to the law

$$q'^a(x'^i) = \left(\frac{\partial x'^a}{\partial x^b}\right)q^b(x^i) \tag{4.11}$$

under a coordinate transformation $x^i \to x'^i$. The x'^i in $q'^a(x'^i)$ in the left hand side and the x^i in $q^a(x^i)$ in the right hand side represent the same event in different coordinate frames.

This definition, of course, reduces to the one used in special relativity when the coordinate transformation $x'^i = L^i_j x^j$ is a Lorentz transformation, for which $(\partial x'^a/\partial x^b) = L^a_b$ are constants independent of the coordinates. But for a general coordinate transformation, the coefficients $(\partial x'^a/\partial x^b)$ will depend on the event x^i. Hence vectors located at different points in the spacetime will transform differently under a general coordinate transformation.

In our discussion of special relativity we pointed out that the coordinate differentials dx^i transform as a four-vector. Under a general coordinate transformation characterized by four functions $x^i = f^i(x'^0, x'^1, x'^2, x'^3)$, the differential dx^i transforms as

$$dx^i = \frac{\partial x^i}{\partial x'^k}dx'^k. \tag{4.12}$$

This is essentially the object we used in defining the four-vector in curved spacetime as well; a tangent vector $v^i = dx^i/d\lambda$ transforms exactly as dx^i since $d\lambda$ is invariant. In the case of Lorentz transformations, *both* x^i and dx^i transform as four-vectors; however, as we stressed during the discussion on page 16, the differential dx^i continues to retain its four-vector status under curvilinear coordinate transformations. In contrast, x^is do not enjoy any special status when we allow for arbitrary coordinate transformations. This is why we have not used x^i to define four-vectors even in the case of special relativity.

In a similar vein, we can use the transformation law of the gradient of a scalar to define the covariant components of a vector. We know, from the normal rules of calculus, that

$$\partial_a' \phi = \frac{\partial x^b}{\partial x'^a} \, \partial_b \phi. \tag{4.13}$$

(As we shall see later partial derivatives of other quantities like vectors, e.g. $\partial_i v^j$, do *not* transform as a tensor under general coordinate transformation, though they do so under Lorentz transformations. The derivative of a *scalar* is an exception.) We will define the covariant components of the vector using Eq. (4.13). That is, A_i represents the covariant components of a vector if it transforms as

$$A_i' = \frac{\partial x^j}{\partial x'^i} \, A_j. \tag{4.14}$$

It follows that, as long as we are dealing with tensorial quantities defined at a given event, we can easily generalize the definitions and algebraic concepts of Chapter 1 by just replacing L_b^a by $(\partial x'^a / \partial x^b)$. For example, let us consider the definition of a second rank contravariant tensor. If we take a specific tensor of the form $T^{ab} = v^a u^b$ then, from the known transformation law for v^a and u^b, we can determine the transformation law for T^{ab}. We find that

$$T'^{ab} \equiv v'^a u'^b = \frac{\partial x'^a}{\partial x^c} \frac{\partial x'^b}{\partial x^d} v^c u^d = \frac{\partial x'^a}{\partial x^c} \frac{\partial x'^b}{\partial x^d} T^{cd}. \tag{4.15}$$

The quantity $T^{ab} = v^a u^b$ is clearly a second rank tensor though it is a special kind of tensor. The transformation law arrived at above, however, is stated entirely in terms of the components T^{ab} and does not depend on the vectors v^a and u^b. It is therefore natural to define a transformation law for a tensor T^{ab} based on the above equation.

The definition can be generalized easily for higher rank tensors. To every index we introduce a factor $(\partial x'^a / \partial x^b)$ in such a way as to relate the component a in the new coordinates to a component b in the old coordinates. For example, a covariant tensor of rank 2 transforms as

$$A_{ik} = \frac{\partial x'^l}{\partial x^i} \frac{\partial x'^m}{\partial x^k} A_{lm}' \tag{4.16}$$

and a mixed tensor transforms as

$$A_k^i = \frac{\partial x^i}{\partial x'^l} \frac{\partial x'^m}{\partial x^k} A_m'^l. \tag{4.17}$$

The rules for multiplying tensors as well as for contracting over indices remain the same as in the special relativistic context. For example, the scalar product of two vectors $A^i B_i$ remain invariant under general coordinate transformations:

$$A^i B_i = \frac{\partial x^i}{\partial x'^l} \frac{\partial x'^m}{\partial x^i} A'^l B'_m = \frac{\partial x'^m}{\partial x'^l} A'^l B'_m = \delta_l^m A'^l B'_m = A'^l B'_l. \qquad (4.18)$$

Note that the Kronecker delta, defined with components $\delta_k^i = 0$ for $i \neq k$ and $\delta_k^i = 1$ for $i = k$, is generally covariant (as is obvious from the relation $A^k \delta_k^i = A^i$) and has the same components in all coordinate systems; i.e. $\delta_m'^l = \delta_m^l$.

Let us next consider the metric tensor and the process of raising and lowering of indices. Treating the metric tensor g_{ik} as a matrix we can define its inverse g^{ik} such that $g^{ik} g_{kl} = \delta_l^i$. The transformation law for g^{ik} can be obtained by noting that the relation $g^{ik} g_{kl} = \delta_l^i$ should hold in all coordinate frames. From the definition of the metric tensor it follows that:

$$g'_{ik} = \frac{\partial x^a}{\partial x'^i} \frac{\partial x^b}{\partial x'^k} g_{ab}. \qquad (4.19)$$

Therefore we must have

$$g'^{ik} = \frac{\partial x'^i}{\partial x^a} \frac{\partial x'^k}{\partial x^b} g^{ab}. \qquad (4.20)$$

Thus g^{ik} is a contravariant second rank tensor, g_{ik} is a covariant tensor and both transform appropriately under coordinate transformations.

In the case of special relativity, we defined the process of raising and lowering of an index using η_{ab} and this procedure allowed us to associate either contravariant or covariant components with a given tensor. In curved space, we can define the operation of 'raising and lowering of an index' of a tensor by using an appropriate form of metric tensor. That is, given T^{ik}, we define two new tensors $T^i{}_k$ and T_{ik} by the relations

$$T^i{}_k \equiv g_{ka} T^{ia}; \quad T^{ia} = g^{ka} T^i{}_k; \quad T_{ik} \equiv g_{ia} g_{kb} T^{ab}. \qquad (4.21)$$

When we transform from x^i to x'^i the quantity $T^i{}_k$ transforms as

$$T'^i{}_k = g'_{ka} T'^{ia} = \frac{\partial x^b}{\partial x'^k} \frac{\partial x^c}{\partial x'^a} g_{bc} \cdot \frac{\partial x'^i}{\partial x^m} \frac{\partial x'^a}{\partial x^n} T^{mn} = g_{bc} T^{mn} \frac{\partial x^b}{\partial x'^k} \frac{\partial x'^i}{\partial x^m} \delta_n^c$$

$$= g_{bn} T^{mn} \frac{\partial x^b}{\partial x'^k} \frac{\partial x'^i}{\partial x^m} = T^m{}_b \frac{\partial x^b}{\partial x'^k} \frac{\partial x'^i}{\partial x^m}. \qquad (4.22)$$

This relation shows that $T^i{}_k$ transforms correctly when the coordinates are changed. Similar result can be obtained for the covariant components for T_{ik}. Thus raising and lowering of indices using g_{ik} is a legitimate tensorial operation. Similarly, the dot products between four-vectors and the norms of four-vectors will be defined using g_{ab} like $A^i B^k g_{ik}$, $A^i A^j g_{ij}$, etc.

All these can be stated more elegantly – but abstractly – in terms of the notation introduced in Section 1.6. At any given event $\mathcal{P}$ in the spacetime, we can define a tangent vector space $T(\mathcal{P})$, the elements of which are the tangent vectors to all the curves passing through the event. The linearly independent set of basis vectors of $T(\mathcal{P})$ will be denoted by $\boldsymbol{e}_i(\mathcal{P})$ and any other vector in $T(\mathcal{P})$ can be expanded as $\boldsymbol{v} = v^i \boldsymbol{e}_i$. (In this notation, introduced in Section 1.6, the superscript i on v^i denotes a component of a vector while the subscript i on $\boldsymbol{e}_i$ denotes *which* vector and *not* which component of a vector. We take the summation convention to be valid over repeated indices of either kind.) If we take the kth component of $\boldsymbol{e}_i$ to be δ_i^k, we also have $\boldsymbol{e}_i \cdot \boldsymbol{e}_j \equiv g_{kl}(\boldsymbol{e}_i)^k(\boldsymbol{e}_j)^l = g_{ij}$. All these operations are local and, to each event $\mathcal{P}$, we have an associated tangent vector space $T(\mathcal{P})$ and its basis vectors. A general coordinate transformation involves rotating the basis vectors at each event to a new set $\boldsymbol{e}_i' = M_i^j(x)\boldsymbol{e}_j$ where the rotation matrix $M_i^j(x)$ will – in general – vary from event to event. A vector $\boldsymbol{v}$, of course, is a geometric object which is independent of the basis chosen to describe it and we have

$$\boldsymbol{v} = v'^i \boldsymbol{e}_i' = v'^i M_i^j(x)\boldsymbol{e}_j = v^j \boldsymbol{e}_j, \tag{4.23}$$

showing that the components at a given event $\mathcal{P}(x)$ transform as $v^j = v'^i M_i^j(x)$. This conforms to our original definition with $M_i^j(x) = (\partial x^j / \partial x'^i)$. Once the vector is defined all contravariant tensors can be defined by expanding them in the direct product space as $\boldsymbol{T} = T^{abc\cdots} \boldsymbol{e}_a \otimes \boldsymbol{e}_b \otimes \boldsymbol{e}_c \otimes \cdots$, etc.

To obtain the covariant components we need to introduce the dual vector space of $T(\mathcal{P})$ which is usually denoted by a star, $T(\mathcal{P})^*$, with the basis vectors $\boldsymbol{w}^k$. The dual basis allows one to define the covariant components of the vector by taking $\boldsymbol{u} = u_i \boldsymbol{w}^i$, etc., for each element of $T(\mathcal{P})^*$. (All these ideas, introduced in Section 1.6, are local and hence continue to be valid in a curved spacetime.) For any two elements $\boldsymbol{u} \in T(\mathcal{P})^*$ and $\boldsymbol{v} \in T(\mathcal{P})$, we define a scalar product $\langle \boldsymbol{u} | \boldsymbol{v} \rangle \equiv u_i v^i$. This shows that a given element in $T(\mathcal{P})^*$ will allow us to associate a real number with any element of $T(\mathcal{P})$. Further, this association is bi-linear in both vectors. (This is, in fact, the mathematical procedure one uses to define a dual vector space.) The bi-linearity of the scalar product shows that we can compute the scalar product between any two vectors in $T(\mathcal{P})$ and $T(\mathcal{P})^*$ once we know the result $\langle \boldsymbol{w}^k | \boldsymbol{e}_j \rangle$ for basis vectors. Given the basis vectors $\boldsymbol{e}_i$ with components $e_i^k = \delta_i^k$, we can choose the components of $\boldsymbol{w}^k$ to be $w_i^k = \delta_i^k$ with $\langle \boldsymbol{w}^k | \boldsymbol{e}_j \rangle = \delta_j^k$. (Such a basis is called *coordinate basis*. It is possible to choose $\boldsymbol{e}_i$ and $\boldsymbol{w}^k$ in a more general manner and yet satisfy $\langle \boldsymbol{w}^k | \boldsymbol{e}_j \rangle = \delta_j^k$; we shall, however, stick to a coordinate basis for our discussion.)

These operations, so far, have not required a metric. But we know that, given two vectors $\boldsymbol{p}$ and $\boldsymbol{v}$ both of which are elements of $T(\mathcal{P})$, one can obtain a real number by taking the dot product $\boldsymbol{p} \cdot \boldsymbol{v} = g_{ij}p^i v^j$. Now let $\boldsymbol{u}$ be an element of $T(\mathcal{P})^*$ such that $\langle \boldsymbol{u} | \boldsymbol{v} \rangle = \boldsymbol{p} \cdot \boldsymbol{v} = g_{ij}p^i v^j$. That is, suppose both these operations lead to the same real number. Using this we can associate with every vector $\boldsymbol{p}$ (which is an element of $T(\mathcal{P})$) another element $\boldsymbol{u}$ of $T(\mathcal{P})^*$ such that the above result holds. Taking components, we immediately see that $u_j = g_{ij}p^i$; i.e. u_j is obtained by the standard procedure of lowering the index of p^i. Since there is a one-to-one correspondence, it makes sense to use the same symbol for both and simply write $p_j = g_{ij}p^i$, etc. This is the origin of the covariant components of a vector. The logic is same as in the case of special relativity with g_{ij} replacing η_{ij}.

The covariant tensor and mixed rank tensors can now be defined as before by taking appropriate direct product spaces: $\boldsymbol{S} = S_{ij}^k \, \boldsymbol{w}^i \otimes \boldsymbol{w}^j \otimes \boldsymbol{e}_k$, etc. We stress that all these operations are local and take place at some given event $\mathcal{P}$.

4.4 Volume and surface integrals

In special relativity, the volume element d^4x provides an invariant integration measure over four dimensions because the Lorentz transformation matrix has a unit determinant. This ensures that $d^4x' = d^4x$ under Lorentz transformations. This is no longer true when we consider arbitrary coordinate transformation and it is necessary to determine an invariant integration measure for the volume. To do this, we note that d^4x transforms under arbitrary curvilinear transformations as

$$d^4x' = \left| \frac{\partial x'}{\partial x} \right| d^4x = J \, d^4x, \tag{4.24}$$

where J is the Jacobian of the transformation and d^4x stands for the product $dx^0 dx^1 \, dx^2 dx^3$. On the other hand, we can consider Eq. (4.19) as a matrix equation and take the determinant on both sides. This gives $g' = J^{-2} g$, where $g = \det g_{ik}$ and $g' = \det g_{ik}'$. Since the signature of the metric is $(-, +, +, +)$, $\det g_{ik}$ will be a negative quantity. Taking this into account, we have $\sqrt{-g'} = (1/J)\sqrt{-g}$. Using this in Eq. (4.24), we get

$$\sqrt{-g'} \, d^4x' = \sqrt{-g} \, d^4x, \tag{4.25}$$

showing that the quantity $\sqrt{-g} \, d^4x$ is generally covariant and can provide the appropriate measure of integration over the four-volume. This result is quite general and is independent of the number of dimensions of the space. In fact, in ordinary vector analysis, one routinely uses the volume element $dV = r^2 \sin\theta \, d\theta \, d\phi \, dr$ in spherical coordinates; this is just $\sqrt{g}d^3x$ in terms of spherical polar coordinates.

Similar considerations apply in defining three-dimensional surface integrals in four-dimensional space. In special relativity, we defined these quantities using the completely antisymmetric four-tensor ϵ_{ijkl} (see Eq. (1.58) and Eq. (1.59)). The components of this tensor were the same in all Lorentz frames, again because the Jacobian of Lorentz transformations is unity. Under general coordinate transformations, the components of this tensor change because the Jacobian is no longer unity. The combination

$$\epsilon_{abcd} = \sqrt{-g}\,[abcd], \qquad (4.26)$$

however, is a generally covariant tensor where the symbol $[abcd]$ now stands for the completely antisymmetric fourth rank object introduced in special relativity. That is, $[abcd] = +1$ if $[abcd]$ is an even permutation of 0123, $[abcd] = -1$ if $[abcd]$ is an odd permutation of 0123 and zero if any two indices are the same. To prove that ϵ_{abcd} is a tensor, we note that the quantity

$$[abcd]\,\frac{\partial x^a}{\partial x'^a}\frac{\partial x^b}{\partial x'^b}\frac{\partial x^c}{\partial x'^c}\frac{\partial x^d}{\partial x'^d} \qquad (4.27)$$

is completely antisymmetric in the primed indices and hence must be proportional to $[a'b'c'd']$. Therefore, this expression must be equal to $\lambda[a'b'c'd']$, where λ is a proportionality constant. If we now put $a'b'c'd' = 0123$, we get

$$\lambda = [abcd]\,\frac{\partial x^a}{\partial x'^0}\frac{\partial x^b}{\partial x'^1}\frac{\partial x^c}{\partial x'^2}\frac{\partial x^d}{\partial x'^3}. \qquad (4.28)$$

But from the first relation in Eq. (1.52) with $(i, k, l, m) = (0, 1, 2, 3)$ we see that the right hand side of Eq. (4.28) is the determinant of the matrix $(\partial x^a / \partial x'^a)$, which is J^{-1}. From our previous result we know that $J^{-1} = \sqrt{g'/g}$; so we have

$$\sqrt{-g}\,[abcd]\,\frac{\partial x^a}{\partial x'^a}\frac{\partial x^b}{\partial x'^b}\frac{\partial x^c}{\partial x'^c}\frac{\partial x^d}{\partial x'^d} = \sqrt{-g'}\,[a'b'c'd'] \qquad (4.29)$$

showing that the ϵ_{abcd} defined by Eq. (4.26) transforms as a tensor. By a similar analysis, we can show that the completely antisymmetric tensor with contravariant indices is given by

$$\epsilon^{abcd} = -\frac{1}{\sqrt{-g}}\,[abcd]. \qquad (4.30)$$

Using the ϵ-tensor, we can define the surface integrals in curved spacetime exactly as in the case of special relativity using the formulas Eq. (1.58) and Eq. (1.59) with the understanding that the ϵ-tensors now contain the $\sqrt{-g}$ factors.

In this connection there is one result which is of some importance, so that we will state it explicitly. Consider a three-dimensional surface S in four-dimensional spacetime given in parametric form as $x^a = x^a(y^\mu)$, where y^μ are three parameters. The vectors $e^a_\mu = (\partial x^a / \partial y^\mu)$ for $\mu = 1, 2, 3$ are tangents to the curves contained in the three-dimensional surface and are orthogonal to the normal n_a of the surface; i.e. $e_\mu^a n_a = 0$. When we restrict ourselves to infinitesimal coordinate displacements on $\mathcal{S}$, we have the result

$$ds_{\mathcal{S}}^2 = g_{ab}\, dx^a\, dx^b\Big|_{\mathcal{S}} = g_{ab}\left(\frac{\partial x^a}{\partial y^\alpha}\, dy^\alpha\right)\left(\frac{\partial x^b}{\partial y^\beta}\, dy^\beta\right) \equiv h_{\alpha\beta}\, dy^\alpha\, dy^\beta. \quad (4.31)$$

The last equality defines the *induced metric* $h_{\alpha\beta} = g_{ab}e_\alpha^a e_\beta^b$ on the surface. The infinitesimal volume element on $\mathcal{S}$ is given by $d^3V = |h|^{1/2}\, d^3y$, where h is the determinant of the induced metric. A directed surface element on $\mathcal{S}$ will be $d^3\sigma_a = n_a\, |h|^{1/2}\, d^3y$. If the unit normal to the surface $\mathcal{S}$ is n_a, then it follows that $n_a e_\alpha^a = 0$. Hence we can also write the induced metric as $h_{\alpha\beta} = h_{ab}e_\alpha^a e_\beta^b$, where $h_{ab} = g_{ab} + \epsilon n_a n_b$ with $\epsilon = +1$ for the spacelike hypersurfaces and $\epsilon = -1$ for a timelike hypersurface. Very often, we will use h_{ab} and $h_{\alpha\beta}$ interchangeably since they carry the same information. Similar definitions can be given for two-dimensional surfaces embedded in a four-dimensional space.

These expressions show that $\det g_{ab} = g$ will play a key role while dealing with curvilinear coordinates. We shall now derive some results related to the differential dg of the determinant g in terms of the components of the metric tensor, which will prove to be useful. To obtain this, we begin with the general result

$$\ln \det M = \mathrm{Tr}\, \ln M \quad (4.32)$$

for any matrix M_{ab}, which is easily proved by diagonalizing the matrix. Taking the differential of the left hand side gives $(\delta \det M/\det M)$. On the right hand side, the operations of taking differentials and trace commute and we have the result $\delta \ln M = M^{-1}\delta M$. We therefore get

$$\frac{\delta \det M}{\det M} = \mathrm{Tr}\, M^{-1}\delta M. \quad (4.33)$$

In our case, $M \to g_{ab}$, $M^{-1} \to g^{ab}$ so that we get the result as $\delta g = g g^{ik}\delta g_{ik}$. It follows that

$$\partial_a g = g g^{ik} \partial_a g_{ik} = -g g_{ik}\partial_a g^{ik}. \quad (4.34)$$

The second equality arises from the fact that $\partial_a(g_{ik}g^{ik}) = 0$. Manipulating this condition, we can also easily show that

$$\partial_m g^{ik} = -g^{ip}g^{lk}(\partial_m g_{pl}). \quad (4.35)$$

Note that the partial derivative operator ∂_m is the same on both sides while the indices on the metric are raised in the usual manner *except for an overall sign*. This result also arises in a different context which deserves comment. Consider an infinitesimal variation of the metric in the form $g_{ab} \to g_{ab} + \delta g_{ab}$. Let the corresponding change in the inverse matrix be $g^{ab} \rightarrow g^{ab} + \delta g^{ab}$. By varying the condition $g_{ab}g^{bd} = \delta_a^d$ we get $g^{bd}\delta g_{ab} = -g_{ab}\delta g^{bd}$. Contracting this equation with g^{ak} we immediately obtain $\delta g^{kd} = -g^{ak}g^{bd}\delta g_{ab}$, which is essentially the same as Eq. (4.35). This result shows that one must *not* think of δg^{kd} as the contravariant components of δg_{ab} since in such a case the minus sign will be absent.

4.5 Geodesic curves

The first nontrivial problem in curved spacetime we will discuss will be the motion of a particle in a given curved spacetime. To do this, let us recall the action principle in special relativity which was used to obtain the equation of motion for the free particle:

$$\mathcal{A} = -m \int d\tau = -m \int \left(-\eta_{ab} \frac{dX^a}{d\tau} \frac{dX^b}{d\tau} \right)^{1/2} d\tau. \tag{4.36}$$

The variation of $\mathcal{A}$ gives $(d^2 X^a/d\tau^2) = (du^a/d\tau) = 0$, where $u^a = (dX^a/d\tau)$ is the four-velocity which is the tangent vector to the trajectory $X^a(\tau)$. This equation can be equivalently written as

$$0 = \frac{du^a}{d\tau} = \left(\frac{dX^b}{d\tau} \right) \left(\frac{\partial u^a}{\partial X^b} \right) = u^b \left(\frac{\partial u^a}{\partial X^b} \right) \equiv u^b \partial_b u^a. \tag{4.37}$$

Since u^b transforms as a vector under coordinate transformation, this equation will retain its form, under arbitrary coordinate transformations, provided the derivative $(\partial u^a/\partial x^b)$ transforms as a tensor.

In special relativity, the derivatives of a vector field $\partial_b u^a$ transform as a mixed tensor of rank 2. This is, however, not true when we consider arbitrary coordinate transformations as can be seen by explicit evaluation in two different coordinate frames:

$$\partial_b' u'^a \equiv \frac{\partial u'^a}{\partial x'^b} = \frac{\partial x^c}{\partial x'^b} \frac{\partial}{\partial x^c} \left\{ \frac{\partial x'^a}{\partial x^d} u^d \right\}$$

$$= \left(\frac{\partial x^c}{\partial x'^b} \frac{\partial x'^a}{\partial x^d} \right) \partial_c u^d + \left[\left(\frac{\partial x^c}{\partial x'^b} \right) \frac{\partial^2 x'^a}{\partial x^c \partial x^d} \right] u^d. \tag{4.38}$$

If $\partial_b u^a$ has to transform as a mixed tensor of rank 2, the second term in the above equation has to vanish. This term vanishes for Lorentz transformations (when x'^i are linear functions of x^i) but not for arbitrary coordinate transformations. Hence $\partial_b u^a$ is not a tensor under general coordinate transformation.

The reason behind this result is easy to understand. The derivative $\partial_b u^a$ involves subtracting $u^a(x^b)$ from $u^a(x^b + \Delta x^b)$. In special relativity, the transformation law for four-vectors is the same all over the spacetime and hence the difference between $u^a(x^b)$ and $u^a(x^b + \Delta x^b)$ will be a four-vector. But in a curved spacetime or when curvilinear coordinates are used, the transformation law for the four-vector involves the quantities $\partial x'^i/\partial x^j$ which are functions of the coordinates. Therefore, the transformation laws for vectors at x^b and $x^b + \Delta x^b$ are different and the subtraction of vectors located at two different events is not a tensorial operation. It is clear that we need to generalize the definition of partial derivative of a vector in order to give it a covariant meaning.

The difficulty has its roots in the fact that the action $\mathcal{A}$ in Eq. (4.36) uses η_{ab} which is a tensor only under Lorentz transformations and not under general coordinate transformations. This problem can be easily tackled as follows. In any small region around an event, one can always choose an inertial coordinate system. In such a small region, the form of the integrand in Eq. (4.36) remains valid. However, in this small region $\eta_{ab} = g_{ab}$ and so we can *also* write the integrand as $(-g_{ab}dx^a dx^b)^{1/2}$, leading to the action in the form

$$\mathcal{A} = -m \int d\tau = -m \int \sqrt{-g_{ab}dx^a dx^b}. \tag{4.39}$$

Since *this* action is made of proper tensorial quantities it will remain valid in any coordinate system. Therefore, the equations derived from this action will provide us with the correct generalization of Eq. (4.37). As a bonus, it will provide us with an insight into the manner in which partial derivatives can be generalized to a covariant notion.

The variation of the action in Eq. (4.39) can be obtained as follows. We first note that

$$\delta d\tau^2 = 2d\tau\, \delta d\tau = \delta(-g_{ik}\, dx^i\, dx^k) = -dx^i\, dx^k\, (\partial_l g_{ik})\, \delta x^l - 2g_{ik}dx^i\, d\delta x^k. \tag{4.40}$$

Substituting for $\delta(d\tau)$ from this expression, we get

$$\delta\mathcal{A} = -m \int \delta d\tau = m \int \left[\frac{1}{2}\frac{dx^i}{d\tau}\frac{dx^k}{d\tau}\,(\partial_l g_{ik})\, \delta x^l + g_{ik}\frac{dx^i}{d\tau}\frac{d\delta x^k}{d\tau} \right] d\tau \tag{4.41}$$

$$= m \int \left[\frac{1}{2}\frac{dx^i}{d\tau}\frac{dx^k}{d\tau}\,(\partial_l g_{ik})\, \delta x^l - \frac{d}{d\tau}\left(g_{ik}\frac{dx^i}{d\tau} \right) \delta x^k \right] d\tau + m g_{ik}\frac{dx^i}{d\tau}\delta x^k \Big|_{\tau_1}^{\tau_2}$$

in which we have integrated one of the terms by parts. As usual, we will first consider variations for which $\delta x^i = 0$ at the end points and obtain the equations of motion. Demanding $\delta\mathcal{A} = 0$ for paths with $\delta x^i = 0$ at the end points, gives the equations of motion:

$$\frac{d}{d\tau}(g_{il}u^i) = \frac{d}{d\tau}(u_l) = \frac{1}{2}(\partial_l g_{ik})\, u^i u^k. \tag{4.42}$$

This equation shows that the rate of change of the *covariant* component of the velocity is driven by the derivatives of the metric. If we multiply the whole equation by the mass m, the left hand side can be interpreted as the rate of change of the covariant components of the momentum $p_l = mu_l$, suggesting that the right hand side can be interpreted as a velocity dependent force. As a quick check, consider the situation that arises with slowly moving particles ($u^0 \approx 1$; $d\tau \approx dt$; $u_\alpha \approx u^\alpha$) in a weak gravitational field, for which we obtained an approximate metric in Eq. (3.72) with:

$$g_{00} \approx -(1 + 2\phi); \quad g_{\alpha\beta} \approx \delta_{\alpha\beta}. \tag{4.43}$$

To this order of accuracy, the spatial component of Eq. (4.42) now becomes $du_\alpha/dt \approx -\partial_\alpha\phi$, which is precisely the equation of motion for a particle in Newtonian gravity.

We will now rewrite Eq. (4.42) in a different and more conventional form. Expanding out the derivative term with $d/d\tau$ we get

$$g_{il}\frac{du^i}{d\tau} + u^i u^k \, \partial_k g_{il} = \frac{1}{2} u^i u^k \, \partial_l g_{ik}. \tag{4.44}$$

This equation can be cast in a more symmetric form by rewriting the second term in the left hand side in an equivalent form as $(1/2)u^i u^k [\partial_k g_{il} + \partial_i g_{kl}]$. Then we get

$$g_{il}\frac{du^i}{d\tau} = \frac{1}{2} \left(\partial_l g_{ik} - \partial_i g_{lk} - \partial_k g_{li}\right) u^i u^k. \tag{4.45}$$

We can eliminate the g_{il} on the left hand side by multiplying the equation by g^{kl} and using $g^{kl}g_{il} = \delta_i^k$. This gives the final result:

$$\frac{du^k}{d\tau} = -\frac{1}{2}g^{kl}\left(-\partial_l g_{ij} + \partial_j g_{il} + \partial_i g_{lj}\right) u^i u^j. \tag{4.46}$$

We will now define two sets of quantities, Γ_{imn} and Γ^i_{mn} called (called *affine connection* or *Christoffel symbols*) in terms of the derivatives of the metric by:

$$\Gamma_{lij} \equiv \frac{1}{2}\left(-\partial_l g_{ij} + \partial_j g_{il} + \partial_i g_{lj}\right) \tag{4.47}$$

and

$$\Gamma^k_{ij} \equiv g^{kl}\Gamma_{lij} = \frac{1}{2}g^{kl}\left(-\partial_l g_{ij} + \partial_j g_{il} + \partial_i g_{lj}\right). \tag{4.48}$$

The Γ^k_{ij} is symmetric in the lower indices while Γ_{lij} is symmetric in i, j; the raising of the index l to get Γ^k_{ij} from Γ_{lij} is done by the standard procedure. With this definition, the equation of motion Eq. (4.46) can be written in the form:

$$\frac{du^i}{d\tau} + \Gamma^i_{mn}u^m u^n = 0. \tag{4.49}$$

Equation (4.49) (called the *geodesic equation*) governs the motion of material particles in a given gravitational field and can be thought of as the generalization of the force equation in Newtonian gravity.

Finally, let us consider the variation $\delta \mathcal{A}$ when the end points are varied by δx^k for a trajectory which satisfies the equation of motion. As we have seen several times by now, this procedure allows us to determine the relation between action and canonical momentum leading to the Hamilton–Jacobi equation. From Eq. (4.41), we find that (when the equation of motion is satisfied, making the term in square brackets vanish) $\delta \mathcal{A} = m g_{ik} u^i \delta x^k$ so that $\partial \mathcal{A}/\partial x^k = m u_k = p_k$. The Hamilton–Jacobi equation can now be obtained from $u_i u^i = -1$ giving

$$g^{ik} \frac{\partial \mathcal{A}}{\partial x^i} \frac{\partial \mathcal{A}}{\partial x^k} = -m^2. \tag{4.50}$$

Often this will provide a fast route to obtaining the trajectory of a particle in a given spacetime.

Exercise 4.2

Two ways of splitting spacetimes into space and time On several occasions, we would like to separate the components of the metric into three sets $(g_{00}, g_{0\alpha}, g_{\alpha\beta})$. This will prove to be useful just as separating a four-vector into its time and space components is useful in certain contexts. In the literature such a (1+3) split is achieved in two different ways. In the first procedure, one writes the metric in the form

$$ds^2 = -M^2 \left(dt - M_\alpha \, dx^\alpha\right)^2 + \gamma_{\alpha\beta} \, dx^\alpha \, dx^\beta, \tag{4.51}$$

where the functions $M^2, M_\alpha, \gamma_{\alpha\beta}$ determine the metric. The second procedure is to write the metric as

$$ds^2 = -N^2 \, dt^2 + g_{\alpha\beta} \left(dx^\alpha + N^\alpha dt\right) \left(dx^\beta + N^\beta \, dt\right), \tag{4.52}$$

where the functions $N^2, N_\alpha, g_{\alpha\beta}$ determine the metric.

(a) Show that, in the first procedure, the metric components are given by

$$g_{00} = -M^2 \; ; \qquad g_{0\alpha} = M^2 M_\alpha \; ; \qquad g_{\alpha\beta} = \gamma_{\alpha\beta} - M^2 M_\alpha M_\beta$$
$$g^{00} = -\left(M^{-2} - M_\gamma M^\gamma\right) \; ; \qquad g^{0\alpha} = M^\alpha \; ; \qquad g^{\alpha\beta} = \gamma^{\alpha\beta}, \tag{4.53}$$

where g^{ab} is the inverse of the matrix g_{ab} with $g^{ab} g_{bc} = \delta_c^a$ and, in the second procedure,

$$g_{00} = -\left(N^2 - N_\gamma N^\gamma\right) \; ; \qquad g_{0\alpha} = N_\alpha \; ; \qquad g_{\alpha\beta} = g_{\alpha\beta}$$
$$g^{00} = -N^{-2} \; ; \qquad g^{0\alpha} = N^{-2} N^\alpha \; ; \qquad g^{\alpha\beta} = g^{\alpha\beta} - N^{-2} N^\alpha N^\beta. \tag{4.54}$$

(b) In the literature, the first procedure is sometimes called 'threading the spacetime' while the second one is called 'slicing the spacetime'. Understand the geometrical structure of the two procedures and convince yourself that this terminology is appropriate.

Exercise 4.3

Hamiltonian form of action – particle in curved spacetime In the case of a special relativistic particle, the equations of motion could be obtained from an alternative Hamiltonian form of the action introduced in Exercise 1.9 and pursued further in Exercise 1.10 and Exercise 2.4. Show that the corresponding Hamiltonian form of the action in curved spacetime is obtained by the replacement of η_{ab} by $g_{ab}(x)$ in the action in Eq. (1.85). That is, show that varying x^a, p^a, C independently in

$$\mathcal{A} = \int_{\lambda_1}^{\lambda_2} d\lambda \left[p_a \dot{x}^a - \frac{1}{2} C \left(\frac{H}{m} + m \right) \right], \tag{4.55}$$

where $H = g_{ab} p^a p^b$ and C is an auxiliary variable (which is set to unity after the variation) leads to the correct equations of motion for a free particle and fixes the parameter λ to be proportional to the proper time.

Exercise 4.4

Gravo-magnetic force In several contexts, a metric can be usefully approximated to a near Newtonian form with $g_{ab} = \eta_{ab} + h_{ab}$, $|h_{ab}| \ll 1$ and $\partial_0 g_{ab} = 0$. This exercise explores one such case with $h_{0\alpha} \neq 0$.

(a) Work out the geodesic equation for a particle in this metric to the lowest order in (v/c) and show that it can be expressed in the form (with c-factors temporarily displayed):

$$\frac{d^2 x^\alpha}{dt^2} \approx \frac{1}{2} c^2 \delta^{\alpha\beta} \partial_\beta h_{00} + c \delta^{\alpha\gamma} \left(\partial_\beta h_{0\gamma} - \partial_\gamma h_{0\beta} \right) v^\beta. \tag{4.56}$$

(b) Write this in a more suggestive form as:

$$\ddot{\boldsymbol{x}} = -\nabla\phi + [\boldsymbol{v} \times (\nabla \times \boldsymbol{A})], \tag{4.57}$$

where $g_{00} \equiv -(1 + 2\phi/c^2)$ and $g_{0\alpha} \equiv -A_\alpha/c$. This allows interpreting $\nabla \times \boldsymbol{A}$ as a 'gravo-magnetic field'.

[Hint. The quickest way to do this is to express the variational principle for the geodesic equation in terms of t as the parameter leading to a Lagrangian

$$L = -m \frac{d\tau}{dt} = -m \sqrt{-g_{ab} \frac{dx^a}{dt} \frac{dx^b}{dt}}, \tag{4.58}$$

which can be expanded to the relevant order of accuracy to give

$$L = \frac{1}{2} v^2 - \phi + \frac{\boldsymbol{A} \cdot \boldsymbol{v}}{c}, \tag{4.59}$$

where the vector operations are done with the flat space metric. This maps the problem to the corresponding one in electromagnetism and the result follows.]

(c) Consider a spinning gyroscope in this spacetime. Show that its spin will precess by an amount $\boldsymbol{\Omega}_{\mathrm{LT}} = -(1/2)\nabla \times \boldsymbol{A}$. This is known as *Lens–Thirring precession*. [Answer. The spin precession formula in electrodynamics is $\dot{\boldsymbol{S}} = \boldsymbol{\mu} \times \boldsymbol{B}$, where $\boldsymbol{\mu} = (e/2m)\boldsymbol{S}$. For gravity, we only need to substitute e by m to get the result.]

(d) The metric of a weakly rotating source (like the spinning Earth or the spinning Sun), located at the origin, with angular momentum $\boldsymbol{J}$ has the form $g_{ab} = \eta_{ab} + h_{ab}$ with $g_{0\alpha} = (2/r^3)\epsilon_{\alpha\beta\gamma} x^\beta J^\gamma$. (We will derive this result in Chapter 6, Section 6.4.) Compute the Lens–Thirring precession for this case. [Answer. Using the result of part (c) we get: $\boldsymbol{\Omega}_{\mathrm{LT}} = (G/r^3)[-\boldsymbol{J} + 3((\boldsymbol{J} \cdot \boldsymbol{x})\boldsymbol{x}/r^2)]$.]

4.5.1 Properties of geodesic curves

We shall now describe several features of the result obtained in the previous section. To begin with, we know that Eq. (4.49) must be generally covariant, since it is derived from a variational principle that is generally covariant. The procedure to obtain this equation is valid in any dimension and in metrics with any signature. In two or three dimensions, for example, the solutions to this equation will give the curves of shortest distance (geodesics) between two points in the space described by a metric g_{ab}.

While the full equation is generally covariant, neither the first term nor the second term in Eq. (4.49) is a generally covariant tensor. The simplest way to gain some insight into this feature is to write Eq. (4.49) in another form. Writing $(du^i/d\tau) = (dx^a/d\tau)\partial_a u^i = u^a\partial_a u^i$ we can transform Eq. (4.49) into the form

$$u^a\left[\partial_a u^i + \Gamma^i_{ja}u^j\right] \equiv u^a\nabla_a u^i = 0, \tag{4.60}$$

where the first equality defines the symbol $\nabla_a u^i$. Since u^a is a tensor, the quantity in the square brackets in the above equation

$$\nabla_b u^a \equiv \partial_b u^a + \Gamma^a_{ib}u^i \tag{4.61}$$

must be a tensor. But we have seen before in Eq. (4.38) that the partial derivative of the vector $\partial_a u^i$ is *not* generally covariant. Therefore the second term in Eq. (4.61) also could not be generally covariant but together they must transform as a second rank tensor. To verify this we need to know how Γ^a_{bc} transforms under general coordinate transformations. Using its definition from Eq. (4.48) and the transformation law for g_{ik}, it can be directly verified that Γ^a_{bc} transforms as

$$\Gamma'^a_{bc} = \frac{\partial x'^a}{\partial x^i}\frac{\partial x^k}{\partial x'^b}\frac{\partial x^m}{\partial x'^c}\Gamma^i_{km} - \frac{\partial x^d}{\partial x'^b}\frac{\partial x^f}{\partial x'^c}\frac{\partial^2 x'^a}{\partial x^d\partial x^f}. \tag{4.62}$$

Comparing the second term in Eq. (4.62) with the second term in Eq. (4.38), we see that the extra terms cancel out in $\nabla_b u^a$ thereby making it transform as a tensor even though neither Γ^a_{bc} nor $\partial_b u^a$ is a tensor. (We stress this fact since the notation involving superscripts and subscripts might give the wrong impression that Γ^a_{bc} is a tensor.)

The above discussion did not make any distinction between a genuinely curved spacetime and the flat spacetime expressed in curvilinear coordinates. Whenever the metric tensor depends on the coordinates – either because we are using curvilinear coordinates in flat spacetime or because spacetime is genuinely curved – we will have nonzero Γ^a_{bc} and the trajectory of a free particle will be determined by Eq. (4.49). This result, again, is easy to understand in flat two-dimensional space. In a flat two-dimensional plane, the shortest distance between any two points (which will be a geodesic) is given by a straight line with the parametric equation $Y(s) = as + b; X(s) = s$ with the arc length s acting as a parameter. Clearly $d^2X/ds^2 = d^2Y/ds^2 = 0$. If we now decide to use the polar coordinates in the plane, then the geodesics are still straight lines but when expressed in polar coordinates they have the form $r = [(as + b)^2 + s^2]^{1/2}; \theta = \tan^{-1}[a + (b/s)]$. Clearly, we no longer have a linear relation between r (or θ) and s and $d^2r/ds^2 \neq 0$, etc. This is precisely what happens when we study the motion of a free particle in spacetime in a curvilinear coordinate system. In the inertial frame, the trajectory (which is a geodesic in the spacetime obtained by extremizing the arc length) satisfies the equation $d^2X^i/d\tau^2 = 0$. But the same straight line motion when viewed from a non-inertial frame (using curvilinear coordinates) will not have $d^2x^i/d\tau^2 = du^i/d\tau \neq 0$. Equation (4.49) shows that $du^i/d\tau$ is expressible in terms of the derivatives of the metric which do not vanish in an accelerated frame. We shall elaborate on this point and provide a geometrical interpretation for the affine connection later on.

We have seen that one can make the first derivatives of the metric tensor vanish around any given event by a suitable choice of coordinates. Since the first derivatives are in one-to-one correspondence with the Christoffel symbols, the vanishing of $\partial_a g_{bc}$ is equivalent to the vanishing of Christoffel symbols at a given event. It is possible to give an explicit coordinate transformation from x^a to x'^a which will achieve this around any event $\mathcal{P}$. Let this event have the coordinates $x^a(\mathcal{P})$ in a given system of coordinates in which the Christoffel symbols have a nonzero value $\Gamma^a_{bc}(\mathcal{P})$ at the same event. Consider a new set of coordinates given by

$$x'^a = x^a - x^a(\mathcal{P}) + \frac{1}{2}\Gamma^a_{bc}(\mathcal{P})\left[x^b - x^b(\mathcal{P})\right][x^c - x^c(\mathcal{P})]. \tag{4.63}$$

This transformation is constructed such that $\partial x'^a/\partial x^b = \delta^a_b$ at the event $\mathcal{P}$ and $\partial^2 x'^a/\partial x^b \partial x^c = \Gamma^a_{bc}(\mathcal{P})$. Using the transformation law in Eq. (4.62), it is easy to see that $\Gamma'^a_{bc} = 0$. This provides an explicit construction of a local inertial frame – which we described at the end of Section 4.2 – around any given event.

Exercise 4.5
Flat spacetime geodesics in curvilinear coordinates (a) Introducing the matrices $L^{a'}_i = (\partial x'^a/\partial x^i)$ and its inverse $L^b_{j'} \equiv (\partial x^b/\partial x'^j)$ we can write Eq. (4.62) as

$$\Gamma'^a_{bc} = L^{a'}_i L^k_{b'} L^m_{c'} \Gamma^i_{km} - L^d_{b'} L^f_{c'} \partial_d L^{a'}_f. \tag{4.64}$$

Invert Eq. (4.64) and show that

$$\Gamma^i_{km} = L^i_{a'} L^{b'}_k L^{c'}_m \Gamma'^{a'}_{b'c'} + L^i_{a'} \partial_k L^{a'}_m. \tag{4.65}$$

(b) Consider a flat spacetime expressed in two coordinate systems: the Cartesian one (ξ^a) with $ds^2 = \eta_{mn}d\xi^m d\xi^n$ and a curvilinear coordinate system $[x^a = x^a(\xi)]$ with a metric g_{ab} obtained by transforming the coordinates. Show that the trajectories of particles, moving with uniform velocity in the inertial coordinates, will satisfy the equation

$$\frac{d^2 x^l}{d\tau^2} + \Gamma^l_{mn}\frac{dx^m}{d\tau}\frac{dx^n}{d\tau} = 0; \quad \Gamma^l_{mn} = \frac{\partial x^l}{\partial \xi^a}\frac{\partial^2 \xi^a}{\partial x^m \partial x^n} \tag{4.66}$$

in the curvilinear coordinates where the parameter τ is the proper time. Explain the result.

Exercise 4.6

Gaussian normal coordinates Geodesics can be used to set up a useful coordinate system around any event. On any given spacelike hypersurface specified by, say, $x^0 = T$, one can introduce a coordinate grid (x^1, x^2, x^3). Let geodesic worldlines emerge from the events on the hypersurface orthogonal to the hypersurface. Give all the events in a given worldline the spatial coordinates of the event $\mathcal{P}_0$ on $\mathcal{S}$ at which it originates and the time coordinate $x^0 = T + \tau$ where τ is the proper time along the worldline (with $\tau = 0$ on the hypersurface). Near the $x^0 = T$ hypersurface there will be one unique geodesic connecting the events on the hypersurface to any other event $\mathcal{P}$ and our procedure assigns a unique set of four numbers to $\mathcal{P}$.

(a) Show that, in this coordinate system, the metric takes the form $ds^2 = -dt^2 + g_{\mu\nu}dx^\mu dx^\nu$. This is called the *synchronous form* of the metric.

[Hint. Start with the general metric $ds^2 = g_{ab}dx^a dx^b$. Since x^α is a constant on any geodesic, argue that $g_{00} = -1$ everywhere. Let $(e_0 = u, e_\mu)$ be the coordinate basis vectors where u is the tangent vector field to the geodesic. Argue that $u \cdot e_\mu = g_{0\mu}$ vanishes on the hypersurface and that its derivative along the geodesic also vanishes. This is enough to show that $g_{0\mu}$ vanishes everywhere.]

(b) Consider a spacetime metric expressed in the form $ds^2 = d\lambda^2 + g_{AB}(\lambda, x^A)dx^A dx^B$ in arbitrary dimension and signature. Show that the coordinate lines of λ (i.e. curves of the form $x^A = $ constant with λ varying) are geodesics in this geometry. Give some familiar examples of such metrics.

4.5.2 Affine parameter and null geodesics

When the geodesic corresponds to the path taken by a material particle, it is natural to use the proper time as a parameter for the curve. This is what was done in arriving at Eq. (4.49) which can be rewritten in terms of the coordinates as

$$\frac{d^2 x^i}{d\tau^2} + \Gamma^i_{jk}\frac{dx^j}{d\tau}\frac{dx^k}{d\tau} = 0. \tag{4.67}$$

But a geodesic in the spacetime, like any other curve $x^i(\lambda)$, is a geometrical entity that is independent of the parameter that is used to describe it. We want to comment on the occurrence of the proper time τ in Eq. (4.49) and some related issues.

Note that the second equality in Eq. (4.39) defining the action $\mathcal{A}$ does not use *any* parameters. One says that the action $\mathcal{A}$ is *re-parametrization invariant*. In proceeding from Eq. (4.40) to Eq. (4.41) we have introduced $d\tau$ by hand which appears in the final equation. Since $\mathcal{A}$ is re-parametrization invariant, and the geodesic is a geometric entity that is independent of the parametrization used to define it, we are allowed to use any other parameter λ, say, to describe the same curve. If the new parameter is expressed in terms of the old by a function $\tau = \tau(\lambda)$, we can convert the derivatives in Eq. (4.67) from λ to τ and – after some simple algebra – arrive at

$$\frac{d^2 x^i}{d\lambda^2} + \Gamma^i_{jk} \frac{dx^j}{d\lambda} \frac{dx^k}{d\lambda} = \left(\frac{d^2\tau/d\lambda^2}{d\tau/d\lambda} \right) \frac{dx^i}{d\lambda} \equiv f(\lambda) \frac{dx^i}{d\lambda}. \tag{4.68}$$

This is a slightly more general form of a geodesic equation which does not restrict the parameter that is used to describe the geodesic. If τ is a linear function of λ, the right hand side vanishes and we get back the same equation as before. A class of parameters related to each other by linear transformations that preserves the form of Eq. (4.67) are called *affine parameters*. Proper time is an affine parameter for timelike geodesics. Incidentally, our analysis also shows that the arc length of the curve is an affine parameter for the spacelike geodesics.

Equation (4.68) can be written as $u^a \nabla_a u^i = f(\lambda) u^i$ where $u^a = dx^a/d\lambda$. This equation has a simple physical interpretation. The left hand side gives the acceleration a^i of the trajectory; we would expect the curve to be non-accelerating when the spatial part of the acceleration vanishes in the instantaneous comoving frame. This can happen if $\boldsymbol{a}$ is in the direction of $\boldsymbol{u}$ which is precisely what the equation indicates.

Given any equation of this form, one can also make the inverse transformation from the parameter λ to another parameter μ, say, so that in terms of the new parameter the geodesic curve $x^i(\mu)$ satisfies the equation $u^a \nabla_a u^i = 0$, where $u^a = dx^a/d\mu$. The explicit form of this transformation is easily determined using Eq. (4.68):

$$\frac{d\mu}{d\lambda} = \exp \left(\int^\lambda f(\lambda') \, d\lambda' \right). \tag{4.69}$$

Therefore, one can always choose an affine parameter to describe a geodesic.

It is also possible to provide several other variational principles from which the geodesic equation with an affine parametrization can be obtained. Consider a modified action principle with

$$\mathcal{A}_1 = \int d\lambda \left[g_{ab}(x) \frac{dx^a}{d\lambda} \frac{dx^b}{d\lambda} \right]. \tag{4.70}$$

This is essentially a classical mechanics problem with λ playing the role of time and x^i playing the role of generalized coordinates. Variation of this action keeping $x^i(\lambda)$ fixed at the end points will lead to the Euler Lagrange equation

$\partial_a L = (d/d\lambda)(\partial L/\partial \dot{x}^a)$, which becomes

$$\frac{d}{d\lambda}\left(g_{ab}\frac{dx^a}{d\lambda}\right) = \frac{1}{2}(\partial_b g_{ij})\frac{dx^i}{d\lambda}\frac{dx^j}{d\lambda}. \tag{4.71}$$

This is identical to Eq. (4.42), provided λ is an affine parameter, showing that the variational principle based on $\mathcal{A}_1$ leads to the same equation of motion. If $Q = (g_{ab}\,\dot{x}^a\,\dot{x}^b)^{1/2}$, where the dot denotes the derivative with respect to an affine parameter, then our original choice in $\mathcal{A}$ was $L = Q$ and the choice in $\mathcal{A}_1$ corresponds to $L = Q^2$. There is, however, one crucial difference between the original action $\mathcal{A}$ and any other choice. The action $\mathcal{A}$, as mentioned before, is invariant under the re-parametrization $\tau \to \lambda = \lambda(\tau)$. It is not necessary to choose an affine parameter with $\mathcal{A}$, but more generalized actions like $\mathcal{A}_1$ will lead to the correct geodesic equation only when the affine parameter is used as a parameter.

It is trivial to verify that if a geodesic is timelike, null or spacelike at a given event, then it will continue to remain so. This arises from the result

$$\frac{d}{d\lambda}(u^a u_a) = u^j \nabla_j(u^a u_a) = 2u_a\,(u^j \nabla_j u^a) = 0, \tag{4.72}$$

where the last equality follows from the fact that u^a satisfies the geodesic equation. This shows that $u^a u_a$ is a constant along the geodesic.

A null geodesic will be the path taken by a light ray in a curved spacetime just as a timelike geodesic describes the path of a material particle. The issue of the parameter that is used to describe a geodesic equation becomes particularly important when the geodesic is null. We cannot use proper time as a valid parameter for a null geodesic since it vanishes along the path of the light ray. Hence null geodesics are, in general, described using an arbitrary parameter. That is, we define the null geodesics, in general, as the integral curves of a vector field $k^a(x)$ which satisfies $k^b \nabla_b k^a = f(x)k^a$ and $k^a k_a = 0$. However, if k^a is a null vector field, then $\mu(x)k^a$ will also be a null vector for an arbitrary function $\mu(x)$. Using this freedom, we can make $f(x) = 0$ in the geodesic equation, thereby describing the null geodesic with an affine parametrization corresponding to $k^b \nabla_b k^a = 0$ and $k^a k_a = 0$. It is usual to choose the parametrization such that k^a is the momentum of the photon, say, which is travelling on the null geodesic – unlike the case of a timelike geodesic in which mu^a gives the momentum. Note that, with this convention, the energy attributed to the particle moving in the trajectory $x^a(\lambda)$ by an observer with four-velocity U^a will be $E = -(dx^a/d\lambda)U_a$ irrespective of whether the trajectory is null or timelike.

An important example in which null geodesics occur with a *non*-affine parameter is in the case of – what are called – conformally related spacetimes. Two metrics are called conformally related if $g'_{ab} = \Omega^2(x)g_{ab}$. It is straightforward to show that the Christoffel symbols for these two metrics are related by

$$\Gamma_{bc}^{a}{}' = \Gamma_{bc}^{a} + \delta_{b}^{a}\,\partial_{c}\ln\Omega + \delta_{c}^{a}\,\partial_{b}\ln\Omega - g_{bc}\partial^{a}\ln\Omega. \tag{4.73}$$

The geodesic equation changes under such a conformal transformation. In general, if $x^{a}(\lambda)$ is a geodesic in the metric g_{ab}, it will not be a geodesic in the metric g_{ab}'. One exception to this rule are null geodesics, which will continue to remain as null geodesics under conformal transformations. But the null geodesic equation will now get modified to the form

$$\frac{d^{2}x^{a}}{d\lambda^{2}} + \Gamma_{bc}^{a}{}'\,\frac{dx^{b}}{d\lambda}\frac{dx^{c}}{d\lambda} = \frac{d^{2}x^{a}}{d\lambda^{2}} + \Gamma_{bc}^{a}\,\frac{dx^{b}}{d\lambda}\frac{dx^{c}}{d\lambda} + 2\frac{dx^{a}}{d\lambda}\frac{d\ln\Omega}{d\lambda} = 0. \tag{4.74}$$

This equation is in the form of Eq. (4.68) showing that $x^{a}(\lambda)$ is still a geodesic but the parameter λ is no longer an affine parameter in the new metric; but a transformation to a new parameter μ with $d\mu/d\lambda = \Omega^{-2}$ will reduce Eq. (4.74) to the geodesic equation with an affine parameter.

In the case of null geodesics in a *static* spacetime (with $g_{0\alpha} = 0$ and all other components independent of $x^{0} = t$), one can introduce another variational principle, which is a generalization of Fermat's principle to curved spacetime. Consider all null *curves* connecting two events $\mathcal{P}$ and $\mathcal{Q}$ in a static spacetime. Each null curve can be described by the three functions $x^{\alpha}(t)$ and will take a particular amount of coordinate time Δt to go from $\mathcal{P}$ to $\mathcal{Q}$. We will now show that the null *geodesic* connecting these two events extremizes Δt. To do this, we shall change the independent variable in Eq. (4.67) from the affine parameter λ to the coordinate time t by using the relation

$$0 = dt^{2} + \left(\frac{g_{\alpha\beta}}{g_{00}}\right)dx^{\alpha}\,dx^{\beta}. \tag{4.75}$$

This gives

$$g_{\beta\gamma}\frac{d^{2}x^{\gamma}}{dt^{2}} + \Gamma_{\beta\gamma\delta}\frac{dx^{\gamma}}{dt}\frac{dx^{\delta}}{dt} - \Gamma_{\beta00}\frac{g_{\gamma\delta}}{g_{00}}\frac{dx^{\gamma}}{dt}\frac{dx^{\delta}}{dt} + \frac{dt^{2}/d\lambda^{2}}{(dt/d\lambda)^{2}}g_{\beta\gamma}\frac{dx^{\gamma}}{dt} = 0. \tag{4.76}$$

Combining this with the zeroth component of the geodesic equation

$$\frac{(d^{2}t/d\lambda^{2})}{(dt/d\lambda)^{2}} = -2\,\Gamma_{0\gamma0}\frac{(dx^{\gamma}/dt)}{g_{00}} \tag{4.77}$$

and using the expression for Christoffel symbols in terms of the metric, one can show, after some straightforward algebra, that

$$H_{\beta\gamma}\frac{d^{2}x^{\gamma}}{dt^{2}} + \frac{1}{2}\,(\partial_{\delta}H_{\beta\gamma} + \partial_{\gamma}H_{\beta\delta} - \partial_{\beta}H_{\gamma\delta})\frac{dx^{\gamma}}{dt}\frac{dx^{\delta}}{dt} = 0, \tag{4.78}$$

where $H_{\alpha\beta} = -(g_{\alpha\beta}/g_{00})$. This is the same as a geodesic equation with an affine parameter t in a three-dimensional space with metric $H_{\alpha\beta}$. It follows that the null geodesics in a static spacetime can be obtained from the extremum principle for coordinate time

$$\delta \int dt = 0, \qquad (4.79)$$

which is the same as *Fermat's principle*.

This result shows that light does not travel in the three-dimensional space along a path of least *length* but travels along a path of least time. The difference arises because the gravitational field acts like a medium with a spatially varying refractive index producing an effective speed of light that is different from unity. To see this explicitly, consider a special case – which arises in several practical examples including all spherically symmetric spacetimes – in which $g_{\alpha\beta} = f^2(x^\alpha)\delta_{\alpha\beta}$ so that the three-dimensional space is conformally flat. In that case, along the light path, $dt = [f/\sqrt{|g_{00}|}]\, dl$, where $dl^2 = \delta_{\alpha\beta}\, dx^\alpha dx^\beta$ is the usual Cartesian line element. The Fermat principle is now equivalent to the statement that such a gravitational field acts like a medium with a refractive index $n(x) = f(x)/\sqrt{|g_{00}(x)|}$. In addition to the bending of light, such an effective refractive index will also lead to a time delay in the propagation of light rays. This delay, called *Shapiro time-delay* has been observationally verified (see Exercise 4.8).

Exercise 4.7
Non-affine parameter: an example Vary the action functional based on the Lagrangian

$$L = \left(-g_{ab}\frac{dx^a}{d\lambda}\frac{dx^b}{d\lambda}\right)^{1/2}, \qquad (4.80)$$

where λ is some *arbitrary* parameter. Show that the resulting equation has the form in Eq. (4.68) and identify $f(\lambda)$.

Exercise 4.8
Refractive index of gravity We will see in Chapter 7 that the spacetime around a star (like the Sun) can be expressed in the form

$$ds^2 = -\left(1 - \frac{2GM}{r}\right)dt^2 + \frac{dr^2}{\left(1 - \dfrac{2GM}{r}\right)} + r^2(d\theta^2 + \sin^2\theta d\phi^2), \qquad (4.81)$$

where M is the mass of the star. This is called the Schwarzschild metric which we will derive in Chapter 7.

(a) Show that one can convert the spatial part of the metric into a conformally flat form by making the coordinate transformation from r to ρ with

$$r = \rho\left[1 + \frac{GM}{2\rho}\right]^2. \qquad (4.82)$$

The ranges of coordinates are $2GM < r < \infty$ and $GM/2 < \rho < \infty$.

(b) Show that, in the new coordinates, the metric has an effective refractive index given by

$$n(x) = \frac{[1 + (GM/2|x|)]^3}{[1 - (GM/2|x|)]}. \tag{4.83}$$

The slowing down of light rays due to this 'refractive index' has been verified using the Cassini satellite when it was in opposition to Earth. A radio wave was sent from Earth (located at the radius r_1) to the satellite (located at the radius r_2) and was bounced back to Earth. Compute the line integral of $2ndl$ along a straightline path in isotropic coordinates and show that the time delay is given by

$$\Delta t = \frac{4GM}{c^3} \ln \left(\frac{4r_1 r_2}{b^2} \right), \tag{4.84}$$

where b is the impact factor of the light ray.

(c) The circumference of a circle of coordinate radius ρ, centred at the origin, in the metric in Eq. (4.81) is $2\pi\rho n(\rho)$. Show that this circumference has a minimum value at $r = 3M$. What does this imply for the existence of circular null geodesics in the spacetime?

Exercise 4.9
Practice with the Christoffel symbols The purpose of this exercise is to compute the Christoffel symbols in flat two-dimensional space in polar coordinates by different methods.

(a) Write down the coordinate transformation from (x, y) to (r, θ) and the metric in both coordinates. The Christoffel symbols vanish in the Cartesian coordinates. Use the explicit transformation law of the Christoffel symbols to obtain them in the polar coordinates.

(b) From the metric in the polar coordinates, compute the Christoffel symbols by direct differentiation.

(c) The geodesics in flat two-dimensional space are just straight lines. Choosing the geodesics judiciously and knowing the geodesic equation, determine the Christoffel symbols.

Exercise 4.10
Vanishing Hamiltonians In the text, we discussed two different variational principles: one based on $L_1 = (-g_{ab} \dot{x}^a \dot{x}^b)^{1/2}$ and the other based on $L_2 = -g_{ab} \dot{x}^a \dot{x}^b$ in obvious notation.

(a) Explain why both Lagrangians work under appropriate conditions. What is the general condition under which two Lagrangians L and $F(L)$ (where F is a monotonic function) will lead to the same equations of motion?

(b) Compute the canonical momenta $p_a = \partial L/\partial \dot{x}^a$ as well as the Hamiltonian $H = p_a \dot{x}^a - L$ for both these Lagrangians.

(c) Show that the Hamiltonian corresponding to L_1 vanishes identically for any trajectory while the Hamiltonian corresponding to L_2 is conserved when the equations of motion are satisfied.

(d) Explain why *any* Lagrangian, like L_1, which is invariant under the re-parametrization of the independent variable ('time') will lead to vanishing Hamiltonian. What does this mean physically?

Exercise 4.11
Transformations that leave geodesics invariant Express the geodesic equation as a differential equation for $x^\alpha(t)$. What is the most general transformation of the Christoffel symbols that will leave these equations invariant?

4.6 Covariant derivative

The analysis of the trajectory of a free particle has also provided us with a means of generalizing the concept of a partial derivative in a generally covariant manner. Taking the cue from Eq. (4.61), we will define the covariant derivative of any vector field $v^i(x)$ by the rule

$$\nabla_b v^a \equiv \partial_b v^a + \Gamma^a_{ib} v^i. \tag{4.85}$$

The fact that $\nabla'_a v'^i = \nabla_a v^i$ can be directly verified using Eq. (4.62) and Eq. (4.38). We will now introduce covariant derivative for several tensorial quantities using this basic definition by assuming that the covariant derivative operator of tensorial quantities should obey the 'chain rule' of differentiation.

To begin with, since the ordinary derivative of a scalar $\partial_a \phi$ does transform like a covariant vector, we define $\nabla_a \phi = \partial_a \phi$ for scalars. Next, since $(u^i u_i)$ is a scalar, we must have $\nabla_b(u^i u_i) = \partial_b(u^i u_i)$. Using the chain rule on both sides and the known form of $\nabla_b u^i$, we find that

$$\nabla_b u_i = \partial_b u_i - \Gamma^k_{ib} u_k. \tag{4.86}$$

This defines the covariant derivative of the covariant components of the vector.

The covariant derivatives of higher rank tensors are defined in a similar fashion by using these results and the chain rule. Consider, for example, the tensor $T^a_b = u^a v_b$. Using

$$\nabla_i T^a_b = \nabla_i(u^a v_b) = (\nabla_i u^a) v_b + u^a(\nabla_i v_b), \tag{4.87}$$

and the known forms for the covariant derivatives of vectors, one gets

$$\nabla_i T^a_b = \partial_i T^a_b + \Gamma^a_{ki} T^k_b - \Gamma^k_{bi} T^a_k. \tag{4.88}$$

This will be the definition for $\nabla_i T^a_b$ for an *arbitrary* T^a_b. A similar procedure is adopted for higher rank tensors.

Studying these expressions, we can provide the following 'rule' for covariant differentiation of any tensor which will be of practical use. (i) First write down the ordinary partial derivative of the tensor. (ii) This expression is to be 'corrected' by a set of terms with one correction term for each index of the tensor. The correction term has a positive sign if the index is a superscript (contravariant) and a negative sign if the index is a subscript (covariant). (iii) Each correction term will be a product of a Γ and the original tensor. The differentiating index will always be taken to be one of the lower indices of Γ. (iv) The index of the tensor that is being corrected will be replaced by a dummy index. One index of the Γ will be a dummy index contracting with the tensor index which is being corrected. There will be only one unique way of placing the remaining indices.

For these operations to be consistent, it is necessary that the covariant derivative of the metric tensor vanishes. This is because we have the relation

$$\nabla_i v_a = g_{ab} \nabla_i v^b = \nabla_i (g_{ab} v^b) = g_{ab} \nabla_i v^b + v^b \nabla_i g_{ab}. \tag{4.89}$$

The first equality must hold if the covariant derivative leads to a tensor so that by lowering the index b of the tensor $\nabla_i v^b$ we can obtain $\nabla_i v_b$. The last equality arises from applying the chain rule. Clearly the term with $\nabla_i g_{ab}$ has to vanish for consistency. Demanding $\nabla_i g_{ab} = 0$ and using our rules for covariant derivative we can write

$$\nabla_l g_{ik} = \partial_l g_{ik} - g_{mk} \Gamma^m_{il} - g_{im} \Gamma^m_{kl} = \partial_l g_{ik} - \Gamma_{kil} - \Gamma_{ikl}. \tag{4.90}$$

Direct substitution of the expression in Eq. (4.47) for the Christoffel symbols shows that this expression vanishes identically. Thus the whole procedure is internally consistent.

Finally, note that the covariant derivative provides the natural generalization of the directional derivative of any tensorial quantity in the direction of a vector. We will often use the notation $\nabla_{\mathbf{v}}(...)$ to indicate the tensorial object $v^i \nabla_i (...)$, where $(...)$ is any tensor.

Exercise 4.12

Accelerating without moving A particle dropped from a height towards the Earth's surface will follow a geodesic and hence will have zero proper acceleration $a^i \equiv u^j \nabla_j u^i = 0$. In contrast, a particle at rest on the Earth's surface will not be moving in a geodesic and hence will have $a^i \neq 0$. [This is in stark contrast with Newtonian terminology!]
 (a) Show that the four-velocity of a particle at rest ($x^\alpha = $ constant) in a static gravitational field will be $u^a = N^{-1} \delta^a_0$ where $g_{00} = -N^2$.
 (b) Show that its four-acceleration will be $a^i = (0, a^\alpha)$ with $a_\alpha = -\partial_\alpha N$. Compute its magnitude.

4.6.1 Geometrical interpretation of the covariant derivative

The concept of the covariant derivative and affine connection introduced in the previous sections has a very simple geometrical origin. While the geometrical meaning is not relevant for manipulation of expressions (which can be simply based on the 'rules' introduced in the previous section), it provides a deeper insight and is necessary to understand different geometrical structures in a more general manner. We

Metric tensor, geodesics and covariant derivative

shall first describe the geometrical meaning using the simple example of curvilinear coordinates in two-dimensional flat spacetime since the generalization to higher
dimensions is completely obvious.

Consider the two-dimensional flat space described in Cartesian as well as polar
coordinates with the line element

$$ds^2 = dx^2 + dy^2 = dr^2 + r^2 d\theta^2 \tag{4.91}$$

with $x = r\cos\theta$, $y = r\sin\theta$. The coordinate systems are shown in Fig. 4.1 with
two representative points P_1 and P_2. At P_1, we have the two unit vectors e_x and e_y
along the x- and y-axes. Let us assume that there is a vector field $v(x, y)$ defined
in the two-dimensional plane that could vary from point to point. (For example,
it could be an electric field which exists on the plane.) At P_1 the vector v can
be expanded in terms of the basis vectors as $v(P_1) = v^x(P_1)e_x + v^y(P_1)e_y$
which defines the two components $v^i(P_1)$, $i = (x, y)$ at the point P_1. At a different point P_2, we can do a similar expansion in terms of the basis vectors at that
point: $v(P_2) = v^x(P_2)e_x + v^y(P_2)e_y$. The Cartesian basis vectors do not vary on
the plane (i.e., the basis vector e_x, say, points to the same direction everywhere)
and hence the e_x, e_y in these two expansions can be taken to be the same. If we
now want to consider how the vector varies with respect to the x coordinate, say,
we can compute the derivative

$$\frac{\partial v}{\partial x} = \frac{\partial v^x}{\partial x} e_x + \frac{\partial v^y}{\partial x} e_y. \tag{4.92}$$

In this differentiation, e_x, e_y remain as constants.

Let us next consider the corresponding situation in curvilinear coordinates. The
unit vectors along the r and θ directions are indicated by e_r, e_θ. At P_1 the vector v can be expanded in terms of the basis vectors as $v(P_1) = v^r(P_1)e_r(P_1) +
v^\theta(P_1)e_\theta(P_1)$, which defines the two components $v^i(P_1)$, $i = (r, \theta)$ at the point
P_1. At a different point P_2, we can do a similar expansion in terms of the basis vectors at that point: $v(P_2) = v^r(P_2)e_r(P_2) + v^\theta(P_2)e_\theta(P_2)$. The crucial difference
between the Cartesian and curvilinear coordinates is that the basis vectors e_r, e_θ
point to different directions at P_1 and P_2 (i.e. the e_r at P_2 is not parallel to e_r at
P_1, etc.) as is obvious from Fig. 4.1. Therefore, the *components* of a vector v in the
curvilinear coordinates can vary from place to place due to two reasons. First, the
vector may be intrinsically different at two points. (For example, the electric field –
represented by a vector field – could be varying from place to place.) Second, the
components might reflect the change due to the basis vectors being different from
place to place. To see the second effect clearly, let us consider a vector v which is
a constant everywhere. This statement is intuitively obvious in flat space if we use
Cartesian coordinates. If the Cartesian components of a vector field v^x and v^y do
not depend on the coordinates x and y, we can call it a constant vector field, almost

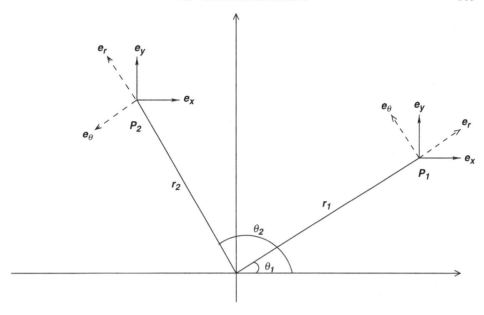

Fig. 4.1. Basis vectors at two points P_1 and P_2 in Cartesian and polar coordinates in the plane.

by definition. Pictorially, such a vector field will be described by a set of 'arrows' on the plane that are parallel to each other. But if the same constant vector field is expanded in terms of the basis vectors in polar coordinates, then it will *not* have constant components. When we write $v = v^r e_r + v^\theta e_\theta$ the components v^r and v^θ have to change in such a manner as to cancel the effect of e_r and e_θ varying from place to place, thereby keeping the vector v pointing in the same direction.

To make this idea a bit more formal and to discuss the general case of the N-dimensional spacetime, let us write the vector as $v = v^i e_i$, where e_i now denotes the basis vectors in some system of coordinates and v^i denotes the components. We now want to introduce an operator ∇_j which acts on any tensorial object to produce a new tensorial object and has the following properties: (i) it obeys the chain rule; (ii) it acts linearly; and (iii) it reduces to the ordinary partial derivative ∂_j while acting on functions. Then, operating on a vector field, it produces

$$\nabla_j v = \nabla_j (v^i e_i) = (\partial_j v^i) e_i + v^i \nabla_j e_i. \tag{4.93}$$

(We are using the symbol ∇ in anticipation of the fact that it will turn out to be the covariant derivative operator introduced in the previous section.) The second equality arises from using the chain rule and the fact that $v^i(x)$ are just functions as far as the operator ∇_j is concerned. In the final expression, the first term is due to intrinsic variation of the vector field from place to place and will exist even in flat spacetime Cartesian coordinates. The second term describes the effect due to the variation of the basis vectors from place to place. This term is absent in flat spacetime if we use a Cartesian coordinate system but will be present in any curvilinear coordinate system in flat spacetime and, of course, in curved spacetime. Since $\nabla_j \boldsymbol{e}_i$ is another vector we can expand it in the same basis and write:

$$\nabla_j \, \boldsymbol{e}_i \equiv \Gamma_{ij}^k \, \boldsymbol{e}_k, \tag{4.94}$$

which defines a set of quantities Γ_{ij}^k in this approach. (Again, we use the symbol Γ in anticipation of the next result.) Substituting this into Eq. (4.93) and changing the dummy index in the first term to k we get

$$\nabla_j \boldsymbol{v} = (\partial_j v^k + \Gamma_{ij}^k \, v^i) \, \boldsymbol{e}_k = (\nabla_j v^k) \, \boldsymbol{e}_k. \tag{4.95}$$

We can now think of the left hand side as a vector whose components are the covariant derivatives we introduced earlier with the identification of the symbols Γ_{ij}^k with the Christoffel symbols.

The rest of the formalism generalizes easily. Given the basis vectors $\boldsymbol{e}_i$ with components $e_i^k = \delta_i^k$, we can introduce a dual basis (see Section 1.6 and Section 4.3) $\boldsymbol{w}^k$ with the components $w_i^k = \delta_i^k$ and a scalar product between the original vector space and its dual by $\langle \boldsymbol{w}^k, \boldsymbol{e}_j \rangle = \delta_j^k$. The dual basis allows one to define the covariant components of the vector by taking $\boldsymbol{v} = v_i \boldsymbol{w}^i$, etc. Taking the covariant derivative of this expression and using Eq. (4.94), we find that

$$\nabla_j \boldsymbol{w}^i = -\Gamma_{kj}^i \boldsymbol{w}^k. \tag{4.96}$$

This ensures that $\nabla_i u_a = \partial_i u_a - \Gamma_{ai}^k u_k$ for the covariant components of a vector.

Once we have the set $\boldsymbol{e}_i, \boldsymbol{w}^j$, any tensorial object can be expanded in the direct product basis to get the components. For example, a third rank tensor with two contravariant and one covariant index can be defined through the relation:

$$\boldsymbol{T} = T^{ab}_{c} \, \boldsymbol{e}_a \otimes \boldsymbol{e}_b \otimes \boldsymbol{w}^c. \tag{4.97}$$

Here the left hand side is treated as a geometric object that has the components in a basis as shown in the right hand side. The action of ∇_j on this object can again be computed by the standard chain rule, in terms of $\nabla_j \boldsymbol{e}_i$ and $\nabla_j \boldsymbol{w}^i$. This will give

$$\begin{aligned}
\nabla_j \boldsymbol{T} &= \partial_j T^{ab}_{c} \boldsymbol{e}_a \otimes \boldsymbol{e}_b \otimes \boldsymbol{w}^c + T^{ab}_{c} (\nabla_j \boldsymbol{e}_a) \otimes \boldsymbol{e}_b \otimes \boldsymbol{w}^c \\
&\quad + T^{ab}_{c} \boldsymbol{e}_a \otimes (\nabla_j \boldsymbol{e}_b) \otimes \boldsymbol{w}^c + T^{ab}_{c} \boldsymbol{e}_a \otimes \boldsymbol{e}_b \otimes (\nabla_j \boldsymbol{w}^c) \\
&= \left(\partial_j T^{ab}_{c} + T^{ib}_{c} \Gamma_{ij}^a + T^{ai}_{c} \Gamma_{ij}^b - T^{ab}_{i} \Gamma_{cj}^i \right) \boldsymbol{e}_a \otimes \boldsymbol{e}_b \otimes \boldsymbol{w}^c, \tag{4.98}
\end{aligned}$$

clearly showing the origin of the 'rules' we gave in the previous section.

To summarize, the covariant derivative takes into account the fact that basis vectors will, in general, vary from point to point on a manifold. To define a generally covariant concept of a derivative of a tensorial quantity we need to take into account the variation of its components as well as the variation of the basis vectors. In a flat spacetime, the variation of basis vectors arises when we use curvilinear coordinates for some mathematical or physical reason. (For example, in three dimensions, a particular problem might be easier to tackle in spherical polar coordinates rather than in Cartesian coordinates; in spacetime, we may want to work in a coordinate system which is appropriate to an accelerated observer, etc.) In this case, the dependence of the metric on the coordinates – leading to nonzero values for Γ^a_{bc} – arises due to our choice and one can, if needed, go back to Cartesian coordinates. But we have seen earlier that, in curved spacetime, it is impossible to choose the coordinates such that the metric will reduce to the Cartesian form globally. So, in a curved spacetime, the basis vectors will – of necessity – vary from point to point in the manifold and we cannot make Γ^a_{bc} vanish globally. In this case, we need to use a covariant derivative instead of an ordinary derivative to maintain the covariance of the equations under arbitrary coordinate transformations.

4.6.2 Manipulation of covariant derivatives

The covariant derivative operator ∇_j will replace the ordinary derivative operator ∂_j when we generalize many of the concepts from special relativity in inertial coordinates to either curvilinear coordinates or curved spacetime. We shall describe several mathematical aspects of this operator and the resulting expressions in this section.

The derivatives of the metric tensor can be expressed in terms of the Christoffel symbols. From the definition, it follows that

$$\partial_l g_{ik} = \Gamma_{kil} + \Gamma_{ikl}, \tag{4.99}$$

which expresses the partial derivatives in terms of Christoffel symbols. From the equation $\nabla_l g^{ik} = 0$, we can also express the derivatives of g^{ik} in terms of Christoffel symbols

$$\partial_l g^{ik} = -\Gamma^i_{ml} g^{mk} - \Gamma^k_{ml} g^{im}. \tag{4.100}$$

The calculation of the covariant derivatives can be simplified by using two identities which we now derive. From the definition of Γ^i_{km} it follows that

$$\Gamma^a_{ba} = \frac{1}{2} g^{ad} \partial_b g_{ad} = \frac{1}{2g} \partial_b g = \partial_b(\ln \sqrt{-g}), \tag{4.101}$$

where we have used Eq. (4.34). Similarly,

$$g^{bc}\Gamma^a_{bc} = g^{bc} g^{ad} \left(\partial_c g_{db} - \frac{1}{2} \partial_d g_{bc} \right) = -\frac{1}{\sqrt{-g}} \partial_b(\sqrt{-g}\, g^{ab}). \tag{4.102}$$

These results allow us to express covariant derivatives of antisymmetric and symmetric tensors in a simple form. For an antisymmetric tensor Q^{ab}, we have

$$\nabla_b Q^{ab} = \partial_b Q^{ab} + \Gamma^a_{db} Q^{db} + \Gamma^b_{db} Q^{ad} = \partial_b Q^{ab} + \Gamma^b_{db} Q^{ad} \tag{4.103}$$

since $\Gamma^a_{db} Q^{db} = 0$ for an antisymmetric Q^{db}. Using Eq. (4.101), we get

$$\nabla_b Q^{ab} = \frac{1}{\sqrt{-g}} \partial_b(\sqrt{-g} Q^{ab}). \tag{4.104}$$

This shows that one can compute the covariant derivatives of an antisymmetric tensor directly from the metric without first having to compute the Christoffel symbols. The symmetry of the Christoffel symbols in the lower two indices also shows that, for any vector field V_a,

$$\nabla_b V_a - \nabla_a V_b = \partial_b V_a - \partial_a V_b \tag{4.105}$$

since the term involving Γs cancels out in this expression. So the antisymmetric part of the covariant derivative of any vector field is the same as the ordinary derivative. In fact, these results generalize to any completely antisymmetric tensor $Q^{abc\cdots} = Q^{[abc\cdots]}$. This is because, in the computation of $\nabla_a Q^{abc\cdots}$, only one term with Christoffel symbols (the one with $\Gamma^a_{ja} Q^{jbc\cdots}$) will survive since all others will involve the contraction of two symmetric indices of the Christoffel symbols with antisymmetric indices in the tensor.

Consider next a symmetric tensor T^{ab} for which we have the result,

$$\nabla_k T^k_i = \partial_k T^k_i + \Gamma^k_{lk} T^l_i - \Gamma^l_{ik} T^k_l = \frac{1}{\sqrt{-g}} \partial_k \left(\sqrt{-g} T^k_i \right) - \Gamma^l_{ki} T^k_l. \tag{4.106}$$

Expanding out Γ^l_{ki} and using the symmetry of T^{kl}, we get

$$\nabla_k T^k_i = \frac{1}{\sqrt{-g}} \partial_k \left(\sqrt{-g} T^k_i \right) - \frac{1}{2} (\partial_i g_{kl}) T^{kl}. \tag{4.107}$$

This expression is not as simple as the one for antisymmetric tensors but is still easier to use than the basic definition because we do not need to compute the Christoffel symbols.

We shall now take up the generalization of the concept of covariant divergence of a vector field to curved spacetime. In special relativity, in Cartesian coordinates, the covariant divergence is defined as $\partial_i A^i$, which generalizes to a curvilinear coordinates as $\nabla_i A^i$. From the definition of the covariant derivative we get

$$\nabla_i A^i = \partial_i A^i + \Gamma^i_{ci} A^c = \partial_i A^i + A^c \partial_c (\ln \sqrt{-g}) = \frac{1}{\sqrt{-g}} \partial_i(\sqrt{-g} A^i), \tag{4.108}$$

where we have used Eq. (4.101). This structure is identical to the covariant divergence for an antisymmetric tensor obtained in Eq. (4.104). Further, if A_i was a gradient of some function ϕ, so that $A_i = \partial_i \phi$, $A^i = g^{ik} \partial_k \phi \equiv \partial^i \phi$, then $\nabla_a A^a$ will represent the covariant Laplacian $\nabla_a \nabla^a \phi$ of the scalar ϕ. We see that

$$\nabla_i \nabla^i \phi = \frac{1}{\sqrt{-g}} \partial_i \left(\sqrt{-g} g^{ik} \partial_k \phi \right). \tag{4.109}$$

The above considerations remain valid in any dimension and for metrics with any signature with $\sqrt{-g}$ interpreted as $\sqrt{|g|}$. In fact, these results are routinely used in three-dimensional vector analysis in the context of spherical polar coordinates. If the metric is taken to be

$$ds^2 = dr^2 + r^2 (d\theta^2 + \sin^2 \theta \, d\phi^2) \tag{4.110}$$

the Laplacian operator in Eq. (4.109) becomes

$$\nabla_\alpha \nabla^\alpha f = \frac{1}{r^2 \sin \theta} \partial_\alpha \left(r^2 \sin \theta \, g^{\alpha\beta} \partial_\beta f \right)$$
$$= \frac{1}{r^2} \partial_r (r^2 \partial_r f) + \frac{1}{r^2 \sin \theta} \partial_\theta (\sin \theta \, \partial_\theta f) + \frac{1}{r^2} \partial_\phi^2 f, \tag{4.111}$$

which should be a result familiar from standard vector analysis. (The same is true as regards Eq. (4.108) except that the θ and ϕ components of a vector in spherical polar coordinates (v^θ, v^ϕ) are usually defined with extra factors of r and $r \sin \theta$ by convention; such a complication does not arise in the defintion of a Laplacian.)

Using the definition of divergence, we can generalize the Gauss theorem to the curved spacetime in a straightforward manner. Consider a region of spacetime $\mathcal{V}$ with a boundary $\partial \mathcal{V}$. Since the proper volume element is now $\sqrt{-g} d^4 x$, we have the result

$$\int_\mathcal{V} \sqrt{-g} \, d^4 x \, (\nabla_i J^i) = \int_{\partial\mathcal{V}} |h|^{1/2} d^3 y \, (n_i J^i), \tag{4.112}$$

where h is the determinant of the induced metric (see Eq. (4.31)) on the surface $\partial \mathcal{V}$, which is given in parametric form, as $x^i = x^i(y^\alpha)$, and n_i is the normal to the surface.

As we mentioned in the context of special relativity (see page 28), the above result holds even if the J^is are not components of a four-vector but just a set of four functions. In that case, the integral over the coordinates $d^4 x$ (without the $\sqrt{-g}$ factor) of the quantity $\sqrt{-g} \, J^i$ will be given by $\sqrt{-g} \, n_i J^i$ evaluated on the boundary of region of integration. Of course, when the J^is are not components of a vector, the result of integration will not be generally covariant.

Exercise 4.13

Covariant derivative of tensor densities The objects of the form $\sqrt{-g} \, \boldsymbol{T}$, where $\boldsymbol{T}$ is tensor of arbitrary rank, are called a *tensor densities*. Show that ϵ_{abcd} and $\sqrt{-g}$ have vanishing covariant derivatives. Therefore one has, for example, $\nabla_a(\sqrt{-g}\, v^i) = \sqrt{-g}\, \nabla_a v^i$, etc. When one thinks of $\sqrt{-g}\, v^i$ as a single entity, it is sometimes useful to write the expansion of $\nabla_a(\sqrt{-g}\, v^i)$ in terms of $\partial_a(\sqrt{-g}\, v^i)$ and correction terms. Find a suitable rule for calculating the covariant derivatives of such tensor densities in terms of the partial derivative and correction terms.

In later chapters, we will need the notion of the spacetime being foliated by a family of hypersurfaces. One simple example of this could be a sequence of spacelike surfaces parametrized by a variable t that can generalize the notion of constant time slice in special relativity. We can define the normal to this sequence of hypersurfaces everywhere in spacetime, thereby obtaining a vector field $n_i(x)$. Such vector fields are called *hypersurface-orthogonal vector fields.*

We can show that such a hypersurface-orthogonal vector field, $n_i(x)$, will satisfy the condition $n_{[l}\nabla_n n_{m]} = 0$. If a vector field n_a is orthogonal to a family of hypersurfaces given by $\Phi(x) = $ constant, it can be expressed in the form

$$n_a = -\mu \partial_a \Phi \tag{4.113}$$

with some function μ. Differentiating this relation gives $\nabla_b n_a = -\mu \nabla_b \nabla_a \Phi - \partial_a \Phi \partial_b \mu$. We now construct the combination:

$$n_{[c}\nabla_b n_{a]} \equiv \frac{1}{3!}\left(n_c \nabla_b n_a + n_b \nabla_a n_c + n_a \nabla_c n_b - n_c \nabla_a n_b - n_b \nabla_c n_a - n_a \nabla_b n_c\right) \tag{4.114}$$

which, on explicit evaluation using Eq. (4.113) and the result $\nabla_b \nabla_a \Phi = \nabla_a \nabla_b \Phi$, will vanish. Thus we conclude that any vector field which is hypersurface orthogonal must satisfy the condition

$$n_{[c}\nabla_b n_{a]} = 0 \tag{4.115}$$

This result and its converse, which is also true, are called the *Frobenius theorem.*

4.7 Parallel transport

In Cartesian coordinates, a vector field will be considered as constant if its components with respect to the coordinates are the same everywhere. In this case, the vector field can be thought of as being 'parallel to itself' at every point in space. This is a consistent definition in Cartesian coordinates because the basis vectors of the Cartesian coordinates themselves are parallel to the axes everywhere.

If we use curvilinear coordinates in flat spacetime, the basis vectors change from point to point and the notion of a constant vector is more difficult to define; however, in this case, one can circumvent the problem by first introducing the Cartesian coordinates globally, defining a constant vector using the Cartesian components and transforming back to the curvilinear coordinates. In general, to define a constant vector, we need a well-defined procedure to move a vector 'parallel to itself' from one point to another along a curve. When Cartesian coordinates can be introduced globally, this can be achieved by moving the vector keeping its components constant with respect to the Cartesian coordinates. This agrees with our intuitive notion of a parallel transport.

But in a curved spacetime, it will not be possible to introduce Cartesian coordinates globally. Therefore, the above procedure for parallel transport of a vector fails. It is necessary to generalize this concept in a manner that will remain valid in a curved spacetime.

The covariant derivative, defined in the previous section, can be used to provide a natural definition for parallel transport which is valid in an arbitrary spacetime. To arrive at this, let us consider a parametrized curve $x^i(\lambda)$ passing through an event $\mathcal{P}$ in the spacetime. Let the coordinates of $\mathcal{P}$ be $x^i(0)$. If $v^j(x)$ is a vector field, then we can interpret the quantity

$$\left(\frac{dx^b}{d\lambda}\right)\nabla_b v^a = \left(\frac{dx^b}{d\lambda}\right)\frac{\partial v^a}{\partial x^b} + \Gamma^a_{ib}v^i\left(\frac{dx^b}{d\lambda}\right) = \frac{dv^a}{d\lambda} + \Gamma^a_{ib}v^i\left(\frac{dx^b}{d\lambda}\right) \quad (4.116)$$

as the generalization of the *directional derivative* of the vector field along a particular direction, specified by the tangent vector $(dx^b/d\lambda)$. Once the curve $x^i(\lambda)$ is specified, we can compute Γ^a_{ib} at any point on the curve and treat it as a function of λ; that is, along the curve, $\Gamma^a_{ib}(x) = \Gamma^a_{ib}(\lambda)$. Suppose we are given a vector k^a at one event $\mathcal{P}$ and some curve $x^a(\lambda)$ passing through $\mathcal{P} = x^a(0)$. We can now solve the differential equation

$$\frac{dv^a}{d\lambda} + \Gamma^a_{ik}(\lambda)\frac{dx^i}{d\lambda}v^k = 0 \quad (4.117)$$

with the boundary condition $v^a(\lambda = 0) = k^a$, say, and obtain a vector field $v^a(\lambda)$ everywhere on the curve. (Since Eq. (4.117) is a first order differential equation, it only requires the value of the vector field at $\lambda = 0$ as the initial condition.)

The solution $v^a(\lambda)$ thus allows us to *define* a vector *field* along the curve $x^a(\lambda)$, given a vector k^a only at *one point*. Equation (4.117) is equivalent to demanding that the directional derivative of the vector vanishes along the tangent vector to a given curve. If the spacetime is actually flat, then this construction is equivalent to 'moving' the vector from event to event maintaining the same Cartesian components; i.e. the vector is moved 'parallel' to itself. Therefore, the above construction generalizes the notion of 'parallel transport' to curved spacetime. It is easy to verify from the chain rule that parallel transport maintains the values of the dot products:

$$\nabla_u(\mathbf{a}\cdot\mathbf{b}) = \mathbf{a}\cdot\nabla_u\mathbf{b} + \mathbf{b}\cdot\nabla_u\mathbf{a} = 0, \quad (4.118)$$

since $\nabla_u\mathbf{b} = \nabla_u\mathbf{a} = 0$.

The idea of parallel transport gives an alternative interpretation to the geodesic Eq. (4.60) as determining a 'straight' path between two events. In flat space, a 'straight' line can be defined in two ways: (i) it is the shortest distance between two points; or (ii) it is that curve for which the tangent vector always points in the same direction: i.e. the tangent vector moves parallel to itself as we move along the curve. The generalization of (i) to curved space was obtained in Eq. (4.39) to Eq. (4.49) where we extremized the interval between the events; we found that the 'straight line' in curved space satisfies the Eq. (4.49). To generalize (ii) to curved space, we should demand that 'straight' paths in a curved space are those curves $x^i(\lambda)$ for which the tangent vector $u^i = (dx^i/d\lambda)$ should be parallel transported along the same curve. This requires $u^a \nabla_a u^i = 0$, which is indeed the same as Eq. (4.49) showing that both the definitions (i) and (ii) coincide in this formalism. In our approach, we have started from a metric $g_{ab}(x)$ and have defined Γ^k_{ij} through Eq. (4.48). More formally, one can consider manifolds in which a set of Γ^k_{ij} are given without any metric having been specified. In that case, the idea of parallel transport based on criterion (ii) can still be introduced since the equation $u^a \nabla_a u^i = 0$ requires only the Christoffel symbols. To define the length of a curve, we need the notion of a metric and criterion (i) can be used only when the metric is introduced in a manifold. This distinction, however, is irrelevant for us because of the physical, metric based approach we have adopted.

One can also understand the geodesic equation with a general parametrization (see Eq. (4.68)) in terms of the notion of parallel transport. This provides an alternative way of understanding the origin of the term $f(\lambda)u^i$ in the general parametrization. Consider two neighbouring points P and P' on the geodesic curve corresponding to the parametric values λ and $\lambda + d\lambda$. The tangent vectors at P and P' are given by $(dx^i/d\lambda)$ and $[dx^i/d\lambda + (d^2x^i/d\lambda^2)d\lambda]$. On the other hand, if we parallel transport the tangent vector $u^i = (dx^i/d\lambda)$ from P to P', then it will change by an amount $\delta u^i = -\Gamma^i_{jk}u^k u^j d\lambda$. The curve will be a geodesic only if the components of this vector are parallel to the tangent vector at P'. Clearly, the proportionality factor will differ from unity by a quantity which is first order in $d\lambda$ which we can write in the form $(1 + f(\lambda)d\lambda)$. Hence we demand,

$$u^m + \frac{du^m}{d\lambda}d\lambda = [1 + f(\lambda)d\lambda][u^m - \Gamma^m_{ps}u^p u^s d\lambda]. \tag{4.119}$$

Simplifying this, one is again led to Eq. (4.68).

Exercise 4.14
Parallel transport on a sphere The vector tangent to the curve $\phi = 0$, located at the north pole of a 2-sphere, is parallel transported along the following curves: (i) along $\phi = 0$ from

the north pole to the point $\theta = \theta_0$; (ii) along the curve $\theta = \theta_0$ from $\phi = 0$ to $\phi = \phi_0$; and (iii) along the curve $\phi = \phi_0$ from $\theta = \theta_0$ to the north pole. Calculate the angle through which the original vector has rotated.

[Note. This result shows that a vector could change when parallel transported around a closed curve. In other words, the vector we obtain at a point $\mathcal{P}$ by parallel transporting a vector from $\mathcal{Q}$, say, will – in general – depend on the curve which is used. We will discuss this feature in detail in the next chapter.]

4.8 Lie transport and Killing vectors

The parallel transport introduced in the previous section allows one to define a vector field at all points on a curve if the vector is known at a given point on the curve. This construction is a natural generalization of the intuitive idea of moving a vector parallel to itself to curved spacetime. To do this, we need to know the Christoffel symbols along the curve.

There is another way of transporting a vector along the curve which does not require the knowledge of Christoffel symbols but requires only another vector field defined on the spacetime. While this concept can also be introduced quite formally, it will be useful to start with a physical context which will lead to this idea.

Let us recall from Eq. (4.42) that the covariant component of a velocity u_l (and the corresponding momentum $p_l = mu_l$) is a constant along the geodesic if the metric tensor does not depend on the coordinate x^l. Such a conserved quantity arises from an underlying symmetry, viz. the translational invariance of the metric along the x^l direction. For example, if the metric tensor is independent of time $t = x^0$, the p_0 is conserved. Such conservation laws will be very useful, for example, in solving the geodesic equation in a given spacetime.

There is, however, one difficulty brought about by the freedom in the choice of (arbitrary) coordinate systems. It is possible that the metric g_{ab} is independent of t in a particular coordinate system; but if one makes an arbitrary coordinate transformation to a new set of coordinates, then the metric will, in general, depend on the new time coordinate. If we are working with the new coordinates it will be difficult to figure out the underlying symmetry. The situation is somewhat like attempting to solve the Kepler problem in, say, an (x, y, z) coordinate system in which the conservation of angular momentum is not apparent. In fact, saying that the metric is independent of a particular coordinate is clearly not a covariant statement. It is important that we arrive at a geometrical concept which can capture the symmetries of the metric tensor (or any other tensor) in a coordinate independent way. It is possible to do this along the following lines.

Let us consider a spacetime in which a vector field $u^i(x)$ is defined in some region. The *integral curves* to this vector field are defined to be a set of curves

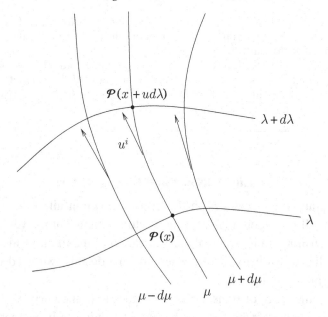

Fig. 4.2. A congruence of curves $x^i(\lambda, \mu)$ and the geometrical notion of sliding along the congruence is illustrated.

$x^i(\lambda, \mu)$ which satisfies the condition $\partial x^i/\partial \lambda = u^i(x)$ at each point. In this notation, the parameter μ specifies the curve we are considering and the parameter λ defines the location along a curve. We will assume that these curves fill a region of spacetime and there is precisely one curve passing through each event in that region. (Such a set of curves is called a *congruence*.) A congruence provides a way of mapping the spacetime onto itself by sliding each event along the integral curve of a vector field to a new point (see Fig. 4.2). For an infinitesimal sliding, an event x^i will move to an event $x^i + u^i d\lambda$ along the curve. Such a sliding of the spacetime does two separate things. First, treated as a coordinate transformation, it will make the components of any tensorial object (like the metric tensor) change by the usual rules. Second, it will attribute different coordinate labels to the events in the spacetime. We can now compare how the *functional* form of a tensorial quantity changes under the action of these two effects by considering this quantity at the same numerical value of the coordinates. If the functional form does not change, then we have a geometrical way of describing its invariance.

For example, consider the notion that the metric tensor is independent of time. Let us introduce a vector field which points at every location of the spacetime along the time axis. One can define a set of integral curves such that this vector field is a tangent to these curves at every event. The intuitive notion of the metric being independent of time can be made precise by introducing a notion which allows the spacetime to slide along these integral curves. The metric will not change during this sliding which will provide a geometrical notion of the symmetry. We will now make these ideas precise.

Consider a curve $\mathcal{C}$ with the parametrization $x^a(\lambda)$ and the tangent vector $u^a = (dx^a/d\lambda)$. Let $\mathcal{P}(x^a)$ and $\mathcal{Q}(x^a + dx^a)$ be two points on $\mathcal{C}$ with infinitesimally different coordinates. Another vector field v^i is defined in the spacetime in the neighbourhood of $\mathcal{C}$. We now introduce an infinitesimal coordinate transformation to a new set of primed coordinates with $x'^a = x^a + u^a d\lambda$. Under this transformation, the vector field v^a changes to

$$v'^a(x') = \frac{\partial x'^a}{\partial x^b} v^b(x) = (\delta_b^a + d\lambda\, \partial_b u^a) v^b(x) = v^a(x) + d\lambda\, v^b(x)\partial_b u^a. \quad (4.120)$$

Since $x'^a = x^a + u^a d\lambda = x^a + dx^a$ can also be interpreted as the coordinates of $\mathcal{Q}$, we can interpret the above equation as

$$v'^a(\mathcal{Q}) = v^a(\mathcal{P}) + d\lambda\, v^b(\mathcal{P})\partial_b u^a. \quad (4.121)$$

On the other hand, the original value of the vector field v^a at $\mathcal{Q}$ can be expressed as

$$v^a(\mathcal{Q}) = v^a(x + dx) = v^a(x) + dx^b \partial_b v^a(x) = v^a(\mathcal{P}) + d\lambda\, u^b \partial_b v^a(\mathcal{P}). \quad (4.122)$$

The crucial point is that $v'^a(\mathcal{Q})$ and $v^a(\mathcal{Q})$ will not – in general – be equal. Their difference can be used to define the *Lie derivative* of the vector v^a along the curve $\mathcal{C}$ as:

$$\pounds_u v^a(\mathcal{P}) \equiv \lim_{d\lambda \to 0} \frac{v^a(\mathcal{Q}) - v'^a(\mathcal{Q})}{d\lambda} \quad (4.123)$$

or

$$\pounds_u v^a \equiv u^b \partial_b v^a - v^b \partial_b u^a = u^b \nabla_b v^a - v^b \nabla_b u^a. \quad (4.124)$$

The second equality follows from the fact that terms involving the Christoffel symbols cancel out in the expression. It also shows $\pounds_u v^a$ is a generally covariant vector. In the abstract, index-free, notation we will denote the Lie derivative by $\pounds_{\boldsymbol{u}}\mathbf{v}$. If $\pounds_{\boldsymbol{u}}\mathbf{v} = 0$, we say that $\mathbf{v}$ is *Lie transported* along the integral curves of $\boldsymbol{u}$.

It is possible to obtain some additional insight into the concept of the Lie derivative by relating it to the commutator of two vector fields. Consider two vector fields $\boldsymbol{u}$ and $\mathbf{v}$ in a spacetime with corresponding integral curves parameterized by λ and μ such that one can set up the correspondence $\boldsymbol{u} \leftrightarrow u^i \partial_i = d/d\mu$ and $\mathbf{v} \leftrightarrow v^i \partial_i = d/d\lambda$. (The idea of associating a directional derivative operator with a vector field was introduced in Section 1.6.) Figure 4.3 shows a region in the neighbourhood of an event $\mathcal{P}_0$. We want to compute the difference in the value of some

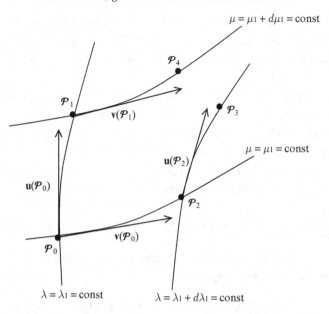

Fig. 4.3. The relation between $\mathcal{L}_{\boldsymbol{u}}\boldsymbol{v}$ and $[\boldsymbol{u}, \boldsymbol{v}]$.

function when evaluated at the events $\mathcal{P}_3$ and $\mathcal{P}_4$. Writing

$$f(\mathcal{P}_4) - f(\mathcal{P}_3) = [f(\mathcal{P}_4) - f(\mathcal{P}_1)] + [f(\mathcal{P}_1) - f(\mathcal{P}_0)]$$
$$- [f(\mathcal{P}_2) - f(\mathcal{P}_0)] - [f(\mathcal{P}_3) - f(\mathcal{P}_2)] \qquad (4.125)$$

we see that the term in the first square bracket can be expressed – correct to quadratic order – as $v^i \partial_i f + (1/2) v^i v^j \partial_i \partial_j f$ evaluated at $\mathcal{P}_1$. (We have not bothered to display the infinitesimal factors $d\lambda, d\mu$ for notational simplicity and will treat the vectors themselves as infinitesimal.) Similarly, the term in the second square bracket is given by $u^i \partial_i f + (1/2) u^i u^j \partial_i \partial_j f$ evaluated at $\mathcal{P}_0$. The term in the third square bracket is given by $v^i \partial_i f + (1/2) v^i v^j \partial_i \partial_j f$ evaluated at $\mathcal{P}_0$, while the term in the fourth square bracket is given by $u^i \partial_i f + (1/2) u^i u^j \partial_i \partial_j f$ evaluated at $\mathcal{P}_2$. Each of these terms has a simple interpretation in terms of sliding of vectors along the congruences as can be seen from Fig. 4.3. Combining them and simplifying we find that

$$f(\mathcal{P}_4) - f(\mathcal{P}_3) = \left(u^b \partial_b v^a - v^b \partial_b u^a \right) \partial_a f = (\mathcal{L}_{\boldsymbol{u}}\boldsymbol{v}) f = [\boldsymbol{u}, \boldsymbol{v}] f. \qquad (4.126)$$

In this relation we are interpreting all vector fields using their correspondence with directional derivatives. The second equality shows that the Lie derivative of the vector fields is closely related to the sliding of the vectors, as indicated in Fig. 4.3; its equality with the commutator of two vector fields confirms this and also shows that geometrically both these quantities indicate the 'gap between the tips of the vectors' when $\mathbf{u}$ is transported in the $\mathbf{v}$ direction and $\mathbf{v}$ is transported in the $\mathbf{u}$ direction.

An immediate consequence of this result, which will be very useful in future discussions, is the following. Consider the integral curves of a vector field which form a congruence. We can define a 'connecting vector field' in such a congruence, as follows. Let $x^i(\lambda, \mu)$ be the congruence with μ specifying the curve and λ specifying the location on the curve. Then $u^i = (\partial x^i/\partial\lambda)$ at a given μ denotes the tangent vector while $k^i = (\partial x^i/\partial\mu)$ at a given λ denotes the 'connecting vector' which connects two infinitesimally separated curves (labelled by μ and $\mu + d\mu$) in the congruence at the same value of λ. Treated as directional derivative operators, $\mathbf{u} \leftrightarrow u^i\partial_i = \partial/\partial\lambda$ and $\mathbf{k} \leftrightarrow k^i\partial_i = (\partial/\partial\mu)$. Obviously,

$$\pounds_{\mathbf{u}}\mathbf{k} = [\mathbf{u}, \mathbf{k}] = [(\partial/\partial\lambda), (\partial/\partial\mu)] = 0. \tag{4.127}$$

This shows that any connecting vector is Lie transported along a congruence.

Given the Lie derivative of the vector, it is easy to obtain the Lie derivative of all other tensorial objects by using exactly the same procedure we followed for the case of parallel transport. We first define the Lie derivative of a scalar to be $\pounds_{\mathbf{u}}f \equiv df/d\lambda = u^a\partial_a f$. From the chain rule for differentiation and the fact that $p_a v^a$ is a scalar, we can obtain the Lie derivative of covariant vectors. We get

$$\pounds_{\mathbf{u}}p_a = u^b\,\partial_b p_a + p_b\,\partial_a u^b = u^b\,\nabla_b p_a + p_b\,\nabla_a u^b. \tag{4.128}$$

By constructing tensors from the direct product of vectors, one can obtain the Lie derivatives of tensors. For example, we have

$$\pounds_{\mathbf{u}}T^a{}_b = u^i\partial_i\,T^a{}_b - T^i{}_b\partial_i u^a + T^a{}_i\partial_b u^i = u^i\nabla_i\,T^a{}_b - T^i{}_b\nabla_i u^a + T^a{}_i\nabla_b u^i. \tag{4.129}$$

Any tensor field $T^{a\cdots}$ is said to be Lie transported along a curve $\mathcal{C}$ if its Lie derivative along the curve vanishes; that is, $\pounds_{\mathbf{u}}T^{a\cdots} = 0$ where $\mathbf{u}$ is the tangent vector to the curve. From our definition it follows that $\pounds_{\mathbf{u}}T^{a\cdots} = 0$ if the transformed value of the tensor field, when the coordinates are infinitesimally shifted, matches with the original value of the tensor field. This is the generally covariant notion of a vector not changing when we 'slide' the spacetime. To give an explicit example, let us suppose we choose the coordinates such that the x^αs are constant along $\mathcal{C}$ and $\lambda = x^0$ so that the tangent vector is $u^a = \delta^a_0$. In this particular coordinate system, derivatives of the tangent vector vanish and we have

$$\pounds_{\mathbf{u}}T^{a\cdots}{}_{b\ldots} = u^i\partial_i\,T^{a\cdots}{}_{b\ldots} = \frac{\partial}{\partial x^0}\,T^{a\cdots}{}_{b\ldots}. \tag{4.130}$$

So the statement that the tensor $T^{a\cdots}{}_{b\ldots}$ is Lie transported along the curve $\mathcal{C}$ is equivalent to saying that the tensor is independent of the time coordinate x^0. We have thus succeeded in giving an invariant meaning to such a statement.

Coming back to the metric tensor, we can say that if the metric tensor is independent of a particular coordinate, say, x^0 in a given coordinate system, then $\mathcal{L}_\xi g_{ab} = 0$, where $\xi^a = \delta^a_0$. Since this is a generally covariant condition, we can write this in any coordinate system as

$$\mathcal{L}_\xi g_{ab} = \nabla_b \xi_a + \nabla_a \xi_b = 0. \tag{4.131}$$

Any vector that satisfies this condition is called a *Killing vector*. This equation provides an operational way of determining the symmetries of the metric tensor. If the spacetime metric is described in some arbitrary coordinate system, then we need to solve the above equation and determine all independent vector fields which satisfy this equation. The integral curves to these vector fields will define the directions of symmetry of the spacetime in the sense that, if we 'slide' the spacetime along the integral curves, the metric and other geometrical features will remain invariant.

Considering the importance of this idea, we shall provide an equivalent, more concrete, description of the same for the case of the metric tensor. Consider an infinitesimal coordinate transformation from x^i to $x^i + \xi^i(x)$. Under such a transformation, the metric g^{ik} changes to

$$g'^{ik}(x'^l) = g^{lm}(x^l) \frac{\partial x'^i}{\partial x^l} \frac{\partial x'^k}{\partial x^m} = g^{lm} \left(\delta^i_l + \partial_l \xi^i \right) \left(\delta^k_m + \partial_m \xi^k \right)$$
$$\approx g^{ik}(x^l) + g^{im} \partial_m \xi^k + g^{kl} \partial_l \xi^i. \tag{4.132}$$

In this equation, $g'^{ik}(x'^i)$ is a function of x'^i, where both x'^i and x^i refer to the same physical event $\mathcal{P}$ in two coordinate systems. The change represented by the last two terms in the above equation is the change in the components at a given location. We want to compute how the *functional dependence* of the metric tensor on the coordinates changes under our coordinate transformation. To do this, we need to compare $g'^{ik}(x^l)$ with $g^{ik}(x^l)$; that is, we need to compare them with the same numerical values for the coordinates x^l. This is easily done by expanding the left hand side of Eq. (4.132) $g'^{ik}(x') = g'^{ik}(x+\xi)$ in a Taylor series in ξ^i retaining up to linear order in ξ. This will lead to the relation

$$g'^{ik}(x^l) = g^{ik}(x^l) - \xi^l \partial_l g^{ik} + g^{il} \partial_l \xi^k + g^{kl} \partial_l \xi^i \equiv g^{ik} + \delta g^{ik}. \tag{4.133}$$

Here we are comparing the functional dependence of g'^{ik} on the coordinates and δg^{ik} gives the change in the functional dependence. It is easy to verify that the three terms in the middle expression of Eq. (4.133) together can be written as $\nabla^k \xi^i + \nabla^i \xi^k$. Therefore the functional change in the metric tensor is given by

$$g'^{ik} = g^{ik} + \delta g^{ik}, \qquad \delta g^{ik} = \nabla^k \xi^i + \nabla^i \xi^k. \tag{4.134}$$

The corresponding changes in the covariant components are

$$g'_{ik} = g_{ik} + \delta g_{ik}, \qquad \delta g_{ik} = -\nabla_k \xi_i - \nabla_i \xi_k. \tag{4.135}$$

We see that when ξ^i is a Killing vector field, we have $\delta g^{ik} = 0$ showing that the functional form of the metric does not change. This is precisely the symmetry characterized by the vanishing of the Lie derivative: $\mathcal{L}_\xi g_{ik} = 0$.

Our original notion of u_l being constant along the geodesic when the metric is independent of a particular coordinate x_l can now be stated in a generally covariant fashion. The symmetry of the metric implies that there exists a Killing vector ξ^a which satisfies Eq. (4.131). In that case, we can easily show that the quantity $u^i \xi_i$ is conserved along the geodesic:

$$\frac{d}{d\lambda} \left(u^a \xi_a \right) = u^b \nabla_b \left(u^a \xi_a \right) = \xi_a u^b \nabla_b u^a + u^a \, u^b \nabla_b \xi_a = 0. \tag{4.136}$$

The last equality arises from using the following facts: (i) $u^b \nabla_b u^a = 0$ along the geodesic and (ii) $u^a u^b \nabla_b \xi_a = 0$ because $\nabla_b \xi_a$ is antisymmetric for a Killing vector. This completes the circle and connects up with our original idea. In a similar vein, one can prove that if T^{ab} is a conserved energy-momentum tensor and ξ_b is a Killing vector, then $P^a \equiv T^{ab} \xi_b$ is also conserved; that is,

$$\nabla_a (T^{ab} \xi_b) \equiv \nabla_a P^a = 0. \tag{4.137}$$

These results will be useful in our future discussions.

Finally, we mention a result which is quite useful, especially when working with action principles. We will then need to compute variations of quantities of the form $Q\sqrt{-g}$, where Q is a scalar, when $x^a \to \bar{x}^a = x^a + \xi^a$. We can easily show that

$$\delta(Q\sqrt{-g}) = -\sqrt{-g} \nabla_a (Q\xi^a). \tag{4.138}$$

To prove this, we note that $\delta Q \equiv \bar{Q}(x) - Q(x) = \bar{Q}(\bar{x} - \xi) - Q(x) = -\xi^a \nabla_a Q$, where we have used the facts $\bar{Q}(\bar{x}) = Q(x)$ and $\partial_a Q = \nabla_a Q$ for any scalar. On combining with $\delta\sqrt{-g} = -(1/2)\sqrt{-g}\, g_{ab} \delta g^{ab}$ and $\delta g^{ab} = \nabla^a \xi^b + \nabla^b \xi^a$, the result in Eq. (4.138) follows immediately.

Exercise 4.15
Jacobi identity Prove that, for any three vector fields $\boldsymbol{u}, \boldsymbol{v}, \boldsymbol{w}$ treated as directional derivative operators, the following relation (called the Jacobi identity) holds:

$$[\boldsymbol{u}, [\boldsymbol{v}, \boldsymbol{w}]] + [\boldsymbol{v}, [\boldsymbol{w}, \boldsymbol{u}]] + [\boldsymbol{w}, [\boldsymbol{u}, \boldsymbol{v}]] = 0. \tag{4.139}$$

Express this identity displaying the indices and provide an interpretation in terms of Lie derivatives.

Exercise 4.16
Understanding the Lie derivative (a) Show that $\mathcal{L}_v \epsilon_{abcd} = -(\nabla_i v^i) \epsilon_{abcd}$. [Hint. This can be worked out somewhat tediously after obtaining the expression for the Lie derivative of a fourth rank tensor. A more physical approach will be to consider the change in the volume of an infinitesimal element of spacetime when it 'flows' along the vector field v. Such an element has the volume $\epsilon_{ijkl} v^i a^j b^k c^l$, where a, b, c are infinitesimal connecting vectors. This implies that Lie derivative of these vectors vanish. Take it from there.]

(b) Show that, for any two vector fields u and v, treated as directional derivative operators, we have: $\mathcal{L}_u \mathcal{L}_v - \mathcal{L}_v \mathcal{L}_u = \mathcal{L}_{[u,v]}$.

Exercise 4.17
Understanding the Killing vectors (a) Solve Eq. (4.131) on the 2-sphere and in the flat Minkowski spacetime and determine all the Killing vectors.

[Hint. In the case of 2-sphere, our knowledge of angular momentum operators in quantum mechanics and the fact that one can set up a one-to-one correspondence between vectors and directional derivative operators will be adequate to determine the Killing vectors. In the Minkowski space, the ten Killing vectors should correspond to four translations, three rotations and three Lorentz boosts. Killing vectors for translation are easy to find. To find the rest, try an ansatz of the form $\xi_a = M_{ab} x^b$, where x^i are standard Cartesian coordinates and determine the conditions on M_{ab}.]

(b) Show that the linear combination of two Killing vector fields is another Killing vector field and that the commutator of two Killing vector fields is also another Killing vector field.

(c) If ξ^i is a timelike Killing vector, then we can introduce a four-velocity in the direction of the Killing vector by $u^i \equiv \xi^i / \sqrt{-\xi^a \xi_a} \equiv \xi^i / N$, where N is the norm of the Killing vector. Show that the four-acceleration corresponding to this four-velocity is $a = \nabla_u u = (1/2) \nabla \ln N$.

(d) Consider a stationary metric with a Killing vector ξ^i. We assume that ξ^i is timelike at spatial infinity but could become spacelike at other events. A test particle with four-momentum p and mass μ moves in this spacetime with a conserved energy $E = -p \cdot \xi$. Find the minimum value of E/μ that the particle can have at any event in terms of the norm $N^2 = -\xi \cdot \xi$.

[Hint. Examine the quantity E/μ in a local orthonormal frame in which the particle has a three velocity v^μ and show that $E/\mu = \gamma(\xi_0 - v^\mu \xi_\mu)$, where $\gamma = (1 - v_\alpha v^\alpha)^{-1/2}$ and the dot products are evaluated in the local Euclidean 3-space. Argue using this result that, if ξ is spacelike, then all values of E/μ are possible: $-\infty < E/\mu < \infty$. But if ξ is timelike then there is a lower bound on E/μ corresponding to $(E/\mu)_{\min} = N$ so that $N < E/\mu < \infty$.]

(e) Let $\xi^a(x)$ be an *arbitrary* vector field in the spacetime. Show that it is always possible to construct a coordinate system such that ξ^a points along one of the coordinate basis vectors. That is, show that one can choose the coordinates in such a manner that $\xi^a \partial_a = \partial/\partial x^M$, where x^M is one of the coordinates. Now show that, in this coordinate system,

$$\nabla_a \xi_b + \nabla_b \xi_a = \xi^i \partial_i g_{ab} = \partial g_{ab}/\partial x^M. \tag{4.140}$$

This generalizes the Killing equation and demonstrates that the Killing equation is satisfied by ξ^i if and only if the metric components are independent of x^M.

(f) A vector $\boldsymbol{v}$ is called homothetic if $\pounds_{\boldsymbol{v}} g_{ab} = c g_{ab}$, where c is a constant. Can you provide a geometric meaning for such a vector field when $c \neq 0$? If $u^i(\tau)$ is the tangent vector to a timelike geodesic, determine how $(\boldsymbol{v} \cdot \boldsymbol{u})$ changes with τ.

(g) A stationary, axisymmetric, spacetime has two Killing vector fields $[\boldsymbol{\xi}_t, \boldsymbol{\xi}_\phi]$ corresponding to translation along t or ϕ directions. A particle of unit mass moving in this spacetime has a four-velocity $\boldsymbol{u} = \gamma[\boldsymbol{\xi}_t + \Omega \boldsymbol{\xi}_\phi]$. (i) Explain why we can interpret this as a particle moving in a circular orbit. (ii) Suppose the particle is acted upon by some dissipative forces making it lose energy by the amount δE and orbital angular momentum by the amount δL and move to another circular orbit of smaller radius. Show that $\delta E = \Omega \delta L$.

Exercise 4.18
Killing vectors for a gravitational wave metric Consider a spacetime with the line element

$$ds^2 = -2du\,dv + a^2(u)dx^2 + b^2(u)dy^2, \tag{4.141}$$

where $u = (t - z), v = (t + z)$, which could represent a gravitational wave propagating along the z direction. (We will discuss gravitational waves in Chapter 9.) Find the complete set of Killing vectors for this metric. [Answer. Three Killing vectors corresponding to translation along the v, x, y directions can be written down by inspection. This spacetime actually has two more Killing vectors which can be found by solving the Killing equation written in the form $\partial_m \xi_n + \partial_n \xi_m = 2\Gamma^l_{mn} \xi_l$. Since the metric is fairly simple, one can, however, follow an alternate route. Consider the explicit form of the infinitesimal transformation of the coordinates with $\delta u = 0, \delta v = \epsilon f, \delta x = \epsilon g, \delta y = \epsilon h$, where f, g and h are arbitrary functions of all the four coordinates and ϵ is an infinitesimal parameter. We can now work out the change in the metric to the lowest order in ϵ and equate it to zero. This results in a set of differential equations for f, g and h to be solved leading to two more Killing vectors

$$\boldsymbol{\xi}_1 = \left(0, y, 0, \int b^{-2} du\right); \qquad \boldsymbol{\xi}_2 = \left(0, x, \int a^{-2} du, 0\right). \tag{4.142}$$

It can be directly verified that these two solve the Killing equation.]

4.9 Fermi–Walker transport

We have so far introduced two different ways of transporting a vector from one event to another in a curved spacetime, viz. parallel transport and Lie transport. Both allow us to define a vector *field* in spacetime given a vector *at one event*. In the case of parallel transport we need to know the Christoffel symbols and the transport is along a specified curve. In the case of Lie transport, we do not need to know the Christoffel symbols but we will require a vector field defined in the spacetime. We will now introduce a third procedure for transporting a vector from event to event, called *Fermi–Walker transport*, which requires knowledge of the metric and the resulting Christoffel symbols. Each of these procedures of transport has its own use and Fermi–Walker transport is particularly useful to set up coordinate systems for different classes of observers in the spacetime.

Consider an observer who is moving along some worldline (which is not necessarily a geodesic) with the four-velocity $\boldsymbol{u}$. Let $(\boldsymbol{e}_0, \boldsymbol{e}_\mu)$ be the tetrad of basis vectors transported along the trajectory by this observer. We are interested in the variation of these basis vectors along the observer's trajectory which is determined by $\nabla_{\boldsymbol{u}}\boldsymbol{e}_a$. Being another vector, this quantity can be expanded in terms of the basis vectors as $\nabla_{\boldsymbol{u}}\boldsymbol{e}_a = -\Omega_a{}^b(\tau)\,\boldsymbol{e}_b$, which defines $\Omega_a{}^b(\tau)$ at any location on the observer's worldline parametrized by proper time τ. We want to study the structure of Ω_a^b which completely determines how the observer's basis vectors change along the worldline.

The first condition we will impose is that the basis should satisfy the standard relations, $\boldsymbol{e}_a \cdot \boldsymbol{e}_b = g_{ab}$. The demand $\nabla_{\boldsymbol{u}}(\boldsymbol{e}_a \cdot \boldsymbol{e}_b) = \nabla_{\boldsymbol{u}} g_{ab} = 0$ leads to the condition $\Omega_{ab} = -\Omega_{ba}$. We can now write the transport equation in the form

$$\nabla_{\boldsymbol{u}}\boldsymbol{e}_a = -\Omega_a{}^b(\tau)\,\boldsymbol{e}_b = -\left(\Omega^{cb}\boldsymbol{e}_b \otimes \boldsymbol{e}_c\right)\cdot \boldsymbol{e}_a \equiv -\boldsymbol{\Omega}\cdot\boldsymbol{e}_a. \tag{4.143}$$

The components Ω^{ab} of $\boldsymbol{\Omega} = \Omega^{cb}\boldsymbol{e}_b \otimes \boldsymbol{e}_c$ can be thought of as the four-dimensional version of the three-dimensional antisymmetric rotation matrix.

The next natural condition on the tetrad is that the observer uses the proper time to define the time coordinate so that $\boldsymbol{e}_0 = \boldsymbol{u}$, which is the four-velocity of the observer. To see its implication, write $\boldsymbol{\Omega}$ in terms of another second rank antisymmetric tensor ω^{ab} and a four-vector v^a as

$$\Omega^{ab} = v^a u^b - u^a v^b + \omega^{ab} \tag{4.144}$$

with the conditions that both are orthogonal to four-velocity (i.e., $\omega^{ab} u_b = 0, v^i u_i = 0$). In the rest frame of the observer, both ω^{ab} and v^a are purely spatial with three components each, which is the same number of independent components in Ω^{cb}. To determine Ω^{cb}, we need to determine these two quantities. Equation (4.143) with $a = 0$ gives the condition $\boldsymbol{\Omega} \cdot \boldsymbol{u} = -\nabla_{\boldsymbol{u}}\boldsymbol{u} \equiv -\boldsymbol{a}$, while Eq. (4.144) gives $\boldsymbol{\Omega}\cdot\boldsymbol{u} = -\boldsymbol{v}$, allowing us to identify $\boldsymbol{v} = \boldsymbol{a}$, the acceleration. Thus our transport law becomes $\nabla_{\boldsymbol{u}}\boldsymbol{e}_a = -\boldsymbol{e}_a\cdot(\boldsymbol{a}\otimes\boldsymbol{u} - \boldsymbol{u}\otimes\boldsymbol{a} + \omega)$ or equivalently:

$$\frac{d\boldsymbol{e}_a}{d\tau} = -[\boldsymbol{a}(\boldsymbol{u}\cdot\boldsymbol{e}_a) - \boldsymbol{u}(\boldsymbol{a}\cdot\boldsymbol{e}_a) + (\omega\cdot\boldsymbol{e}_a)]. \tag{4.145}$$

As for the tensor ω^{ab}, we know that – being orthogonal to the four-velocity – it has only three independent (spatial) components which can be traded off for a vector describing a three-dimensional rotation. That is, we can write

$$\omega^{ab} = \epsilon^{abls}u_l\omega_s; \quad \omega^a = -(1/2)\epsilon^{amls}u_m\omega_{ls} \tag{4.146}$$

by introducing a four-vector ω^a which is orthogonal to u^i. This tensor represents the rotation of the spatial basis vectors of the observer. Such a freedom will always exist because the observer can choose the orientation of the spatial vectors differently at different events along the trajectory.

It is, however, more economical to assume that the tetrad is transported along the worldline of the observer with the spatial vectors remaining non-rotating. This is the same as assuming $\omega^{ab} = 0$. In that case, the transport equation for the basis vectors will read as

$$\frac{d\boldsymbol{e}_a}{d\tau} = [\boldsymbol{u}(\boldsymbol{a} \cdot \boldsymbol{e}_a) - \boldsymbol{a}(\boldsymbol{u} \cdot \boldsymbol{e}_a)]. \tag{4.147}$$

Taking a cue from this, we shall say that any vector v^i is Fermi–Walker transported along a curve if it satisfies the above equation with $\boldsymbol{e}_a$ replaced by $\boldsymbol{v}$. In component form, the condition for Fermi–Walker transport is

$$\frac{dv^m}{d\tau} = u^i \nabla_i v^m = (u^m a^n - u^n a^m) v_n, \tag{4.148}$$

where u^i and a^j are the velocity and acceleration of an observer along the curve.

The Fermi–Walker transport has a very simple physical meaning and, in fact, provides the most natural coordinate basis for an observer moving along an arbitrary worldline in curved spacetime. It ensures that the tetrad remains orthonormal and the time direction coincides with the direction of four-velocity. As one moves along an arbitrary curve, the local Lorentz frames will necessarily be different at different events. Treating Lorentz transformations as generalized rotations, we see that the basis vectors will necessarily be rotating to accommodate this. The factor $(u^m a^n - u^n a^m) v_n$ accounts for this inevitable 'rotation' and Fermi–Walker transport ensures that there is no *additional* rotation of the spatial basis vectors. Roughly speaking, this rotation described by $(u^m a^n - u^n a^m)$ is in the velocity–acceleration plane; if at a given instant acceleration is along the x-axis, the 'rotation' is in the $t–x$ plane corresponding to the appropriate Lorentz boost needed to account for the change in velocity induced by the acceleration.

From the definition, it is easy to show that (a) Fermi–Walker transport along a geodesic curve is the same as parallel transport and (b) the dot product between any two vectors is preserved under Fermi–Walker transport.

Exercise 4.19
Tetrad for a uniformly accelerated observer Consider an observer undergoing uniform accelerated motion along the x-axis who carries with her an orthonormal tetrad $\boldsymbol{e}_i(\tau)$ of basis vectors which keeps changing with her proper time τ. Let the magnitude of the acceleration be κ. Explain why it is reasonable to choose an orthonormal tetrad with the

following components in the instantaneous inertial frame

$$\boldsymbol{e}_0(\tau) = \left(\cosh \frac{\kappa\tau}{c}, \sinh \frac{\kappa\tau}{c}, 0, 0\right), \qquad \boldsymbol{e}_1(\tau) = \left(\sinh \frac{\kappa\tau}{c}, \cosh \frac{\kappa\tau}{c}, 0, 0\right),$$

$$\boldsymbol{e}_2(\tau) = (0, 0, 1, 0), \qquad \boldsymbol{e}_3(\tau) = (0, 0, 0, 1). \tag{4.149}$$

One elementary application of Fermi–Walker transport is in the study of spinning objects in curved spacetime. The spin of an object can be described by a four-vector S^j, which is orthogonal to the four-velocity ($S^j u_j = 0$) so that it has only three spatial components in the local rest frame. Such a spin vector will be Fermi–Walker transported when the spinning object moves along some trajectory. The orthogonality with four-velocity implies that the spin vector is Fermi–Walker transported with the equation $dS^m/d\tau = u^m(a^n S_n)$, which is precisely Eq. (1.152) that we arrived at in special relativity. Being a local equation, it continues to hold in the general relativistic context.

The observer can use the tetrad carried by her to set up a local coordinate system in the neighbourhood of her worldline. Such a coordinate system will, of course, depend on the manner in which the tetrad was transported along her worldline and will be simplest when the tetrad undergoes Fermi–Walker transport. It is, however, useful to study the coordinate system set up by an observer who transports the tetrad in a slightly more general manner through Eq. (4.145) with $\omega^{ij} \neq 0$. To set up the coordinate system, we will proceed as follows. At a given event along the worldline at proper time τ, the observer sends out spatial geodesics characterized by the unit vector $\boldsymbol{n}$ which is orthogonal to the four-velocity ($\boldsymbol{n} \cdot \boldsymbol{u} = 0$). In the neighbourhood of the observer's worldline, each event will be intersected by one (and only one) geodesic. (This will not be true far away from the observer because the geodesics may intersect; as such, the coordinate system remain valid only in the local neighbourhood of the worldline of the observer.) Any event $\mathcal{P}$ in the local neighbourhood can therefore be identified by four numbers (τ, sn^α), where s is the arclength along the geodesic which passes through $\mathcal{P}$. We will now determine the form of the metric in this coordinate system.

By construction, along the worldline, $g_{ab} = \eta_{ab}$ and the worldline has the coordinates $x^\alpha = 0$. The condition that the curves of the form $x^\alpha = sn^\alpha, x^0 = \tau$ are geodesics shows that $\Gamma^a_{\mu\nu} n^\mu n^\nu = \Gamma_{a\mu\nu} = 0$. (This is obtained from the geodesic equation by noting that s, τ are affine parameters and $d^2 x^a/ds^2 = 0$.) Further, comparing the transport law $\nabla_{\boldsymbol{u}} \boldsymbol{e}_a = -\boldsymbol{e}_b \Omega^b_a$ with the definition of Christoffel symbols $\nabla_{\boldsymbol{u}} \boldsymbol{e}_a = \boldsymbol{e}_b \Gamma^b_{a0}$ we find that $\Gamma^b_{a0} = -\Omega^b_a$. Noting that, along the worldline, $u^i = \delta^i_0$,

$a_0 = 0$ we get the Christoffel symbols:

$$\Gamma^0_{00} = \Gamma_{000} = 0; \quad \Gamma^0_{\mu 0} = -\Gamma_{0\mu 0} = +\Gamma_{\mu 00} = \Gamma^\mu_{00} = a^\mu;$$
$$\Gamma^\mu_{\nu 0} = \Gamma_{\mu\nu 0} = -\omega^\alpha \epsilon_{0\alpha\mu\nu}, \tag{4.150}$$

where we have temporarily ignored the positioning of the indices. Together, these conditions on the Christoffel symbols translate to the following conditions on the metric along the worldline:

$$\partial_0 g_{ab} = 0, \quad \partial_\alpha g_{\mu\nu} = 0, \quad \partial_\alpha g_{00} = -2a_\alpha, \quad \partial_\mu g_{0\nu} = -\epsilon_{0\nu\mu\lambda}\omega^\lambda. \tag{4.151}$$

These can be integrated to the lowest order to provide the metric in this coordinate system as

$$ds^2 = -(1 + 2a_\mu x^\mu)dt^2 - 2(\epsilon_{\alpha\kappa\lambda}x^\kappa\omega^\lambda)dt dx^\alpha + \delta_{\mu\nu}dx^\mu dx^\nu, \tag{4.152}$$

which is correct to $\mathcal{O}(x^2)$. The acceleration of the observer shows up in g_{00} in a manner we have already seen in Chapter 3 for the case of motion along the x-axis (see Eq. (3.70)). The rotation of the spatial tetrad leads to a $g_{0\mu}$ term; this term will be absent if the tetrad was Fermi–Walker transported along the worldline.

The effect and interpretation of these terms are transparent when we use this spacetime to study the motion of another test particle. From the geodesic equation, it is easy to see that the spatial location of the test particles will vary as

$$\frac{d^2 x^\alpha}{dt^2} = 2\epsilon^\alpha_{\nu\mu}\omega^\mu \frac{dx^\nu}{dt} - a^\alpha - \frac{d\lambda}{dt}\frac{d}{dt}\left(\frac{dt}{d\lambda}\right)\frac{dx^\alpha}{dt}. \tag{4.153}$$

The time component of the geodesic equation gives the relation

$$\frac{d^2 t}{d\lambda^2} + 2 a_\nu \frac{dt}{d\lambda}\frac{dx^\nu}{d\lambda} = 0, \tag{4.154}$$

which relates t to λ. Combining these two, the spatial part of the geodesic equation can be expressed in three-vector notation (with c-factors temporarily inserted) as

$$\ddot{\boldsymbol{r}} = -\boldsymbol{a} - 2\boldsymbol{\omega} \times \dot{\boldsymbol{r}} + \frac{2(\boldsymbol{a} \cdot \dot{\boldsymbol{r}})\,\dot{\boldsymbol{r}}}{c^2}. \tag{4.155}$$

It is clear that the particle is seen to move with an acceleration a^α and feels the Coriolis acceleration, $(\boldsymbol{\omega} \times \boldsymbol{v})$, caused by the rotation of the spatial components of the tetrad. The last term is a relativistic correction and may be ignored in the limit of $c \to \infty$.

The three different rules for transporting the vectors – parallel transport, Lie transport, Fermi–Walker transport – have different uses as would have been clear from the above discussion. It is interesting to ask what kinds of vector fields in a spacetime automatically transport themselves according to each of these rules. In case of parallel transport, this requires finding vector fields which satisfy the conditions $\nabla_u u = 0$. We know that this equation is satisfied by the tangent vector to any geodesic. Since $\mathcal{L}_u u = 0$ is an identity for any vector field, all vector fields Lie transport themselves. In a similar manner, it can be directly verified that (a) properly normalized four-velocities are Fermi–Walker transported along themselves and (b) the tangent vectors of any congruence of geodesics are also Fermi–Walker transported along themselves.

PROJECTS

Project 4.1
Velocity space metric

In Section 1.4 we derived the metric in the velocity space based on the relativistic addition law for the velocities. The purpose of this project is to investigate several interesting features of this velocity space as an example of a space with a nontrivial metric.

(a) Consider a rocket which changes its four-velocity from some value v_1 to another value v_2, spending the minimum amount of fuel. Show that the path traced by the rocket in the velocity space will be a geodesic. [Hint. This does not require any calculation!]

(b) A rocket changes its velocity from v_1 to v_2, spending the least amount of fuel. What is the minimum speed of the rocket ship with respect to an inertial frame during this operation?

[Hint. This will require solving for geodesics in the velocity space. Note that the velocity space metric in Eq. (1.48) can be mapped to that of a 3-sphere by analytically continuing the rapidity to imaginary values by $\chi \rightarrow i\chi$. This should help in determining the nature of geodesics without much work. Using this approach, argue that the general geodesic in velocity space has the form

$$v = \left[\frac{1 - \beta^2}{1 + l^2}\right]^{1/2} l\,a + \beta\,b, \tag{4.156}$$

where a and b are two mutually perpendicular unit vectors and l parametrizes the geodesics with $-\infty < l < +\infty$ for each choice of β with $\beta < 1$. Take it from there.]

(c) The velocity space metric also helps in the study of a relativistic random walk. Consider a particle which is moving because of a series of Lorentz boosts that are executed in uncorrelated random directions. Each boost introduces a 'step' σ in the velocity space with $\sigma \ll c$ in the instantaneous comoving frame. Find the probability distribution function of the resultant velocity of the particle after a large number (n) of steps.

[Hint. Familiarize yourself with the essential idea by first proving that, in the nonrelativistic case, the probability distribution $P(n, v)$ satisfies the diffusion equation

$$\frac{\partial P}{\partial n} = \frac{\sigma^2}{6}\nabla^2 P \tag{4.157}$$

in the velocity space. Then show that, in the case of a relativistic random walk, the same result holds with ∇^2 replaced by the Laplacian with an appropriate velocity space metric. From this show that the relevant equation to solve is

$$\frac{\partial P}{\partial n} = \frac{\sigma^2}{6}\left[\frac{1}{\sinh^2\chi}\frac{\partial}{\partial\chi}\left(\sinh^2\chi\frac{\partial P}{\partial\chi}\right)\right]. \tag{4.158}$$

Show that the solution is given by

$$P(\mathbf{v}, n) = \left(\frac{1}{4\pi x}\right)^{3/2} \frac{\chi}{\sinh \chi} \exp[-x - (\chi^2/4x)], \tag{4.159}$$

where $x \equiv n\sigma^2/6$.]

(d) Investigate the asymptotic limits of a relativistic random walk and compare it with the non-relativistic case. Can you explain the asymptotic limits from simple physical considerations?

(e) Recall the discussion of Thomas precession in Section 1.10. An electron moving in a circular orbit in real space with speed v will trace a circle of radius v in the velocity space. Compute the rate of change of proper area (using the velocity space metric) swept by the electron as it moves. Do you see any connection with Thomas precession? More generally, can you provide an interpretation of Thomas precession in terms of the velocity space metric?

Project 4.2

Discovering gauge theories

In Exercise 2.15 we introduced the electromagnetic field through the gauge covariant derivative of a scalar field. This idea can be generalized in a useful way to a multi-component field. Consider an N-component field $\phi_A(x)$ with $A = 1, 2, ...N$, which is a vector in some N-dimensional linear vector space (usually called an *internal symmetry* space). A 'rotation' in the internal symmetry space is generated through a matrix transformation of the type

$$\phi' = U\phi; \qquad U = \exp(-i\tau_A \alpha^A), \tag{4.160}$$

where ϕ is treated as a column vector, α^A is a set of N parameters and τ_A are N matrices which satisfy the commutation rules $[\tau_A, \tau_B] = iC_{AB}^J \tau_J$, where C_{AB}^J are constants. [In the standard context, one considers a theory based on a gauge group, say, $SU(N)$, in which case the C_{AB}^J will be the structure constants of the group. You do not need to know any group theory to do this project, though.] Consider now a field theory for ϕ_A based on a Lagrangian of the form

$$L = -\frac{1}{2} \left[\partial_i \phi^\dagger \, \partial^i \phi + \mu^2 \phi^\dagger \, \phi \right], \tag{4.161}$$

where $\phi^\dagger$ is the Hermitian conjugate. This Lagrangian is invariant under the transformations $\phi' = U\phi$ with *constant* parameters α^A.

(a) Consider now the 'local rotations' with $\alpha^A = \alpha^A(x)$ depending on spacetime coordinates. Show that the Lagrangian in Eq. (4.161) is no longer invariant under such transformations.

(b) Introduce now an N-component *gauge field* $A_i^K(x)$ with $i = 0, 1, 2, 3$ being a space-time index and $K = 1, 2, 3, ...N$ being an internal space index. Let the $A_i \equiv \tau_J A_i^J$ denote a set of matrices corresponding to the gauge field. Show that the Lagrangian in Eq. (4.161) can be made invariant under the local rotations if: (i) we replace partial derivatives by *gauge covariant derivatives*

$$D_i = \partial_i + iA_i, \tag{4.162}$$

where the first term is multiplied by a unit matrix, and (ii) we assume that the gauge field transforms according to the rule:

$$A'_i = U A_i U^{-1} - i U \partial_i U^{-1}. \tag{4.163}$$

(c) To complete the picture, we need to write down a Lagrangian for the gauge field itself. Show that

$$F_{ik} = \partial_i A_k - \partial_k A_i + i[A_i, A_k], \tag{4.164}$$

where each term is interpreted as an $N \times N$ matrix, is gauge covariant. That is, when A^i changes as in Eq. (4.163) F_{ik} changes to $U F_{ik} U^{-1}$. This allows us to construct a Lagrangian for the gauge field which is gauge covariant

$$\mathcal{L} = -\frac{1}{2} \operatorname{Tr} \left(F_{ik} F^{ik} \right). \tag{4.165}$$

Make sure you understand the connection with electromagnetic field which occurs for $N = 1$, etc.

(d) It is clear that gauge theories use objects which have both internal space index and spacetime index. This idea can be made more formal as follows. In the text, we introduced a tangent space $T(\mathcal{P})$ and its dual $T^*(\mathcal{P})$ at each event in the spacetime $\mathcal{P}$. These linear vector spaces are closely tied to the spacetime since they were defined through the tangents to curves in the spacetime manifold. But it is possible to have a more general structure in which we associate some other linear vector space $V(\mathcal{P})$ of N dimensions [*internal symmetry space*] with each event in the spacetime. We then postulate that the basis vectors of this space $\mathbf{e}_J$ (with $J = 1, 2, \ldots N$) change from event to event according to the rule $D_i \mathbf{e}_J = A_{iJ}^K \mathbf{e}_K$ with a set of *connections* $A_{iJ}^K(x)$ which carry one spacetime index and two internal space indices. They can be thought of as four matrices A_i, each of which is of $N \times N$ dimension. We can now define gauge covariant derivatives of vectors in the internal space by $D_i \mathbf{v} = \partial_i \mathbf{v} + A_i \mathbf{v}$, where $\mathbf{v}$ is a column vector and A_is are matrices, etc. In this spirit, one should interpret the Christoffel symbols Γ^i_{kj} as the ijth element of a matrix Γ_k and identify the internal space with the tangent space. See[2] how far you can push such a comparison. (We will revisit gauge theories in Chapter 11.)

5

Curvature of spacetime

5.1 Introduction

In this chapter, we will develop further the mathematical formalism required to understand different aspects of curved spacetime, focusing on the description of spacetime curvature. It uses extensively the concepts developed in Chapter 4, especially the idea of parallel transport. Most of the topics described here will be used in the subsequent chapters, except for the ideas discussed in Section 5.6 related to the classification of spacetime curvature which fall somewhat outside the main theme of development.

5.2 Three perspectives on the spacetime curvature

The discussion in the previous chapter used only the fact that the metric tensor depended on the coordinates. This dependence can arise either due to the use of curvilinear coordinates in flat spacetime or due to genuine curvature of the spacetime. The gravitational field generated by matter manifests itself as the curvature of the spacetime and hence, to study the gravitational effects, we need to develop the mathematical machinery capable of describing and analysing the curvature of spacetime. This will be the aim of the current chapter and we shall begin by introducing the concept of spacetime curvature from three different – but closely related – perspectives. At a fundamental level, these three perspectives stem from the same source, viz. behaviour of vectors under parallel transport; however, we will discuss them separately in the next three subsections for greater clarity.

5.2.1 Parallel transport around a closed curve

The first perspective on curvature originates from the changes induced in a vector when it is parallel transported around a small, closed, curve in the spacetime. The covariant derivative introduced in the previous chapter provides a procedure for the parallel transport of vectors in curved spacetime or in curvilinear coordinates. Let $C[x^i(\lambda)]$ be a curve connecting the event P having the coordinates $x^i(\lambda = 0)$ with the event Q having the coordinates $x^i(\lambda = 1)$. If we are given a vector $v^i(0)$ at the event P, we can generate a vector *field* all along C using parallel transport. Let us assume that we obtain a vector $v^i(1; C)$ at the event Q by this process where we have used a notation which explicitly keeps track of the curve C that was used. Consider now another curve S connecting the events P and Q. Let us suppose that we also parallel transport the vector along S from P to Q obtaining, say, the vector $v^i(1; S)$ at Q. We can then ask the question: will the two vectors $v^i(1; C)$ and $v^i(1; S)$ be the same? Equivalently, will a vector change if it is parallel transported from P to Q and back, around a closed loop built by C and S?

If the spacetime was actually flat and the dependence of the metric on the coordinates was only the result of our using curvilinear coordinates, then this process of parallel transport would have moved the vector keeping its components parallel to the *global* Cartesian axes constant – which is just the definition of parallel transport. In that case, we should obtain the same vector at Q by parallel transporting it from P through any curve; and a vector will not change if it is parallel transported around a closed curve. Obviously, this is a test for the flatness of spacetime and we can expect the change in a vector, on being parallel transported around a closed curve, to provide an insight into the curvature of the spacetime. We will now study this effect.

Consider an infinitesimal closed loop $\mathcal{L}$ in spacetime described by the parametrized equation $x^a(\lambda)$. During the parallel transport, the vector v^a changes according to the relation

$$\frac{dv^a}{d\lambda} = -\Gamma^a_{bc} v^b \frac{dx^c}{d\lambda}. \tag{5.1}$$

When we parallel transport the vector from P (corresponding to $\lambda = 0$) to some event at the parameter value λ, it will become

$$v^a(\lambda) = v^a_P - \int_0^\lambda \Gamma^a_{bc} v^b \frac{dx^c}{d\lambda} d\lambda. \tag{5.2}$$

Since the entire loop $\mathcal{L}$ is assumed to be infinitesimally small, we can Taylor expand the integrand in the form

$$\Gamma^a_{bc}(\lambda) = (\Gamma^a_{bc})_P + (\partial_d \Gamma^a_{bc})_P \left[x^d(\lambda) - x^d_P \right] + \cdots$$
$$v^a(\lambda) = v^a_P - (\Gamma^a_{bc})_P v^b_P \left[x^c(\lambda) - x^c_P \right] + \cdots . \tag{5.3}$$

Substituting into Eq. (5.2) we get

$$v^a(\lambda) = v_{\mathcal{P}}^a - (\Gamma_{bc}^a)_{\mathcal{P}} \, v_{\mathcal{P}}^b \int_0^\lambda \frac{dx^c}{d\lambda} d\lambda$$

$$- (\partial_d \Gamma_{bc}^a - \Gamma_{ec}^a \Gamma_{bd}^e)_{\mathcal{P}} \, v_{\mathcal{P}}^b \int_0^\lambda \left(x^d - x_{\mathcal{P}}^d \right) \frac{dx^c}{d\lambda} d\lambda. \qquad (5.4)$$

The second term, as well as the one involving $x_{\mathcal{P}}^d$ in the last term, vanishes when we integrate around a closed loop. So the net change in the vector is

$$\Delta v^a = - (\partial_d \Gamma_{bc}^a - \Gamma_{ec}^a \Gamma_{bd}^e)_{\mathcal{P}} \, v_{\mathcal{P}}^b \oint x^d \, dx^c. \qquad (5.5)$$

This clearly shows that a vector, in general, changes when moved around an infinitesimal closed loop. To write this in a more meaningful form, we will add to it the expression obtained by interchanging the dummy indices c and d and use the fact that the integral of $x^c dx^d + x^d dx^c = d(x^c x^d)$ around a closed loop vanishes. This leads to the result

$$\Delta v^a = -\frac{1}{2} (\partial_c \Gamma_{bd}^a - \partial_d \Gamma_{bc}^a + \Gamma_{ec}^a \Gamma_{bd}^e - \Gamma_{ed}^a \Gamma_{bc}^e)_{\mathcal{P}} \, v_{\mathcal{P}}^b \oint x^c \, dx^d$$

$$\equiv -\frac{1}{2} (R^a{}_{bcd})_{\mathcal{P}} \, v_{\mathcal{P}}^b \oint x^c \, dx^d, \qquad (5.6)$$

where we have defined the four-indexed object

$$R^a{}_{bcd} \equiv \partial_c \Gamma_{bd}^a - \partial_d \Gamma_{bc}^a + \Gamma_{ec}^a \Gamma_{bd}^e - \Gamma_{ed}^a \Gamma_{bc}^e. \qquad (5.7)$$

It is obvious that $R^a{}_{bcd}$ is antisymmetric in c and d. In the case of an infinitesimal loop made of a parallelogram with sides $\delta a^c \delta b^d$, the integral in Eq. (5.6) is equal to the area enclosed by the parallelogram:

$$\oint x^c \, dx^d = \delta a^c \, \delta b^d - \delta a^d \, \delta b^c \equiv \Delta \sigma^{cd}. \qquad (5.8)$$

Therefore, for an infinitesimal loop, we can write

$$\Delta v^a = -\frac{1}{2} R^a{}_{bcd} \, v_{\mathcal{P}}^b \, \Delta \sigma^{cd}, \qquad (5.9)$$

where $\Delta \sigma^{cd}$ can be interpreted as the area enclosed by the loop $\mathcal{L}$. This result shows that the change in the vector when it is moved around a closed loop is related to $R^a{}_{bcd}$ and vanishes when $R^a{}_{bcd} = 0$. Further, Eq. (5.9) shows that $R^a{}_{bcd}$ is a tensor since all other quantities in that expression are tensors, as long as we are dealing with an infinitesimal region. (We will provide a more direct proof of this claim in the next section; see Eq. (5.15) below.)

If the spacetime is actually flat, we can always choose an inertial coordinate system in which the Γs vanish at *all* events. Then the derivatives of the Γs will also vanish making $R^a{}_{bcd} = 0$. But since this is a tensor equation, it will be valid in any other coordinate system. Thus we find that, in flat spacetime, we must have $R^a{}_{bcd} = 0$. This condition provides the first invariant characterization of curvature we were seeking; the tensor $R^a{}_{bcd}$ is called the *curvature tensor*.

5.2.2 Non-commutativity of covariant derivatives

The above calculation can be given a geometrical, though a bit abstract, interpretation. Consider two vector fields X and Y in the spacetime. We can now consider an infinitesimal loop in spacetime built out of the vectors X, Y and $[X, Y]$. The latter is needed because – as we saw in Section 4.8 – the commutator of the vector fields $[X, Y] = \pounds_X Y$ gives the vector which 'closes' the loop built from X, Y by appropriate sliding. (See Fig. 4.3 and Fig. 5.1.) The quantity $\nabla_X \equiv X^i \nabla_i$ is the covariant generalization of the directional derivative along a vector X. So we can interpret $\nabla_X \nabla_Y(v)$ as the effect of parallel transporting the vector v first along Y and then along X; similarly, $\nabla_Y \nabla_X(v)$ gives the effect of parallel transporting the vector first along Y and then along X. Finally, the term $\nabla_{[X,Y]} v$ will give the effect of parallel transport along the 'closing' vector $[X, Y]$. So, from our result in Eq. (5.9), we will expect

$$\left([\nabla_X, \nabla_Y] - \nabla_{[X,Y]}\right) v^a = -\frac{1}{2} R^a{}_{bcd} \, v^b \, \Delta\sigma^{cd} = -\frac{1}{2} R^a{}_{bcd} \, v^b \left(X^d Y^c - X^c Y^d\right)$$
$$= R^a{}_{bcd} \, v^b \, X^c Y^d. \tag{5.10}$$

The second equality follows from the fact that, for a sufficiently small loop, the area $\Delta\sigma^{cd}$ of a (near) parallelogram with sides X and Y is $(X^c Y^d - X^d Y^c)$. (The fact that the parallelogram does not close only leads to a higher order term, which can be ignored here.) The third equality arises from the fact that $R^a{}_{bcd}$ is antisymmetric in c and d.

In particular, if we take X and Y to be along the coordinate basis vectors e_i and e_j, then the commutator $[X, Y] = [\partial_i, \partial_j] = 0$, leading to the result

$$[\nabla_i, \nabla_j] v^a = R^a{}_{bij} v^b. \tag{5.11}$$

That is, unlike ordinary derivatives, covariant derivatives do not commute – which provides our second perspective on curvature.

It is easy to verify this fact directly. Starting from the covariant derivative $\nabla_b A^a$ of a vector field

$$\nabla_b A^a = \partial_b A^a + \Gamma^a_{mb} A^m, \tag{5.12}$$

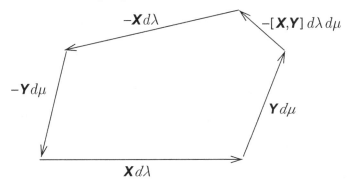

Fig. 5.1. A vector **v** is parallel transported around an infinitesimal closed loop made of vector fields, **X** and **Y**. The parameters $d\lambda$ and $d\mu$ are infinitesimal.

and differentiating once again, treating $\nabla_b A^a$ as a second rank tensor, we get

$$\nabla_c \nabla_b A^a = \partial_c \left(\nabla_b A^a\right) + \Gamma^a_{nc}\nabla_b A^n - \Gamma^m_{bc}\nabla_m A^a \tag{5.13}$$

which reduces to

$$\nabla_c \nabla_b A^a = \partial_c \partial_b A^a + \partial_c \left(\Gamma^a_{bm}A^m\right) + \Gamma^a_{cn}\left(\partial_b A^n + \Gamma^n_{mb}A^m\right)$$
$$- \Gamma^m_{bc}\left(\partial_m A^a + \Gamma^a_{nm}A^n\right). \tag{5.14}$$

Computing the difference $\left(\nabla_c \nabla_b A^a - \nabla_b \nabla_c A^a\right)$ we find that the terms with first derivatives of A^a vanish giving

$$\nabla_c \nabla_b A^a - \nabla_b \nabla_c A^a = - \left(\partial_b \Gamma^a_{ic} - \partial_c \Gamma^a_{ib} + \Gamma^a_{kb}\Gamma^k_{ic} - \Gamma^a_{kc}\Gamma^k_{ib}\right) A^i \equiv -R^a{}_{ibc} A^i. \tag{5.15}$$

This result also provides a very direct proof that $R^a{}_{ibc}$ is a tensor since the left hand side is a tensor and A^i is a vector.

The two perspectives on curvature obtained above – in terms of the change in vector on transport around a loop and as a measure of non-commutativity of covariant derivatives – are closely related to the integrability conditions on the parallel transport equation. We have seen that the result of parallel transporting a vector from event to event will – in general – depend on the curve that is used for this purpose. On the other hand, if we can find a unique solution to the equation

$$\partial_c v^a = -\Gamma^a_{bc}(x)v^b \tag{5.16}$$

for the four functions $v^a(x)$ independent of any curve in the spacetime, we can satisfy Eq. (4.117) identically for any curve and the result of the parallel transport will be independent of the curve chosen. We know that, in general, this will not be possible. This implies that Eq. (5.16) cannot – in general – be integrated to give a unique vector field $v^a(x)$. This equation will have an acceptable, unique, solution only if the integrability conditions

$$\partial_b \partial_c v^a = \partial_c \partial_b v^a \qquad (5.17)$$

are satisfied. This is the same as demanding $dv^a = -\Gamma^a_{bc}(x)v^b(x)dx^c$ to be an exact differential. Differentiating Eq. (5.16) with respect to x^d and using Eq. (5.16) again, we find

$$\partial_c \partial_d v^a = -\left[\partial_d \Gamma^a_{lc} - \Gamma^a_{bc}\Gamma^b_{dl}\right]v^l. \qquad (5.18)$$

Using Eq. (5.18) in Eq. (5.17) we find that the integrability condition is equivalent to $R^a_{\ ldc}v^l = 0$. If this result should hold for an arbitrary v^a, then $R^a_{\ ldc}$ has to vanish. In general, this quantity will *not* vanish and hence parallel transport will not give a unique vector except in flat spacetime. In flat spacetime one can always choose a system of coordinates such that $\Gamma^i_{kl} = 0$. In such a coordinate system R^i_{klm} will vanish. Since R^i_{klm} is a tensor, it follows that it will vanish in any coordinate system in flat spacetime and parallel transport will be unique.

Let us now consider the converse of the above result. Suppose we are working in coordinate system in which the metric tensor is a function of coordinates. But we find, on computation, that $R^a_{\ bcd} = 0$. Does this imply that the spacetime is flat? That is, can we now construct a coordinate transformation which will lead from the given metric g_{ab} to η_{ab} globally? This is indeed true as can be seen by the following argument.

We choose four mutually orthonormal vectors $\boldsymbol{a}, \boldsymbol{b}, \boldsymbol{c}, \boldsymbol{d}$ (one timelike, three spacelike) at some event $\mathcal{P}$ with $\boldsymbol{a} \cdot \boldsymbol{a} = -1, \boldsymbol{b} \cdot \boldsymbol{b} = 1; \boldsymbol{c} \cdot \boldsymbol{c} = 1$ and $\boldsymbol{d} \cdot \boldsymbol{d} = 1$ and all other dot products vanishing. We next parallel transport them to all other events, thereby producing four-vector *fields* in the spacetime $[\boldsymbol{w}^{(0)}(x), \boldsymbol{w}^{(1)}(x), \boldsymbol{w}^2(x), \boldsymbol{w}^{(3)}(x)]$ from $\boldsymbol{a}, \boldsymbol{b}, \boldsymbol{c}, \boldsymbol{d}$. Since the curvature tensor vanishes, this procedure is unique and is independent of the curve used to parallel transport the vectors. Each of these vector fields satisfy the analogue of Eq. (5.16) for the covariant components of the vector field:

$$\partial_c v_a = \Gamma^b_{ac}(x)v_b \qquad (5.19)$$

with $\boldsymbol{v} = \boldsymbol{w}^{(j)}$, where $j = 0, 1, 2, 3$. From the symmetry of Christoffel symbols in b, c in the above equation, we get $\partial_j v_i = \partial_i v_j$, etc., so that we can express the covariant components of each of these vectors as a gradients: $(\boldsymbol{w}^{(i)})_j = \partial_j \phi^{(i)}, i = 0, 1, 2, 3$ of four functions $\phi^{(j)}$. Since the dot product of the vectors is preserved under parallel transport, the relation $\boldsymbol{w}^{(i)} \cdot \boldsymbol{w}^{(j)} = \eta^{ij}$ translates to

$$g^{ij}\left(\frac{\partial\phi^{(k)}}{\partial x^i}\right)\left(\frac{\partial\phi^{(l)}}{\partial x^j}\right) = \eta^{kl}. \qquad (5.20)$$

This shows that a coordinate transformation $x'^i(x) \equiv \phi^{(i)}(x)$ transforms the metric to Cartesian form globally. Thus the vanishing of $R^i{}_{klm}$ is both necessary and sufficient for the spacetime to be flat.

In summary, we have now arrived at a mathematical characterization of a curved spacetime. Given a metric g_{ab} one can compute Christoffel symbols and from them obtain $R^a{}_{bcd}$ using Eq. (5.7). If this quantity vanishes identically, then the spacetime is flat and one will be able to transform to standard Cartesian coordinates. There can be no real gravitational field in such a spacetime and any nontrivial effect can be attributed to the choice of curvilinear coordinates. If any of the components of $R^a{}_{bcd}$ is nonzero then the spacetime is curved and describes a genuine gravitational field.

In obtaining the expression for curvature, we used the effect of $[\nabla_i, \nabla_j]$ on a vector but it is clear that similar results also hold for higher rank tensors. We conclude this section with a description of the effect of $[\nabla_i, \nabla_j]$ on a general tensor. For example, explicit computation will give, for a second rank tensor, the result:

$$[\nabla_b, \nabla_a]S^{mn} = \nabla_b \nabla_a S^{mn} - \nabla_a \nabla_b S^{mn} = R^m{}_{rba}S^{rn} + R^n{}_{rba}S^{mr}. \quad (5.21)$$

The generalization to a tensor of arbitrary rank should be obvious. It can be explicitly computed by working in a local inertial frame in which the Γ^a_{bc} vanish but their derivatives do not. We then have, for a tensor of arbitrary rank, in a local inertial frame:

$$\nabla_k \nabla_n (T_{a\ldots}{}^{b\ldots}) = \partial_k \left(\partial_n T_{a\ldots}{}^{b\ldots} + T_{a\ldots}{}^{s\ldots}\Gamma^b_{sn} + \cdots - T_{s\ldots}{}^{b\ldots}\Gamma^s_{an} - \cdots \right)$$
$$= \partial_k \partial_n T_{a\ldots}{}^{b\ldots} + T_{a\ldots}{}^{s\ldots}\partial_k \Gamma^b_{sn} + \cdots - T_{s\ldots}{}^{b\ldots}\partial_k \Gamma^s_{an}. $$
$$(5.22)$$

In the same frame we also have $R^l{}_{cnk} = \partial_n \Gamma^l_{ck} - \partial_k \Gamma^l_{cn} = -2\partial_{[k}\Gamma^l_{n]c}$ so that we can write Eq. (5.22) as

$$[\nabla_k, \nabla_n]T_{a\ldots}{}^{b\ldots} = T_{a\ldots}{}^{s\ldots}R^b{}_{skn} - \cdots - T_{s\ldots}{}^{b\ldots}R^s{}_{akn} + \cdots. \quad (5.23)$$

The 'rule' here is similar to that for the covariant derivative. If the tensor has N indices, then the commutator $[\nabla_k, \nabla_n]$ has N 'correction' terms, each made of a product of original tensor and curvature tensor with index placement $R^a{}_{bkn}$. The term which 'corrects' a contravariant index has a plus sign and that which 'corrects' a covariant index has a minus sign. When a particular index is corrected, that index is moved to the curvature tensor and is replaced by a dummy index. There is only one way of doing this.

Exercise 5.1

Curvature in the Newtonian approximation Consider the metric g_{ab} with $g_{00} = -(1+2\phi)$

and $g_{\alpha\beta} = \delta_{\alpha\beta}$. Compute the Christoffel symbols and the curvature tensor for this metric to the lowest order. [Answer. $R^{\alpha}{}_{0\beta 0} = \partial^{\alpha}\partial_{\beta}\,\phi$.]

5.2.3 *Tidal acceleration produced by gravity*

We shall now take up the most important effect of curvature, viz. how it acts as a manifestation of real gravity by changing the relative separation between the nearby geodesics, thereby providing a third perspective on curvature.

The importance of this effect can be understood from an elementary consideration of the principle of equivalence. Consider a small, closed, box located near the surface of Earth (where the acceleration due to gravity is g) and another similar box located in interstellar space which is moving with an acceleration g. The principle of equivalence states that the mechanical experiments involving gravity will yield identical results in both the contexts. This is true at the lowest order of accuracy in which we can treat the acceleration due to the gravity of Earth as uniform over the size of the box. If the scale over which the acceleration due to gravity is varying is L and the size of the box is l, then we need to ensure that $l \ll L$. If ϕ is the gravitational potential, then $g \approx \partial\phi$ and $L^{-2} \approx \phi^{-1}\partial^2\phi$. Our condition, therefore, requires $(l/L)^2 \approx \phi^{-1}[l^2\partial^2\phi] \ll 1$. When this condition is satisfied, the Earth's gravitational field is indistinguishable from the one produced by an accelerated frame. The latter, we know, is merely a coordinate effect and does not represent genuine curvature of spacetime. It follows that the genuine gravitational effects cannot be revealed in a region of size l, where $(l/L)^2 \ll 1$. This is also obvious from the fact that treating g as constant and studying its effects is equivalent to working with approximately constant Γ^a_{bc} without introducing any effects due to derivatives of the Christoffel symbols. (Recall the correspondence $\phi \Leftrightarrow g_{ab}$; $\nabla\phi \Leftrightarrow \Gamma^a_{bc}$; $\nabla\nabla\phi \Leftrightarrow \partial\Gamma^a_{bc}$, etc.) As we have seen in the previous chapter, nonzero Christoffel symbols can arise either due to curvilinear coordinates or due to curvature of spacetime, and the nonzero Christoffel symbols alone cannot imply the existence of genuine gravitational effects.

To probe genuine gravitational effects, we need to proceed to the next order. In the context of the simple example discussed above, let us consider an experiment in which two different masses are dropped from a height within the box. In reality, since the Earth is spherical and not flat, both these particles will be moving towards the centre of the Earth. Hence the separation between the two particles will decrease as they fall towards the floor. This effect arises owing to the $\partial_i\partial_j\phi$ terms and will be absent if we treat the gravitational field as uniform over the box. (In particular, this effect will be absent in the box accelerating in interstellar space.) It is this change in the separation between the trajectories of two nearby particles which probes the genuine gravitational effects due to spacetime curvature. In relativity, the trajectories of free particles are geodesics and hence the acceleration of the deviation vector between two nearby geodesics should contain information about the curvature of the spacetime. We shall now obtain the relation between *geodesic deviation* and curvature.

The effect of curvature on the relative acceleration of geodesics is easy to see even in two-dimensional examples. Consider a flat two-dimensional space in which two geodesics (which are straight lines) start from some point A with an angle θ between them. We know that this angle does not change as we move along the geodesics and – obviously – the geodesics will not intersect again. The separation between geodesics grows linearly with distance and the relative acceleration vanishes. Consider next a similar situation on the surface of a sphere: two geodesics (which are great circles) start from some point A of the surface and makes an angle θ at A. These geodesics, however, will *not* maintain the same angle between them. In fact they will eventually intersect again. This is yet another manifestation of the curvature and we will expect the curvature tensor to determine the rate at which the separation between the nearby geodesics changes.

To set the stage, let us first consider a bunch of particles in the non-relativistic, Newtonian, gravity moving along trajectories $x^\alpha(t, n)$, where n labels the trajectory and t is the absolute Newtonian time. The separation between any two nearby trajectories is given by the vector $n^\alpha \equiv (\partial x^\alpha/\partial n)$. We are interested in the relative acceleration of two nearby trajectories, which is given by:

$$\frac{\partial^2 n^\alpha}{\partial t^2} = \frac{\partial^2}{\partial t^2}\left(\frac{\partial x^\alpha}{\partial n}\right) = \frac{\partial}{\partial n}\left(\frac{\partial^2 x^\alpha}{\partial t^2}\right) = \frac{\partial}{\partial n}\left(-\frac{\partial \phi}{\partial x^\alpha}\right) = -n^\beta \frac{\partial}{\partial x^\beta}\left(\frac{\partial \phi}{\partial x^\alpha}\right)$$
$$= -n^\beta \frac{\partial^2 \phi}{\partial x^\beta \partial x^\alpha}, \tag{5.24}$$

where we have temporarily ignored the positioning of the indices. We see that this quantity is proportional to the second derivative of the gravitational potential while the usual gravitational force is proportional to the first derivative. It is precisely this quantity which measures those effects of gravity that cannot be made to vanish by transforming to a freely falling frame.

Let us now study the same question in relativistic gravity. Consider a bunch of particles moving along geodesics in a spacetime. Let $x^i = x^i(\tau, v)$ denote the family of geodesics, where τ represents an affine parameter along the geodesics and v labels a particular geodesic. The vector $n^i = (\partial x^i/\partial v)\delta v \equiv v^i \delta v$ denotes the deviation between two neighbouring geodesics parametrized by the values v and $v + dv$. From the definition of the covariant derivative and the relation $(\partial u^i/\partial v) = \partial v^i/\partial \tau$ (where u^i is the tangent vector to the geodesic), it follows that

$$v^k \nabla_k u^i = u^k \nabla_k v^i. \tag{5.25}$$

Consider now the variation of the separation vector v^i between two neighbouring geodesics. We have

$$\frac{D^2 v^i}{D\tau^2} \equiv u^l \nabla_l \left(u^k \nabla_k v^i \right) = u^l \nabla_l \left(v^k \nabla_k u^i \right) = u^l v^k \nabla_l \nabla_k u^i + u^l \nabla_l v^k \nabla_k u^i. \tag{5.26}$$

Using Eq. (5.25) in the second term and changing the order of covariant derivatives in the first term we get

$$\frac{D^2 v^i}{D\tau^2} = v^k \nabla_k \left(u^l \nabla_l u^i \right) + u^m R^i{}_{mkl} u^k v^l. \tag{5.27}$$

The first term vanishes since $u^l \nabla_l u^i = 0$ along the geodesics and we are left with

$$\frac{D^2 v^i}{D\tau^2} = R^i{}_{klm} u^k u^l v^m. \tag{5.28}$$

This result shows how the curvature generates the geodesic deviation and is called the *geodesic deviation equation*. In the Newtonian approximation, we can set $u^i \approx \delta^i_0$ and replace the derivative on the left hand side by $\partial^2/\partial t^2$. This gives

$$\frac{\partial^2 v^i}{\partial t^2} = R^i{}_{00m} v^m = -R^i{}_{0m0} v^m. \tag{5.29}$$

Comparing with Eq. (5.24), we find that, in the Newtonian, non-relativistic approximation the only non-vanishing component of the curvature that is relevant is

$$R^\alpha{}_{0\beta 0} = \partial^\alpha \partial_\beta \phi, \tag{5.30}$$

which agrees with the result of Exercise 5.1.

It is instructive to obtain the result in Eq. (5.28) using index-free analysis. In this case, the deviation vector corresponds to the operator $\boldsymbol{v} = (\partial/\partial v)$ while the tangent vector is $\boldsymbol{u} = (\partial/\partial s)$. Therefore $[\boldsymbol{v}, \boldsymbol{u}] = 0$ implying $\mathcal{L}_{\boldsymbol{v}} \boldsymbol{u} = 0$ leading to

$$0 = [\boldsymbol{v}, \boldsymbol{u}] = \mathcal{L}_{\boldsymbol{v}} \boldsymbol{u} = \nabla_{\boldsymbol{v}} \boldsymbol{u} - \nabla_{\boldsymbol{u}} \boldsymbol{v}. \tag{5.31}$$

From Eq. (5.10) we have

$$R^a{}_{bcd} u^b u^c v^d = \nabla_{\boldsymbol{u}} \nabla_{\boldsymbol{v}} \, \boldsymbol{u} - \nabla_{\boldsymbol{v}} \nabla_{\boldsymbol{u}} \, \boldsymbol{u} - \nabla_{[\boldsymbol{u},\boldsymbol{v}]} \, \boldsymbol{u} = \nabla_{\boldsymbol{u}} \nabla_{\boldsymbol{u}} \, \boldsymbol{v}. \tag{5.32}$$

To get the second equality we have used $[\boldsymbol{v}, \boldsymbol{u}] = 0$ as well as the geodesic equation $\nabla_{\boldsymbol{u}} \boldsymbol{u} = 0$ and Eq. (5.31). The economy and transparency achieved in the index-free notation is obvious.

The geodesic deviation in Eq. (5.28) can also be given an alternate interpretation in the case of rigid bodies in free fall. In such a case, the worldline of the centre of mass of the body will be on a timelike *geodesic*. But some other point on the rigid body is constrained to move on a curve which is at a fixed spatial distance from the centre of mass. In general, such a point cannot be on a geodesic. It is necessary that internal stresses must act on the body to keep all the parts together in a rigid shape. In this particular case, the acceleration of the geodesic deviation vector in Eq. (5.28) will be equal to the internal forces (per unit mass) which are needed to keep a point moving on a curve at a fixed distance from the geodesic followed by the center of mass. Mathematically, these stresses can be calculated by first constructing the tensor $S^m{}_n \equiv R^m{}_{srn} u^s u^r$ and then determining its eigenvalues by solving the equation $S^m{}_n v^n = \lambda v^m$. This equation has one zero eigenvalue with $v^i = u^i$ and three other nontrivial eigenvalues which give the principal stresses on the body.

Exercise 5.2
Non-geodesic deviation Show that when the family of curves are not geodesics, the deviation vector satisfies the equation

$$\frac{D^2 v^k}{D\tau^2} = R^k{}_{bcd} u^b u^c v^d + v^j \nabla_j a^k, \tag{5.33}$$

where $a^k = u^i \nabla_i u^k$ is the acceleration.

Exercise 5.3
Measuring the curvature tensor Explain how the geodesic deviation equation can be used to measure the components of the curvature tensor. Give careful thought to the issue that geodesic deviation gives $R^a{}_{bcd} u^b n^c n^d$ only for a spacelike vector n^c and a timelike vector u^a.

Exercise 5.4
Spinning body in curved spacetime Show that the equations of motion governing the spin of a non-spherical spinning body in an inhomogeneous gravitational field is

$$u^i \nabla_i S^k = \epsilon^{kbam} u_m u^s u^l (Q_{bn} R^n{}_{sal}), \tag{5.34}$$

where: (i) Q_{bn} is the quadrupole moment tensor orthogonal to four-velocity such that in the rest frame it reduces to

$$Q^{\mu\nu} = \int d^3 x \rho \left[x^\mu x^\nu - \frac{1}{3} r^2 \delta^{\mu\nu} \right], \tag{5.35}$$

and (ii) the curvature is assumed to be approximately constant over the size of the body.

[Hint. Work everything out in the comoving frame of the centre of mass of the body which is following a geodesic. Express the relative acceleration of a mass element at position x^μ using a geodesic equation and hence show that the ν component of the torque per unit volume is $-\rho \epsilon^\nu{}_{\lambda\rho} x^\lambda x^\kappa R^\rho{}_{0\kappa 0}$. Integrating over the body, one can obtain dS^ν/dt. Use the symmetries of the curvature tensor and define Q_{ij} appropriately to obtain the result.]

5.3 Properties of the curvature tensor

The curvature tensor will play a crucial role in all our future discussions. So, before studying its role in gravitational physics, we will first describe several important mathematical properties of this tensor. Writing $R_{iklm} = g_{in}R^n{}_{klm}$ and expanding it out in terms of metric and Christoffel symbols, we get

$$R_{iklm} = \frac{1}{2}\left(\partial_k\partial_l g_{im} + \partial_i\partial_m g_{kl} - \partial_k\partial_m g_{il} - \partial_i\partial_l g_{km}\right) + g_{np}\left(\Gamma^n_{kl}\Gamma^p_{im} - \Gamma^n_{km}\Gamma^p_{il}\right).$$
(5.36)

To analyse its structure, it is convenient to work in a local inertial frame with $g_{ij} = \eta_{ij}$; $\Gamma^i_{jk} = 0$. In such a frame, the curvature tensor becomes

$$R_{abcd} = \frac{1}{2}(\partial_c\partial_b g_{ad} + \partial_d\partial_a g_{bc} - \partial_d\partial_b g_{ac} - \partial_c\partial_a g_{bd}).$$
(5.37)

This expression allows us to read off all the required symmetries and properties of the curvature tensor.

5.3.1 Algebraic properties

To begin with, we have $R_{abcd} = -R_{abdc}$ which follows directly from our definition. It reflects the fact that the last two covariant indices in $R^a{}_{bcd}$ arise from the $[\nabla_c, \nabla_d]$ operation, which is antisymmetric. Next, we see that

$$R_{abcd} = -R_{bacd}.$$
(5.38)

This is a consequence of the fact that $\nabla_i g_{ab} = 0$. In fact, we can see this directly from

$$0 = [\nabla_c, \nabla_d]\, g_{ab} = R^l{}_{acd}g_{lb} + R^l{}_{bcd}g_{al} = (R_{abcd} + R_{bacd}).$$
(5.39)

Third, we can show that the cyclic sum on the three indices is zero:

$$R_{i[klm]} = R_{iklm} + R_{imkl} + R_{ilmk} = 0.$$
(5.40)

This can be understood by applying $[\nabla_c, \nabla_d]$ to a gradient $\nabla_b\phi$ of a scalar ϕ. We have $\nabla_{[c}\nabla_d\nabla_{b]}\,\phi = 0$ for any scalar, implying

$$R^l{}_{[bcd]}\nabla_l\phi = 0.$$
(5.41)

Since this is true for all gradients, taking $\phi(x) = x^k$, say with $\nabla_i\phi = \delta^k_i$, immediately gives the result. It is also obvious from Eq. (5.37) that the curvature tensor is symmetric under the pair exchange

$$R_{abcd} = R_{cdab}.$$
(5.42)

These symmetries significantly reduce the number of independent components of the curvature tensor. In view of the future applications, we shall work out the number of independent components of the curvature tensor in N dimensions. Originally, in the absence of any symmetry, the curvature tensor has N^4 independent components in N dimensions. The antisymmetry in ab and cd in R_{abcd} shows that there are $M = (1/2)N(N-1)$ ways of choosing independent pairs among this. But since the tensor is symmetric under the exchange of these pairs, there are $(1/2)M(M+1)$ independent ways of choosing the combination $abcd$. The last symmetry we need to take into account is the cyclic identity in Eq. (5.40). The pair symmetry guarantees that $R_{a[bcd]} = 0$ is trivial unless $abcd$ are all distinct. Hence the number of independent constraints from this equation is the same as the number of combinations of four objects which can be chosen from N objects. This is given by $^{N}C_4 = [N!/4!(N-4)!]$ with the interpretation that $^{N}C_4 = 0$ for $N < 4$. This condition correctly takes into account the fact that in dimensions less than four, Eq. (5.40) does not lead to any new constraint. Therefore, the number of independent components of the curvature tensor is

$$\frac{1}{2}M(M+1) - \frac{N!}{(N-4)!\,4!} = \frac{N^2(N^2-1)}{12}. \tag{5.43}$$

We have seen in Section 4.2 (see Eq. (4.7)) that the number of independent components in the second derivatives of the metric $\partial_a\partial_b g_{cd}$ that cannot be made to vanish at an event by a suitable choice of coordinates is also given by $N^2(N^2-1)/12$. Equation (5.43) shows that the number of independent components of the curvature tensor matches exactly with the number of independent terms involving second derivatives of the metric tensor which survive after we have made the coordinate transformation to a local inertial frame at a given event.

In fact, we can use this fact to improve upon the coordinate system we described in Section 4.2 by specifying the metric up to quadratic order in the coordinates, along the following lines. Let us assume that we have transformed to a locally inertial coordinate system at some event $\mathcal{P}$ so that $g_{ab} = \eta_{ab}$ and $\Gamma^i_{jk} = 0$. We will now make a coordinate transformation

$$x^n = x'^n + \frac{1}{6}D^n_{pqr}\,x'^p x'^q x'^r,$$

$$L^n_i = \frac{\partial x^n}{\partial x'^i} = \delta^n_i + \frac{1}{2}D^n_{pqi}x'^p x'^q. \tag{5.44}$$

Since the new coordinates differ from the old only at cubic order, this transformation will not change either the metric or Christoffel symbols at the event $\mathcal{P}$ (which is taken to be the origin of both coordinate systems). Let us consider how it affects the first derivatives of the Christoffel symbols, which are given by

$$\partial_n\Gamma^m_{ab} = \partial_n\left(\Gamma'^r_{ik}L^m_r L^i_a L^k_b - L^i_a L^k_b\,\partial_k L^m_i\right) = \partial_n\Gamma'^m_{ab} - D^m_{abn}, \tag{5.45}$$

where the second equality comes from using the fact that $\Gamma^a_{bc} = 0$ at $\mathcal{P}$. The coefficients D^m_{abn} are completely symmetric in the lower three indices whereas $\partial_n \Gamma^a_{bc}$ has no such symmetry. But by making the choice

$$D^m_{abn} = \frac{1}{3} \left(\partial_n \Gamma'^m_{ab} + \partial_b \Gamma'^m_{na} + \partial_a \Gamma'^m_{bn} \right) \tag{5.46}$$

we can ensure that the fully symmetric part vanishes:

$$\partial_n \Gamma^m_{ab} + \partial_b \Gamma^m_{na} + \partial_a \Gamma^m_{bn} = 0. \tag{5.47}$$

Further, in this frame the curvature tensor is $R^r{}_{msq} = \partial_s \Gamma^r_{mq} - \partial_q \Gamma^r_{ms}$. Hence we get, on using Eq. (5.47):

$$R^r{}_{msq} + R^r{}_{smq} = -3\partial_q \Gamma^r_{ms}. \tag{5.48}$$

Relating the first derivatives of Γ^i_{jk} to second derivatives of g_{ab} in this result and using the symmetries of the curvature tensor, we get:

$$\partial_n \partial_b g_{ia} = g_{mi} \partial_n \Gamma^m_{ab} + g_{ma} \partial_n \Gamma^m_{ib} = \frac{1}{3} \left(R_{iban} + R_{abin} \right). \tag{5.49}$$

To find the metric correct to quadratic order in the coordinates, we can integrate this equation, treating the right hand side as constant to the required order of accuracy. This gives the metric in these coordinates to be:

$$g_{ia} = \eta_{ia} - \frac{1}{3} \left(R_{iban} + R_{abin} \right) x^b x^n. \tag{5.50}$$

This result shows explicitly that one can set up a coordinate system at some event $\mathcal{P}$ such that $g_{ab} = \eta_{ab}$, $\Gamma^i_{jk} = 0$ with the second derivatives of the metric traded off for the components of curvature tensor. (Such a coordinate system is sometimes called the *Riemann normal coordinates*.) The deviation of the metric from η_{ij} is at quadratic order in coordinates and is determined by the curvature tensor.

Exercise 5.5

Explicit transformation to the Riemann normal coordinates The transformations in Eq. (5.44) give an implicit relation between a general coordinate system and the Riemann normal coordinates. Often, it is convenient to have an explicit transformation from general coordinates z^a to the Riemann normal coordinates x^a around a given event accurate to quadratic order of accuracy in the metric. Show that such a transformation and its inverse are given by

$$x^m = z^m + \frac{1}{2} \Gamma^m_{ab} z^a z^b + \frac{1}{6} \left(\Gamma^m_{at} \Gamma^t_{br} + \partial_r \Gamma^m_{ab} \right) z^a z^b z^r + \cdots \tag{5.51}$$

$$z^m = x^m - \frac{1}{2} \Gamma^m_{ab} x^a x^b + \frac{1}{6} \left(\Gamma^m_{at} \Gamma^t_{br} - \partial_r \Gamma^m_{ab} \right) x^a x^b x^r + \cdots . \tag{5.52}$$

Exercise 5.6
Curvature tensor in the language of gauge fields In Project 4.2 (assuming you did it!), we saw that the field tensor for a gauge field can be given a matrix representation as in Eq. (4.164). Show that this equation also applies to R^l_{mik} if we treat l and m as matrix indices and introduce the matrices $(\Gamma_k)^a_b = \Gamma^i_{kb}$ as described in Project 4.2. That is, prove that the curvature tensor can be written in the form

$$R_{ik} = \partial_i \Gamma_k - \partial_k \Gamma_i + [\Gamma_i, \Gamma_k], \tag{5.53}$$

where each term is interpreted as a matrix.

5.3.2 Bianchi identity

Another important property of the curvature tensor is not an algebraic one but a differential identity which it satisfies. This relation – which will play a key role in the formulation of field equations of gravity in Chapter 6 – is called the *Bianchi identity* and states that:

$$\nabla_m R^n_{ikl} + \nabla_l R^n_{imk} + \nabla_k R^n_{ilm} = 0. \tag{5.54}$$

Once again, this is most easily proved in the local inertial frame. Differentiating the curvature tensor and then using $\Gamma^i_{jk} = 0$ leads to the result, valid in the local inertial frame around an event,

$$\nabla_m R^n_{ikl} = \partial_m R^n_{ikl} = \partial_m \partial_k \Gamma^n_{il} - \partial_m \partial_l \Gamma^n_{ik}. \tag{5.55}$$

From the form of this expression, it is easy to verify by inspection that Eq. (5.54) holds.

The Bianchi identity is also important in the mathematical problem of determining a metric given the form of the curvature tensor. In four dimensions, this problem reduces to the following. We are given a tensor R^a_{bcd} with the symmetries of a curvature tensor and the 20 independent components of this tensor are specified as functions of coordinates. We need to determine whether this set is made of the components of a valid curvature tensor that arises from a metric or not. This will require solving a set of 20 second order differential equations for the 10 metric functions g_{ab}. Obviously, such a system is over-constrained and will not, in general, possess consistent solutions. One needs to satisfy some additional integrability conditions for these equations which should involve the third derivatives of the metric tensor. It turns out that the Bianchi identity is necessary and sufficient for a given tensor to be derivable as a curvature tensor of some metric.

As an interesting example of a curvature tensor which *cannot* be obtained from a spacetime metric, let us consider the question of representing the geodesic deviation in Newtonian gravity in terms of a possible metric tensor. In Newtonian theory, we use an absolute time coordinate t and Cartesian space coordinates x^α with a gravitational potential ϕ. The equation of motion $\ddot{\boldsymbol{x}} = -\nabla\phi$ can be written in four-dimensional form as

$$\frac{d^2 t}{d\lambda^2} = 0; \qquad \frac{d^2 x^\alpha}{d\lambda^2} + \partial^\alpha \phi \left(\frac{dt}{d\lambda}\right)^2 = 0, \tag{5.56}$$

with λ being an affine parameter. This can arise from a geodesic equation with the Christoffel symbol $\Gamma^\alpha_{00} = \partial^\alpha \phi$ with all other Christoffel symbols vanishing. From this, we find that the curvature tensor should have the form

$$R^\alpha{}_{0\beta 0} = -R^\alpha{}_{00\beta} = \partial^\alpha \partial_\beta \phi, \tag{5.57}$$

with all other components vanishing. The question is whether these Christoffel symbols and the curvature tensor can arise from a metric. If so, we must have the component $R_{\alpha 0 \beta 0} = g_{\alpha\delta} R^\delta{}_{0\beta 0}$ to be nonzero since the right hand side is, in general, non-vanishing. But

$$R_{0\alpha\beta 0} = g_{0\delta} R^\delta{}_{\alpha\beta 0} = 0 \neq -R_{\alpha 0 \beta 0}. \tag{5.58}$$

This clearly violates the symmetry property of the curvature tensor and hence there is no spacetime metric from which we can obtain the Newtonian gravity *exactly*. While this result is somewhat of a curiosity, it has the mathematical significance that the geometrization of Newtonian gravity cannot be based on a *spacetime* with a metric.

5.3.3 Ricci tensor, Weyl tensor and conformal transformations

From the curvature tensor, we can obtain a second rank tensor by contracting on a pair of indices. Since R_{abcd} is antisymmetric in ab and cd, the only nontrivial contraction which is possible is between a and c (or between a and d which merely changes the sign). This way we arrive at the *Ricci tensor* defined by

$$R_{ik} = g^{lm} R_{limk} = R^l{}_{ilk} = \partial_l \Gamma^l_{ik} - \partial_k \Gamma^l_{il} + \Gamma^l_{ik} \Gamma^m_{lm} - \Gamma^m_{il} \Gamma^l_{km}. \tag{5.59}$$

This tensor is clearly symmetric. Contracting once again, we can obtain the *Ricci scalar* or *scalar curvature* defined by

$$R = g^{ik} R_{ik} = g^{il} g^{km} R_{iklm}. \tag{5.60}$$

From the Bianchi identity in Eq. (5.54) we can obtain a corresponding identity for the Ricci scalar obtained by contracting the Bianchi identity on ik and lm. This gives

$$\nabla_i \left(R^i_k - \frac{1}{2}\delta^i_k R\right) = 0. \tag{5.61}$$

The combination that occurs inside the bracket in the above expression is of special importance and is called the *Einstein tensor*:

$$G_{ik} \equiv R_{ik} - \frac{1}{2}g_{ik}R. \tag{5.62}$$

In view of the key role it will play later on, we will provide a direct geometric interpretation for this tensor. Consider an observer moving with a four-velocity u^i through the spacetime. The projection tensor $h^i_k = \delta^i_k + u^i u_k$ to the space orthogonal to the four-velocity of the observer can be used to obtain components of the curvature tensor which are purely spatial in the instantaneous rest frame of the observer. These are given by the equation

$$\mathcal{R}_{ijkl} = h^a_i h^b_j h^c_k h^d_l R_{abcd}. \tag{5.63}$$

It should be stressed that $\mathcal{R}_{ijkl}$ is *not* the curvature tensor $^{(3)}R_{ijkl}$ of the 3-space orthogonal to u^i but is just a projection defined as above. (The relation between $\mathcal{R}_{ijkl}$ and $^{(3)}R_{ijkl}$ is discussed in Chapter 12; see Eq. (12.25).) We can build a scalar out of $\mathcal{R}_{ijkl}$ by contracting on i and k and j and l leading to

$$\begin{aligned} \mathcal{R} = h^{ac}h^{bd}R_{abcd} &= (g^{ac} + u^a u^c)(g^{bd} + u^b u^d)R_{abcd} \\ &= (g^{ac} + u^a u^c)(R_{ac} + u^b u^d R_{abcd}) \\ &= R + 2u^a u^c R_{ac} = 2u^a u^c G_{ac}. \end{aligned} \tag{5.64}$$

It follows that the scalar built out of the projection of curvature components orthogonal to a world line, as measured by an observer with four-velocity u^i is proportional to $G_{ac}u^a u^c$. It measures the local scalar curvature of the spatially projected curvature tensor.

The contractions of $R^a{}_{bcd}$ have led to nonzero results in terms of R_{ij} and R. This suggests the possibility that one may be able to subtract from $R^a{}_{bcd}$ a suitable fourth rank tensor built from R_{ij}, g_{ab} and R so that the resulting tensor is tracefree on all indices; i.e. the contraction on any two indices of such a tensor will give zero. This is indeed true and this tensor, called the *Weyl tensor*, is defined by

$$C_{abcd} = R_{abcd} - \frac{2}{N-2}\left(g_{a[c}R_{d]b} - g_{b[c}R_{d]a}\right) + \frac{2}{(N-1)(N-2)}R\,g_{a[c}\,g_{d]b} \tag{5.65}$$

in $N \geq 3$-dimension. In four dimensions, this reduces to the form

$$C_{iklm} = R_{iklm} - R_{l[i}\,g_{k]m} + R_{m[i}\,g_{k]l} + \frac{1}{3}R\,g_{l[i}\,g_{k]m} \tag{5.66}$$

$$\begin{aligned} = R_{iklm} &- \frac{1}{2}R_{il}g_{km} + \frac{1}{2}R_{im}g_{kl} + \frac{1}{2}R_{kl}g_{im} - \frac{1}{2}R_{km}g_{il} \\ &+ \frac{1}{6}R(g_{il}g_{km} - g_{im}g_{kl}). \end{aligned}$$

It is fairly straightforward to verify that this tensor has the same algebraic symmetries as the curvature tensor. Furthermore, the contraction of this tensor on any two indices vanishes. The above relation allows us to express the curvature tensor as a sum of a Weyl tensor and a tensor built from R_{ij}, g_{ab} and R. Since R_{ab} has ten independent components in four dimension while $R^a{}_{bcd}$ has 20 it follows that the Weyl tensor has ten independent components.

Exercise 5.7

Conformal transformations and curvature Consider a conformal transformation of the form $g_{ab} \to \bar{g}_{ab} \equiv \Omega^2(x)g_{ab}$. (In general relativity, conformal transformations are defined through this modification of the metric – and related quantities – with no change in the coordinates. Sometimes, in flat spacetime, the scaling transformation $x^i \to \lambda x^i$ is called conformal transformation. In this book, conformal transformations are defined as the change in the metric.)

(a) We have seen earlier that under such a transformation the Christoffel symbols change according to Eq. (4.73). Using this, obtain the transformation properties of curvature tensor, Ricci tensor and scalar curvature in an N-dimensional space. Show that

$$\bar{R}^a{}_{bcd} = R^a{}_{bcd} - 2\left(\delta^a_{[c}\delta^e_{d]}\delta^f_b - g_{b[c}\delta^e_{d]}g^{af}\right)\frac{\nabla_e\nabla_f\Omega}{\Omega}$$

$$+ 2\left(2\delta^a_{[c}\delta^e_{d]}\delta^f_b - 2g_{b[c}\delta^e_{d]}g^{af} + g_{b[c}\delta^a_{d]}g^{ef}\right)\frac{(\nabla_e\Omega)(\nabla_f\Omega)}{\Omega^2}, \tag{5.67}$$

$$\bar{R}_{bc} = R_{bc} + \left[(N-2)\delta^f_b\,\delta^e_c + g_{bc}g^{ef}\right]\frac{\nabla_e\nabla_f\Omega}{\Omega}$$

$$- \left[2(N-2)\delta^f_b\,\delta^e_c - (N-3)g_{bc}g^{ef}\right]\frac{(\nabla_e\Omega)(\nabla_f\Omega)}{\Omega^2}, \tag{5.68}$$

$$\bar{R} = \frac{R}{\Omega^2} + 2(N-1)g^{ef}\frac{\nabla_e\nabla_f\Omega}{\Omega^3} + (N-1)(N-4)g^{ef}\frac{(\nabla_e\Omega)(\nabla_f\Omega)}{\Omega^4}. \tag{5.69}$$

(b) Consider a scalar field in a four-dimensional curved spacetime described by the action

$$\mathcal{A} = -\frac{1}{2}\int d^4x\,\sqrt{-g}\,\partial_i\phi\,\partial^i\phi. \tag{5.70}$$

Assume that, under the conformal transformation, ϕ changes as $\phi \to \Omega^n\phi$ with some n. Show that this action cannot be made conformally invariant with any possible choice for n. On the other hand, show that the action in

$$\mathcal{A} = -\int d^4x\,\sqrt{-g}\,\left(\frac{1}{2}\partial_i\phi\,\partial^i\phi + \frac{1}{6}R\phi^2\right), \tag{5.71}$$

where R is the scalar curvature, will lead to a field equation $[\Box - (1/6)R]\phi = 0$, which is conformally invariant if ϕ is also transformed as $\phi \to \Omega^{-1}\phi$.

(c) Show that all two-dimensional space(time)s are locally conformally flat irrespective of their signature. That is, show that given any two-dimensional metric $g_{ab}(x)$ one can find a coordinate transformation from x^i to x'^i such that $g'_{ab}(x') = \Omega^2(x')\,\eta_{ab}$ [Hint. Assume that the new coordinates are given by the functions $x'^1 = \alpha(x^1, x^2)$, $x'^2 = \beta(x^1, x^2)$. Show that the condition $g'^{12} = 0$ can be satisfied by the choice $\partial_k\alpha = \kappa(x)\,\epsilon_{ab}\,g^{bc}\,\partial_c\beta$, where κ is an arbitrary function. Determine the second condition suitably so as to reduce the metric to conformally flat form. Make sure that you cover all possible signatures.]

Exercise 5.8

Splitting the spacetime and its curvature Consider a D-dimensional space with coordinates $x^A = (x^a, x^\mu)$ where the Latin letters go over $1, 2...n$ and the Greek letters go over $(n+1), (n+2), \ldots D$. Further assume that the metric g_{AB}, when expressed as a matrix, is block diagonal with g_{ab} depending only on the coordinates x^a, and $g_{\alpha\beta}$ depending only on the coordinates x^μ with no off-diagonal terms like $g_{a\mu}$. Show that the curvature tensor also gets factorized in an exactly similar manner.

The Weyl tensor has another useful property which is straightforward (but tedious) to prove.[1] From the results of Exercise 5.7 for the transformation of curvature tensor, etc., under a conformal transformation $g_{ab} \rightarrow \Omega^2(x)g_{ab}$, one can directly verify that the Weyl tensor transforms as $C_{abcd} \rightarrow \Omega^2(x)C_{abcd}$ while $C^a{}_{bcd}$ is conformally invariant in any dimension N. From this it follows that the Weyl tensor vanishes in any spacetime which is conformally flat. Therefore, in any conformally flat spacetime with a metric of the form $g_{ab} = f^2(x)\eta_{ab}$, the curvature tensor can be expressed entirely in terms of R_{ij}, g_{ab} and R. The converse is also true for all spacetimes with $N > 3$; that is, in four and higher dimensions, the vanishing of the Weyl tensor ensures that the spacetime is conformally flat. The situation is special for $N = 2, 3$. All $N = 2$ spaces are conformally flat (see Exercise 5.8) and the Weyl tensor is undefined for $N = 2$. In $N = 3$ the Weyl tensor vanishes identically but all three-dimensional spaces are *not* conformally flat. We will briefly describe the reason for this peculiarity for the sake of completeness. If a space is indeed conformally flat with $g_{ab} = f^2(x)\eta_{ab}$, then one should be able to make a conformal transformation from g_{ab} to η_{ab}. Since we can determine how the curvature tensor changes under the conformal transformation, demanding that the new curvature tensor should vanish will lead to a differential equation for $f(x)$. The integrability condition for this equation to possess solutions turns out to be identically satisfied if the Weyl tensor vanishes in $N > 3$. But in $N = 3$, one term in the relevant equation (which has a coefficient $(N - 3)$) vanishes and the integrability requires an additional condition to be satisfied:

$$C^{\alpha\beta} \equiv 2\epsilon^{\alpha\gamma\delta}\nabla_\delta[R^\beta_\gamma - \frac{1}{4}\delta^\beta_\gamma R] = 0. \tag{5.72}$$

This tensor $C^{\alpha\beta}$ is called the Cotton–York tensor.

Exercise 5.9

Matrix representation of the curvature tensor From the symmetries of the curvature tensor, it is clear that one can write it in an unambiguous form as $R^{ab}_{cd} \equiv R^{ab}{}_{cd}$. Associate an index $I = 1, 2, ...6$ with each of the pairs $01, 02, 03, 23, 31, 12$ of independent values ab and cd

can take. Hence show that, in a suitable basis, the curvature tensor can be expressed in the form of a 6×6 matrix corresponding to

$$
M_J^I = \begin{array}{c} \\ 01 \\ 02 \\ 03 \\ 23 \\ 31 \\ 12 \end{array}
\begin{array}{c}
\overset{\displaystyle 01 \quad 02 \quad 03 \qquad 23 \quad 31 \quad 12}{
\left(
\begin{array}{ccc|ccc}
& E & & & H & \\
& & & | & & \\
- & - & - & - & - & - \\
& & & | & & \\
& -H^T & & & F & \\
& & & | & &
\end{array}
\right)}
\end{array}, \tag{5.73}
$$

where E, F, H are 3×3 matrices with $E = E^T, F = F^T$ and $\mathrm{Tr}\, H = 0$. The superscript T denotes the transpose of a matrix.

Exercise 5.10
Curvature in synchronous coordinates Using the freedom in the choice of coordinates, one can locally impose the condition $g_{00} = -1$, $g_{0\alpha} = 0$ thereby reducing the line element to the form $ds^2 = -dt^2 + h_{\alpha\beta}dx^\alpha dx^\beta$. Such a coordinate system is called synchronous. Show that the mixed components of the Ricci tensor can be expressed in the form

$$
R_0^0 = -\frac{1}{2}\frac{\partial}{\partial t}\Omega_\alpha^\alpha - \frac{1}{4}\Omega_\alpha^\beta\Omega_\beta^\alpha, \tag{5.74}
$$

$$
R_\alpha^0 = \frac{1}{2}\left(\nabla_\beta\Omega_\alpha^\beta - \nabla_\alpha\Omega_\beta^\beta\right), \tag{5.75}
$$

$$
R_\alpha^\beta = -\,^{(3)}R_\alpha^\beta - \frac{1}{2\sqrt{h}}\frac{\partial}{\partial t}\left(\sqrt{h}\,\Omega_\alpha^\beta\right), \tag{5.76}
$$

where $\Omega_{\alpha\beta} = \partial_0 h_{\alpha\beta}$ and $^{(3)}R_\beta^\alpha$ is the Ricci tensor made from $h_{\alpha\beta}$. All the raising and lowering, as well as covariant differentiation, are carried out with the three-dimensional metric $h_{\alpha\beta}$.

5.4 Physics in curved spacetime

The description of a gravitational field in terms of g_{ik} assumes that all the laws of physics can be suitably generalized from flat spacetime to curved spacetime. The procedure for doing this is straightforward and we have, in fact, used it in the previous chapter without explicitly stating it. Around any event $\mathcal{P}$ we choose a locally inertial frame in which the laws of special relativity are valid. By writing these laws in a covariant manner using suitable tensors and covariant derivatives, we will have a description that is valid in any arbitrary coordinate system around $\mathcal{P}$. Since the form of the equations is expected to be the same (locally) in an accelerated frame and in the presence of gravity, the above procedure allows us to describe physics in curved spacetime. This was the procedure we followed earlier to describe the motion of a particle in a curved spacetime.

Note that this provides a prescription for coupling gravity to any other physical system using the fact that laws of special relativity should be valid in the local inertial frame. One can consider this as a version of the principle of equivalence. While this procedure is consistent and natural, it is not unique. For example, it cannot be used to rule out (or determine) explicit coupling to the curvature. Hence this procedure is sometimes called *minimal coupling*. We shall comment on this issue, when relevant, in the ensuing discussion. In this section, we shall adopt this procedure to indicate how several physical phenomena described in Chapters 1 and 2 are generalized to the curved spacetime.

5.4.1 Particles and photons in curved spacetime

The simplest example is that of a particle moving in a gravitational field, which has already been explored in Section 4.5 by generalizing the action principle to the curved spacetime. In the weak field limit of gravity, the resulting equation of motion reduces to the form obtained in Eq. (4.57).

It is possible to extend these ideas and obtain some general results whenever the spacetime is stationary; that is, when all the metric components are independent of the time coordinate x^0. It is convenient, in this case, to introduce the notation

$$N^2 \equiv -g_{00}, \qquad g_\alpha \equiv -\left(\frac{g_{0\alpha}}{N^2}\right). \tag{5.77}$$

It can be easily verified that the line interval on the three space x^0 = constant is given by

$$dl^2 = (g_{\alpha\beta} - N^2 g_\alpha g_\beta) dx^\alpha dx^\beta \equiv \gamma_{\alpha\beta} dx^\alpha dx^\beta. \tag{5.78}$$

The inverse of the three metric $\gamma_{\alpha\beta}$ is given by $\gamma^{\alpha\beta} = g^{\alpha\beta}$ (see Exercise 4.2).

Since the metric is independent of time, we know that the time component of the momentum $p_i = m u_i = m g_{ik} u^k$ is conserved. Using the form of the line element $ds^2 = -N^2 (dx^0)^2 + dl^2$, the conserved energy can be expressed as

$$E = m N^2 \frac{dx^0}{d\tau} = m N^2 \frac{dx^0}{\sqrt{N^2 (dx^0)^2 - dl^2}}. \tag{5.79}$$

Introducing a velocity

$$v = \frac{dl}{d\tau} = \frac{dl}{N \, dx^0} \tag{5.80}$$

as measured by an observer located at the given point, we can express the energy in the form

$$E = \frac{mN}{\sqrt{1 - v^2}}, \tag{5.81}$$

which is conserved during the motion of the particle.

It is also possible to obtain a general result for the redshift of a photon propagating in this spacetime. Suppose an observer who is at rest (x^α = constant) in this spacetime measures the energy of a photon with four-momentum p^i. The measured energy, which will be proportional to the measured frequency, will be $E = -p_i u^i$, where u^i is the four-velocity of the observer. Taking $u^i = N^{-1}(1, \mathbf{0})$, in which the N^{-1} factor ensures that the velocity is properly normalized with $g_{ij}u^i u^j = -1$, the observed energy of the photon will be $E = -p_0/N$. Further, we know that p_0 is conserved during the propagation of the photon leading to the result that the observed frequency of the photon will vary as $\omega \propto (-g_{00})^{-1/2} \propto N^{-1}$. In many cases, the spacetime will be asymptotically flat with $g_{00} \to -1$ at spatial infinity. We can, therefore, write $\omega(\boldsymbol{x}) = \omega_\infty/N(\boldsymbol{x})$ or $\omega_\infty = \omega(\boldsymbol{x})N(\boldsymbol{x})$. This shows that light emitted from a region of strong gravitational field (where $N \ll 1$) and received at large distances, will be strongly redshifted.

More generally, suppose two observers A and B in the spacetime measure the frequency of a light ray when it intersects their world lines. The ratio of the frequencies as measured by them (which defines the relative red shift factor, usually denoted by z) will be $(1 + z) = (\boldsymbol{p} \cdot \boldsymbol{u}_A/\boldsymbol{p} \cdot \boldsymbol{u}_B)$. On the other hand A and B can also determine the frequencies by measuring the proper time it takes for a wave train to cross them and using $(1 + z) = \Delta\tau_A/\Delta\tau_B$. It is interesting to verify – as an important consistency check – that these two expressions are the same. To do this, let us assume that the 'head' of the wave train is described by a null geodesic corresponding to the phase $\theta = \theta_0$ of the wave and the 'tail' corresponds to the phase $\theta = \theta_0 + \Delta\theta$. The null geodesic has a tangent vector k^a which defines the ray associated with the wave train. Since the ray is normal to the surfaces of constant phase, we have $\boldsymbol{k} = \nabla\theta$. Using the fact that k^a is a null vector we have

$$0 = \boldsymbol{k} \cdot \boldsymbol{k} = \boldsymbol{k} \cdot \nabla\theta = \nabla_{\boldsymbol{k}}\theta, \tag{5.82}$$

showing that the phase is constant along the rays. The phase difference between the head and the tail of the wave train, as measured by either observer, will be $\Delta\theta = (\Delta\tau)(d\theta/d\tau) = (\Delta\tau)(\boldsymbol{u} \cdot \boldsymbol{k})$. The constancy of $\Delta\theta$ implies that $(\Delta\tau_A)/(\Delta\tau_B) = (\boldsymbol{u}_B \cdot \boldsymbol{k})/(\boldsymbol{u}_A \cdot \boldsymbol{k})$, thereby demonstrating the equivalence of the two ways of defining the redshift.

5.4.2 Ideal fluid in curved spacetime

To proceed from single particle dynamics to a system of particles, we have used the concept of the distribution function in Chapter 1. This idea can be generalized to curved spacetime with the conservation law $\partial_a T^a_b = 0$ now being modified to the relation $\nabla_a T^a_b = 0$. This equation will lead to the equations of motion for a fluid

and will be a generalization of Eq. (1.118) obtained in Chapter 1. However, before
we do that, it is important to clarify a crucial difference regarding the interpretation
of the equation $\nabla_a T^{ab} = 0$ in contrast with the equation $\partial_a T_b^a = 0$.

The equation $\partial_a T_b^a = 0$ in special relativity was interpreted as a conservation
law because it could be transformed into a form demonstrating the conservation
of a particular quantity. Integrating this equation over a spacetime volume $\mathcal{V}$ and
using the Gauss theorem gives:

$$0 = \int_{\mathcal{V}} d^4x \, (\partial_a T_b^a) = \int_{\partial \mathcal{V}} d^3\sigma_a \, T_b^a = \int_{t_2} d^3x \, T_b^0 - \int_{t_1} d^3x \, T_b^0. \qquad (5.83)$$

The second equality arises from the Gauss theorem and the third arises from the
contributions on the surfaces $t = $ constant. As usual, we have ignored the contribution from a surface at spatial infinity assuming that the energy-momentum tensor
of the system is confined to a finite region in 3-space. This shows that there exist
four quantities

$$Q_b(t) = \int_t d^3x \, T_b^0(t, \boldsymbol{x}) \qquad (5.84)$$

which are conserved in the sense that $dQ_b/dt = 0$. It is natural to identify this
as the covariant components of the four-momentum of the system with energy-
momentum tensor T_b^a.

Consider now placing the above physical system in a gravitational field
described by the metric tensor $g_{ab}(x)$. (For example, T_b^a could then describe the
energy-momentum tensor of a gas of particles in the Earth's atmosphere.) The gravitational field will influence the system and can, in general, change its energy and
momentum. It is unlikely that, in this case, we will have a conserved quantity for
the system because it is interacting with an external field. The equation $\partial_k T_i^k = 0$
will now be modified to $\nabla_k T_i^k = 0$. Using Eq. (4.107) we can write this as

$$\frac{1}{\sqrt{-g}} \partial_k \left(\sqrt{-g} \, T_i^k \right) = \frac{1}{2} \left(\partial_i g_{kl} \right) T^{kl}. \qquad (5.85)$$

Integrating both sides over the proper volume $\sqrt{-g} \, d^4x$ we get the result

$$\int_{\mathcal{V}} d^4x \partial_k \left(\sqrt{-g} T_i^k \right) = \int_{t_2} d^3x \sqrt{-g} T_i^0 - \int_{t_1} d^3x \sqrt{-g} T_i^0 \equiv Q_i(t_2) - Q_i(t_1)$$

$$= \frac{1}{2} \int_{\mathcal{V}} d^4x \sqrt{-g} \, T^{kl}(\partial_i g_{kl}). \qquad (5.86)$$

The first equality follows from the standard application of the Gauss theorem. This
shows that the difference $Q_i(t_2) - Q_i(t_1)$ is nonzero and is related to an integral
over a term containing $\partial_i g_{kl}$. As long as $\partial_i g_{kl}$ is nonzero, the right hand side, in

general, will not vanish and $Q_i(t)$ is no longer conserved. As we said before, this is exactly what we will expect in the presence of an external gravitational field.

The analysis also reveals another expected consequence. If g_{kl} is independent of a particular coordinate x^M, say, then Q_M will be conserved. In particular, if g_{ik} is independent of time, the total energy of the system represented by Q_0 will be conserved. This result agrees with what we had already seen, in terms of Killing vectors as the connection between translational invariance of the metric and a conservation law. More formally, given a Killing vector ξ^a in the spacetime, Eq. (4.137) shows that $P^i = T^i_j \xi^j$ is conserved with $\nabla_i P^i = 0$. Using Eq. (4.108) for the covariant derivative of the vector and integrating over $\sqrt{-g}\, d^4x$ we get

$$
0 = \int_V \sqrt{-g}\, d^4x \nabla_i P^i = \int_V d^4x\, \partial_i \left(\sqrt{-g}\, P^i\right)
$$
$$
= \int_{t_2} d^3\mathbf{x} \sqrt{-g} P^0 - \int_{t_1} d^3\mathbf{x} \sqrt{-g} P^0, \tag{5.87}
$$

which shows that the integral over all space of $\sqrt{-g}\, P^0$ is conserved. A comparison between Eq. (4.107) and Eq. (4.108) clearly shows the structural difference between the expanded forms of equations $\nabla_i T^{ik} = 0$ for a symmetric tensor and $\nabla_i P^i = 0$ for a vector. The former cannot be converted to a pure surface term by using the Gauss theorem (so that the right hand side of Eq. (5.86) is nonzero) while in the latter case we do get a pure surface term. So a vector equation like $\nabla_i P^i = 0$ is a genuine conservation law in curved spacetime but a tensor equation like $\nabla_i T^{ik} = 0$ is not. A Killing vector allows us to convert a tensor equation to a vector equation, thereby leading to a conservation law.

We shall now discuss some explicit examples of the equation $\nabla_i T^{ik} = 0$ starting with the energy-momentum tensor for a single particle. In special relativity, one could have taken the expression to be

$$
T^{mn} = m \int \delta_D[x^a - z^a(\tau)]\, u^m u^n\, d\tau, \tag{5.88}
$$

where $z^a(\tau)$ is the trajectory of the particle. In this expression, $u^m u^n d\tau$ is generally covariant but not the Dirac delta function. It is, however, easy to see from the relation

$$
1 = \int \delta_D(x^a)\, d^4x = \int \frac{\delta_D(x^a)}{\sqrt{-g}} \sqrt{-g}\, d^4x \tag{5.89}
$$

that $\delta_D(x)/\sqrt{-g}$ is a scalar. Therefore, the generally covariant definition of the energy-momentum tensor for a single particle is given by

$$T^{mn} = m \int \delta_D[x^a - z^a(\tau)] \frac{u^m u^n}{\sqrt{-g}} \, d\tau$$

$$= \frac{m}{\sqrt{-g}} \int \delta_D[x^a - z^a(\tau)] \, u^m u^n \, d\tau, \tag{5.90}$$

where we have pulled out the $\sqrt{-g}$ factor which depends only on x. Using

$$\nabla_m T^{mn} = \frac{1}{\sqrt{-g}} \partial_m \left(\sqrt{-g} \, T^{mn} \right) + \Gamma^n_{sm} T^{ms} \tag{5.91}$$

we can express the relation $\nabla_m T^{mn} = 0$ in the form

$$\int u^m u^n \partial_m \delta^4 \left(x - z(\tau) \right) d\tau + \Gamma^n_{sm} \int u^m u^s \delta^4 \left(x - z(\tau) \right) d\tau = 0. \tag{5.92}$$

Since the Dirac delta function depends only on the difference of the coordinates, we can replace $(\partial/\partial x^m)$ by $-(\partial/\partial z^m)$. Using

$$u^m \frac{\partial}{\partial z^m} \delta^4 \left(x - z(\tau) \right) = \frac{d}{d\tau} \delta^4 \left(x - z(\tau) \right) \tag{5.93}$$

we get

$$-\int u^n \frac{d}{d\tau} \delta^4 \left(x - z(\tau) \right) d\tau + \Gamma^n_{sm} \int u^m u^s \delta^4 \left(x - z(\tau) \right) d\tau = 0. \tag{5.94}$$

Doing an integration by parts on the first term, this reduces to

$$\int \left[\frac{du^n}{d\tau} + \Gamma^n_{sm} u^m u^s \right] \delta^4 \left(x - z(\tau) \right) d\tau = 0. \tag{5.95}$$

The vanishing of this integral requires the expression within the square brackets to vanish along the trajectory of the particle, which is identical to the geodesic equation for the particle. Thus, for a single particle, the condition $\nabla_m T^{mn} = 0$ is equivalent to a geodesic equation. As we have already seen, the geodesic equation is capable of encoding the effect of external gravitational field on a material particle and – in general – will not lead to any conservation law. It follows that the equation $\nabla_m T^{mn} = 0$ describes the way material systems are influenced by external gravity and, of course, is not a conservation law either.

Let us next consider an ideal fluid with the energy-momentum tensor $T^{mn} = (\rho + p)u^m u^n + pg^{mn}$. In this case it is faster – and instructive – to use the index-free notation, in which we start with the energy-momentum tensor written as $\boldsymbol{T} = (\rho + p)\boldsymbol{u} \otimes \boldsymbol{u} + p\boldsymbol{g}$. Taking the divergence of this equation and equating to zero we get

$$0 = \nabla \cdot \boldsymbol{T} = [\nabla(\rho + p) \cdot \boldsymbol{u}] \, \boldsymbol{u} + [(\rho + p)\nabla \cdot \boldsymbol{u}] \, \boldsymbol{u} + [(\rho + p)\boldsymbol{u}] \cdot \nabla \boldsymbol{u} + (\nabla p) \cdot \boldsymbol{g}$$

$$= [\nabla_{\boldsymbol{u}}\rho + \nabla_{\boldsymbol{u}}p + (\rho + p)\nabla \cdot \boldsymbol{u}] \, \boldsymbol{u} + (\rho + p)\nabla_{\boldsymbol{u}}\boldsymbol{u} + \nabla p, \tag{5.96}$$

where we have used $\nabla \boldsymbol{g} = 0$ and $\boldsymbol{g} \cdot \nabla p = \nabla p$. (The notation of dot products like $\nabla \cdot \boldsymbol{T}$ is unambiguous for symmetric tensors like $\boldsymbol{T}$.) This is equivalent to four equations, one for each value of the free index, in $\nabla \cdot \boldsymbol{T} \rightarrow \nabla_i T^{ij}$. This can be cast into a useful form by separating these four equations into: (i) one obtained by taking the dot product with $\boldsymbol{u}$ and (ii) three others obtained by projecting it to a space orthogonal to $\boldsymbol{u}$. The dot product with $\boldsymbol{u}$ gives

$$0 = \boldsymbol{u} \cdot (\nabla \cdot \boldsymbol{T}) = -\left[\nabla_{\boldsymbol{u}}\rho + \nabla_{\boldsymbol{u}}p + (\rho + p)\nabla \cdot \boldsymbol{u}\right] + \nabla_{\boldsymbol{u}}p$$
$$= -\nabla_{\boldsymbol{u}}\rho - (\rho + p)\nabla \cdot \boldsymbol{u}, \tag{5.97}$$

where we have used $\boldsymbol{u} \cdot \nabla_{\boldsymbol{u}}\boldsymbol{u} = (1/2)\nabla_{\boldsymbol{u}}(\boldsymbol{u}^2) = 0$. Projection orthogonal to the velocity requires contraction with the projection tensor $\boldsymbol{P} = \boldsymbol{g} + \boldsymbol{u} \otimes \boldsymbol{u}$. This leads – after some simple algebra – to the Euler equation

$$(\rho + p)\nabla_{\boldsymbol{u}}\boldsymbol{u} = -\boldsymbol{P} \cdot (\nabla p) \equiv -\left[\nabla p + (\nabla_{\boldsymbol{u}}p)\boldsymbol{u}\right]. \tag{5.98}$$

In component form these equations are the same as Eq. (1.120) with partial derivatives replaced by covariant derivatives as to be expected. The operator $u^i\nabla_i = d/d\tau$ represents derivative along the integral curves of u^i ('stream lines') with τ as the parameter. Explicitly, we can write Eq. (5.97) and Eq. (5.98) as

$$u^i\nabla_i\rho = \frac{d\rho}{d\tau} = -(\rho + p)\nabla_i u^i \tag{5.99}$$

$$(\rho + p)u^j\nabla_j u^i = (\rho + p)\frac{du^i}{d\tau} = -\nabla^i p - u^i u^j \nabla_j p. \tag{5.100}$$

In the special case of pressureless fluid, made of a collection of particles, $\boldsymbol{T} = \rho \boldsymbol{u} \otimes \boldsymbol{u}$ and $\nabla \cdot \boldsymbol{T} = 0$ lead to

$$\nabla \cdot (\rho \boldsymbol{u}) = 0; \qquad \nabla_{\boldsymbol{u}}\boldsymbol{u} = 0. \tag{5.101}$$

The first equation is the conservation mass energy and the second is the geodesic equation for each particle.

In the case of an ideal fluid, it is possible to interpret the equation of motion as the conservation of entropy per baryon along the fluid streamlines. The conservation of baryons implies that $\nabla \cdot (n\boldsymbol{u}) = 0$, where n is the number of baryons per unit volume. Using this to determine $\nabla \cdot \boldsymbol{u}$ and combining with Eq. (5.99) we get the equation

$$\frac{d\rho}{d\tau} = \frac{(p + \rho)}{n}\frac{dn}{d\tau}. \tag{5.102}$$

To see that this equation implies conservation of specific entropy, let us consider the first law of thermodynamics in curved spacetime applied to a volume V containing $N \equiv nV$ baryons. The energy is $E = \rho V = (N\rho/n)$, the volume is $V = N/n$

and the entropy is $S \equiv Ns$. The first law is identical in form to the non-relativistic equation $d(\rho V) = TdS - pdV$ except that ρ now includes rest mass energy. So we have

$$d\left(\frac{\rho}{n}\right) = -pd\left(\frac{1}{n}\right) + Tds. \tag{5.103}$$

This is equivalent to

$$d\rho = (\rho + p)\frac{dn}{n} + nTds. \tag{5.104}$$

Comparing Eq. (5.102) with Eq. (5.104), we find that $(ds/d\tau) = \nabla_{\boldsymbol{u}}s = 0$ which shows that the entropy per baryon is conserved along the flow lines as expected.

One can easily verify that Eq. (5.100) reduces to the equations describing a fluid in a Newtonian gravitational field in the appropriate limit, characterized by the following conditions. For the energy density, we take $\rho = \rho_0(1+\epsilon)$, where $\rho_0 = nm_B$ denotes the rest mass energy of a collection of baryons, each of mass m_B. The ϵ will represent the *specific* internal energy of the fluid. In the Newtonian approximation the rest energy will be the dominant contribution and we can assume $\epsilon \ll 1$. For the metric we will take $g_{00} = -(1+2\phi)$ with $|\phi| \ll 1$. Finally, non-relativistic motion will require $(p/\rho_0) \ll 1$ and $v^2 \ll 1$. To obtain a Newtonian approximation with these conditions, we write the spatial components of Eq. (5.100) as:

$$\rho_0\left(1 + \epsilon + \frac{p}{\rho_0}\right)(u^n\partial_n u_\alpha - \Gamma^a_{na}u_a u^n) = -\partial_\alpha p - u_\alpha\frac{dp}{d\tau}. \tag{5.105}$$

For the four-velocity $u^i = (u^0, v^\alpha)$, we can use the approximations $u^0 \approx 1$, $u_0 \approx -1$ to obtain:

$$\rho_0\left(\frac{dv_\alpha}{d\tau} + \Gamma^0_{0\alpha}\right) \approx -\partial_\alpha p. \tag{5.106}$$

But since $\Gamma^0_{0\alpha} \approx -\Gamma_{00\alpha} = -(1/2)\partial_\alpha g_{00} \approx \partial_\alpha \phi$, we obtain the standard equation for a fluid moving under the action of the pressure gradient and gravity given by

$$\frac{dv_\alpha}{d\tau} = -\partial_\alpha\phi - \frac{1}{\rho_0}\partial_\alpha p \tag{5.107}$$

as expected.

Exercise 5.11
Pressure gradient needed to support gravity Consider a perfect fluid at rest (for which 'flow' lines are just x^α = constant) in a static gravitational field. Show that the Euler equation now gives

$$\frac{\partial p}{\partial x^0} = 0, \quad \frac{\partial p}{\partial x^\alpha} = -(\rho + p)\frac{\partial \ln\sqrt{-g_{00}}}{\partial x^\alpha}. \tag{5.108}$$

Use this to show that an ultra-relativistic fluid with $p = (1/3)\,\rho$ in hydrostatic equilibrium in a gravitational field cannot have a free surface (i.e. a surface where ρ goes to zero.)

[Hint. Integrate the Euler equation to show that $\rho \propto (-g_{00})^{-2}$; as long as g_{00} is finite, ρ cannot vanish.]

Exercise 5.12

Thermal equilibrium in a static metric Show that if a fluid is at rest in a static metric then its temperature satisfies the relation $T\sqrt{-g_{00}} = $ constant. We will need this result in several later chapters. [Hint. An interesting route to this result is the following: we know that, for photons in a stationary metric, $\omega\sqrt{-g_{00}}$ is a constant. Energy can be exchanged between two different regions of the fluid by emission and absorption of radiation but in thermal equilibrium such an exchange should not change anything. Take it from there.]

Exercise 5.13

Weighing the energy A container with a substance having an energy-momentum tensor T^{ab} is kept on a scale pan. Express the weight shown by the scale pan in terms of the components of the energy-momentum tensor if it is located in a static, uniform, gravitational field. If two different containers have identical densities for mass energy but different internal stresses, will they weigh the same?

[Hint. First show that the metric of a static, uniform gravitational field has the generic form $ds^2 = g_{tt}dt^2 + g_{zz}dz^2 + dx^2 + dy^2$, where the metric components depend only on z. Next show that the weight shown by the scale pan at $z = 0$ is $W(0)$ where

$$W(z) = \int dx\,dy\,(g_{zz}\,T^{zz}). \tag{5.109}$$

Now comes the hard part. Manipulate the conservation law $\nabla_a T^{ab} = 0$ taking careful account of surface terms to show that $W(z)$ satisfies the differential equation

$$\partial_z\left(|g_{tt}|^{1/2}\,W\right) = \left(\partial_z|g_{tt}|^{1/2}\right)\int g_{tt}T^{tt}\,dxdy. \tag{5.110}$$

Integrate this with appropriate boundary conditions.]

Exercise 5.14

General relativistic Bernoulli equation (a) An ideal gas is undergoing adiabatic, stationary flow in a stationary gravitational field with a four-velocity $\boldsymbol{u}$. Show that, along the flow lines, we must have $u_0 \propto n/(\rho + p)$ where n is the baryon number density.

[Hint. Since the gravitational field is stationary, there is a timelike Killing vector field ξ. Take the dot product of this with the Euler equation, use the stationarity of the flow and show that $(\rho + p)\,(du^0/d\tau) = u^0(dp/d\tau)$. For adiabatic flow of ideal fluids, the first law of thermodynamics gives $d\rho = (\rho + p)dn/n$. Combining these should lead to the required result.]

(b) Show that, for slow velocities and weak gravitational fields, this equation reduces to the standard Bernoulli equation.

5.4.3 Classical field theory in curved spacetime

We shall next consider the modifications which arise for the scalar field theory and electrodynamics (discussed in Chapter 2) in curved spacetime. In the case of the scalar field, we need to modify the action in Eq. (3.72) making it generally covariant. This is easily done by changing d^4x to $\sqrt{-g}\,d^4x$ and $\partial_i\phi \to \nabla_i\phi$. The second change is purely cosmetic because, for scalars, $\nabla_i\phi$ is same as $\partial_i\phi$. This leads to the action

$$\mathcal{A} = -\int \sqrt{-g}\,d^4x \left[\frac{1}{2}\nabla_i\phi\,\nabla^i\phi + V\right]. \tag{5.111}$$

Varying ϕ in the action leads to the field equation

$$\Box\phi \equiv \nabla_i\nabla^i\phi = \frac{1}{\sqrt{-g}}\partial_i\left(\sqrt{-g}\,g^{ik}\partial_k\phi\right) = \frac{\partial V}{\partial\phi}, \tag{5.112}$$

where we have used Eq. (4.109). Unlike the special relativistic case one cannot, in general, introduce plane wave solutions to this equation or reduce the equation to an algebraic one in the Fourier space. One needs to solve the equations on a case-by-case basis depending on the symmetries of the metric. Other than such mathematical complications, the generalization of the scalar field theory to curved spacetime poses no conceptual issues.

Let us next consider the generalization of electrodynamics to curved spacetime. To do this, we need to write down the second and third terms of Eq. (2.111) in a generally covariant manner. The generalization of the current vector to curved spacetime is straightforward. If the determinant of the spatial metric on the $t = $ constant hypersurface is given by h then the proper three-volume element is $\sqrt{h}\,dx^1dx^2dx^3 \equiv \sqrt{h}dV$. The charge contained in this volume will be $dq = \rho\sqrt{h}dV$. Using the same procedure that was adopted to arrive at Eq. (2.73) we now get

$$dq\,dx^i = \rho\,dx^i\sqrt{h}\,dx^1dx^2dx^3 = \frac{\rho}{\sqrt{|g_{00}|}}\sqrt{-g}\,d^4x\frac{dx^i}{dx^0}, \tag{5.113}$$

where we have used the relation $g = hg_{00}$. Since $\sqrt{-g}d^4x$ is invariant, this allows us to determine the current vector in the curved spacetime to be

$$J^i = \frac{\rho}{\sqrt{|g_{00}|}}\frac{dx^i}{dx^0}. \tag{5.114}$$

For a set of point charges in curved spacetime, the current four-vector will be

$$J^i = \sum_A \frac{q_A}{\sqrt{-g}}\,\delta(\boldsymbol{x} - \boldsymbol{x}_A)\frac{dx^i}{dx^0}, \tag{5.115}$$

which is consistent with the modification of the Dirac delta function in the curved spacetime obtained in Eq. (5.89). Given the current vector, the interaction term in Eq. (2.111) can be generalized to curved spacetime by just replacing d^4x by $\sqrt{-g}\,d^4x$.

As regards the action for the field, we first note that the definition of F_{ik} carries over to curved spacetime because of the identity Eq. (4.105). So we get

$$F_{ik} = \nabla_i A_k - \nabla_k A_i = \partial_i A_k - \partial_k A_i. \tag{5.116}$$

As for the action, we only need to replace d^4x by $\sqrt{-g}\,d^4x$ in the last term of Eq. (2.111). This leads to the full action of the form

$$\mathcal{A} = -\sum \int m\,d\tau + \int \sqrt{-g}\,d^4x\,A_i J^i - \frac{1}{16\pi}\int F_{ab}F^{ab}\sqrt{-g}d^4x. \tag{5.117}$$

Varying the trajectory of the particle in this action leads to the equation of motion

$$mu^j\,\nabla_j u^i = m\left(\frac{du^i}{d\tau} + \Gamma^i_{kl}u^k u^l\right) = qF^{ik}u_k \tag{5.118}$$

for each particle of charge q and mass m. This shows that the geodesic acceleration of the particle is provided by the electromagnetic field, as expected. Varying A_i in the action leads to the generalization of Maxwell's equations in curved spacetime given by

$$\nabla_k F^{ik} = \frac{1}{\sqrt{-g}}\,\partial_k(\sqrt{-g}\,F^{ik}) = -4\pi J^i, \tag{5.119}$$

where the first equality comes from using the identity Eq. (4.104). In the absence of current, the free electromagnetic field is described by the equations

$$\nabla_k F^{ik} = \frac{1}{\sqrt{-g}}\,\partial_k(\sqrt{-g}\,F^{ik}) = 0. \tag{5.120}$$

Equations (5.116) and (5.120) describe the influence of the gravitational field on the electromagnetic phenomena. As in the case of scalar field, we can no longer obtain plane wave solutions and these equations need to be solved in each spacetime on a case-by-case basis.

The second pair of Maxwell's equations is already incorporated in the definition of F_{ik} in terms of A_i in Eq. (5.116). It can also be stated as

$$\nabla_l F_{ik} + \nabla_k F_{li} + \nabla_i F_{kl} = \frac{\partial F_{ik}}{\partial x^l} + \frac{\partial F_{li}}{\partial x^k} + \frac{\partial F_{kl}}{\partial x^i} = 0. \tag{5.121}$$

The first equality can be easily verified for the antisymmetric tensor F_{ik}.

Just as in the case of special relativity, the field equations, as well as the action, are invariant under the gauge transformation $A_i \rightarrow A_i + \partial_i f$. We saw in Chapter 2 that gauge invariance is intimately related to conservation of current. This conservation was expressed as $\partial_i J^i = 0$ in special relativity and will generalize to

$$\nabla_i J^i = \frac{1}{\sqrt{-g}}\, \partial_i \left(\sqrt{-g}\, J^i \right) = 0 \tag{5.122}$$

in curved space. This equation is consistent with both field equation as well as gauge invariance. Writing the field equations Eq. (5.119) as

$$\partial_k (\sqrt{-g}\, F^{ik}) = -4\pi \sqrt{-g}\, J^i \tag{5.123}$$

it follows that

$$\nabla_i J^i = -\frac{1}{4\pi} \frac{1}{\sqrt{-g}}\, \partial_i \partial_k (\sqrt{-g}\, F^{ik}) = 0, \tag{5.124}$$

where the last equality follows from the antisymmetry of F^{ik}. Note that the structure of covariant derivatives in Eq. (4.104) and Eq. (4.108) has played an important role in maintaining the consistency. On the other hand, when we perform the gauge transformation $A^i \to A^i + \partial_i f$ the middle term in Eq. (5.117) picks up an additional contribution which can be written as

$$\int_{\mathcal{V}} \sqrt{-g}\, d^4x\, J^i \partial_i f = \int_{\mathcal{V}} d^4x\, \partial_i [f \sqrt{-g}\, J^i] - \int_{\mathcal{V}} \sqrt{-g}\, d^4x\, (f\, \nabla_i J^i), \tag{5.125}$$

where we have performed an integration by parts and used Eq. (4.108). The first term on the right hand side can be converted to a surface term which vanishes when we choose f such that it vanishes on $\partial \mathcal{V}$. Gauge invariance now requires $\nabla_i J^i = 0$, which is consistent with our previous results.

Exercise 5.15

Conformal invariance of electromagnetic action Consider the conformal transformations $g_{ab} \to \Omega^2(x) g_{ab}$, $g^{ab} \to \Omega^{-2}(x) g^{ab}$ with A_i and F_{ik} remaining unchanged. Show that the free field electromagnetic action is invariant under these transformations. What does it imply for the solutions of Maxwell's equations in a spacetime with a metric of the form $g_{ab} = f^2(x) \eta_{ab}$?

Exercise 5.16

Gravity as an optically active media Maxwell's equations can be expressed in a suggestive form in a curved spacetime. Take the metric to be

$$ds^2 = g_{ik} dx^i dx^k = g_{00} dt^2 + 2g_{0\alpha} dt dx^\alpha + g_{\alpha\beta} dx^\alpha dx^\beta \tag{5.126}$$

and introduce the notation: $N^2 \equiv -g_{00}$; $g_\alpha \equiv -(g_{0\alpha}/N^2)$ and $\gamma_{\alpha\beta} \equiv [g_{\alpha\beta} - N^2 g_\alpha g_\beta]$. Also define the electric and magnetic fields by $E_\alpha \equiv F_{0\alpha}$ and $B^\alpha \equiv -(1/2\sqrt{\gamma})\epsilon^{\alpha\beta\gamma} F_{\beta\gamma}$, where γ is the determinant of the metric of three-dimensional space.

(a) Work out various components of Maxwell's equations in the (1+3) form. Show that Eq. (5.121) becomes

$$\nabla \cdot \boldsymbol{B} = 0; \qquad \nabla \times \boldsymbol{E} = -\frac{1}{\sqrt{\gamma}} \frac{\partial}{\partial t} \left(\sqrt{\gamma} \boldsymbol{B} \right). \tag{5.127}$$

The three-dimensional vector operations are carried out using the metric $\gamma_{\alpha\beta}$.

(b) Next show that the equations Eq. (5.120) can be written as

$$\nabla \cdot \boldsymbol{D} = 0; \qquad \nabla \times \boldsymbol{H} = \frac{1}{\sqrt{\gamma}} \frac{\partial}{\partial t} \left(\sqrt{\gamma} \boldsymbol{D} \right), \tag{5.128}$$

where

$$\boldsymbol{D} = \frac{\boldsymbol{E}}{N} + \boldsymbol{H} \times \boldsymbol{g}; \qquad \boldsymbol{B} = \frac{\boldsymbol{H}}{N} + \boldsymbol{g} \times \boldsymbol{E}. \tag{5.129}$$

(c) In static gravitational fields with $\boldsymbol{g} = 0$ and other metric components independent of time, these equations simplify to

$$\nabla \cdot \boldsymbol{B} = 0; \quad \nabla \times \boldsymbol{E} = -\frac{\partial \boldsymbol{B}}{\partial t}; \quad \nabla \cdot \left(\frac{\boldsymbol{E}}{N} \right) = 0; \quad \nabla \times (N\boldsymbol{B}) = \frac{1}{N} \frac{\partial \boldsymbol{E}}{\partial t}. \tag{5.130}$$

Interpret this result.

Exercise 5.17

Curvature and Killing vectors (a) Show that any Killing vector $\boldsymbol{\xi}$ satisfies the equation

$$\nabla_b \nabla_a \, \xi_c = R_{ibac} \, \xi^i. \tag{5.131}$$

Hence (or otherwise) show that

$$\Box \xi^a \equiv \nabla_b \nabla^b \, \xi^a = -R^a{}_c \xi^c. \tag{5.132}$$

[Hint. Start with the general result $[\nabla_m, \nabla_p]\xi_s = R^l{}_{spm}\xi_l$. Permute the three lower indices of the curvature tensor and use the cyclic identity Eq. (5.40) to obtain an identity involving second derivatives of the vector field. Now use the fact that $\boldsymbol{\xi}$ satisfies the Killing equation.]

(b) Show that a Killing vector is an acceptable solution as a vector potential satisfying Maxwell's equations in a curved spacetime if $R_{ij} = 0$. What are the electromagnetic fields which correspond to the Killing vectors of the Minkowski spacetime? Are they physically meaningful?

Exercise 5.18

Christoffel symbols and infinitesimal diffeomorphism Show that, under an infinitesimal coordinate transformation $x^i \to x^i + \xi^i(x)$ the Christoffel symbols change by

$$\delta\Gamma^d_{bc} = -\frac{1}{2} \nabla_{(b} \nabla_{c)} \xi^d + \frac{1}{2} R^d{}_{(bc)i} \xi^i, \tag{5.133}$$

where we use the notation $Q^{(ij)} \equiv Q^{ij} + Q^{ji}$. Hence (or otherwise) show that $\delta\Gamma^d_{bc} = 0$ when ξ^a is a Killing vector.

Exercise 5.19

Conservation of canonical momentum In the presence of an electromagnetic field, the canonical four-momentum $\boldsymbol{P}$ of a particle picks up a term which depends on the vector potential: $\boldsymbol{P} = \boldsymbol{p} + q\boldsymbol{A}$. Suppose $\boldsymbol{\xi}$ is a Killing vector in the spacetime with $\mathcal{L}_{\boldsymbol{\xi}} A^i = \mathcal{L}_{\boldsymbol{\xi}} g_{ab} = 0$. Show that $\boldsymbol{P} \cdot \boldsymbol{\xi}$ is conserved along the trajectory of the particle.

5.4.4 Geometrical optics in curved spacetime

The wave solutions to Maxwell's equations describe the propagation of light and, in the limit of high frequencies (small wavelengths), the wave propagation reduces to conventional ray optics of light rays. In this limit, light rays in a spacetime are supposed to travel along null geodesics. In fact, we have repeatedly used this feature in several previous chapters in discussing light rays and photon trajectories. We shall now derive this geometrical optics limit systematically, starting from the wave optics in curved spacetime.

To arrive at the geometrical optics limit, we start from the assumption that the wave front is a good approximation to a plane wave over a length scale L and assume that the wavelength $\lambda \ll L$. In curved spacetime, we also need to assume that the wavelength is small compared with the typical length scale associated with the curvature of the spacetime. The latter could be taken as $\mathcal{R}^{-1/2}$, where $\mathcal{R}$ is the magnitude of a typical component of the curvature tensor. Under these conditions, we can make a systematic expansion of the solution to the wave equation in the small parameter λ/L_0 where $L_0 = \min \left[\mathcal{R}^{-1/2}, L \right]$. To do this, we will first express Maxwell's equations in terms of the vector potential by substituting $F_{ij} = \nabla_i A_j - \nabla_j A_i$ into $\nabla_i F^{ij} = 0$, obtaining

$$0 = -\nabla_i \nabla^i \, A^j + \nabla_i \nabla^j \, A^i = -\nabla_i \nabla^i \, A^j + \nabla^j (\nabla_i \, A^i) + R^j_{\ i} \, A^i. \quad (5.134)$$

In arriving at the second equality, we have interchanged the covariant derivatives thereby bringing in the Ricci tensor. Imposing the Lorenz gauge condition $\nabla_i A^i = 0$, the equation for vector potential reduces to

$$\Box A^j - R^j_{\ i} \, A^i = 0. \quad (5.135)$$

To solve this equation, we shall use an ansatz of the form

$$A_i = \text{Re} \left[(a_i + \epsilon \, b_i + \cdots) \, e^{i\psi/\epsilon} \right], \quad (5.136)$$

where ϵ is a bookkeeping parameter used to define the different orders of approximation. We take $\epsilon = \mathcal{O}(\lambda/L_0)$ to define the order-by-order approximation. We will also define $k_a = \partial_a \psi$ (which will eventually be identified with the wave vector), a scalar amplitude $a = \sqrt{a_i \bar{a}^i}$, where $\bar{a}^i$ is the complex conjugate of a^i, and a polarization vector $f_i = a_i/a$. Substituting the expansion in Eq. (5.136) into the

Lorenz gauge condition gives

$$0 = \nabla_j A^j = \mathrm{Re}\left[\left(i\frac{k_i}{\epsilon}(a^i + \epsilon\, b^i + \cdots) + \nabla_i(a^i + \epsilon\, b^i + \cdots)\right)e^{i\psi/\epsilon}\right].$$
(5.137)

Equating the coefficients of similar powers of ϵ, we get different conditions on the physical parameters. The leading term gives $k_a f^a = 0$, implying that the polarization vector is perpendicular to the wave vector. At the next order, we have $\mathbf{k} \cdot \mathbf{b} = i\nabla \cdot \mathbf{a}$, which determines the propagation of b^j in terms of the leading order. Next we substitute Eq. (5.136) into Eq. (5.135) to get

$$0 = \mathrm{Re}\left[\left(\frac{1}{\epsilon^2}k^j k_j(a^i + \epsilon\, b^i + \cdots) - 2\frac{i}{\epsilon}k^j\nabla_j(a^i + \epsilon\, b^i + \cdots)\right.\right.$$
$$\left.\left. - \frac{i}{\epsilon}\nabla_j k^j(a^i + \epsilon\, b^i + \cdots) + R^i{}_j(a^j + \cdots)\right)e^{i\psi/\epsilon}\right].$$
(5.138)

To the leading order, this gives $k^a k_a = 0$ showing that k^i is a null vector. Expressing the wave vector in terms of the phase as $k_i = \partial_i\psi$ we get the *ikonal* equation

$$g^{ij}\,\partial_i\psi\,\partial_j\psi = 0.$$
(5.139)

We also find that $\nabla_i(k^j k_j) = 2\,k_j\nabla_i k^j = 0$ from which, on using $k_i = \partial_i\psi$ and noting $\partial_i\partial_j\psi = \partial_j\partial_i\psi$, we get the result

$$k^j\,\nabla_j\,k_i = 0.$$
(5.140)

This shows that the wave vector, at the leading order, satisfies the geodesic equation. Since it is also null, we have the result that, in the ray optics limit, Maxwell's equations imply that light travels along null geodesics. At the next order, $\mathcal{O}(\epsilon^{-1})$, on using $k_a k^a = 0$ we get

$$k^j\,\nabla_j a^i = -\frac{1}{2}\,a^i\,\nabla_j k^j.$$
(5.141)

On substituting $a^j = a f^j$ this equation gets transformed to

$$2ak^j\partial_j a = 2ak^j\nabla_j a = k^j\nabla_j(a^2) = k^j\nabla_j(a_i\bar{a}^i)$$
$$= \bar{a}^i k^j\nabla_j a_i + a_i k^j\nabla_j\bar{a}^i = -\frac{1}{2}\nabla_j k^j(\bar{a}^i a_i + a_i\bar{a}^i)$$
(5.142)

so that

$$k^j\partial_j a = -\frac{1}{2}\,a\,\nabla_j k^j.$$
(5.143)

Using Eq. (5.141) again in the form

$$0 = k^j \nabla_j(af^i) + \frac{1}{2} af^i \nabla_j k^j$$

$$= ak^j \nabla_j f^i + f^i \left(k^j \nabla_j a + \frac{1}{2} a \nabla_j k^j \right) = ak^j \nabla_j f^i, \qquad (5.144)$$

where we have used Eq. (5.143). This equation shows that the polarization vector is parallel propagated along the null geodesics. We can also rewrite Eq. (5.143) in the form $(k^j \nabla_j)a^2 + a^2 \nabla_j k^j = 0$ leading to

$$\nabla \cdot (a^2 \mathbf{k}) = \nabla_i(a^2 k^i) = 0. \qquad (5.145)$$

This is in the form of a conservation law for the vector $a^2 \mathbf{k}$, which is a conserved current. This conservation law has no simple interpretation in classical optics but when light is treated as a bunch of photons, it translates directly into conservation of the number of photons.

It is possible to obtain another interesting result in this context. If the cross-sectional area of a congruence of null rays is $\mathcal{S}$ (see Exercise 1.8) then the conservation of the intensity of light will imply that $k^j \partial_j (a^2 \mathcal{S}) = 0$. Expanding this equation and using Eq. (5.145) we get the propagation equation for the cross-sectional area of a bundle of rays

$$k^j \partial_j \mathcal{S} = \partial_{\mathbf{k}} \mathcal{S} = (\nabla \cdot \mathbf{k}) \, \mathcal{S}. \qquad (5.146)$$

If we define a characteristic 'radius' $\mathcal{R} = \mathcal{S}^{1/2}$, then the above equation implies that the radius evolves along the rays as

$$\frac{d\mathcal{R}}{d\lambda} = \frac{1}{2}(\nabla \cdot \mathbf{k})\mathcal{R}, \qquad (5.147)$$

where λ is an affine parameter. Further, if we think of $\mathcal{R}$ as the characteristic separation between two extreme rays in the bundle of rays, one can again introduce the relative acceleration by the quantity $d^2\mathcal{R}/d\lambda^2$ which is analogous to the acceleration of the geodesic deviation vector in the case of a timelike congruence (see Section 5.5). Differentiating Eq. (5.147) with respect to λ, we get

$$\frac{d^2\mathcal{R}}{d\lambda^2} = \frac{1}{2}\mathcal{R}k^b \nabla_b(\nabla \cdot \mathbf{k}) + \frac{1}{2}(\nabla \cdot \mathbf{k})\frac{d\mathcal{R}}{d\lambda} = \frac{1}{2}\mathcal{R}k^b \nabla_b(\nabla_a k^a) + \frac{1}{4}(\nabla_m k^m)^2 \mathcal{R}.$$
$$(5.148)$$

The first term on the right hand side can be rewritten in the form

$$k^b \nabla_b(\nabla_a k^a) = k^b(\nabla_a \nabla_b k^a + R^a{}_{nba}k^n)$$
$$= \nabla_a(k^b \nabla_b k^a) - (\nabla_b k^a)(\nabla_a k^b) - R_{ab}k^a k^b. \qquad (5.149)$$

The first term vanishes because the rays are propagating along the geodesics thereby allowing us to write

$$\frac{d^2 \mathcal{R}}{d\lambda^2} = -\left[\frac{1}{2}\left(\nabla_b k_a\right)\left(\nabla^b k^a\right) - \frac{1}{4}(\nabla_m k^m)^2 + \frac{1}{2}R_{ab}k^a k^b\right]\mathcal{R}, \qquad (5.150)$$

which is usually called the *focusing equation*. We shall see latter (see Eq. (5.161)) that similar results can be obtained for any congruence of geodesics and that this is a special case. Working in a local Lorentz frame, it is easy to show that the quantity

$$\frac{1}{2}\nabla_b k_a \nabla^b k^a - \frac{1}{4}(\nabla_m k^m)^2 = \frac{1}{4}(\partial_x k_x - \partial_y k_y)^2 + (\partial_y k_x)^2 \qquad (5.151)$$

is positive semi-definite. This is essentially the shear of a bundle of rays and Eq. (5.150) shows that the shear always focuses a bundle of rays $(d^2\mathcal{R}/d\lambda^2 < 0)$ while spacetime curvature will also focus the bundle if $R_{ab}k^a k^b > 0$. This result plays a crucial role in the study of black hole physics.

Exercise 5.20
Energy-momentum tensor and geometrical optics Show that the energy-momentum tensor of the electromagnetic wave in the ikonal approximation, averaged over several wavelengths, is

$$\langle T^{mn}\rangle = \frac{1}{8\pi}\, a^2\, k^m\, k^n. \qquad (5.152)$$

Show that $\nabla_n \langle T^{nm}\rangle = 0$ follows from the equations for the ikonal approximations derived in the text.

Exercise 5.21
Ray optics in Newtonian approximation Consider the ikonal equation in the metric with $g_{00} = -(1 + 2\phi)$, $g_{\alpha\beta} = (1 - 2\phi)\delta_{\alpha\beta}$. Show that it has a solution in the form $\psi = S(\boldsymbol{x} - \omega kt)$ with

$$(\nabla S)^2 = n^2\omega^2, \qquad n = 1 - 2\phi, \qquad (5.153)$$

where the gradient is in three-dimensional space. Interpret this relation physically.

5.5 Geodesic congruence and Raychaudhuri's equation

The deviation of geodesics from one another – discussed in Section 5.2.3 – manifests itself in a different way when we study a system of particles (or fluids) moving in a spacetime. Considering its use in different contexts, we shall obtain these result in a general form.

5.5.1 Timelike congruence

Let us study a fluid made of particles which move along a set of timelike curves with a velocity field $u^i(x)$. We will call this set of curves a *congruence* in a particular region of spacetime if there is only one curve passing through each event in this region. (Congruence can be defined more precisely but this is adequate for our purposes.) If all the curves in the congruence are geodesics, we call it a geodesic congruence. The deviation vector $\boldsymbol{\xi}$ between any two particles, travelling along any two curves in the congruence, is parallel transported along the congruence and will satisfy the condition $\mathcal{L}_{\boldsymbol{u}}\boldsymbol{\xi} = [\boldsymbol{u}, \boldsymbol{\xi}] = 0$. This implies that

$$u^b \nabla_b \xi^a = \xi^b \nabla_b u^a = \xi^b Q^a{}_b, \tag{5.154}$$

showing that the quantity $Q^a{}_b \equiv \nabla_b u^a$ determines the manner in which spatial separation between nearby particles changes with time as the fluid moves. The fact that $u_a u^a = $ constant implies that $u_a Q^a{}_b = 0$. In the case of a geodesic congruence, we also have $u^b \nabla_b u^a = 0$ ensuring $u^b Q^a{}_b = 0$. So, in the rest frame of the fluid element in a geodesic congruence, this tensor $Q^a{}_b$ is purely spatial.

For many applications, it is convenient to rewrite $\nabla_j u_i$ by the following procedure. First we will separate out from $\nabla_j u_i$ the part $a_i u_j$, where a^i is the acceleration vector. This is convenient since this term will vanish for a geodesic congruence which is what we will deal with most of the time. The remaining second rank tensor can be separated out into symmetric and antisymmetric parts. The symmetric part can be further split into a trace term proportional to δ^i_j and a tracefree part. For our purpose it is also convenient to concentrate on all the physical quantities projected onto the 3-space orthogonal to u^i. This gives the decomposition:

$$\nabla_j u_i = \omega_{ij} + \sigma_{ij} + \frac{1}{3}\theta P_{ij} - a_i u_j, \tag{5.155}$$

where $P^i_j = \delta^i_j + u^i u_j$ is the projection tensor, $a_i = u^j \nabla_j u^i$ is the acceleration of the fluid, $\theta = \nabla_i u^i$ is called *expansion*, ω_{ij} is called *rotation* given by

$$\omega_{ij} \equiv \frac{1}{2}\left(P^m_j \nabla_m u_i - P^m_i \nabla_m u_j\right), \tag{5.156}$$

and σ_{ij} is called the *shear*

$$\sigma_{ij} \equiv \frac{1}{2}\left(P^m_j \nabla_m u_i + P^m_i \nabla_m u_j\right) - \frac{1}{3}\theta P_{ij}. \tag{5.157}$$

From the definitions, it is easy to verify the following facts: (i) ω_{ij} is antisymmetric while σ_{ij} is symmetric and traceless; (ii) both shear and rotation are orthogonal to the four-velocity u^i.

This separation also has another useful feature. In Section 4.6.2, we proved the Frobenius theorem (see Eq. (4.115)) for any hypersurface orthogonal vector field. If the congruence is made of geodesics, and u^i is hypersurface orthogonal, one can prove another useful result. Whenever a geodesic congruence is hypersurface orthogonal, it must have $\omega_{ab} = 0$. To prove this, we first note that the hypersurface orthogonality implies that the vector field has the form $u_i(x) = -\mu \partial_i \Phi$ for two functions $\mu(x), \Phi(x)$. From this we have the relation $\nabla_b u_a = -\mu \nabla_b \nabla_a \Phi - \partial_a \Phi \partial_b \mu$, which implies $\omega_{ab} = -\partial_{[a} \Phi \partial_{b]} \mu$. We now use the condition that $u^a \omega_{ab} = 0$. Using $\omega_{ab} = -\partial_{[a} \Phi \partial_{b]} \mu$ in this equation and using Eq. (4.113) we get the condition $\partial_a(\mu) = -u_a(u^b \partial_b \mu)$. Substituting this back into the expression for ω_{ab} we find that it vanishes identically for a geodesic congruence.

While all these quantities (expansion, shear and rotation) will evolve as the fluid moves through spacetime, we will find that it is particularly useful to follow the evolution of θ. So we want to obtain an equation for determining how the expansion $\theta \equiv \nabla \cdot \boldsymbol{u}$ of the congruence changes along the flow lines. To keep it general, we shall assume that the flow lines are made of arbitrary timelike curves which are not necessarily geodesics. Direct differentiation of θ gives

$$\frac{d\theta}{d\tau} = u^b \nabla_b(\nabla_a u^a) = u^b \nabla_b \nabla_a u^a = u^b(\nabla_a \nabla_b u^a - R^a{}_{bac} u^c)$$

$$= u^b \nabla_a \nabla_b u^a - R_{bc} u^b u^c. \tag{5.158}$$

We can write

$$u^b \nabla_a \nabla_b u^a = \nabla_a(u^b \nabla_b u^a) - \nabla_b u^a \nabla_a u^b = \nabla_a a^a - \left(\omega^a_b + \sigma^a_b + \frac{1}{3}\theta P^a_b - a^a u_b\right)$$

$$\times \left(\omega^b_a + \sigma^b_a + \frac{1}{3}\theta P^b_a - a^b u_a\right), \tag{5.159}$$

where we have decomposed $Q^a{}_b$ into expansion, shear and rotation along the lines described above. Using the symmetry properties of ω_{ab}, σ_{ab} and P_{ab} and the fact that all three tensors are orthogonal to $\boldsymbol{u}$ we get, after some elementary algebra,

$$\frac{d\theta}{d\tau} = \nabla_a a^a + 2\omega^2 - 2\sigma^2 - \frac{1}{3}\theta^2 - R_{ab} u^a u^b, \tag{5.160}$$

where $2\sigma^2 \equiv \sigma_{ab}\sigma^{ab}$ and $2\omega^2 \equiv \omega_{ab}\omega^{ab}$. If the congruence is made of geodesics $(a^a = 0)$, this equation reduces to the form

$$\frac{d\theta}{d\tau} = -\frac{1}{3}\theta^2 + 2\omega^2 - 2\sigma^2 - R_{ab} u^a u^b, \tag{5.161}$$

which is known as *Raychaudhuri's equation*.

The importance of this equation arises in the following context. If the timelike geodesics are orthogonal to some hypersurface so that $u_a = f_1(x)\partial_a f_2(x)$, where f_1 and f_2 are scalar functions, then we saw earlier that $\omega_{ab} = 0$. Since the shear σ_{ab} is orthogonal to the four-velocity, it is purely spatial and $\sigma_{ab}\sigma^{ab} \geq 0$. If the spacetime is such that $R_{ab}u^a u^b \geq 0$, then the right hand side of Eq. (5.161) is negative definite showing $d\theta/d\tau \leq -(1/3)\theta^2$. This immediately implies that $\theta^{-1}(\tau) \geq \theta(0)^{-1} + (1/3)\tau$. Therefore, if the congruence was initially converging $[\theta(0) < 0]$, then $\theta^{-1}(\tau) \to 0^-$ and $\theta(\tau) \to -\infty$ within a proper time $\tau \leq (3/|\theta(0)|)$. This is a crucial result which shows that in any spacetime with $R_{ab}u^a u^b \geq 0$ a set of timelike, converging geodesics that are hypersurface orthogonal will focus a singular point – with all the geodesics converging to a point. This result can be used to predict the existence of singularities in a class of spacetime with very minimal physical assumptions.

The quantity θ can be interpreted as the fractional rate of change of the cross-sectional volume δV of the congruence; that is,

$$\theta = \frac{1}{\delta V}\frac{d}{d\tau}\delta V. \tag{5.162}$$

A formal proof can be provided along the following lines (though the result is intuitively obvious from the defintion of θ). We start with some fiducial geodesic $\mathcal{G}$ and choose an event P on it determined by some parameter value τ_P. In an infinitesimal neighbourhood around P, we can choose a set of points such that $\tau = \tau_P$ for each of the geodesics passing through the points in this neighourhood. Roughly speaking, we are constructing the $\tau = $ constant three-dimensional submanifold $\Sigma(P)$, which can be called the cross-section of the geodesic $\mathcal{G}$ around the event P. We are interested in the volume δV of this hypersurface. We can introduce coordinates on $\Sigma(P)$ by just choosing the labels y^α which are assigned to each of the geodesics in the congruence. This construction provides the relation $x^a = x^a(\tau, y^\alpha)$ defining the congruence; each value of y^α defines a unique geodesic (in a local region) and τ represents the affine parameter along the geodesic. The tangent vector u^a is given by $u^a = (\partial x^a/\partial\tau)_{y^\alpha}$ and, more importantly, we now have the vectors

$$e_\alpha^a = \left(\frac{\partial x^a}{\partial y^\alpha}\right)_\tau \tag{5.163}$$

which are tangents to the cross-section. It follows that $\boldsymbol{u} \cdot \boldsymbol{e}_\alpha = 0$ on $\mathcal{G}$ and $\pounds_{\boldsymbol{u}}\boldsymbol{e}_\alpha = 0$. The induced metric on $\Sigma(\tau)$ is given by

$$h_{\alpha\beta} = g_{ab}e_\alpha^a e_\beta^b \tag{5.164}$$

(see Eq. (4.31)). It is also clear that $h_{\alpha\beta} = h_{ab}e_\alpha^a e_\beta^b$ on $\mathcal{G}$, where $h_{ab} = g_{ab} + u_a u_b$ is the induced metric in the four-dimensional form. The inverse metric is given by

$$h^{ab} = h^{\alpha\beta}e_\alpha^a e_\beta^b. \tag{5.165}$$

The cross-sectional volume is $\delta V = \sqrt{h}d^3 y$, where $h = \det(h_{\alpha\beta})$. We are interested in the change in this volume as we vary τ. Since the coordinates y^α are comoving, d^3y does not change and the change in the volume arises only from the change in $\sqrt{h}$. Therefore,

$$\frac{1}{\delta V}\frac{d}{d\tau}\delta V = \frac{1}{\sqrt{h}}\frac{d}{d\tau}\sqrt{h} = \frac{1}{2}h^{\alpha\beta}\frac{dh_{\alpha\beta}}{d\tau}. \tag{5.166}$$

The quantity $dh_{\alpha\beta}/d\tau$ can be computed directly and shown to be equal to

$$\frac{dh_{\alpha\beta}}{d\tau} \equiv u^m\nabla_m\left(g_{ab}e_\alpha^a e_\beta^b\right) = e_\alpha^a e_\beta^b\nabla_a u_b + e_\alpha^a e_\beta^b\nabla_b u_a = (Q_{ab} + Q_{ba})\,e_\alpha^a e_\beta^b, \tag{5.167}$$

where we have used the relation $\pounds_{\boldsymbol{u}}\boldsymbol{e}_\alpha = 0$ to express covariant derivatives of $\boldsymbol{e}_\alpha$ in terms of covariant derivatives of $\boldsymbol{u}$. It follows that

$$h^{\alpha\beta}\frac{dh_{\alpha\beta}}{d\tau} = (Q_{ab} + Q_{ba})\left(h^{\alpha\beta}e_\alpha^a e_\beta^b\right) = 2Q_{ab}h^{ab} = 2Q_{ab}g^{ab} = 2\theta \tag{5.168}$$

which, together with Eq. (5.166), leads to Eq. (5.162).

5.5.2 Null congruence

It is possible to obtain very similar relationships in case of a congruence of null geodesics. However, since some new features emerge in this case, it is worth repeating the above analysis explicitly for the null geodesics. In this case we are dealing with a set of null curves $x^a(\lambda)$ with the tangent vector $\boldsymbol{k}$ and a deviation vector $\boldsymbol{\xi}$ which satisfy the relations

$$k^a k_a = 0, \qquad k^b\nabla_b k^a = 0, \qquad \xi^b\nabla_b k^a = k^b\nabla_b\xi^a, \qquad k^a\xi_a = 0. \tag{5.169}$$

As before, we will be interested in the submanifold defined by a constant value of λ which can be thought of as the cross-section of the null congruence. The new features arise because the cross-section is now *two*-dimensional rather than *three*-dimensional. As an example, consider the flat spacetime expressed in null coordinates $u = t - x$, $v = t + x$ with the metric

$$ds^2 = -dudv + dx^2 + dy^2. \tag{5.170}$$

The null surface $u = $ constant, say, is now two-dimensional and not three-dimensional – which is a general feature of any null surface. We cannot now define a metric on this null surface by $q_{ab} = g_{ab} + k_a k_b$ because $k^a(g_{ab} + k_a k_b) = k_b \neq 0$ when $k^2 = 0$ and hence q_{ab} is not transverse to k^a. To tackle this problem we need to introduce another null vector field l_a such that, say, $l_a k^a = -1$. For example, if $k_a = -\partial_a u$ in the local inertial frame, we can take $l_a = -(1/2)\partial_a v$. We can then define the transverse metric by

$$h_{ab} = g_{ab} + k_a l_b + l_a k_b, \tag{5.171}$$

which satisfies the conditions

$$h_{ab}k^b = h_{ab}l^b = 0, \qquad h_a^a = 2, \qquad h_m^a h_b^m = h_b^a, \qquad k^s \nabla_s h_{ab} = 0. \quad (5.172)$$

These relations also show that h_b^a acts as a projection tensor to the transverse space. The conditions $l^2 = 0$ and $\mathbf{k} \cdot \mathbf{l} = -1$, however, do not uniquely determine $\mathbf{l}$; so the transverse metric in this case is not unique but we will see that physically relevant quantities will be independent of the choice made for l.

To study a null congruence, we will again consider the rate of change of a deviation vector V^m in the orthogonal subspace and define $Q_{ab} = \nabla_a k_b$. We then have, as before,

$$\frac{DV^m}{d\lambda} \equiv k^n \nabla_n V^m = Q_n^m V^n. \quad (5.173)$$

It is, however, easy to prove that only the two-dimensional projection $\widehat{Q}_n^m = h_a^m h_n^b Q_b^a$ of the tensor Q_{ab} is relevant in this equation. This follows from straightforward manipulation:

$$\frac{DV^m}{d\lambda} = k^n \nabla_n V^m = k^n \nabla_n (h_r^m V^r) = h_r^m k^n \nabla_n V^r$$
$$= h_r^m Q_n^r V^n = h_r^m Q_n^r h_s^n V^s = \widehat{Q}_s^m V^s, \quad (5.174)$$

where we have used the last relation in Eq. (5.172) and Eq. (5.173). We now decompose the projected deviation tensor $\widehat{Q}_{mn}$ into expansion, shear and rotation in the standard manner:

$$\widehat{Q}_{mn} = \frac{1}{2}\theta h_{mn} + \widehat{\sigma}_{mn} + \widehat{\omega}_{mn}. \quad (5.175)$$

The rest of the analysis proceeds exactly as before. We now get

$$\frac{d\theta}{d\lambda} = -\frac{1}{2}\theta^2 - \sigma^{ab}\sigma_{ab} + \omega^{ab}\omega_{ab} - R_{ab}k^a k^b. \quad (5.176)$$

The main difference compared with the case of timelike geodesic congruence is the change in the numerical factor $(1/3)$ to $(1/2)$ which arises because the orthogonal space to the null surface is two-dimensional. This result has the same physical content as Eq. (5.150). To make this correspondence explicit, we shall now prove that, in the case of a null congruence, we can interpret θ as the fractional rate of change of the congruences cross-sectional *area*. That is,

$$\theta = \frac{1}{\delta A}\frac{d}{d\lambda}\delta A. \quad (5.177)$$

The interpretation here is somewhat more involved than the case of timelike congruence because the transverse space is two-dimensional. We can, as before, pick one null geodesic $\mathcal{G}$ and an event P on it parameterized by a given value $\lambda = \lambda_P$. Next we introduce another set of auxiliary curves for which l^a are the tangent vectors and label them using a parameter μ such that μ is a constant on the null geodesics. Let us now consider a small neighbourhood of P such that, through each of the events in this neighbourhood, there passes a geodesic from the congruence and another auxiliary curve (with fixed values for λ and μ) through each of the events in this neighbourhood. This defines a two-dimensional surface with a coordinate system θ^A [with $A = (1, 2)$]. These coordinates can again be chosen so that they remain constant on each of the geodesics. We thus have a four-dimensional coordinate system defined by the relation $x^a = x^a(\lambda, \mu, \theta^A)$. We can now define the vectors

$$k^a = \left(\frac{\partial x^a}{\partial \lambda}\right)_{\mu, \theta^A} ; \qquad e_A^a = \left(\frac{\partial x^a}{\partial \theta^A}\right)_{\lambda, \mu}. \tag{5.178}$$

As before, we also have the relations $\pounds_k e_A = 0$ and on the geodesic $\mathcal{G}$ the relations $k \cdot e_A = l \cdot e_A = 0$. The induced metric on the two surface can now be taken as

$$\sigma_{AB} = g_{ab} e_A^a e_B^b. \tag{5.179}$$

The rest of the analysis proceeds as before and one can easily show that θ represents the fractional rate of change of the cross-sectional area.

The analogue of Raychaudhuri's equation obtained above (Eq. (5.176)) assumes that the null geodesics are affinely parametrized. If this is not the case, then the expansion gets modified to $\theta = \nabla_i k^i - \kappa$, where κ is defined through the geodesic equation (without affine parametrization), $k^b \nabla_b k^a = \kappa k^a$. In this case, it is easy to see that Eq. (5.176) picks up an addition term on the right hand side and is now given by

$$\frac{d\theta}{d\lambda} = \kappa\theta - \frac{1}{2}\theta^2 - \sigma^{ab}\sigma_{ab} + \omega^{ab}\omega_{ab} - R_{ab}k^a k^b, \tag{5.180}$$

where λ is not an affine parameter. We will need this equation in Chapter 8 while discussing the black hole physics.

5.5.3 Integration on null surfaces

The concepts introduced above also allows us to perform integration over the cross-section transverse to a congruence, including the case in which the congruence is null. When the congruence is made of timelike geodesics, this is completely straightforward and we can take the directed element of three-volume to be $d\Sigma_a = n_a \sqrt{h} d^3 y$ where the congruence is parametrized through the relations $x^a = x^a(\lambda, y^\alpha)$. In the case of a null congruence, the situation is slightly more complicated. We first note that if k^a is the normal to the null surface, then the directed surface element has the natural parametrization

$$d\Sigma_m = k^n dS_{mn} d\lambda \tag{5.181}$$

with

$$dS_{mn} = \epsilon_{mnbc}e_2^b e_3^c \, d^2\theta \tag{5.182}$$

representing the two-dimensional surface element. Since $\epsilon_{mnbc}e_2^b e_3^c$ is orthogonal to e_A^i, it is contained in the space spanned by k^a and l^a. So it can be expressed as

$$\epsilon_{mnbc}e_2^b e_3^c = 2fk_{[m}l_{n]} = f(k_m l_n - l_m k_n) \tag{5.183}$$

with some function f. By working in the local inertial frame, we see that $f = \sqrt{\sigma}$ so that we can also express the surface element in Eq. (5.182) as

$$dS_{ab} = 2k_{[a}l_{b]}\sqrt{\sigma} \, d^2\theta. \tag{5.184}$$

Using Eq. (5.184) in Eq. (5.181) we get

$$d\Sigma_a = -k_a\sqrt{\sigma} \, d^2\theta \, d\lambda, \tag{5.185}$$

which is the same as the volume element $d\Sigma_a = n_a\sqrt{h}d^3y$ in the non-null case with the choice of coordinates $y^\alpha = (\lambda, \theta^A)$ except for a sign that results from the convention we have chosen for l^a.

Exercise 5.22
Expansion and rotation of congruences Consider the vector field

$$u^i = \frac{1}{\sqrt{1 - 2M/r}}\delta_0^i + \frac{\sqrt{M/r^3}}{\sqrt{1 - 3M/r}}\delta_\theta^i \tag{5.186}$$

in the Schwarzschild spacetime described by the metric in Eq. (4.81). (i) Show that this vector field is a tangent to a set of timelike geodesics and determine the geodesic curves. (ii) Calculate the expansion $\nabla_i u^i$ of this congruence. Is the expansion singular anywhere? Does it have the same sign everywhere? (iii) Compute the rotation tensor ω_{ab} for this congruence and show that its square is given by

$$\omega^{ab}\omega_{ab} = \frac{M}{8r^3}\left(\frac{1 - 6M/r}{1 - 3M/r}\right)^2. \tag{5.187}$$

5.6 Classification of spacetime curvature

The curvature tensor characterizes the key geometrical features of spacetime and is directly relevant to the study of the gravitational field. Given this fact, the classification of generic types of the curvature tensor is of importance to understand the properties of gravitational fields they represent. Though the physically relevant case corresponds to $N = 4$ dimensions, we will provide a description of the curvature tensors in $N = 2, 3$ as well since they are of interest in some special contexts and in Chapter 15.

5.6.1 Curvature in two dimensions

In two dimensions, the curvature tensor has $(1/12)N^2(N^2 - 1) = 1$ independent component which could be conveniently taken as R_{1212}. In this case, we can think of the two-dimensional surface to be embedded in the usual three-dimensional Euclidean space and introduce, at any given point P on the surface, a tangent plane. We will choose the Cartesian coordinates (x, y) on the tangent plane with P being the origin. The deviation $z(x, y)$ of the curved surface from the tangent plane can describe the local properties of the curved surface. In our coordinate system, the curvature effects will arise owing to second derivatives of $z(x, y)$ and hence it is adequate to use a quadratic function $z(x, y) = (1/2)[ax^2 + 2bxy + cy^2]$. Diagonal-izing the quadratic function by a linear transformation, from the coordinates (x, y) to (ξ, η), we can reduce the equation to the surface to the form

$$z = \frac{1}{2}\left(\kappa_1\xi^2 + \kappa_2\eta^2\right) \equiv \frac{1}{2}\left(\frac{\xi^2}{\rho_1} + \frac{\eta^2}{\rho_2}\right), \tag{5.188}$$

where κ_1 and κ_2 are called the two *principal curvatures* and ρ_1 and ρ_2 are called *principal radii* of curvature. The metric on the two-dimensional surface can be found from the metric on the Euclidean 3-space by expressing z as a function of the coordinates (ξ, η). This gives

$$ds^2 = dz^2 + d\xi^2 + d\eta^2 = [(\kappa_1\xi d\xi + \kappa_2\eta d\eta)^2 + (d\xi^2 + d\eta^2)] \equiv \gamma_{ab}dx^a dx^b. \tag{5.189}$$

From this metric it is easy to show that the scalar curvature of the two-dimensional surface has the general form

$$R = \frac{2R_{1212}}{\gamma}, \qquad \gamma = \det|\gamma_{ab}| = \gamma_{11}\gamma_{22} - (\gamma_{12})^2. \tag{5.190}$$

The product $\kappa_1\kappa_2 = (\rho_1\rho_2)^{-1}$ can be expressed entirely in terms of intrinsic mea-surements on the surface without any reference to the external embedding space. One possible way of doing this will be as follows. Start from some point P on the surface and proceed along a geodesic on the surface for a proper distance ϵ thereby arriving at point Q. By repeating this with geodesics starting off in different direc-tions, one can obtain a continuous set of points $Q_1, Q_2, Q_3...$, etc., all of which are at a proper distance ϵ from P. This is a natural defintion of a 'circle' on the sur-face centred at P, the circumference of which can also be measured intrinsically. A simple calculation using the metric in Eq. (5.189) shows that

$$\lim_{\epsilon\to0}\frac{6}{\epsilon^2}\left(1 - \frac{\text{circumference}}{2\pi\epsilon}\right) = \kappa_1\kappa_2 = \frac{1}{\rho_1\rho_2} = \det\begin{pmatrix} a & b \\ b & c \end{pmatrix}. \tag{5.191}$$

This quantity $\kappa_1\kappa_2$ is called the *intrinsic curvature* (or *Gaussian curvature*) of the surface and is related to the scalar curvature by $R = 2(\kappa_1\kappa_2)$. In contrast, one can also define a quantity called *extrinsic curvature* by $(\kappa_1 + \kappa_2)$. The difference between these two could be understood by considering, for example, a plane sheet of paper (with $\kappa_1 = \kappa_2 = 0$) rolled in the form of a cylinder of radius r which will be a curved two-dimensional surface embedded in three-dimensional space. For the cylindrical surface, we have $\kappa_1 = (1/r)$, $\kappa_2 = 0$. The intrinsic curvature $\kappa_1\kappa_2 = 0$ and retains its value for the sheet of paper; but the extrinsic curvature, which indicates how the surface is embedded in the three-dimensional space, changes from 0 to a nonzero value $(\kappa_1 + \kappa_2) = (1/r)$.

Curvature of two-dimensional surfaces can also occur at a single point on the surface and the most important example of this occurs in the case of a cone. If a sector of a circle with a central angle α is rolled to make the radii at the edges of the sector meet each other, one obtains a conical surface embedded in 3-space. The metric on the surface of the cone is the same as the metric on the original plane sheet of paper, $ds^2 = dr^2 + r^2 d\theta^2$ but now the range of θ is limited to $0 \leq \theta < \alpha$. It is easy to see that such a surface has a *conical singularity* with infinite curvature at the apex of the cone. This is an example of a two-dimensional space which is locally flat but has concentrated curvature at one point.

Exercise 5.23
Euler characteristic of two-dimensional spaces Consider a two-dimensional manifold $\mathcal{V}$ which is compact (i.e. boundaryless) and has a scalar curvature R. The integral

$$\chi(\mathcal{V}) = \frac{1}{4\pi} \int_{\mathcal{V}} R\sqrt{|g|}\, d^2x \qquad (5.192)$$

is called the Euler characteristic of the manifold and depends only on the topology of the manifold. In general, it is given by $\chi(\mathcal{V}) = 2(1 - k)$, where k is the genus of the manifold. Show that $k = 0$ for a sphere and evaluate it for a torus.

5.6.2 Curvature in three dimensions

Let us next consider three dimensions in which the curvature tensor has $(1/12) \times 3^2 \times (3^2 - 1) = 6$ independent components, which is the same as the number of independent components of the Ricci tensor $R_{\alpha\beta}$ in three dimensions. From the linear relation $R_{\alpha\beta} = g^{\gamma\delta} R_{\gamma\alpha\delta\beta}$, it follows that all the components of curvature tensor can be expressed in terms of those of the Ricci tensor. Given the linearity, we can assume a form

$$R_{\alpha\beta\gamma\delta} = A_{\alpha\gamma}\gamma_{\beta\delta} - A_{\alpha\delta}\gamma_{\beta\gamma} + A_{\beta\delta}\gamma_{\alpha\gamma} - A_{\beta\gamma}\gamma_{\alpha\delta}, \qquad (5.193)$$

which satisfies the symmetry requirements of a curvature tensor with $A_{\alpha\beta}$ being some symmetric tensor. Contracting on α and γ we get $R_{\alpha\beta} = A\gamma_{\alpha\beta} +$

$A_{\alpha\beta}$, $A_{\alpha\beta} = R_{\alpha\beta} - (1/4)R\gamma_{\alpha\beta}$. This leads to the expression

$$R_{\alpha\beta\gamma\delta} = R_{\alpha\gamma}\gamma_{\beta\delta} - R_{\alpha\delta}\gamma_{\beta\gamma} + R_{\beta\delta}\gamma_{\alpha\gamma} - R_{\beta\gamma}\gamma_{\alpha\delta} + \frac{R}{2}(\gamma_{\alpha\delta}\gamma_{\beta\gamma} - \gamma_{\alpha\gamma}\gamma_{\beta\delta}). \quad (5.194)$$

As we mentioned earlier in Section 5.3.3, the Weyl tensor vanishes identically in three-dimensions.

5.6.3 Curvature in four dimensions

In four dimensions, we have already seen that the curvature tensor has 20 independent components while the Ricci tensor has ten components. From Eq. (5.66) it is clear that the remaining ten components are contained in the Weyl tensor C_{abcd}. In any spacetime with $R_{ik} = 0$ (which, as we shall see in Chapter 6 corresponds to vacuum solutions to Einstein's equations) the classification of curvature tensor is reduced to classification of Weyl tensor. Further, we saw in Section 4.2 (see Eq. (4.6)) that even after transforming to the local inertial frame at a given event, we have the freedom of three rotations of the axes and three Lorentz boosts, which allows us to impose six more conditions. This reduces the number of independent components we need to deal with, in general, to four. This problem of classifying the independent generic forms of the Weyl tensor can be tackled in a general manner and the resulting classification is called *Petrov classification*. The key idea is to use the fact that curvature tensor can be thought of as a 6×6 matrix (see Exercise 5.9), which further separates into four blocks. By reducing these matrices to canonical form, we can determine the general categories of curvature tensors.[2]

The structure of the 6×6 matrix representation for the curvature tensor is based on separating out the components of R_{abcd} into three sets $(R_{0\alpha0\beta}, R_{0\beta\gamma\delta}, R_{\gamma\delta\lambda\mu})$. The first of these can easily be thought of as a 3×3 matrix in the indices α and β. In the other two we shall use three-dimensional $\epsilon_{\alpha\beta\gamma}$ to trade off two antisymmetric indices for a single vector index. This allows us to introduce three 3×3 matrices:

$$A_{\alpha\beta} = R_{0\alpha0\beta}, \qquad B_{\alpha\beta} = \frac{1}{2}\epsilon_{\alpha\gamma\delta}R^{\gamma\delta}_{0\beta}, \qquad C_{\alpha\beta} = \frac{1}{4}\epsilon_{\alpha\gamma\delta}\epsilon_{\beta\lambda\mu}R^{\gamma\delta\lambda\mu}. \quad (5.195)$$

The condition $R_{ab} = 0$ is equivalent to the relation:

$$A_{\alpha\alpha} = 0, \qquad B_{\alpha\beta} = B_{\beta\alpha}, \qquad A_{\alpha\beta} = -C_{\alpha\beta}. \quad (5.196)$$

To proceed further, we will define a symmetric complex tensor by

$$D_{\alpha\beta} = \frac{1}{2}(A_{\alpha\beta} + 2iB_{\alpha\beta} - C_{\alpha\beta}) = A_{\alpha\beta} + iB_{\alpha\beta}. \quad (5.197)$$

(This may be thought of as analogous to a combination $\boldsymbol{E} + i\boldsymbol{B}$ in electrodynamics.) Because of the doubling of degrees of freedom in moving to the complex plane, we can think of the four-dimensional transformations of the curvature tensor as equivalent to three-dimensional complex rotations of $D_{\alpha\beta}$. The task of classifying the curvature tensor is now reduced to finding the different canonical structures for this matrix which – in turn – depends on the nature of eigenvalues of this matrix. In the eigenvalue equation $D_{\alpha\beta}\, n_\beta = \lambda\, n_\alpha$ the complex eigenvalues $\lambda = \lambda_R + i\lambda_I$ satisfy the condition $\lambda^{(1)} + \lambda^{(2)} + \lambda^{(3)} = 0$ since $D_\alpha^\alpha = 0$. The classification of the matrix now depends on the number of independent eigenvectors and there are essentially three cases to consider, leading to what are called Petrov Types I, II and III. For the sake of completeness, we shall briefly describe their structure.

Petrov Type I
In this case there are three independent eigenvectors and diagonalizing $D_{\alpha\beta}$ and taking its real and imaginary parts we get

$$A_{\alpha\beta} = \text{dia}\left(\lambda_R^{(1)},\ \lambda_R^{(2)},\ -(\lambda_R^{(1)} + \lambda_R^{(2)})\right)$$

$$B_{\alpha\beta} = \text{dia}\left(\lambda_I^{(1)},\ \lambda_I^{(2)},\ -(\lambda_I^{(1)} + \lambda_I^{(2)})\right). \tag{5.198}$$

With straightforward computation, one can show that the eigenvalues can be expressed in terms of the scalars

$$I_1 = \frac{1}{48}\left(R_{iklm}\, R^{iklm} - i\, R_{iklm}\, (*R)^{iklm}\right)$$

$$I_2 = \frac{1}{96}\left(R_{iklm}\, R^{lmpr}\, R_{pr}^{ik} + i\, R_{iklm}\, R^{lmpr}\, (*R)_{pr}^{ik}\right), \tag{5.199}$$

where $(*R)_{iklm} = (1/2)\epsilon_{ikpr} R_{lm}^{pr}$ is the dual of the curvature tensor. The eigenvalues are related to the curvature invariants by

$$I_1 = \frac{1}{3}\left(\lambda^{(1)2} + \lambda^{(2)2} + \lambda^{(3)2}\right), \qquad I_2 = \frac{1}{3}\lambda^{(1)}\lambda^{(2)}\left(\lambda^{(1)} + \lambda^{(2)}\right). \tag{5.200}$$

The special case in which $\lambda^{(1)} = \lambda^{(2)}$ is usually called Type D.

Petrov Type II
This corresponds to the situation with two independent eigenvectors so that $n_\alpha n^\alpha$ vanishes for one of them. This means we cannot rotate the coordinate axis and align a coordinate in its direction but we can always choose this vector to lie in the x^1–x^2 plane so that $n_2 = in_1$ and $n_3 = 0$. The eigenvalue equations now read $D_{11} + iD_{12} = \lambda$, $D_{22} - iD_{12} = \lambda$ so that we can write $D_{11} = \lambda - i\mu$, $D_{2} = \lambda + i\mu$, $D_{12} = \mu$. Since the quantity μ can be taken to be real without any loss of generality, we obtain the following canonical form of the matrices

$$A_{\alpha\beta} = \begin{pmatrix} \lambda_R & \mu & 0 \\ \mu & \lambda_R & 0 \\ 0 & 0 & -2\lambda_R \end{pmatrix}, \qquad B_{\alpha\beta} = \text{dia}\left(\lambda_R - \mu,\ \lambda_R + \mu,\ -2\mu\right), \tag{5.201}$$

with just two invariants λ_R and λ_I so that $I_1 = \lambda^2$ and $I_2 = \lambda^3$. In this spacetime, we have the condition $I_1^3 = I_2^2$. The special case when $\lambda = 0$ corresponds to a situation in which both the curvature invariances vanish and this type is usually referred to as Type N.

Petrov Type III

In this case there is only one eigenvector with $n_\alpha n^\alpha = 0$ and all eigenvalues are identically zero. The eigenvalue equation has the solution $D_{11} = D_{22} = D_{12} = 0$, $D_{13} = \mu$, $D_{23} = i\mu$. The matrices $A_{\alpha\beta}$ and $B_{\alpha\beta}$ can be easily determined to be

$$A_{\alpha\beta} = \begin{pmatrix} 0 & 0 & \mu \\ 0 & 0 & 0 \\ \mu & 0 & 0 \end{pmatrix}, \qquad B_{\alpha\beta} = \begin{pmatrix} 0 & 0 & 0 \\ 0 & 0 & \mu \\ 0 & \mu & 0 \end{pmatrix}. \tag{5.202}$$

In this case, the curvature tensor has no nonzero invariants even though the spacetime is curved. This is analogous to the situation in electrodynamics in which both $E^2 - B^2$ and $E \cdot B$ vanish though both E and B are nonzero.

The above analysis started with Eq. (5.195) in which we have set some indices of R_{abcd} to the time component in order to define spatial vectors. A more formal and covariant procedure is to use an arbitrary timelike vector u^i to define the four-dimensional quantities

$$A_{ab} = R_{iajb} u^i u^j,$$

$$B_{ab} = \frac{1}{2} u^i u^j \epsilon_{iacd} R_{jb}^{cd} = \frac{1}{2} u^i u^j (*R)_{iajb},$$

$$C_{ab} = \frac{1}{4} u^i u^j \epsilon_{iamn} \epsilon_{jbpq} R^{mnpq} = \frac{1}{4} u^i u^j (**R)_{iajb}, \tag{5.203}$$

where $(*R)_{abcd}$ is the dual of R_{abcd} on two of the antisymmetric indices and $(**R)_{abcd}$ is the double dual of R_{abcd} on both pairs of antisymmetric indices. In the rest frame of the timelike vector u^i, these definitions reduce to the ones in Eq. (5.195). In general, all these tensors are orthogonal to u^i. Repeating our analysis with these definitions, one can show that the results are independent of the four-vector u^i used in the definition.

PROJECTS

Project 5.1
Parallel transport, holonomy and curvature

It is possible to provide a formal but interesting solution to the parallel transport equation which is of some interest in different contexts. This project discusses several features of such a solution. Consider Eq. (5.1), which parallel transports the vector v^a along some curve $\mathcal{C}$. Given the curve, we can define the quantities $A^m{}_r \equiv -\Gamma^m_{sr}(dx^s/d\lambda)$ and rewrite Eq. (5.1) as $dv^m/d\lambda = A^m{}_r v^r$. This suggests defining a *parallel propagator* $P^m{}_r(\lambda, \lambda_0; \mathcal{C})$ which depends on the initial and final points as well as the curve chosen for the parallel transport, through the definition

$$v^m(\lambda) = P^m{}_r(\lambda, \lambda_0)v^r(\lambda_0). \tag{5.204}$$

Clearly, the parallel propagator satisfies the differential equation

$$\frac{d}{d\lambda}P^m{}_r(\lambda, \lambda_0) = A^m{}_s P^s{}_r(\lambda, \lambda_0). \tag{5.205}$$

(a) Show that Eq. (5.205) is equivalent to the integral equation

$$P^m{}_r(\lambda, \lambda_0) = \delta^m_r + \int_{\lambda_0}^{\lambda} A^m{}_s(\eta) P^s{}_r(\eta, \lambda_0) d\eta. \tag{5.206}$$

Iterating the process, show that the parallel propagator can be expressed as

$$P^m{}_r(\lambda, \lambda_0) = \delta^m_r + \int_{\lambda_0}^{\lambda} A^m{}_r(\eta) d\eta + \int_{\lambda_0}^{\lambda}\int_{\lambda_0}^{\eta} A^m{}_s(\eta) A^s{}_r(\eta') d\eta' d\eta + \cdots. \tag{5.207}$$

(b) To proceed further, we need to rewrite the integral with the same range of integration in each. This can be done by introducing a *path ordering symbol* which is defined as follows: the expression $\mathcal{P}[A(\eta_n)A(\eta_{n-1})\cdots A(\eta_1)]$ is defined to be the product of n matrices $A(\eta_i)$ ordered in such a way that the largest value of η is on the left and each subsequent value is less than or equal to the previous value. Show that

$$\int_{\lambda_0}^{\lambda}\int_{\lambda_0}^{\eta_n}\cdots\int_{\lambda_0}^{\eta_2} A(\eta_n)A(\eta_{n-1})\cdots A(\eta_1) d^n\eta \tag{5.208}$$

$$= \frac{1}{n!}\int_{\lambda_0}^{\lambda}\int_{\lambda_0}^{\lambda}\cdots\int_{\lambda_0}^{\lambda} \mathcal{P}[A(\eta_n)A(\eta_{n-1})\cdots A(\eta_1)] d^n\eta.$$

Hence show that the parallel propagator can be expressed in the form

$$P^m{}_n(\lambda, \lambda_0) = \mathcal{P}\exp\left(-\int_{\lambda_0}^{\lambda}\Gamma^m_{sv}\frac{dx^s}{d\eta} d\eta\right), \tag{5.209}$$

where the exponential is defined through its power series expansion.

(c) When the path is a loop starting and ending at the same point, the resulting parallel propagator matrix must be just a Lorentz transformation at the point in question. Explain what happens to it if the loop is infinitesimally small. [Note. The resulting transformation is called the holonomy of the loop. It turns out that knowledge of the holonomy of every possible loop is equivalent to knowing the metric.[3]]

Project 5.2
Point charge in the Schwarzschild metric

Consider a spacetime described by the metric in Eq. (4.81). Suppose a test (point) charge q is located at $r = r', \theta = 0$; that is, on the z-axis at a distance r' from the origin.

(a) Solve Maxwell's equations assuming that only A_0 is nonzero and show that the resulting potential is

$$A_0\left(r,\theta\right) = \frac{q[(r-M)(r'-M) - M^2\cos\theta)]}{rr'\sqrt{(r-M)^2 + (r'-M)^2 - 2(r-M)(r'-M)\cos\theta - M^2\sin\theta}}.$$

(5.210)

Compute the resulting electric field and plot the field lines for different valuse of r', particularly when $r' \gg 2M$ and when $r' \to 2M$. Can you explain the nature of field lines in terms of a 'refractive index' obtained in Exercise 4.8?

(b) Obtain the vector potential and the electric field due to a charge at rest in the Rindler frame introduced in Chapter 3. [Hint. You can get this from the result in part (a) by taking suitable limits.]

(c) A charge at rest in the Rindler frame is accelerating uniformly with respect to the inertial frame. Transform the fields obtained in part (b) to the inertial frame and determine the radiation field of the charge.

(d) A charge accelerating through the inertial frame will feel a radiation reaction. If we transform to a non-inertial frame in which the charge is at rest, where does the radiation reaction come from? (This approach[4] provides an interesting insight into the radiation reaction.)

6

Einstein's field equations and gravitational dynamics

6.1 Introduction

We begin this chapter by introducing the action functional for gravity and obtaining Einstein's equations. We then describe the general properties of Einstein's equations and discuss their weak field limit. The action functional and its properties will play a crucial role in Chapters 12, 15 and 16, while the linearized field equations will form the basis of our discussion of gravitational waves in Chapter 9.

6.2 Action and gravitational field equations

Let us recall that we studied the scalar and electromagnetic fields in Chapter 2 in two steps. First, in Sections 2.3.1 and 2.4.1, we considered the effect of the field (scalar or electromagnetic) on other physical systems (like material particles). In the second step, in Sections 2.3.2 and 2.7, we studied the dynamics of the field itself, by adding a new term to the action principle. This new term depended on the field and on its first derivatives; by varying the field in the total action we could obtain the dynamical equations governing the field.

In the case of gravity, we have already completed the corresponding first step in the previous two chapters. We have seen that the effect of gravity on any other physical system (particles, fluids, electromagnetic field, ...) can be incorporated by modifying the action functional for the physical system by changing d^4x to $\sqrt{-g}\, d^4x$, partial derivatives by covariant derivatives and replacing η_{ab} by g_{ab}. That is, the action for a material system interacting with an externally specified gravitational field will have the form

$$\mathcal{A}_m = \int d^4x \sqrt{-g}\; L_m(\phi, \nabla\phi; g_{ab}), \qquad (6.1)$$

where ϕ is a generic symbol to denote any matter variable (position of a particle, electromagnetic vector potential, ...) that is to be varied to obtain the equations

239

of motion *for matter* in the presence of an externally specified gravitational field described by the metric tensor g_{ab}. (We saw examples of this in Eq. (5.111) and in Eq. (5.117).) We now have to take the second step, viz. to determine the dynamics of the gravitational field itself.

To do this, we have to add to the matter action $\mathcal{A}_m$ another term $\mathcal{A}_g[g_{ab}]$ which is a functional of the metric. Varying the total action, $\mathcal{A}_{\rm tot} = \mathcal{A}_m + \mathcal{A}_g$ with respect to the metric tensor g_{ab}, we should be able to get the field equations for gravity. This is precisely the procedure we have followed in the case of scalar and electromagnetic fields. Just as in those examples, we may expect the gravitational action to be an integral over a Lagrangian L_g which depends on the metric and its first derivatives:

$$\mathcal{A}_g = \int d^4x \sqrt{-g}\, L_g(g_{ab}, \partial_c g_{ab}). \tag{6.2}$$

Since L_g contains only up to the first derivatives of the metric, we will obtain field equations that are second order in the derivatives of the metric, which is precisely what we will expect for a dynamical system.

The above program runs into an unexpected difficulty which makes gravity completely different from all other fundamental interactions described through an action principle. Considering its importance (and a somewhat inadequate discussion of it in the standard text books), we will provide a detailed discussion of this issue and its possible resolution.

The difficulty arises from the fact that Eq. (6.2) requires us to find a generally covariant scalar L_g built out of the metric and its first derivatives, if $\mathcal{A}_g$ is to be a generally covariant scalar. However, no nontrivial scalar quantity L_g can be constructed from the metric and its first derivatives alone. This can be proved as follows. Consider any scalar quantity $f(g_{ik}, \partial_l g_{ik})$ which depends only on the metric tensor and its first derivatives. In a small region around any event in spacetime we can construct a locally inertial coordinate system in which $g_{ik} = \eta_{ik}$ and $(\partial_l g_{ik}) = 0$. Hence the function f will become a constant around any event in the locally inertial frame. (In fact, one can construct a locally inertial frame with $\Gamma^i_{kl} = 0$ all along any geodesic in a spacetime. Hence the scalar can be reduced to a constant all along a curve in the spacetime.) If the function is a scalar, then its value will remain the same in any other coordinate system. In other words, any covariant scalar function built out of the metric tensor and its derivatives will be a trivial constant. This result, of course, can be verified explicitly.

There are two ways out of this difficulty. The first is to look for a generally covariant scalar $L_g[g, \partial g, \partial^2 g]$ which depends not only on the metric and the first derivatives but also on the second derivatives but *depends on the second derivatives only linearly*. When we integrate the Lagrangian over a spacetime volume $\mathcal{V}$ to construct the action, such a second derivative term will give a contribution only on the boundary $\partial\mathcal{V}$. If we consider variations in which this contribution is fixed on $\partial\mathcal{V}$, then we are effectively using a Lagrangian that is quadratic in the first derivatives of the metric and we will get reasonable field equations. A simple example from classical mechanics illustrating this procedure will be a Lagrangian of the form $L = \ddot{q} + (1/2)\dot{q}^2$. The action resulting from this Lagrangian will be

$$\mathcal{A} = \int_{t_1}^{t_2} dt \left[\ddot{q} + \left(\frac{1}{2}\right)\dot{q}^2\right] = \dot{q}(t_2) - \dot{q}(t_1) + \int_{t_1}^{t_2} dt \left(\frac{1}{2}\right)\dot{q}^2. \tag{6.3}$$

If we now assume that *both* $\dot{q}$ and q are held fixed at the end points, then $\dot{q}(t_2) - \dot{q}(t_1)$ will not contribute when we vary $q(t)$ and we will get the standard equations of motion from the $(1/2)\dot{q}^2$ term. There are, however, several difficulties in an approach in which one attempts to freeze *both* q and $\dot{q}$ at the boundary. The first problem is that, for arbitrary values of q and $\dot{q}$ at both $t = t_1$ and $t = t_2$, we may not have a classical solution satisfying the boundary conditions. Further, one would like the action principle to obey the composition rule of the following kind. We expect $\mathcal{A}(1 \to 2 \to 3) = \mathcal{A}(1 \to 2) + \mathcal{A}(2 \to 3)$, where the paths connecting (q_1, t_1) and (q_3, t_3) are decomposed at an intermediate time t_2 with $t_1 < t_2 < t_3$. The paths are expected to be continuous at $t = t_2$ but need not be smooth at $t = t_2$. This requires leaving $\dot{q}$ at $t = t_2$ arbitrary in the action principle. Finally, the action principle has its roots in quantum theory and freezing q and $\dot{q}$ simultaneously will require specifying the values of both coordinate and momentum at a given time, which is inappropriate in the quantum theory.

The second way out of the difficulty is just to drop the second derivative terms from $L_g(g, \partial g, \partial^2 g)$ and work with the remaining part of the Lagrangian. (The 'dropping' of the second derivatives can be done either unceremoniously or by adding an extra term to the action such that its variation cancels the unwanted terms; either way the consequences are the same.) Now we need only to fix the metric on the boundary and we will get the same field equations. The price we pay is that the remaining part of the resulting Lagrangian is not a covariant scalar – though its variation will be covariant, so that the field equations will be generally covariant. We shall broadly adopt this strategy. Our approach will be to identify a generally covariant scalar which depends on the metric, its first and second derivatives, but is linear in the second derivatives. We will then separate out the second derivative part of this scalar and use the remaining terms as the action for gravity.

Since R_{abcd} is linear in the second derivatives, a scalar formed out of it will satisfy this criterion. We have seen in Section 5.3.3 that only one such scalar exists, viz. the Ricci scalar R. It is, of course, possible to add a constant term to it and multiply by another constant. Thus one can take the Lagrangian to have the form

$$L_g = \frac{1}{16\pi\kappa}(R - 2\Lambda), \tag{6.4}$$

where κ and Λ are constants. The particular choice of factors and signs are dictated by future convenience; with this choice, we will see that κ will turn out to be the Newtonian gravitational constant and Λ can be identified with what is usually called the cosmological constant. To obtain the total action we will have to add L_g to the matter Lagrangian L_m. We can, therefore, also think of the constant term $(-\Lambda/8\pi\kappa)$ in Eq. (6.4) as a constant added to the matter Lagrangian L_m, which is anyway unspecified at this stage. So, for the moment, we will ignore the constant term in L_g and study the properties of the Lagrangian $L_g = R/16\pi\kappa$. The key features of the gravitational action are therefore encoded in R, which we shall now study more closely.

6.2.1 Properties of the gravitational action

We want to rewrite this Lagrangian $L_g = (16\pi\kappa)^{-1}R$ separating out the quadratic part and a total derivative term. There are several ways of obtaining this result and we will choose an approach which will turn out to be useful later on in Chapters 15 and 16 when we discuss higher dimensional theories. We begin by expressing the Lagrangian as:

$$16\pi\kappa L_g = R \equiv Q_a{}^{bcd} R^a{}_{bcd} = Q^{cd}_{ab} R^{ab}_{cd} \equiv \delta^{cd}_{ab} R^{ab}_{cd}, \tag{6.5}$$

where

$$Q_a{}^{bcd} = \frac{1}{2}(\delta^c_a g^{bd} - \delta^d_a g^{bc}); \quad Q^{cd}_{ab} = \delta^{cd}_{ab} = \frac{1}{2}(\delta^c_a \delta^d_b - \delta^d_a \delta^c_b). \tag{6.6}$$

The tensor $Q_a{}^{bcd}$ is the only fourth rank tensor that can be constructed from the metric (alone) that has all the symmetries of the curvature tensor. In addition it has zero divergence on all the indices, $\nabla_a Q^{abcd} = 0$, etc. The δ^{cd}_{ab} is the alternating tensor ('determinant' tensor, discussed in Section 1.5) obtained by lowering one index of $Q_a{}^{bcd}$. This also shows that $\partial_i Q^{cd}_{ab} = 0$; i.e. it can be treated as a tensor with constant components. (Because of the symmetries of the curvature tensor, a notation like Q^{ab}_{cd}, R^{ab}_{cd}, etc., is unambiguous.) Expressing $R^a{}_{bcd}$ in terms of Γ^i_{jk} and using the antisymmetry of $Q_a{}^{bcd}$ in c and d, we can write

$$\sqrt{-g}Q_a{}^{bcd} R^a{}_{bcd} = 2\sqrt{-g}Q_a{}^{bcd}(\partial_c \Gamma^a_{db} + \Gamma^a_{ck}\Gamma^k_{db})$$

$$= 2\sqrt{-g}Q_a{}^{bcd}\Gamma^a_{ck}\Gamma^k_{db} + 2\partial_c\left[\sqrt{-g}Q_a{}^{bcd}\Gamma^a_{db}\right] - 2\Gamma^a_{db}\partial_c[\sqrt{-g}Q_a{}^{bcd}]$$

$$= 2\sqrt{-g}Q_a{}^{bcd}\Gamma^a_{ck}\Gamma^k_{db} + 2\partial_c\left[\sqrt{-g}Q_a{}^{bcd}\Gamma^a_{db}\right]$$

$$- 2\sqrt{-g}\Gamma^a_{db}\partial_c Q_a{}^{bcd} - 2\sqrt{-g}\Gamma^a_{db}\Gamma^j_{cj}Q_a{}^{bcd}. \tag{6.7}$$

In arriving at the last equality we have used Eq. (4.101). We now want to re-express the derivative term $\partial_c Q_a{}^{bcd}$ in terms of $Q_a{}^{bcd}$ itself. This can be done by writing $\partial_c Q_a{}^{bcd}$ in terms of $\nabla_c Q_a{}^{bcd}$ and using the fact that $\nabla_c Q_a{}^{bcd} = 0$. This gives the required relation:

$$\Gamma^a_{db}\partial_c Q_a{}^{bcd} = -\Gamma^a_{db}\Gamma^b_{kc}Q_a{}^{kcd} + \Gamma^a_{db}\Gamma^k_{ac}Q_k{}^{bcd} - \Gamma^a_{db}\Gamma^c_{kc}Q_a{}^{bkd}. \qquad (6.8)$$

Substituting Eq. (6.8) into Eq. (6.7) we notice that two *pairs* of the terms cancel out leading to a rather simple result

$$\sqrt{-g}\,Q_a{}^{bcd}R^a{}_{bcd} = 2\sqrt{-g}\,Q_a{}^{bcd}\Gamma^a_{dk}\Gamma^k_{bc} + 2\partial_c\left[\sqrt{-g}\,Q_a{}^{bcd}\Gamma^a_{bd}\right] \equiv \sqrt{-g}\,L_{\text{quad}} + L_{\text{sur}}, \qquad (6.9)$$

where we have separated out the expression into one term $[L_{\text{quad}}]$ which is quadratic in Γ^a_{bc} (and hence quadratic in the first derivatives of the metric) and another term $[L_{\text{sur}}]$ which is a total divergence that could lead to a surface term on integration. (Note that, in arriving at Eq. (6.9) we have only used the condition $\nabla_a Q^{abcd} = 0$ and its symmetries but not the explicit form of Q^{abcd}; we will need this fact in Chapters 15 and 16.)

When Q^{ab}_{cd} is given by Eq. (6.6), which is the case of our current interest, the explicit expression for the quadratic part is

$$L_{\text{quad}} = 2Q_a{}^{bcd}\Gamma^a_{dk}\Gamma^k_{bc} = g^{ab}\left(\Gamma^i_{ja}\Gamma^j_{ib} - \Gamma^i_{ab}\Gamma^j_{ij}\right). \qquad (6.10)$$

We can express Christoffel symbols in terms of the derivatives of the metric as

$$\Gamma^a_{bc} = \frac{1}{2}g^{ak}(-\partial_k g_{bc} + \partial_b g_{ck} + \partial_c g_{kb}) = \frac{1}{2}(-g^{ap}\delta^q_b\delta^r_c + g^{ar}\delta^p_b\delta^q_c + g^{aq}\delta^p_c\delta^r_b)\partial_p g_{qr}. \qquad (6.11)$$

Using this in Eq. (6.10) and simplifying the terms, we can write L_{quad} displaying explicitly the quadratic dependence on the derivatives of the metric:

$$L_{\text{quad}} = \frac{1}{4}M^{abcijk}\partial_a g_{bc}\partial_i g_{jk}, \qquad (6.12)$$

where

$$M^{abcijk} = \left[g^{ai}g^{bc}g^{jk} - g^{ai}g^{bj}g^{ck} + 2g^{ak}g^{bj}g^{ci} - 2g^{ak}g^{bc}g^{ij}\right]. \qquad (6.13)$$

This is the quadratic part of the gravitational action.

Let us next look at the total divergence term L_{sur} which can be written as

$$L_{\text{sur}} = 2\partial_c\left[\sqrt{-g}\,Q_a{}^{bcd}\Gamma^a_{bd}\right] = 2\partial_c\left[\sqrt{-g}\,Q^{cd}_{ak}g^{bk}\Gamma^a_{bd}\right] \equiv \partial_c\left[\sqrt{-g}\,V^c\right], \qquad (6.14)$$

where we have defined a four component object V^c (which is *not* a four-vector) by:

$$V^c \equiv \left(g^{ik}\Gamma^c_{ik} - g^{ck}\Gamma^m_{km}\right) = (g^{ia}g^{cj} - g^{aj}g^{ci})\partial_i g_{aj} = -\frac{1}{g}\partial_b(gg^{bc}). \quad (6.15)$$

The first equality arises from using the definition of Q^{cd}_{ak} in Eq. (6.6) and the second from expressing Christoffel symbols in terms of derivatives of the metric. The last equality arises from using the identities in Eq. (4.101) and Eq. (4.102).

The result in Eq. (6.9) shows that the Lagrangian $L_g \propto R$ can be separated into a quadratic part and a four-divergence. We will use the quadratic part as the Lagrangian to derive the field equations. That is, we take the total action for the system made of gravity and matter to be:

$$\mathcal{A}_{\text{tot}} = \mathcal{A}_{\text{quad}} + \mathcal{A}_m = \frac{1}{16\pi\kappa}\int_{\mathcal{V}} \sqrt{-g}\, d^4x L_{\text{quad}}[g, \partial g] + \int_{\mathcal{V}} \sqrt{-g}\, d^4x L_m[\phi, \nabla\phi; g],$$
$$(6.16)$$

where ϕ symbolically denotes the variables pertaining to the matter. We will assume for the moment that the constant term in Eq. (6.4) is also absorbed into L_m and comment on it at the end. Varying this action $\mathcal{A}_{\text{tot}}$ with respect to the metric g^{ab} should lead to the field equations of gravity. We shall now work out the variations of the two terms $\mathcal{A}_{\text{quad}}$ and $\mathcal{A}_m$ separately.

6.2.2 Variation of the gravitational action

It is possible to find the variation of $\mathcal{A}_{\text{quad}}$ directly from its explicit form in Eq. (6.12). However, it is difficult to re-express the resulting equations in a nice tensorial form. Hence we shall approach this variation somewhat indirectly. We first write Eq. (6.9) as $\sqrt{-g}\, L_{\text{quad}} = \sqrt{-g}\, R - L_{\text{sur}}$ and evaluate the required variation as

$$\delta(\sqrt{-g}\, L_{\text{quad}}) = \delta\left(\sqrt{-g}\, g^{ab} R_{ab}\right) - \delta L_{\text{sur}} \quad (6.17)$$
$$= \sqrt{-g}\, G_{ab}\delta g^{ab} + \sqrt{-g}\, g^{ab}\delta R_{ab} - \delta L_{\text{sur}},$$

where we have used the definition of the Einstein tensor $G_{ab} \equiv R_{ab} - (1/2)g_{ab}R$ (see Eq. (5.62)) and the result (see Eq. (4.34))

$$\delta\sqrt{-g} = -\frac{1}{2\sqrt{-g}}\delta g = -\frac{1}{2}\sqrt{-g}\, g_{ik}\delta g^{ik}. \quad (6.18)$$

To evaluate the quantity $g^{ik}\delta R_{ik}$, it is convenient to work in a local inertial frame in which $\Gamma^i_{kl} = 0$ and g_{ab} is a constant. Using the definition of R_{ik}, we find

$$g^{ik}\delta R_{ik} = g^{ik}\left\{\partial_l\delta\Gamma^l_{ik} - \partial_k\delta\Gamma^l_{il}\right\} = g^{ik}\partial_l\delta\Gamma^l_{ik} - g^{il}\partial_l\delta\Gamma^k_{ik} \equiv \partial_l w^l \quad (6.19)$$

with

$$w^l \equiv g^{ik}\delta\Gamma^l_{ik} - g^{il}\delta\Gamma^k_{ik}. \quad (6.20)$$

This result is valid in the local inertial frame in which g_{ab} are constants and hence can be moved through the partial derivatives. We now use the fact that, even though Γ^i_{kl} is not a tensor, $\delta\Gamma^i_{kl}$ is a tensor. To see this, we have only to note that $\Gamma^i_{kl}A^k dx^l$ is the change in a vector under parallel displacement between two infinitesimally separated points; hence $\delta\Gamma^i_{kl}A^k dx^l$ is the difference between two vectors obtained by two parallel displacements (one with the connection Γ^i_{kl} and the other with the connection $(\Gamma^i_{kl} + \delta\Gamma^i_{kl})$) between two infinitesimal points. The difference between two vectors at the same point is a vector and hence $\delta\Gamma^i_{kl}$ is a tensor. (This result can also be obtained by using the explicit transformation law for Christoffel symbols given by Eq. (4.62) by noting that the second term in Eq. (4.62) cancels out in the transformation of $(\Gamma^i_{kl} + \delta\Gamma^i_{kl}) - \Gamma^i_{kl}$.) Since $\delta\Gamma^i_{kl}$ is a tensor we can write Eq. (6.19) in any coordinate system by replacing $\partial_l w^l$ by $\nabla_l w^l$. This gives

$$g^{ik}\delta R_{ik} = \frac{1}{\sqrt{-g}}\partial_l\left(\sqrt{-g}\,w^l\right) \tag{6.21}$$

and consequently

$$\sqrt{-g}\,g^{ik}\delta R_{ik} = \partial_l\left(\sqrt{-g}\,w^l\right) = \partial_l\left(\sqrt{-g}\,(g^{ik}\delta\Gamma^l_{ik} - g^{il}\delta\Gamma^k_{ik})\right)$$
$$= 2\partial_c\left[\sqrt{-g}g^{bk}Q^{cd}_{ak}\delta\Gamma^a_{bd}\right], \tag{6.22}$$

where the last equality can be verified by using the definition of Q^{cd}_{ak} in Eq. (6.6).

We will next compute δL_{sur} in order to complete the evaluation of the right hand side of Eq. (6.17). Using the definition of L_{sur} in Eq. (6.14), we find that

$$\delta L_{\text{sur}} = 2Q^{cd}_{ak}\partial_c\left[\sqrt{-g}g^{bk}\delta\Gamma^a_{bd} + \Gamma^a_{bd}\delta(\sqrt{-g}g^{bk})\right]. \tag{6.23}$$

Using

$$\delta(\sqrt{-g}g^{bk}) = \sqrt{-g}[\delta^b_l\delta^k_m - \frac{1}{2}g^{bk}g_{lm}]\delta g^{lm} \equiv \sqrt{-g}B^{bk}_{lm}\delta g^{lm}, \tag{6.24}$$

with the last equality defining B^{bk}_{lm}, we see that the second term in Eq. (6.23) can be written as

$$2Q^{cd}_{ak}\partial_c\left[\Gamma^a_{bd}\delta(\sqrt{-g}g^{bk})\right] = 2Q^{cd}_{ak}\partial_c\left[\sqrt{-g}\Gamma^a_{bd}B^{bk}_{lm}\delta g^{lm}\right] \equiv \partial_c[\sqrt{-g}M^c{}_{lm}\delta g^{lm}] \tag{6.25}$$

where we have defined the 3-index *non*-tensorial object,

$$M^c{}_{lm} = 2Q^{cd}_{ak}B^{bk}_{lm}\Gamma^a_{bd} = \Gamma^c_{lm} - \Gamma^d_{ld}\delta^c_m - \frac{1}{2}g_{lm}V^c \tag{6.26}$$

for ease of notation. (Also recall that $\partial_i Q^{cd}_{ak} = 0$ so that it can be moved through the partial derivatives.) So

$$\delta L_{\text{sur}} = \partial_c\left[2\sqrt{-g}g^{bk}Q^{cd}_{ak}\delta\Gamma^a_{bd} + \sqrt{-g}M^c{}_{lm}\delta g^{lm}\right]. \tag{6.27}$$

We are interested in the combination $\sqrt{-g}\,g^{ab}\delta R_{ab} - \delta L_{\text{sur}}$ which occurs in the right hand side of Eq. (6.17). Comparing the results in Eq. (6.22) and Eq. (6.27), we find that the terms involving $\delta\Gamma$ cancel out in $(\sqrt{-g}\,g^{ab}\delta R_{ab} - \delta L_{\text{sur}})$ so that Eq. (6.17) becomes:

$$\delta(\sqrt{-g}\,L_{\text{quad}}) = \sqrt{-g}\,G_{ik}\delta g^{ik} - \partial_c\left[\sqrt{-g}M^c_{\ lm}\delta g^{lm}\right]. \tag{6.28}$$

Hence the variation of the action arising from L_{quad} is given by

$$\delta\mathcal{A}_{\text{quad}} = \frac{1}{16\pi\kappa}\int_{\mathcal{V}}d^4x\,\delta\left(\sqrt{-g}\,L_{\text{quad}}\right) \tag{6.29}$$

$$= \frac{1}{16\pi\kappa}\int_{\mathcal{V}}d^4x\,\sqrt{-g}\,G_{ik}\delta g^{ik} - \frac{1}{16\pi\kappa}\int_{\partial\mathcal{V}}d^3x\,\sqrt{h}\,(n_c M^c_{\ ik})\,\delta g^{ik},$$

where we have integrated the four-divergence using Gauss's theorem to obtain the surface term. (The induced metric on $\partial\mathcal{V}$ is h_{ab} with determinant h and n_c is the normal to the surface.) As usual, we shall consider variations in which $\delta g^{ik} = 0$ on the boundary $\partial\mathcal{V}$. For such variations the surface term vanishes and we get

$$\delta\mathcal{A}_{\text{quad}} = \frac{1}{16\pi\kappa}\int_{\mathcal{V}}d^4x\,\sqrt{-g}\,G_{ik}\,\delta g^{ik} \qquad \left(\text{when }\delta g^{ik} = 0\text{ on }\partial\mathcal{V}\right). \tag{6.30}$$

We see that the variation of the action based on L_{quad} leads to a generally covariant expression even though L_{quad} itself is not a scalar. The reason is clear from the expression in Eq. (6.29). If we consider an infinitesimal coordinate transformation of the form $x^a \to x^a + \xi^a$ with ξ^a and its derivatives vanishing on $\partial\mathcal{V}$, then δg^{ik} will vanish on $\partial\mathcal{V}$. In Eq. (6.29), the second term vanishes and the first term will have an integrand proportional to $G_{ik}\nabla^i\xi^k$. Using the Bianchi identity in Eq. (5.61) this integrand can be expressed as $\nabla^i(G_{ik}\xi^k)$, which on integration will vanish if ξ^a vanishes on $\partial\mathcal{V}$. It follows that $\delta\mathcal{A}_{\text{quad}} = 0$ for infinitesimal coordinate transformations on the bulk $x^a \to x^a + \xi^a$ with $\xi^a = 0$, $\nabla_a\xi_b = 0$ on $\partial\mathcal{V}$. Hence $\mathcal{A}_{\text{quad}}$ remains invariant under such transformations.

We have seen in Chapter 2 that the variation of the action with respect to the dynamical variables can also be used to determine the canonical momenta of the system. From Eq. (6.29) it is clear that $M^c_{\ ik}$ is related to the canonical momenta of the gravitational field. This is explored in greater detail in Project 6.3.

Exercise 6.1

Palatini variational principle There is an alternative procedure for obtaining gravitational field equations that avoids the problem of second derivatives. We start with the action in Eq. (6.4) (with $\Lambda = 0$ for simplicity) and write $\sqrt{-g}\,R = \sqrt{-g}\,g^{ab}R_{ab} = \sqrt{-g}\,g^{ab}R^l_{\ alb}$. We now use the fact that the quantity $R^l_{\ alb}$ – with one contravariant and three covariant indices – can be expressed entirely in terms of Γ^i_{jk} and its derivatives *without the metric appearing anywhere.* This allows us to treat both g^{ab} and Γ^i_{jk} as independent variables in this action, forgetting for the moment the relationship between them.

(a) Vary g^{ab} and Γ^i_{jk} independently in the action and show that

$$\delta A_g = \frac{1}{16\pi\kappa} \int_{\mathcal{V}} \left[\left(R_{ab} - \frac{1}{2} g_{ab} R \right) \delta g^{ab} + g^{ab} \left(\nabla_l \delta \Gamma^l_{ab} - \nabla_b \delta \Gamma^l_{al} \right) \right] \sqrt{-g} \, d^4 x. \quad (6.31)$$

(b) Assume that both δg^{ab} and $\delta \Gamma^i_{jk}$ vanish on $\partial \mathcal{V}$. Manipulate the term involving the variation of Christoffel symbols and show that demanding the coefficient of $\delta \Gamma^i_{jk}$ to vanish leads to the equation

$$\nabla_l g_{bc} = \partial_l g_{bc} - g_{cs} \Gamma^s_{bl} - g_{bs} \Gamma^s_{cl} = 0. \quad (6.32)$$

Hence prove that the Christoffel symbols are related to the derivatives of the metric in the usual way. The remaining term in the variation in Eq. (6.31) matches with what we have found in Eq. (6.30).

Exercise 6.2
Connecting Einstein gravity with the spin-2 field Write the metric in Eq. (6.12) as $g_{ab} = \eta_{ab} + h_{ab}$ and expand L_{quad} to quadratic order in the perturbations. Show that the resulting Lagrangian is the same as that of a spin-2 field discussed in Chapter 3.

6.2.3 A digression on an alternative form of action functional

Before we proceed further, we want to comment on an alternative form of action functional which will lead to the same variation as in Eq. (6.30) when the metric is held fixed at the boundaries. To obtain this, let us inspect the surface term arising from integrating L_{sur} in Eq. (6.14) more closely. Integrating L_{sur} over a four-dimensional region leads to the surface term in the action:

$$16\pi\kappa A_{\text{sur}} = \int_{\mathcal{V}} d^4 x L_{\text{sur}} = \int_{\mathcal{V}} d^4 x \partial_c [\sqrt{-g} V^c]. \quad (6.33)$$

Obviously, A_{sur} is *not* generally covariant. Let us, as usual, choose the boundary $\partial \mathcal{V}$ to be made of two spacelike surfaces at $t = t_1$ and $t = t_2$ and one timelike surface at spatial infinity and make the usual assumption that the fields do not contribute at the spatial infinity. In that case, the integral in Eq. (6.33) arises only from the two $t = $ constant surfaces. (A more general boundary condition is explored in Exercise 6.3.) From any one of them, we will have the contribution:

$$16\pi\kappa A_{\text{sur}} = \int_{t = \text{const}} d^3 x \sqrt{-g} \, V^0. \quad (6.34)$$

To facilitate further computations, it is convenient to separate the metric into the $(g_{00}, g_{0\alpha}, g_{\alpha\beta})$ components at this stage and write the line interval as

$$ds^2 = -(N dt)^2 + h_{\mu\nu} [dx^\mu + N^\mu dt][dx^\nu + N^\nu dt] \quad (6.35)$$

so that $g^{00} = -1/N^2, g^{0\mu} = N^\mu/N^2, g^{\mu\nu} = h^{\mu\nu} - N^\mu N^\nu/N^2, g = -N^2 h$. We take the normal to the $t = $ constant surface as $n_a = -N\delta_a^0$ with $n^0 = 1/N, n^\mu = -N^\mu/N$. Let us now compare $\mathcal{A}_{\text{sur}}$ with another surface term obtained by integrating a quantity $-K \equiv \nabla_a n^a$ over this surface. (We will see in Chapter 12 that K, called the *trace of the extrinsic curvature*, has a geometric meaning.) We will define this new surface term $\mathcal{A}'_{\text{sur}}$ by:

$$16\pi\kappa\mathcal{A}'_{\text{sur}} \equiv -2\int_{\partial\mathcal{V}} d^3x\sqrt{h}K = -2\int_{t=\text{const}} d^3x\frac{\sqrt{-g}}{N}K$$

$$= -2\int_{t=\text{const}} d^3x\sqrt{-g}(n^0 K). \tag{6.36}$$

The difference between the two actions $\mathcal{A}_{\text{sur}}$ and $\mathcal{A}'_{\text{sur}}$ is

$$16\pi\kappa(\mathcal{A}_{\text{sur}} - \mathcal{A}'_{\text{sur}}) = \int_{t=\text{const}} d^3x\sqrt{-g}\,(V^0 + 2Kn^0). \tag{6.37}$$

We can compute the difference in the integrand by straightforward algebra. From the definition of V^c given in Eq. (6.15), we have

$$V^0 = -\frac{1}{g}\partial_a(gg^{0a}) = -\frac{1}{N^2}\partial_a(N^2 g^{0a}) - g^{0a}\partial_a \ln h. \tag{6.38}$$

On the other hand, $2Kn^0 = -(2/N)\nabla_a n^a$ which can be expanded to:

$$2Kn^0 = -\frac{2}{N}\frac{1}{N\sqrt{h}}\partial_a(N\sqrt{h}n^a) = \frac{2}{N^2\sqrt{h}}\partial_a(N^2\sqrt{h}g^{0a})$$

$$= \frac{2}{N^2}\partial_a(N^2 g^{0a}) + 2g^{0a}\partial_a \ln\sqrt{h}. \tag{6.39}$$

Adding Eq. (6.38) and Eq. (6.39), we get:

$$V^0 + 2Kn^0 = \frac{1}{N^2}\partial_a(N^2 g^{0a}) = \frac{1}{N^2}\partial_\mu(N^2 g^{0\mu}) = \frac{\partial_\mu N^\mu}{N^2}. \tag{6.40}$$

Therefore we can write:

$$16\pi\kappa(\mathcal{A}_{\text{sur}} - \mathcal{A}'_{\text{sur}}) = \int_t d^3x\sqrt{h}\left[\frac{\partial_\mu N^\mu}{N}\right]. \tag{6.41}$$

This result allows us to draw several important conclusions.

Consider the *variation* of these two actions when the metric is varied by δg_{ab}. For a general variation, $(\delta\mathcal{A}_{\text{sur}} - \delta\mathcal{A}'_{\text{sur}}) \neq 0$. But if we consider variations with g^{ab} held fixed on $\partial\mathcal{V}$, then N and h will not vary on $\partial\mathcal{V}$; further, if the metric is fixed everywhere on $\partial\mathcal{V}$, then the *spatial* derivative $\partial_\mu N^\mu$ is also fixed everywhere on $\partial\mathcal{V}$ and cannot contribute to the variation. So we find that

$$\delta\mathcal{A}_{\text{sur}} = \delta\mathcal{A}'_{\text{sur}} \quad \text{(when } \delta g^{ab} = 0 \text{ on } \partial\mathcal{V}). \tag{6.42}$$

This allows us to obtain an alternative form of action for gravity. From our previous discussion we know that $\mathcal{A}_g$ can be expressed as the sum:

$$\mathcal{A}_g \equiv \frac{1}{16\pi\kappa} \int_{\mathcal{V}} d^4x \sqrt{-g}\, R = \mathcal{A}_{\mathrm{quad}} + \mathcal{A}_{\mathrm{sur}}. \tag{6.43}$$

Consider now a new action $\mathcal{A}_{\mathrm{new}}$ defined by:

$$\mathcal{A}_{\mathrm{new}} \equiv \mathcal{A}_g - \mathcal{A}'_{\mathrm{sur}} = \mathcal{A}_{\mathrm{quad}} + (\mathcal{A}_{\mathrm{sur}} - \mathcal{A}'_{\mathrm{sur}}). \tag{6.44}$$

Varying this action $\mathcal{A}_{\mathrm{new}}$ and using Eq. (6.42), we find that

$$\delta\mathcal{A}_{\mathrm{new}} = \delta\mathcal{A}_{\mathrm{quad}} = \frac{1}{16\pi\kappa} \int_{\mathcal{V}} d^4x \sqrt{-g}\, G_{ik}\, \delta g^{ik}. \tag{6.45}$$

In other words, the variation of the newly defined action

$$\mathcal{A}_{\mathrm{new}} \equiv \mathcal{A}_g - \mathcal{A}'_{\mathrm{sur}} = \frac{1}{16\pi\kappa} \int_{\mathcal{V}} d^4x \sqrt{-g}\, R + \frac{1}{8\pi\kappa} \int_{\partial\mathcal{V}} d^3x \sqrt{h}\, K \tag{6.46}$$

is the same as the variation of the quadratic part of the action. Therefore one can also use $\mathcal{A}_{\mathrm{new}}$ defined by Eq. (6.46) as a valid form of action for the gravitational field. The extra term that is added is called the Gibbons–Hawking–York counterterm.

To avoid possible misunderstanding, we should stress several caveats in this procedure. To begin with, the difference $(\mathcal{A}_{\mathrm{sur}} - \mathcal{A}'_{\mathrm{sur}})$ is in general *nonzero*. If the coordinates are chosen such that the surface $\partial\mathcal{V}$ corresponds to $x^M = $ constant, then $(\mathcal{A}_{\mathrm{sur}} - \mathcal{A}'_{\mathrm{sur}}) = 0$ only for the coordinate choice in which the metric has no off-diagonal terms with respect to the coordinate labelled by M; in the case of constant *time* foliation, this requires $g^{0\mu} = 0$. It is important to note that at the level of actions $\mathcal{A}_{\mathrm{sur}} \neq \mathcal{A}'_{\mathrm{sur}}$ and $\mathcal{A}_{\mathrm{quad}} \neq \mathcal{A}_{\mathrm{new}}$ in general.

Second, even the variations of the two actions do not match *in general*. The result $\delta\mathcal{A}_{\mathrm{sur}} = \delta\mathcal{A}'_{\mathrm{sur}}$ holds only for variations in which the metric is fixed at the boundary.

Third, note that the action $\mathcal{A}_{\mathrm{new}}$ can be expressed as an integral over a local Lagrangian in the form:

$$16\pi\kappa L_{\mathrm{new}} = R + 2\nabla_i(Kn^i) = R - 2\nabla_i(n^i\nabla_j n^j). \tag{6.47}$$

It may appear at first sight that we now have a generally covariant Lagrangian that we can use. But this Lagrangian L_{new} depends not only on the metric but also on an arbitrary vector field n_i and hence is conceptually no better than the non-covariant Lagrangian L_{quad}. In fact, any non-covariant expression can be written in a generally covariant manner if one is allowed to introduce extra vector fields. For example, a component of a tensor say T_{00}, is not generally covariant. But a quantity $\rho \equiv T_{ab}u^a u^b$ is a generally covariant scalar which will reduce to T_{00} in a local frame in which $u^a = (1, 0, 0, 0)$. It is appropriate to say that ρ is generally covariant but foliation dependent. The $\mathcal{A}_{\text{new}}$ uses the normal vector n^i of the boundary in a similar manner.

We will see later on (in Chapters 15 and 16) that the surface term in the action has several curious and important properties. But, for the moment, we will proceed with the derivation of the field equations.

Exercise 6.3

Action with Gibbons–Hawking–York counterterm Let $h_{ab} = g_{ab} \pm n_a n_b$ be the induced metric on the boundary $\partial\mathcal{V}$ of a spacetime region $\mathcal{V}$ where n_a is the normal to the boundary. (The sign $\pm$ is chosen based on whether the part of $\partial\mathcal{V}$ under consideration is spacelike or timelike.)

  (a) When V^a is defined through Eq. (6.33), prove that

$$V^a n_a = 2K + 2h^{ab}\partial_b n_a - n^m h^{ns}\partial_n g_{sm}. \tag{6.48}$$

Use this to generalize the result in Eq. (6.41) for an arbitrary surface $\partial\mathcal{V}$. [Hint. Supply the necessary logic for the equalities in the following string of equations:

$$n_a V^a = n_a \left(g^{mn}\Gamma^a_{mn} - g^{am}\Gamma^n_{mn} \right) \tag{6.49}$$

$$= \left(n^s h^{mn} - n^m h^{ns} \right)\Gamma_{smn} = n_a h^{mn}\Gamma^a_{mn} - \frac{1}{2}n_a g^{ma}h^{ns}\partial_m g_{ns}$$

$$= 2n_a h^{mn}\Gamma^a_{mn} - n^m h^{ns}\partial_n g_{sm} = 2K + 2h^{ab}\partial_b n_a - n^m h^{ns}\partial_n g_{sm}.$$

Take it from there.]

  (b) Consider the action given in Eq. (6.46). Vary this explicitly, assuming that $\delta g_{ab} = 0$ on the boundary and show that you recover the correct result. [Hint. When $\delta g_{ab} = 0$ everywhere on the boundary, the variation of its derivatives tangential to the boundary will also vanish. First show that variation of Einstein–Hilbert action under such conditions will only leave a term with the integrand $-\sqrt{|h|}\, h^{ab}n^m\partial_m\delta g_{ab}$. Next, show that δK can be expressed as

$$-\delta K = \delta\left(h^{ab}(\partial_a n_b - \Gamma^m_{ab}n_m) \right) = -h^{ab}n_m\delta\Gamma^m_{ab} = \frac{1}{2}h^{ab}n^m\partial_m\delta g_{ab}. \tag{6.50}$$

That should be adequate.]

6.2.4 Variation of the matter action

After this digression, we will now resume our main discussion based on the action principle $\mathcal{A}_{\text{tot}} = \mathcal{A}_m + \mathcal{A}_{\text{quad}}$. We have already computed $\delta\mathcal{A}_{\text{quad}}$ in Eq. (6.30) and, to obtain the field equations, we also need to find the variation of the matter action $\mathcal{A}_m$ when the metric is varied. This is our next task.

The explicit form of this variation can be computed only when the form of matter action is given. But, quite generally, we know that the variation has the following structure

$$\delta \mathcal{A}_m \equiv -\frac{1}{2} \int T_{ik} \delta g^{ik} \sqrt{-g} \, d^4 x. \tag{6.51}$$

This equation, in fact, defines the second rank symmetric tensor T_{ik} and we need to understand its properties. We will argue that this tensor can be identified with the energy-momentum tensor of matter. (Our notation T_{ik}, of course, anticipates this identification.)

The quantity T_{ik} defined through Eq. (6.51) already has three of the properties required of an energy-momentum tensor: it is (i) of second rank, (ii) symmetric and (iii) will occur in the field equations of gravity as a source. The fourth key property we require is that it must satisfy the condition $\nabla_i T^i{}_k = 0$, irrespective of the detailed form of the matter Lagrangian. (See the discussion in Section 5.4.2.) It is important to be able to prove this result for a general matter Lagrangian for the consistency of the theory.

This can be done as follows. Consider an infinitesimal coordinate transformation from x^i to $x'^i = x^i + \xi^i(x)$, where $\xi^i(x)$ are considered to be infinitesimal quantities. We know from Eq. (4.135) that under this transformation the metric tensor changes to $g'^{ik} = g^{ik} + \delta g^{ik}$ with $\delta g^{ik} = \nabla^k \xi^i + \nabla^i \xi^k$. Consider now the variation in the matter action $\mathcal{A}_m$ when the coordinates are changed by an infinitesimal amount. The change of coordinates will induce certain variation $\delta \phi$ in the matter variables and a variation $\delta g^{ik} = \nabla^k \xi^i + \nabla^i \xi^k$ in the metric. The net change in the matter action is

$$\delta \mathcal{A}_m = \left(\frac{\delta \mathcal{A}_m}{\delta \phi} \right)_g \delta \phi + \left(\frac{\delta \mathcal{A}_m}{\delta g^{ik}} \right)_\phi \delta g^{ik}. \tag{6.52}$$

However, when the equations of motion for the matter variables are satisfied, we know that $(\delta \mathcal{A}/\delta \phi)_g = 0$. Further, since the matter Lagrangian L_m is a scalar, we know from Eq. (4.138) that the total variation $\delta \mathcal{A}_m$ is given by

$$\delta \mathcal{A}_m = \delta \int L_m \sqrt{-g} \, d^4 x = -\int \nabla_a (L_m \xi^a) \sqrt{-g} \, d^4 x. \tag{6.53}$$

This integral can be converted to one over the boundary of the region under consideration. We will choose ξ^a to vanish at the boundary of the region of integration, obtaining $\delta \mathcal{A}_m = 0$. Hence we must have

$$0 = \delta \mathcal{A}_m = -\frac{1}{2} \int T_{ik} \delta g^{ik} \sqrt{-g} \, d^4 x \quad (\text{when } \delta g^{ik} = \nabla^k \xi^i + \nabla^i \xi^k). \tag{6.54}$$

We can write this as

$$0 = \frac{1}{2} \int T_{ik} \left(\nabla^k \xi^i + \nabla^i \xi^k\right) \sqrt{-g}\, d^4x = \int T_{ik} \nabla^k \xi^i \sqrt{-g}\, d^4x$$
$$= \int \nabla_k \left(T_i^k \xi^i\right) \sqrt{-g}\, d^4x - \int (\nabla_k T_i^k) \xi^i \sqrt{-g}\, d^4x, \tag{6.55}$$

which should hold for any ξ^i that vanishes at the boundary. The first term can be converted into an integral over the boundary, which again vanishes allowing us to conclude that

$$\nabla_k T_i^k = 0. \tag{6.56}$$

This result shows that Eq. (6.51) allows us to construct a symmetric second rank tensor T^{ik} with zero covariant divergence for *any* matter action when the matter equations of motion are satisied.

All these features strongly suggest that the tensor T_{ab}, defined through Eq. (6.51), can be identified with the energy-momentum tensor of any physical system. As we emphasized on page 91 in Section 2.9, any relativistic theory for gravity must provide a suitable definition for the energy-momentum tensor and this is precisely what we have achieved. Given any action functional for matter, Eq. (6.51) allows us to determine the energy-momentum tensor.

Exercise 6.4

Electromagnetic current from varying the action Our definition of the source of gravitational field, T^{ab}, through variation of matter action with respect to g^{ab} has a direct analogy in the case of electromagnetism. One can similarly define the source of electromagnetic field, J^i, by a variation of matter action with respect to A^i. Let the action for an electromagnetic field interacting with charged matter be $\mathcal{A} = \mathcal{A}_f + \mathcal{A}_m$, where $\mathcal{A}_f$ is the last term in Eq. (5.117) and $\mathcal{A}_m$ is some action which describes the matter that is coupled to the electromagnetic field. Show that we can define the current J^i via the variation

$$\delta \mathcal{A}_m = \int J^i \delta A_i \sqrt{-g}\, d^4x \tag{6.57}$$

and obtain the standard electromagnetic field equations given by Eq. (5.119). Equation (6.57) is analogous to Eq. (6.51). Assume that the matter action $\mathcal{A}_m$ is gauge invariant in the sense that it does not change under the transformations $A_i \rightarrow A_i + \partial_i f$. Show that this leads to the condition $\nabla_i J^i = 0$. This is similar to the derivation of Eq. (6.56) from the general covariance of matter action. (However, as we have explained in Section 5.4.2 the condition in Eq. (6.56) does not lead to a conservation law.)

Let us now consider a few simple physical systems and determine their T_{ab} to convince ourselves that we get reasonable results. As a first example, consider a single point particle of mass m moving along some trajectory $z^a(\tau)$. The action for such a particle is given by

$$\mathcal{A}_m = -m \int d\tau = -m \int \sqrt{-g_{ab}\, dz^a\, dz^b}. \qquad (6.58)$$

If we vary the metric in this action, we get

$$\delta\mathcal{A}_m = m \int \frac{1}{2} u^a u^b \delta g_{ab}\, d\tau = -\frac{1}{2} \int m\, u_a\, u_b\, \delta g^{ab}\, d\tau, \qquad (6.59)$$

where $u^a = dz^a/d\tau$ and we have used the result that $\delta g^{ab} = -g^{ai} g^{bj} \delta g_{ij}$, which introduces a change of sign (see the discussion just after Eq. (4.35)). We need to express this result as an integral over $\sqrt{-g}\, d^4x$ in order to identify the energy-momentum tensor. This can be done by using the analogue of Eq. (2.11) in curved spacetime which allows us to write

$$\delta\mathcal{A}_m = -\frac{1}{2} \int \rho\, u_a\, u_b\, \delta g^{ab}\, \sqrt{-g}\, d^4x, \qquad (6.60)$$

where ρ is the spacetime density of the particle defined using a Dirac delta function

$$\rho(x) = m \int \frac{d\tau}{\sqrt{-g}} \delta_D[x^a - z^a(\tau)]. \qquad (6.61)$$

Comparing with Eq. (6.51) we find that the energy-momentum tensor in this case is given by $T_{ab} = \rho u_a u_b$ – which agrees with the result we obtained previously in Eq. (5.90).

As a second example, let us work out the explicit form of T_{ik} for the electromagnetic field. In this case, the matter action is:

$$\mathcal{A} = -\frac{1}{16\pi} \int F_{ab}F^{ab}\sqrt{-g}d^4x = -\frac{1}{16\pi} \int F_{ab}F_{dc}g^{ad}g^{cb}\sqrt{-g}d^4x. \qquad (6.62)$$

On varying g_{ab}, we get

$$\delta\mathcal{A} = -\frac{1}{16\pi} \int d^4x \left[2F_{ab}F_{dc}g^{ad}\delta g^{bc}\sqrt{-g} + F_{mn}F^{mn}\delta(\sqrt{-g}) \right]$$

$$= \frac{1}{8\pi} \int d^4x\sqrt{-g}\delta g^{bc} \left[-F_{ab}F^a{}_c + \frac{1}{4}F_{mn}F^{mn}g_{bc} \right] \qquad (6.63)$$

so that

$$4\pi T_{bc} = F_{ab}F^a{}_c - \frac{1}{4}F_{mn}F^{mn}g_{bc}, \qquad (6.64)$$

which is precisely the expression for the energy-momentum tensor of the electromagnetic field obtained in Section 2.9 (Eq. (2.141)). Similar results are obtained in the case of a scalar field with the action

$$\mathcal{A}_\phi = \int \left[-\frac{1}{2}g^{mn}\nabla_m\phi\nabla_n\phi - V(\phi) \right] \sqrt{-g}\, d^4x. \qquad (6.65)$$

Varying the action with respect to the metric, we get

$$\delta A_\phi = \int d^4x \left[\sqrt{-g} \left(-\frac{1}{2} \delta g^{mn} \nabla_m \phi \nabla_n \phi \right) + \delta\sqrt{-g} \left(-\frac{1}{2} \nabla_m \phi \nabla^m \phi - V(\phi) \right) \right]$$

$$= \int d^4x \sqrt{-g}\, \delta g^{mn} \left[-\frac{1}{2} \nabla_m \phi \nabla_n \phi + \left(-\frac{1}{2} g_{mn} \right) \left(-\frac{1}{2} \nabla_r \phi \nabla^r \phi - V(\phi) \right) \right].$$

$$(6.66)$$

This allows us to identify the energy-momentum tensor as

$$T_{mn} = \nabla_m \phi \nabla_n \phi - g_{mn} \left[\frac{1}{2} \nabla_r \phi \nabla^r \phi + V(\phi) \right], \tag{6.67}$$

which is the same expression obtained in Chapter 2, Eq. (2.28).

Finally, let us go back to the issue of the constant term $\rho_0 \equiv \Lambda/(8\pi\kappa)$ in the Lagrangian in Eq. (6.4). We said that we can think of this constant term as a part of the matter Lagrangian. If we do that, it will contribute an action of the form

$$A_\Lambda = -\frac{\Lambda}{8\pi\kappa} \int \sqrt{-g}\, d^4x \equiv - \int \rho_0 \sqrt{-g}\, d^4x. \tag{6.68}$$

Varying this action we get

$$\delta A_\Lambda = \frac{1}{2} \int \rho_0\, g_{ab}\, \delta g^{ab} \sqrt{-g}\, d^4x, \tag{6.69}$$

which corresponds to the energy-momentum tensor of the form

$$T_{ab} = -\rho_0 g_{ab}. \tag{6.70}$$

Comparing this with the energy-momentum tensor for an ideal fluid, $T_{ab} = (\rho + p) u_a u_b + p g_{ab}$, we find that this corresponds to a 'fluid' with an equation of state $p = -\rho_0$. One must necessarily have either the pressure or the energy density being negative for such a system. While this might appear bizarre, there are two specific situations in which such an energy-momentum tensor can arise quite naturally. The first case corresponds to shifting the potential energy function $V(\phi)$ by a constant in the action for the scalar field in Eq. (6.65). If we change $V(\phi) \to V(\phi) + \rho$, the energy-momentum tensor in Eq. (6.67) shifts by the amount $-\rho g_{ab}$, which effectively has an equation of state $p = -\rho$. The second situation corresponds to choosing the action functional for gravity itself to be the one in Eq. (6.4) with $\Lambda \neq 0$. We will have occasion to discuss these issues in Chapters 10 and 14.

In all the examples of T_{ab} obtained above, by varying the matter action with respect to the metric tensor, it should be noted that the matter action depended only on the metric but not on its derivatives. This is obvious in the case of Eq. (6.70) and for a massive particle. For the scalar field, since $\nabla_a \phi = \partial_a \phi$, no Christoffel symbols arise. In the case of the electromagnetic field again, we know that

$F_{ij} = \nabla_{[i}A_{j]} = \partial_{[i}A_{j]}$, thereby making Christoffel symbols disappear. In general, however, one can imagine a matter action which can depend on both the metric and the Christoffel symbols. In that case, the variation of the matter action with respect to g_{ab} can be more complicated. However, the physical systems that we are usually interested in do not seem to have a matter action which depends on the Christoffel symbols.

Our definition of the energy-momentum tensor and the derivation of Eq. (6.56) actually allows us to prove a much more general result. Consider any functional ('action') defined as

$$\mathcal{A} = \int d^4x \sqrt{-g}\, C[g_{ab}, R_{abcd}, \ldots], \tag{6.71}$$

where C is a generally covariant scalar that is a general functional of the metric tensor g_{ab} and all its derivatives. It can depend, as indicated, on R_{abcd}, its derivatives, etc., in a completely arbitrary manner. We can now define a second rank symmetric tensor Q_{ik} by varying this functional with respect to g_{ik}:

$$\delta\mathcal{A} = \int d^4x \sqrt{-g}\, Q_{ik}\delta g^{ik}, \tag{6.72}$$

in a way similar to the manner in which we defined the energy-momentum tensor T_{ik} in Eq. (6.51). The argument leading to Eq. (6.56) was independent of the form of the functional and hence will continue to hold for any functional constructed from a generally covariant scalar C. Therefore, we conclude that $\nabla_i Q^{ik} = 0$ for all such functionals. In other words we can construct a whole family of divergence-free symmetric second rank tensors by starting with generally covariant scalars and constructing the variation of the functional built out of them. When the scalar is R, the resulting tensor is proportional to G_{ik} and the above argument leads to the Bianchi identity $\nabla_i G^{ik} = 0$. When the scalar is L_m, the resulting tensor is proportional to T^{ik} and we get $\nabla_i T^{ik} = 0$, etc. (We see that, from this perspective, the Bianchi identity is related to the general covariance of the scalar R.)

Finally, we would like to comment on the rather subtle role played by general covariance and the principle of equivalence in the above discussion that is not often appreciated. While studying physical systems within the framework of special relativity, one routinely uses spherical polar coordinates (r, θ, ϕ) which are obtained from the Cartesian coordinates by a highly nonlinear coordinate transformation. The action principle for the electromagnetic field, say, will depend on the spatial metric functions when such a curvilinear coordinate system is used in flat spacetime. In a completely analogous manner, one could write down the action functional or the field equations for any physical system in arbitrary curvilinear coordinates in *four*-dimensional spacetime. This will certainly involve changing ordinary derivatives ∂_i to ∇_i, changing d^4x to $\sqrt{-g}\, d^4x$, etc. Such changes constitute mere coordinate relabelling and do not have any more fundamental significance than using spherical polar coordinates instead of Cartesian coordinates; no major issue of principle is involved here. It is, however, an entirely different matter to postulate that such a modification (like changing ∂_i to ∇_i, or changing d^4x to $\sqrt{-g}\,d^4x$, etc.) will lead to the correct field equations in a genuinely *curved* spacetime. By postulating this, we have uniquely determined the manner in which gravity couples to matter fields – which is a nontrivial issue. Determining the energy-momentum tensor by varying the metric in the matter action is valid *only if* the action functional for the matter field is assumed to be valid in all spacetimes with an arbitrary metric g_{ab}. This is because a *general* variation of the form $g_{ab} \to g_{ab} + \delta g_{ab}$ will take us from a flat spacetime to a curved spacetime even if the original metric g_{ab} represented flat spacetime in curvilinear coordinates. This postulate, in turn, is equivalent to assuming the validity of laws of special relativity in the local inertial frames (which could be thought of as a possible way of interpreting the principle of equivalence) and demanding that the description of physics in terms of a metric in curvilinear coordinates should carry over, in the same form, to curved spacetime (which could be thought of as a version of the principle of general covariance).

Exercise 6.5
Geometrical interpretation of the spin-2 field Consider a field $\phi_A(x^a)$ described by a Lagrangian density $L(\phi_A, \partial \phi_A, \eta_{ab})$ in *flat* spacetime, in the Cartesian coordinates in which the metric is $\eta_{ab} = \text{dia}\,(-1, 1, 1, 1)$. The index A formally denotes all the indices the field carries depending on its spin. We now modify this action to $A_0 = A(\phi_A, \nabla \phi_A, \gamma_{ab})$ in a curvilinear coordinate system with a metric γ_{ab} and *postulate* that the functional form of this action continues to be valid for *any* metric γ_{ab} including those representing genuine curved spacetime. We can now define a tensor K_{ab} by

$$\delta A_0 = \frac{1}{2} \int d^4x \, \sqrt{-\gamma}\, K^{ab} \delta \gamma_{ab}; \quad K^{ab}(x) \equiv \left[\frac{2}{\sqrt{-\gamma}} \frac{\delta A_0}{\delta \gamma_{ab}(x)} \right]_{\gamma=\eta}. \tag{6.73}$$

(a) Let H_{ab} be a second rank tensor field which couples to K_{ab} with the lowest order coupling being obtained by changing the action from A_0 to $A_{\leq 1} \equiv A_0 + A_1$ where:

$$\delta A_1 = \frac{\lambda}{2} \int d^4x_1 \sqrt{-\gamma}\, K^{ab}(x_1) \delta H_{ab}(x_1) = \lambda \int d^4x_1 \left[\frac{\delta A_0}{\delta \gamma_{ab}(x_1)} \right]_{\gamma=\eta} \delta H_{ab}(x_1), \tag{6.74}$$

where λ is a coupling constant. Integrate this to obtain A_1.

(b) The addition of this coupling will, however, change the definition of K^{ab}, since the second term A_1 contributes to K^{ab} via a relation similar to Eq. (6.73). To take this into account, we need to add a term A_2 in a manner similar to what we did in Eq. (6.74); that is, we need to choose A_2 such that

$$\delta A_2 = \lambda \int d^4x_2 \left[\frac{\delta A_1}{\delta \gamma_{cd}(x_2)} \right]_{\gamma=\eta} \delta H_{cd}(x_2). \tag{6.75}$$

Obviously, the process needs to be repeated as an infinite iteration. Show that the final action obtained by such a procedure is the same as the one obtained by replacing γ_{ab} by $g_{ab} \equiv \gamma_{ab} + \lambda H_{ab}$ *followed by* taking the limit $\gamma_{ab} \to \eta_{ab}$. That is, prove

$$A_\infty = A_0(\gamma_{ab} + \lambda H_{ab})\big|_{\gamma=\eta}, \tag{6.76}$$

where the dependence on matter variables is not shown. Explain why there is a subtle difference between Eq. (6.76) and the expression obtained by replacing η_{ab} by $(\eta_{ab}+\lambda H_{ab})$ in the original action.[1]

(c) Argue that this result shows that a consistent coupling of a spin-2 field to K_{ab} to all orders leads to a geometrical interpretation for the spin-2 field as far as coupling to matter fields are concerned.

Exercise 6.6

Conditions on the energy-momentum tensor If we take the energy-momentum tensor to be defined through Eq. (6.51), then its detailed properties will depend on the nature of matter action. We are only assured that the tensor is symmetric and has zero covariant derivative. Very often, to draw tangible conclusions, one would require some further conditions on the energy-momentum tensor related to the positivity of the energy density, etc. The purpose of this exercise is to list a series of energy conditions and invite you to examine them for an ideal fluid and a scalar field.

Consider a class of vectors v^a and postulate that $T_{ab}v^a v^b > 0$. The demand that this should hold for all timelike v^a is called the weak energy condition (WEC) while the condition that it holds for all null v^a is called the null energy condition (NEC). If we postulate, in addition to the weak energy condition, that $T^{ab}v_b$ should be a non-spacelike vector, it is called the dominant energy condition (DEC). A corresponding statement with the null energy condition is called the null dominant energy condition (NDEC). Finally, the condition $T_{ab}v^a v^b \geq (1/2)T_a^a v^b v_b$ for all timelike v^a is called the strong energy condition (SEC).

(a) Express each of these conditions in the ρ–p plane for an ideal fluid, graphically indicating the regions that are selected out by these conditions.

(b) Does the SEC imply the WEC? Does it imply the NEC?

Exercise 6.7

Pressure as the Lagrangian for a fluid Consider an action for a scalar field ϕ of the form

$$S = -\int p(X,\phi)\sqrt{-g}\, d^4x, \tag{6.77}$$

where $X \equiv (1/2)g^{ab}\partial_a\phi\partial_b\phi$ and p is an arbitrary function of its arguments. Obtain the energy-momentum tensor for this system and show that it can be expressed in the form $T_b^a = (\rho + p)u^a u_b + p\delta_b^a$ with $u_a = (2X)^{-1/2}\partial_a\phi$ and $\rho = 2X(\partial p/\partial X) - p$. Interpret this result.

Exercise 6.8

Generic decomposition of an energy-momentum tensor Let T_{ab} be the energy-momentum tensor of some system and u^i be a four-velocity. Show that we can always decompose T_{ab} in the form

$$T_{ab} = \rho u_a u_b + pP_{ab} + 2u_{(a}q_{b)} + \pi_{ab}, \tag{6.78}$$

where P_{ab} is the projection tensor orthogonal to u^a and

$$\rho = T_{ab}u^a u^b; \quad p = \frac{1}{3}P^{ab}T_{ab}; \quad q_n = -P_n^r u^m T_{mr}; \quad \pi_{mn} = T_{rs}\left(P_m^r P_n^s - \frac{1}{3}P^{rs}P_{mn}\right).$$

(6.79)

Decompose the electromagnetic energy-momentum tensor in the above form using $u^i = \delta_0^i$ and expressing all the quantities in terms of the electric and magnetic fields.

6.2.5 *Gravitational field equations*

Having found the variations of both the matter action and the gravitational action we shall now combine them to obtain the field equations. Setting the total variation $\delta A_{\text{quad}} + \delta A_m$ to zero and using Eq. (6.30) and Eq. (6.51), we obtain the field equations for gravity:

$$G_{ik} = R_{ik} - \frac{1}{2}g_{ik}R = 8\pi\kappa T_{ik}.$$

(6.80)

This shows that T^{ik} is the source of the gravitational field in Einstein's theory.

The above result corresponds to the action functional in Eq. (6.4) with $\Lambda = 0$. If we had kept a nonzero Λ, we would have obtained the equation

$$R_b^a - \frac{1}{2}\delta_b^a R + \Lambda\delta_b^a = 8\pi\kappa T_b^a.$$

(6.81)

But, as we explained in the previous section, this is mathematically the same as adding to the right hand side of the equation an energy-momentum tensor for a fluid with the equation of state $p = -\rho$. For most of the remaining part of this chapter we shall set $\Lambda = 0$ but the issue of a cosmological constant will occupy our attention in later chapters, especially in Chapters 10 and 14.

The procedure by which we have derived Eq. (6.80) – Einstein's equation for the gravitational field – hides its geometrical interpretation. It can be given an elegant geometrical meaning along the following lines. Consider an observer with a four-velocity u^i. The energy density measured in the rest frame of the observer is $\rho = T_{ik}u^i u^k$. Let us next consider the spatial curvature $\mathcal{R}_{ijkl}$ obtained by projecting the curvature tensor R_{abcd} orthogonal to the four-velocity u^i (see Eq. (5.63)). The scalar constructed out of this tensor $\mathcal{R} = \mathcal{R}_{ij}^{ij} = h^{ac}h^{bd}R_{abcd}$ has the value $2G_{ik}u^i u^k$ (see Eq. (5.64)). This quantity measures the curvature of the spatial sections orthogonal to the observer with a four-velocity u^i. The geometrical content of Einstein's equation can, therefore, be stated as follows:

$$\left\{\begin{array}{l} \text{Scalar curvature of the spatial} \\ \text{sections as measured by any} \\ \text{observer with four-velocity } u^i \end{array}\right\} = 16\,\pi\,\kappa \left\{\begin{array}{l} \text{Energy density as} \\ \text{measured by the observer} \end{array}\right\}$$

$$\mathcal{R} = \mathcal{R}^{ij}_{ij} = h^{ac}h^{bd}R_{abcd} = 16\,\pi\,\kappa\,T_{ab}u^a u^b = 16\,\pi\,\kappa\,\rho. \qquad (6.82)$$

This is equivalent to $[G_{ik} - 8\pi\kappa T_{ik}]u^i u^k = 0$ for all timelike unit vectors u^i. It is easy to see by using local Lorentz transformations that this relation can hold for all such u^i only if the term in the square bracket vanishes. This geometrical interpretation shows that every observer in the spacetime can think of the spatial scalar curvature to be generated by the locally measured energy density. (To avoid possible confusion, we stress that $\mathcal{R}$ is *not* the three-dimensional scalar curvature ^3R; see the comment on page 205.)

We saw in Section 5.3.3 (see Eq. (5.61)) that G^a_b satisfies the identity $\nabla_a G^a_b = 0$. From Eq. (6.80), this requires the source energy-momentum tensor to satisfy the equation $\nabla_a T^a_b = 0$. We have already seen that this condition is satisfied when the equations of motion for the matter hold and hence we find that the formalism is consistent. On the other hand, one can also say that when the gravitational field equations hold, the energy momentum tensor of the source must satisfy this equation which, in turn, will imply the field equations. In such an interpretation, one takes the point of view that gravitational field equations imply the equations of motion for the source. In the approach we have followed where one begins from an action principle for the full system, there doesn't seem to be any special advantage in choosing one interpretation over the other.

Taking the trace of Eq. (6.80) we get $-R = (8\pi\kappa)T$. Using this to eliminate R from the left hand side, Eq. (6.80) can also be written in terms of the mixed tensors as

$$R^i_k = 8\pi\kappa[T^i_k - \frac{1}{2}\delta^i_k T]. \qquad (6.83)$$

This form of Einstein's equations will often be useful.

Our next task will be to obtain the non-relativistic limit of Einstein's equations and determine κ by comparing it with Newtonian gravity. To do this, consider a source made of a collection of particles with the energy-momentum tensor $T_{ab} = \rho u_a u_b$. In the non-relativistic limit, the dominant term in the energy-momentum tensor is $T_{00} = -T^0_0 = \rho$. We also find, to the same order of accuracy, $T = -\rho$. Hence, the right hand side of Eq. (6.83) has only the 00 component that is nonzero with $T^0_0 - (1/2)T = -(1/2)\,\rho$. On the other hand, we found in Exercise 5.1 that, in the non-relativistic limit with $g_{00} = -(1 + 2\phi)$, $g_{\alpha\beta} = \delta_{\alpha\beta}$, the only nonzero curvature component is given by $R^\alpha_{0\beta 0} \approx \partial^\alpha \partial_\beta \phi$. Hence the dominant component of R_{ab} is $R_{00} \approx \nabla^2 \phi = -R_0^0$. So, in this limit, Eq. (6.83) reduces to $\nabla^2 \phi = 4\pi\kappa\rho$. Comparing with the standard equation in Newtonian gravity, we see that $\kappa = G_N$, the Newtonian gravitational constant.

Our analysis has assumed that the lowest order solution is flat spacetime with the metric η_{ab}. This requires that, at the lowest order, we ignore T_b^a. We saw earlier that in the presence of the cosmological constant the situation can be different. If we think of the cosmological constant as some kind of energy-momentum tensor with $T_{ab} = -\rho_0 g_{ab}$, then our approximation of $g_{ab} = \eta_{ab} + h_{ab}$ requires us to ignore ρ_0 at the lowest order and treat the cosmological constant also as a perturbation, acting as a source for h_{ab}. On the other hand, one may consider the cosmological constant as *not* a part of the energy-momentum tensor but as a separate term Λg_{ab} in the left hand side of Einstein's equations. In this case, the solution to Einstein's equations in the absence of matter is *not* the flat spacetime but is given by a solution to $G_{ab} = -\Lambda g_{ab}$. The perturbation theory now needs to be developed around this background and the perturbation h_{ab} will be sourced by the matter energy-momentum tensor T_{ab}.

Exercise 6.9

Something to think about: disaster if we vary g_{ab} rather than g^{ab}? In the left hand side of Einstein's equations, Eq. (6.80), the second term $(-1/2)g_{ik}R$ has its origin in the variation of $\sqrt{-g}$ in Eq. (6.18). Suppose that, instead of writing $\sqrt{-g}\,R = \sqrt{-g}\,g^{ab}R_{ab}$, we had chosen to write $\sqrt{-g}\,R = \sqrt{-g}\,g_{ab}R^{ab}$ and had varied g_{ab} instead of g^{ab}. Now, instead of the minus sign in the right hand side of Eq. (6.18), we would have got a plus sign because $g_{ik}\delta g^{ik} = -g^{ik}\delta g_{ik}$. It appears that this will lead to the appearance of $R^{ab} + (1/2)g^{ab}R$ in the left hand side of Einstein's equations. That will be a disaster (inconsistency with $\nabla_a T^{ab} = 0$, wrong Newtonian approximation ...). Find the flaw in the above argument and convince yourself that you get the correct equations even if you vary g_{ab}. This is important because all the textbooks (including this one!) get away with it by choosing g^{ab} as the quantity to vary and do not discuss this case.

Exercise 6.10

Newtonian approximation with cosmological constant Show that in the presence of a cosmological constant, which is treated as an energy-momentum tensor and hence a first order correction to flat spacetime, the Newtonian approximation leads to a differential equation for the gravitational potential given by $\nabla^2 \phi = 4\pi\kappa\rho - \Lambda c^2$ with c factors reintroduced. Show that the potential ϕ produced by a point particle located at the origin is now given by

$$\phi = -\frac{\kappa M}{r} - \frac{\Lambda c^2 r^2}{6}. \tag{6.84}$$

Interpret this result.

6.3 General properties of gravitational field equations

In the coming chapters, we will be concerned with different solutions to Einstein's equations and their properties. The purpose of this section is to describe several general features related to the structure of Einstein's equations without confining ourselves to any specific solution.

In Newtonian theory, the key equation connecting the gravitational field to its source is $\nabla^2\phi = 4\pi\kappa\rho$. Given a distribution of matter, $\rho(t, x)$, this equation determines $\phi(t, x)$ subject to standard boundary conditions at infinity. Since the equation is linear, its solutions can be superposed and one can, in fact, write down the most general solution for any $\rho(t, x)$ as an integral

$$\phi(t, x) = -\kappa \int d^3 y\, \frac{\rho(t, y)}{|x - y|}. \tag{6.85}$$

The situation is quite different in the case of Einstein's theory. The left hand side of Einstein's equation is a nonlinear function of the metric tensor g_{ab} and hence we cannot superpose solutions. What is more, even the source T_{ab} will – in general – depend on the metric g_{ab} and its functional form cannot be independently specified. Thus one needs to solve the equations for each source separately, which makes the problem mathematically quite intractable. This is one of the reasons we have very few exact solutions in general relativity.

The source for the gravitational field is the energy-momentum tensor of matter, T_{ab} and, in the absence of any source, the metric satisfies the equation $G_{ab} = 0$. While the flat spacetime is indeed a solution to this equation, it is also possible to have genuinely curved spacetimes satisfying this equation. The situation is somewhat analogous to the electromagnetic theory in which the electromagnetic field F_{ab} depends on the current J_b through $\partial_a F^a{}_b = -4\pi J_b$. In the absence of the current, Maxwell's equations reduce to $\partial_a F^a{}_b = 0$, which does possess electromagnetic wave solutions. In the case of gravity, we have analogous gravitational wave solutions (see Chapter 9) as well as more complicated configurations.

We have seen that both sides of Einstein's equations have identically vanishing covariant divergences. Thus the ten equations in Eq. (6.80) are constrained by the four identities in Eq. (5.61) leaving six independent equations. On the other hand, among the ten variables g^{ik}, four can be assigned specific values by a suitable choice of coordinates. Thus there are only six independent functions we need to solve for, which is the same as the number of independent equations available.

Structurally, Einstein's equations are ten second order partial differential equations for g^{ab}. Thus, *a priori*, one would have thought that the values of the metric tensor g^{ik} and its first time derivatives $(\partial g^{ik}/\partial t)$ need to be specified as the initial conditions. This is, however, not true because of certain peculiar features in Einstein's equations which we shall now discuss.

It is clear from the symmetries of R_{iklm} that the second derivatives with respect to time are contained only in the components $R_{0\alpha 0\beta}$ in which they enter through the term $\ddot{g}_{\alpha\beta}$ (see Eq. (5.36)). This shows that the second derivatives of the metric components $g_{0\alpha}$ and g_{00} do not appear in R_{iklm} and hence in Einstein's equations. Further, even the second derivatives of $g_{\alpha\beta}$ appear only in the space–space part of Einstein's equations. The time–time part and the time–space part contain time derivatives *only up to first order*. To prove this claim, note that the Bianchi identity implies $\nabla_k G_i^k = 0$, which can be written in the expanded form as

$$\nabla_0 \left(R_i^0 - \frac{1}{2}\delta_i^0 R \right) = -\nabla_\alpha \left(R_i^\alpha - \frac{1}{2}\delta_i^\alpha R \right),\qquad (6.86)$$

where $i = 0, 1, 2, 3$ and $\alpha = 1, 2, 3$. The highest time derivative appearing on the right hand side of these equations is a second derivative; since one time derivative occurs explicitly in ∇_0 in the left hand side, it follows that the quantities within the bracket on the left hand side can only contain first order time derivatives. Hence the time–time and time–space components G_0^0 and G_α^0 contain only the first time derivatives of the metric tensor.

Furthermore, the space–time and time–time equations do not contain the first derivatives $\dot{g}_{0\alpha}$ and $\dot{g}_{00}$ but only $\dot{g}_{\alpha\beta}$. This is because among all the Christoffel symbols, only $\Gamma_{\alpha,00}$ and $\Gamma_{0,00}$ contain these quantities; but these appear only in the components $R_{0\alpha 0\beta}$ which drop out from the time–time and time–space part of Einstein's equations.

The above considerations show the following. (i) The space–time and the time–time parts of Einstein's equations only involve $\dot{g}_{\alpha\beta}$ as the highest order time derivatives. Hence the time–time and the space–time parts of Einstein's equations are constraint equations. (ii) The space–space part contains $\ddot{g}_{\alpha\beta}$. (iii) The time derivatives of g_{00} and $g_{0\alpha}$ do not appear in Einstein's equations.

It is, therefore, possible to assign as initial conditions the functions $g_{\alpha\beta}$ and $\dot{g}_{\alpha\beta}$ at some time $t = t_0$. The space–time and time–time parts of Einstein's equation will then determine the initial values of $g_{0\alpha}$ and g_{00}. (The full set, of course, is not freely specifiable and is inter-related by the constraint equations.) The initial values of $\dot{g}_{0\alpha}, \dot{g}_{00}$ remain arbitrary. These are the valid initial data for integrating Einstein's equations. The evolution of $g_{\alpha\beta}$ can be accomplished through the space–space components of Einstein's equations, which are six in number. As already mentioned, this is adequate since one can impose four conditions on the metric tensor by a suitable choice of coordinates, thereby reducing the number of independent variables that need to be solved for to six. A convenient coordinate choice (called the *harmonic gauge* [2]) is the one in which the coordinates satisfy the condition

$$\frac{1}{\sqrt{-g}}\partial_a \left(\sqrt{-g}\, g^{ab}\partial_b \right) x^m \equiv \Box x^m = 0.\qquad (6.87)$$

(The name 'harmonic' originates from the fact that this equation makes the coordinates harmonic functions.) Simplifying, we get the condition

$$\partial_l \left(\sqrt{-g} \, g^{lm} \right) = 0. \tag{6.88}$$

Differentiating this equation with respect to t gives

$$\frac{\partial^2}{\partial t^2} \left(\sqrt{-g} \, g^{0m} \right) = -\partial_\alpha \left[\frac{\partial}{\partial t} \left(\sqrt{-g} \, g^{\alpha m} \right) \right]. \tag{6.89}$$

This gives a second order differential equation for evolving the components g^{0n}, given the initial data. Thus, once the harmonic gauge is chosen, we can use the space–space part of Einstein's equations to evolve $g^{\alpha\beta}$ and Eq. (6.89) to evolve g^{0m} thereby completely determining the metric.

If one uses the liberty in the choice of coordinate system, it is possible to show that the gravitational field has only two genuine degrees of freedom per event. This can be seen as follows. We first use the four coordinate transformations $x^i \to x'^i$ to arrange that $g_{00} = 1$ and $g_{0\alpha} = 0$ in the neighbourhood of any event. This reduces the metric to the form

$$ds^2 = -dt^2 + g_{\alpha\beta} dx^\alpha dx^\beta, \tag{6.90}$$

leaving six components of $g_{\alpha\beta}$ nonzero. Consider now the infinitesimal coordinate transformation $t \to t' = t + f$, $x^\alpha \to x'^\alpha = x^\alpha + \lambda^\alpha$. This will change the metric coefficients to

$$g'_{00} = -(1 + 2\dot{f}), \quad g'_{0\alpha} = \partial_\alpha f + g_{\alpha\beta}\dot{\lambda}^\beta, \quad g'_{\alpha\beta} = g_{\alpha\beta} + g_{\gamma\beta}\partial_\alpha\lambda^\gamma + g_{\alpha\gamma}\partial_\beta\lambda^\gamma. \tag{6.91}$$

Demanding that the coordinate transformation should not change the conditions $g_{00} = 1$, $g_{0\alpha} = 0$ will lead to the constraints

$$\dot{f} = 0; \quad \partial_\alpha f = -g_{\alpha\beta}\dot{\lambda}^\beta. \tag{6.92}$$

The first equation implies that f is a function of space alone: $f = f(\boldsymbol{x})$. Using this, the second equation can be integrated to give

$$\lambda^\alpha = p^\alpha(\boldsymbol{x}) - \partial_\beta f \int^t dt \, g^{\alpha\beta}, \tag{6.93}$$

where $p^\alpha(\boldsymbol{x})$ are arbitrary functions. Thus on *any given spatial hypersurface*, we have the freedom to choose the four functions $f(\boldsymbol{x})$ and $p^\alpha(\boldsymbol{x})$ in order to bring four out of the six components of $g^{\alpha\beta}(t, \boldsymbol{x})$ to pre-assigned values. Thus, only two components of $g_{\alpha\beta}$ remain arbitrary to be propagated forward in time.

In Chapter 5, we argued that the real characteristic of a gravitational field is the curvature of spacetime represented by the curvature tensor R_{abcd}. It is interesting to ask how Einstein's equations constrain the curvature when the source T_{ab} is specified. We saw in Section 5.3.3 that, out of the 20 independent components in the curvature tensor, 10 are in R_{ab} and the other 10 are in the Weyl tensor C_{abcd}. The 10 degrees of freedom in R_{ab} are directly determined in terms of T_{ab} through Eq. (6.83). As regards the Weyl tensor, it is easy to prove (using the Bianchi identity and Eq. (5.65)) that C_{abcd} satisfies the identity

$$\nabla_r C^r{}_{smn} = 2\frac{(D-3)}{(D-2)}\left(\nabla_{[m}R_{n]s} + \frac{1}{2(D-1)}g_{s[m}\nabla_{n]}R\right) \qquad (6.94)$$

in D dimensions. In four dimensions this gives

$$\nabla^r C_{rsmn} = \nabla_{[m}R_{n]s} + \frac{1}{6}g_{s[m}\nabla_{n]}R = 8\pi\kappa\left(\nabla_{[m}T_{n]s} + \frac{1}{3}g_{s[m}\nabla_{n]}T\right), \qquad (6.95)$$

where we have used Einstein's equations. This equation shows that the Weyl tensor is determined 'non-locally' by the energy-momentum tensor while R_{ab} is determined locally (i.e. algebraically) by the energy-momentum tensor. Equation (6.95) is analogous to Maxwell's equations written in the form $\nabla_a F^{ab} = -4\pi J^b$. In this spirit, one can think of the degrees of freedom contained in the Weyl tensor as being similar to radiative degrees of freedom. In any spacetime that is a vacuum solution to Einstein's equation, the Weyl tensor determines the curvature and thus the geodesic deviation.

In a similar manner, one can obtain a differential equation directly connecting the curvature tensor to the matter energy-momentum tensor but this equation is fairly complicated to allow useful, further manipulation. Considering its formal importance – and its application in the case of gravitational waves – we shall outline its derivation. We start with the Bianchi identity written with a slightly unusual placement of indices as

$$\nabla_a R_{mnbc} - \nabla_b R_{acmn} + \nabla_c R_{abmn} = 0. \qquad (6.96)$$

Contracting on a and m and using Einstein's equations, we get the result

$$\nabla_a R^a{}_{nbc} = 8\pi\kappa\left(\nabla_b \bar{T}_{cn} - \nabla_c \bar{T}_{bn}\right); \qquad \bar{T}_{ij} \equiv T_{ij} - \frac{1}{2}g_{ij}T. \qquad (6.97)$$

To proceed further, we work out the expression $[\nabla_a, \nabla_b]R^a{}_{cmn}$ and use the symmetries of the curvature tensor repeatedly. This leads to the relation

$$\nabla_a\nabla_b R^a{}_{cmn} = 8\pi\kappa\nabla_b\bar{T}_{cmn} + 8\pi\kappa\bar{T}_{sb}R^s{}_{cmn}$$
$$- R^s{}_{cab}R^a{}_{smn} - R^s{}_{mab}R^a{}_{csn} - R^s{}_{nab}R^a{}_{cms}, \qquad (6.98)$$

where $\bar{T}_{nbc} = \nabla_b\bar{T}_{cn} - \nabla_c\bar{T}_{bn}$. Next, taking the covariant derivative of the Bianchi identity it is easy to show that

$$\Box R_{bcmn} = \nabla_a\nabla_b R^a{}_{cmn} - \nabla_a\nabla_c R^a{}_{bmn}. \qquad (6.99)$$

Putting all these together we get the 'wave equation' for the curvature tensor given by

$$\Box R_{bcmn} = 8\pi\kappa(\nabla_b \bar{T}_{cmn} - \nabla_c \bar{T}_{bmn}) + 8\pi\kappa \bar{T}_{sb} R^s{}_{cmn} - 8\pi\kappa \bar{T}_{sc} R^s{}_{bmn}$$
$$- R^s{}_{cab} R^a{}_{smn} - R^s{}_{mab} R^a{}_{csn} - R^s{}_{nab} R^a{}_{cms}$$
$$+ R^s{}_{bac} R^a{}_{smn} + R^s{}_{mac} R^a{}_{bsn} + R^s{}_{nac} R^a{}_{bms}. \tag{6.100}$$

This result shows that the source for curvature is essentially the energy-momentum tensor and its derivatives along with nonlinear self-coupling terms. In the linear limit, when the curvature is small (so that terms quadratic in the curvature can be ignored) and the source energy-momentum tensor is a first order perturbation (so that TR type terms can also be ignored), this equation reduces to

$$\Box R_{bcmn} = 8\pi\kappa \left[\partial_b \left(\partial_m \bar{T}_{nc} - \partial_n \bar{T}_{mc} \right) - \partial_c \left(\partial_m \bar{T}_{nb} - \partial_n \bar{T}_{mb} \right) \right]. \tag{6.101}$$

This equation will be of use in the study of gravitational radiation.

Exercise 6.11
Wave equation for F_{mn} in curved spacetime Show that the analogue of Eq. (6.100) for the electromagnetic field in curved spacetime is given by

$$\Box F_{mn} = -4\pi \left(\nabla_m J_n - \nabla_n J_m \right) + R^i{}_m F_{in} - R^i{}_n F_{im} + \left(R_{imtn} - R_{intm} \right) F^{ti}. \tag{6.102}$$

Explain the origin of the 'source terms' on the right hand side of the equation which are independent of the current vector.

It is also possible to obtain an exact equation for the geodesic acceleration, very similar to the one for gravitational acceleration in Newtonian theory. Recall that the genuine gravitational effects lead to an acceleration of the relative displacement of two geodesics. In Newtonian gravity, the divergence of the acceleration can be related to the mass density. Similarly, we would expect a suitably defined divergence of the acceleration of the geodesic separation to be related to the source T_{ab}. To do this, let us choose a coordinate system that is locally inertial along the trajectory of a given observer. In such a frame, the observer's four-velocity will be $u^i = \delta_0^i$ and the spatial component of the geodesic deviation equation, Eq. (5.28), becomes

$$g^\alpha \equiv \frac{D^2 v^\alpha}{Ds^2} = R^\alpha{}_{abi} u^a u^b v^i = R^\alpha{}_{00\beta} v^\beta. \tag{6.103}$$

(Notice that $R^a{}_{b00} = 0$ due to antisymmetry in the last two indices and hence only the spatial part of v^β contributes on the right hand side.) Taking the divergence of g^α with respect to the separation vector v^α and using Einstein's equations, we get

$$\frac{\partial g^\alpha}{\partial v^\alpha} \equiv \nabla_v \cdot g = R^\alpha{}_{00\alpha} = -R_{00} = -8\pi\kappa \left(T_{00} + \frac{1}{2}T \right). \qquad (6.104)$$

In the locally inertial coordinates we are using $(T_{00} + (1/2)T) = (1/2)(\rho + T^\alpha_\alpha)$, where α is summed over the spatial indices. In the case of an ideal fluid, with $T_{ab} = (\rho + p)u_a u_b + g_{ab}p$, this equation becomes

$$\nabla \cdot g = -4\pi\kappa(\rho + 3p). \qquad (6.105)$$

This equation – which does *not* require any approximation regarding the gravitational field and is always valid in the frame of a freely falling observer – shows that the effective gravitational mass in the frame of a freely falling observer that is leading to the geodesic acceleration is $(\rho + 3p)$. This has two interesting consequences. First, radiation with an equation of state $p = (1/3)\rho$ has $\rho + 3p = 2\rho$. Therefore, radiation produces a gravitational acceleration of the geodesic separation which is twice as strong as that of a material particle with the same energy density. (Also see Project 3.3(c).) Second, any fluid which has $\rho + 3p < 0$ will behave as though the sign of κ is reversed. That is, such a fluid will make the geodesic curves diverge from each other giving the semblance of a repulsive gravity. We saw earlier in Section 6.2.4 that the cosmological constant behaves like a fluid with $p = -\rho$ so that $\rho + 3p = -2\rho < 0$ if $\rho > 0$. This will mimic repulsive gravity and will play a role in our future discussions in Chapters 10 and 14.

Finally we would like to comment on the connection between the gravitational action principle $\mathcal{A}_m + \mathcal{A}_g$ obtained in this chapter and the theory of the spin-2 field discussed in Chapter 2. To relate the two we need to introduce a spin-2 field H_{ab} such that, in some appropriate limit, the $(\mathcal{A}_g + \mathcal{A}_m)$ reduces to the action in Eq. (3.38). Since H_{ab} has the dimensions of inverse length, we can attempt to express the metric as $g_{ab} = \eta_{ab} + lH_{ab}$, where l is a constant with dimensions of length. In the matter action, performing a Taylor series expansion in l and using Eq. (6.51), we get

$$\mathcal{A}_m[q_A; g_{ab}] = \mathcal{A}_m[q_A; \eta_{ab}] - \frac{l}{2} \int T_{ik} H^{ik} d^4 x, \qquad (6.106)$$

where we have denoted the matter degrees of freedom by q_A. This matches with the interaction term in Eq. (3.38). As for the gravitational action, we have two terms, arising from L_{quad} and L_{sur}. Expanding both in Taylor series in l and choosing $l^2 = 16\pi\kappa$, we find that the action functional becomes

$$\mathcal{A} \equiv \frac{1}{16\pi\kappa} \int d^4 x \sqrt{-g}\, R \approx \mathcal{A}_{\text{quad}} + \mathcal{A}_{\text{sur}}, \qquad (6.107)$$

where

$$\mathcal{A}_{\text{quad}} = \frac{1}{4} \int d^4x \, M^{abcijk}(\eta^{mn}) \partial_a H_{bc} \partial_i H_{jk} + \mathcal{O}(l) \tag{6.108}$$

and

$$\mathcal{A}_{\text{sur}} = \frac{1}{4l} \int d^4x \, \partial_a \partial_b [H^{ab} - \eta^{ab} H^i_i] + \mathcal{O}(1). \tag{6.109}$$

Comparing with Eq. (3.38) we see that the $\mathcal{A}_{\text{quad}}$ matches exactly with the action for the spin-2 field. However, the surface term – which is usually ignored – is *non-analytic* in the coupling constant. This result shows that it will be impossible to obtain $\mathcal{A}_{\text{sur}}$ by starting from the action for the spin-2 field ($\mathcal{A}_{\text{quad}}$) and doing a perturbative expansion in the coupling constant l. (One cannot obtain Eq. (6.46) either by this procedure. As mentioned in the discussion following Eq. (6.46), the terms involving K in Eq. (6.46) do *not* cancel the surface term in Einstein–Hilbert action and only their variations cancel under specific circumstances.)

In fact, this result can be obtained from fairly simple considerations related to the algebraic structure of the curvature scalar. In terms of a spin-2 field, the final metric arises as $g_{ab} = \eta_{ab} + l \, H_{ab}$, where $l \propto \sqrt{\kappa}$ has the dimension of length and H_{ab} has the correct dimension of (length)$^{-1}$ in natural units with $\hbar = c = 1$. We now want to obtain the full action for nonlinear gravity by a suitable iteration in powers of l, starting from the zeroth order Lagrangian $L_0 \simeq (\partial h)^2$ for a spin-2 field, which has the dimension of (length)$^{-4}$. Since the scalar curvature has the structure $R \simeq (\partial g)^2 + \partial^2 g$, substitution of $g_{ab} = \eta_{ab} + l \, h_{ab}$ gives, to the lowest order:

$$L_{EH} \propto \frac{1}{l^2} R \simeq (\partial h)^2 + \frac{1}{l} \partial^2 h. \tag{6.110}$$

Thus the full Einstein–Hilbert Lagrangian is non-analytic in l. *It is not possible, by starting from* $(\partial h)^2$ *and doing a proper iteration on* l, *to obtain a piece which is non-analytic in* l. At best, one can hope to get the quadratic part of L_{EH} which gives rise to the Γ^2 action but not the four-divergence term involving $\partial^2 g$. In other words, it is not possible to obtain the Einstein–Hilbert *action* by starting from the *action* for the spin-2 field in flat spacetime, coupled to the energy-momentum tensor of matter, and iteratively coupling it to its own energy-momentum tensor.[3] Such an approach will involve doing a perturbation series in l and will never lead to a term that is non-analytic in l.

The above comments, of course, are applicable only to the task of obtaining the form of the *action* by a series expansion on the coupling constant. It is certainly possible to obtain the field equations – to which $\mathcal{A}_{\text{sur}}$ does not contribute – by such an iteration, provided suitable additional assumptions are introduced.

Exercise 6.12
Structure of the gravitational action principle We saw in the text that only six out of the ten components of Einstein's equations contain second time derivatives of the metric and that even these only contain $\ddot{g}_{\alpha\beta}$. Obtain this result more directly by examining the explicit form of the quadratic action given in Eq. (6.12). Separate out this Lagrangian in terms of $g_{00}, g_{0\alpha}$ and $g_{\alpha\beta}$ as well as in terms of the spatial and time derivatives. Studying the resulting structure, convince yourself of the results regarding the occurrence of different time derivatives in Einstein's equations in a direct manner.

6.4 The weak field limit of gravity

As we mentioned before, the nonlinear nature of gravitational field equations makes solving them a difficult task. Hence it is important to understand the nature of these equations and their solutions in the case of a weak gravitational field. In addition to the practical utility – arising from the fact that the gravitational fields encountered in many physical systems are indeed weak – the formalism that we develop here will allow us to study gravitational waves in Chapter 9.

A weak gravitational field is characterized by a metric of the form $g_{ab} = \eta_{ab} + h_{ab}$ with $|h_{ab}| \ll 1$. In this situation, one can think of the spacetime as being nearly flat or, equivalently, the spacetime being exactly flat with h_{ab} representing a second rank tensor field propagating in this flat spacetime. While these two interpretations are philosophically quite different, they would lead to the same observable predictions at the lowest order of accuracy. The second point of view also has an additional technical advantage. In our notation, $\eta_{ab} = \text{dia}\,(-1, +1, +1, +1)$, which implies that we are using a Cartesian coordinate system. This is, of course, unnecessary and we can use any convenient coordinate system (for example, a spherical polar coordinate system) to describe the flat background spacetime. In this case, one could treat h_{ab} as a second rank tensor and transform it appropriately. One should also replace η_{ab} by the appropriate metric in the curvilinear coordinates g_{ab}^{flat} and replace ordinary derivatives by covariant derivatives corresponding to the Christoffel symbols obtained from this metric. While this is technically feasible (and is also convenient when studying, for example, spherically symmetric perturbations in flat spacetime), we shall develop the formalism using the Cartesian coordinates for the background metric as it simplifies the mathematics. Transformation to the curvilinear coordinate in flat spacetime is straightforward.

To obtain the linearized version of Einstein's equations, we need to compute G_{ab} for the metric of the form $g_{ab} = \eta_{ab} + h_{ab}$, retaining terms that are linear in h_{ab}. This is straightforward and we get, first, for the Christoffel symbols:

$$\Gamma^s_{nm} = \frac{1}{2}\eta^{sr}\left(\partial_n h_{rm} + \partial_m h_{rn} - \partial_r h_{mn}\right) = \frac{1}{2}\left(\partial_n h^s_m + \partial_m h^s_n - \partial^s h_{mn}\right).$$

$$(6.111)$$

In these expressions as well as in what follows, the raising and lowering of indices is performed using η_{ab}. Next, the curvature tensor is:

$$R^s{}_{mnr} = \frac{1}{2}\partial_n\left(\partial_r h^s_m + \partial_m h^s_r - \partial^s h_{mr}\right) - \frac{1}{2}\partial_r\left(\partial_n h^s_m + \partial_m h^s_n - \partial^s h_{mn}\right)$$

$$= \frac{1}{2}\left(\partial_n\partial_m h^s_r + \partial_r\partial^s h_{mn} - \partial_n\partial^s h_{mr} - \partial_r\partial_m h^s_n\right). \tag{6.112}$$

Contracting on two of the indices, we get the Ricci tensor and the scalar to be:

$$R_{mn} = -\frac{1}{2}\left(\partial_n\partial_m h + \Box h_{mn} - \partial_n\partial_r h^r_m - \partial_r\partial_m h^r_n\right), \tag{6.113}$$

$$R = R^m_m = \eta^{mn}R_{mn} = -\Box h + \partial_r\partial_m h^{mr}. \tag{6.114}$$

Finally, combining these together, we can obtain G_{mn} and write Einstein's equations in the form:

$$\partial_n\partial_m h + \Box h_{mn} - \partial_n\partial_r h^r_m - \partial_r\partial_m h^r_n - \eta_{mn}\left(\Box h - \partial_r\partial_s h^{sr}\right) = -16\pi\,\kappa T_{mn}. \tag{6.115}$$

To proceed further, it is convenient to change the variables from h_{mn} to $\bar{h}_{mn}$ defined by the relation

$$\bar{h}_{mn} \equiv h_{mn} - \frac{1}{2}\eta_{mn}h. \tag{6.116}$$

Notice that, taking the 'bar' once again reproduces h_{mn}. That is, $\bar{\bar{h}}_{mn} = h_{mn}$. In terms of $\bar{h}_{mn}$ Eq. (6.115) becomes

$$\Box\bar{h}_{mn} + \eta_{mn}\partial_r\partial_s\bar{h}^{rs} - \partial_n\partial_r\bar{h}^r_m - \partial_m\partial_r\bar{h}^r_n = -16\pi\,\kappa T_{mn}. \tag{6.117}$$

This equation is identical to Eq. (3.40) describing a spin-2 field in flat spacetime. (This, of course, is to be expected because we have already seen that L_{quad} in Eq. (6.12) reduces the corresponding Lagrangian for spin-2 field in Eq. (3.26); see Exercise 6.2.) Hence most of our discussion in Chapter 3 will be applicable here. In particular, Eq. (6.117) is invariant under the gauge transformation:

$$h'_{mn} = h_{mn} - \partial_m\xi_n - \partial_n\xi_m. \tag{6.118}$$

This gauge transformation leaves the curvature tensor $R^s{}_{mnr}$ invariant as can be verified directly. The curvature tensor in Eq. (6.112) can be written more concisely as

$$R_{ambn} \approx g_{al}\left(\partial_b\Gamma^l_{mn} - \partial_n\Gamma^l_{mb}\right) \approx 2\left(\partial_b\Gamma_{amn} - \partial_n\Gamma_{amb}\right). \tag{6.119}$$

By explicit substitution we see that under the gauge transformation this goes to

$$R_{ambn} \to R_{ambn} - \xi_{a[,n,b],m} + \xi_{m[,n,b],a} - \xi_{n[,a,b],m} + \xi_{n[,m,b],a}. \tag{6.120}$$

Since the partial derivatives commute with each other, the additional terms cancel out and the curvature tensor remains invariant under the gauge transformation as it should. (In this sense, the curvature tensor is analogous to the field tensor F_{ab} in electromagnetism which is also gauge invariant.)

One possible way of understanding this gauge freedom is to note that, under a coordinate shift $x^a \rightarrow x'^a = x^a + \xi^a(x)$, the metric changes by the amount $-(\partial_a \xi_b + \partial_b \xi_a)$. This was derived in Section 4.8, Eq. (4.135), in an arbitrary curved spacetime; in the flat spacetime background, we need only replace the covariant derivatives by ordinary partial derivatives. In our approach, we have defined the tensor field h_{ab} through the equation $g_{ab} = \eta_{ab} + h_{ab}$. Thus a change of coordinates in the form of an infinitesimal coordinate transformation is equivalent to changing h_{mn} to h'_{mn} in Eq. (6.118). Thus starting from the full nonlinear theory, which is invariant under general coordinate transformations, and introducing the linear approximation, we can understand *why* the linear theory is invariant under Eq. (6.118). This physical origin of the gauge invariance is more obscure if we think of h_{ab} as a spin-2 field propagating in flat spacetime (as we did in Chapter 3) since – in that context – we have only Lorentz invariance as the symmetry.

The corresponding change in $\bar{h}^{mn}$ is given by

$$\bar{h}'^{mr} = \bar{h}^{mr} - \partial^m \xi^r - \partial^r \xi^m + \eta^{mr} \partial_s \xi^s. \tag{6.121}$$

Clearly, one can choose the four functions ξ_n to put four conditions on the tensor field h_{mn}. One convenient choice for this gauge condition is

$$\partial_r \bar{h}'^{mr} = 0. \tag{6.122}$$

As we saw in Chapter 3, this choice is always possible by choosing ξ^m to be the solution of the equation $\Box \xi^m = \partial_r \bar{h}^{mr}$. (In fact, we know that this condition does not completely fix ξ^m; one can always add to it any solution to the equation $\Box \xi^m = 0$.) The gauge condition in Eq. (6.122) is precisely the harmonic gauge condition in Eq. (6.88) evaluated for a linearized metric. As we have described in Section 6.3 this leads to a well-defined evolutionary equation.

When the gauge condition in Eq. (6.122) is imposed, several terms in Eq. (6.117) drop out and it reduces to a wave equation of the form

$$\Box \bar{h}^{mn} = -16\pi \kappa T^{mn}. \tag{6.123}$$

We have not bothered to put a prime on h_{mn} and have assumed that the gauge condition is imposed on $\bar{h}^{mn}$. This is the basic equation of linearized gravity which relates the metric perturbation to the source energy-momentum tensor.

The gauge condition in Eq. (6.122) now requires $\partial_m T^{mn} = 0$. This shows that, to the lowest order in h_{mn} at which we are working, the source energy-momentum tensor is conserved and the back reaction of the gravitational field on the source is ignored. We know, however, that for *any* source $\nabla_m T^{mn} = 0$. Expanding out the covariant derivative, we have the structure $\nabla T \sim \partial T + \Gamma T$. The field equation Eq. (6.123) tells us that h is of the order of T; therefore, ΓT is of the order of h^2, which is ignored in our approximation making everything consistent.

Since Eq. (6.123) is a wave equation in the flat spacetime background, one can immediately write down its general solution (see Eq. (2.162)) which corresponds to the standard retarded boundary conditions as:

$$\bar{h}^{mn}(t, \boldsymbol{x}) = 4\kappa \int \frac{T^{mn}(t - |\boldsymbol{x} - \boldsymbol{y}|, \boldsymbol{y})}{|\boldsymbol{x} - \boldsymbol{y}|} d^3 \boldsymbol{y}. \tag{6.124}$$

It is obvious that, in the linear limit, gravitational influences propagate at the speed of light and $\bar{h}_{mn}$ bears the same relationship to T_{mn} as the vector potential A_k in electromagnetism bears to the current J_k. A time dependent source will lead to the emission of gravitational waves just as accelerating charges will lead to electromagnetic radiation. We shall explore these features in detail in Chapter 9. In this section we shall confine our attention to some simple applications of this linearized theory for *stationary* sources.

6.4.1 Metric of a stationary source in linearized theory

The simplest example of linearized gravity occurs in the case of stationary sources in which T^{mn} (and consequently $\bar{h}^{mn}$ produced by it) are independent of time. In that case, the solution in Eq. (6.124) simplifies to

$$\bar{h}_{mn}(\boldsymbol{x}) = \frac{4G}{c^4} \int \frac{T_{mn}(\boldsymbol{y})}{|\boldsymbol{x} - \boldsymbol{y}|} d^3 \boldsymbol{y}, \tag{6.125}$$

where we have set $\kappa = G/c^4$ and reintroduced the c-factors. If we further assume that the source is non-relativistic, then the dominant terms in the energy momentum tensor will be

$$T_{00} = \rho c^2, \qquad T_{0\alpha} = -c\rho u_\alpha, \qquad T_{\alpha\beta} = \rho u_\alpha u_\beta. \tag{6.126}$$

It is obvious that, to the lowest order, $|T^{\alpha\beta}|/|T^{00}| \approx (v^2/c^2)$ so that – to the leading order – we can set $T^{\alpha\beta} \approx 0$ but retain $T^{0\alpha}$. (Since $T^{00} \gg T^{0\alpha} \gg T^{\alpha\beta}$ it is a matter of choice, depending on the physical context, whether to retain $T^{0\alpha}$ or to ignore it with respect to T^{00}.) Solving for the remaining components of the metric perturbations we get

$$\bar{h}_{00} = -\frac{4\Phi}{c^2}, \qquad \bar{h}_{0\alpha} = \frac{A_\alpha}{c}, \qquad \bar{h}_{\alpha\beta} = 0, \tag{6.127}$$

where

$$\Phi(\boldsymbol{x}) \equiv -G \int \frac{\rho(\boldsymbol{y})}{|\boldsymbol{x} - \boldsymbol{y}|} d^3 \boldsymbol{y}; \qquad A_\alpha(\boldsymbol{x}) \equiv \frac{4G}{c^2} \int \frac{T_{0\alpha}(\boldsymbol{y})}{|\boldsymbol{x} - \boldsymbol{y}|} d^3 \boldsymbol{y}. \tag{6.128}$$

Given $\bar{h}_{mn}$, it is easy to determine h_{mn}. We find that

$$h_{00} = h_{11} = h_{22} = h_{33} = -\frac{2\Phi}{c^2}, \qquad g_{0\alpha} = h_{0\alpha} = \frac{A_\alpha}{c}. \qquad (6.129)$$

Hence, the linearly perturbed metric produced by a stationary, non-relativistic source is given by

$$ds^2 = -\left(1 + \frac{2\Phi}{c^2}\right) c^2 dt^2 + \frac{2A_\alpha}{c} c \, dt dx^\alpha + \left(1 - \frac{2\Phi}{c^2}\right) dl^2, \qquad (6.130)$$

where $dl^2 = dx^2 + dy^2 + dz^2$. We notice that: (i) Φ is just the Newtonian gravitational potential due to the source distribution ρ, and (ii) the form of the metric, to leading order, agrees with the Newtonian metric we have used previously in several contexts. The perturbation A_α arises because of the motion of the source and can be interpreted as a vector potential produced due to a mass current. In fact, the analogy between (Φ, A_α) in linearized gravity and the electrostatic and vector potentials in electromagnetic theory is quite precise. To see this, note that Eq. (6.128) can be expressed in the differential form as

$$\nabla^2 \Phi_g = 4\pi G \rho; \qquad \nabla^2 A_g = \frac{16\pi G}{c^2} j; \qquad j \equiv \rho v. \qquad (6.131)$$

If we now define the 'gravo-electric' and 'gravo-magnetic' fields by $E_g = -\nabla \Phi_g$; $B_g = \nabla \times A_g$, then the linearized field equations (in the stationary case) take the suggestive form

$$\nabla \cdot E_g = -4\pi G \rho, \qquad \nabla \cdot B_g = 0,$$

$$\nabla \times E_g = 0, \qquad \nabla \times B_g = -\frac{16\pi G}{c^2} j. \qquad (6.132)$$

These are quite similar to Maxwell's equations except for two obvious differences. The change of sign in the term involving ρ arises because 'like charges' in gravity attract rather than repel. The extra factor of 4 in the term involving j is a reflection of the fact that h_{ab} is a spin-2 field unlike the electromagnetic field potential which is a spin-1 field. We have seen earlier (in Exercise 4.4) that a particle moving in this metric indeed experiences an acceleration $E_g + (v \times B_g)$ thereby completing the analogy with the electromagnetic field.

As a first example of the above formalism, consider a point mass located at the origin. The gravitational potential due to this mass is $\Phi(r) = -GM/r$ and $A_g = 0$. The metric now becomes

$$ds^2 = -\left(1 - \frac{2GM}{c^2 r}\right) c^2 \, dt^2 + \left(1 + \frac{2GM}{c^2 r}\right) [dr^2 + r^2 d\theta^2 + r^2 \sin^2 \theta d\phi^2].$$

$$(6.133)$$

This form of the metric has also been used in several examples earlier. It represents the metric at large distances produced by a body of mass M located at the origin.

As a second example we will consider the metric at large distances from a spherical body which is rigidly rotating about the z-axis with a constant angular velocity. We will see that such a body produces a nonzero gravo-magnetic field. Let us assume that the body is centred at the origin and has a radius R. We are interested in the metric at large distances from the body. In Eq. (6.125), the y ranges over a region of size R and when $x \gg R$, we can expand the denominator in the right hand side as

$$\frac{1}{|\boldsymbol{x} - \boldsymbol{y}|} = \frac{1}{r} + \frac{x_\beta}{r^2}\frac{y^\beta}{r} + \cdots. \qquad (6.134)$$

To the leading order, the h_{00} will be produced by the gravitational potential of the body $\Phi = -GM/r$ where

$$M = \int T^{00} d^3 y. \qquad (6.135)$$

We are more interested in the off-diagonal term. Using Eq. (6.134) in Eq. (6.128), we find that the integral can be expressed in terms of the angular momentum of the body

$$S_\alpha = \epsilon_{\alpha\mu\nu} \int y^\mu T^{\nu 0} d^3 y \qquad (6.136)$$

leading to the final result

$$A_\alpha = 2G\epsilon_{\alpha\beta\gamma}\frac{x^\beta S^\gamma}{r^3} = -\frac{2G}{r^3}(\boldsymbol{S} \times \boldsymbol{x})_\alpha. \qquad (6.137)$$

The resulting line element has the form

$$ds^2 = -\left(1 - \frac{2GM}{c^2 r}\right)dt^2 + \left(1 + \frac{2GM}{c^2 r}\right)dl^2 - 4G\epsilon_{\alpha\beta\gamma}S^\beta\frac{x^\gamma}{r^3}dt dx^\alpha. \qquad (6.138)$$

This form of the metric was used earlier in Exercise 4.4.

Using this result, one can show that a rotating mass 'drags' the coordinate frames with it. This 'dragging of inertial frames' is a general feature which occurs in all stationary spacetimes with $g_{0\phi} \neq 0$. When the metric is independent of time and the ϕ coordinate, any particle which moves in such a metric will possess two conserved quantities: p_ϕ and p_t. The corresponding contravariant components are given by:

$$p^\phi = g^{\phi m}p_m = g^{\phi t}p_t + g^{\phi\phi}p_\phi; \qquad p^t = g^{tm}p_m = g^{tt}p_t + g^{t\phi}p_\phi. \qquad (6.139)$$

Suppose the particle was 'dropped radially' from a large distance with zero angular momentum $p_\phi = 0$. Substituting $p_\phi = 0$ and the relations

$$p^t \propto \frac{dt}{d\lambda}, \qquad p^\phi \propto \frac{d\phi}{d\lambda}, \qquad (6.140)$$

where λ is an affine parameter, in Eq. (6.139), we find that

$$\frac{d\phi}{dt} = \frac{p^\phi}{p^t} = \frac{g^{t\phi}}{g^{tt}} \equiv \omega(r, \theta). \tag{6.141}$$

So we see that a particle dropped 'radially' $[p_\phi = 0]$ from infinity will acquire a nonzero angular velocity in the same direction as the rotation of the source, as it approaches it! Recall that inertial frames are conventionally defined as the ones in which a free particle will move with uniform velocity. If we want to think of the particle described above to be at rest in a some locally inertial frame at every (r, θ, ϕ), then such frames should also be rotating with the angular velocity $\omega(r, \theta)$; hence we say that inertial frames are dragged by the rotating source.

Another situation involving a rotating mass that is of interest is the following. Consider a rigid spherical *shell* of radius R and total mass M (distributed uniformly on the shell) which is rotating slowly with a constant angular velocity Ω about the z-axis. We are interested in determining the metric *inside* the shell far away from its edges. We first note that the Newtonian gravitational potential Φ due to such a spherical shell is constant inside the shell; therefore g_{00} and $g_{\alpha\beta}$ are constants. Hence, by rescaling the coordinates we can reset them to the Cartesian values. Any effect of such a rotating shell is encoded purely in the off-diagonal term $h_{0\alpha}$.

To compute this, let us concentrate on the component $\bar{h}_{0y}$ which will be generated by T_{0y}. (We will set $c = 1$ to simplify the notation.) For a spherical shell rotating with constant angular velocity, the latter is given by

$$T_{0y} = r\Omega\rho \sin\theta \cos\phi. \tag{6.142}$$

Solving Eq. (6.125) for this source is facilitated by noting that the combination $\sin\theta \cos\phi$ is proportional to the component of the spherical harmonic $Y_{11}(\theta, \phi)$. Hence in the differential version $\nabla^2 \bar{h}_{0y} = -16\pi G T_{0y}$ of Eq. (6.125), we can take the angular dependence on both sides to be proportional to $Y_{11}(\theta, \phi)$ because spherical harmonics are solutions to the angular part of the Laplacian. Taking $\bar{h}_{0y} = f(r) \sin\theta \cos\phi$, the radial dependence is determined by the equation

$$\frac{1}{r^2} \frac{d}{dr}\left[r^2 \frac{d}{dr} f(r)\right] - \frac{2f(r)}{r^2} = 16\pi G r\Omega\rho. \tag{6.143}$$

(The factor 2 in the second term on the left hand side arises from the standard factor $l(l+1)$ for $l = 1$.) When the source is made of a spherical shell, the density becomes $\rho = (M/4\pi R^2)\delta_D(r - R)$. For this case, it is easy to integrate the above equation by elementary techniques and we get the solution to be

$$f(r) = -\frac{4GM\Omega}{3} \times \begin{bmatrix} r/R & (\text{for} \quad r < R) \\ (R/r)^2 & (\text{for} \quad r > R) \end{bmatrix}. \tag{6.144}$$

Therefore the metric perturbation inside the shell is given by

$$g_{0y} \approx h_{0y} = \bar{h}_{0y} = -\frac{4GM\Omega}{3R} r \sin\theta \cos\phi = -\frac{4GM\Omega}{3R} x. \qquad (6.145)$$

There is also a corresponding g_{0x} component:

$$g_{0x} \approx +\frac{4GM\Omega}{3R} y. \qquad (6.146)$$

In spherical polar coordinates, this result corresponds to adding a $g_{0\phi}$ component. Transforming the coordinates, we find that

$$g_{0\phi} = -\frac{4GM\Omega}{3R} r^2 \sin^2\theta = -\frac{4GM\Omega}{3R} \bar{r}^2, \qquad (6.147)$$

where $\bar{r} = r \sin\theta$ is the radial distance in the x–y plane.

To understand the meaning of this result, let us consider the flat spacetime metric in a rotating coordinate system. If we start with the flat spacetime metric in the cylindrical coordinates $(t, \bar{r}, \phi, z)$ and make a transformation to the rotating frame by $\phi \to \phi - \omega t$, we obtain

$$ds^2 = -\left(1 - \bar{r}^2\omega^2\right) dt^2 + 2\omega\,\bar{r}^2 dt d\phi + d\bar{r}^2 + \bar{r}^2 d\phi^2 + dz^2. \qquad (6.148)$$

To the linear order in the angular velocity ω, this metric becomes

$$ds^2 \simeq -dt^2 + d\bar{r}^2 + dz^2 + \bar{r}^2 d\phi^2 + 2\omega\,\bar{r}^2 dt d\phi \qquad (6.149)$$

so that $g_{0\phi} = \omega\bar{r}^2$. Comparing with Eq. (6.147) we see that the effect of the rotating shell is to mimic inside it the effects obtained by transforming to a rotating frame of reference with

$$\omega = -\frac{4\Omega GM}{3Rc^2} \qquad (6.150)$$

(to linear order in ω) where we have reintroduced the c-factor.

This result has some implications for an idea usually called *Mach's Principle* which can be interpreted – in this context – as follows. Suppose we choose to use a rotating coordinate system in flat spacetime. In such a frame, we shall see the rest of the matter as rotating with respect to us. We will also notice centrifugal and Coriolis forces in the rotating frame. Suppose, instead, we stay in the inertial frame but manage to rotate the rest of the universe with the same angular velocity in the opposite direction. The question arises as to whether such a rotating universe will lead to the same pseudo-forces. If this is true, then we can claim that *all* motion is relative; that is, transforming to an accelerated frame is the same as accelerating the rest of the universe. We see in Eq. (6.150) a hint in this direction. If the total mass and radius of the rest of the universe satisfy the condition $(GM/Rc^2) \approx 1$, then we have $\omega \approx -\Omega$ within factors of order unity. Unfortunately, it is impossible to proceed further from this approximate result (obtained in perturbation theory to linear order in ω, assuming the rest of the universe is a rotating single shell) to the fully nonlinear theory in any sensible manner. Hence it is not possible to realize the naive version of Mach's principle in full general relativity.

The results in Eq. (6.146) and Eq. (6.145) can be written more concisely as $A_g = (1/2)(x \times B_g)$ with $B_g = \nabla \times A_g = -(8/3)(M\Omega/R)$. The A_g and B_g represent the gravo-magnetic potential and the gravo-magnetic field due to a rotating shell on the inside. We have seen earlier (see Exercise 4.4) that a gravo-magnetic field B_g induces a precession on a gyroscope by the amount $\Omega_{\mathrm{LT}} = -(1/2)B_g$. Therefore, in this case, we will expect a gyroscope inside the rotating shell to precess with the angular velocity in Eq. (6.150).

6.4.2 Metric of a light beam in linearized theory

The two examples discussed in the previous section are fairly standard and are of some practical utility. As a final example in the use of linearized field equations, we will consider a situation which is more of an interesting theoretical curiosity.[4]

Consider two parallel beams of laser light travelling in the same direction. We assume that the first beam has a much stronger intensity than the second. In that case, we can think of the first beam of light as producing a spacetime curvature around it and the second beam of light will be propagating in this spacetime as a test beam. Since both beams of light have certain energy density, which should be equivalent to some mass density, one would have expected a gravitational attraction between the two beams of light. Curiously enough, this is *not* the case and the two laser beams do not exert any force on each other. The result is easy to prove when the gravitational field produced by the first beam is weak and one can use linearized theory. (The result is actually exact but the general proof is more involved.)

To see this, we begin by recalling that, for a beam of light moving along the x-direction, the only nonzero components of the energy-momentum tensor are given by $T_{00} = T_{xx} = -T_{0x}$. Therefore, in the linear theory, the only nonzero components of the metric perturbations are $\bar{h}_{00} = \bar{h}_{xx} = -\bar{h}_{0x}$. Since the electromagnetic energy-momentum tensor is tracefree, the trace of $\bar{h}_{ij}$ is zero implying $\bar{h}_{ij} = h_{ij}$; so the above relations hold for h_{ij} as well; that is, $h_{00} = h_{xx} = -h_{0x}$. We will now be able to show that the second beam of light, treated as a bunch of photons travelling along the x-axis, does not feel any transverse acceleration, say, towards the y-axis.

For a photon moving along the x-axis (with $dt = dx$), the coordinate acceleration in the y-direction is obtained from the geodesic equation

$$\frac{d^2y}{d\lambda^2} = -\Gamma^y_{ab}\frac{dx^a}{d\lambda}\frac{dx^b}{d\lambda} = -\left(\frac{dt}{d\lambda}\right)^2(\Gamma^y_{00} + \Gamma^y_{xx} + 2\Gamma^y_{0x}), \qquad (6.151)$$

where λ is an affine parameter. Using the nonzero components of the metric perturbations we find that the right hand side vanishes:

$$2\Gamma^y_{0x} + \Gamma^y_{00} + \Gamma^y_{xx} = -\frac{1}{2}(\partial_y h_{00} + \partial_y h_{xx} + 2\partial_y h_{0x}) = 0. \qquad (6.152)$$

This shows that two parallel beams of light moving in the same direction do not attract each other in the linearized theory. (The same result can be obtained using the spin-2 field theory; see Project 3.3.) This example illustrates that one can obtain results which are quite counter-intuitive compared with the Newtonian theory in spite of the fact that linearized theory appears to be very similar to Newtonian gravity.

Exercise 6.13
Deflection of light in the Newtonian approximation The purpose of this exercise is to study the trajectory of a light ray in a metric of the form

$$ds^2 = -\left(1 - \frac{2GM}{r}\right)dt^2 + \left(1 + \frac{2GM}{r}\right)(dx^2 + dy^2 + dz^2) \qquad (6.153)$$

and show that it gives a deflection which is twice that obtained in Newtonian theory. Assume that the light ray ('photons') is moving in the x–y plane with an unperturbed trajectory $x = t, y = b$, where b is called the impact parameter. The geodesic equation for the light ray is $(dp^a/d\lambda) + \Gamma^a_{bc}p^bp^c = 0$, where λ is an affine parameter and p^a is the tangent vector to the photon's null geodesic interpreted as its momentum.
   (a) Evaluate the Christoffel symbols in the $z = 0$ plane and show that, when $|p^y| \ll p^0 \approx p^x$, the equations reduce to

$$\frac{dp^y}{d\lambda} = -\frac{2GMb}{(x^2 + b^2)^{3/2}}p^x\frac{dx}{d\lambda}, \qquad p^x = \text{const}\left[1 + \mathcal{O}\left(\frac{GM}{b}\right)\right]. \qquad (6.154)$$

   (b) Integrate this equation assuming that $p^y = 0$ at $x = -\infty$ and obtain

$$p^y(x = +\infty) = -\frac{4GM}{b}p^x. \qquad (6.155)$$

Hence show that the deflection of the light through an angle $4GM/b = 4GM/c^2b$ in conventional units with $c \neq 1$.
   (c) Solve the same problem in purely Newtonian theory. That is, compute the deflection for a particle moving initially with a speed c and impact parameter b in the x–y plane under the influence of a point particle of mass M at the origin and show that it gives half the above value. Can you explain the origin of the extra factor 2?

Exercise 6.14
Metric perturbation due to a fast moving particle We saw in the text that the metric perturbation $\bar{h}^{ab}(\boldsymbol{x})$ due to a particle of mass m at rest at the location $\boldsymbol{x}_0$ is given by

$\bar{h}^{00} = 4Gm/|\boldsymbol{x} - \boldsymbol{x}_0|$ with all other components vanishing. We want to use this result to find the metric perturbation due to a fast moving particle.

(a) Let $\boldsymbol{k}$ be a null vector from the field event P to the source point P_0 and let $\boldsymbol{u}$ be the four-velocity of the particle producing the metric perturbation. Show that $|\boldsymbol{x} - \boldsymbol{x}_0| = \boldsymbol{k} \cdot \boldsymbol{u}$. Hence show that the metric perturbation can be written in an invariant manner as

$$\bar{h}^{ab} = 4G\frac{p^a p^b}{\boldsymbol{k} \cdot \boldsymbol{p}}; \qquad \boldsymbol{p} = m\boldsymbol{u}. \tag{6.156}$$

(b) Consider a test particle at rest at the origin with the world line $x^a = (t, \boldsymbol{0})$. Let another particle of mass m and energy E fly past this test particle with an impact parameter b so that its trajectory is given by $x = vt$, $y = b$, $z = 0$. Compute $\bar{h}_{ab}$ near the test particle due to the moving particle. Using this in the geodesic equation and integrating, show that the moving particle gives an impulse to the test particle in the y-direction and the latter acquires a net velocity in the y-direction given by

$$v_y = \frac{2Gm\gamma}{vb}\left(1 + v^2\right); \qquad \gamma = (1 - v^2)^{-1/2}. \tag{6.157}$$

Hence show that, in the limit of $m \to 0$ with $m\gamma = E$ finite, the induced velocity is $v_y = 4GE/b$. Compare this with the Newtonian result and Exercise 6.13.

Exercise 6.15

Metric perturbation due to a non-relativistic source In the text we obtained the metric perturbation for a slowly rotating spherical source in two different contexts. The purpose of this exercise is to generalize this result for any stationary, non-relativistic, source not necessarily spherically symmetric.

(a) Consider a source with the proper density $\rho(\boldsymbol{y})$ and a momentum density $p^\alpha(\boldsymbol{y}) = \rho(\boldsymbol{y})u^\alpha(\boldsymbol{y})$. Far away from the source, show that the metric perturbation is given to the lowest order by

$$\bar{h}^{00}(\boldsymbol{x}) = -\frac{4GM}{c^2|\boldsymbol{x}|} + \mathcal{O}\left(\frac{1}{|\boldsymbol{x}|^2}\right); \qquad \bar{h}^{0\alpha}(\boldsymbol{x}) = -\frac{2G}{c^3|\boldsymbol{x}|^3}x_\beta J^{\alpha\beta} + \mathcal{O}\left(\frac{1}{|\boldsymbol{x}|^3}\right), \tag{6.158}$$

where the total mass and the angular momentum tensor are defined by

$$M = \int_V \rho(\boldsymbol{y})\,d^3\boldsymbol{y} \quad \text{and} \quad J^{\alpha\beta} = \int_V \left[y^\alpha p^\beta(\boldsymbol{y}) - y^\beta p^\alpha(\boldsymbol{y})\right]d^3\boldsymbol{y}. \tag{6.159}$$

The coordinate system is chosen such that the centre of momentum of the source is at the origin. [Hint. For a stationary source, we have $\partial_0 T^{0\alpha} = 0$. Using this show that

$$\int_V \left(T^{0\alpha}y^\beta + T^{0\beta}y^\alpha\right)d^3\boldsymbol{y} = 0 \tag{6.160}$$

when the integral is over a volume enclosing the source. This result could prove to be useful.]

(b) Rewrite Eq. (6.158) in terms of the gravo-electric and gravo-magnetic potentials as

$$\Phi_g(\boldsymbol{x}) = -\frac{GM}{|\boldsymbol{x}|}; \qquad \boldsymbol{A}_g(\boldsymbol{x}) = -\frac{2G}{c^2|\boldsymbol{x}|^3}\,\boldsymbol{S} \times \boldsymbol{x}, \tag{6.161}$$

where $\boldsymbol{S}$ is the total angular momentum of the source.

(c) Compute the gravo-electric and gravo-magnetic fields due to these potentials and obtain

$$E_g(\boldsymbol{x}) = -\frac{GM}{|\boldsymbol{x}|^2}\,\hat{\boldsymbol{x}} \quad \text{and} \quad B_g(\boldsymbol{x}) = \frac{2G}{c^2|\boldsymbol{x}|^3}\left[\boldsymbol{S} - 3\left(\boldsymbol{S}\cdot\hat{\boldsymbol{x}}\right)\hat{\boldsymbol{x}}\right], \tag{6.162}$$

where $\hat{\boldsymbol{x}}$ is a unit vector in the radial direction.

(d) Consider a particle moving under the influence of the metric derived above. In particular, assume that the particle is moving in a circular orbit of radius r in the equatorial plane. Show that the angular velocity of the particle, accurate to first order in S, is given by

$$\omega^2 = \frac{GM}{r^3} \mp \frac{2GS}{c^2 r^4}\sqrt{\frac{GM}{r}}, \tag{6.163}$$

where the two signs correspond to prograde and retrograde orbits. Clearly, the retrograde orbit has a shorter period than the prograde orbit. Can you provide a qualitative explanation for this result?

6.5 Gravitational energy-momentum pseudo-tensor

In the study of scalar and electromagnetic fields in Chapter 2, we could write down an explicit expression for the energy-momentum tensor of the field. In the case of an electromagnetic field interacting with charged particles, we showed that the sum of the energy-momentum tensor for the field and those for the charged particles is conserved. Such a conservation law allows us to define the total four-momentum of the system in a sensible manner.

When such a physical system (material particles, scalar field, electromagnetic field, etc.) is placed in a given external gravitational field, then the conservation law $\partial_a T^{ab} = 0$ becomes modified to the relation $\nabla_a T^{ab} = 0$. As we explained in detail in Section 5.4.2, page 211, the latter equation is *not* a conservation law for anything; in fact, we should not expect the energy and momentum of the physical system to be conserved by themselves when it is acted upon by an external gravitational field.

One might feel that this situation could be tackled if we include the energy-momentum tensor for the gravitational field in the total energy-momentum budget. Just as in the case of electromagnetism, in which it is only the sum of the energies of the source and the field that is conserved, one might feel that a formulation of energy conservation will be possible if one adds a suitably defined energy-momentum tensor for the gravitational field to the matter energy-momentum tensor.

It is, however, impossible to do this in a generally covariant manner. To see this, we have only to note that a conservation law can be obtained only from an equation involving partial derivatives and not with covariant derivatives. If the energy-momentum tensor for the gravitational field is denoted by t_{ab}, then the conservation law for the total energy would require an equation of the type

$$\partial_n T_{\text{tot}}^{mn} = \partial_n (T^{mn} + t^{mn}) = 0. \tag{6.164}$$

Such an equation cannot be generally covariant for any transformation property of t^{mn}.

There is a fundamental reason why we cannot formulate a generally covariant conservation law for a total energy-momentum tensor. Consider a small region of spacetime around any given event $\mathcal{P}$. By a suitable choice of coordinates, we can make the first derivatives of the metric vanish in this small region. The energy-momentum tensor for gravity – which is expected to be quadratic in the first derivatives of the metric tensor – will vanish in this coordinate system. If it is a genuine tensor, it has to vanish in all coordinate systems around the event $\mathcal{P}$. Therefore no generally covariant, local, energy-momentum tensor for gravity can exist.

It is, however, possible to define a useful concept of an energy-momentum *pseudo-tensor*, t_{ab}, for gravity if we give up the requirement that it should be generally covariant (hence the nomenclature, *pseudo-tensor*). Such a definition cannot be unique since it is inherently coordinate dependent but it can be useful in some specific situations in which a natural choice of coordinates is available. What is more, by defining suitable integrals of t_{ab} over spacetime, one can arrive at the definition of total momentum for the physical system made of a source and the gravitational field. These integrated expressions will be meaningful in any coordinate system in which $g_{ab} \rightarrow \eta_{ab}$ at spatial infinity. We shall first describe how to obtain such a pseudo-tensor and then discuss its properties.

It is convenient to start from the definition of integrated conserved quantities and then work towards a local definition. In electrodynamics, the conservation of total charge is guaranteed by Maxwell's equations. What is more the total charge Q producing a given electromagnetic field F_{ab} can be expressed as an integral over the field along the following lines:

$$Q = \int J^0 d^3x = \frac{1}{4\pi} \int \partial_n F^{0n} d^3x = \frac{1}{4\pi} \int \partial_\beta F^{0\beta} d^3x = \frac{1}{4\pi} \int_S F^{0\beta} d^2S_\beta, \tag{6.165}$$

where S is a 2-sphere at spatial infinity. The last equality arises from the Gauss theorem. This expression gives the familiar result that the flux of the electric field through a sphere at large distances is a measure of the total charge Q. We only need to know the asymptotic form of E to determine the total charge.

The corresponding result in gravity will be an expression for the total four-momentum expressed as an integral over metric variables. To obtain such a result and understand what is involved, let us begin with the linearized theory. If we can express Eq. (6.117) in the form

$$2G^{mn} = \partial_b \partial_a H^{manb} = 16\pi\, T^{mn} \tag{6.166}$$

(we are using units with the Newtonian gravitational constant G set to unity for notational simplicity), with a suitably defined H^{manb} which is antisymmetric in n and b, then the expression for the total four-momentum of the source can be expressed as

$$P^m = \int T^{m0}\, d^3x = \frac{1}{16\pi} \int \partial_b \partial_a H^{ma0b}\, d^3x = \frac{1}{16\pi} \int \partial_\beta \partial_a H^{ma0\beta}\, d^3x$$

$$= \frac{1}{16\pi} \int_{\mathcal{S}} \partial_a H^{ma0\beta}\, d^2 S_\beta. \tag{6.167}$$

The third equality follows from the fact that H^{manb} is antisymmetric in n and b and the last equality follows from the Gauss theorem. This result shows that the total four-momentum of the source can be expressed as an integral at spatial infinity of a suitably defined field component just as in electromagnetism we could express the total charge as an integral at spatial infinity of the electromagnetic field.

This is easy to do in the linearized theory. Examining the structure of Eq. (6.117), one can identify the form of H^{manb} to be

$$H^{manb} \equiv -\left(\bar{h}^{mn}\eta^{ab} + \eta^{mn}\bar{h}^{ab} - \bar{h}^{an}\eta^{mb} - \bar{h}^{mb}\eta^{an}\right). \tag{6.168}$$

This tensor is not only antisymmetric in n and b but it also has all the symmetries of the curvature tensor:

$$H^{manb} = H^{nbma} = H^{[ma][nb]}, \qquad H^{m[anb]} = 0. \tag{6.169}$$

The symmetries also lead to the identity

$$\partial_n T^{mn} = \frac{1}{16\pi}\partial_n \partial_b \partial_a H^{manb} = 0, \tag{6.170}$$

which is analogous to the equation $\partial_a J^a \propto \partial_a \partial_b F^{ab} = 0$ in electromagnetism and is consistent with the result obtained earlier using Eq. (6.122) and Eq. (6.123).

Using the integral expression for P^m in Eq. (6.167), we can express the total energy in a somewhat simplified form in terms of the full metric $g_{ab} = \eta_{ab} + h_{ab}$. Simple algebra shows that

$$P^0 = \frac{1}{16\pi} \int_{\mathcal{S}} \left[\partial_\gamma g^{\beta\gamma} - \partial^\beta\left(\eta^{\mu\nu}g_{\mu\nu}\right)\right] d^2 S_\beta. \tag{6.171}$$

A somewhat more lengthy calculation allows us to obtain a similar expression for the total angular momentum of the source as well (see Exercise 6.18):

$$J^{mn} = \int \left(x^m T^{n0} - x^n T^{m0} \right) d^3x \tag{6.172}$$

$$= \frac{1}{16\pi} \int_{\mathcal{S}} \left(x^m \partial_a H^{na0\beta} - x^n \partial_a H^{ma0\beta} + H^{m\beta 0n} - H^{n\beta 0m} \right) d^2 S_\beta.$$

Having settled these issues in the case of linearized theory, we now return to the full theory. The idea now is to express the full Einstein field equations in the form

$$\partial_b \partial_a H^{manb} = 16\pi \, T^{mn}_{\text{tot}}, \tag{6.173}$$

where the H^{manb} has to be defined in terms of $h_{ab} \equiv g_{ab} - \eta_{ab}$ but *without* making any linear approximations. The expression for the total four-momentum will now become

$$P^m = \frac{1}{16\pi} \int_{\mathcal{S}} \partial_a H^{ma0\beta} d^2 S_\beta = \frac{1}{16\pi} \int \partial_\beta \partial_a H^{ma0\beta} d^3x$$

$$= \frac{1}{16\pi} \int \partial_b \partial_a H^{ma0b} d^3x = \int T^{m0}_{\text{tot}} d^3x. \tag{6.174}$$

The explicit form of T^{m0}_{tot} can be computed by recalling that the quantity $\partial_b \partial_a H^{manb}$ is a linearized approximation to $2G^{mn}$. This suggests defining a quantity

$$16\pi t^{mn} \equiv \partial_b \partial_a H^{manb} - 2G^{mn} \tag{6.175}$$

in terms of which we have the result

$$\partial_b \partial_a H^{manb} = 16\pi t^{mn} + 2G^{mn} = 16\pi(t^{mn} + T^{mn}). \tag{6.176}$$

It immediately follows that:

$$\partial_n T^{mn}_{\text{tot}} = \partial_n \left(T^{mn} + t^{mn} \right) = 0, \tag{6.177}$$

which is what we needed to achieve. This allows us to identify t^{mn} as the energy-momentum pseudo-tensor. To explicitly calculate it, we need to use the definition in Eq. (6.175) and evaluate the expression on the right hand side. Here H^{manb} has to be expressed in terms of $h_{ab} = g_{ab} - \eta_{ab}$ without any linearization. The calculation is tedious but straightforward.

There is, however, one basic difficulty with any such procedure. The expression one obtains for t^{mn} is not unique and there are no additional criteria by which any one particular expression can be chosen in preference to others. Hence, several different expressions for t^{mn} exist in the literature all of which allow the definition of a conserved total four-momentum along the lines described above. For the sake of completeness, we shall quote one such expression which is often used in literature called the *Landau–Lifshitz pseudo-tensor*. This differs slightly from the analysis presented above in the placement of $\sqrt{-g}$ factors but follows exactly the same philosophy.

To obtain this result, we recall that, in the local inertial frame in which the Christoffel symbols vanish and the metric tensor as well as $\sqrt{-g}$ are constants, the equation $\nabla_k T^{ik} = 0$ reduces to $\partial_k T^{ik} = 0$. Writing $T^{ik} = (8\pi)^{-1} G^{ik}$ and expressing the right hand side in the local inertial frame in terms of second derivatives of the metric, it is easy to show that

$$T^{ik} = \partial_l \left\{ \frac{1}{16\pi} \frac{1}{(-g)} \, \partial_m \left[(-g)(g^{ik} g^{lm} - g^{il} g^{km}) \right] \right\}, \tag{6.178}$$

which makes $\partial_k T^{ik} = 0$ an identity. This is equivalent to the relation

$$\partial_l \partial_m \lambda^{iklm} = (-g) T^{ik}, \tag{6.179}$$

where

$$\lambda^{iklm} = \frac{1}{16\pi} (-g)(g^{ik} g^{lm} - g^{il} g^{km}). \tag{6.180}$$

(Note that $\lambda^{iklm} = (8\pi)^{-1}(-g) Q^{ilkm}$, where Q^{ilkm} is defined in Eq. (6.6).) This result, in Eq. (6.179), of course will not hold when we go from a coordinate system (in which the metric tensor is a constant) to an arbitrary coordinate system. But we can use the difference between $\partial_l \partial_m \lambda^{iklm}$ and $(-g) T^{ik}$ to define a quantity $(-g) t_{LL}^{ik}$ where the subscript LL is 'Landau–Lifshitz'. That is, we take

$$(-g)(T^{ik} + t_{LL}^{ik}) = \partial_l \partial_m \lambda^{iklm}. \tag{6.181}$$

(Comparing with Eq. (6.176) we see that the key difference is in the placement $\sqrt{-g}$ factors.) We thus get the identity from which we can determine t^{ik}. It is easy to show that the expression for λ^{manb} reduces to H^{manb} in the weak field limit so that it will lead to the same integrated four-momentum as H^{manb}. Explicit computation now gives the result (in normal units):

$$
\begin{aligned}
t_{LL}^{ik} = \frac{c^4}{16\pi G} \Big\{ &\left(2\Gamma_{lm}^n \Gamma_{np}^p - \Gamma_{lp}^n \Gamma_{mn}^p - \Gamma_{ln}^n \Gamma_{mp}^p \right) \left(g^{il} g^{km} - g^{ik} g^{lm} \right) \\
&+ g^{il} g^{mn} \left(\Gamma_{lp}^k \Gamma_{mn}^p + \Gamma_{mn}^k \Gamma_{lp}^p - \Gamma_{np}^k \Gamma_{lm}^p - \Gamma_{lm}^k \Gamma_{np}^p \right) \\
&+ g^{kl} g^{mn} \left(\Gamma_{lp}^i \Gamma_{mn}^p + \Gamma_{mn}^i \Gamma_{lp}^p - \Gamma_{np}^i \Gamma_{lm}^p - \Gamma_{lm}^i \Gamma_{np}^p \right) \\
&+ g^{lm} g^{np} \left(\Gamma_{ln}^i \Gamma_{mp}^k - \Gamma_{lm}^i \Gamma_{np}^k \right) \Big\}.
\end{aligned}
\tag{6.182}
$$

Since, by construction we have the result $\partial_k [(-g)(T^{ik} + t_{LL}^{ik})] = 0$ we have a conserved four-momentum given by

$$P^i = \int (-g)(T^{ik} + t_{LL}^{ik}) \, d\Sigma_k = \int \partial_l h^{ikl} \, d\Sigma_k = \int_{\mathcal{S}} h^{i0\alpha} \, dS_\alpha, \tag{6.183}$$

where $h^{ikl} \equiv \partial_m \lambda^{iklm}$ and the last result follows when we work with a $t =$ constant hypersurface. This result expresses the total momentum as an integral over the 3-surface.

Given a conserved energy-momentum tensor, we can also obtain a conservation law for angular momentum defined through

$$M^{ik} = \int (x^i dP^k - x^k dP^i) = \int \left[x^i (T^{kl} + t^{kl}_{LL}) - x^k (T^{il} + t^{il}_{LL}) \right] (-g) d\Sigma_l$$

$$= \int \left(x^i \partial_m \partial_n \lambda^{klmn} - x^k \partial_m \partial_n \lambda^{ilmn} \right) d\Sigma_l. \tag{6.184}$$

Simplifying this expression and using the results

$$\lambda^{ilkn} - \lambda^{klin} = \lambda^{ilnk}, \qquad \lambda^{inlk} = -\lambda^{ilnk} \tag{6.185}$$

one can manipulate this expression to express the angular momentum as

$$M^{ik} = \int \left(x^i h^{k\,0\alpha} - x^k h^{i\,0\alpha} + \lambda^{i\,0\alpha k} \right) dS_\alpha. \tag{6.186}$$

Several other aspects of energy-momentum pseudo-tensors are explored in the following exercises.

Exercise 6.16
Landau–Lifshitz pseudo-tensor in the Newtonian approximation Consider the Newtonian approximation in which the metric takes the form

$$ds^2 = -(1 + 2\phi) \, dt^2 + (1 - 2\phi) \, \delta_{\beta\gamma} \, dx^\beta dx^\gamma. \tag{6.187}$$

If the source is slowly varying, so that the time derivatives of ϕ can be ignored in comparison with spatial derivatives, show that t^{mn}_{LL} is given by

$$t^{00}_{LL} = -\frac{7}{8\pi} \partial^\beta \phi \, \partial_\beta \phi \, ; \qquad t^{0\beta}_{LL} = 0 \, ; \qquad t^{\beta\gamma}_{LL} = \frac{1}{4\pi} \left(\partial^\beta \phi \, \partial^\gamma \phi - \frac{1}{2} \delta^{\beta\gamma} \partial_\delta \, \phi \partial^\delta \phi \right). \tag{6.188}$$

Exercise 6.17
More on the Landau–Lifshitz pseudo-tensor An alternative expression for the Landau–Lifshitz pseudo-tensor t^{mn}_{LL} is given by

$$(-g)t^{ab}_{LL} = \frac{1}{16\pi} \left\{ \partial_l \mathfrak{g}^{ab} \partial_m \mathfrak{g}^{lm} - \partial_l \mathfrak{g}^{al} \partial_m \mathfrak{g}^{bm} + \frac{1}{2} g^{ab} g_{lm} \partial_r \mathfrak{g}^{ln} \partial_n \mathfrak{g}^{rm} \right.$$

$$- \left(g^{al} g_{mn} \partial_r \mathfrak{g}^{bn} \partial_l \mathfrak{g}^{mr} + g^{bl} g_{mn} \partial_r \mathfrak{g}^{an} \partial_l \mathfrak{g}^{mr} \right) + g_{lm} g^{nr} \partial_n \mathfrak{g}^{al} \partial_r \mathfrak{g}^{bm}$$

$$\left. + \frac{1}{8} \left(2g^{al} g^{bm} - g^{ab} g^{lm} \right) \left(2g_{nr} g_{st} - g_{rs} g_{nt} \right) \partial_l \mathfrak{g}^{nt} \partial_m \mathfrak{g}^{rs} \right\}, \tag{6.189}$$

where $\mathfrak{g}^{ab} \equiv \sqrt{-g}\, g^{ab}$. Though not significantly better than the expression given in the text, it has some advantages in manipulation. Define a quantity $\bar{h}^{ab}$ by the definition $\mathfrak{g}^{ab} \equiv \eta^{ab} - \bar{h}^{ab}$. We have shown in the text that Einstein's field equations are equivalent to

$$\partial_b \partial_a H^{manb} = \partial_b \partial_a \left[\mathfrak{g}^{mn}\mathfrak{g}^{ab} - \mathfrak{g}^{an}\mathfrak{g}^{mb} \right] = 16\pi(-g)\left(T^{mn} + t^{mn}_{\mathrm{LL}}\right). \tag{6.190}$$

(a) Introduce the harmonic gauge by the condition $\partial_b \mathfrak{g}^{ab} = 0$. Show that the field equations can now be expressed as

$$\mathfrak{g}^{ab}\partial_b \partial_a \mathfrak{g}^{mn} = 16\pi(-g)\left(T^{mn} + t^{mn}_{\mathrm{LL}}\right) + \partial_b \mathfrak{g}^{an}\partial_a \mathfrak{g}^{mb}. \tag{6.191}$$

(b) Show that, in terms of $\bar{h}^{mn}$, the same equation becomes:

$$\eta^{ab}\partial_b \partial_a \bar{h}^{mn} = -16\pi(-g)\left(T^{mn} + t^{mn}_{\mathrm{LL}}\right) - \partial_b \bar{h}^{an}\partial_a \bar{h}^{mb} + \bar{h}^{ab}\partial_a \partial_b \bar{h}^{mn}, \tag{6.192}$$

with t^{mn}_{LL} having the same form as before except for the replacement of $\partial_l \mathfrak{g}^{ab}$ by $-\partial_l \bar{h}^{ab}$. [The equations obtained so far are exact. If we assume that $\bar{h}^{ab}$ is a small perturbation, then the above equation can be manipulated iteratively order by order in $\bar{h}^{ab}$.]

Exercise 6.18

Integral for the angular momentum Show that the field equations can be expressed in the form

$$16\pi x^m T^{n0} = \partial_\gamma \left(x^m \partial_a H^{na0\gamma} \right) - \partial_\beta H^{n\beta0m} - \partial_0 H^{n00m}. \tag{6.193}$$

Use this result and the Gauss theorem to prove Eq. (6.172).

Exercise 6.19

Several different energy-momentum pseudo-tensors In the text, we discussed the Landau–Lifshitz pseudo-tensor for the gravitational field. You will find in the literature several other choices for this energy-momentum pseudo-tensor and this exercise introduces a few of them.

(i) Einstein's energy-momentum pseudo-tensor is defined through the relations

$$t^{ik} = \frac{g^{il}}{16\pi}\, \partial_m \chi_l^{km}; \qquad \chi_i^{km} = \frac{g_{in}}{\sqrt{-g}}\, \partial_p \left[-g\left(g^{kn}g^{mp} - g^{mn}g^{kp}\right)\right]. \tag{6.194}$$

(ii) Bergmann–Thomson's energy-momentum pseudo-tensor is given by

$$t^{ik} = \frac{1}{16\pi}\, \partial_m \left[g^{il}\chi_l^{km}\right]. \tag{6.195}$$

(iii) Moller's energy-momentum pseudo-tensor is given by

$$t^n_m = \frac{1}{8\pi}\, \partial_a\, \rho_m^{na}; \qquad \rho_m^{na} = \sqrt{-g}\, \left[\partial_c g_{mb} - \partial_b g_{mc}\right] g^{nc}g^{ab}. \tag{6.196}$$

(iv) Papapetrou's energy-momentum pseudo-tensor is given by

$$t^{mn} = \frac{1}{16\pi}\, \partial_b \partial_a\, \lambda^{mnab};$$
$$\lambda^{mnab} = \sqrt{-g}\, \left(g^{mn}\eta^{ab} - g^{ma}\eta^{nb} + g^{ab}\eta^{mn} - g^{nb}\eta^{ma}\right). \tag{6.197}$$

(v) Landau–Lifshitz's energy-momentum pseudo-tensor discussed in the text is given by

$$t^{ij} = \frac{1}{16\pi} \partial_l \partial_k \lambda^{ikjl}; \quad \lambda^{ikjl} = -g \left(g^{ij} g^{kl} - g^{il} g^{kj} \right). \tag{6.198}$$

Compute the energy density and the total four-momentum of the following metric

$$ds^2 = -f(r)dt^2 + \frac{dr^2}{f(r)} + r^2 d\Omega^2 \tag{6.199}$$

using each of these energy-momentum pseudo-tensors.

Finally we want to describe another procedure for obtaining integral expressions for the conserved quantities in a spacetime, in the presence of Killing vectors. Consider a solution to Einstein's equation that admits one or more Killing vectors ξ^a. We can now form a current vector by $Q^a = \xi_b R^{ab}$ which has zero divergence:

$$\nabla_a Q^a = R^{ab} \nabla_a \xi_b + \xi^b \nabla_a R_b^a = \frac{1}{2} \xi^b \partial_b R = 0. \tag{6.200}$$

The second equality arises from the fact that for a Killing vector $\nabla_a \xi_b = -\nabla_b \xi_a$ and that $\nabla_a G_b^a = 0$; the third equality arises from noting that the directional derivative of all geometrical scalars in the direction of ξ must be zero if ξ is a Killing vector. Thus we find that Q^a is a conserved vector and the integral of Q^0 over all space is a conserved quantity. More formally, the integral

$$I \equiv -\frac{1}{4\pi} \int d^3 \Sigma_m \xi^n R_n^m = -2 \int d^3 \Sigma_m \xi^n [T_n^m - \frac{1}{2} \delta_n^m T] \tag{6.201}$$

is a conserved quantity where the integral is over a spacelike hypersurface. The physical meaning of this conserved quantity will, of course, depend on the nature of the Killing vector used in its definition. We will now show that, for the Killing vectors which represent time translation invariance and rotational invariance about an axis, these integrals can be related to the mass and the angular momentum. To do this, we first rewrite I in Eq. (6.201) as

$$I = -\frac{1}{4\pi G} \int R_n^m \xi^n \, d^3 \Sigma_m = \frac{1}{4\pi G} \int \nabla_n \nabla^n \xi^m \, d^3 \Sigma_m = \frac{1}{8\pi G} \int_S \nabla^n \xi^m d^2 \Sigma_{mn}. \tag{6.202}$$

The second equality follows from the identity in Eq. (5.132) while the last equality arises from the Gauss theorem. The final result allows us to express I as an integral over a two-dimensional surface located at the spatial infinity even though the source itself has a finite extent. Using

$$d^2 \Sigma_{mn} = \frac{1}{2!} \epsilon_{mnab} \frac{\partial(x^a, x^b)}{\partial(\theta, \phi)} d\theta \, d\phi = \epsilon_{mn\theta\phi} r^2 \sin\theta \, d\theta \, d\phi \tag{6.203}$$

we find that the relevant integration measure is:

$$\nabla^n \xi^m d^2 \Sigma_{mn} = 2\nabla^r \xi^t r^2 \sin\theta \, d\theta \, d\phi. \tag{6.204}$$

For explicit evaluation of the integral we need to know: (i) which Killing vector we are using and (ii) the asymptotic form of the metric. The latter is given by the weak field expression:

$$ds^2 = -\left(1 - \frac{2GM}{r}\right)dt^2 - \frac{4GJ\sin^2\theta}{r}dtd\phi + \left(1 + \frac{2GM}{r}\right)$$
$$\times \left(dr^2 + r^2 d\theta^2 + r^2 \sin^2\theta d\phi^2\right). \tag{6.205}$$

As for the Killing vector let us first consider the one corresponding to time translation invariance: $\xi^a = \delta^a_0$. Then, in the asymptotic region, to first order accuracy in metric coefficients, we get:

$$\nabla^r \xi^t \approx \nabla_r \xi^t = \partial_r \xi^t + \Gamma^t_{ar}\xi^a = \Gamma^t_{tr} \approx -\frac{1}{2}\partial_r g_{tt} = \frac{GM}{r^2}. \tag{6.206}$$

Substituting into the integral we get the result

$$I = \frac{1}{8\pi G} \int \frac{2GM}{r^2} r^2 \sin\theta \, d\theta \, d\phi = M. \tag{6.207}$$

In other words we have proved that

$$I \equiv -2\int d^3\Sigma_m \xi^n [T^m_n - \frac{1}{2}\delta^m_n T] = -2\int d^3x \sqrt{-g}\, [T^0_0 - \frac{1}{2}T] = M \tag{6.208}$$

for any source which is confined to a finite region in space and the metric perturbations are small at large distances. For an ideal fluid the integral is over $(\rho + 3p)$ rather than over ρ; pressure also contributes to the conserved mass.

In the case of axial symmetry, the relevant Killing vector is $\xi^a = \delta^a_\phi$ and we get:

$$\nabla^r \xi^t \approx \partial_r \xi^t + \Gamma^t_{ar}\xi^a = \Gamma^t_{\phi r} = g^{tt}\Gamma_{t\phi r} + g^{t\phi}\Gamma_{\phi\phi r} = -3GJ\frac{\sin^2\theta}{r^2} \tag{6.209}$$

so that the integral becomes

$$S = \frac{1}{8\pi G}\int 6GJ \sin^3\theta \, d\theta \, d\phi = 2J. \tag{6.210}$$

While this gives an answer that is proportional to angular momentum, the result has an extra factor 2. (Alternatively, if we define the conserved integral to get J for $\xi^a = \delta^a_\phi$, the corresponding result for $\xi^a = \delta^a_0$ will be $M/2$. There is no simple explanation for this factor of 2.) It is obvious from the nature of the derivation that the results will hold even in spacetimes which have a vector field that satisfies the Killing equation only in the asymptotic region.

Exercise 6.20

Alternative expressions for the mass Consider a spacetime which is asymptotically flat with the metric perturbation $h_{ab} \equiv g_{ab} - \eta_{ab} \approx \mathcal{O}(1/r)$ as $r \to \infty$ in a coordinate system which is Cartesian in the asymptotic region. Let the source T_{ab} be static.

(a) Show that Eq. (6.117) can now be reduced to the form

$$T_{00} = \frac{1}{16\pi G} \partial_\alpha \left(\partial_\beta h^{\alpha\beta} - \partial^\alpha h^\beta_\beta \right). \tag{6.211}$$

[Hint. Use the 00 component as well as the trace of Eq. (6.117).] Hence show that the mass can also be expressed in the form

$$M = \frac{1}{16\pi G} \oint_\infty dS_\alpha \left(\partial_\beta h^{\alpha\beta} - \partial^\alpha h^\beta_\beta \right), \tag{6.212}$$

where the integral is over a 2-surface at infinity.

(b) Show that Eq. (6.117) also implies

$$\partial_\alpha \left(\partial_\beta h^{\alpha\beta} - \partial^\alpha h^\beta_\beta \right) = 2\nabla^2 h_{00}. \tag{6.213}$$

(c) Prove that $g^{\alpha\beta}\Gamma^0_{0\beta} \approx -(1/2)\partial^\alpha h_{00} + \mathcal{O}(r^{-3})$. Hence obtain

$$M = \frac{1}{4\pi G} \oint_\infty dS_\alpha g^{\alpha\beta}\Gamma^0_{0\beta}. \tag{6.214}$$

Finally, expressing $\nabla_i \xi^0$ of the Killing vector in terms of the Christoffel symbols in the asymptotic region, prove the equivalence between the above formula and the one obtained in the text.

PROJECTS

Project 6.1

Scalar tensor theories of gravity

One possible way of generalizing Einstein's theory is to introduce a scalar field into the action functional for gravity. While such a theory is unlikely to have anything to do with nature, it is a convenient test bed for exploring several mathematical features and hence is quite popular. The purpose of this project is to explore several aspects of such theories.

(a) As a crucial warm up, show that the variation $g^{ab}\delta R_{ab}$ computed in the text can be expressed in the form

$$g^{ab}\delta R_{ab} = \nabla_i \nabla^i (g_{ab}\delta g^{ab}) - \nabla_a \nabla_b (\delta g^{ab}). \tag{6.215}$$

(b) Consider now an action principle for gravity of the form

$$\mathcal{A}_g = \int d^4x \sqrt{-g}\, f(\phi)R + \mathcal{A}_{(m)}, \tag{6.216}$$

where $\mathcal{A}_{(m)}$ is the matter action. Vary this Lagrangian with respect to g^{ab} and show that the resulting equations can be written in the form

$$f(\phi)G_{mn} = \left(\frac{1}{2}T_{mn}^{(M)} + \nabla_m\nabla_n f - g_{mn}\Box f\right).$$ (6.217)

[Hint. When you perform the variation, you will get a term with $f(\phi)g^{ab}\delta R_{ab}$. Using the result of part (a) above, you can do an integration by parts on this term. Assume that all surface terms can be ignored.]

(c) Consider now a conformal transformation to a new metric (denoted by a tilde on top) through the relation

$$\tilde{g}_{mn} = 16\pi G\, f(\phi)\, g_{mn}.$$ (6.218)

Using the transformation properties of scalar curvature under conformal transformations, show that the action becomes

$$\mathcal{A}_g = \int d^4x\sqrt{-g}\, f(\phi)R = \int d^4x\sqrt{-\tilde{g}}\,\frac{1}{16\pi G}\left[\tilde{R} - \frac{3}{2}\tilde{g}^{rs}f^{-2}\left(\frac{df}{d\phi}\right)^2\left(\tilde{\nabla}_r\phi\right)\left(\tilde{\nabla}_s\phi\right)\right].$$
(6.219)

This result shows that, after the conformal transformation is made, the resulting field equations will look like Einstein's equations for a somewhat complicated matter action. [In the literature, the original metric g_{ab} is called the metric in the *Jordan frame* while the metric $\tilde{g}_{ab}$ is called the metric in the *Einstein frame*.]

(d) In particular, if we make the choice $f(\phi) = \exp(\phi/\sqrt{3})$, then the action in the Einstein frame corresponds to standard gravity coupled to a massless scalar field. How would you interpret this result?

(e) For further generalization, add to the action in part (b) another piece which represents the action for the scalar field ϕ itself, given by

$$\mathcal{A}_\phi = \int d^4x\sqrt{-g}\left[-\frac{1}{2}h(\phi)g^{mn}(\partial_m\phi)(\partial_n\phi) - U(\phi)\right].$$ (6.220)

Show that the previous results continue to hold and that the action in the Einstein frame now has the form

$$\mathcal{A}_{\text{tot}} = \int d^4x\sqrt{-\tilde{g}}\left[\frac{\tilde{R}}{16\pi G} - \frac{1}{2}K(\phi)\tilde{g}^{rs}\left(\tilde{\nabla}_r\phi\right)\left(\tilde{\nabla}_s\phi\right) - \frac{U(\phi)}{(16\pi G)^2 f^2(\phi)}\right],$$
(6.221)

where

$$K(\phi) = \frac{1}{16\pi G f^2}\left[fh + 3(f')^2\right].$$ (6.222)

The action in Eq. (6.221) can be expressed in a more conventional form by redefining the field by

$$\psi = \int K^{1/2}d\phi$$ (6.223)

in terms of which we have

$$\mathcal{A}_{\text{tot}} = \int d^4x\sqrt{-\tilde{g}}\left[\frac{\tilde{R}}{16\pi G} - \frac{1}{2}\tilde{g}^{rs}\left(\tilde{\nabla}_r\psi\right)\left(\tilde{\nabla}_s\psi\right) - V(\psi)\right]$$ (6.224)

with

$$V(\psi) = \frac{U(\phi(\psi))}{(16\pi G)^2 f^2(\phi(\psi))}. \tag{6.225}$$

Hence a much wider class of theories can be mapped to Einstein's gravity coupled to a scalar field.

(f) Another possible extension of the Einstein gravity is to use a Lagrangian $L_g = F(R)$, which is an arbitrary function of the scalar curvature instead of a function that is linear in R. Work out the resulting field equations in this particular case.

(g) Show that the field equations obtained in part (f) above are identical to one derived in an Einstein–scalar field system with a scalar field $\phi = \sqrt{3/2} \ln F'(R)$ if we choose the potential for the scalar field to be

$$V(\phi) = \frac{RF'(R) - F(R)}{F'^2(R)}. \tag{6.226}$$

It is assumed that the right hand side is expressed in terms of ϕ by inverting the relation $\phi = \sqrt{3/2} \ln F'(R)$.

(h) In the above analysis we have ignored the surface terms in the action to derive the field equations. We saw in the text that, in case of Einstein–Hilbert action, adding an integral over the boundary of $2K$ will make the variation well-defined. Determine the corresponding counterterms to be added to the action in the cases considered above.

Project 6.2

Einstein's equations for a stationary metric

Consider a metric with the notation $g_{00} = -N^2$, $g_{0\alpha} = -N^2 g_\alpha$ and the definition of the gravo-magnetic field

$$f_{\alpha\beta} = \partial_\alpha g_\beta - \partial_\beta g_\alpha. \tag{6.227}$$

Assume that the metric is independent of time. Show that the components of the Ricci tensor are given by the relation

$$R_{00} = -N \nabla_\alpha \nabla^\alpha (N) + \frac{N^4}{4} f_{\alpha\beta} f^{\alpha\beta} \tag{6.228}$$

$$R_0^\alpha = \frac{N^2}{2} \nabla_\beta f^{\alpha\beta} - \frac{3N}{2} f^{\alpha\beta} \nabla_\beta (N) \tag{6.229}$$

$$R^{\alpha\beta} = {}^{(3)}R^{\alpha\beta} + \frac{N^2}{2} f^{\alpha\gamma} f_\gamma^\beta - \frac{1}{N} \nabla^\beta \nabla^\alpha (N), \tag{6.230}$$

where all computations are performed in a three-dimensional space with the metric

$$\gamma_{\alpha\beta} = \left(g_{\alpha\beta} - \frac{g_{0\alpha} g_{0\beta}}{g_{00}} \right). \tag{6.231}$$

The ∇_α denotes covariant derivative calculated using $\gamma_{\alpha\beta}$; ${}^{(3)}R^{\alpha\beta}$ is the three-dimensional Ricci scalar constructed from $\gamma_{\alpha\beta}$. See how far you can proceed in determining the nature of time independent spacetimes using these equations. A sample of questions to ask yourself will be the following. If there is no source, what kind of solutions will emerge? When is it possible to set $f_{\alpha\beta}$ to zero? If you perform a Taylor series expansion in the metric perturbation, what kind of corrections emerge due to the nonlinear nature of gravity at the lowest order?

Project 6.3

Holography of the gravitational action

The purpose of this project is to prove a curious and remarkable relation[5] between L_{quad} and L_{sur} in Eq. (6.9) and understand its implications.[6] These results will be used and discussed in Chapters 15 and 16.

(a) As a warm up, consider a simple problem in classical mechanics. Given a Lagrangian $L_q(\dot{q}, q)$, one can obtain the standard Euler–Lagrange equations by varying q in the action functional with the condition $\delta q = 0$ at the boundary. Consider now a different Lagrangian defined as

$$L_p(\ddot{q}, \dot{q}, q) \equiv L_q(\dot{q}, q) - \frac{d}{dt}\left(q\frac{\partial L_q}{\partial \dot{q}}\right). \tag{6.232}$$

This Lagrangian, unlike L_q contains $\ddot{q}$. Vary the action resulting from this Lagrangian but – instead of demanding $\delta q = 0$ at the boundary – demand that $\delta p = 0$ at the boundary, where $p(\dot{q}, q) \equiv (\partial L_q/\partial \dot{q})$ is the momentum. Show that we now obtain the same equations of motion as the one obtained from L_q. *That is, even though L_p contains the second derivatives of q, it leads to second order differential equations for q (rather than third order) if we fix p at the boundary.*

(b) If you are familiar with path integral formulation of quantum mechanics, show that L_p can be used to define a path integral which will propagate the momentum space wave function from $t = t_1$ to $t = t_2$.

(c) Generalize this result[7] to a situation in which we keep some given function $C(q, \dot{q})$ fixed at the end points. That is, find a Lagrangian of the form

$$L_C(q, \dot{q}, \ddot{q}) = L_q(q, \dot{q}) - \frac{df(q, \dot{q})}{dt} \tag{6.233}$$

such that it will lead to the same equations of motion as $L_q(q, \dot{q})$ when we vary q keeping $C(q, \dot{q})$ fixed. [Hint. Invert the relation $C = C(q, \dot{q})$ to determine $\dot{q} = \dot{q}(q, C)$ and express $p(q, \dot{q}) \equiv \partial L_q/\partial \dot{q}$ in terms of (q, C) obtaining the function $p = p(q, C)$. Then, f is given by

$$f(q, C) = \int p(q, C)dq + F(C), \tag{6.234}$$

where the integration is with constant C and F is an arbitrary function.]

(d) Generalize these results to a multi-component field $q_A(x^a)$ in a spacetime with coordinates x^a, where A collectively denotes the tensor indices. Suppose a Lagrangian $L_q(q_A, \partial q_A)$ gives the field equations when the action is varied keeping q_A fixed at the boundary $\partial \mathcal{V}$ of a spacetime region $\mathcal{V}$. We now want to add to the action a four-divergence $\partial_a V^a$ such that the same equations are obtained when the action is varied keeping some given functions $U_A^a(q_A, \partial q_A)n_a$ fixed at the boundary where n_a is the normal to $\partial \mathcal{V}$. Show that the Lagrangian we are looking for is

$$L_U(\partial^2 q_A, \partial q_A, q_A) = L_q(q_A, \partial q_A) - \partial_a V^a(q_A, \partial q_A) \tag{6.235}$$

with

$$V^j(q_A, U_A^b) = \int \pi_A^j(q_A, U_A^b)dq^A + F^j(U_A^b). \tag{6.236}$$

(e) Remarkably enough, the Lagrangian $\sqrt{-g}\,R$ has exactly the same structure as L_p introduced above. Prove this by showing that $\sqrt{-g}\,R$ can be expressed in the form

$$\sqrt{-g}\,R = \sqrt{-g}\,L_{\text{quad}} - \partial_c\left[g_{ab}\frac{\partial\sqrt{-g}L_{\text{quad}}}{\partial(\partial_c g_{ab})}\right].\tag{6.237}$$

The straightforward (but tedious) way to do this is to use Eq. (6.12) and work out everything explicitly. This is not a bad idea since it will train you well in index gymnastics.

(f) A shorter procedure is to use some of the scaling relations that are available in L_{quad} and argue as follows.

(i) For a Lagrangian written as $L(q_A, \partial_i q_A)$ – which depends on a set of dynamical variables q_A, where A could denote a collection of indices (in the case of gravity $q_A \to g_{ab}$ with A denoting a pair of indices) – we define the Euler–Lagrange function as:

$$F^A \equiv \frac{\partial L}{\partial q_A} - \partial_i\left[\frac{\partial L}{\partial(\partial_i q_A)}\right].\tag{6.238}$$

Assume that L is a homogeneous function of degree μ in q_A and a homogeneous function of degree λ in $\partial_i q_A$. Taking the contraction $q_A F^A$ and manipulating the terms show that:

$$q_A F^A = (\lambda + \mu)L - \partial_i\left[q_A\frac{\partial L}{\partial(\partial_i q_A)}\right].\tag{6.239}$$

(ii) In the case of gravity, show that $F^A = -(R^{ab} - (1/2)g^{ab}R)\sqrt{-g}$ and $q_A F^A = R\sqrt{-g}$. Further show that, in this case $\mu = -1, \lambda = +2$. Complete the proof of Eq. (6.237) using these facts.

[Hint. In the case of gravity, if we change $g_{ab} \to f g_{ab}$ then $g^{ab} \to f^{-1}g^{ab}, \sqrt{-g} \to f^2\sqrt{-g}$. If the first derivatives $g_{ab,c}$ are held fixed, the above scaling will change $\sqrt{-g}L_{\text{quad}}$ in Eq. (6.12) by the factor $f^2 f^{-3} = f^{-1}$ showing that $\sqrt{-g}L_{\text{quad}}$ is of degree $\mu = -1$ in g_{ab}. When g_{ab} is held fixed and $\partial_c g_{ab}$ is changed by a factor f, $\sqrt{-g}L_{\text{quad}}$ changes by a factor f^2; so $\sqrt{-g}L_{\text{quad}}$ is of degree $\lambda = +2$ in the derivatives.]

(g) In the text, we obtained an expression for $\delta(\sqrt{-g}\,L_{\text{quad}})$ in Eq. (6.28). Rewrite this expression for the variation δg_{ab} of covariant components of the metric using $\delta g^{ab} = -g^{ai}g^{bk}\delta g_{ik}$. Using further the fact that $M^c_{\ ik}g^{ik} = -V^c$ (see Eq. (6.14) and Eq. (6.26)), show that the resulting variation can be written in the form

$$\delta(\sqrt{-g}\,L_{\text{quad}})+\delta[\partial_c(\sqrt{-g}\,V^c)] = \delta(R\sqrt{-g}) = -\sqrt{-g}\,G^{ab}\delta g_{ab}-\partial_c[g_{ik}\delta(\sqrt{-g}\,M^{cik})],\tag{6.240}$$

where

$$M^{cik} = g^{il}g^{km}\Gamma^c_{lm} - g^{il}g^{ck}\Gamma^d_{ld} - \frac{1}{2}g^{ik}V^c = \frac{1}{2}M^{cikpqr}\partial_p g_{qr}\tag{6.241}$$

in the notation of Eq. (6.12). It is now obvious that one can devise a variational principle for gravity based on R provided we keep $\sqrt{-g}\,M^{cik}$ fixed at the boundary. From our previous discussion we know that this is equivalent to a momentum space description for gravity. This leads to

$$\delta\int_{\mathcal{V}} d^4x\,\sqrt{-g}\,R = -\int_{\mathcal{V}} d^4x\,\sqrt{-g}\,G^{ab}\delta g_{ab} - \int_{t=\text{const}} d^3x\,g_{ik}\delta(\sqrt{-g}\,M^{0ik})\tag{6.242}$$

on the spacelike boundary surface, allowing us to identify M^{0ik} as the canonical momenta associated with g_{ik}. Investigate the properties of M^{0ik}.

7

Spherically symmetric geometry

7.1 Introduction

In this chapter we will obtain the simplest of the exact solutions to Einstein's equations, which are the ones with spherical symmetry. The chapter also discusses the orbits of particles and photons in these spacetimes and the tests of general relativity. All of this will be used in the study of black holes in the next chapter.

7.2 Metric of a spherically symmetric spacetime

One of the simplest – but fortunately very useful – class of solutions to Einstein's equations is obtained when the source T_{ik} and the resulting metric possess spherical symmetry. We shall first obtain the general form of the metric in the spherically symmetric context and then use Einstein's equations to relate the metric to the source. While the form of the spherically symmetric metric (given in Eq. (7.12) below) can be obtained almost 'by inspection', we shall perform a rather formal analysis in order to illustrate a useful technique.

If the spacetime exhibits a particular symmetry which can be characterized by the action of an element of a group, then the functional change in the form of the metric under the action of this element of the group should vanish. In the case of spherical symmetry, the relevant group is the group of rotations under which the Cartesian coordinates will change by $x^a \to x^a + \xi^a$, where ξ^a has the components

$$\xi^0 = 0, \qquad \xi^\alpha = \epsilon^{\alpha\beta} x_\beta. \tag{7.1}$$

Here $\epsilon^{\alpha\beta} = -\epsilon^{\beta\alpha}$ is a set of arbitrary infinitesimal constants (with only three elements being independent due to the antisymmetry) which describe infinitesimal rotations. We already know that the change in the functional form of the metric under an infinitesimal coordinate transformation is given by Eq. (4.133). Using the form of ξ^a in Eq. (7.1) in Eq. (4.133) and setting $\delta g_{ik} = 0$ leads to a set of equations for different combinations of i and k which constrains the form of the metric.

When $i = k = 0$, we get the condition $(\partial_\alpha g_{00}) \, \epsilon^{\alpha\beta} x_\beta = 0$; that is,

$$(\partial_\alpha \, g_{00}) \, x_\beta = (\partial_\beta \, g_{00}) \, x_\alpha, \qquad (7.2)$$

which requires $g_{00} = g_{00}(t, r)$. Next, in the case of $i = 0, k \neq 0$, we get the condition

$$g_{0\alpha} \, \epsilon^\alpha{}_\beta + (\partial_\alpha g_{0\beta}) \, \epsilon^{\alpha\gamma} \, x_\gamma = 0, \qquad (7.3)$$

which shows that

$$g_{0\beta} \, dx^\beta \, dt = a_1(r, t) \, x^\beta \, dx_\beta \, dt = f(r, t) dr \, dt, \qquad (7.4)$$

where a_1 and f are arbitrary functions. This fixes the form of the off-diagonal term. When $i \neq 0$ and $k \neq 0$, we get the condition

$$g_{\alpha\gamma} \, \epsilon^\gamma_\beta + g_{\delta\beta} \, \epsilon^\delta_\alpha + (\partial_\delta g_{\alpha\beta}) \, \epsilon^{\delta\gamma} \, x_\gamma = 0, \qquad (7.5)$$

which requires that

$$g_{\alpha\beta} \, dx^\alpha dx^\beta = \left[a_2(r, t) \delta^{\alpha\beta} + a_3(r, t) x^\alpha x^\beta \right] dx_\alpha dx_\beta \qquad (7.6)$$

with a_2 and a_3 being arbitrary functions. Using $\delta^{\alpha\beta} dx_\alpha dx_\beta = |d\mathbf{x}|^2$ and $x^\alpha dx_\alpha = r dr$ and combining with the previous results, we conclude that the line element can be written in the form

$$ds^2 = -A(t, r) \, dt^2 + B(t, r) \, dtdr + C(t, r) \, dr^2 + D(t, r) \left(d\theta^2 + \sin^2 \theta \, d\phi^2 \right), \qquad (7.7)$$

where A, B, C, D are arbitrary functions of r and t. This result is intuitively obvious but the technique based on the group of motions induced by the relevant Killing vectors has a wider domain of applicability.

One can simplify this metric further by using the freedom in the choice of coordinates. First, one can transform to a new radial coordinate $\bar{r}$ using $\bar{r}^2 \equiv D(t, r)$. This will change the explicit forms of A, B and C but, since they are arbitrary, we will continue to denote the new functions by the same symbols. This reduces the form of the line element to

$$ds^2 = -A(t, \bar{r}) \, dt^2 + B(t, \bar{r}) \, dtd\bar{r} + C(t, r) \, d\bar{r}^2 + \bar{r}^2 \left(d\theta^2 + \sin^2 \theta \, d\phi^2 \right). \qquad (7.8)$$

Next, we can remove the cross term $dtd\bar{r}$ by transforming to a new time coordinate $\bar{t}$ determined by the equation

$$d\bar{t} = \Phi(t, \bar{r}) \left[A(t, \bar{r}) \, dt - \frac{1}{2} B(t, \bar{r}) \, d\bar{r} \right]. \qquad (7.9)$$

The integrating factor Φ on the right hand side ensures that $\bar{t}$ is an exact differential and hence such a coordinate transformation is always possible (at least locally). Squaring the above expression and simplifying, we find that

$$A \, dt^2 - B \, dt \, d\bar{r} = \frac{1}{A\Phi^2} \, d\bar{t}^2 - \frac{B}{4A} \, d\bar{r}^2, \tag{7.10}$$

which shows that the cross term has been successfully eliminated. Defining two new functions

$$\bar{A} = \frac{1}{A\Phi^2}; \qquad \bar{B} = C + \frac{B}{4A} \tag{7.11}$$

we can express the line element with only g_{tt} and g_{rr} being nontrivial. At this stage, we shall drop the 'bar' on all the functions and coordinates for notational convenience and take the metric to be

$$ds^2 = -A(t,r) \, dt^2 + B(t,r) \, dr^2 + r^2 \left(d\theta^2 + \sin^2 \theta \, d\phi^2 \right). \tag{7.12}$$

In this metric, the two-dimensional surface with $r = $ constant, $t = $ constant has the standard line element $dl^2 = r^2 d\Omega^2$ of the 2-sphere with the proper area $\mathcal{A} = 4\pi r^2$. We can therefore identify θ and ϕ with the standard angular coordinates on a 2-sphere and think of r as being given by $r = (\mathcal{A}/4\pi)^{1/2}$, where $\mathcal{A}$ is the proper area of the $t = $ constant, $r = $ constant surface. Thus three of the coordinates r, θ and ϕ have natural interpretations.

There is a caveat to the naturalness of the interpretation of the r-coordinate. For the moment we are assuming that $A > 0$ and $B > 0$ so that $t = $ constant surfaces are spacelike and $r = $ constant surfaces are timelike. When this fails, the interpretations of the coordinates r and t will become more complicated; we will have occasion to comment on this later in Chapter 8.

So far we have used only the spherical symmetry to determine the form of the metric. Our next task is to compute R_{ik} for the above metric in terms of A and B, equate it to $8\pi G[T_{ik} - (1/2)g_{ik}T]$ and solve the resulting equations to determine A and B for different kinds of sources.

Exercise 7.1

A reduced action principle for spherical geometry The spherically symmetric metric in Eq. (7.12) has two free functions $A(t,r)$ and $B(t,r)$. Introduce two other functions $a(t,r) = (1/B)$ and $b(t,r) = \sqrt{AB}$, thereby reducing the metric to the form

$$ds^2 = -a(r,t)b(r,t)^2 dt^2 + \frac{dr^2}{a(r,t)} + r^2 d\Omega_{D-2}^2 \tag{7.13}$$

where, keeping future applications in mind, we are working in a D-dimensional spacetime with $d\Omega_{D-2}^2$ giving the line element on the $D - 2$-dimensional sphere.

(a) Express the Einstein–Hilbert action (with a D-dimensional cosmological constant λ)

$$I = \int d^D x \sqrt{-g}(R - 2\lambda) \tag{7.14}$$

in terms of a and b. Show that the resulting action is equivalent to the one which can be expressed in the form

$$I = (D-2) \int_0^\infty dr \, r^{D-3} \left(a - 1 + \frac{2\lambda r^2}{D-2} \right) b', \tag{7.15}$$

where the prime denotes derivatives with respect to r and suitable integration of second derivative terms have been performed.

(b) Vary this action to obtain the equations of motion for a and b and show that the solutions have the form

$$a(r) = 1 - \frac{2M}{r^{D-3}} - \frac{2\lambda}{D-2} r^2; \qquad b(r) = \text{constant} \tag{7.16}$$

for $D > 3$ while one gets $a = k - 2\lambda r^2$, $b = \text{constant}$ for $D = 3$. Here M and k are integration constants.

(c) In the next section we shall write down the full Einstein equations for the spherically symmetric metric and solve them. It turns out that the solution is identical to the one obtained above in which the action is first 'reduced' and then varied. Do you think this procedure is always justified? If not, when do you expect the procedure to work? [Hint. The following expression for the scalar curvature may be of use

$$R = a'' + \frac{2ab''}{b} + \frac{3a'b'}{b} + 2(D-2)\frac{1}{r}\left(a' + \frac{ab'}{b} \right) + (D-2)(D-3)\frac{1}{r^2}(a-1)$$

$$+ \frac{1}{a^2 b^2}\left(\ddot{a} - 2\frac{\dot{a}^2}{a} - \frac{\dot{a}\dot{b}}{b} \right) \tag{7.17}$$

in obtaining the form of action.]

7.2.1 Static geometry and Birkoff's theorem

The first case we want to study corresponds to a static, spherically symmetric source for which the metric coefficients A and B are independent of time. So, we take $A = A(r)$ and $B = B(r)$.

To compute R_{ik} for this metric, we first need to compute the Christoffel symbols. When this is done rather blindly, several of the Christoffel symbols will turn out to be zero after laborious computation. A simpler way to obtain the relevant *nonzero* Christoffel symbols is to proceed as follows. We saw in Section 4.5.2 (see Eq. (4.70)) that the square of the standard Lagrangian $L^2 = -g_{ab}\dot{x}^a\dot{x}^b$ will lead to the same geodesic equation as L when proper time is used as the affine parameter. Since the geodesic equation has the form $\ddot{x}^a = -\Gamma^a_{bc}\dot{x}^b\dot{x}^c$, one can read off the nonzero Christoffel symbols from the geodesic equation. Explicitly,

$$L^2 = -g_{mn}\dot{x}^m\dot{x}^n = A\dot{t}^2 - B\dot{r}^2 - r^2\left(\dot{\theta}^2 + \sin^2\theta\,\dot{\phi}^2\right) \qquad (7.18)$$

with the Euler–Lagrange equation:

$$\frac{d}{d\tau}\left(\frac{\partial L^2}{\partial\dot{x}^m}\right) - \frac{\partial L^2}{\partial x^m} = 0. \qquad (7.19)$$

Consider, for example, the case of $m = 0$ when this equation reduces to $d(A\dot{t})/d\tau = 0$; this immediately gives the result $\Gamma^t_{tr} = \Gamma^t_{rt} = A'/2A$ with other $\Gamma^t_{ab} = 0$. Working out the other components, using the form of L^2 in Eq. (7.18), we find that the non-vanishing Christoffel symbols are given by

$$\Gamma^r_{rr} = \frac{B'}{2B}, \quad \Gamma^r_{tt} = \frac{A'}{2B}, \quad \Gamma^r_{\theta\theta} = -\frac{r}{B}, \quad \Gamma^r_{\phi\phi} = -\frac{r\sin^2\theta}{B},$$

$$\Gamma^\theta_{\theta r} = \Gamma^\phi_{\phi r} = \frac{1}{r}, \quad \Gamma^t_{tr} = \frac{A'}{2A}, \quad \Gamma^\theta_{\phi\phi} = -\sin\theta\cos\theta, \quad \Gamma^\phi_{\phi\theta} = \cot\theta, \qquad (7.20)$$

where a prime denotes differentiation with respect to r. Given the Christoffel symbols, we can directly calculate the Ricci tensor R_{mn} using the formula

$$R_{mn} = R^l_{mln} = \partial_l\Gamma^l_{mn} - \partial_n\Gamma^l_{ml} + \Gamma^l_{lr}\Gamma^r_{mn} - \Gamma^l_{nr}\Gamma^r_{ml}. \qquad (7.21)$$

Once again, it is easy to prove from symmetry considerations that several of the components will vanish. For example, the metric is invariant under $\theta \to -\theta$ or $\phi \to -\phi$. On the other hand, a coordinate transformation $\theta \to -\theta$ will change the sign of $R_{r\theta}$, requiring that $R_{r\theta} = 0$. By similar reasoning, we can easily conclude that

$$R_{rt} = R_{r\theta} = R_{r\phi} = R_{t\theta} = R_{t\phi} = R_{\theta\phi} = 0, \qquad (7.22)$$

etc. Consider next a coordinate transformation on the 2-sphere given by $(\theta, \phi) \to (\theta', \phi')$. The line interval on the unit sphere changes by

$$d\theta^2 + \sin^2\theta\,d\phi^2 = \left[\left(\frac{\partial\theta}{\partial\theta'}\right)^2 + \sin^2\theta\left(\frac{\partial\phi}{\partial\theta'}\right)^2\right]d\theta'^2 + \cdots. \qquad (7.23)$$

The line element will remain invariant if we satisfy the condition

$$\left(\frac{\partial\theta}{\partial\theta'}\right)^2 + \sin^2\theta\left(\frac{\partial\phi}{\partial\theta'}\right)^2 = 1. \qquad (7.24)$$

On the other hand, $R_{\theta\theta}$ transforms as $R_{\theta'\theta'} = (\partial\theta/\partial\theta')^2\,R_{\theta\theta} + (\partial\phi/\partial\theta')^2\,R_{\phi\phi}$; so demanding $R_{\theta'\theta'} = R_{\theta\theta}$ and using the condition Eq. (7.24) we get

$$R_{\theta\theta} = R_{\theta\theta}\left(1 - \sin^2\theta\left(\frac{\partial\phi}{\partial\theta'}\right)^2\right) + \left(\frac{\partial\phi}{\partial\theta'}\right)^2 R_{\phi\phi}, \qquad (7.25)$$

which requires

$$R_{\phi\phi} = \sin^2\theta\, R_{\theta\theta}. \tag{7.26}$$

Thus, we only need to compute the components R_{tt}, R_{rr} and $R_{\theta\theta}$, which is a fairly simple job. We get

$$R_{tt} = \frac{A''}{2B} - \frac{A'}{4B}\left(\frac{A'}{A} + \frac{B'}{B}\right) + \frac{A'}{rB}$$

$$R_{rr} = -\frac{A''}{2A} + \frac{A'}{4A}\left(\frac{A'}{A} + \frac{B'}{B}\right) + \frac{B'}{rB}$$

$$R_{\theta\theta} = 1 - \frac{1}{B} - \frac{r}{2B}\left(\frac{A'}{A} - \frac{B'}{B}\right). \tag{7.27}$$

To proceed further, we have to specify the form of the source energy-momentum tensor T_{ab}. This tensor should, of course, be consistent with the symmetries of R_{ik} obtained above which requires the source to be static and respect the spherical symmetry. Let us first consider the most important case in which the T_{ab} of the source is nonzero only within a radius $r = R$ so that we can assume $T_{ab} = 0$ for $r > R$. We will first begin by determining the form of the metric *outside* the source where $R_{ik} = 0$. Using Eq. (7.27) we notice that:

$$BR_{tt} + AR_{rr} = \frac{1}{rB}\left(A'B + B'A\right) = \frac{1}{rB}\frac{d}{dr}(AB). \tag{7.28}$$

The vanishing of the left hand side requires AB to be a constant everywhere. At very large distances from the source, we expect the spacetime to approach the flat spacetime so that $A \to 1$ and $B \to 1$ as $r \to \infty$. Hence this constant AB has the value unity. Therefore,

$$B(r) = \frac{1}{A(r)}. \tag{7.29}$$

The explicit form of A or B can be easily determined from any of the other equations in Eq. (7.27); for example, demanding $R_{\theta\theta} = 0$ leads to $A - 1 + rA' = 0$ which has the solution

$$A(r) = 1 + \frac{C}{r}, \tag{7.30}$$

where C is a constant related to the properties of the source located at $r < R$. But we know that, far away from the source, where the metric is close to that of flat spacetime, $g_{00} \to -(1 - 2GM/r)$ with M being the total mass energy of the system. Therefore we must have $C = -2GM$. We conclude that the metric outside a static spherically symmetric source, located within a radius $r = R$, is given by

$$ds^2 = -\left(1 - \frac{2GM}{r}\right)dt^2 + \left(1 - \frac{2GM}{r}\right)^{-1}dr^2 + r^2d\Omega^2 \quad (\text{for } r \geq R). \tag{7.31}$$

This metric, called the *Schwarzschild metric*, describes the spacetime in the empty region around a spherically symmetric source.

We note that the above metric can be expressed in the form

$$ds^2 = -f(r)dt^2 + \frac{dr^2}{f(r)} + r^2 d\Omega^2 \quad \text{(for } r \geq R\text{)}, \tag{7.32}$$

where $f(r)$ has a simple zero at $r = 2GM$ and has the explicit form $f(r) = [1 - (2GM/r)]$. It turns out that (see, for example, Exercise 7.3 and Exercise 7.4) a wide class of spherically symmetric solutions to Einstein's equations has this generic form with $g_{tt}(r) = -1/g_{rr}(r)$ with the function $f(r)$ having simple zeros at one or more values of r. All these metrics share several common properties which we will be studying in Chapter 8. To save the algebraic effort, we will very often work with Eq. (7.32) and will introduce the specific form of $f(r)$ only when required. In this manner many of the results that we obtain in this chapter are generalizable to a much wider class of spacetimes.

As a simple example, consider the question of expressing the metric in Eq. (7.32) such that the spatial form of the metric is conformal to the flat spatial line element $d\rho^2 + \rho^2 d\Omega^2 = |d\boldsymbol{x}|^2$. If we make a transformation from the coordinate r to a coordinate ρ through a function $r = r(\rho)$, then the spatial part of the metric in Eq. (7.32) becomes $(r'^2/f)d\rho^2 + r^2(\rho)d\Omega^2$, where a prime denotes differentiation with respect to ρ. If we choose $r(\rho)$ such that $r'/\sqrt{f} = r/\rho$ then the spatial part of the metric will become $(r/\rho)^2[d\rho^2 + \rho^2 d\Omega^2]$, which is conformal to the flat spatial line element. This equation can be easily integrated to give

$$\ln\left(\frac{\rho}{\rho_0}\right) = \int_{r_0}^{r} \frac{d\bar{r}}{\bar{r}\sqrt{f(\bar{r})}}, \tag{7.33}$$

where ρ_0 and r_0 are arbitrary constants. In the case of the Schwarzschild metric, this integrates to give

$$r = \rho\left(1 + \frac{GM}{2\rho}\right)^2 \tag{7.34}$$

with the integration constant chosen such that $\rho \to r$, at spatial infinity. The metric now becomes

$$ds^2 = -\left[\left(1 - \frac{GM}{2\rho}\right)^2\left(1 + \frac{GM}{2\rho}\right)^{-2}\right]dt^2 + \left(1 + \frac{GM}{2\rho}\right)^4(d\rho^2 + \rho^2 d\Omega^2). \tag{7.35}$$

We have already encountered this result in Exercise 4.8 (see Eq. (4.81)). Such a spatial coordinate system is called isotropic.

It is clear from Eq. (7.31) that the nature of the Schwarzschild metric depends crucially on the length scale (called the Schwarzschild radius or *gravitational radius*), given by $r_g \equiv 2GM/c^2 \approx 3\,\mathrm{km}(M/M_\odot)$ in normal units. The signs of g_{tt}, g_{rr} change at $r = r_g$. Note, however, that the Schwarzschild metric is valid only in the region with $T_{ab} = 0$ – that is, for the region outside the spherical source at $r > R$. If $r_g < R$, then the Schwarzschild metric in Eq. (7.31) is not valid at $r = r_g$ and we need not worry about these issues. For most of the discussion in this chapter (except when stated to the contrary), we will assume that this is the case; i.e. $r_g < R$ which, for example, is very well satisfied for a star like the Sun. When the size of the source is smaller than its gravitational radius, new physical phenomena are indicated which will be discussed in Chapter 8. Similar comments apply to the metric in Eq. (7.32) whenever $f(r) = 0$ at some $r = a$.

The metric in Eq. (7.31) was obtained by making three assumptions: (i) spherical symmetry, (ii) time independence and (iii) $T_{ab} = 0$. Interestingly, it turns out that the form of the metric remains valid even if we *do not* assume time independence but make the other two assumptions. We shall briefly describe how this comes about, leaving the details to Exercise 7.5 below.

Spherical symmetry alone leads us to the form of the metric in Eq. (7.12). If we assume that the metric is given by Eq. (7.12) with A and B depending on both r and t, then the equations $G_{ik} = 0$ reduces to (see Eq. (7.51) in Exercise 7.5 below):

$$\frac{1}{B}\left(\frac{A'}{rA} + \frac{1}{r^2}\right) - \frac{1}{r^2} = 0, \qquad \frac{1}{B}\left(\frac{B'}{rB} - \frac{1}{r^2}\right) + \frac{1}{r^2} = 0, \qquad \dot{B} = 0. \quad (7.36)$$

The condition $\dot{B} = 0$ (which arises from the form of $R^r_t = \dot{B}/Br$) shows that B depends only on r. Adding the other two equations, it is easy to see that $\partial_r(AB) = 0$. Therefore, $A = q(t)/B(r)$ where q is a function of t alone and the metric becomes:

$$ds^2 = -(q(t)/B(r))\,dt^2 + B(r)\,dr^2 + r^2\,(d\theta^2 + \sin^2\theta\,d\phi^2). \quad (7.37)$$

But note that we are still allowed to make a change of variables from t to any other $t' = F(t)$ which is an arbitrary function of t. If we take $dt' = \sqrt{q}dt$ the metric reduces to

$$ds^2 = -(1/B(r))\,dt'^2 + B(r)\,dr^2 + r^2\,(d\theta^2 + \sin^2\theta\,d\phi^2). \quad (7.38)$$

That is, we can set $q(t) = 1$ and $A = 1/B(r)$ (making A also independent of t) even in this case. Using this result in any of Eq. (7.36) shows that $A(r) = (1+C/r)$ as before so that we recover the metric in Eq. (7.31).

This result, known as *Birkoff's theorem*, shows that the metric we obtained in Eq. (7.31) is valid even for a *time dependent* but spherically symmetric source. As an example, consider a star which is expanding and contracting radially in a spherically symmetric manner thereby undergoing radial pulsations. The outside metric will still be static and given by Eq. (7.31).

There is an interesting corollary to this result. Consider a spherically symmetric distribution of matter with T_{ab} being nonzero only in an annular region $R_1 < r < R_2$. The metric at $r > R_2$ is, of course, given by the one in Eq. (7.31) as demonstrated above. But the region inside the spherical shell $r < R_1$ is also spherically symmetric and empty. Hence Birkoff's theorem, leading to Eq. (7.30), should be applicable in this region as well with some constant C. For the region outside the shell, we considered the $r \to \infty$ limit to determine the value of C. For the region inside the shell, if the metric has to be finite and well behaved at the origin, we must have $C = 0$. For the same reasons explained earlier the same result holds even when the shell is moving in a spherically symmetric manner; the result only depends on spherical symmetry and $R_{ik} = 0$ and we need not separately assume that metric is independent of time.

Thus we get the – rather surprising – result that the region inside a spherically symmetric shell is flat spacetime. This result is true in Newtonian gravity as well but there are two significant differences between the general relativistic result and the corresponding one in Newtonian theory. First, in Newtonian theory it is possible to have very nontrivial distributions of matter (not necessarily spherically symmetric) that will also lead to a vanishing gravitational field within a cavity. This does not seem to be the case in general relativity. Second, the Newtonian result for a spherical shell is often proved by using the fact that the Newtonian force at an inside point due to an element of shell, at a distance l, varies as $(1/l^2)$ while the amount of matter dm in a region in the shell, subtending a solid angle $d\Omega$ at a point, increases as l^2. While the latter result $(dm \propto l^2 d\Omega)$ continues to hold in general relativity for spherically symmetric spacetime, it is certainly not true that the force falls strictly as $(1/r^2)$. In addition, superposition of forces due to the material located at different regions of the shell is not allowed in general relativity. In view of all these, the general relativistic result is somewhat intriguing. We will see later that several such peculiar features – in which Newtonian results carry over to general relativity – arise in the case of spherically symmetric geometries.

Exercise 7.2

Superposition in spherically symmetric spacetimes We explained in the text that Einstein's equations are nonlinear and solutions cannot be superposed. A surprising exception to this result is in the case of spherically symmetric spacetimes of a particular kind. This exercise explores this feature.

(a) Consider a metric of the form

$$ds^2 = -f(r)dt^2 + f(r)^{-1}dr^2 + r^2(d\theta^2 + \sin^2\theta d\phi^2) \tag{7.39}$$

with a general function $f(r)$ that needs to be determined via Einstein's equations. Show that this metric will satisfy Einstein's equations provided the the source energy-momentum tensor has the form

$$T_t^t = T_r^r = -\frac{\epsilon(r)}{8\pi G}; \quad T_\theta^\theta = T_\phi^\phi = -\frac{\mu(r)}{8\pi G}. \tag{7.40}$$

Equation (7.40) also defines the functions $\epsilon(r)$ and $\mu(r)$.
   (b) Show that the Einstein equations now reduce to:

$$\frac{1}{r^2}(1-f) - \frac{f'}{r} = \epsilon; \quad \nabla^2 f = -2\mu. \tag{7.41}$$

The remarkable feature about the metric in (7.39) is that Einstein's equations become linear in $f(r)$ so that solutions for different $\epsilon(r)$ can be superposed.
   (c) Given any $\epsilon(r)$, integrate the Eq. (7.41) to determine the solution to be

$$f(r) = 1 - \frac{a}{r} - \frac{1}{r}\int_a^r \epsilon(r)r^2\,dr \tag{7.42}$$

with a being an integration constant chosen such that $f = 0$ at $r = a$ and $\mu(r)$ is fixed by $\epsilon(r)$ through

$$\mu(r) = \epsilon + \frac{1}{2}r\epsilon'(r). \tag{7.43}$$

Exercise 7.3

Reissner–Nordstrom metric A simple generalization of the Schwarzschild metric occurs in the case of a spherically symmetric source carrying a spherically symmetric distribution of electric charge Q on it. Though the charge and mass are confined to a region $r < R$, the electric field due to the charge and its *energy density* will exist all over the place.
   (a) Show that, in the notation of the previous problem, this will correspond to $\epsilon = (Q^2/r^4)$ due to the electric field in suitable units. Show that, in this case,

$$f(r) = \left(1 - \frac{2GM}{r} + \frac{GQ^2}{r^2}\right) \tag{7.44}$$

which gives the relevant metric called the *Reissner–Nordstrom metric*.
   (b) Also show that, in this case, $\mu = -(Q^2/r^4)$. Explain the origin of this term from the electromagnetic energy-momentum tensor.
   (c) The above analysis shows that one could superpose the solutions due to mass and charge in Eq. (7.41). But to do this, we have assumed that $g_{00} = -(1/g_{rr})$. Relax this assumption and start with the more general spherically symmetric metric Eq. (7.12). Find the general form of an electromagnetic field which is static and spherically symmetric and show that it can be described with a single nonzero component F_{tr}. Solve the Maxwell equations as well as the Einstein equations in this spacetime and show that the metric is indeed given by Eq. (7.39) with Eq. (7.44). This is the generalization of Birkoff's theorem to a source with charge Q. Also show that $F_{tr} \propto (1/r^2)$.

Exercise 7.4

Spherically symmetric solutions with a cosmological constant Consider Einstein's equations with $\Lambda \neq 0$ but with $T_{ab} = 0$.

(a) In the spherically symmetric case, show that this is equivalent to the situation in Exercise 7.2 with $\epsilon = \Lambda =$ constant. Also show that this requires $\mu = \Lambda$ and $T_b^a = -(\Lambda/8\pi G)\delta_b^a$.

(b) Integrate Eq. (7.41) to obtain the metric to be

$$ds^2 = -\left(1 - \frac{1}{3}\Lambda r^2\right)dt^2 + \left(1 - \frac{1}{3}\Lambda r^2\right)^{-1}dr^2 + r^2(d\theta^2 + \sin^2\theta d\phi^2). \qquad (7.45)$$

This spacetime (for $\Lambda > 0$) is called the *de Sitter spacetime* and it will play a crucial role in our discussions in Chapters 10 and 14.

(c) Combine with the results of the previous problem and obtain the metric due to a charged, massive, spherically symmetric source to be of the form in Eq. (7.39) with

$$f(r) = \left(1 - \frac{2GM}{r} + \frac{GQ^2}{r^2} - \frac{1}{3}\Lambda r^2\right). \qquad (7.46)$$

Such is the power of superposition!

(d) The Reissner–Nordstrom metric can also be obtained by varying a 'reduced action' just as in the case of Exercise 7.1. We will again work in D dimensions for generality. One has to add to the action the term coming from electromagnetic Lagrangian, $-(1/16\pi)F_{ik}F^{ik}$, evaluated for a vector potential $A_i = [\phi(r), 0, 0, 0, \ldots]$. Show that the reduced action in this case is equivalent to

$$I = \int_0^\infty dr \left[r^{D-3}\left(a - 1 + \frac{2\lambda}{D-2}\right)b' - \frac{1}{2b}r^{D-2}\phi'^2\right]. \qquad (7.47)$$

Vary the action with respect to ϕ, a, b to obtain the equations of motion. Show that the solutions can be expressed in the form

$$\phi \propto \frac{1}{r^{D-3}}; \qquad a(r) = 1 - \frac{2M}{r^{D-3}} - \frac{2\lambda}{D-2}r^2 + \frac{Q^2}{r^{2D-6}}; \qquad b = \text{constant}, \qquad (7.48)$$

where M and Q are integration constants.

Exercise 7.5
Time dependent spherically symmetric metric Consider a metric of the form

$$ds^2 = -e^\nu dt^2 + e^\lambda dr^2 + r^2(d\theta^2 + \sin^2\theta d\phi^2), \qquad (7.49)$$

where λ and ν are functions of both t and r.

(a) Work out the nonzero components of G_{ik} and verify Eq. (7.36).

[Answer. For the sake of reference we give the Christoffel symbols and the components of the Einstein tensor. The non-vanishing Christoffel symbols are

$$\Gamma_{11}^1 = \frac{\lambda'}{2}, \quad \Gamma_{10}^0 = \frac{\nu'}{2}, \quad \Gamma_{33}^2 = -\sin\theta\cos\theta,$$

$$\Gamma_{11}^0 = \frac{\dot\lambda}{2}e^{\lambda-\nu}, \quad \Gamma_{22}^1 = -re^{-\lambda}, \quad \Gamma_{00}^1 = \frac{\nu'}{2}e^{\nu-\lambda},$$

$$\Gamma_{12}^2 = \Gamma_{13}^3 = \frac{1}{r}, \quad \Gamma_{23}^3 = \cot\theta, \quad \Gamma_{00}^0 = \frac{\dot\nu}{2},$$

$$\Gamma_{10}^1 = \frac{\dot\lambda}{2}, \quad \Gamma_{33}^1 = -r\sin^2\theta\,e^{-\lambda}. \qquad (7.50)$$

Here the prime denotes the derivative with respect to r and the dot denotes differentiation with respect to t. The components of G^i_k can be computed in a straightforward manner and we find

$$G^1_1 = e^{-\lambda}\left(\frac{\nu'}{r} + \frac{1}{r^2}\right) - \frac{1}{r^2},$$

$$G^2_2 = G^3_3 = \frac{1}{2}e^{-\lambda}\left(\nu'' + \frac{\nu'^2}{2} + \frac{\nu' - \lambda'}{r} - \frac{\nu'\lambda'}{2}\right) - \frac{1}{2}e^{-\nu}\left(\ddot{\lambda} + \frac{\dot{\lambda}^2}{2} - \frac{\dot{\lambda}\dot{\nu}}{2}\right)$$

$$G^0_0 = e^{-\lambda}\left(\frac{1}{r^2} - \frac{\lambda'}{r}\right) - \frac{1}{r^2},$$

$$G^1_0 = e^{-\lambda}\frac{\dot{\lambda}}{r}. \tag{7.51}$$

All other components vanish identically.]

(b) We showed in Exercise 5.9 how the curvature tensor R^{ab}_{cd} can be represented in a matrix form. Work out the structure of this matrix for this metric. [Hint. Make full use of the symmetries.]

Exercise 7.6

Schwarzschild metric in a different coordinate system Show that the following metric is actually the Schwarzschild metric in a different coordinate system by providing an explicit coordinate transformation to the standard Schwarzschild form of the metric:

$$ds^2 = -c^2 dT^2 + \frac{4}{9}\left[\frac{9GM}{2(R - cT)}\right]^{2/3} dR^2 + \left[\frac{9GM}{2}(R - cT)^2\right]^{2/3} d\Omega^2. \tag{7.52}$$

Consider observers located at fixed spatial coordinates. Show that they are on a free fall trajectory starting with zero velocity at infinite distances and that T represents their proper time. This provides a simple interpretation of this coordinate system.

7.2.2 Interior solution to the Schwarzschild metric

The Schwarzschild metric in Eq. (7.31) and its properties will occupy most of our attention in the later sections of this chapter. However, before we do that, we will complete the picture by analysing the form of the metric in the region occupied by the source. The simplest kind of source one can assume is an ideal fluid with the energy-momentum tensor given by $T_{mn} = (\rho + p)u_m u_n + p g_{mn}$. Since the metric is diagonal and the off-diagonal components of the Ricci tensor vanish, it follows that $u_0 u_\alpha = 0$, requiring $u_\alpha = 0$. Hence the spatial three-velocity of the fluid must vanish identically, as to be expected for a source which is static and in a state of hydrostatic equilibrium. Combining with $u_a u^a = -1$, it follows that the four-velocity has the components $[u_m] = \sqrt{A}\,(1, 0, 0, 0)$. We can now evaluate the form of T_{mn} explicitly and write down the Einstein equations. Simple calculation shows that the nontrivial equations are

$$R_{tt} = 4\pi A(\rho + 3p); \qquad R_{rr} = 4\pi B(\rho - p); \qquad R_{\theta\theta} = 4\pi r^2(\rho - p). \tag{7.53}$$

Using the components in Eq. (7.27) and manipulating these equations, it is easy to obtain

$$\frac{d}{dr}\left[r\left(1-\frac{1}{B}\right)\right]=8\pi\,r^2\,\rho. \tag{7.54}$$

Integrating this, we find that $B(r)$ is given by

$$B(r)=\left[1-\frac{2Gm(r)}{r}\right]^{-1}, \tag{7.55}$$

where we have defined

$$m(r)=4\pi\int_0^r \rho(\bar{r})\,\bar{r}^2\,d\bar{r} \tag{7.56}$$

with the boundary condition $m(0)=0$, which is needed to ensure that $B(0)$ is finite. This determines $B(r)$ in terms of the energy density distribution of the source. This equation can also be written in the form

$$\frac{dm}{dr}=4\pi r^2\rho(r) \tag{7.57}$$

with the boundary condition $m(0)=0$.

The following feature needs to be noted regarding the integral in Eq. (7.56). The proper three-volume element in the $t=$ constant surface is *not* $r^2\sin\theta dr d\theta d\phi$ but $\sqrt{B(r)}\,r^2\sin\theta dr d\theta d\phi$. Therefore, the *proper* integrated mass contained in a sphere of radius r is

$$\epsilon(r)\equiv 4\pi\int_0^r \rho(\bar{r})\sqrt{B(\bar{r})}\,\bar{r}^2\,d\bar{r}=4\pi\int_0^r \rho(\bar{r})\left[1-\frac{2Gm(\bar{r})}{\bar{r}}\right]^{-1/2}\bar{r}^2\,d\bar{r}, \tag{7.58}$$

which is different from $m(r)$ defined in Eq. (7.56); in fact, $m(r)<\epsilon(r)$. If the source extends to a region of radius R then $B(R)=1-2GM/R$ with $m(R)=M$ being identified with the mass which appears in the external Schwarzschild metric so that the $B(r)$ in the external and internal metrics match at $r=R$. Obviously, $E\equiv\epsilon(R)$ will not be equal to M. The difference $\Delta E\equiv E-M$ can be interpreted as the gravitational binding energy of the source.

The equation for A, obtained by substituting the form of $B(r)$ in any of Eq. (7.53), can be simplified to give

$$\frac{A'}{A}=-\frac{2p'}{\rho+p}. \tag{7.59}$$

Given $\rho(r)$ and $p(r)$, Eq. (7.59), Eq. (7.57) and Eq. (7.55) can be used to determine the metric coefficients $A(r)$ and $B(r)$. For practical purposes, however, it is better

Spherically symmetric geometry

to rewrite these equations as equations for $\rho(r)$ and $p(r)$. This is easily achieved as follows. The equation for the $\theta\theta$ component has the explicit form

$$\frac{1}{B} - 1 + \frac{r}{2B}\left(\frac{A'}{A} - \frac{B'}{B}\right) = -4\pi(\rho - p)r^2. \tag{7.60}$$

Eliminating A'/A and B'/B using Eq. (7.59) and Eq. (7.55) and replacing $m'(r)$ using Eq. (7.57) we can obtain, after some simple algebra, the equation (called the *Oppenheimer–Volkoff equation*):

$$\frac{dp}{dr} = -\frac{G[\rho + p][m + 4\pi r^3 p]}{r[r - 2Gm]} = -\frac{G[\rho + (p/c^2)][m + (4\pi r^3 p/c^2)]}{r[r - (2Gm/c^2)]}, \tag{7.61}$$

where we have introduced the c-factors in the last part. The non-relativistic limit, obtained by taking $c \to \infty$ is the standard equation for pressure balance:

$$\frac{dp}{dr} \approx -\frac{G\rho m}{r^2}. \tag{7.62}$$

In the same limit, Eq. (7.56) can be thought of relating the mass density with total mass; Eq. (7.56), Eq. (7.62) and an equation of state $p = p(\rho)$ provide three equations for the three unknowns $p(r)$, $\rho(r)$ and $m(r)$.

Substituting Eq. (7.61) into Eq. (7.59), we can rewrite the latter as

$$\frac{A'}{A} = \frac{2G[m + 4\pi r^3 p]}{r[r - 2Gm]}. \tag{7.63}$$

The Eq. (7.57), Eq. (7.61) and Eq. (7.63) and an equation of state $p = p(\rho)$ provide four equations to determine the four functions $A(r), \rho(r), p(r), m(r)$ in the relativistic case; the $B(r)$ is determined purely algebraically in terms of $m(r)$ by Eq. (7.55). The initial condition for $m(r)$ must be $m = 0$ at $r = 0$; for Eq. (7.61), the initial condition can be specified as either the central pressure $p_c = p(0)$ or equivalently the central density. The situation regarding A is a bit more complicated since it is constrained by the boundary condition $A(R) = 1 - 2GM/R$ at $r = R$. To tackle these mixed boundary conditions, one can proceed as follows. We start by assuming some fiducial value for $A_c = A(0)$. Given p_c, A_c (and $m_c = 0$), we can integrate the equations forward from $r = 0$ until a point $r = R$, say, at which p vanishes, which should be taken as the surface of the source. (Since $dp/dr < 0$, it is clear that the pressure will decrease as we proceed outwards from the origin; the equation of state should be so chosen such that it decreases to zero at a *finite* radius.) The value of $m(R) \equiv M$ determines the total mass and fixes the value of $B(r)$ at the surface of the source to be $B(R) = (1 - 2GM/R)^{-1}$. The integration of the A'/A equation will lead to some value for A at the surface $r = R$. But we notice that Eq. (7.63) is invariant under scaling of A by a constant; that is, if $A(r)$ is a solution, so is $kA(r)$ where k is a constant. Using this freedom, we can ensure that $A(R) = (1 - 2GM/R)$. This will make the metric coefficients of the interior solution match smoothly with the Schwarzschild solution for $r > R$.

To illustrate the above procedure, we will consider an unrealistic but simple model for a source with a constant density ρ. This is unrealistic in the sense that matter has to be infinitely stiff to have constant density, which violates the principle of special relativity. The speed of sound in any ideal fluid is given by $c_s = (dp/d\rho)^{1/2}$; a constant density fluid will have $c_s \to \infty$ so that information can be transmitted across the fluid with infinite speed to maintain the constant density. Since relativity requires $c_s < c$, the model is clearly unrealistic; but it can be thought of as an approximation to more realistic models. The mathematical simplicity certainly makes it worth discussing.

When the density is constant, the mass function $m(r)$ is given by

$$m(r) = \begin{cases} \frac{4}{3}\pi\rho r^3 & \text{for} \quad r \le R \\ \frac{4}{3}\pi\rho R^3 \equiv M & \text{for} \quad r > R \end{cases}, \tag{7.64}$$

where R is the radius of the constant density 'star'. Equation (7.61) now becomes

$$\frac{dp}{dr} = -\frac{4\pi G}{3} r\,(\rho + p)(\rho + 3p)\left(1 - \frac{8\pi G}{3}\rho r^2\right)^{-1}. \tag{7.65}$$

This can be integrated by elementary means to give $p(r)$:

$$\frac{\rho + 3p}{\rho + p} = \frac{\rho + 3p_c}{\rho + p_c}\left(1 - \frac{8\pi G}{3}\rho r^2\right)^{1/2}, \tag{7.66}$$

where $p_c = p(0)$ is the central pressure. As we said before, the radius R of the star is determined by the condition $p(R) = 0$. This gives the radius in terms of the central pressure to be

$$R^2 = \frac{3}{8\pi G\rho}\left[1 - \left(\frac{\rho + p_c}{\rho + 3p_c}\right)^2\right]. \tag{7.67}$$

This equation, combined with Eq. (7.64), allows us to express the central pressure as well as the pressure at any radius in terms of the total mass M and the radius R of the star. We have:

$$p_c = \rho\,\frac{1 - (1 - 2GM/R)^{1/2}}{3(1 - 2GM/R)^{1/2} - 1}, \tag{7.68}$$

and

$$p(r) = \rho\,\frac{(1 - 2GMr^2/R^3)^{1/2} - (1 - 2GM/R)^{1/2}}{(1 - 2GM/R)^{1/2} - 3(1 - 2GMr^2/R^3)^{1/2}} \quad \text{(for } r \le R). \tag{7.69}$$

The expression for the central pressure in Eq. (7.68) diverges when $2GM/R \to 8/9$ so that we need to assume $2GM/R < 8/9$ for the solution to exist. (As we shall see below, this result can be proved more generally and is *not* an artefact of our assumption $\rho = $ constant.) As we explained earlier, we are already assuming that $2GM/R < 1$ so that the metric is well behaved at the gravitational radius but $2GM/R < (8/9)$ is a marginally tighter constraint. So, if the amount of mass inside a region of radius R is increased above a critical value, we cannot obtain a static interior solution in general relativity. This is in sharp contrast to Newtonian gravity in which, by increasing the pressure support, we can balance arbitrarily strong gravitational fields. The mathematical reason is clear from the structure of Eq. (7.61) which shows that – in general relativity – pressure also contributes on the right hand side as a source for the gravitational force. When we increase the pressure to support gravity, the gravitational force in Eq. (7.61) will also increase and – beyond a critical value – it becomes counterproductive.

To complete the solution, we have to determine $A(r)$ and $B(r)$ such that they match smoothly with the external Schwarzschild metric. This leads to the solutions

$$B(r) = \left(1 - \frac{2GMr^2}{R^3}\right)^{-1}; \quad A(r) = \frac{1}{4}\left[3\left(1 - \frac{2GM}{R}\right)^{1/2} - \left(1 - \frac{2GMr^2}{R^3}\right)^{1/2}\right]^2.$$
(7.70)

So Eq. (7.70), Eq. (7.64) and Eq. (7.69) provide the complete description of a constant density star.

We showed that, in the case of the constant density star, its mass and radius must satisfy the condition $2GM/R < 8/9$. This result continues to hold in a more general situation even for stars with varying density and we shall briefly describe a proof of this claim. Consider an interior solution for which: (a) $\rho(r) = 0$ for $r > R$; (b) the total mass obtained by integrating $4\pi r^2 \rho(r)$ between 0 and R has a fixed value M; (c) the metric coefficient $B(r)$ is nowhere singular so that $m(r) < r/2G$; and (d) $\rho'(r) \leq 0$ so that the density does not increase outwards. Using these conditions, we can show that the pressure will remain finite everywhere only if there is a bound on M/R.

To do this, we will require an equation that allows $A(r)$ to be calculated from $\rho(r)$ without having to determine $p(r)$. This can be obtained by manipulating the equations in a straightforward manner and eliminating $p(r)$. We get

$$A'' - \frac{A'}{2}\left(\frac{B'}{B} + \frac{A'}{A} + \frac{2}{r}\right) = \frac{A}{rB}\left[3B' - 16\pi G\rho r B^2\right]. \quad (7.71)$$

Substituting $A \equiv C^2$, this equation can be linearized in the form

$$\frac{d}{dr}\left[\frac{1}{r}\left(1 - \frac{2Gm(r)}{r}\right)^{1/2}\frac{dC(r)}{dr}\right] = G\left(1 - \frac{2Gm(r)}{r}\right)^{-1/2}\left(\frac{m(r)}{r^3}\right)'C(r). \quad (7.72)$$

The outer boundary conditions on $C(r)$ are

$$C(R) = \left[1 - \frac{2GM}{R}\right]^{1/2}; \qquad C'(R) = \frac{GM}{R^2}\left[1 - \frac{2GM}{R}\right]^{-1/2}. \qquad (7.73)$$

We can now obtain an upper bound on $C(0)$. The quantity $3m(r)/4\pi r^3$, which is the mean density within the radius r, cannot increase with r if $\rho(r)$ does not. Further, since $C(r)$ is positive, the right hand side of Eq. (7.72) is negative, leading to:

$$\frac{d}{dr}\left[\frac{1}{r}\left(1 - \frac{2Gm(r)}{r}\right)^{1/2}\frac{dC(r)}{dr}\right] \leq 0. \qquad (7.74)$$

The equality can be achieved only for a constant density configuration. Integrating Eq. (7.74) we get

$$C'(r) \geq \frac{GMr}{R^3}\left(1 - \frac{2Gm(r)}{r}\right)^{-1/2}. \qquad (7.75)$$

Integrating once again from $r = R$ to $r = 0$ and using the boundary condition in Eq. (7.73) on $C'(R)$, leads to

$$C(0) \leq \left[1 - \frac{2GM}{R}\right]^{1/2} - \frac{GM}{R^3}\int_0^R \frac{r\,dr}{[1 - (2Gm(r)/r)]^{1/2}}. \qquad (7.76)$$

The right hand side will increase when $m(r)$ decreases. For a fixed M and R and with a decreasing density distribution, the minimum value for $m(r)$ is obtained when $\rho(r)$ is a constant – for which $m(r) = Mr^3/R^3$. Using this in the integral, we get the bound

$$C(0) \leq \frac{3}{2}\left[1 - \frac{2GM}{R}\right]^{1/2} - \frac{1}{2}. \qquad (7.77)$$

But on the other hand, $C(0)$ must be positive. The condition $C(0) > 0$ gives

$$\frac{2GM}{R} < \frac{8}{9}, \qquad (7.78)$$

which is precisely the result obtained earlier for a uniform density star.

Exercise 7.7
Variational principle for pressure support Show that Eq. (7.61) can be derived by extremizing the expression

$$m(r) = 4\pi \int_0^r \rho r'^2\,dr' \qquad (7.79)$$

with respect to an adiabatic, Eulerian, variation in which the number of baryons

$$N = \int_0^R 4\pi r^2 n(r) \left(1 - \frac{2Gm(r)}{c^2 r}\right)^{-1/2} dr \qquad (7.80)$$

remains constant. [Hint. (i) Vary $m + \lambda N$ where λ is a Lagrangian multiplier and use the adiabatic condition $(\delta n/\delta \rho) = nc^2/(p + \rho c^2)$ after proving it. (ii) Solve for λ in terms of other variables in the resulting equation. (iii) Since λ is a constant, set its derivative to zero.]

Exercise 7.8

Internal metric of a constant density star Consider a star of mass M, radius R and constant density. We derived the form of the interior solution in the text. Show that the spatial line element on a $t =$ constant surface of this metric can be expressed in the form

$$dL^2 = \frac{Rc^2}{2GM} \left[d\chi^2 + \sin^2 \chi (d\theta^2 + \sin^2 \theta d\phi^2)\right]. \qquad (7.81)$$

Do you recognize the space described by the metric within the square bracket? Provide a physical reason for this result.

Exercise 7.9

Clock rates on the surface of the Earth Assume that Earth is a rigidly rotating perfect fluid in equilibrium. Suppose clocks are kept at different locations on the Earth's surface and their rates are compared with a standard clock. The Doppler shift due to the Earth's rotation and the redshift due to the Earth's gravitational field (taking into account the rotational deformation of the Earth's surface) will affect the clock rate. Calculate the clock rate at any given location on the surface of the Earth compared with the standard clock. [Hint. This is the easiest exercise in this book!]

Exercise 7.10

Metric of a cosmic string Consider a static, infinitely long, cylindrical configuration of matter described in cylindrical polar coordinates (t, r, ϕ, z).

(a) Show that the spacetime due to such a cylindrical distribution of matter can be described by a metric of the form

$$ds^2 = -dt^2 + dr^2 + b^2(r)d\phi^2 + dz^2 \qquad (7.82)$$

with a source energy-momentum tensor given by

$$[T^{mn}] = \text{diag}\,(\rho, 0, 0, -\rho). \qquad (7.83)$$

Note that the source has negative pressure along the cylindrical axis.

(b) Compute G_{ab} for the metric and show that the Einstein equation reduces to the condition

$$\frac{d^2 b}{dr^2} = -8\pi G \rho b. \qquad (7.84)$$

(c) A cosmic string is expected to have uniform density across the string so that $\rho(r) = \rho_0$ for $r < r_0$ and zero otherwise. Solve the Einstein equations and show that the metric that is well behaved everywhere is given by

$$ds^2 = -dt^2 + dr^2 + \left(\frac{\sin \lambda r}{\lambda}\right)^2 d\phi^2 + dz^2, \qquad (r \leq r_0) \qquad (7.85)$$

$$ds^2 = -dt^2 + dr^2 + \left[\frac{\sin \lambda r_0}{\lambda} + (r - r_0)\cos \lambda r_0\right]^2 d\phi^2 + dz^2, \qquad (r > r_0) \quad (7.86)$$

where $\lambda^2 = 8\pi G \rho_0 c^2$.

(d) The physically relevant case probably corresponds to $\lambda r_0 \ll 1$. In this case, show that the metric outside the string has the form

$$ds^2 = -dt^2 + dr^2 + \left(1 - \frac{8G\mu}{c^2}\right) r^2 d\phi^2 + dz^2, \qquad (7.87)$$

where $\mu = \rho_0 \pi r_0^2$ is the mass per unit length of the string. Is this spacetime curved or flat? Can you provide a physical interpretation for this metric?

Exercise 7.11
Static solutions with perfect fluids A spacetime is static if there exists a timelike Killing vector field $\boldsymbol{\xi}$ which is hypersurface orthogonal. Show that, if a static metric is generated by a perfect fluid, then the fluid four-velocity must be parallel to $\boldsymbol{\xi}$.

7.2.3 Embedding diagrams to visualize geometry

One way of visualizing the geometry of a spherical star described in the previous section is to depict it using an *embedding diagram*, which we will now describe. Since the spacetime is static, the geometry at any $t = $ constant slice is completely representative of the solution. Similarly, spherical symmetry implies that the geometry of all planar sections passing through the origin will be identical so that we can conveniently work with the $\theta = \pi/2$ equatorial plane. The two-dimensional geometry of the $t = $ constant $\theta = \pi/2$ surface is described by the metric:

$$ds^2 = [1 - 2Gm(r)/r]^{-1}dr^2 + r^2 d\phi^2 = \frac{dr^2}{f(r)} + r^2\,d\phi^2. \qquad (7.88)$$

We now want to construct a two-dimensional surface embedded in the ordinary three-dimensional flat space such that it is described by the above metric. Such a construction will help in visualizing the geometry.

We see that, if $m(r) = 0$ (that is $f = 1$) then the geometry in Eq. (7.88) is flat and we can think of it as the $z = 0$ plane in the standard (r, ϕ, z) cylindrical coordinates. Let us suppose that, when $m \neq 0$, the above metric represents a surface of revolution given by $z = z(r), 0 < \phi \leq 2\pi$. The induced metric on such a two-dimensional surface, embedded in three dimensions, is given by:

$$ds^2 = dz^2 + dr^2 + r^2 d\phi^2 = \left[1 + \left(\frac{dz(r)}{dr}\right)^2\right] dr^2 + r^2 \, d\phi^2. \qquad (7.89)$$

Comparing with Eq. (7.88), we find that the equation to the surface is related to $f(r)$ by

$$\left(\frac{dz(r)}{dr}\right)^2 + 1 = \frac{1}{f(r)}. \qquad (7.90)$$

Integrating this equation, we get

$$z(r) = \int_0^r dx \left[\frac{1 - f(x)}{f(x)}\right]^{1/2} = \int_0^r dx \left[\frac{x}{2Gm(x)} - 1\right]^{-1/2}, \qquad (7.91)$$

which allows us to construct the embedded surface for any model.

The actual form of the embedded surface in the interior of the star depends on the form of $f(r)$ but its behaviour outside the star is easy to determine. When $r > R$ in the above integral, we find that

$$z(r) = [8M(r - 2M)]^{1/2} + \text{constant} \qquad \text{(outside the star)}. \qquad (7.92)$$

At large radius, $z \propto r^{1/2}$. One can also determine its general form near the centre of the star, where we can approximate $m(r)$ as $(4\pi/3)\rho_c r^3$. We can integrate the equation for $z(r)$ in this case as well; introducing a variable $a = (3c^2/8\pi G\rho_c)^{1/2}$ (in normal units) we find that the surface is a segment of a sphere of radius a near the centre of the star:

$$[a - z(r)]^2 + r^2 = a^2; \qquad \text{(for } r \ll a). \qquad (7.93)$$

This result is exact for a constant density star and holds approximately near the origin for any other model.

Such an embedding diagram is shown in Fig. 7.1. For any realistic density and pressure, the geometry will be similar. It will open upwards and outwards like a bowl and will flatten out only asymptotically as $r \to \infty$. At the surface of the star, the density will (usually) drop discontinuously to zero but the interior and exterior geometries depicted in the embedding diagram will join smoothly since dz/dr occurring in Eq. (7.90) is continuous.

Exercise 7.12

Model for a neutron star When a neutron star has a density in the range 10^{13} gm cm$^{-3} <$ $\rho < 10^{16}$ gm cm^{-3}, its equation of state can be approximated by $p = A\rho^{5/3}$, where $A = (3^{2/3}\pi^{4/3}/5)(\hbar^2/m^{8/3})$ and m is the mass of a neutron. Integrate the equations of

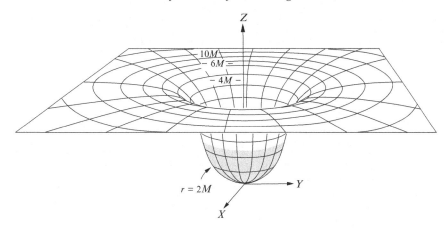

Fig. 7.1. Embedding diagram showing the curvature around a spherically symmetric source distribution.

stellar structure numerically and determine the total mass and the total radius of the star for a sequence of central densities in the range $(10^{14}-10^{16})$ gm cm^{-3}.

7.3 Vaidya metric of a radiating source

There is a simple and interesting generalization of the Schwarzschild metric called the *Vaidya metric*, which can be interpreted as a spacetime with an outgoing spherically symmetric radiation of massless particles. For example, consider a spherically symmetric body ('star') that emits a continuous stream of photons with each photon travelling radially outwards. The metric will now have as its source both the energy-momentum tensor of the star as well as the energy-momentum tensor of the radially outgoing null rays. The Vaidya metric is capable of describing this situation and provides an interesting model for a time dependent spherically symmetric metric.

To obtain this metric, we shall first make a coordinate transformation from the Schwarzschild time coordinate t to another coordinate u such that

$$dt = du + \frac{dr}{(1 - 2GM/r)}. \tag{7.94}$$

The coordinate u has a simple physical meaning. From the above transformation we find that $u = $ constant corresponds to curves which satisfy the equation $dr/dt = +(1 - 2GM/r)$. The radial null lines in the Schwarzschild geometry will correspond to the curves with $d\theta = 0$, $d\phi = 0$ and $ds = 0$. This immediately gives $(dr/dt)^2 = (1 - 2GM/r)^2$. The positive root of this equation will represent a radially outgoing null ray. (The negative root will give a radially ingoing null ray.) Thus we see that $u = $ constant along the radially outgoing null rays. With this transformation, the Schwarzschild metric becomes

$$ds^2 = -\left(1 - \frac{2GM}{r}\right) du^2 - 2\,du\,dr + r^2(d\theta^2 + \sin^2\theta\,d\phi^2). \qquad (7.95)$$

Since we are interested in the metric due to radially outgoing photons, we will now generalize the above line element making M a function of u; i.e. $M = M(u)$. Computing the Christoffel symbols, we find that the nonzero ones are given by

$$\Gamma^r_{uu} = \frac{GM}{r^2}\left(1 - \frac{2GM}{r}\right) - \frac{GM'}{r}, \qquad \Gamma^r_{\theta\theta} = -r\left(1 - \frac{2GM}{r}\right),$$

$$\Gamma^r_{\phi\phi} = -\left(1 - \frac{2GM}{r}\right) r\sin^2\theta, \qquad \Gamma^r_{ur} = \frac{GM}{r^2}, \qquad \Gamma^\theta_{r\theta} = \frac{1}{r},$$

$$\Gamma^\theta_{\phi\phi} = -\sin\theta\,\cos\theta, \qquad \Gamma^\phi_{r\phi} = \frac{1}{r}, \qquad \Gamma^\phi_{\theta\phi} = \cot\theta,$$

$$\Gamma^u_{uu} = -\frac{GM}{r^2}, \qquad \Gamma^u_{\theta\theta} = r, \qquad \Gamma^u_{\phi\phi} = r\sin^2\theta, \qquad (7.96)$$

where the prime denotes differentiation with respect to u. From these, we find that the only nonzero component of the Ricci tensor is

$$R_{uu} = -\frac{2GM'}{r^2} \qquad (7.97)$$

so that the source for the metric in Eq. (7.95) with $M = M(u)$ must have only one component of energy-momentum tensor nonzero: $T_{uu} = -(M'/4\pi r^2)$. This is precisely what we will get for a stream of photons moving radially outwards with a four-momentum $k_a = \nabla_a u$. The energy-momentum tensor will be $T^{ab} = -(M'/4\pi r^2)\,k^a k^b$. Hence we can interpret the metric as being due to a spherical source which is losing mass through radiation of photons in a spherically symmetric manner.

The procedure we adopted to find this solution is worth noting. It involved guessing at a reasonable form of a line element, evaluating the left hand side of Einstein's equations and thus identifying the relevant T_{ab}. The procedure will be successful only when the resulting T_{ab} is physically reasonable – which often is *not* the case. It worked in this situation, again, because of spherical symmetry.

7.4 Orbits in the Schwarzschild metric

The metric in Eq. (7.31) describes the gravitational field around any massive body (like the Sun), which may be approximated as spherically symmetric. The behaviour of particles and light rays in this metric will show characteristic features of general relativity which are absent in Newtonian gravity – and, in fact, most of the crucial tests of general relativity are based on this fact. We shall now study the motion of particles and light rays in this metric.

The motion of a particle of mass m in a metric g^{ik} can be most easily determined by using the relativistic Hamilton–Jacobi equation and we will adopt exactly the same strategy and the formalism used in Section 2.5 to study the motion of a particle in the Coulomb field. We know that the momentum and energy of a particle can be expressed in terms of the derivatives of the action $\mathcal{A}(x^i)$ of a particle treated as a function of the end points. In the relativistic situation, this relation has the form $p_i = (\partial \mathcal{A}/\partial x^i)$. Since $p_i p^i = -m^2$ the relativistic Hamilton–Jacobi equation becomes

$$g^{ik} \frac{\partial \mathcal{A}}{\partial x^i} \frac{\partial \mathcal{A}}{\partial x^k} + m^2 = 0. \tag{7.98}$$

The initial position and velocity of the particle defines a plane and spherical symmetry ensures that gravitational acceleration cannot have a component in the direction normal to the plane. Hence the motion will be confined to a plane which we take to be the one at $\theta = \pi/2$. Then, for the Schwarzschild metric with $(\partial \mathcal{A}/\partial \theta) = 0$, the Hamilton–Jacobi equation becomes:

$$\frac{1}{f(r)} \left(\frac{\partial \mathcal{A}}{\partial t} \right)^2 - f(r) \left(\frac{\partial \mathcal{A}}{\partial r} \right)^2 - \frac{1}{r^2} \left(\frac{\partial \mathcal{A}}{\partial \phi} \right)^2 = m^2, \tag{7.99}$$

where $f(r) = [1 - (2GM/r)]$ for the Schwarzschild metric but we shall keep it general whenever possible so that the expressions are applicable in other cases. The time independence and the rotational symmetry ensure that there are two conserved quantities $\mathcal{E}$ and L allowing the action to be separated out into the form:

$$\mathcal{A} = -\mathcal{E}t + L\phi + \mathcal{A}_r(r), \tag{7.100}$$

where L and $\mathcal{E}$ denote the angular momentum and energy of the particle. Substituting into the Hamilton–Jacobi equation and solving we find

$$\mathcal{A} = -\mathcal{E}t + L\phi + \int \sqrt{\frac{\mathcal{E}^2}{f^2} - \left(m^2 + \frac{L^2}{r^2} \right) \frac{1}{f}} \, dr. \tag{7.101}$$

As usual, the trajectory of the particle is determined by the equation $(\partial \mathcal{A}/\partial L) = $ constant while the time dependence is determined by $(\partial \mathcal{A}/\partial \mathcal{E}) = $ constant (see e.g., Section 2.5 where we used the same procedure). This gives $\phi(r)$ and $r(t)$ to be

$$\phi = \int dr \left(\frac{L}{r^2} \right) \left[\mathcal{E}^2 - \left(m^2 + \frac{L^2}{r^2} \right) f(r) \right]^{-1/2} \tag{7.102}$$

and

$$t = \frac{\mathcal{E}}{m} \int dr \frac{1}{f} \left[\left(\frac{\mathcal{E}}{m} \right)^2 - \left(1 + \frac{L^2}{m^2 r^2} \right) f \right]^{-1/2} . \qquad (7.103)$$

From Eq. (7.103) we get

$$\frac{1}{f} \frac{dr}{dt} = \frac{1}{\mathcal{E}} \left[\mathcal{E}^2 - V_{\text{eff}}^2(r) \right]^{1/2} \qquad (7.104)$$

with

$$V_{\text{eff}}^2(r) = m^2 f(r) \left(1 + \frac{L^2}{m^2 r^2} \right) = m^2 \left(1 - \frac{2GM}{r} \right) \left(1 + \frac{L^2}{m^2 r^2} \right), \qquad (7.105)$$

where the last expression is valid for the Schwarzschild metric. Similarly, from Eq. (7.102) we can determine $(d\phi/dr)$ which, when combined with the expression for (dr/dt), gives

$$r^2 \dot{\phi} = \frac{L}{\mathcal{E}} f(r) = \left(\frac{L}{\mathcal{E}} \right) \left(1 - \frac{2GM}{r} \right) . \qquad (7.106)$$

The left hand side of Eq. (7.104) has a direct physical meaning. Since the metric is independent of time, the covariant component of the four-velocity $u_0 = -\mathcal{E}/m$ is conserved. Therefore,

$$u^0 = \frac{dt}{d\tau} = g^{00} u_0 = \frac{\mathcal{E}/m}{f(r)} = \frac{\mathcal{E}/m}{(1 - 2GM/r)} . \qquad (7.107)$$

This shows that the left hand side of Eq. (7.104) is actually $(m/\mathcal{E})(dr/d\tau)$ and that the equation can be re-written as:

$$\frac{dr}{d\tau} = \frac{1}{m} \left[\mathcal{E}^2 - V_{\text{eff}}^2(r) \right]^{1/2} . \qquad (7.108)$$

Similarly, since the metric is independent of ϕ we know that $mu_\phi = L$ is conserved. So

$$u^\phi = \frac{d\phi}{d\tau} = g^{\phi\phi} \frac{L}{m} = \frac{L}{mr^2} \qquad (7.109)$$

and $\dot{\phi} = u^\phi/u^0 = (L/\mathcal{E}r^2)(1 - 2GM/r)$, which is the same as Eq. (7.106).

The effective potential equation, Eq. (7.104), can be given the following intuitive interpretation. Consider the trajectory of a particle in *special relativity* under the action of a central force. The angular momentum $L = r \times p$ is still conserved but the momentum is now given by $p = \gamma m v$ with $\gamma \equiv (1 - v^2)^{-1/2}$. So the relevant conserved component of the angular momentum is $L = mr^2(d\theta/d\tau) = \gamma mr^2(d\theta/dt)$ rather than $mr^2(d\theta/dt)$. (This, incidentally, means that Kepler's second law regarding the areal velocity does not hold in the case of special relativistic motion in a central force when we use the coordinate time. It does hold in terms of the proper time.) Consider now the motion of a *free* particle described in *polar* coordinates. The standard relation $E^2 = p^2 + m^2$ can be manipulated to give the equation

$$E^2 \left(\frac{dr}{dt}\right)^2 = E^2 - \left(\frac{L^2}{r^2} + m^2\right). \tag{7.110}$$

This is still the description of a *free* particle moving in a straight line but in the polar coordinates. Since special relativity must hold around any event in the Schwarzschild metric, we can obtain the corresponding equation for general relativistic motion by simply replacing dr, dt by the proper quantities $\sqrt{|g_{11}|}dr = dr/f(r)$, $\sqrt{|g_{00}|}dt = f(r)dt$ and the energy E by $E/\sqrt{|g_{00}|} = E/\sqrt{f(r)}$ (which is just the redshift factor for the energy) in this equation. This should give the equation for the orbit of a particle of mass m, energy E and angular momentum L around a body of mass M. With some simple manipulation, this equation can be written in the form of Eq. (7.104).

The analysis so far assumed that $m \neq 0$. But the original Hamilton–Jacobi equation as well as the subsequent results are equally applicable for massless particles like photons. Either by repeating the analysis, or – more simply – by taking the $m \to 0$ limit in the final expressions Eq. (7.102) and Eq. (7.103), we see that the photons are described by the equations

$$\phi = \int dr \left(\frac{L}{\mathcal{E}r^2}\right) \left[1 - \frac{L^2}{\mathcal{E}^2 r^2} f(r)\right]^{-1/2} = \int dr \left(\frac{b}{r^2}\right) \left[1 - \frac{b^2}{r^2} f(r)\right]^{-1/2}, \tag{7.111}$$

where $b \equiv L/\mathcal{E}$ and

$$t = \int \frac{dr}{f} \left[1 - \left(\frac{L^2}{\mathcal{E}^2 r^2}\right) f(r)\right]^{-1/2} = \int \frac{dr}{f} \left[1 - \left(\frac{b^2}{r^2}\right) f(r)\right]^{-1/2}. \tag{7.112}$$

The corresponding effective potential equation can now be taken as:

$$\frac{b^2}{r^2} \left(\frac{dr}{r d\phi}\right)^2 = 1 - V_{\text{eff}}^2; \qquad V_{\text{eff}}^2(r) = \frac{b^2}{r^2} f(r) = \frac{b^2}{r^2} \left(1 - \frac{2GM}{r}\right). \tag{7.113}$$

When $M = 0$ (i.e. in flat spacetime) this equation has the solution $r \cos \phi = x = b$ showing that b can be thought of as the impact parameter of the photon trajectory which is now, of course, a straight line.

These equations describe the motion of a particle in the Schwarzschild metric and the major new features occur for orbits close to $r = 2GM$. In this chapter, however, we have taken the point of view that the Schwarzschild metric is the external metric of source with radius $R > 2GM$. So we shall first concentrate on corrections to the Newtonian theory which arises even when the orbital radius is large compared with $2GM$. A more complete analysis of orbits, as a prelude to discussing black holes in the next chapter, will be taken up in Section 7.5.

7.4.1 Precession of the perihelion

In Newtonian theory, the orbit of a particle in the gravitational field of a spherical source is a conic section and the bound orbit will be an ellipse. We shall begin by calculating the corrections to these orbits which provide a sensitive test of the theory.

To do this it is best to start from the equation for the orbit, Eq. (7.102) with $f(r) = [1 - (2GM/r)]$, and change the variables to $u = 1/r$. Simple algebra gives:

$$\left(\frac{du}{d\phi}\right)^2 = \frac{1}{L^2}\left[\frac{\mathcal{E}^2}{c^2} - (m^2c^2 + L^2u^2)\left(1 - \frac{2GMu}{c^2}\right)\right], \qquad (7.114)$$

where c-factors have been reintroduced. Differentiating this equation with respect to ϕ, we get

$$\frac{d^2u}{d\phi^2} + u = \frac{GMm^2}{L^2} + \frac{3GM}{c^2}u^2. \qquad (7.115)$$

The first term on the right hand side is purely Newtonian and the second term is the correction from general relativity. The ratio of these two terms is $(L/mrc)^2 \approx (v/c)^2$, where r and v are the typical radius and speed of the particle. This correction term changes the nature of the orbits in several ways. To begin with, it changes the relationship between the parameters of the orbit and the energy and angular momentum of the particle. More importantly, it makes the elliptical orbit of Newtonian gravity precess slowly which is of greater observational importance.

The exact solution to Eq. (7.115) (which is the same as the exact integral resulting from Eq. (7.114)) can be given only in terms of elliptic functions and hence is not very useful. An approximate solution to Eq. (7.115), however, can be obtained fairly easily in two different contexts. The first context is when the orbit has a very low eccentricity and is nearly circular. Then the lowest order solution will be $u = u_0 = $ constant and one can find the next order correction by perturbation theory. This can be done *without* assuming that $2GMu_0 = 2GM/r_0$ is small, so that the result is valid even for orbits close to the Schwarzschild radius, as long as the orbit is nearly circular. The second context is when the orbital radius is large compared with $2GM$ but the eccentricity need not be small. Then the zeroth order solution is the ellipse in the Newtonian theory and we can find the perturbative corrections using the small parameter $2GM/a$, where a is the major axis of the ellipse. We shall begin with the case of nearly circular orbits.

If the original orbit has low eccentricity (i.e. nearly circular, as most planetary orbits are), then precession of the orbit can be determined as follows. Let the radius of the circular orbit be r_0 for which $u = (1/r_0) \equiv k_0$. For the actual orbit, $u = k_0 + u_1$ where we expect the second term to be a small correction. Changing the variables from u to u_1, where $u_1 = u - k_0$, Eq. (7.115) can be written as

$$u_1'' + u_1 + k_0 = \frac{GMm^2}{L^2} + \frac{3GM}{c^2}\left(u_1^2 + k_0^2 + 2u_1k_0\right).$$ (7.116)

We now choose k_0 to satisfy the condition

$$k_0 = \frac{GMm^2}{L^2} + k_0^2\frac{3GM}{c^2},$$ (7.117)

which determines the radius $r_0 = 1/k_0$ of the original circular orbit in terms of the other parameters. Now the equation for u_1 becomes

$$u_1'' + \left(1 - \frac{6k_0GM}{c^2}\right)u_1 = \frac{3GM}{c^2}u_1^2.$$ (7.118)

This equation is exact. We shall now use the fact that the deviation from circular orbit, characterized by u_1, is small and ignore the right hand side of equation Eq. (7.118). Solving Eq. (7.118), with the right hand side set to zero, we get

$$u_1 \cong A\cos\left[\left(1 - \frac{6GM}{c^2r_0}\right)^{1/2}\phi\right]$$ (7.119)

so that the solution, given by $u = u_0 + u_1$, is

$$\frac{1}{r} = \frac{1}{r_0} + A\cos\left[\left(1 - \frac{6GM}{c^2r_0}\right)^{1/2}\phi\right].$$ (7.120)

(We can determine the A in terms of other parameters by using the first order equation Eq. (7.114) but we will not need its explicit form.) We see that r does not return to its original value at $\phi = 0$ when $\phi = 2\pi$, indicating a precession of the orbit. We encountered the same phenomenon in the case of motion in a Coulomb field in Section 2.5 (see Eq. (2.83)). As described there, the precession angle can be determined by finding when the argument of the cosine function becomes 2π so that r returns to the original value. This occurs at an angle

$$\phi_c \approx 2\pi[1 - (6GM/c^2r_0)]^{-1/2}$$ (7.121)

which gives the precession $(\phi_c - 2\pi)$ per orbit.

Note that this result does not require relativistic effects to be small and relies only on the assumption that the original orbit was nearly circular. If we *further* assume that $r_0 \gg (GM/c^2)$, then we get from Eq. (7.117) that $k_0 \approx (GMm^2/L^2)$ and the precession is given by

$$\Delta\phi = 2\pi\left(1 - \frac{6G^2M^2m^2}{c^2L^2}\right)^{-1/2} - 2\pi \cong 6\pi\left(\frac{GMm}{Lc}\right)^2. \tag{7.122}$$

This gives the angle by which the major axis rotates per orbit.

Let us next consider a different situation in which the original Newtonian orbit was an ellipse and the eccentricity may not be negligible. In this case, the zeroth order solution to Eq. (7.115) is obtained by ignoring the second term in the right hand side (that is, we take $c \to \infty$ limit) leading to

$$\frac{d^2u_0}{d\phi^2} + u_0 = \frac{GM}{h^2}, \tag{7.123}$$

where $h = L/m$ is the angular momentum per unit mass. This has a solution

$$u_0 = \frac{GM}{h^2}\left(1 + e\cos\phi\right) = \frac{1}{p}\left(1 + e\cos\phi\right), \tag{7.124}$$

where p is the latus rectum and e is the eccentricity of the ellipse. In the limit of $c \to \infty$, Eq. (7.114) becomes $u'^2 + u^2 = (2mE_N/L^2) + (2GMu/h^2)$ where $E_N = \mathcal{E} - mc^2$ is the energy in the Newtonian approximation. Substituting the solution into this equation we can determine the eccentricity in terms of the energy as $e^2 - 1 = 2E_N L^2/G^2M^2m^3$, which completely solves the Newtonian problem. The semi-major axis is given by

$$a = \frac{h^2}{GM(1 - e^2)}. \tag{7.125}$$

We shall now solve Eq. (7.115) perturbatively by taking the solution to be $u = u_0 + u_1$. Substituting into Eq. (7.115), we get an equation correct to the lowest order in u_1:

$$\frac{d^2u_1}{d\phi^2} + u_1 = A\left(1 + e^2\cos^2\phi + 2e\cos\phi\right), \tag{7.126}$$

where

$$A = \frac{3G^3M^3}{c^2h^4} = \frac{3GM}{c^2a}\frac{1}{a(1 - e^2)}. \tag{7.127}$$

For reasonable eccentricities, $Aa = \mathcal{O}(GM/c^2a)$ which we have assumed to be small. A straightforward integration of Eq. (7.126) will lead to terms like $\phi\sin\phi$, which grow without bound with ϕ, invalidating the approximation that $u_1 \ll u_0$. This issue can be handled by choosing an ansatz which uses only bounded functions like $B\cos[(1 - C)\phi]$, substituting into Eq. (7.126) and determining B, C

when $C \ll 1$. This way we get a suitable solution:

$$u = u_0 + u_1 = \frac{1}{p} + \frac{\mu}{p}\left(1 + \frac{e^2}{3}\right) + \frac{e}{p}\cos[(1-\mu)\phi] + \frac{\mu e^2}{3p}\sin^2\phi, \quad (7.128)$$

where we have defined the dimensionless constant $\mu \equiv 3(GM/hc)^2 \ll 1$. The constant second term and the last term (with period 2π) do not lead to observable effects. But the $\cos[(1-\mu)\phi]$ term shows that r does not get back to the value at $\phi = 0$ when $\phi = 2\pi$. Repeating the previous analysis we find that the rate of precession is

$$\Delta\phi = \frac{2\pi}{1-\mu} - 2\pi \approx 2\pi\mu = \frac{6\pi(GM)^2}{h^2c^2} = \frac{6\pi GM}{a(1-e^2)c^2}. \quad (7.129)$$

On using Eq. (7.125) and $h = L/m$, we see that this result matches with the one in Eq. (7.122).

For Mercury, the precession is about $43''$ per century, and for the Earth the precession is about $3.8''$ per century. The result for Mercury has been one of the crucial tests of the general theory of relativity. In Chapter 3, we calculated the same effect using a spin-2 field theory of gravity and found that it does not give this (observed) result. Since the spin-2 field theory is identical in structure to the linear limit of general relativity, we can conclude that the linear limit does not reproduce the observed value for perihelion precession; we are testing the lowest order nonlinear corrections in this situation.

We conclude by pointing out another curious result in the case of Schwarzschild metric for circular orbits. In this case, $u = u_0 = $ constant and Eq. (7.115) allows us to determine the angular momentum to be $L^2 = (m^2 GM/u_0)(1 - 3GMu_0)^{-1}$. From this we find that

$$\left(1 + \frac{L^2 u_0^2}{m^2}\right) = \frac{(1 - 2GMu_0)}{(1 - 3GMu_0)}. \quad (7.130)$$

Using Eq. (7.130) and the condition $dr/dt = 0$ in Eq. (7.104), we get the energy to be $\mathcal{E}^2 = m^2 f^2/(1 - 3GMu_0)$. Hence we get, for circular orbits, the ratio

$$\left(\frac{\mathcal{E}}{L}\right)^2 = \frac{u_0}{GM}(1 - 2GMu_0)^2. \quad (7.131)$$

On the other hand Eq. (7.106) gives the angular velocity $\Omega = \dot{\phi}$ for the circular orbit to be

$$\Omega^2 = \left(\frac{d\phi}{dt}\right)^2 = \left(\frac{L}{\mathcal{E}}\right)^2 u_0^4(1 - 2GMu_0)^2 = \frac{GM}{r_0^3}, \quad (7.132)$$

which is precisely the result in Newtonian gravity that leads to Kepler's third law!

Exercise 7.13

Exact solution of the the orbit equation in terms of elliptic functions Write down the equation governing the orbit of a massive particle in the Schwarzschild metric in the form

$$\left(\frac{du}{d\phi}\right)^2 = f(u) = 2(u - u_1)(u - u_2)(u - u_3), \tag{7.133}$$

where $u = (GM/r)$ and $u_1 < u_2 < u_3$.

(a) Show that $u_1 + u_2 + u_3 = 1/2$ and that the aphelion and perihelion of the orbit are at $r_1 = M/u_1$ and $r_2 = M/u_2$.

(b) Express the energy and angular momentum of the particle in terms of u_1, u_2 and u_3.

(c) Integrate Eq. (7.133) to obtain the solution to be

$$u = u_1 + (u_2 - u_1)\text{sn}^2\left[\alpha\phi|\ \beta\right] \tag{7.134}$$

with $\alpha = (u_3 - u_1)^{1/2}/\sqrt{2}$; $\beta = (u_2 - u_1)/(u_3 - u_1)$ and $\text{sn}(z|\mu)$ is the Jacobi elliptic function with modulus μ. Use this to plot the trajectory for several different cases discussed in the text.

Exercise 7.14

Contribution of nonlinearity to perihelion precession Consider a metric of the form in Eq. (7.12) with $A(r)$ and $B(r)$ given parametrically by an expansion of the form

$$B(r) = 1 + 2\gamma\frac{GM}{r} + \cdots; \qquad A^{-1}(r) \simeq 1 + \frac{2GM}{r} + \frac{2(2 - \beta + \gamma)G^2M^2}{r^2} + \cdots. \tag{7.135}$$

Here γ and β are two arbitrary parameters; in standard general relativity $\gamma = 1$, $\beta = 1$. We have used an expansion for $1/A(r)$ rather than for $A(r)$ since it simplifies calculations somewhat. Compute the perihelion precession of an orbit in this metric and show that

$$\Delta\phi = \left(\frac{6\pi GM}{L}\right)\left(\frac{2 - \beta + 2\gamma}{3}\right). \tag{7.136}$$

[Note. The linear approximation will correspond to $\gamma = 1$, $\beta = 3$ which gives an incorrect result to the perihelion precession. In contrast, we saw in Exercise 6.13 that the linear approximation can correctly predict the bending of light.]

Exercise 7.15

Perihelion precession for an oblate Sun If the Sun has a quadrupole moment, then a perihelion shift to the orbit of Mercury will arise even in Newtonian theory. In such a case, the gravitational potential due to the Sun can be expressed in the form

$$U = \frac{M_\odot}{r}\left(1 - J_2\frac{R_\odot^2}{r^2}\frac{3\cos^2\theta - 1}{2}\right) \tag{7.137}$$

in units with $G = c = 1$ and J_2 is a constant which characterizes the quadrupole moment of the Sun. Show that, in this case, the total precession per orbit is given by

$$\delta\phi = \frac{6\pi M_\odot}{a(1 - e^2)} + J_2\frac{3\pi R_\odot^2}{a^2(1 - e^2)^2}, \tag{7.138}$$

where a and e are the semi-major axis and the eccentricity of the original elliptic orbit, respectively. The first term is the same as the result derived in the text while the second term arises due to the oblateness of the Sun. What is the value for J_2 if these two terms are comparable? Does this value appear reasonable?

7.4.2 Deflection of an ultra-relativistic particle

The correction discussed above is for a bound elliptical orbit in Newtonian theory. A different effect in general relativity arises when an ultra-relativistic particle passes near a central mass, as – for example – in the case of a light ray travelling near the Sun. We are interested in the deflection of the particle from a straight line trajectory due to the gravitational attraction of the central mass. When this deflection is small, it can be computed by perturbing the straight line trajectory.

The starting point is again the equation for the orbit given by Eq. (7.115). (The first term on the right hand side will vanish for photons since $m \to 0$; we retain it because we are interested in an ultra-relativistic particle.) The undeflected straight line solution to this equation is given by $u_0 = b^{-1} \cos \phi$, which is obtained when both terms on the right hand side are neglected. (Here, b is the impact parameter of the orbit and the particle is travelling parallel to the y-axis at $x = b$.) To find the deflection due to the gravitational field, we try a solution to Eq. (7.115) of the form $u = b^{-1} \cos \phi + v(\phi)$ with $bv \ll 1$. Substituting this ansatz into Eq. (7.115) and retaining only the leading order terms, we get

$$v'' + v \cong \frac{GMm^2}{L^2} + \frac{3GM}{2c^2b^2} \left[\cos 2\phi + 1 \right]. \tag{7.139}$$

The first term on the right hand side describes the deflection due to purely Newtonian attraction and the second term gives the general relativistic correction. This equation is identical to that of a harmonic oscillator perturbed by an external force and can be easily solved. Solving the equation and substituting the solution in our ansatz for u, we get

$$u = \frac{1}{b} \cos \phi - \frac{GM}{c^2b^2} \cos^2 \phi + \frac{GMm^2}{L^2} + \frac{2GM}{c^2b^2}. \tag{7.140}$$

The first term is the undeflected trajectory and the other three terms give the deflection due to the gravitational field of the central mass. To find the angle of deflection when the particle moves past the central mass, we set $u = 0$ (which corresponds to $r = \infty$) and solve for ϕ. To the leading order, we can ignore the $\cos^2 \phi$ term and obtain

$$-\cos \phi \cong \frac{GMm^2b}{L^2} + \frac{2GM}{c^2b} \equiv q. \tag{7.141}$$

To the same order of accuracy, the solution to this equation is given by $\phi = \pm[(\pi/2) + q]$ with $q \ll 1$, and the net deflection will be $\delta\phi \approx 2q$. To relate q to parameters of the particle at large distance, we note that the angular momentum L is given by bp_∞, where $p_\infty = \gamma m v_\infty$ is the momentum of the particle at large distances. Using these relations in the definition of q we find that the deflection is given by

$$\delta\phi = 2q = \frac{2GM}{bv_\infty^2}\left(1 + \frac{v_\infty^2}{c^2}\right). \tag{7.142}$$

When $c \to \infty$, this reduces to the Newtonian deflection of $(2GM/bv_\infty^2)$. The second term in the bracket gives the correction due to general relativity. For photons, with $v_\infty = c$, this leads to a deflection of $\delta\phi \approx (4GM/c^2 b)$ which is twice the Newtonian value. This was another prediction from general relativity which has been successfully tested.

Note that the deflection in Eq. (7.142) has a very different velocity dependence compared with the deflection of a *relativistic* charged particle moving in the Coulomb field (see Chapter 2, Section 2.5, Eq. (2.94)) even though both results have the same velocity dependence in the non-relativistic limit (that is, when $c \to \infty$). But as $v \to c$, the electromagnetic deflection vanishes while the gravitational deflection becomes twice the non-relativistic value. This difference is due to the fact that the electromagnetism is based on a vector field while gravity is based on a second rank tensor field.

The deflection of light by a gravitational field also leads to an important phenomenon called *gravitational lensing* which will be discussed in Section 10.7.

Exercise 7.16
Angular shift of the direction of stars What is actually observed in the observations involving the deflection of light by the Sun is the shift in the angular position of a star. Show that this shift is given by

$$\delta\alpha = \frac{2GM_\odot}{c^2 r_{ES}}\cot\left(\frac{\alpha}{2}\right), \tag{7.143}$$

where α is the unperturbed direction of the star when viewed from the Earth and r_{ES} is the distance between the centres of the Earth and the Sun. Estimate this quantity $\delta\alpha$ when the direction of the ray of light varies from $\alpha = \pi$ (opposite to the Sun's direction) through $\alpha = \pi/2$ (perpendicular to the Earth–Sun line) to the grazing incidence touching the rim of the Sun. Stellar positions can now be measured to an accuracy of 10^{-3} arcsec. Does one have to take into account general relativistic effects in these measurements?

Exercise 7.17
Time delay for photons An alternative way of obtaining the radar time delay discussed previously in Exercise 4.8 is as follows. Using the fact that the undeflected part of a light ray is given by $r\sin\phi = b$ in the equatorial plane, where b is the impact factor, obtain an

expression for $r^2 d\phi^2$. Inserting this into the equation $ds^2 = 0$ in the Schwarzschild metric evaluated on the equatorial plane, get the relation

$$cdt \simeq \pm \frac{r}{\sqrt{r^2 - b^2}} \left(1 + \frac{2M}{r} - \frac{b^2 M}{r^3} \right) dr \tag{7.144}$$

which is correct to $\mathcal{O}(M^2/r^2)$. Integrate this to obtain the time taken for light to travel from the point of closest approach to a distance r to give

$$c\Delta t \cong \sqrt{r^2 - b^2} + 2M \log \left(\frac{r}{b} + \sqrt{\frac{r^2}{b^2} - 1} \right) - \frac{M}{r} \sqrt{r^2 - b^2}. \tag{7.145}$$

The round trip time will involve four such pieces. Show that the final result agrees with the one obtained in Exercise 4.8.

Exercise 7.18
Deflection of light in the Schwarzschild–de Sitter metric In the presence of a cosmological constant the Schwarzschild metric gets modified to the form discussed in Exercise 7.4 with $Q = 0$. The metric is now no longer asymptotically flat and one cannot define the deflection angle for a light ray by taking the $r \to \infty$ limit which has led to some controversy in the literature. But if $GM/c^2 \ll \Lambda^{-1/2}$ then one should be able to study the trajectories of light rays and obtain a correction to the result in the text due to the presence of a cosmological constant. Work out this deflection properly and show that it is given by

$$\delta\phi \approx \frac{4GM}{b} \left(1 - \frac{\Lambda b^4}{24 G^2 M^2} \right). \tag{7.146}$$

Interpret this result.

Exercise 7.19
Solar corona and the deflection of light by the Sun One difficulty faced by the precision tests of relativity is to ensure that all *other* effects are taken into account properly. This exercise discusses the effect of the variable refractive index of the solar corona on the bending of light.

(a) Let $\boldsymbol{x}(s)$ be the trajectory of a light ray with s denoting the arc length along the trajectory. Show that, if the refractive index of the medium is $n(\boldsymbol{x})$, then the trajectory is determined by the equation

$$\frac{d}{ds} \left(n \frac{d\boldsymbol{x}}{ds} \right) = \nabla n. \tag{7.147}$$

(b) Let the electron density in the solar corona be $N_e(r)$. Given the electron density, the refractive index is given by $n = \sqrt{1 - (\omega_p^2/\omega^2)}$, where $\omega_p^2 = (4\pi q^2/m_e)N_e$. Show that the net deflection due to the corona can be estimated to be

$$\Delta\phi \simeq \int_{-\infty}^{\infty} n' \left(\sqrt{x^2 + b^2} \right) \frac{b}{\sqrt{x^2 + b^2}} dx, \tag{7.148}$$

where $n' = (dn/dr)$ and b is the impact parameter of the photon.

(c) Assume that $N_e(r)$ is well approximated by the fitting formula

$$N_e(r) = \frac{A}{(r/R_\odot)^6} + \frac{B}{(r/R_\odot)^2}, \tag{7.149}$$

where $A = 10^8$ cm^{-3} and $B = 10^6$ cm^{-3}. Make a rough estimate of $\Delta\phi$. Can this have a bearing on the experiments to measure the deflection of light by the Sun?

7.4.3 Precession of a gyroscope

The spin angular momentum of a body will also be affected by the curvature of the spacetime and – in particular – a spinning gyroscope orbiting a massive body will undergo precession of the spin vector. This can serve as a sensitive test of general relativity. We will now describe the spin precession in the case of the Schwarzschild geometry.

We have seen in Section 1.10 and in Section 4.9, page 183, that the spin of a gyroscope can be described by a four-vector s^a that is orthogonal to the four-velocity u^a of the gyroscope so that $u^a s_a = 0$. We will consider a gyroscope which is moving in a circular orbit on the equatorial plane in the Schwarzschild metric. For such an orbit, $u^r = u^\theta = 0$ and $u^\phi/u^t = \Omega = (GM/r^3)^{1/2}$ is the angular velocity of the circular orbit (see Eq. (7.132)). Using $u^i u_i = -1$, we get

$$u^t = \frac{dt}{d\tau} = \left(1 - \frac{3GM}{r}\right)^{-1/2}; \qquad u^\phi \equiv u^t \Omega = u^t \left(\frac{GM}{r^3}\right)^{1/2}. \qquad (7.150)$$

The condition $u^a s_a = 0$ reduces to the equation

$$\left(1 - \frac{2GM}{r}\right) s^t u^t - r^2 s^\phi u^\phi = 0. \qquad (7.151)$$

Since $u^\phi/u^t = \Omega$ is the angular velocity of the orbit, we get the relation

$$s^t = \frac{\Omega r^2}{(1 - 2GM/r)} s^\phi. \qquad (7.152)$$

We next use the fact that the spin vector is parallel transported along the trajectory so that

$$\frac{ds^m}{d\tau} + \Gamma^m_{nl} s^n u^l = 0. \qquad (7.153)$$

Using the known form of Christoffel symbols for the Schwarzschild metric, direct computation gives the equations

$$\frac{ds^r}{d\tau} - \frac{r\Omega}{u^t} s^\phi = 0, \qquad \frac{ds^\theta}{d\tau} = 0, \qquad \frac{ds^\phi}{d\tau} + \frac{u^t \Omega}{r} s^r = 0. \qquad (7.154)$$

We now convert the derivatives with respect to proper time τ to the derivatives with respect to coordinate time t using $u^t = dt/d\tau$. We can also eliminate s^ϕ from the first equation using the third equation thereby obtaining the set

$$\frac{d^2 s^r}{dt^2} + \left(\frac{\Omega}{u^t}\right)^2 s^r = 0, \qquad \frac{ds^\theta}{dt} = 0, \qquad \frac{ds^\phi}{dt} + \frac{\Omega}{r} s^r = 0. \qquad (7.155)$$

These equations can be integrated once we know the initial condition for the spin vector $s(0)$. For the sake of simplicity, we shall assume that the initial direction of the spin vector was radial so that $s^\theta(0) = s^\phi(0) = 0$. The solution to the above set of equations with this initial condition is given by

$$s^r(t) = s^r(0) \cos \Omega' t, \qquad s^\theta(t) = 0, \qquad s^\phi(t) = -\frac{\Omega}{r\Omega'} s^r(0) \sin \Omega' t, \quad (7.156)$$

where

$$\Omega' = \frac{\Omega}{u^t} = \Omega \left(1 - \frac{3GM}{r}\right)^{1/2}. \qquad (7.157)$$

The solution shows that an s^ϕ component is generated in the negative ϕ direction and both the s^r and s^ϕ components rotate relative to the radial direction with an angular speed Ω' in the negative ϕ direction. What is relevant, of course, is the difference between the rotation of the radial direction at an angular velocity Ω and the rotation of the spin components. Since one revolution of the gyroscope is completed in the time interval $t - (2\pi/\Omega)$, the direction of the spin changes by an amount $\alpha = (2\pi/\Omega)(\Omega - \Omega')$. Therefore, the *geodesic precession*, as it is called, of a spin vector during one orbit of the gyroscope is given by the amount

$$\alpha = 2\pi \left[1 - \left(1 - \frac{3GM}{r}\right)^{1/2}\right]. \qquad (7.158)$$

For a gyroscope in a near Earth orbit, this effect is about $8''$ per year and should be measurable in a satellite experiment.

In general, the spin precession of a gyroscope will have two more components; one due to the rotation of the central body (if it is nonzero) and another due to Thomas precession (see Section 1.10). When the effects are small, they add linearly and the total precession can be written, in a somewhat general form as

$$\mathbf{\Omega} = -\frac{1}{2}\left(\nabla \times \mathbf{A}_g\right) + \frac{1}{2}\left(\mathbf{a} \times \mathbf{v}\right) + \frac{3}{2}\left(\nabla \phi \times \mathbf{v}\right). \qquad (7.159)$$

The first term, $-(1/2)(\nabla \times \mathbf{A}_g)$ arises due to the rotation of the central body which causes the inertial frames to be 'dragged along' with respect to the fixed stars. This has already been worked out in Section 6.4.1. The second term is purely a special relativistic effect (Thomas precession) which has been derived in Section 1.10. The last term arises for two separate reasons. Part of this term $(1/2)(\nabla \phi \times \mathbf{v})$ arises due to precession in the gravo-magnetic field while another part $(\nabla\phi \times v)$ arises from the curvature of space (see Exercise 7.20).

These three terms lead to different amounts of total precession for different objects. The first term in Eq. (7.159) is of purely geometrical origin and is independent of the orbital elements of the gyroscope. The second term will contribute whenever there is a non-gravitational source of acceleration but will vanish for a gyroscope that is in free fall orbit around a massive body. Such an orbiting gyroscope will respond only to the first and third terms. For a gyroscope at rest on the surface of the Earth, say, there is partial cancellation between the second and third terms (since $a = -\nabla\phi$) leading to a net precession of $\nabla\phi \times v$.

Exercise 7.20
General expression for relativistic precession Obtain the general formula in Eq. (7.159) for the precession of the spin for a gyroscope orbiting a massive body taking all three effects mentioned above into account along the following lines. The equation of motion for the spin four-vector s^i has a simple form relative to an instantaneous local Lorentz frame. During an infinitesimal increment $d\tau$ of the proper time, any spatial component s^α will vary by the amount $ds^\alpha = \Gamma^\alpha_{ij}s^j u^i d\tau = \Gamma^\alpha_{\beta 0}s^\beta d\tau$. Since in the local Lorentz frame $u^i = (1,0)$ and $s^j = (0,S)$, the precession of the spatial component of the spin vector can be obtained using the expression for $\Gamma^\alpha_{\beta 0}$. Compute this to the lowest nontrivial order of accuracy for a metric

$$ds^2 = -(1+2\phi)dt^2 + (1-2\phi)\delta_{\alpha\beta}dx^\alpha dx^\beta - 2A_\alpha dx^\alpha dt \tag{7.160}$$

in the appropriate frame of reference and show that

$$\Gamma_{\alpha,\beta 0} = \frac{3}{2}\left(v_\beta\partial_\alpha\phi - v_\alpha\partial_\beta\phi\right) + \frac{1}{2}\left(\partial_\alpha A_\beta - \partial_\beta A_\alpha\right) + \frac{1}{2}\left(a_\alpha v_\beta - a_\beta v_\alpha\right), \tag{7.161}$$

where v is the three velocity and a is the three acceleration of the spinning object. Using this expression, the equation of motion $(ds^\alpha/d\tau) = \Gamma^\alpha_{\beta 0}s^\beta$ reduces to the form $(ds/d\tau) = \Omega \times s$, where Ω is given by Eq. (7.159).

Exercise 7.21
Hafele–Keating experiment Consider two clocks circumnavigating the Earth in two aeroplanes, one flying in the western direction and the other flying in the eastern direction. Let the aeroplanes fly at a constant altitude h and constant speed u (with $u > 0$ for the east bound plane and $u < 0$ for the west bound plane) and let R and Ω be the Earth's radius and angular velocity, and g be the Earth's acceleration due to gravity. Let $\delta\tau$ be the proper time measured by one of the clocks and $\delta\tau_0$ be the proper time measured by a clock located on the ground ($h = u = 0$). Show that the fractional time difference is

$$f \equiv \frac{\delta\tau - \delta\tau_0}{\delta\tau_0} = \frac{gh}{c^2} - \frac{u^2 + 2R\Omega u}{2c^2}. \tag{7.162}$$

Hence show that there will be a difference in the time recorded by the two clocks. Estimate the time difference in nanoseconds. (Current atomic clocks have confirmed this difference to about 20 per cent accuracy.)

7.5 Effective potential for orbits in the Schwarzschild metric

So far we have concentrated on corrections to Newtonian results that arise from general relativity. But, as we said earlier, the really important new features arise at $r \gtrsim 2GM$. To discuss this issue properly we need to study what happens to a source when its radius is smaller than the gravitational radius, which will occupy our attention in the next chapter in the study of black holes. However, as a prelude, we shall now describe several peculiar features of the orbits in the Schwarzschild metric without worrying about the nature of the source that is producing it. In other words, we will assume in this section that the Schwarzschild metric is valid throughout the spacetime and analyse the nature of the orbits.

The qualitative features of the orbit can be understood from the study of the effective potential introduced through the equation (see Eq. (7.105)):

$$\left(1 - \frac{2GM}{r}\right)^{-1} \frac{dr}{dt} = \frac{1}{\mathcal{E}} \left[\mathcal{E}^2 - V_{\text{eff}}^2(r)\right]^{1/2} \tag{7.163}$$

with

$$V_{\text{eff}}^2(r) = m^2\left(1 - \frac{2GM}{r}\right)\left(1 + \frac{L^2}{m^2 r^2}\right). \tag{7.164}$$

To understand the nature of relativistic orbits, we will first determine the maxima and minima of $V_{\text{eff}}^2(r)$ using the dimensionless variables $u \equiv GM/c^2 r$, and $\bar{L} = L/mM = Lc/GMm$. The extrema of the function $V_{\text{eff}}^2(u)$ occur at

$$u_m \equiv \frac{1 \pm \sqrt{1 - 12/\bar{L}^2}}{6} \tag{7.165}$$

with the maximum value of the potential being

$$V_{\text{max}}^2(\bar{L}) = m^2(1 - 2u_m)(1 + \bar{L}^2 u_m^2). \tag{7.166}$$

Clearly, for $\bar{L} > \sqrt{12}$ (which corresponds to $L > 2\sqrt{3}GMm$) the function $V_{\text{eff}}(r)$ has one maximum and one minimum. The two extrema merge for $L = 2\sqrt{3}GMm$ and there are no extrema (i.e. the function $V_{\text{eff}}(r)$ becomes monotonic) for $L < 2\sqrt{3}GMm$. The maximum value of the potential reaches the value m when $L = 4GMm$. Figure 7.2 gives a plot of (V_{eff}/m) against $(rc^2/GM) = r/M$ (in units with $G = c = 1$) for different values of L. Several important aspects of the motion can be deduced from this figure.[1]

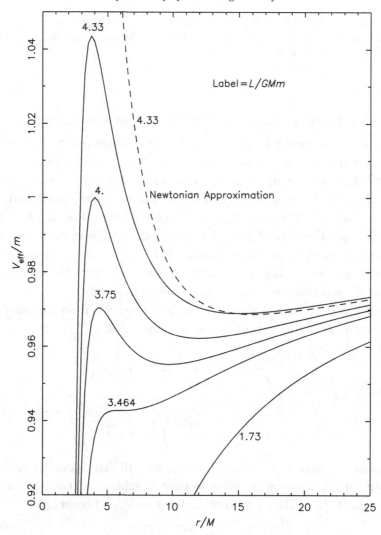

Fig. 7.2. The effective potential for particle motion in the Schwarzschild metric.

(i) For a given value of L and $\mathcal{E}$ the nature of the orbit will, in general, be governed by the turning points in r, determined by the equation $V_{\text{eff}}^2(r) = \mathcal{E}^2$. If $L > 4GMm$, then the value of $V_{\max}$ is greater than m. Further, $V_{\text{eff}} \to m$ as $r \to \infty$ for all values of L. If the energy $\mathcal{E}$ of the particle is lower than m (and $L > 4GMm$) then there will be two turning points. The particle will orbit the central body with a perihelion and aphelion undergoing the precession described in Section 7.4.1. This is similar to elliptic orbits in Newtonian gravity.

(ii) If $m < \mathcal{E} < V_{\text{max}}(L)$ there will be only one turning point. The particle will approach the central mass from infinity, reach a radius of closest approach and will travel back to infinity. This is similar to the hyperbolic orbits in Newtonian gravity.

(iii) If $\mathcal{E} = V_{\text{max}}(L)$, then the orbit will be circular at some radius r_0 determined by the condition $V'(r_0) = 0, V(r_0) = \mathcal{E}$. Solving these equations simultaneously, we find that the radii of circular orbits and their energies are given by

$$\frac{r_0}{2GM} = \frac{L^2}{4G^2M^2m^2} \left[1 \pm \sqrt{1 - \frac{12G^2M^2m^2}{L^2}} \right] ; \quad \mathcal{E}^2 = \frac{L^2}{GMr_0} \left(1 - \frac{2GM}{r_0} \right)^2 .$$
$$(7.167)$$

The upper sign refers to stable orbit and the lower sign refers to the unstable orbit. The stable orbit closest to the centre has the parameters $r_0 = 6GM, L = 2\sqrt{3}GMm$ and $\mathcal{E} = (\sqrt{(8/9)})m \approx 0.943m$. When a particle falls into the black hole from the stable circular orbit closest to the centre, it can release a fraction 0.057 of its energy in radiation. This fact is of some interest in the case of accretion disks around black holes and we will see in Chapter 8 that much higher efficiencies can be achieved in the case of rotating black holes.

(iv) If $\mathcal{E} > V_{\text{max}}(L)$, the particle 'falls to the centre'. This behaviour is in sharp contrast to Newtonian gravity in which a particle with nonzero angular momentum can never reach $r = 0$.

(v) As the angular momentum is lowered, V_{max} decreases and for $L = 4GMm$ the maximum value of the potential is at $V_{\text{max}} = m$. In this case all the particles (with $\mathcal{E} > m$) moving inward from spatial infinity will fall to the origin. Particles with $\mathcal{E} < m$ will still have two turning points and will have bound orbits.

(vi) When L is reduced still further, the maxima and minima of $V(r)$ approach each other. For $L \leq 2\sqrt{3}GMm$, there are no turning points in the $V(r)$ curve. Particles with lower angular momentum will fall to the origin irrespective of the energy.

Exercise 7.22

Exact solution of the orbital equation Equation (7.115) is valid for both $m = 0$ and $m \neq 0$ and – interestingly enough – it is possible to find an exact solution to this equation for both these cases.

(a) Show that the following function solves the equation:

$$u = \frac{1}{6GM} + \frac{2\alpha^2}{3GM} - \frac{2\alpha^2}{GM \cosh^2(\alpha\phi)}, \quad (7.168)$$

where

$$\alpha^2 = \pm \frac{1}{4} \sqrt{\left(1 - \frac{12G^2M^2m^2}{L^2} \right)}. \quad (7.169)$$

Explain in algebraic terms why such an analytic solution exists.

(b) Plot the solution and describe its features for different values of α^2, especially for $\alpha^2 = 1/4, 1/8, 0$.

Exercise 7.23

Effective potential for the Reissner–Nordstrom metric In Exercise 7.3, we derived the Reissner–Nordstrom metric which represents the spacetime of a charged spherically symmetric source.

(a) Write down the Hamilton–Jacobi equation for an uncharged particle moving in this spacetime and obtain the effective potential. Determine the radii of stable circular orbits.

(b) Consider an observer in a stable circular orbit in the Reissner–Nordstrom metric. What is the magnetic field measured by this observer?

Exercise 7.24

Horizons are forever Consider a Reissner–Nordstrom black hole with $Q_i < M_i$ so that it has a well-defined event horizon. A small charged sphere of mass m and charge q is dropped on to the black hole, changing the total charge and mass of the black hole to $Q_f = Q_i + q$ and $M_f = M_i + m$. Show that we necessarily will have $Q_f < M_f$ so that one cannot use this process to destroy the horizon of the Reissner–Nordstrom black hole.

As an application of the formalism of effective potential, let us consider the absorption cross-section for particles in the Schwarzschild metric. We have just seen that particles with $\mathcal{E} > V_{\max}$ will fall towards the centre. For a given value of energy, the limiting angular momentum for which this will occur can be found by equating $\mathcal{E}^2 = V_{\max}^2$. Using Eq. (7.165), the expression for $V_{\max}$ in Eq. (7.166) can be expressed in an equivalent form as

$$\frac{V_{\max}^2}{m^2} = \frac{36 + \bar{L}^2 + (\bar{L}^2 - 12)(1 - 12/\bar{L}^2)^{1/2}}{54}. \tag{7.170}$$

Let us first consider the $\bar{L} \gg 1$, $\mathcal{E} \gg m$ limit. Then Eq. (7.170) has the asymptotic behaviour for large values of $\bar{L}$ given by:

$$\frac{V_{\max}^2}{m^2} \approx \frac{\bar{L}^2 + 9}{27} \tag{7.171}$$

so that the condition $\mathcal{E}^2 = V_{\max}^2$, which gives the critical value of the angular momentum, becomes $\bar{L}_{\text{crit}}^2 = 27(\mathcal{E}/m)^2 - 9$. If the momentum of the particle at infinity was $p = (\mathcal{E}^2 - m^2)^{1/2}$, then the corresponding critical impact parameter is given by $b_{\text{crit}} = L_{\text{crit}}/p$. The particle will be captured for $b < b_{\text{crit}}$ so that the capturing cross-section is given by

$$\sigma_{\text{cap}} = \pi b_{\text{crit}}^2 \approx 27\pi (GM)^2 \left(1 + \frac{2m^2}{3\mathcal{E}^2}\right). \tag{7.172}$$

These results are valid for high energies with $\mathcal{E} \gg m$. In the opposite extreme when $\mathcal{E}/m \approx 1$, we can write $(\mathcal{E}/m)^2 \approx (1 + \beta^2)$ and repeat the above analysis to the lowest order in β^2. The equation $\mathcal{E}^2 = V_{\max}^2$ now becomes

$$18 + 54\beta^2 \approx \bar{L}^2 + \frac{(\bar{L}^2 - 12)^{3/2}}{\bar{L}} \tag{7.173}$$

which gives, to the lowest order,

$$\bar{L}_{\text{crit}}^2 = 16(1 + 2\beta^2) + \mathcal{O}(\beta^4). \tag{7.174}$$

In this case, the corresponding capture cross-section is given by

$$\sigma_{\text{cap}} = \pi b_{\text{crit}}^2 \approx \frac{16\pi G^2 M^2}{\beta^2}. \tag{7.175}$$

We stress that these phenomena, in which a particle with *nonzero* angular momentum is captured by a central source, have no analogies in Newtonian gravity.

The orbits of the photons can be analysed in exactly the same manner as we did for the material particles but with the effective potential given in Eq. (7.113). This potential has just one maximum at $r = 3GM$ and $V_{\max} = b/(3\sqrt{3}GM)$. It follows that the photons with impact parameter $b > 3\sqrt{3}GM$ which move inwards from infinity are 'reflected back' by the potential. These correspond to rays that are deflected by the geometry and go back to $r = \infty$. When $b \gg 3\sqrt{3}GM$, the orbit is almost a straight line and the deflection is just $4GM/b$. Photons with impact parameter $b < 3\sqrt{3}GM$ which fall in from infinity spiral down to $r = 0$.

Exercise 7.25

Redshift of the photons A spaceship on a circular orbit at radius r in a Schwarzschild metric emits a photon with the rest frame frequency ν_0 at an angle α outward from the tangential direction of the motion, in the plane of the orbit. What is the frequency of the photon as seen by a stationary observer at a large distance? [Answer. The frequency observed at infinity will be

$$\nu_\infty = \gamma \nu_0 \left(1 + v \cos\alpha\right) \left(1 - \frac{2GM}{r}\right)^{1/2}, \tag{7.176}$$

where $\gamma = (1 - v^2)^{-1/2}$, $v^2 = (GM/r)(1 - 2GM/r)^{-1}$.]

Exercise 7.26

Going into a shell Consider a spherical thin shell of mass M and radius R. A photon travels along a radial trajectory from a radius $r = R_i > R$ towards the centre of the shell. (Assume that a small hole is drilled on the shell to let this happen or that the shell is transparent to the photon.) Show that the observer at the origin of the shell will find the photon frequency to have been blue shifted by the amount

$$\frac{\nu_0}{\nu_E} = \left(\frac{1 - 2GM/R_i}{1 - 2GM/R}\right)^{1/2}. \tag{7.177}$$

Exercise 7.27
You look fatter than you are   The effect of gravity on light makes massive objects look bigger than they are. Consider a light ray grazing the surface of a massive spherical star of radius R and reaching $r \to \infty$ with an impact parameter b. Show that

$$b = R\left(1 - \frac{2GM}{c^2 R}\right)^{-1/2}. \tag{7.178}$$

This effect increases the apparent diameter of the Sun by about 3 km.

Exercise 7.28
Capture of photons by a Schwarzschild black hole   Consider a photon travelling at an angle Ψ to the radial direction in a Schwarzschild metric.

   (a) Show the following. (i) If the initial location of the photon was at $r < 3GM$ and the motion was inwardly directed, it will be captured by the black hole; (ii) if the initial location was at $r > 3GM$ and was outwardly directed, the photon will escape to infinity.

   (b) If the photon was at $r < 3GM$ and was moving outward, then the condition for the escape is

$$\sin \Psi < \frac{3\sqrt{3}\,GM}{r}\sqrt{1 - \frac{2GM}{r}}. \tag{7.179}$$

(c) If the photon was at $r > 3GM$ and was moving inwards, the condition for the escape is

$$\sin \Psi > \frac{3\sqrt{3}\,GM}{r}\sqrt{1 - \frac{2GM}{r}}. \tag{7.180}$$

Exercise 7.29
Twin paradox in the Schwarzschild metric?   Let us consider two observers A and B in the Schwarzschild metric. Observer A is on a circular orbit at the radius $r = 4GM$; observer B starts from a radius $r < 4GM$, moves radially outward to a maximum radius and falls back to $r = 4GM$. The orbits are so arranged that observer A completes exactly N orbits during the time interval taken by the observer B to cross $r = 4GM$ in the onward and the return trips and that A and B meet when their orbits cross. Both the observers are travelling on geodesics in the Schwarzschild metric and will experience no gravitational force in their respective frames. If their clocks were synchronized during the first meeting, how much will they differ when they meet for the second time? Is there a 'twin paradox' in this case, since each observer can consider himself to be located in an inertial, freely falling, frame?

7.6 Gravitational collapse of a dust sphere

We saw in Section 7.2.2 that static, spherically symmetric, solutions may not exist for all values taken by the parameters describing the source. By and large, if the gravitational field becomes too strong, the source will undergo a gravitational collapse which cannot be halted. We will study the result of such a collapse in detail in the next chapter but will describe one simple case of gravitational collapse in this section. This corresponds to the collapse of a sphere of pressureless dust. While it is not surprising that such a body collapses under self-gravity in the absence of any pressure gradient to prevent the collapse, the model has two advantages. First, it can be solved analytically; second the general features of the collapse that we describe here appear even in more realistic collapse scenarios.

To describe the geometry produced by a sphere of spherically symmetric dust with the energy-momentum tensor $T_{ab} = \rho u_a u_b$ we will use a line element of the form

$$ds^2 = -d\tau^2 + e^{\lambda(\tau, R)} dR^2 + r^2(\tau, R)(d\theta^2 + \sin^2\theta d\phi^2). \tag{7.181}$$

This is an example of what is known as a synchronous coordinate system in which the time coordinate τ gives the proper time of the particles at rest with fixed spatial coordinate values. Such a form is obtained from Eq. (7.12) by trading off the functional freedom in g_{00} for $g_{\theta\theta}$ and $g_{\phi\phi}$. More generally, a synchronous coordinate system is obtained by using the choice of four coordinates to set $g_{00} = -1$ and $g_{0\alpha} = 0$. It may not be possible to do this globally, in general, but in this case of spherical symmetry one can introduce such a coordinate system.

Calculating the components of the Ricci tensor for this metric and equating it to the energy-momentum tensor of dust, we get the following set of equations (we are temporarily choosing units with $G = 1$):

$$-e^{-\lambda} r'^2 + 2r\ddot{r} + \dot{r}^2 + 1 = 0 \tag{7.182}$$

$$-\frac{e^{-\lambda}}{r} \left(2r'' - r'\lambda'\right) + \frac{\dot{r}\dot{\lambda}}{r} + \ddot{\lambda} + \frac{\dot{\lambda}^2}{2} + \frac{2\ddot{r}}{r} = 0 \tag{7.183}$$

$$-\frac{e^{-\lambda}}{r^2} \left(2rr'' + r'^2 - rr'\lambda'\right) + \frac{1}{r^2} \left(r\dot{r}\dot{\lambda} + \dot{r}^2 + 1\right) = 8\pi\rho \tag{7.184}$$

$$2\dot{r}' - \dot{\lambda}r' = 0, \tag{7.185}$$

where the prime denotes differentiation with respect to r and the dot denotes differentiation with respect to τ. Integrating Eq. (7.185) we get

$$e^{\lambda} = \frac{r'^2}{1 + f(R)}, \tag{7.186}$$

where $f(R)$ is an arbitrary function with $(1 + f) > 0$. On using this in Eq. (7.182) and simplifying one can get the first integral

$$\dot{r}^2 = f(R) + \frac{F(R)}{r}, \tag{7.187}$$

where $F(R)$ is another arbitrary function. Integrating this, we can obtain the solution $r(R, \tau)$ in parametric form (in terms of the parameter η) as

$$r = \frac{F}{2f}(\cosh \eta - 1), \qquad \tau_0(R) - \tau = \frac{F}{2f^{3/2}}(\sinh \eta - \eta) \qquad (\text{for} \quad f > 0)$$
(7.188)

$$r = \frac{F}{-2f}(1 - \cos \eta), \qquad \tau_0(R) - \tau = \frac{F}{2(-)f^{3/2}}(\eta - \sin \eta) \qquad (\text{for} \quad f < 0)$$
(7.189)

$$r = \left(\frac{9F}{4}\right)^{1/3} [\tau_0(R) - \tau]^{2/3} \qquad (\text{for} \quad f = 0),$$
(7.190)

where $\tau_0(R)$ is again an arbitrary function. Finally, using these results to eliminate f we can obtain an expression for energy density

$$8\pi\rho = \frac{F'}{r'r^2}.$$
(7.191)

The above equations determine the complete solution through the functions $\lambda(\tau, R)$, $r(\tau, R)$ and $\rho(\tau, R)$ and three arbitrary functions $f(R), F(R), \tau_0(R)$; but in the metric we are allowed to transform R to any other function of itself $R \to R'(R)$. Because of this freedom, there are only two independent functions in the solutions which correspond to the freedom in the choice of distribution of density and radial velocity of the matter at some initial instant and that need to be given as an initial condition.

There is one interesting feature of this solution which is worth mentioning. In the synchronous frame, each dust particle has a fixed value of R and its motion is determined by the function $r(\tau, R)$ with $\dot{r}$ being the radial velocity. If we now fix the form of the arbitrary functions which appear in the solutions in the interval $(0 < R < L)$, then we completely determine the behaviour of a sphere of radius L *independent* of the behaviour of the functions at $R > L$. This is again reminiscent of the situation in Newtonian gravity and arises because of the high degree of symmetry. The total mass contained in the sphere of size L is given by

$$m = 4\pi \int_0^{r(t,L)} \rho r^2 \, dr = 4\pi \int_r^L \rho r^2 r' \, dR.$$
(7.192)

Using Eq. (7.191) we find that

$$m = \frac{F(L)}{2}.$$
(7.193)

This shows that the function $F(R)$ is directly related to the distribution of mass.

The solutions given in Eq. (7.188), Eq. (7.189) and Eq. (7.190) are capable of describing both expanding and contracting spheres depending on the range of values taken by η. We are interested in the collapse scenario in which the solutions are parametrized such that the instant $\tau = \tau_0(R)$ corresponds to matter at a given radial coordinate R collapsing to the origin. Close to the time of the collapse, all these solutions tend to the same asymptotic forms given by

$$r \approx \left(\frac{9F}{4}\right)^{1/3}[\tau_0 - \tau]^{2/3}, \qquad e^{\lambda/2} \approx \left(\frac{2F}{3}\right)^{1/3}\frac{\tau_0'}{\sqrt{1+f}}[\tau_0 - \tau]^{-1/3}.$$
(7.194)

The matter density becomes infinite, clearly indicating a singularity:

$$8\pi\rho \approx \frac{2F'}{3F\tau_0'(\tau_0 - \tau)}.$$
(7.195)

A very special kind of collapse – in which all the particles reach the origin simultaneously – is obtained for the choice $\tau_0(R) = $ constant. In this case, the asymptotic behaviour is given by

$$r \approx \left(\frac{9F}{3}\right)^{1/3}[\tau_0 - \tau]^{2/3}, \qquad e^{\lambda/2} \approx \left(\frac{2}{3}\right)^{1/3}\frac{F'}{2F^{2/3}\sqrt{1+f}}[\tau_0 - \tau]^{2/3},$$
(7.196)

and

$$8\pi\rho \approx \frac{4}{3(\tau_0 - \tau)^2}.$$
(7.197)

These are the simplest kind of gravitational collapse solutions.

Exercise 7.30
Spherically symmetric collapse of a scalar field Consider a massless scalar field $\phi(t, r)$ which is acting as a source for Einstein's equations in the context of spherical symmetry. Take the metric to be of the form:

$$ds^2 = -b^2(t, r)dt^2 + a^2(t, r)dr^2 + r^2 d\Omega^2.$$
(7.198)

Show that the scalar wave equation can be reduced, in terms of the variables $\Phi \equiv \partial_r\phi$, $\Pi = (a/b)\partial_t\phi$, to the first order form:

$$\partial_t\Phi = \partial_r[(a/b)\Pi]; \quad \partial_t\Pi = \frac{1}{r^2}\partial_r[(a/b)\Phi],$$
(7.199)

while two of the Einstein's equations become:

$$\frac{1}{b}\frac{\partial a}{\partial r} + \frac{a^2 - 1}{2r} - 2\pi r[\Pi^2 + \Phi^2] = 0; \quad \frac{1}{b}\frac{\partial b}{\partial r} + \frac{1}{a}\frac{\partial a}{\partial r} - \frac{a^2 - 1}{2r} = 0.$$
(7.200)

(The numerical integration of these equations leads to an interesting critical phenomenon in the formation of a black hole in such a system.)

PROJECTS

Project 7.1

Embedding the Schwarzschild metric in six dimensions

It is possible to think of the Schwarzschild metric as the induced metric[2] on a four-dimensional hypersurface embedded in a *flat* six-dimensional spacetime with the line element

$$ds_6^2 = -dZ_1^2 + dZ_2^2 + dZ_3^2 + dZ_4^2 + dZ_5^2 + dZ_6^2. \tag{7.201}$$

(a) Prove that the following functions will do the embedding by computing explicitly the induced metric and showing it to be the Schwarzschild metric

$$Z_1 = 4GM\sqrt{1 - \frac{2GM}{r}}\,\sinh\left(\frac{t}{4GM}\right), \qquad Z_2 = 4GM\sqrt{1 - \frac{2GM}{r}}\,\cosh\left(\frac{t}{4GM}\right),$$

$$Z_3 = \pm\int\left[\frac{2GM}{r} + \left(\frac{2GM}{r}\right)^2 + \left(\frac{2GM}{r}\right)^3\right]^{1/2}dr,$$

$$Z_4 = r\sin\theta\cos\phi, \qquad Z_5 = r\sin\theta\sin\phi, \qquad Z_6 = r\cos\theta. \tag{7.202}$$

It is assumed that $Z_3 > 0$ for $r > 2GM$ and $Z_3 = 0$ for $r = 2GM$.

(b) In spite of the fact that six-dimensional space is difficult to visualize, one should be able to use this embedding to understand the Schwarzschild metric better. Plot different cross-sections of the six-dimensional space, making full use of the symmetries of the Schwarzschild metric, and see whether you can understand this embedding. How do various physical phenomena (for example, the particle motion) appear when viewed in the six-dimensional space? See whether you can obtain deeper insights into the Schwarzschild metric using the embedding scheme.

Project 7.2

Poor man's approach to the Schwarzschild metric

It is possible to obtain the Schwarzschild metric by a remarkably simple, but strange, line of reasoning. This project explores this approach and investigates related issues.[3]

(a) Consider a point P at a distance r from the origin where the Newtonian gravitational potential due to a mass M at the origin is $\phi_N = -GM/r$. If we consider a small box around P which is freely falling towards the origin, then the metric in terms of the coordinates used by a freely falling observer in the box will be just that of special relativity:

$$ds^2 = -c^2 dt_{\text{in}}^2 + dr_{\text{in}}^2. \tag{7.203}$$

(The subscript 'in' is for inertial frame.) Transform the coordinates from the inertial frame to a frame (T, r) which will be used by observers who are at rest around the point P using *Galilean transformation* with $dt_{\text{in}} = dT$, $dr_{\text{in}} = dr - vdT$, where $v(r) = -\hat{r}\sqrt{2GM/r}$. Show that the metric in the new coordinates is

$$ds^2 = -\left[1 - \frac{2GM}{c^2 r}\right]c^2 dT^2 + 2\sqrt{(2GM/r)}\,drdT + dr^2. \tag{7.204}$$

(b) Incredibly enough, the metric in Eq. (7.204) turns out to be the Schwarzschild metric.

Prove this by transforming to a new time coordinate:

$$ct = c \int dT + \frac{1}{c^2} \int dr \frac{\sqrt{(2GM/r)}}{(1 - \frac{2GM}{c^2 r})}$$

$$= cT - \left[\sqrt{\frac{8GMr}{c^2}} - \frac{4GM}{c^2} \tanh^{-1} \sqrt{\frac{2GM}{c^2 r}} \right], \qquad (7.205)$$

and showing that in the (t, r, θ, ϕ) coordinates the metric has the standard Schwarzschild form.

(c) This approach, of course, raises more questions than it answers and you are invited to explore them. Can you justify the use of (i) a Galilean transformation (note that a Lorentz transformation will give nothing new) and (ii) the form of the velocity field obtained from Newtonian considerations? Is this yet another 'accidental' correspondence between Newtonian and general relativistic results in the case of spherical symmetry or is there something deeper?

(d) In Exercise 7.3 and Exercise 7.4, we worked out several other spherically symmetric solutions like the Reissner–Nordstrom spacetime, de Sitter spacetime, etc. See whether you can obtain all these spacetimes by a similar reasoning.

(e) Consider a spacetime with the line element

$$ds^2 = -dt^2 + (d\boldsymbol{x} - v dt)^2, \qquad (7.206)$$

where $v = v(t, \boldsymbol{x})$. Compute G_{ab} for this metric and determine what kind of source can produce such a spacetime geometry.

Project 7.3

Radiation reaction in curved spacetime

In Chapter 2 we obtained the expressions for radiation and radiation reaction in the context of special relativity; see Eq. (2.174) and Eq. (2.193). The situation becomes much more complicated in the case of curved spacetime. One cannot write down in closed form the Green function for the wave equation in an arbitrary curved spacetime and hence one cannot write down a formula analogous to Eq. (2.174) in curved spacetime. This project explores several related issues.

(a) In flat spacetime, the electromagnetic field produced at an event x^i will depend on the properties of the worldline of the charged particle at an event y^i, where $(x - y)^2 = 0, x^0 > y^0$. This implies that the field produced by a charge reacts back on the charge only through the self-force at the event of emission, so to speak. Show that, in curved spacetime, the radiation can 'back-scatter' from the curvature, making this result invalid. Convince yourself that a charged particle can now feel the effect of scattered radiation which was originally emitted by the same particle earlier on. Explain qualitatively the origin of non-locality. [Hint. The result of Exercise 6.11 might help.]

(b) Consider a naive generalization of Eq. (2.193) to a generally covariant expression by using generally covariant forms of acceleration $(a^i = u^j \nabla_j u^i)$ and its time derivative $(u^j \nabla_j a^i)$. Do you think this is a valid generalization? If this is the correct expression, will a charged particle orbiting a mass in the Schwarzschild metric feel any radiation reaction? Will it radiate energy?

(c) A really challenging problem is to obtain the radiation reaction force acting on a charged particle in curved spacetime taking into account the non-locality arising from the back scattering. Carry this out for a particle moving in a circular orbit in the Schwarzschild metric, with help from the literature[4] and resolve the question raised in part (b) above.

8

Black holes

8.1 Introduction

This chapter covers several aspects of black hole physics, concentrating mostly on the Schwarzschild and Kerr black holes. It also introduces some important new concepts like the *maximal extension* of a manifold, Penrose–Carter diagrams to visualize the causal structure of the spacetime and the geometrical description of horizons as null surfaces. A derivation of the zeroth law of black hole mechanics and illustrations of first and second laws of black hole mechanics are also provided. The material developed here will be required for Chapters 13, 15 and 16. We will use units with $G = c = 1$ in this chapter.[1]

8.2 Horizons in spherically symmetric metrics

The Schwarzschild metric due to a mass M, located at the origin, has the form

$$ds^2 = -f(r)dt^2 + \frac{dr^2}{f(r)} + r^2 d\Omega^2; \quad f(r) = \left(1 - \frac{2M}{r}\right), \tag{8.1}$$

which depends on the characteristic length scale $r_g \equiv 2M$. The g_{tt} component vanishes and g_{rr} diverges at $r = r_g$ showing that this radius must play a key role in the description of physics in the Schwarzschild metric. It must, however, be remembered that the Schwarzschild metric describes the *empty* region outside a spherical source and is not applicable in the region occupied by the source where the metric will be quite different. If the size of the source is L, then the singular behaviour of the metric coefficients at $r = r_g$ is not of any concern as long as $r_g < L$. But if the size of the source is such that $L < r_g$, then several new features come into play due to the singular behaviour of the metric at $r = r_g$.

A physical context in which such a situation can arise, for example, is in the gravitational collapse of matter. Consider, as an idealized example, a sphere of pressureless particles ('dust') described by a energy-momentum tensor $T_b^a = \rho u^a u_b$

which was occupying a spherical region of radius $r = L_1$ at time $t = t_1$. Let the total mass of the dust sphere be such that $L_1 > r_g$. At $t = t_1$, the metric at $r > L_1$ is given by the Schwarzschild metric which is everywhere well behaved because of our assumption that $L_1 > r_g$. This initial configuration, however, cannot be static because there is no pressure gradient to balance the gravitational attraction. We expect the dust sphere to collapse under its own gravitational force thereby decreasing its size. When the matter contracts to a radius less than r_g, the metric outside becomes ill behaved at $r = r_g$ and we need to consider several new features introduced by such a collapse.

In fact, such a situation can arise even in the presence of pressure gradients. We saw in the previous chapter that, under very reasonable conditions, we can obtain a bound $2GM/c^2R < 8/9$ for a static spherically symmetric solution to exist. (We have temporarily reintroduced the G and c factors.) If we start with an initial, spherically symmetric, configuration for which $8/9 < 2GM/c^2R < 1$, then there arises the possibility that the gravitational collapse will contract the mass to a radius smaller than $(2GM/c^2)$. We need to understand the physics and geometry of the resulting configuration.

The singular behaviour of the metric (in which a component vanishes or diverges in a given coordinate system) can arise owing to two reasons. First, if the spacetime geometry itself is singular at that event then the metric will reflect the singular behaviour of the geometry. Second, the metric can appear to be singular just because of a bad choice of coordinates. For example, the metric on the surface of the 2-sphere can be expressed in terms of the coordinates $\mu \equiv \sin\theta$ and ϕ as:

$$dl^2 = \frac{d\mu^2}{1 - \mu^2} + \mu^2 d\phi^2. \tag{8.2}$$

This metric is clearly singular at $\mu = \pm 1$ but we know that the geometry of the 2-sphere is everywhere well behaved. In this case, one will be able to make a coordinate transformation to a new set of coordinates in which the metric is non-singular.

One way to distinguish between these two situations is to construct the scalar quantities built out of the curvature tensor and inspect whether they diverge at the event in question. A scalar quantity will have the same numerical value in all coordinates and if it diverges in a given coordinate system, it will do so in all coordinate frames. Such a pathology will indicate that the geometry itself is singular at that event. If none of the scalar invariants built from the curvature tensor diverges at a given event, then the geometry is not singular and one should – in principle – be able to transform to a new coordinate system in which the metric is well behaved.

This is precisely what happens in the case of the surface $r = r_g$ in the Schwarzschild metric. None of the scalar invariants made from the curvature tensor diverges on this surface. For example, the scalar built out of the curvature tensor has the form

$$I \equiv R_{abcd} R^{abcd} = \frac{48 M^2}{r^6}, \tag{8.3}$$

which is well behaved at $r = 2M$ showing that the geometry does not exhibit singular behaviour. (It is also clear from the above expression that the curvature blows up at $r = 0$ showing it is a location of genuine singularity.)

Given this background, our first task is to introduce a new set of coordinates in which the metric is non-singular at $r = r_g$ and examine the physics in these new coordinates. This study will show that the surface $r = r_g$ in the Schwarzschild metric – though not singular – has several peculiar properties and acts as a *horizon* preventing signals from being transmitted from $r < r_g$ to $r > r_g$. Most of these features related to the existence of the horizon are quite generic and occur in a wider class of metrics than the Schwarzschild metric. For example, we have seen in the previous chapter (see Exercise 7.3) that the metric of a charged source is again given by Eq. (8.1) with

$$f(r) = \left(1 - \frac{2M}{r} + \frac{Q^2}{r^2} \right). \tag{8.4}$$

In this case $f(r) = 0$ at two values $r = r_\pm$ with $r_\pm = M \pm \sqrt{M^2 - Q^2}$ for $M^2 > Q^2$. The larger of the two radii r_+ will act as a horizon in this case. Similarly, we found that (see Exercise 7.4) vacuum Einstein's equations with cosmological constant has the solution in the form of Eq. (8.1) with

$$f(r) = \left(1 - \frac{1}{3}\Lambda r^2 \right) \equiv \left(1 - H^2 r^2 \right). \tag{8.5}$$

Here $f(r) = 0$ at $r = \sqrt{3/\Lambda} = H^{-1}$ and this surface will act as a horizon. In view of all these, it will prove to be useful to keep our discussion somewhat more general and deal with an arbitrary function $f(r)$ in Eq. (8.1) to the extent possible.

We shall, therefore, consider metrics of the form in Eq. (8.1) with the condition that $f(r)$ has a simple zero at some $r = r_H$ with $f'(r_H) \neq 0$. We will see that all such metrics will have a horizon at $r = r_H$ and will share many of the physical features in the Schwarzschild metric. (If the function $f(r)$ has more than one simple zero, the situation becomes more complicated but most of our discussion can be generalized in a straightforward manner.) In such a case, the metric near $r = r_H$ becomes:

$$ds^2 \approx -f'(r_H)(r - r_H)dt^2 + \frac{dr^2}{f'(r_H)(r - r_H)} + dL_\perp^2, \tag{8.6}$$

where $dL_\perp^2$ denotes the metric on the t =constant, r =constant surface. Introducing a new variable $\kappa \equiv (1/2)f'(r_H)$ and a new coordinate $l \equiv (r - r_H)$ in place of r, the metric becomes

$$ds^2 \approx -2\kappa l \, dt^2 + \frac{dl^2}{2\kappa l} + dL_\perp^2. \tag{8.7}$$

This is precisely the metric of the Rindler frame encountered in Chapter 3 with κ being the parameter characterizing the acceleration (see Exercise 3.5). The horizon is now located at $l = 0$. We see that, when the flat spacetime is described in the Rindler coordinates, the metric in Eq. (8.7) is singular at $l = 0$; but since the underlying spacetime is flat, we know that the geometry has no real singularity at $l = 0$ and the peculiar behaviour of the metric in Eq. (8.1) must be due to the bad choice of coordinates. This is a clear indication that, for metrics of the form in Eq. (8.1), the spacetime geometry at $r = r_H$ is well-defined. As we shall see below, the coordinates used in Eq. (8.1) are similar to the Rindler coordinates and we will be able to introduce a new set of coordinates – analogous to the inertial coordinates of the flat spacetime – in which the singular behaviour disappears. We shall now describe how one can transform to a new set of coordinates in which the metric is non-singular and well behaved at the horizon.

8.3 Kruskal–Szekeres coordinates

As a warm up to achieving this, we shall first take a closer look at the Rindler metric in Eq. (8.7) and the transformation to the inertial coordinates. We shall discuss some key geometrical features of this transformation in this context before taking up the case of more general metrics like the Schwarzschild metric.

In the case of flat spacetime, the coordinate transformation between the inertial coordinates (T, X, Y, Z) and the Rindler coordinates (t, l, y, z) in Eq. (8.7) is given by $Y = y, Z = z$ and

$$\kappa T = \sqrt{2\kappa l} \sinh(\kappa t); \quad \kappa X = \pm\sqrt{2\kappa l} \cosh(\kappa t) \tag{8.8}$$

for the region $|X| > |T|$. We choose the positive sign in the right wedge $\mathcal{R}$ (with $X > 0$) and the negative sign in the left wedge $\mathcal{L}$ (with $X < 0$) (see Fig. 8.1). In Chapter 3, we considered only the region $\mathcal{R}$ but it can be easily verified that the transformation in Eq. (8.8) with the negative sign works in $\mathcal{L}$. In fact, we can introduce a similar set of transformations for the region $|X| < |T|$. These are

$$\kappa T = \pm\sqrt{-2\kappa l} \cosh(\kappa t); \quad \kappa X = \sqrt{-2\kappa l} \sinh(\kappa t) \tag{8.9}$$

with $l < 0$; the positive sign is chosen in the future light cone ($\mathcal{F}$) through the origin ($|X| < T, T > 0$) and the negative sign is chosen in the past light cone ($\mathcal{P}$)

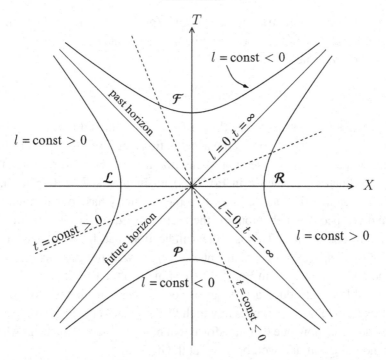

Fig. 8.1. The global manifold with different coordinate systems in the four quadrants. See text for discussion.

through the origin ($|X| < T, T < 0$). The inverse transformations are

$$l = \frac{1}{2}\kappa(X^2 - T^2); \qquad \frac{T}{X} = \pm \left\{ \begin{array}{c} \tanh(\kappa t) \\ \coth(\kappa t) \end{array} \right. , \qquad (8.10)$$

in each of the four sectors.

We will now discuss some key features which arise in the context of these coordinate transformations (from Rindler coordinates to inertial coordinates) that will prove to be valuable in the study of the Schwarzschild metric as well. We note that the Rindler coordinates (t, l) are analogous to the Schwarzschild metric coordinates (t, r) and the surface $r = r_g$ is analogous to the surface $l = 0$. In the (t, l) coordinates, t is timelike where $l > 0$ and spacelike where $l < 0$ (see Eq. (8.7)), which matches with the behaviour of the t coordinate in the Schwarzschild metric for $r > 2M$ and $r < 2M$. Further, in the case of Rindler coordinates, it is obvious from the light cone structure of the flat spacetime that the surface $l = 0$ acts as a 'one-way membrane'; signals can go from $l > 0$ to $l < 0$ but not the other way around. (When we talk of an $l = 0$ surface as a horizon, we often have this interpretation in mind.) We will see soon that this is indeed the case with the $r = r_H$ surface in all the metrics of the form Eq. (8.1).

But the most important feature that arises in the transformation from Rindler to inertial coordinates is that the transformation has 'doubled up' the spacetime region. This is obvious in the inverse transformations Eq. (8.10) which shows that the two distinct events (T, X) and $(-T, -X)$ are mapped to the same value of (t, l) in the Rindler frame. The original Rindler coordinates cover only the right wedge ($\mathcal{R}$) and the future wedge ($\mathcal{F}$). But when we use the transformations in Eq. (8.8) and Eq. (8.9) to go back to the inertial coordinates, a given value of (t, l) gets mapped to a *pair* of points in $\mathcal{R}$ and $\mathcal{L}$ for $l > 0$ and to a *pair* of points in $\mathcal{F}$ and $\mathcal{P}$ for $l < 0$. (Figure 8.1 shows the geometrical features of the coordinate systems.) In the inertial coordinates, we now have two copies of the region originally covered by the Rindler coordinates, but with different values for the new coordinates (T, X). This shows that the original system of coordinates was inadequate in revealing the complete structure of the spacetime manifold under consideration (which is just the flat spacetime in this case) but the coordinate transformation to the inertial coordinates has *analytically extended* the coordinates to the full manifold. This is, of course, because the coordinate transformations in Eq. (8.8) and Eq. (8.9) give the same coordinate label (t, l) to two distinct events in the spacetime thereby hiding the information about half the spacetime when the Rindler coordinates are used.

We shall see that similar features arise in the case of the Schwarzschild metric – as well as in a wide class of metrics described by Eq. (8.1). Introducing a good set of coordinates on the manifold is analogous to transforming from the Rindler coordinates to inertial coordinates; we will then find that a given event with coordinates (t, r) gets mapped to two distinct points in the new coordinate system. That is, the full manifold will have two copies of the region described by the coordinates used in the metric in Eq. (8.1). We will now show how one can obtain a similar set of 'good' coordinates – called *Kruskal–Szekeres coordinates* – for the more general case of metrics of the form in Eq. (8.1) with arbitrary $f(r)$ but for the conditions $f(r_H) = 0, \kappa = (1/2) f'(r_H) \neq 0$.

We will first introduce a new radial coordinate r^*, called the *tortoise coordinate*, by

$$r^* \equiv \int \frac{dl}{f(l)}. \tag{8.11}$$

Its exact form depends on the function $f(r)$ but near the horizon, with suitable choice for the integration constants, we have the relations:

$$r^* \simeq \frac{1}{2\kappa} \ln(2\kappa(r - r_H)) \simeq \frac{1}{2\kappa} \ln(2\kappa l); \quad f(r) \approx 2\kappa(r - r_H) \approx \exp(2\kappa r^*). \tag{8.12}$$

It is clear that the region outside the horizon $l = (0, \infty)$ is mapped to $r^* = (-\infty, \infty)$. Near the horizon, r^* changes very slowly – logarithmically – compared with r, which explains the name *tortoise coordinate*. We first transform the line element in Eq. (8.7) from (t, r) to (t, r^*) and then to the null coordinates $u = (t - r^*), v = (t + r^*)$, thereby obtaining:

$$ds^2 = f(r^*)(-dt^2 + dr^{*2}) + dL_\perp^2 = -f(v - u)dudv + dL_\perp^2. \tag{8.13}$$

Near the horizon, $f \approx \exp[2\kappa r^*] = \exp[(\kappa(v - u)]$, which is still singular as $r^* \to -\infty$. This singularity can be removed by the transformation to two new null coordinates (U, V) with

$$\kappa V = \exp[\kappa v], \qquad \kappa U = -\exp[-\kappa u] \tag{8.14}$$

in terms of which we have the metric in the form:

$$ds^2 = -f(v - u)dudv + dL_\perp^2 = \frac{f(U, V)}{\kappa^2 UV} dU dV + dL_\perp^2. \tag{8.15}$$

We see that the metric in the U, V coordinates is now perfectly regular at the horizon. Near the horizon, $f \approx \exp[(\kappa(v - u)] = -\kappa^2 UV$ so that (f/UV) is finite at the horizon. From the null coordinates U, V we can transform to standard timelike (T) and spacelike (X) coordinates by $U \equiv (T - X), V = (T + X)$. Putting it all together, we get the required transformation to a non-singular coordinate system, for the general case of the metric in Eq. (8.1), to be

$$\kappa X = e^{\kappa r^*} \cosh \kappa t; \quad \kappa T = e^{\kappa r^*} \sinh \kappa t \tag{8.16}$$

with r^* given by Eq. (8.11). The transformation in Eq. (8.16) is valid in the $|X| > |T|$ region. The other ranges can be covered by changes of signs in this transformation exactly as in the case of Eq. (8.8) and Eq. (8.9). (We will display these explicitly for the Schwarzschild metric in Eq. (8.20) and Eq. (8.21) below.) The metric in terms of (T, X) coordinates has the form

$$ds^2 = \frac{f}{\kappa^2(X^2 - T^2)}(-dT^2 + dX^2) + dL_\perp^2, \tag{8.17}$$

where f needs to be expressed in terms of (T, X) using the coordinate transformations.

In general, this metric will be quite complicated and will *not* even be static (that is, the metric will depend on the final time coordinate T) but we can easily obtain some general properties. Since $f \cong -\kappa^2 UV$ near the horizon $\mathcal{H}$, the horizon at $f = 0$ corresponds to the light cones $T^2 - X^2 = 0$ in these coordinates and $[f(T, X)/\kappa^2(T^2 - X^2)]$ is finite on the horizon by construction. Thus the (T, X) coordinates are, in fact, the locally inertial coordinates near $\mathcal{H}$. The transformations in Eq. (8.16) show that $(X^2 - T^2)$ is purely a function of r (or l) while (X/T) is a function of t. Thus $t = $ constant curves are radial lines through the origin with the $T = 0$ plane coinciding with $t = 0$ plane. Curves of $r = $ constant are hyperbolas (see Fig. 8.1). These properties are true in general irrespective of the form of $f(r)$.

Let us now carry out this procedure explicitly in the case of the Schwarzschild metric. The tortoise coordinate is given by

$$r^* = \int \frac{dr}{f(r)} = \int \frac{rdr}{r - r_g} = r + r_g \ln \left| \left(\frac{r}{r_g} - 1 \right) \right| \tag{8.18}$$

with a suitable choice for the integration constant. Further, for the Schwarzschild metric, we have $\kappa = (1/2) f'(r_H) = 1/2r_g$. Therefore, the prefactor $\exp(\kappa r^*)$ in Eq. (8.16) is given by

$$\exp(\kappa r^*) = e^{r/2r_g} \left| \left(\frac{r}{r_g} - 1 \right) \right|^{1/2}. \tag{8.19}$$

Using this, we find that the transformations to the so-called *Kruskal–Szekeres coordinates* (T, X, θ, ϕ) from the Schwarzschild coordinates (t, r, θ, ϕ) are given by

$$\left.\begin{aligned} X &= \left(\frac{r}{r_g} - 1 \right)^{1/2} e^{r/2r_g} \cosh(t/2r_g) \\ T &= \left(\frac{r}{r_g} - 1 \right)^{1/2} e^{r/2r_g} \sinh(t/2r_g) \end{aligned}\right\} \quad \text{(for } r > r_g\text{)}, \tag{8.20}$$

$$\left.\begin{aligned} X &= \left(1 - \frac{r}{r_g} \right)^{1/2} e^{r/2r_g} \sinh(t/2r_g) \\ T &= \left(1 - \frac{r}{r_g} \right)^{1/2} e^{r/2r_g} \cosh(t/2r_g) \end{aligned}\right\} \quad \text{(for } r < r_g\text{)}. \tag{8.21}$$

We have rescaled the coordinates to make T and X dimensionless. In these coordinates the Schwarzschild line element is given by

$$ds^2 = \frac{4r_g^3}{r} e^{-r/r_g} \left(-dT^2 + dX^2 \right) + r^2 \left(d\theta^2 + \sin^2 \theta d\phi^2 \right), \tag{8.22}$$

with r being given as an implicit function of X and T determined by the relation

$$\left(\frac{r}{r_g} - 1 \right) e^{r/r_g} = X^2 - T^2. \tag{8.23}$$

One major advantage of these coordinates is that radial $(d\theta = d\phi = 0)$ light rays obeying $ds^2 = 0$ are given by $dT = \pm dX$. Thus light cones are made of 45 degree lines in the X–T coordinates just like in the flat spacetime. Curves of constant r become hyperbolas in Kruskal–Szekeres coordinates and curves of constant t becomes straight lines passing through the origin. The line $X = T$ corresponds (simultaneously) to the horizon $r = r_g$ and $t = \infty$. Since light cones are made of 45 degree lines, it is obvious that the region $r < r_g$ cannot send signals to the region with $r > r_g$. The surfaces with $t =$ constant, $r =$ constant are 2-spheres with finite area in the case of Schwarzschild coordinates; for example, the horizon at $r = 2M$ has the area $16\pi M^2$. These are mapped to surfaces with $T =$ constant, $X =$ constant in the Kruskal–Szekeres coordinates.

There is a striking similarity between the transformation from the Rindler coordinates to inertial coordinates on the one hand and the transformation from the Schwarzschild coordinates to the Kruskal–Szekeres coordinates on the other hand. In particular, we now see that the full Schwarschild manifold has four regions which are mapped to two regions in the Schwarzschild coordinates. This is completely analogous to the four regions of the flat spacetime (in inertial coordinates) getting mapped to two regions of the Rindler coordinates. The usual Schwarzschild metric covers the $\mathcal{R}$ and $\mathcal{F}$ regions and attributes the *same* Schwarzschild coordinates (t, r) to the *pair* of events (T, X) and $(-T, -X)$ in the Kruskal–Szekeres coordinate system. The $r = 2M$ surface now gets mapped to the two surfaces $X = \pm T$. There are two exterior regions $r > 2M$, one corresponding to the right wedge $\mathcal{R}$ and another corresponding to the left wedge $\mathcal{L}$. Even the asymptotically flat region gets doubled in the process. The coordinate transformation has analytically extended the region of the spacetime manifold under consideration.

We saw earlier that the Schwarzschild metric describes a manifold with a physical singularity at $r = 0$ which corresponds to the curve $T^2 - X^2 = 1$ in the Kruskal–Szekeres coordinates. Again, because of the two-to-one mapping, the $r = 0$ gets mapped to two curves $T = \pm[X^2 + 1]^{1/2}$ in the Kruskal–Szekeres coordinates. Thus a worldline $X =$ constant, for example, crosses the horizon and hits the singularity in finite T. The region $T^2 - X^2 > 1$ is treated as physically irrelevant in the manifold.

The form of the metric in Eq. (8.22) shows that the metric is *not* static in the Kruskal–Szekeres coordinates while it was independent of t in the Schwarzschild coordinates. There is a nontrivial, time dependent, dynamics in this manifold which is not easy to see in the Schwarzschild coordinates but is obvious in the Kruskal–Szekeres coordinates (see Exercise 8.1). Because of these features, the (t, l) Schwarzschild coordinate system has an intuitive appeal that the Kruskal–Szekeres coordinate system lacks, in spite of the mathematical fact that the Kruskal–Szekeres coordinate system is analogous to the inertial coordinate system while the Schwarzschild coordinate system is like the Rindler coordinate system.

These features certainly appear a bit bizarre and one may wonder whether they have any physical relevance. If we treat the Schwarzschild metric as a family of exact solutions to vacuum Einstein equations, parametrized by M, then the resulting manifold is indeed represented by the Kruskal–Szekeres coordinates. (Such a solution is usually called an eternal black hole.) But, in most physical contexts one is interested in a solution in which matter collapses to a region within $r = r_g$. In that case the surface of the matter distribution will follow a timelike worldline in the Kruskal–Szekeres coodinates (see Fig. 8.3). This worldline will start in the right wedge $\mathcal{R}$ at early times, cross the $X = T$ horizon and move into the future wedge $\mathcal{F}$ later on. The Schwarzschild metric is, of course, valid only to the right of this worldline in the Kruskal–Szekeres coordinates since the region to the left of this worldline is occupied by collapsing matter and is not empty. So, in the case of a black hole formed by collapsing matter, the Kruskal–Szekeres coordinates are not valid in the regions $\mathcal{L}$ and $\mathcal{P}$ thereby solving the issue of 'doubled up' space-time. In any collapse scenario we need only concern ourselves with the two sectors $\mathcal{R}$ and $\mathcal{F}$.

There are, however, several counterintuitive physical phenomena taking place even in these two sectors $\mathcal{R}$ and $\mathcal{F}$ and we will next try to understand these by studying the simple physical processes in both.

Exercise 8.1

The weird dynamics of an eternal black hole The Schwarzschild metric depends explicitly on r, which is a *timelike* coordinate inside the horizon. This already indicates that a manifold described entirely by the Schwarzschild metric (with no interior solution for collapsing matter, etc.) should have a nontrivial time dependent evolution! This feature is more obvious in the Kruskal–Szekeres coordinates in which the metric explicitly depends on the time coordinate T. The purpose of this exercise is to explore some features of this dynamics.

(a) In Section 7.2.3 we worked out the embedding diagram for a spherically symmetric star with a Schwarzschild exterior and some general interior metric. In the same manner, you can work out the embedding diagram for a manifold described by the Schwarzschild metric for the entire range of coordinates with no other interior solution. Show that the relevant two-dimensional surface embedded in Euclidean three-dimensional space, having this geometry, is generated by rotating the curve

$$z(r) = \sqrt{8M(r - 2M)} + \text{constant} \qquad (8.24)$$

about the z-axis. Plot this surface. (The region connecting the two asymptotically flat regions is called the *Einstein–Rosen bridge*.)

(b) Show that the above description of the geometry can be translated directly to the geometry of $T = \text{constant} = 0$ surface in Kruskal–Szekeres coordinates.

(c) Consider now what happens to the geometry as T changes through the ranges $0 < T < 1$, $T = 1$ and $T > 1$. Show that the Einstein–Rosen bridge narrows and the two regions join at the $r = 0$ singularity pinching off the bridge at $T = 1$. Further show that for $T > 1$ the two sectors, each containing a singularity at $r = 0$, are completely separate.

8.3.1 Radial infall in different coordinates

To understand the geometrical features of the Schwarszchild manifold in the two coordinate systems, we will start by considering the radial trajectory of a particle of dust located at the surface of the dust sphere. In the absence of pressure, this particle follows a radial geodesic in the Schwarzschild metric. To determine the trajectory, we need to integrate Eq. (7.104) and Eq. (7.108) with $V_{\text{eff}}(r)$ evaluated for $L = 0$. From Eq. (7.104) we find that, with $\bar{E} = \mathcal{E}/m$,

$$\left(\frac{dr}{d\tau}\right)^2 = \left[\bar{E}^2 - (1 - 2M/r)\right] = \frac{2M}{r} - \frac{2M}{R}, \tag{8.25}$$

where we have assumed that $(dr/d\tau) = 0$ at $r = R$. (Incidentally, this is *precisely* the same equation which occurs for a particle in Newtonian theory, moving radially under the attraction of mass M – which is another peculiar feature arising from spherical symmetry.) The trajectory of the particle in terms of its proper time τ is obtained by the integral

$$\tau = \int \frac{dr}{[(2M/r) - (2M/R)]^{1/2}} = \left(\frac{R^3}{8M}\right)^{1/2} \left[2\left(\frac{r}{R} - \frac{r^2}{R^2}\right)^{1/2}\right.$$
$$\left. + \cos^{-1}\left(\frac{2r}{R} - 1\right)\right]. \tag{8.26}$$

The solution is usually expressed in parametric form as:

$$r = \frac{R}{2}(1 + \cos\eta), \qquad \tau = \frac{R}{2}\left(\frac{R}{r_g}\right)^{1/2}(\eta + \sin\eta) \tag{8.27}$$

with $r_g = 2M$. This equation, which describes the trajectory of the dust particles in terms of the the proper time τ measured by a clock carried by a particle, shows that the total *proper time* for the dust particle to fall from $r = R$ to $r = 0$ is finite and given by

$$\tau = \frac{\pi}{2}R\left(\frac{R}{r_g}\right)^{1/2}. \tag{8.28}$$

(This is, again, identical to the corresponding result in the Newtonian gravity.) Since the dust particle is at the surface of a sphere, the entire body will collapse to a point in finite proper time as measured by clocks located at the surface of the body. As the particle crosses $r = r_g$, no peculiar effects will be noticed by an observer located on the surface. The trajectory $r(\tau)$ is smooth throughout the motion.

The situation, however, is quite different when viewed by an outside observer using the time coordinate t. This can be found by integrating Eq. (7.104) but it is more convenient to use the above result and proceed as follows. Using Eq. (7.107),

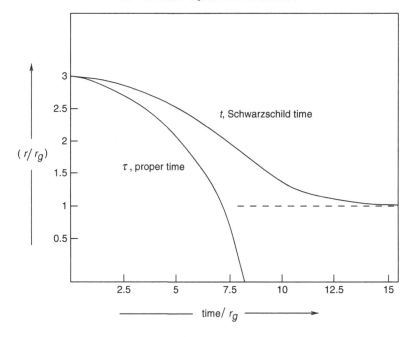

Fig. 8.2. The variation of coordinate time and proper time as a particle moves in a radial trajectory in the Schwarzschild metric.

we have

$$t = \left(1 - \frac{2M}{R}\right)^{1/2} \int \left(\frac{R^3}{8M}\right)^{1/2} \frac{(1 + \cos\eta)d\eta}{1 - 4M[R(1 + \cos\eta)]^{-1}}. \tag{8.29}$$

The integration is elementary but a bit tedious, leading to the result:

$$\frac{t}{2M} = \log\left| \frac{\left(\frac{R}{2M} - 1\right)^{1/2} + \tan\left(\frac{\eta}{2}\right)}{\left(\frac{R}{2M} - 1\right)^{1/2} - \tan\left(\frac{\eta}{2}\right)} \right| + \left(\frac{R}{2M} - 1\right)^{1/2}\left[\eta + \frac{R}{4M}(\eta + \sin\eta)\right]. \tag{8.30}$$

Once again the constant of integration is chosen such that $r = R$ at $t = 0$. Equation (8.27) along with Eq. (8.30) implicitly give the trajectory of the particle $r(t)$ in terms of the time coordinate t of an observer located at a large distance from the origin in the Schwarzschild metric.

This function $r(t)$ has a very different behaviour compared to $r(\tau)$. From the expressions Eq. (8.27) and Eq. (8.30) it is easy to see that $t \to \infty$ as $r \to r_g$. That is, it takes infinite *coordinate time* for the particle at the surface of the dust sphere to reach the Schwarzschild radius $r = r_g$. An outside observer can conclude that

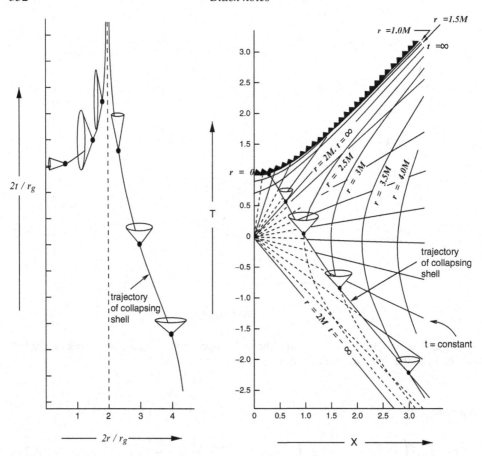

Fig. 8.3. A collapsing spherical region is shown in the Schwarzschild coordinates (left panel) as well as in the Kruskal–Szekeres coordinates (right panel). The trajectory of a particle located on the surface is shown along with the nature of light cones on the path in both the cases. In the Schwarzschild coordinate, there is a discontinuity at $r = r_g$; the surface reaches $r = r_g$ only asymptotically as $t \to \infty$. On the other hand, the collapse is continuous when viewed in the Kruskal–Szekeres coordinate system.

it takes infinite time for the body to collapse to the size $r = r_g$ even though it collapses to $r = 0$ within a finite proper time as determined by the clocks on the surface. The asymptotic form of the trajectory, $r(t)$, as r approaches r_g is easy to determine. We find that

$$r - r_g = \text{constant } e^{-(t/r_g)}. \tag{8.31}$$

The behaviour is shown in Fig. 8.2. This is an extreme example of the gravitational field affecting the clock rate.

Figure 8.3 shows a collapsing spherical region in the X–T coordinates and compares it with the corresponding region in the Schwarzschild coordinate. This figure also shows that, in the case of a collapsing body, only two out of the four regions in the Kruskal–Szekeres coordinates are physically relevant. This is because collapsing matter – which should be described by a different interior solution – will occupy the region to the left of the trajectory shown in Fig. 8.3.

Our explicit solution used the collapse of a pressureless dust. However, even if the material has pressure, any spherical mass distribution that collapses to $r < r_g$ will not be able to communicate with the region at $r > r_g$. This is obvious from the nature of light cones in the Kruskal–Szekeres coordinates. In the case of more complicated collapse scenarios, the trajectories of the collapsing material may be quite complex, but once the surface contracts below r_g, no light signal can escape to infinity.

In the case of a particle following a geodesic, we have demonstrated that it will hit the $r = 0$ singularity within finite proper time. The question arises as to whether a particle can cross to the $r < 2M$ region but can still avoid hitting the $r = 0$ singularity by following a *non-geodesic* timelike trajectory. For example, one can consider a spaceship with a sufficient amount of fuel that crosses to the $r < 2M$ region and tries to avoid hitting the $r = 0$ singularity by firing its rockets appropriately. We will now show that this is not possible and that any particle which crosses $r = 2M$ will hit the singularity in a proper time which is less than πM. To prove this, we will start with the condition that the trajectory of the particle should be timelike so that its four-velocity should satisfy the condition $u^i u_i = -1$. In expanded form this condition reads as

$$
1 = \left(1 - \frac{2M}{r}\right)\left(\frac{dt}{d\tau}\right)^2 - \left(1 - \frac{2M}{r}\right)^{-1}\left(\frac{dr}{d\tau}\right)^2 - r^2\left(\frac{d\theta}{d\tau}\right)^2
$$
$$
- r^2\left(\frac{d\theta}{d\tau}\right)^2 - r^2 \sin^2\theta\left(\frac{d\phi}{d\tau}\right)^2. \tag{8.32}
$$

When $r < 2M$, all the terms in the right hand side, except the one involving $(dr/d\tau)^2$, are negative. This gives us the inequality

$$
\left(\frac{2M}{r} - 1\right)^{-1}\left(\frac{dr}{d\tau}\right)^2 > 1. \tag{8.33}
$$

Writing this as $dr < -(2M/r - 1)^{1/2}\, d\tau$ and integrating from $r = 2M$ to $r = 0$, we find that $\tau < \tau_{\max}$, where

$$
\tau_{\max} = -\int_{2M}^{0}\left(\frac{2M}{r} - 1\right)^{-1/2} dr
$$
$$
= \left[r^{1/2}(2GM - r)^{1/2} + M \cos^{-1}\left(\frac{r}{M} - 1\right)\right]_{2M}^{0} = \pi M. \tag{8.34}
$$

Hence any particle that crosses $r = 2M$ will hit the singularity within a time πM.

The question arises as to the ultimate fate of the material which collapses in the above mentioned manner. It can be easily verified that the curvature of the Schwarzschild metric becomes infinitely large at $r = 0$. In other words, $r = 0$ is a genuine singularity in spacetime with an infinite gravitational field (in contrast to the event horizon at $r = r_g$ at which all geometrical quantities are finite). According to our considerations, the collapsing matter should hit the singularity at $r = 0$ in finite proper time. It is very likely that Einstein's equations become invalid near $r = 0$ where the curvature is arbitrarily strong, possibly owing to quantum gravitational effects. We, however, stress that as long as the causal structure is preserved by the quantum gravitational effects, say, the phenomena at $r < 2M$ are irrelevant to classical physics at $r > 2M$. Currently no theory is available to describe this situation and hence the issue of the final state of matter that collapses to form a black hole is, at present, unsettled.

The collapse also affects the communication between an observer (A, say) located on the surface of the dust sphere and a distant, stationary observer (B). Let us suppose that A is sending light signals to B at periodic intervals. Since the time of propagation for the radial light signals is given by $dt = dr/(1 - r_g/r)$, the time taken by the light signal to propagate from some point r to $r_0 > r$ is given by

$$\Delta t = \int_r^{r_0} \frac{dr}{(1 - r_g/r)} = r_0 - r + r_g \ln \frac{r_0 - r_g}{r - r_g}, \qquad (8.35)$$

which diverges as $r \to r_g$. Hence the signals take progressively a longer time to reach the distant observer and the signal sent when the dust sphere crosses $r = r_g$ (as indicated by the proper time on a clock that is on the surface of the collapsing sphere) reaches the distant observer only after an infinite duration of time. The signals sent by the observer on the collapsing sphere, after it has contracted through $r = r_g$, do not reach the outside observer at all. It follows that the region inside $r = r_g$ cannot communicate with or influence the outside region. The fact that part of the spacetime region gets cut off from the rest is a special feature of Einstein's gravity.

As the body collapses, any light emitted from its surface will be progressively redshifted. Consider an electromagnetic wave with a wave vector k^a propagating radially from a radius r to infinity. We know that $k_a \xi^a = k_0$ is conserved as the wave propagates because of the existence of the Killing vector $\xi^a = (1, 0, 0, 0)$ (see Eq. (4.136)). An observer who is stationary at r will have the four-velocity $u^a = (-g_{00})^{-1/2}(1, 0, 0, 0)$. (The g_{00} factor is needed to ensure the normalization $u^a u_a = -1$.) The frequency of the electromagnetic wave, as measured by this observer, will be $\omega(r) = k^a u_a = k_0 u^0(r)$. The ratio of frequencies measured by observers at infinity and at r will be $(\omega(\infty)/\omega(r)) = (u^0(\infty)/u^0(r)) = \sqrt{-g_{00}(r)}$.

Therefore,

$$\omega(\infty) = \omega(r) \left(1 - \frac{r_g}{r}\right)^{1/2}. \tag{8.36}$$

As the dust sphere approaches r_g, the $\omega(\infty) \to 0$ and the wave suffers infinite redshift.

Exercise 8.2
Dropping a charge into the Schwarzschild black hole The fact that a radially infalling particle reaches $r = 0$ in finite proper time but infinite coordinate time raises some intriguing questions, one of which is the following. Consider a test particle of mass m and charge q that is dropped into a Schwarzschild black hole of mass M. This would increase the total mass of the black hole to $M + m$ and its charge to q. As a consequence, we expect the metric outside the black hole to become the Reissner–Nordstrom metric (see Exercise 7.3). To an observer A falling with the charge, it reaches $r = 0$ in finite proper time and the electric field outside should be directed radially outwards from the origin once the charge has reached $r = 0$. But the outside observer B will argue that the charge only goes up to $r = 2M$ even as $t \to \infty$. Hence observer B might expect to see – at late times – an electric field which is radial outwards from a point displaced from the origin by a distance $r = 2M$. But this experiment can be carried out (in principle) and the electric field outside the horizon can be measured at all times. It seems we cannot simultaneously satisfy the expectations of observer A and observer B. Your task is to reconcile this conflict.

(a) Since the problem of a moving charge in the Schwarzschild metric is intractable, consider instead the field of a charged particle kept at a the spatial location $(0, 0, z)$ along the z-axis of the Schwarzschild metric. If you have done Project 5.2, you already know the electric field in this case. If not, calculate the electric field in this case and determine how it changes as the charge is moved closer and closer to the horizon.

(b) Show that as the position of the charge approaches the horizon (that is, as $z \to 2M$), the asymptotic electric field does look as though it is pointing radially outwards from $r = 0$ and not from the location of the charge! In other words, gravity also 'bends' the field lines of the electric field.

(c) Give a physical interpretation for the above feature and settle the differences between observer A and observer B.

Exercise 8.3
Painlevé coordinates for the Schwarzschild metric It is possible to construct several different coordinate systems which are all well behaved at $r = 2M$ and the Kruskal–Szekeres coordinates are just one of them. Another coordinate system that is well behaved at $r = 2M$ was described in Project 7.2. In case you did not do it, here is a slightly simplified version.

(a) Consider a massive particle falling in the Schwarzschild metric along a radial trajectory starting from rest at infinity. Let its four-velocity be $u_a(x)$ when it is at an event x^a. Show that this four-velocity can be expressed as a gradient of the scalar in the form $u_a = \partial_a T$ where

$$T = t + \int^r \left(\frac{2M}{r'}\right)^{1/2} \left(1 - \frac{2M}{r}\right)^{-1} dr'. \tag{8.37}$$

(b) This suggests the use of T as a new time coordinate. Transform from the Schwarzschild coordinates to (T, r, θ, ϕ) and show that the resulting metric has the form

$$ds^2 = -dT^2 + \left(dr + \sqrt{\frac{2M}{r}} \, dT \right)^2 + r^2(d\theta^2 + \sin^2 \theta d\phi^2). \qquad (8.38)$$

(c) Describe the geometry in terms of the new coordinates by plotting lines of constant t and constant r in the (T, r) coordinate system and vice versa. How does the metric behave near the horizon?

Exercise 8.4

Redshifts of all kinds Consider an observer who is freely falling radially in the Schwarzschild metric starting from rest at infinity. She emits a photon which travels in the outgoing radial direction. This photon is detected by another observer who is stationary at a large distance.

(a) Compute the ratio of the wavelengths of the photon at emission and absorption as a function of the radial coordinate r at which the emission takes place. [Hint. This is different from the emission and absorption by two *static* observers. You should now get $(\lambda_{\rm rec}/\lambda_{\rm em}) \propto [1 - (2M/r_{\rm em})]^{-1}$ without the square root in Eq. (8.36).]

(b) Show that the time of emission of the photon and the reception of the photon are given by

$$t_{\rm em} = -2M \ln \left[1 - \frac{2M}{r_{\rm em}} \right] + \text{constant}; \quad t_{\rm rec} = -4M \ln \left[1 - \frac{2M}{r_{\rm em}} \right] + \text{constant}.$$
$$(8.39)$$

Hence show that $(\lambda_{\rm rec}/\lambda_{\rm em}) \propto \exp(t_{\rm rec}/4M)$.

8.3.2 General properties of maximal extension

Most of what we have seen in the case of the Schwarzschild metric is applicable in any spherically symmetric spacetime with a horizon. We shall now briefly describe a general class of such metrics and their physical properties.

While it is possible to have different kinds of solutions to Einstein's equations with horizons, some of the solutions have attracted significantly more attention than others. Table 8.1 summarizes the features related to three of these solutions. In each of these cases, the metric can be expressed in the form Eq. (8.7) with different forms of $f(l)$ given in the table. All these cases have only one horizon at some surface $l = l_H$ and the parameter κ is well defined. The coordinates (T, X) are well behaved near the horizon while the original coordinate system (t, l) is singular at the horizon. Table 8.1 describes all three cases of horizons in which we are interested, with suitable definition for the coordinates.

In all the cases the horizon at $l = l_H$ corresponds to the light cones through the origin $(T^2 - X^2) = 0$ in the freely falling coordinate system. It is conventional to call the $T = X$ surface the *future horizon* and the $T = -X$ surface the *past horizon*. Also note that the explicit transformations to (T, X) given in Table 8.1

Table 8.1. *Properties of Rindler, Schwarzschild and de Sitter metrics*

Metric	Rindler	Schwarzschild	de Sitter
$f(l)$	$2\kappa l$	$\left[1 - \dfrac{2M}{l}\right]$	$(1 - H^2 l^2)$
$\kappa = \frac{1}{2} f'(l_H)$	κ	$\dfrac{1}{4M}$	$-H$
r^*	$\dfrac{1}{2\kappa} \ln \kappa l$	$l + 2M \ln\left[\dfrac{l}{2M} - 1\right]$	$\dfrac{1}{2H} \ln\left(\dfrac{1 - Hl}{1 + Hl}\right)$
κX	$\sqrt{2\kappa l}\,\cosh \kappa t$	$e^{\frac{l}{4M}}\left[\dfrac{l}{2M} - 1\right]^{1/2} \cosh\left[\dfrac{t}{4M}\right]$	$\left(\dfrac{1 - Hl}{1 + Hl}\right)^{1/2} \cosh Ht$
κT	$\sqrt{2\kappa l}\,\sinh \kappa t$	$e^{\frac{l}{4M}}\left[\dfrac{l}{2M} - 1\right]^{1/2} \sinh\left[\dfrac{t}{4M}\right]$	$\left(\dfrac{1 - Hl}{1 + Hl}\right)^{1/2} \sinh Ht$

correspond to $l > 0$ and the right wedge, $\mathcal{R}$. Changing l to $-l$ in these equations with $l < 0$ will take care of the left wedge, $\mathcal{L}$. The future and past regions will require interchange of the sinh and cosh factors. These are direct generalization of the transformations in Eq. (8.8) and Eq. (8.9).

The simplest case corresponds to flat spacetime in which (T, X) are the Minkowski coordinates and (t, l) are the Rindler coordinates. The range of all the coordinates is $(-\infty, \infty)$. The g_{00} does *not* go to (-1) at spatial infinity in (t, l) coordinates and the horizon is at $l = 0$. The metric is static in both coordinates. The Killing vector corresponding to translation in the Rindler time coordinate corresponds to the Killing vector for the Lorentz boost in the (T, X) plane in the Minkowski coordinates.

The second case is that of a Schwarzschild black hole. The full manifold is described in the (T, X) Kruskal–Szekeres coordinates but the metric is *not* static in terms of the Kruskal–Szekeres time T. The horizon at $X^2 = T^2$ divides the black hole manifold into the four regions $\mathcal{R}, \mathcal{L}, \mathcal{F}, \mathcal{P}$. In terms of the Schwarzschild coordinates, the metric is independent of t and the horizon is at $l = 2M$, where M is the mass of the black hole. The standard Schwarzschild coordinates (t, l) is a 2-to-1 map from the Kruskal–Szekeres coordinates (T, X). The region $l > 2M$ which describes the exterior of the black hole corresponds to both $\mathcal{R}$ and $\mathcal{L}$, and the region $0 < l < 2M$ that describes the interior of the black hole, corresponds to both $\mathcal{F}$ and $\mathcal{P}$. The transverse coordinates are now (θ, ϕ) and the surfaces $t = $ constant, $l = $ constant are 2-spheres.

In the case of a black hole formed due to gravitational collapse, the Schwarzschild solution is applicable only to the region outside the collapsing matter, if the collapse is spherically symmetric. The surface of the collapsing matter will be a timelike curve cutting through $\mathcal{R}$ and $\mathcal{F}$, making the whole of $\mathcal{L}$, $\mathcal{P}$ (and part of $\mathcal{R}$ and $\mathcal{F}$) irrelevant since they will be inside the collapsing matter (see Fig. 8.3). In this case, the past horizon does not exist and we are only interested in the future horizon. Similar considerations apply whenever the actual solution corresponds only to part of the full manifold.

The third spacetime listed in Table 8.1 is the de Sitter spacetime (see Exercise 7.3) which, again, admits a Schwarzschild type coordinate system and a Kruskal–Szekeres type coordinate system. The horizon is now at $l = H^{-1}$ and the spacetime is *not* asymptotically flat. There is also a reversal of the roles of 'inside' and 'outside' of the horizon in the case of de Sitter spacetime. If the Schwarzschild type coordinates are used on the black hole manifold, an observer at large distances $(l \rightarrow \infty)$ from the horizon $(l = 2M)$ will be stationed at nearly flat spacetime and will be confined to $\mathcal{R}$. The corresponding observer in the de Sitter spacetime is at $l = 0$, which is again in $\mathcal{R}$. Thus the nearly inertial observer in the de Sitter manifold is near the origin, 'inside' the horizon, while the nearly inertial observer in the black hole manifold is at a large distance from the horizon and is 'outside' the horizon; but both are located in the region $\mathcal{R}$ in Fig. 8.1 making this figure to be of universal applicability to all these three metrics. The transverse dimensions are compact in the case of the de Sitter manifold as well.

8.4 Penrose–Carter diagrams

One of the major advantages of the Kruskal–Szekeres coordinates compared with Schwarzschild coordinates is that the light-cones in the Kruskal–Szekeres coordinates in the X–T plane are 45 degree lines, just as in the flat spacetime. This makes the causal relationships between any two events easy to understand. We will now describe a simple extension of this idea, leading to what are called *Penrose–Carter diagrams*, which are of considerable help in the study of the causal structure in general and horizons in particular.

To set the stage, let us recall that any two-dimensional section of the spacetime can be described by a conformally flat metric of the form

$$ds^2 = \Omega^2(t, x)(-dt^2 + dx^2) = -\Omega^2(u, v)\,du\,dv, \qquad (8.40)$$

where $u = t - x, v = t + x$ (see Section 5.3.2). Such a coordinate system will have light cones described by the 45 degree lines $x = \pm t + \text{constant}$. The Kruskal–Szekeres metric in Eq. (8.22), for example, has this form in the X–T plane. However, the range of coordinates in the line element in Eq. (8.40) will – in general – be infinite. To study the causal structure of the spacetime, it is often convenient to map the infinite range of coordinates to a finite interval without, of course, losing the property that the light cones should remain 45 degree lines. This will require making a transformation preserving the conformal flatness of the line element but at the same time mapping the infinite range of coordinates to the finite interval. Since any transformation of the form $u \to f_1(u), v \to f_2(v)$ preserves the conformal structure of Eq. (8.40), a simple choice is given by the transformations

$$v \to v' = \tan \frac{1}{2}v; \quad u \to u' = \tan \frac{1}{2}u. \tag{8.41}$$

These transformations make use of the fact that when z varies over $(-\infty, \infty)$ the function $\tan^{-1} z$ varies over the finite range $(-\frac{\pi}{2}, \frac{\pi}{2})$. If we now use $(1/2)(v'+u')$ and $(1/2)(v'-u')$ as the new coordinates in place of t and x, we would have achieved our objective. Let us illustrate this procedure with a couple of examples.

As a first example, let us consider the flat spacetime itself expressed in the spherical polar coordinates (t, r, θ, ϕ) with the range of t and r coordinates being $-\infty < t < +\infty$ and $0 < r < \infty$. We now make a transformation to the null coordinates $t \pm r$ and then use the $\tan^{-1}$ function to obtain a new set of coordinates (T, R) with a finite range. Explicitly, the coordinate transformation is given by

$$t + r = \tan \frac{1}{2}(T + R) ; \quad t - r = \tan \frac{1}{2}(T - R) \tag{8.42}$$

$$ds^2 = \frac{-dT^2 + dR^2}{4 \cos^2 \frac{1}{2}(T + R) \cos^2 \frac{1}{2}(T - R)} + r^2(T, R) \left(d\theta^2 + \sin^2 \theta \, d\phi^2\right). \tag{8.43}$$

The structure of the spacetime in the (T, R) coordinates is shown in Fig. 8.4 along with the results for Cartesian coordinates and the Rindler coordinates.

It is obvious that light cones are still 45 degree lines in this Penrose–Carter diagram, as it is called. What is more, points and regions located at infinite distances in the original coordinates have now been brought to a finite distance. The figure indicates a set of different kinds of 'infinity' that is useful in the discussion of physical phenomena. The following list gives the definition of each of these.

$$i^+ = \text{future timelike infinity } (T = \pi, \, R = 0)$$
$$i^0 = \text{spatial infinity } (T = 0, \, R = \pi)$$
$$i^- = \text{past timelike infinity } (T = -\pi, \, R = 0)$$
$$\mathcal{I}^+ = \text{future null infinity } (T = \pi - R, \, 0 < R < \pi)$$
$$\mathcal{I}^- = \text{past null infinity } (T = -\pi + R, \, 0 < R < \pi)$$

For example, outgoing radiation described by a function $f(t - r)$ will eventually reach future null infinity.

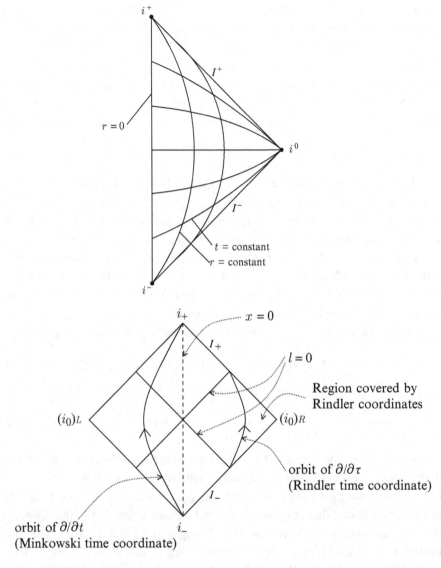

Fig. 8.4. The Penrose–Carter diagram for flat spacetime. The spatial coordinates in the top figure are standard (t, r, θ, ϕ) coordinates while the bottom figure shows the corresponding structure in (t, x, y, z) coordinates concentrating on the t–x plane. Also shown are the Rindler coordinates with (τ, l, y, z) in the τ–l plane. Note the clear difference between the causal structure when we confine ourselves to part of the spacetime in the right wedge $\mathcal{R}$.

As a second example, let us consider the Schwarzschild metric in the Kruskal–Szekeres coordinates. These coordinates are already in conformally flat form and hence we need only to make their ranges compact. This can be done using the

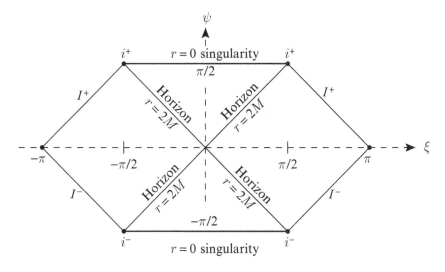

Fig. 8.5. The Penrose–Carter diagram for the Schwarzschild spacetime.

coordinate transformations

$$T + X = \tan \frac{1}{2}(\psi + \xi) \, ; \qquad T - X = \tan \frac{1}{2}(\psi - \xi). \qquad (8.44)$$

These transform the metric into the form

$$ds^2 = \frac{32M^3}{r} \frac{e^{r/2M}(-d\psi^2 + d\xi^2)}{4\cos^2 \frac{1}{2}(\psi + \xi) \cos^2 \frac{1}{2}(\psi - \xi)} + r^2 \left(d\theta^2 + \sin^2 \theta \, d\phi^2 \right), \qquad (8.45)$$

where r is now an implicit function of ψ and ξ given by

$$\left(1 - \frac{r}{2M} \right) e^{r/2M} = T^2 - X^2 = \tan \frac{1}{2}(\psi + \xi) \tan \frac{1}{2}(\psi - \xi). \qquad (8.46)$$

The corresponding conformal diagram is now more complicated and is shown in Fig. 8.5.

It is clear from Fig. 8.5 that the future null infinity contains the future end points of all outgoing null geodesics while the past null infinity contains the past end points of all ingoing null geodesics. The intersection of the future null infinity and the past null infinity defines the spacelike infinity. Similarly, the future end points of all timelike geodesics that do not terminate at $r = 0$ define the future timelike infinity and the past end points of all timelike geodesics that do not originate from $r = 0$ define the past timelike infinity. (It may appear from the Penrose–Carter diagram for the Schwarzschild manifold that $i^{\pm}$ intersects with the $r = 0$ line. This is, however, somewhat misleading since there are several timelike curves which do not go to $r = 0$.) Clearly, the Penrose–Carter diagram can be used to provide a

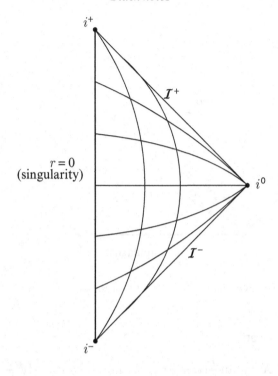

Fig. 8.6. Penrose–Carter diagram for a Reissner–Nordstrom black hole with $Q^2 > M^2$.

simple geometrical interpretation of the *event horizon*. We define the causal past of $\mathcal{I}^+$ as the set of all events in the past light cones emanating from all the events on $\mathcal{I}^+$. Then the boundary of the causal past of $\mathcal{I}^+$ will define the future event horizon. It is now obvious that one needs to know the entire time evolution of the geometry before one can decide which surface is an event horizon. From this point of view, it is very doubtful whether the event horizon can play a fundamental role in any local, causal, process in spite of the considerable importance given to it in the literature. (We will have occasion to comment on this point again in Chapter 16.)

Another feature of the Kruskal–Szekeres manifold representing an eternal black hole is the existence of a past singularity that is visible from $\mathcal{R}$ and in fact from $\mathcal{I}^+$. Such a singularity is called *naked* and, since it can influence future events, it is generally believed that naked singularities cannot arise in normal physical processes.

As another example, let us consider the Penrose–Carter diagram for the Reissner–Nordstrom metric. In this case, $f(r)$ vanishes at two points given by

$$r_\pm = M \pm \sqrt{M^2 - Q^2} \tag{8.47}$$

and obviously the nature of the spacetime depends on the relative values of M^2 and Q^2. When $M^2 < Q^2$, the roots are imaginary, showing that there are no horizons. But $r = 0$ is still a singularity and the Penrose–Carter diagram looks like the one shown in Fig. 8.6. The singularity at $r = 0$ is visible from infinity and hence is naked. As in the case of the past singularity in the Kruskal–Szekeres manifold, it is believed that realistic collapse of charged matter cannot lead to such a situation.

Let us next consider the case with $M^2 > Q^2$. In this case, $f(r)$ has the correct sign (that is, positive) both at large r and at small r but becomes negative in the range $r_- < r < r_+$. The coordinate singularities at both $r = (r_+, r_-)$ can be removed by standard transformation to Kruskal–Szekeres like coordinates. Both the surfaces are null and hence both are event horizons. If a particle falls into the black hole from a large distance, it will first cross $r = r_+$; this is analogous to a particle crossing $r = 2M$ in the Schwarzschild metric and takes place in finite proper time and infinite coordinate time. Such a particle has to move towards $r = r_-$ after which the motion can be controlled in the usual manner as in the region $r > r_+$. There is no compulsion for the particle to hit the $r = 0$ singularity. More interestingly, one can decide to reverse one's motion while at $r < r_-$ and cross through the null surface $r = r_-$. Now r will again be a timelike coordinate but with a reversed orientation so that the particle will be forced to move in the direction of increasing r and cross through $r = r_+$. Such a trajectory can wind back and forth several times. The Penrose–Carter diagram in this case makes this clearer and shows that there are infinite copies of the same basic structure repeated in this case (see Fig. 8.7).

This situation can actually arise in the case of, say, the collapse of a spherical dust ball in which the dust particles have a charge. Using the technique similar to the one adopted in the case of Schwarzschild metric, one can obtain an effective potential for the motion of the particles. The electrostatic repulsion will halt the collapse at some finite radius and the surface of the dust sphere will follow a trajectory like the one shown in Fig. 8.8. In this case there is no singularity that is visible from infinity. However, there are timelike curves that connect completely different asymptotically flat universes like I and I' and it is not clear whether this is physically reasonable.

Exercise 8.5

Extreme Reissner–Nordstrom solution Consider a Reissner–Nordstrom metric for which $Q^2 = M^2$. (Such a solution is called an extreme RN solution.) Show that, using the coordinate transformation from r to $\rho = r - Q$, the RN metric can be written in an isotropic form as

$$ds^2 = -\left(\frac{1}{1 + (Q/\rho)}\right)^2 dt^2 + \left(1 + \frac{Q}{\rho}\right)^2 \left(d\rho^2 + \rho^2 \left(d\theta^2 + \sin^2\theta \, d\phi^2\right)\right). \quad (8.48)$$

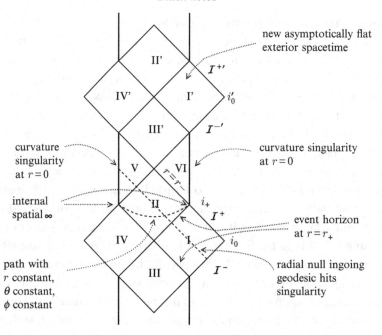

Fig. 8.7. Penrose–Carter diagram for a Reissner–Nordstrom black hole with $Q^2 < M^2$.

The horizon is now mapped to $\rho = 0$. Show that the proper spatial distance to the horizon diverges. Further show that, near $\rho = 0$, the metric can be expressed in the form

$$ds^2 = -e^{-2z}dt^2 + dz^2 + Q^2\left(d\theta^2 + \sin^2\theta\, d\phi^2\right), \tag{8.49}$$

where $z = Q\ln\rho$. [Note. This metric factorizes into a two-dimensional anti-de Sitter space and a 2-sphere.]

Exercise 8.6
Multisource extreme black hole solution Consider a metric of the form

$$ds^2 = -H^{-2}(\boldsymbol{x})dt^2 + H^2(\boldsymbol{x})\left[dx^2 + dy^2 + dz^2\right]. \tag{8.50}$$

Show that this metric is a solution to the Einstein–Maxwell equations if $\nabla^2 H = 0$, where the Laplacian is evaluated in flat spacetime. Determine the electrostatic potential corresponding to this solution. In particular, this equation possesses a solution of the form

$$H = 1 + \sum_{a=1}^{N} \frac{M_a}{|\boldsymbol{x} - \boldsymbol{x}_a|}, \tag{8.51}$$

which corresponds to N charged particles located at $\boldsymbol{x}_a$ with $a = (1, 2, ..., N)$. Show that each of these charges is extreme in the sense that $Q_a = M_a$. Construct the Penrose–Carter

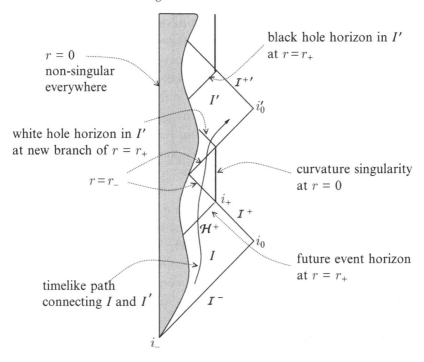

Fig. 8.8. The Penrose–Carter diagram for the Reissner–Nordstrom metric generated by a collapsing charged dust ball. Part of the region occupied by the dust ball is shown by the shaded region.

diagram for this metric when $N = 2$. This is probably the most remarkable exact solution to Einstein's equations.

8.5 Rotating black holes and the Kerr metric

The Schwarzschild metric represents the simplest possible black hole solution to Einstein's equations characterized by a single parameter M, which can be interpreted as the mass of the black hole. A somewhat more complicated black hole solution can be obtained – still within the context of spherical symmetry – if we assume the source has both charge and mass. The spacetime is now described by the Reissner–Nordstrom metric and requires two parameters M and Q. A different type of black hole solution is obtained when one relaxes spherical symmetry to axial symmetry. In this case, vacuum Einstein's equations admit a solution that depends on three parameters M, Q and J, which can be interpreted as the mass, charge and angular momentum of the black hole. This solution is called the Kerr–Newman metric and a special case of this solution corresponding to $Q = 0$ is called the Kerr metric. The Kerr metric represents an axially symmetric vacuum solution to Einstein's equations characterized by two parameters (M and J) which can be interpreted as the mass and angular momentum of the black hole.[2]

There is, however, one crucial conceptual difference between the spherically symmetric black hole solutions (like the Schwarzschild metric or the Reissner–Nordstrom metric) and the axially symmetric Kerr solutions. In the case of the former, it is possible to construct an interior solution made of some kind of matter that will appropriately match with the vacuum exterior solution. As the matter collapses to a singularity, the external region can be interpreted as a spacetime describing, say, a Schwarzschild black hole. In a similar manner, one would like to consider matter, rotating with a certain angular momentum and collapsing under self-gravity, to act as a source for the Kerr metric. We would then expect to find an interior solution made of matter with angular momentum appropriately matched to an external Kerr metric which could then be interpreted as a spacetime of a rotating black hole. Unfortunately, no such interior solution is known. In the absence of such a solution – even an idealized one – we do not know what is the source for the Kerr metric; or even whether such a metric can be realized in nature under appropriate circumstances. It is generally believed that this is purely a technical issue involving the usual difficulties of solving coupled nonlinear partial differential equations and that the Kerr metric does represent the external metric of a source with mass and angular momentum. But it should be stressed that, at present, this is only a hope and conjecture.

Given the form of the Kerr metric, it is a straightforward (but algebraically intensive) exercise to verify that it satisfies the equations $R_{ab} = 0$. In the absence of an interior solution, one cannot provide a more complete picture of the dynamical origin of the Kerr metric through the collapse of rotating matter. In view of the above, we shall not bother to provide a complete derivation of the Kerr metric and will merely describe many of the properties that distinguish it from the spherically symmetric black hole solutions.

In a suitable coordinate system (called *Boyer–Lindquist coordinates*), the Kerr metric can be expressed in the form

$$ds^2 = -\left(1 - \frac{2\mu r}{\rho^2}\right) dt^2 - \frac{4\mu a r \sin^2 \theta}{\rho^2} dt d\phi + \frac{\rho^2}{\Delta} dr^2 + \rho^2 d\theta^2$$
$$+ \left(r^2 + a^2 + \frac{2\mu r a^2 \sin^2 \theta}{\rho^2}\right) \sin^2 \theta d\phi^2, \tag{8.52}$$

where μ and a are constants and

$$\rho^2 \equiv r^2 + a^2 \cos^2 \theta, \qquad \Delta \equiv r^2 - 2\mu r + a^2. \tag{8.53}$$

It is obvious that the metric coefficients are independent of t and ϕ respecting time translational invariance and axial symmetry. Of the two parameters, μ represents the mass and a represents the angular momentum. When $a = 0$ the metric reduces to the Schwarzschild metric, allowing us to identify $\mu = M$. In the limit of $\mu \to 0$, we will expect the metric to represent flat spacetime if the above interpretation is correct. In this limit, the metric reduces to the form

$$ds^2 = -dt^2 + \frac{\rho^2}{r^2 + a^2}dr^2 + \rho^2 d\theta^2 + (r^2 + a^2)\sin^2\theta d\phi^2, \tag{8.54}$$

which is actually the flat spacetime expressed in some form of spheroidal coordinates. The following coordinate transformation

$$x = \sqrt{r^2 + a^2}\sin\theta\cos\phi \; ; \qquad y = \sqrt{r^2 + a^2}\sin\theta\sin\phi \; ; \qquad z = r\cos\theta \tag{8.55}$$

will reduce the metric in Eq. (8.54) to the standard Cartesian form. This feature also indicates that, in the $\theta = \pi/2$ plane, $r = 0$ corresponds to a *disk* of radius a and hence the r coordinate has properties quite different from the usual radial coordinate we are accustomed to. To interpret the parameter a it is better to write the Kerr metric in a different form that is often used:

$$ds^2 = -\frac{\Delta - a^2\sin^2\theta}{\rho^2}dt^2 - \frac{4\mu a r\sin^2\theta}{\rho^2}dtd\phi + \frac{\rho^2}{\Delta}dr^2 + \rho^2 d\theta^2 + \frac{\Sigma^2\sin^2\theta}{\rho^2}d\phi^2$$

$$= -\frac{\rho^2\Delta}{\Sigma^2}dt^2 + \frac{\Sigma^2\sin^2\theta}{\rho^2}(d\phi - \omega dt)^2 + \frac{\rho^2}{\Delta}dr^2 + \rho^2 d\theta^2, \tag{8.56}$$

where

$$\Sigma^2 \equiv (r^2 + a^2)^2 - a^2\Delta\sin^2\theta; \qquad \omega \equiv \frac{2\mu r a}{\Sigma^2}. \tag{8.57}$$

The second line in Eq. (8.56) presents the metric in a form suggestive of a rotating object. In fact, taking the $r \to \infty$ limit and comparing the asymptotic form of $g_{t\phi}$ with the asymptotic metric due to a rotating source in Eq. (6.205) we can identify a as the angular momentum per unit mass. In these coordinates, the contravariant components of the metric are given by

$$g^{rr} = \frac{\Delta}{\rho^2}, \quad g^{\theta\theta} = \frac{1}{\rho^2}, \quad g^{tt} = -\frac{\Sigma^2}{\rho^2\Delta}, \quad g^{t\phi} = -\frac{2\mu a r}{\rho^2\Delta}, \quad g^{\phi\phi} = \frac{a^2\sin^2\theta - \Delta}{\rho^2\Delta\sin^2\theta}. \tag{8.58}$$

This can be presented more compactly through the operator (with $r_g \equiv 2\mu$)

$$g^{ik}\frac{\partial}{\partial x^i}\frac{\partial}{\partial x^k} = -\frac{1}{\Delta}\left(r^2 + a^2 + \frac{r_g r a^2}{\rho^2}\sin^2\theta^2\right)\left(\frac{\partial}{\partial t}\right)^2 + \frac{\Delta}{\rho^2}\left(\frac{\partial}{\partial r}\right)^2 \tag{8.59}$$

$$+ \frac{1}{\rho^2}\left(\frac{\partial}{\partial\theta}\right)^2 - \frac{1}{\Delta\sin^2\theta}\left(1 - \frac{r_g r}{\rho^2}\right)\left(\frac{\partial}{\partial\phi}\right)^2 - \frac{2r_g\,r a}{\rho^2\,\Delta}\frac{\partial}{\partial\phi}\frac{\partial}{\partial t}.$$

Hereafter we shall denote the parameter μ by M. Occasionally we shall also use $J = Ma$ to denote the total angular momentum.

Exercise 8.7

A special class of metric Consider a metric of the form $g_{ab} = \eta_{ab} + 2Hl_a l_b$, where η_{ab} is the usual Lorentzian metric in Cartesian coordinates and l_a are null vectors.
  (a) Find an expression for g^{ab}. Show that $\eta^{ab} l_a l_b = 0$.
  (b) Consider a metric of this form with

$$H = \frac{Mr}{r^2 + a^2 \cos^2 \theta}; \qquad l_a = \left(1, \frac{rx + ay}{r^2 + a^2}, \frac{ry - ax}{r^2 + a^2}, \frac{z}{r} \right), \tag{8.60}$$

with $z = r \cos \theta$ and where r is defined through the equation

$$\frac{x^2 + y^2}{r^2 + a^2} + \frac{z^2}{r^2} = 1. \tag{8.61}$$

Show that this metric satisfies Einstein's equations. Can you identify this metric?

8.5.1 Event horizon and infinite redshift surface

Since the Kerr metric is stationary and axially symmetric, it allows two obvious Killing vector fields given by

$$\boldsymbol{\xi}_{(t)} \equiv \left(\frac{\partial}{\partial t} \right)_{r,\theta,\phi} ; \qquad \boldsymbol{\xi}_{(\phi)} \equiv \left(\frac{\partial}{\partial \phi} \right)_{t,r,\theta}. \tag{8.62}$$

The norms of these Killing vectors, as well as their dot product, are scalar quantities with a coordinate-free, geometrical interpretation. This allows us to represent three of the metric coefficients in the form

$$\boldsymbol{\xi}_{(t)} \cdot \boldsymbol{\xi}_{(t)} = g_{tt} = -\left(\frac{\Delta - a^2 \sin^2 \theta}{\rho^2} \right) \tag{8.63}$$

$$\boldsymbol{\xi}_{(t)} \cdot \boldsymbol{\xi}_{(\phi)} = g_{t\phi} = \frac{a \sin^2 \theta (\Delta - r^2 - a^2)}{\rho^2} \tag{8.64}$$

$$\boldsymbol{\xi}_{(\phi)} \cdot \boldsymbol{\xi}_{(\phi)} = g_{\phi\phi} = \frac{[(r^2 + a^2)^2 - \Delta a^2 \sin^2 \theta] \sin^2 \theta}{\rho^2}. \tag{8.65}$$

We shall see that most of the geometrical features of the Kerr metric are closely related to the behaviour of these two Killing vector fields.

From the form of the Kerr metric in Eq. (8.52), it is obvious that the metric becomes ill-defined at $\rho = 0$ and at $\Delta = 0$. The calculation of scalar invariants of the curvature tensor shows that $\rho = 0$ is indeed a physical singularity. The condition $\rho = 0$ corresponds to

$$\rho^2 = r^2 + a^2 \cos^2 \theta = 0, \tag{8.66}$$

which can only be satisfied with $\theta = \pi/2$ and $r = 0$. However, we saw earlier that $r = 0$ in the equatorial plane corresponds to an outer edge of a disk and hence we have a ring-like singularity in the case of a Kerr metric. While this is quite peculiar, it is difficult to provide an intuitive understanding of this feature in the absence of a valid interior solution.

The curvature invariants are well behaved at $\Delta = 0$. Since $\Delta = 0$ corresponds to $r = 2\mu = 2M$ in the limit of $a \to 0$, we suspect that it is probably a coordinate artefact. But unlike in the case of the Schwarzschild metric, we now find that the surfaces $g_{tt} = 0$ and $g^{rr} = 0$ do not coincide. We will see that these two surfaces have different properties and lead to new physical phenomena which are absent in the case of spherically symmetric spacetimes. We shall now discuss these two surfaces and their physical features.

It is possible to provide simple geometrical interpretations to both conditions $g^{rr} = 0$ and $g_{tt} = 0$. Suppose we choose the coordinate system in such a way that one of the coordinates, r, remains constant on the horizon (without attributing any other property to this coordinate). The normal to a general $r = $ constant surface will be proportional to $\partial_a r$. Hence the condition that the normal to the $r = $ constant surface becomes null is given by $g^{ab}\partial_a r \partial_b r = g^{rr} = 0$. Hence this condition selects the horizon as a null surface.

The surface corresponding to $g^{rr} = 0$ is given by a solution to the quadratic equation $\Delta = 0$ with the two roots

$$r_{\text{hor}} = \frac{r_g}{2} \pm \sqrt{\left(\frac{r_g}{2}\right)^2 - a^2}; \qquad r_g \equiv 2M. \tag{8.67}$$

Both represent null surfaces in the spacetime and hence can act as one-way membranes. Unlike in the case of the Schwarzschild metric, we now have two horizons, usually called the inner and outer horizons, corresponding to the solutions with negative and positive signs in Eq. (8.67). From the point of view of an external observer, the one that is relevant is the solution with the positive sign in Eq. (8.67). We shall hereafter take the horizon to be the outer horizon at

$$r_h = \frac{r_g}{2} + \sqrt{\left(\frac{r_g}{2}\right)^2 - a^2} = M + \sqrt{M^2 - a^2}. \tag{8.68}$$

We also note that, in the limit of $a \to 0$, this goes over to $r = 2M$.

It is also obvious from Eq. (8.67) that the horizon exists only if $a^2 < M^2$. If this condition is violated, $\Delta > 0$ everywhere and the singularity at $\rho = 0$ will be visible from the outside. As we have said before, it is generally believed that such solutions cannot arise in nature and that physically relevant solutions will have $a^2 < M^2$.

Let us next consider the surface defined by the quadratic equation $g_{tt} = 0$. To understand the significance of the condition $g_{tt} = 0$, consider a class of observers with four-velocity $\boldsymbol{u}$ in the direction of the timelike Killing vector $\boldsymbol{\xi}$. Since the four-velocity has to be normalized, we have the relation $\boldsymbol{\xi} = \mathcal{R}\boldsymbol{u}$ where $\mathcal{R}^2 = -\boldsymbol{\xi} \cdot \boldsymbol{\xi}$. For any photon with four-momentum $\boldsymbol{p}$ propagating in this spacetime, the observer with four-velocity $\boldsymbol{u}$ will attribute a frequency

$$\omega = -\boldsymbol{p} \cdot \boldsymbol{u} = -\frac{1}{\mathcal{R}}(\boldsymbol{p} \cdot \boldsymbol{\xi}) = \frac{E}{\mathcal{R}}, \tag{8.69}$$

where $E = -\boldsymbol{p} \cdot \boldsymbol{\xi}$ is the *conserved* energy of the photon. Two different observers will attribute to the photon the wavelengths related by $\lambda_1/\lambda_2 = \mathcal{R}_1/\mathcal{R}_2$. (Hence $\mathcal{R}$ is called the redshift factor.) It follows that the surface with $\mathcal{R} = 0$ corresponds to infinite redshift. Since $\xi^a = \delta_t^a$ the vanishing of $\xi^a \xi_a$ occurs on the surface $g_{tt} = 0$. In general, this surface will be distinct from the event horizon defined through $g^{rr} = 0$.

For the Kerr metric, the equation $g_{tt} = 0$ also has two solutions, $r = r_\pm$, given by

$$r_\pm = \frac{r_g}{2} \pm \sqrt{\left(\frac{r_g}{2}\right)^2 - a^2 \cos^2 \theta} = M \pm \sqrt{M^2 - a^2 \cos^2 \theta}. \tag{8.70}$$

We find that, in the limit of $a \to 0$, one of the solutions r_+ goes to $2M$ while the solution r_- vanishes. Further, the surface $r = r_-$ coincides with the ring singularity in the equatorial plane and is completely contained by the inner horizon (except at the poles where the two surfaces touch). On the other hand, the surface $r = r_+$ completely encloses the outer horizon (except at the poles where they touch).

From the point of view of an outside observer, the structure of the Kerr geometry is therefore characterized by the following two surfaces (see Fig. 8.9). First, we have the surface

$$r_+ = \frac{r_g}{2} + \sqrt{\left(\frac{r_g}{2}\right)^2 - a^2 \cos^2 \theta} = M + \sqrt{M^2 - a^2 \cos^2 \theta} \tag{8.71}$$

on which g_{tt} vanishes. Physically, this corresponds to a surface of infinite redshift usually called an *ergosurface*. As we we proceed inwards, we reach the horizon given by Eq. (8.68) which is a null surface and acts as a one-way membrane. Inside this surface lie two more surfaces as we have discussed above and a ring-like singularity. But since $r = r_h$ acts as a horizon, the internal structure does not affect the physics outside.

The crucial new feature in the Kerr geometry – compared with spherically symmetric black hole solutions – is the separation of the infinite redshift surface $r = r_0$ (on which g_{11} diverges) from the event horizon $r = r_{\text{hor}}$ (on which g_{00} vanishes).

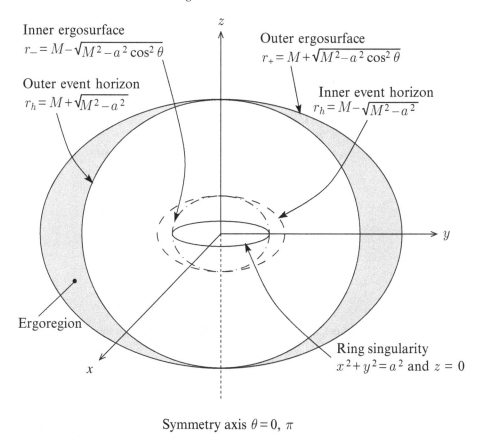

Inner ergosurface
$$r_- = M - \sqrt{M^2 - a^2 \cos^2 \theta}$$

Outer ergosurface
$$r_+ = M + \sqrt{M^2 - a^2 \cos^2 \theta}$$

Outer event horizon
$$r_h = M + \sqrt{M^2 - a^2}$$

Inner event horizon
$$r_h = M - \sqrt{M^2 - a^2}$$

Ergoregion

Ring singularity
$x^2 + y^2 = a^2$ and $z = 0$

Symmetry axis $\theta = 0, \pi$

Fig. 8.9. Schematic picture showing the geometrical structure of the Kerr spacetime.

The region between theses two surfaces is called the *ergosphere* and we shall now describe the interesting new phenomena the existence of ergosphere leads to. Figure 8.10 shows the Penrose–Carter diagram for the Kerr metric.

Exercise 8.8
Closed timelike curves in the Kerr metric Prove that the Kerr metric, when analytically extended to negative values of r, contains closed timelike curves travelling along which you can go back to your past.

[Hint. Consider curves that wind around in ϕ keeping θ and t constant and having a small negative value for r. Show that the line interval along such a path is

$$ds^2 \approx a^2 \left(1 + \frac{2M}{r} \right) d\phi^2, \tag{8.72}$$

which is negative for small negative r.]

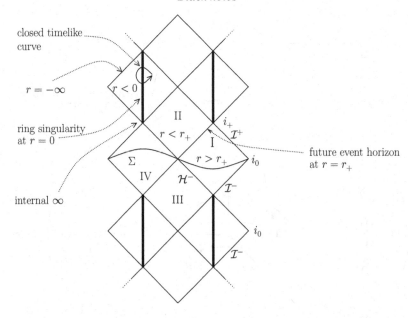

Fig. 8.10. The Penrose–Carter diagram for the Kerr spacetime for the case $\mu > a$.

8.5.2 Static limit

An observer in the Kerr spacetime moving with a constant angular velocity and having fixed values for r, θ will see the geometry to be unchanging. Such observers are called stationary observers. In addition to having fixed values for r, θ if the observers also have a fixed value for ϕ, they are called static observers located at fixed spatial coordinates. In the case of spherically symmetric black holes, we know that static observers can exist only outside the event horizon and – once inside the event horizon – they will move towards the singularity. In the case of a Kerr black hole the situation becomes more complicated, as we will now discuss.

To begin with, we know that geodesic observers will experience the effects of frame dragging due to the existence of a nonzero $g_{t\phi}$ term in the metric (see Section 6.4.1). But non-geodesic observers could possibly remain static in such a metric provided the frame dragging effect is weak. As one approaches the Kerr black hole, one would expect this effect to become stronger and the question arises as to whether an observer can remain static all the way up to the horizon. We will show that static observers can exist only outside the ergosurface and that inside the ergosurface they must necessarily rotate in the same direction as the black hole.

Consider an observer with an angular velocity Ω in the Kerr spacetime with

$$\Omega \equiv \frac{d\phi}{dt} = \frac{d\phi/d\tau}{dt/d\tau} = \frac{u^\phi}{u^t}. \tag{8.73}$$

Such an observer has a four-velocity which can be expressed as

$$\boldsymbol{u} = u^t \left(\frac{\partial}{\partial t} + \Omega \frac{\partial}{\partial \phi} \right) = \frac{\boldsymbol{\xi}_{(t)} + \Omega \boldsymbol{\xi}_{(\phi)}}{|\boldsymbol{\xi}_{(t)} + \Omega \boldsymbol{\xi}_{(\phi)}|} = \frac{\boldsymbol{\xi}_{(t)} + \Omega \boldsymbol{\xi}_{(\phi)}}{(-g_{tt} - 2\Omega g_{t\phi} - \Omega^2 g_{\phi\phi})^{1/2}}. \tag{8.74}$$

Given the two Killing vectors $\boldsymbol{\xi}_{(t)}$ and $\boldsymbol{\xi}_{(\phi)}$, it is obvious that the four-velocity of the observer will be in the direction of $\boldsymbol{\xi}_{(t)} + \Omega \boldsymbol{\xi}_{(\phi)}$. But to be a valid four-velocity corresponding to a timelike trajectory we need to ensure that $\boldsymbol{u}^2 = -1$. This leads to the normalization factor in the denominator of Eq. (8.74). Obviously we can satisfy this condition only if

$$g_{tt} + 2\Omega g_{t\phi} + \Omega^2 g_{\phi\phi} < 0. \tag{8.75}$$

This condition leads to limits in the range of values allowed for the angular velocity to be

$$\Omega_{\min} < \Omega < \Omega_{\max}, \tag{8.76}$$

where

$$\Omega_{\min} = \omega - \sqrt{\omega^2 - (g_{tt}/g_{\phi\phi})}, \qquad \Omega_{\max} = \omega + \sqrt{\omega^2 - (g_{tt}/g_{\phi\phi})}, \tag{8.77}$$

with

$$\omega \equiv \frac{1}{2}(\Omega_{\min} + \Omega_{\max}) = -\frac{g_{\phi t}}{g_{\phi\phi}} = \frac{2Mar}{(r^2 + a^2)^2 - \Delta a^2 \sin^2 \theta}. \tag{8.78}$$

The following features can immediately be obtained from this result. First, far away from the black hole, we have $r\Omega_{\min} = -1$ and $r\Omega_{\max} = +1$, which correspond to the standard result that motion should be at a speed less than that of light. Second, as one moves closer to the black hole, $\Omega_{\min}$ increases due to the dragging of the inertial frames. Eventually, $\Omega_{\min}$ reaches zero at the surface on which $g_{tt} = 0$, which is the ergosurface. Therefore, inside the ergosphere, all stationary observers must orbit the black hole with $\Omega > 0$ and hence static observers can exist only outside the ergosurface. Finally, as one crosses the ergosurface and moves towards the event horizon, the allowed range of angular velocities become ever more positive with the allowed range narrowing down. At the event horizon, where $\omega^2 = g_{tt}/g_{\phi\phi}$, the $\Omega_{\min}$ and $\Omega_{\max}$ coincide and all timelike worldlines point inwards. The limiting angular velocity is given by

$$\Omega_H \equiv \omega(r_h, \theta) = \frac{a}{2Mr_h} = \frac{a}{r_h^2 + a^2}. \tag{8.79}$$

This limiting angular velocity is sometimes called the angular velocity of the horizon. Using this we can form a new Killing vector

$$\boldsymbol{\xi}_{(K)} \equiv \boldsymbol{\xi}_{(t)} + \Omega_H \boldsymbol{\xi}_{(\phi)}, \tag{8.80}$$

which is a linear combination of $\boldsymbol{\xi}_{(t)}$ and $\boldsymbol{\xi}_{(\phi)}$. From the above discussion, it is clear that this Killing vector becomes null (that is, $\boldsymbol{\xi}_{(K)} \cdot \boldsymbol{\xi}_{(K)} = 0$) on the event horizon r_{hor} and is timelike outside the horizon.

Exercise 8.9
Zero angular momentum observers (ZAMOs) Consider an observer in the Kerr geometry moving along the trajectory $\phi = \omega t$, $(r, \theta) = \text{constant}$.
   (a) Show that the observer has the four-velocity $u_a = -N\partial_a t$, where $-N^2 = g_{00}$. Hence show that ZAMO worldlines are orthogonal to the $t = \text{constant}$ hypersurface.
   (b) Show that the acceleration of a ZAMO is given by $a_j = \partial_j \ln N$. Compute the magnitude of the acceleration and show that, when the observer approaches the horizon $r = r_h$, the magnitude of the acceleration goes to κ/N, where

$$\kappa = \frac{r_h - M}{2Mr_h} = \frac{\sqrt{M^2 - a^2}}{2M(M + \sqrt{M^2 - a^2})}. \tag{8.81}$$

Also show that, in this limit, the four-velocity $\boldsymbol{u}$ satisfies the limit $N\boldsymbol{u} \to \boldsymbol{\xi}_{(K)}$.
   (c) Show that $\nabla_{\boldsymbol{\xi}} \boldsymbol{\xi} = \kappa \boldsymbol{\xi}$ on the horizon for the vector $\boldsymbol{\xi} = \boldsymbol{\xi}_{(K)}$.

8.5.3 Penrose process and the area of the event horizon

On several occasions in the past, we have seen that the energy of a particle can be expressed as a time derivative of the action functional describing the motion of a particle in a given spacetime evaluated along the trajectory. If the time derivative is taken with respect to the proper time, we obtain $E_{\text{proper}} = -(\partial \mathcal{A}/\partial \tau)$, which is the energy as measured in the instantaneous rest frame of the particle. This quantity is always positive. On the other hand, when a particle is moving in a stationary spacetime, the *conserved* energy E is given by $E = -(\partial \mathcal{A}/\partial t)$, where the derivative is taken with respect to the *coordinate* time. This quantity is the same as the covariant component of the four-momentum $p_0 = mg_{0i}(dx^i/d\tau)$. In black hole spacetimes the coordinate time t does not have the temporal character throughout the manifold and so there is no guarantee that this quantity will be positive definite. In the ergosphere of the Kerr metric, g_{tt} changes sign and hence this conserved energy

$$E = m(g_{tt}\, u^t + g_{t\phi}\, u^\phi) = m\left(g_{tt}\frac{dt}{d\tau} + g_{t\phi}\frac{d\phi}{d\tau}\right) \tag{8.82}$$

can become negative. Because this quantity should necessarily be positive far away from the black hole, it is not possible for a particle that is either entering the ergosphere from outside or leaving the ergosphere to have $E < 0$. But it is perfectly admissible for a particle which originates within the ergosphere and falls into the event horizon to have $E < 0$. One way of producing such a particle will be to allow a particle (with $E > 0$) to enter the ergosphere, and break it up into two parts such that one part has $E < 0$ and another has $E > 0$. It is now possible to arrange the trajectories such that the part with $E < 0$ falls into the black hole and the other part with $E > 0$ escapes to infinity. In this process one can decrease the rest energy and the rotational energy of the Kerr black hole and transfer them to infinity. We shall now describe in some detail how this can be done.

Since the Kerr metric has a timelike Killing vector field $\boldsymbol{\xi}_{(t)}$ any particle moving on a geodesic will have a conserved energy given by

$$E = -\boldsymbol{p} \cdot \boldsymbol{\xi}_{(t)} = -p_t. \tag{8.83}$$

We want first to ascertain when we can have $E < 0$. We note that the Kerr metric can be expressed in a generic form as

$$ds^2 = -e^{2\nu}dt^2 + e^{2\psi}(d\phi - \omega dt)^2 + e^{2\mu_1}dr^2 + e^{2\mu_2}d\theta^2 \tag{8.84}$$

with the contravariant metric components given by

$$g^{tt} = -e^{-2\nu}\,;\; g^{t\phi} = -\omega e^{-2\nu};\; g^{\phi\phi} = e^{-2\psi} - \omega^2 e^{-2\nu};\; g^{rr} = e^{-2\mu_1};\; g^{\theta\theta} = e^{-2\mu_2}. \tag{8.85}$$

Consider now a particle of mass m moving in this spacetime. It will satisfy the condition

$$-m^2 = \boldsymbol{p} \cdot \boldsymbol{p} = -e^{-2\nu}p_t^2 - e^{-2\nu}2\omega p_t p_\phi + \left(e^{-2\psi} - e^{-2\nu}\omega^2\right)$$
$$\times p_\phi^2 + e^{-2\mu_1}p_r^2 + e^{-2\mu_2}p_\theta^2, \tag{8.86}$$

which can be solved for the conserved energy $E = -p_t$ to give

$$E = \omega p_\phi + \left[e^{2\nu-2\psi}p_\phi^2 + e^{2\nu}\left(e^{-2\mu_1}p_r^2 + e^{-2\mu_2}p_\theta^2 + m^2\right)\right]^{1/2}. \tag{8.87}$$

The sign of the square root necessarily has to be positive in order to maintain $E = +m$ for a particle which is at rest at infinity. Now if we want E to be negative, we must necessarily have: (i) p_ϕ negative as well as (ii) satisfy the condition

$$\left[e^{2\nu-2\psi}p_\phi^2 + e^{2\nu}\left(e^{-2\mu_1}p_r^2 + e^{-2\mu_2}p_\theta^2 + m^2\right)\right]^{1/2} < -\omega p_\phi. \tag{8.88}$$

The boundary of the region in which E can be negative can be found by making the left hand side as small as possible; this is achieved by taking $p_r = p_\theta = 0$ and taking the limit $m \to 0$. This immediately leads to the condition $\exp[2(\nu - \psi)]$ $< \omega^2$, which is equivalent to $g_{tt} > 0$. So clearly the orbit must be inside the ergosphere if the energy has to be negative.

This result can be used to extract energy from the Kerr black hole in several ways, of which the simplest one is the following. Consider, for example, a particle A moving in the ergosphere which breaks into two particles B and C. We let particle B to fall into the black hole and let particle C escape to infinity. All this can be done using suitable timelike trajectories. The conservation of four-momentum requires that $\boldsymbol{p}_A = \boldsymbol{p}_C + \boldsymbol{p}_B$ so that the energies satisfy the relation

$$E_A = E_C + E_B. \tag{8.89}$$

Since the particle A can fall into the ergosphere from infinity, we have $E_A > m$. We can arrange the trajectory of B such that it moves sufficiently fast and with $p_\phi < 0$ thereby making $E_B < 0$. It immediately follows that $E_C > E_A$. When the particle C goes back to infinity, it will have more energy than the original particle had. Thus, using the existence of negative energy orbits in the ergosphere and the local conservation of energy for processes taking place in the ergo region, one can extract energy from the black hole.

The Penrose process decreases both the mass and the angular momentum of the Kerr black hole by an amount equal to the (negative of) the energy and the angular momentum of the particle B that falls into the black hole. We shall now show that the resulting changes in the mass δM and in the angular momentum δJ are related by the inequality $\delta M > \Omega_H \delta J$. To do this, consider the dot product of the four-momentum $\boldsymbol{p}$ with the Killing vector $\boldsymbol{\xi}_{(K)}$ defined in Eq. (8.80). Since $\boldsymbol{\xi}_{(K)}$ is timelike outside the horizon and we want particle B to fall into the horizon, it is necessary that this dot product is negative. Using $\boldsymbol{p} \cdot \boldsymbol{\xi}_{(t)} = -E$, $\boldsymbol{p} \cdot \boldsymbol{\xi}_{(\phi)} = L$, where E and L are the conserved energy and angular momentum of particle B, we get the condition

$$\boldsymbol{p} \cdot (\boldsymbol{\xi}_{(t)} + \Omega_H \boldsymbol{\xi}_{(\phi)}) = -(E - \Omega_H L) < 0, \tag{8.90}$$

which gives the bound $L < (E/\Omega_H)$. Since $E < 0$, it follows that $L < 0$.

When the particle B falls into the black hole, the angular momentum and mass of a Kerr black hole will change by $\delta J = L$ and $\delta M = E$. Hence the above bound translates into the result

$$\delta M > \Omega_H \, \delta J = \frac{a \delta J}{r_h^2 + a^2}. \tag{8.91}$$

This result can be expressed in a more suggestive form. To do this, let us consider the surface area of the event horizon of a Kerr black hole. The metric on the event horizon is given by

$$ds^2 = (r_h^2 + a^2 \cos^2 \theta)d\theta^2 + \frac{(r_h^2 + a^2)^2 \sin^2 \theta d\phi^2}{(r_h^2 + a^2 \cos^2 \theta)}, \tag{8.92}$$

using which we can obtain the area to be

$$A = \int \int \sqrt{\sigma} \, d\theta d\phi = \int \int (r_h^2 + a^2) \sin \theta d\theta d\phi = 4\pi(r_h^2 + a^2). \tag{8.93}$$

We shall next introduce the concept of an irreducible mass M_{irr} through the equation $M_{\text{irr}}^2 = A/(16\pi)$. One can think of the irreducible mass as the mass of the Schwarzschild black hole with the same area for the event horizon as the Kerr black hole. Using Eq. (8.93), one can express the irreducible mass in the form

$$M_{\text{irr}}^2 = \frac{1}{2} \left(M^2 + \sqrt{M^4 - J^2} \right); \qquad J \equiv Ma. \tag{8.94}$$

An equivalent form of the relation is

$$M^2 = M_{\text{irr}}^2 + \left(\frac{J}{2M_{\text{irr}}} \right)^2, \tag{8.95}$$

which shows that the total mass energy of the Kerr black hole can be expressed as a Pythagorean sum of the irreducible mass and a term that can be interpreted as the rotational energy. Direct differentiation of Eq. (8.94) gives the result

$$\delta M_{\text{irr}} = \frac{a}{4M_{\text{irr}}\sqrt{M^2 - a^2}} \left(\Omega_H^{-1}\delta M - \delta J \right). \tag{8.96}$$

Our bound in Eq. (8.91) can now be stated in terms of the irreducible mass or – equivalently – in terms of the area of the event horizon as

$$\delta M_{\text{irr}} > 0 ; \qquad \delta A > 0. \tag{8.97}$$

That is, the area of the event horizon increases during the Penrose process. It turns out that this result is far more general; no classical physical process involving black holes can decrease their total horizon area.

With future applications in mind, we shall write the bound on the change in the area in a different form. Using the explicit expression in Eq. (8.93), it is easy to see that

$$\delta A = 8\pi \frac{a}{\Omega_H \sqrt{M^2 - a^2}} \left(\delta M - \Omega_H \delta J \right). \tag{8.98}$$

This can be manipulated to read

$$\delta M = \frac{\kappa}{8\pi G} \delta A + \Omega_H \delta J, \tag{8.99}$$

where

$$\kappa = \frac{\sqrt{M^2 - a^2}}{2M(M + \sqrt{M^2 - a^2})} \tag{8.100}$$

is the surface gravity (see Eq. (8.9)). We shall see later in Chapters 13 and 16 that this relation can be given a thermodynamic interpretation with $\kappa/(2\pi)$ acting as the temperature of the black hole and $A/4$ acting as the entropy of the black hole.

8.5.4 Particle orbits in the Kerr metric

The orbits of particles in the Kerr metric can be studied, in principle, by the same techniques we have used in the case of the Schwarzschild metric. However, lack of spherical symmetry makes the nature of the orbits very complicated and analytic solutions are impossible to find. It is clear that radial motion will now be possible only along the axis of symmetry and even planar motion will be possible only in the equatorial plane. Though it is not possible to obtain significant insights into the nature of orbits analytically, we shall briefly describe – for the sake of completeness – how the relevant equations can be obtained.

As usual, we shall start with the Hamilton–Jacobi equation for the action $\mathcal{A}$, given by

$$g^{ab}\partial_a \mathcal{A}\partial_b \mathcal{A} = -m^2, \tag{8.101}$$

where m is the mass of the particle. Given the axial symmetry and stationarity of the spacetime, it is obvious that one can separate this partial differential equation into t and ϕ dependences. It turns out that one can actually do better and, in fact, separate the equation even as regards the r and θ dependence. (This is related to the fact that Kerr metric possesses a Killing *tensor*; see Exercise 8.11.) Therefore we use the ansatz

$$\mathcal{A} = -\mathcal{E}_0 t + L\phi + \mathcal{A}_r(r) + \mathcal{A}_\theta(\theta). \tag{8.102}$$

Substituting into the Hamilton–Jacobi equations we find that $\mathcal{A}_r(r)$ and $\mathcal{A}_\theta(\theta)$ satisfy the ordinary differential equations:

$$\left(\frac{d\mathcal{A}_\theta}{d\theta}\right)^2 + \left(a\mathcal{E}_0 \sin\theta + \frac{L}{\sin\theta}\right)^2 + a^2 m^2 \cos^2\theta = K \ ;$$

$$\left(\frac{d\mathcal{A}_r}{d\theta}\right)^2 + \frac{1}{\Delta}\left[(r^2 + a^2)\mathcal{E}_0 - aL\right]^2 + m^2 r^2 = -K, \tag{8.103}$$

where K is a separation constant. This reduces the problem to quadrature. Differentiating the action with respect to $m^2, \mathcal{E}_0, L$ and K and equating to new constants will lead to three equations for the trajectories. Manipulating these equations, one can express the final equations of motion in the form:

$$m\frac{dt}{d\tau} = -\frac{r_g r a}{\rho^2 \Delta}L + \frac{\mathcal{E}_0}{\Delta}\left(r^2 + a^2 + \frac{r_g r a^2}{\rho^2}\sin^2\theta\right) \tag{8.104}$$

$$m\frac{d\phi}{d\tau} = \frac{L}{\Delta\sin^2\theta}\left(1 - \frac{r_g r}{\rho^2}\right) + \frac{r_g r a}{\rho^2 \Delta}\mathcal{E}_0 \tag{8.105}$$

$$m^2\left(\frac{dr}{d\tau}\right)^2 = \frac{1}{\rho^4}\left[(r^2 + a^2)\mathcal{E}_0 - aL\right]^2 - \frac{\Delta}{\rho^4}(K + m^2 r^2) \tag{8.106}$$

$$m^2\left(\frac{d\theta}{d\tau}\right)^2 = \frac{1}{\rho^4}(K - a^2 m^2 \cos^2\theta) - \frac{1}{\rho^4}\left(a\mathcal{E}_0\sin\theta - \frac{L}{\sin\theta}\right)^2. \tag{8.107}$$

These are the equations that govern the particle trajectories in the Kerr metric.

The existence of stable circular orbits in the equatorial plane is of some practical interest in astrophysics. It is generally believed that the matter in the accretion disks around astrophysical black holes will be able to move towards the black hole in a series of approximately circular orbits in the equatorial plane. In that case, the radius of the stable circular orbit closest to the black hole and its energy are of interest in the astrophysics of accretion disks. For the motion in the equatorial plane one can introduce an effective potential as in the case of the Schwarzschild metric along the following lines. We put $\theta = \pi/2$ in Eqs. (8.104)–(8.107) and

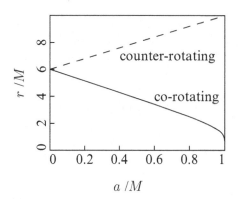
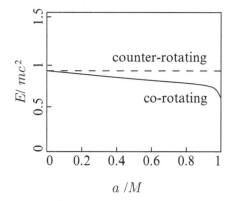

Fig. 8.11. The radius and the binding energy of the innermost stable circular orbit in the equatorial plane of the Kerr geometry. The results for both counter-rotating and co-rotating cases are given as a function of the rotation parameter a of the Kerr metric. The right panel gives the binding energy of the orbit as a function of the rotating parameter.

use the equation $(d\theta/d\tau) = 0$ to express the constant K in terms of $\mathcal{E}_0$ and L. The resulting equation for the radial motion is

$$m^2 \left(\frac{dr}{d\tau} \right)^2 = \frac{1}{r^4} \left[(r^2 + a^2)\mathcal{E}_0 - aL \right]^2 - \frac{\Delta}{r^4} \left[(a\mathcal{E}_0 - L)^2 + m^2 r^2 \right]. \quad (8.108)$$

We can now define an effective $U(r)$ such that the right hand side of the above equation vanishes when $\mathcal{E}_0 = U$. Hence the effective potential is the solution to the equation:

$$\left[(r^2 + a^2)U(r) - aL \right]^2 - \Delta \left[(aU(r) - L)^2 + r^2 m^2 \right] = 0. \quad (8.109)$$

The radii of stable circular orbits are determined by the minima of $U(r)$; that is by the simultaneous solution to the equations $\mathcal{E}_0 = U(r), U'(r) = 0$. Among all the stable circular orbits, we are interested in the innermost one. Fairly lengthy but straightforward calculation shows that this orbital radius is the solution to the quartic equation

$$r^2 - 6Mr - 3a^2 \mp 8a\sqrt{Mr} = 0, \quad (8.110)$$

where the upper (lower) sign corresponds to the counter-rotating (co-rotating) orbit. Figure 8.11(left) plots the radius of the innermost stable circular orbit as a function of the rotation parameter a of the Kerr metric. When $a = 0$ (which corresponds to the Schwarzschild metric) we get the standard result that $r = 6M$ (see page 331, item (iii)); as the rotation parameter increases, the radius of the circular orbit decreases for the co-rotating orbit which is the one that is probably the relevant one for the accretion disk. This shows that one can have stable circular orbits very close to the black hole in the case of rotating black holes. The binding energy E/mc^2 of the orbit is plotted against the rotation parameter on the right of Fig. 8.11. The quantity $(mc^2 - E)/mc^2$ represents the fraction of the rest energy that can be released when a particle falls from the innermost stable circular orbit into the black hole. In the extreme case of $a/M = 1$, this fraction is $1 - (1/\sqrt{3})$ which is about 42 per cent, while the corresponding value for orbits in the Schwarzschild metric is only about 5.7 per cent. This higher efficiency could be of use in certain astrophysical scenarios.

Exercise 8.10
Circular orbits in the Kerr metric Consider a particle moving in a circular orbit in the equatorial plane of the Kerr metric and show that its angular velocity Ω is related to the radius of the orbit r by

$$\Omega^2 = \frac{M}{(aM^{1/2} \pm r^{3/2})^2}, \quad (8.111)$$

where the two signs correspond to direct (co-rotating) and retrograde (counter-rotating) orbits. (For the Schwarzschild metric, this reduces to the relation $\Omega^2 = M/r^3$ obtained earlier in Eq. (7.132). The Newtonian coincidence does not carry over to the Kerr metric!)

[Hint. For circular motion in the equatorial plane, the four-velocity has the form $\boldsymbol{u} = (u^t, 0, 0, u^\phi)$ with both the components remaining constant. Use this in the geodesic equation for the r component of the acceleration and show that one gets a quadratic equation for $\Omega = d\phi/dt$ given by

$$\Omega^2 \left(\frac{Ma^2}{r^2} - r \right) - \frac{2Ma}{r^2}\Omega + \frac{M}{a^2} = 0. \tag{8.112}$$

Solving this will give the required result.]

Exercise 8.11

Killing tensor A symmetric second rank tensor K_{ab} is called a Killing tensor if it obeys the condition $\nabla_{(c}K_{ab)} = 0$. The existence of such Killing tensors can be connected with extra conservation laws which are explored in this exercise. Start with the action for a particle in a curved geometry expressed in the Hamiltonian form as

$$\mathcal{A}[p, x, C] = \int d\lambda \left\{ p_m \dot{x}^m - \frac{1}{2}C\left[m^{-1}g^{mn}(x)p_m p_n + m\right] \right\}, \tag{8.113}$$

where p^a, x^a and C are treated as independent variables (see Exercise 4.3).

(a) Suppose there is a Killing tensor K_{ab} in the spacetime. Consider the transformations

$$\delta x^m = \epsilon K^{mn} p_n \qquad \delta p_m = -\frac{1}{2}\epsilon p_r p_s \partial_m K^{rs}, \tag{8.114}$$

where ϵ is an infinitesimal parameter of first order. Show that the action $\mathcal{A}$ is invariant under the transformation and that the corresponding conserved quantity is $K_{ab}p^a p^b$. One simple example of a Killing tensor is g_{ab} itself, which leads to a trivial constant of motion m^2.

(b) Show that $\xi_a \xi_b$ is a Killing tensor if ξ_a is a Killing vector.

(c) Show that, in the Kerr geometry, there is a Killing tensor given by

$$\sigma_{mn} = 2\rho^2 l_{(m}n_{n)} + r^2 g_{mn}, \tag{8.115}$$

where the vectors $\boldsymbol{l}$ and $\boldsymbol{n}$ are given by

$$l^m = \frac{1}{\Delta}\left(r^2 + a^2, \Delta, 0, a\right) \qquad n^m = \frac{1}{2\rho^2}\left(r^2 + a^2, -\Delta, 0, a\right). \tag{8.116}$$

Show that $\boldsymbol{l} \cdot \boldsymbol{l} = 0$, $\boldsymbol{n} \cdot \boldsymbol{n} = 0$ and $\boldsymbol{n} \cdot \boldsymbol{l} = -1$. What is the conserved quantity corresponding to this Killing tensor?

8.6 Super-radiance in Kerr geometry

We saw earlier that it is possible to extract energy out of a Kerr black hole by making use of the negative energy orbits that exist in the ergosphere. There is an analogous process which works in the case of fields that propagate in the Kerr metric. This process is of particular importance since it is closely related to the particle production by black holes, which we will describe in Chapter 14. In this section we shall show how net energy could be radiated to infinity when a scalar field is propagating in the Kerr geometry.

Consider a massless scalar field Φ in the Kerr geometry described by the wave equation

$$\Box\Phi = (-g)^{-1/2}\partial_b\left[(-g)^{1/2}g^{ab}\partial_a\Phi\right] = 0. \tag{8.117}$$

Expanding this out, the equation can be written as

$$0 = \left[-\frac{(r^2+a^2)^2}{\Delta}+a^2\sin^2\theta\right]\frac{\partial^2\Phi}{\partial t^2} - \frac{arM}{\Delta}\frac{\partial^2\Phi}{\partial t\partial\phi} + \left(\frac{1}{\sin^2\theta}-\frac{a^2}{\Delta}\right)\frac{\partial^2\Phi}{\partial\phi^2}$$
$$+ \frac{\partial}{\partial r}\left(\Delta\frac{\partial\Phi}{\partial r}\right) + \frac{1}{\sin\theta}\frac{\partial}{\partial\theta}\left(\sin\theta\frac{\partial\Phi}{\partial\theta}\right). \tag{8.118}$$

Given the stationary nature and axial symmetry of the Kerr geometry, one can separate out the t and ϕ dependence of the wave equation by using a factor $\exp(-i\omega t + im\phi)$. It turns out that one can also separate out r and θ dependences of the wave equation. (This is similar to the fact that one can separate all the variables in the case of Hamilton–Jacobi equations in the Kerr metric and arises for the same reason – viz., that the Kerr geometry admits a Killing tensor.) Using an ansatz

$$\Phi(t,r,\theta,\phi) = e^{-i\omega t}e^{im\phi}R(r)S(\theta) \tag{8.119}$$

in Eq. (8.118) we find that the equations for S and R separate out into the form

$$\frac{1}{\sin\theta}\frac{d}{d\theta}\left(\sin\theta\frac{dS}{d\theta}\right) - \left(a^2\omega^2\sin^2\theta + \frac{m^2}{\sin^2\theta} - A\right)S = 0 \tag{8.120}$$

and

$$\frac{d}{dr}\left(\Delta\frac{dR}{dr}\right) + \left[\frac{\omega^2(r^2+a^2)^2 - 4arM\omega m + a^2m^2}{\Delta} - A\right]R = 0, \tag{8.121}$$

where A is a separation constant. Equation (8.120) can be thought of as an eigenvalue equation for A and the solutions can be expressed in terms of spheroidal harmonics. We are more concerned with the radial function, the equation for which can be simplified by introducing the tortoise coordinate with the definition $(dr^*/dr) = (r^2+a^2)/\Delta$. As usual, the tortoise coordinate ranges from $-\infty$ to $+\infty$ when r varies from the horizon r_h to infinity and thus covers the region outside the horizon. Equation (8.121) now becomes

$$\frac{d^2R}{dr^{*2}} + \frac{2r\Delta}{(r^2+a^2)^2}\frac{dR}{dr^*} + \left[\omega^2 + \frac{a^2m^2 - 4arM\omega m - \Delta A}{(r^2+a^2)^2}\right]R = 0. \tag{8.122}$$

To study the solutions to this equation which describe the radial dependence of the waves which propagate in the Kerr metric, it is convenient to begin with the asymptotic limits. When $r \to \infty$, this equation reduces to

$$\frac{d^2R}{dr^{*2}} + \frac{2}{r}\frac{dR}{dr^*} + \omega^2 R \approx 0, \tag{8.123}$$

which is just the radial part of the spherical wave equation in flat spacetime and has the solution

$$R \sim r^{-1}\exp(\pm i\omega r^*). \tag{8.124}$$

The two signs correspond to ingoing and outgoing waves near spatial infinity and the $(1/r)$ factor in the amplitude arises from the fact that areas of spherical surfaces increase as r^2 in the asymptotic, flat, spacetime limit. At the other extreme, when $\Delta \to 0$, the Eq. (8.122) reduces to

$$\frac{d^2R}{dr^{*2}} + \left[\omega^2 - \frac{am\omega}{Mr_h} + \frac{a^2m^2}{(2Mr_h)^2}\right]R = \frac{d^2R}{dr^{*2}} - (\omega - m\Omega_H)^2 R = 0; \quad \Omega_H \equiv \frac{a}{2Mr_h}. \tag{8.125}$$

This has the solution

$$R \sim \exp\left[\pm i(\omega - m\Omega_H)r^*\right]. \tag{8.126}$$

Next, we need to identify the wave mode that is ingoing at the horizon. To do this, let us choose an observer just outside the horizon who will be rotating in the positive ϕ direction with the angular velocity $\Omega = \Omega_H$ just outside the horizon. This observer will see the local wave mode to be

$$\Phi = S(\theta)\,\exp(-i\omega t)\,\exp(im\phi)\,\exp\left[\pm i(\omega - m\Omega_H)r^*\right], \tag{8.127}$$

which can be expressed as

$$\Phi = S(\theta)\,\exp\left[-i(\omega - m\Omega_H)t\right]\,\exp\left[\pm i(\omega - m\Omega_H)r^*\right]\exp(im\tilde{\phi}) \tag{8.128}$$

where $\tilde{\phi} = \phi - \Omega_H t$. We therefore conclude that the mode with the dependence $\exp[-i(\omega - m\Omega_H)r^*]$ (with the minus sign in the exponent) represents a wave which is ingoing at the horizon. The energy-momentum tensor corresponding to this wave mode is given by

$$4\pi T_{ab} = \partial_{(a}\Phi\,\partial_{b)}\Phi^* - \frac{1}{2}g_{ab}\left|\partial_c\Phi\partial^c\Phi\right|, \tag{8.129}$$

where the modulus signs and complex conjugation are required because we are using complex functions to represent the wave. The energy flux vector is given by $P_b = -T_{ab}\xi^a_{(t)}$, where $\boldsymbol{\xi}_{(t)}$ is the Killing vector corresponding to time translation.

The integral of the radial component of **P** over the 2-surface $r = r_h$ gives the energy flux:

$$\frac{dE}{dt} = \int T_t^r \, |g|^{1/2} \, d\theta \, d\phi. \tag{8.130}$$

Simple computation using Eq. (8.129) now shows that

$$4\pi T_t^r = \mathrm{Re}\,(\partial_t \Phi \, \partial^r \Phi^*) = \mathrm{Re}\left(\partial_t \Phi \, \partial_{r^*} \Phi^* \frac{r^2 + a^2}{\Sigma}\right)$$

$$= \omega(\omega - m\Omega_H) S^2(\theta) \left(\frac{2Mr_h}{\Sigma}\right) \tag{8.131}$$

so that the flux of energy through the horizon can be expressed in the form

$$\frac{dE}{dt} = \omega(\omega - m\Omega_H)\frac{Mr_h}{2\pi}\int S^2(\theta)\sin\theta \, d\theta d\phi = C_1 \, \omega(\omega - m\Omega_H), \tag{8.132}$$

where C_1 is a constant.

This result shows that the sign of (dE/dt) depends on the sign of $(\omega - m\Omega_H)$. In particular, $(dE/dt < 0)$ – indicating a flow of energy out of the horizon to infinity – if $(\omega - m\Omega_H) < 0$. Therefore, all the modes of a wave with frequency in the range $0 < \omega/m < \Omega_H$ transport more energy out of the Kerr black hole than into the black hole when it scatters off the black hole. These modes are called *super-radiant* modes and this phenomena is called *super-radiance*. One can also similarly compute the amount of angular momentum which is falling into the horizon by using the Killing vector $\xi_{(\phi)}$. The result in this case will be

$$\frac{dJ}{dt} = C_1 \, m(\omega - m\Omega_H). \tag{8.133}$$

This shows that the process of scattering the super-radiant modes of the wave off the black hole reduces both the mass and the angular momentum of the black hole. That is, when a wave in the relevant frequency range scatters off the black hole, it extracts energy and angular momentum from the Kerr black hole.

One can also compute the quantity $(dE/dt) - \Omega_H(dJ/dt)$ for this process using the above expressions. We find that

$$\frac{dE}{dt} - \Omega_H \frac{dJ}{dt} = C_1\omega(\omega - m\Omega_H) - \Omega_H \omega m(\omega - m\Omega_H) = C_1(\omega - m\Omega_H)^2. \tag{8.134}$$

Using Eq. (8.99) and noting that $dM = dE$, we can write this relation as

$$\frac{\kappa}{8\pi G}\frac{dA}{dt} = \frac{dE}{dt} - \Omega_H \frac{dJ}{dt} = C_1(\omega - m\Omega_H)^2 > 0. \tag{8.135}$$

This shows that the area of the event horizon increases during the scattering of the wave just as in the Penrose process studied earlier. This is another illustration of the general fact that the area of the event horizon increases during any classical process.

8.7 Horizons as null surfaces

The properties of horizons discussed in the specific cases of the Schwarzschild metric, the Kerr metric and even the Rindler metric correspond to special cases of a general class of horizons called *Killing horizons*. While each of the specific examples will have *other* features which may not be shared in a general context, several aspects of horizons can be presented in a broad unified framework using the behaviour of suitably defined Killing vectors. In this section, we shall introduce some of these general features and illustrate them using the specific examples we have studied in the previous sections.

In the case of the Schwarzschild spacetime, the metric in the Kruskal–Szekeres null coordinates is given by (see Eq. (8.22))

$$ds^2 = -\frac{32M^2}{r} e^{-r/2M} dU dV + r^2 d\Omega^2 \equiv -C^2(U, V) dU dV + dL_\perp^2, \quad (8.136)$$

where

$$U = -e^{-u/4M}, \qquad V = e^{v/4M} \quad (8.137)$$

with $u = t - r^*$, $v = t + r^*$, where r^* is the tortoise coordinate

$$r^* = r + 2M \ln \left| \frac{r - 2M}{2M} \right|. \quad (8.138)$$

All these expressions have a direct generalization for a metric of the form in Eq. (8.1) with κ replacing $(1/4M)$, the appropriate conformal factor $C(U, V)$ replacing the specific term in the Schwarzschild metric and $f(r)$ replacing $(1 - 2M/r)$, etc. (Some of these examples are given in Table 8.1.)

The Killing vector, $\xi = \partial/\partial t$, representing time translation invariance in the Schwarzschild like coordinates can be expressed in the Kruskal–Szekeres null coordinates as

$$\xi = \frac{\partial}{\partial t} = \frac{\partial X^a}{\partial t} \frac{\partial}{\partial X^a} = \kappa \left(V \frac{\partial}{\partial V} - U \frac{\partial}{\partial U} \right). \quad (8.139)$$

The norm of this vector is given by

$$\xi^2 = g_{UV} \xi^U \xi^V = C^2 \kappa^2 UV. \quad (8.140)$$

From the nature of the transformations, we know that the product UV is negative in $\mathcal{R}$ and positive in $\mathcal{F}$. It vanishes on the horizon, $\mathcal{H}$, given by $T = \pm X$, which corresponds to $UV = 0$ where C is finite. It follows that ξ^2 vanishes on the horizon and switches sign there. In the case of the Schwarzschild metric, for example, we have

$$\xi^2 = -\left(1 - \frac{2M}{r} \right). \quad (8.141)$$

The Killing vector ξ itself becomes

$$\xi\Big|_{\mathcal{H}} = \left(\frac{\partial}{\partial t}\right)_{\mathcal{H}} = \kappa V \frac{\partial}{\partial V} = \frac{\partial}{\partial v} \tag{8.142}$$

on the future horizon $U = 0$. Therefore, the Killing vector ξ is both normal and tangential to the horizon surface which, of course, is possible only because the horizon is a null surface. Similar conclusions apply on the past horizon $V = 0$.

Given a Killing vector ξ, its integral curves are called the orbits of ξ. We see that the orbits of ξ are hyperbolas in $\mathcal{R}$ and $\mathcal{L}$; they degenerate to straight lines on the horizons with the origin $U = 0, V = 0$ being a fixed point. Just like any other event in the UV plane, the origin also represents a 2-sphere with coordinates θ and ϕ on it and is called a *bifurcation 2-sphere*. To understand these properties of the horizon better, we shall begin with a general discussion of null surfaces and their properties.

Consider a surface S given by an equation $S(x^a) = 0$. The covariant normal vector n_a to that surface is in the direction of $\partial_a S$. Let t^a be another vector such that $t^a n_a = 0$. If $t^a = dx^a/d\lambda$ for some curve $x^a(\lambda)$, then the condition $t^a n_a = 0$ implies that $(dx^a/d\lambda)\partial_a S = 0$, showing S does not change along the curve. Hence it follows that t^a is tangential to the surface. When the norm of $\partial_a S$ is nonzero, one can normalize this vector so that its norm is ± 1. We will, however, be interested in null surfaces for which the normal vector is a null vector and $g^{ab}\partial_a S \partial_b S = 0$. With this motivation in mind, we will consider the normal vector to be given by $l_a = \mu(x)\partial_a S$, where $\mu(x)$ is an arbitrary function at the moment. A null surface is then defined by the condition $l_a l^a = 0$. From the comments made above, it follows that l^a is simultaneously normal and tangential to the surface S – which is the first peculiar feature of a null surface.

The vector l^a can be expressed in the form $l^a = dx^a/d\lambda$, where $x^a(\lambda)$ is a null curve on the surface S. In fact, we can prove that $x^a(\lambda)$ is a null *geodesic* on S. For any general $l_a = \mu(x)\partial_a S$ we can easily verify that

$$l^a \nabla_a l^m = \left[\frac{d}{d\lambda}(\ln \mu)\right] l^m + \frac{1}{2}\partial^m l^2 - l^2 \partial^m \ln \mu. \tag{8.143}$$

On the null surface, $l^2 = 0$ and the last term vanishes. Further, since l^2 is constant on S, we know that $t^a \partial_a l^2 = 0$ for any vector t^a tangent to S. Since $l^a l_a = 0$ we can take $\partial_m l^2$ to be in the direction of l_m. It therefore follows that the right hand side of Eq. (8.143) is in the direction of l^m and Eq. (8.143) reduces to the form

$$l^a \nabla_a l^m = F l^m \tag{8.144}$$

for some function F. This is a geodesic equation with a non-affine parametrization. (See Eq. (4.68) and the discussion surrounding it.) We can now use the freedom in the choice of the function μ to change the parametrization to affine parametrization so that l^a is the tangent vector to an affinely parametrized null geodesic. It is usual to call the affinely parametrized null geodesics $x^a(\lambda)$, with $(dx^a/d\lambda)$ being normal to a null hypersurface, the *generators* of $\mathcal{S}$.

As an example, consider the normal to the null surface U = constant in the Kruskal–Szekeres coordinates. This is clearly in the direction of V and can be taken to be

$$l = -\frac{\mu}{C^2}\frac{\partial}{\partial V} = -\frac{\mu r}{32M^3}\,e^{r/2M}\frac{\partial}{\partial V}, \tag{8.145}$$

where the first expression is valid in general and the second expression is for the Schwarzschild metric. On the horizon, it has the value $l = -(\mu e/16M^2)(\partial/\partial V)$. If we now choose $\mu = -16M^2/e$, it follows that $l = \partial/\partial V$ on the horizon. It follows that V is the appropriate affine parameter for the generator of this null surface.

A special kind of null surface is a *Killing horizon* $\mathcal{H}$ which arises if a Killing vector field $\boldsymbol{\xi}$ is normal to the null surface. If l^a is an affinely parametrized normal to $\mathcal{H}$, then $(l \cdot \boldsymbol{\nabla})l = 0$ and $\boldsymbol{\xi} = \mu l$ for some function μ. It follows that, on the Killing horizon, we have the relation

$$(\boldsymbol{\xi} \cdot \boldsymbol{\nabla})\boldsymbol{\xi} = \kappa\boldsymbol{\xi}, \qquad \kappa = \boldsymbol{\xi} \cdot \boldsymbol{\nabla}\ln|\mu|. \tag{8.146}$$

The notation anticipates the fact that κ will indeed be the surface gravity introduced earlier. This can be explicitly verified. We know that, on the horizon, $\boldsymbol{\xi}$ is given by Eq. (8.142) while the affinely parametrized normal is $\partial/\partial V$. It follows that the function μ is given by $\mu = \kappa V$. Therefore,

$$(\boldsymbol{\xi} \cdot \boldsymbol{\nabla})\log\mu = \kappa V\frac{\partial\log|V|}{\partial V} = \kappa. \tag{8.147}$$

This justifies the notation.

Since any constant multiple of a Killing vector is also another Killing vector, Eq. (8.146) does not uniquely specify the scaling of κ; it can be changed by a constant factor by rescaling the Killing vector. If the spacetime is asymptotically flat, then one can normalize the Killing vector at the spatial infinity thereby obtaining a unique value for the surface gravity – which is what we usually do in the case of black hole spacetimes. When the spacetime is not asymptotically flat, some other alternative prescription will be required to determine the normalization of κ.

To avoid any misunderstanding, we would like to point out the following peculiarity of the null surface. Consider the vector $\boldsymbol{m} = \partial/\partial r$ in the Schwarzschild like coordinates. This vector clearly points along the radial direction. Transforming to the null Kruskal–Szekeres coordinates, we find that its components are given by

$$\boldsymbol{m} = \frac{\partial}{\partial r} = \frac{\partial X^a}{\partial r}\frac{\partial}{\partial X^a} = \frac{\partial X^a}{\partial r^*}\frac{1}{f}\frac{\partial}{\partial X^a} = \frac{\kappa}{f}\left(U\frac{\partial}{\partial U} + V\frac{\partial}{\partial V}\right). \tag{8.148}$$

On the future horizon, f diverges but $f\mathbf{m}$ has the limit:

$$f\mathbf{m}\big|_{\mathcal{H}} = \kappa \left(U\frac{\partial}{\partial U} + V\frac{\partial}{\partial V} \right)_{\mathcal{H}} = \kappa \left(V\frac{\partial}{\partial V} \right)_{\mathcal{H}} = \left(\frac{\partial}{\partial t} \right)_{\mathcal{H}} = \boldsymbol{\xi}\big|_{\mathcal{H}}. \qquad (8.149)$$

Therefore, in a limiting sense, we can think of $f\mathbf{m} = \boldsymbol{\xi}$ on the horizon. This is important because we would expect the relation $\xi^a \nabla_a \xi^b = \kappa \xi^b$ to hold in *all* coordinate systems including the original Schwarzschild coordinate system in which $\xi^a = \delta^a_t$. Explicit computation now gives

$$\xi^a \nabla_a \xi^b = \xi^c \xi^a \Gamma^b_{ca} = \Gamma^b_{00} = \frac{1}{2} f f' \delta^b_r, \qquad (8.150)$$

showing that $\xi^a \nabla_a \xi^b$ is in the direction of r while we expect it to be in the direction of t if it has to be equal to $\kappa \xi^b$. This conflict is resolved by noting that $f \partial/\partial r = \partial/\partial t$ on the horizon and identifying $\kappa = (1/2) f'(r_H)$. This was, of course, the original definition of κ which we have introduced. Intuitive concepts like radial and tangential directions have to be used with care on the horizon (also see Exercise 8.12).

As another illustration, let us consider the Kerr metric for which the relevant Killing vector is $\boldsymbol{\xi}_{(K)} = \boldsymbol{\xi}_{(t)} + \Omega_H \boldsymbol{\xi}_{(\phi)}$, which has the norm

$$\xi^b \xi_b = \frac{\Sigma^2 \sin^2 \theta}{\rho^2} (\Omega_H - \omega)^2 - \frac{\rho^2 \Delta}{\Sigma^2} \qquad (8.151)$$

where $\omega = -g_{\phi t}/g_{\phi\phi}$ is defined in Eq. (8.78). Differentiating this expression and using the conditions $\omega = \Omega_H$ and $\Delta = 0$ on the horizon, we get

$$\nabla_a \left(-\xi^b \xi_b \right) = \frac{\rho^2}{\Sigma} \partial_a \Delta. \qquad (8.152)$$

Further, using $\partial_a \Delta = 2(r_h - M)\partial_a r$ and $\xi_a = (1 - a\Omega_H \sin^2 \theta)\partial_a r$ on the horizon, we can easily evaluate the surface gravity of the Kerr metric to be

$$\kappa = \frac{r_h - M}{r_h^2 + a^2} = \frac{\sqrt{M^2 - a^2}}{r_h^2 + a^2}. \qquad (8.153)$$

It is easy to verify that this expression matches with the corresponding quantity denoted by the same symbol κ in Eq. (8.100).

Using the Killing vector ξ_a, one can obtain an explicit and useful expression for κ. Any congruence of null generators is necessarily hypersurface orthogonal and hence will satisfy the Frobenius theorem given by Eq. (4.115). Therefore, we necessarily have $\xi_{[c} \nabla_b \xi_{a]} = 0$. This relation, when expanded out and simplified using the Killing equation, gives

$$\xi_c \nabla_b \xi_a + \xi_b \nabla_a \xi_c + \xi_a \nabla_c \xi_b = 0. \qquad (8.154)$$

On contracting this relation with $\nabla^b \xi^a$ we get

$$\xi_c \, \nabla^b \xi^a \nabla_b \xi_a = -\xi^b \, \nabla_a \xi_c \nabla_b \xi^a + \xi^a \nabla_c \xi_b \nabla_a \xi^b = -\kappa \xi^a \nabla_a \xi_c + \kappa \xi^b \nabla_c \xi_b = 2\kappa^2 \xi_c. \tag{8.155}$$

This allows us to identify

$$\kappa^2 = -\frac{1}{2} \nabla^b \xi^a \, \nabla_b \xi_a, \tag{8.156}$$

which provides an alternate definition of the surface gravity.

Using the expression for surface gravity given in Eq. (8.156), it is easy to show that the surface gravity is constant on the orbits of ξ. If t^a is a tangent vector on $\mathcal{H}$, then from Eq. (8.156) we get that

$$t^a \partial_a \kappa^2 = -(\nabla^m \xi^n) t^r \nabla_r \nabla_m \xi_n \big|_{\mathcal{H}} = -(\nabla^m \xi^n) t^r R^s{}_{nmr} \xi_s, \tag{8.157}$$

where we have used Eq. (5.131). But since ξ is also a tangent vector on $\mathcal{H}$, the variation of κ^2 along ξ is given by

$$\xi^a \partial_a \kappa^2 = -(\nabla^m \xi^n) R_{nmrs} \xi^r \xi^s = 0, \tag{8.158}$$

where the last result follows from the antisymmetry of curvature tensor. In fact, it can be shown that κ is a constant on the *entire horizon* $\mathcal{H}$; that is, its value does not change from orbit to orbit. (Later on, when we study quantum field theory in black hole spacetimes in Chapter 14, we will see that one can associate a temperature to the horizon that is proportional to κ and an entropy that is proportional to the area. The constancy of κ on the horizon is therefore called the *zeroth law of black hole mechanics*, in analogy with the zeroth law of thermodynamics.) This result is quite general but it is somewhat easier to prove if a bifurcation 2-sphere $\mathcal{B}$ exists in the spacetime on which $\xi^a = 0$. We already know that κ^2 is constant on each orbit of ξ. The value of this constant is equal to the value of κ^2 at the limit of this orbit on $\mathcal{B}$. Therefore κ^2 is a constant on the horizon $\mathcal{H}$ if it is constant on $\mathcal{B}$. But

$$t^a \partial_a \kappa^2 = -(\nabla^m \xi^n) t^r R^s{}_{nmr} \xi_s \big|_{\mathcal{H}} = 0 \tag{8.159}$$

on $\mathcal{B}$ since $\xi = 0$ on $\mathcal{B}$. Since t^a can be any tangent on $\mathcal{B}$, it follows that κ is constant on $\mathcal{B}$ and hence on $\mathcal{H}$. A more general proof is given in Exercise 8.13.

Exercise 8.12

Practice with null surfaces and local Rindler frames This exercise explores several features of bifurcation horizons and the associated vectors using the Rindler horizon as a prototype. The ideas generalize in a natural fashion to any bifurcation horizon and also allow the introduction of the notion of a local Rindler frame (which will be useful in Chapters 15 and 16.)

(a) Introduce the Rindler coordinates of different kinds and verify the following relations

$$ds^2 = -dT^2 + dX^2 = -dU\,dV = -e^{\kappa(v-u)}du\,dv = -2\kappa l\,dt^2 + (2\kappa l)^{-1}dx^2 \quad (8.160)$$

by working out explicit coordinate transformations. In particular, note that the null coordinates in the two frames are related by $U = T - X = -\kappa^{-1}e^{-\kappa u}$, $V = T + X = \kappa^{-1}e^{\kappa v}$. We now introduce several closely related vectors and explore their properties.

(b) Let k^a be a future directed null vector with components proportional to $(1,1)$ in the inertial frame. Show that the affinely parametrized null curve can be taken to be $x^a = X(1,1)_I$ with X being the affine parameter and with subscript I indicating the components in the inertial frame.

(c) Next define the Killing vector ξ^a corresponding to translations in the Rindler time coordinate. Show that $\xi^a = (1,0)_R = \kappa(X,T)_I$ and $\xi^a\xi_a = -2\kappa l = -N^2$. This shows that the bifurcation horizon $\mathcal{H}$ is at the location where $\xi^a\xi_a = 0$. Compute the 'acceleration' of this Killing vector and show that $a^i = \xi^b\nabla_b\xi^i = \kappa^2(T,X)_I$ and that, on the horizon, $a^i = \kappa\xi^i$. Also show that, on the horizon, $\xi^a \to \kappa X k^a$.

(d) Let the four-velocity of observers moving along the orbits of the Killing vector ξ^a be u^a. Show that, on $\mathcal{H}$, $Nu^a \to \kappa X k^a$. Also define the unit normal r_a to $l = $ constant surface. Show that $Nr^a \to \kappa X k^a$ when we approach the horizon. Hence conclude that Nu^i, Nr^i, a^i and ξ^i all tend to vectors proportional to k^i on the horizon.

(e) Show that, on any spacelike 2-surface, orthogonal to the plane, one can use the integration measure $d\Sigma_{ab} = (n_a u_b - n_b u_a)dA$, where dA is an infinitesimal area element.

(f) Explain how these concepts can be used to characterize geometrically a local Rindler frame in the neighbourhood of an event $\mathcal{P}$ provided a timelike vector field ξ^a satisfying the Killing equation is available around $\mathcal{P}$. [Hint. Choose a local inertial frame at $\mathcal{P}$ and a null vector k^a and choose the coordinates such that k^a is in the T–X plane. Boost to a local Rindler frame (LRF) with acceleration κ along the X-direction and let ξ^a be the local Killing vector corresponding to the Rindler time.]

The Killing vectors also provide a general procedure for defining the energy and angular momentum of a stationary black hole and obtaining several relationships among them. In the case of the Kerr black hole, one is, of course, dealing with a solution to the vacuum Einstein's equations. In that case, the mass and angular momentum attributed to the spacetime using asymptotic Killing vectors at spatial infinity will correspond to the conserved quantities for the source. But it is possible to think of more general black hole configurations in the presence of matter and one would like to obtain expressions for the mass and angular momentum in this case which can be associated with the black hole. To do this, we shall use the two Killing vectors $\xi^b_{(t)}$ and $\xi^b_{(\phi)}$ corresponding to time translation invariance and axial symmetry and define the total mass and angular momentum along the lines of the discussion in Section 6.5. There we obtained an expression for the conserved quantities in terms of the Killing vectors in Eq. (6.202). The expressions for mass and angular momentum, given in terms of the relevant Killing vectors, are

$$M = -\frac{1}{8\pi}\oint_S \nabla^a\xi^b_{(t)}\,dS_{ab}\,, \qquad J = \frac{1}{16\pi}\oint_S \nabla^a\xi^b_{(\phi)}\,dS_{ab}. \quad (8.161)$$

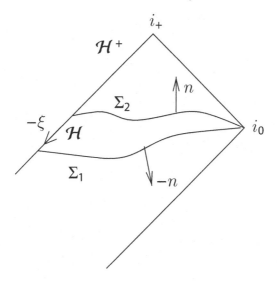

Fig. 8.12. Penrose–Carter diagram showing the relevant region involved in the integrals which define the mass and angular momentum associated with a horizon. Note that $(-\xi)$ is the outward directed normal to $\mathcal{H}^+$.

However, we will now consider these integrals over a spacelike surface which extends from the horizon to infinity (see Fig. 8.12). Then the application of the Gauss theorem will give a contribution on the horizon as well and we will get (instead of Eq. (6.201))

$$M = M_H + 2 \int_\Sigma \left(T_{ab} - \frac{1}{2} T g_{ab} \right) n^a \xi^b_{(t)} \sqrt{h} \, d^3 y \qquad (8.162)$$

$$J = J_H - \int_\Sigma \left(T_{ab} - \frac{1}{2} T g_{ab} \right) n^a \xi^b_{(\phi)} \sqrt{h} \, d^3 y, \qquad (8.163)$$

where the horizon contributions are defined as surface integrals over the horizon:

$$M_H = -\frac{1}{8\pi} \oint_\mathcal{H} \nabla^a \xi^b_{(t)} \, dS_{ab} \, , \quad J_H = \frac{1}{16\pi} \oint_\mathcal{H} \nabla^a \xi^b_{(\phi)} \, dS_{ab}. \qquad (8.164)$$

The expressions in Eq. (8.162) and Eq. (8.163) can be interpreted as giving the total amount of the conserved quantity (mass or angular momentum) as a sum over the contribution from the black hole and a contribution from possible matter existing outside the horizon.

The contributions M_H and J_H from the black hole given by Eq. (8.164) satisfy a curious identity usually called the generalized *Smarr formula*. Straightforward manipulation of the expressions for M_H and J_H in Eq. (8.164) gives

$$M_H - 2\Omega_H J_H = -\frac{1}{8\pi} \oint_{\mathcal{H}} \nabla^a \left(\xi^b_{(t)} + \Omega_H \xi^b_{(\phi)} \right) dS_{ab} = -\frac{1}{8\pi} \oint_{\mathcal{H}} \nabla^a \xi^b_{(K)} dS_{ab}.$$

(8.165)

We shall now use the results of Section(5.5.3), especially Eq. (5.184), to perform the integration over the horizon, treating it as a null surface with the normal $\xi^b_{(K)}$ and introducing an auxiliary null vector field l^a. We then get

$$M_H - 2\Omega_H J_H = -\frac{1}{4\pi} \oint_{\mathcal{H}} \xi^{(K)}_a l_b \nabla^a \xi^b_{(K)} dS$$

$$= -\frac{1}{4\pi} \oint_{\mathcal{H}} \kappa \xi^b_{(K)} l_b \, dS = \frac{\kappa}{4\pi} \oint_{\mathcal{H}} dS,$$

(8.166)

where we have used the facts that $\xi^a_{(K)} l_a = -1$ and that κ is a constant over the horizon. The final integration gives the area A of the horizon so that we have the relation

$$M_H = 2\Omega_H J_H + \frac{\kappa A}{4\pi}.$$

(8.167)

The expressions for the mass and angular momentum given by Eq. (8.162) and Eq. (8.163) allow us to consider processes in which the black hole exchanges mass and angular momentum with the surroundings. In such a process a stationary black hole with the parameters M, J and area of the event horizon A can change to another stationary black hole with parameters $M + \delta M, J + \delta J$ and $A + \delta A$. Let us assume that such a change takes place when the black hole is perturbed by the infall of an infinitesimal amount of matter with energy-momentum tensor T_{ab}. Then, the changes in mass and angular momentum can be expressed in the form

$$\delta M = -\int_H T^a_b \xi^b_{(t)} \, d\Sigma_a \,, \qquad \delta J = \int_H T^a_b \xi^b_{(\phi)} \, d\Sigma_a.$$

(8.168)

We have seen earlier examples of such processes in the case of the Penrose process and in super-radiance. We will now prove that, in general, changes in the parameters describing the black hole in such a process will be related by the equation

$$\delta M = \frac{\kappa}{8\pi} \delta A + \Omega_H \delta J.$$

(8.169)

To prove this relation, we first note that an element of the spacelike surface can be expressed in the form $d\Sigma_a = -\xi^{(K)}_a dS dv$, where $dS = \sqrt{\sigma} d^2 x$ is the integration measure on the 2-surface. Using this result, we get

$$\delta M - \Omega_H \delta J = \int_H T_{ab} \left(\xi^b_{(t)} + \Omega_H \xi^b_{(\phi)} \right) \xi^a_{(K)} dS dv = \int dv \oint_{\mathcal{H}} T_{ab} \xi^a_{(K)} \xi^b_{(K)} dS.$$

(8.170)

To evaluate this integral, it is convenient to use the Raychaudhuri equation ignoring terms which are quadratic in θ and σ_{ab}. This approximation is valid since these quantities are already first order in T_{ab} and we are working to the lowest order in the perturbations caused by the energy-momentum tensor. In this limit, using Einstein's equation, the Raychaudhuri equation obtained in Section 5.5.2 (see Eq. (5.180)) becomes

$$\frac{d\theta}{dv} = \kappa\theta - 8\pi T_{ab}\,\xi^a_{(K)}\,\xi^b_{(K)}. \tag{8.171}$$

Using this we can express $\delta M - \Omega_H \delta J$ in the form

$$\delta M - \Omega_H \delta J = -\frac{1}{8\pi} \int dv \oint_{\mathcal{H}} \left(\frac{d\theta}{dv} - \kappa\theta \right) dS$$

$$= -\frac{1}{8\pi} \oint_{\mathcal{H}} \theta dS \Bigg|_{-\infty}^{\infty} + \frac{\kappa}{8\pi} \int dv \oint_{\mathcal{H}} \theta dS. \tag{8.172}$$

Since the black hole was assumed to be stationary before and after the perturbation, the first term in the above equation vanishes. In the second term, we use the fact that θ is the fractional rate of change of the cross-sectional area of the congruence (see Eq. (5.177)), so that we can write

$$\delta M - \Omega_H \delta J = \frac{\kappa}{8\pi} \int dv \oint_{\mathcal{H}} \left(\frac{1}{dS} \frac{d}{dv} dS \right) dS$$

$$= \frac{\kappa}{8\pi} \oint_{\mathcal{H}} dS \Bigg|_{-\infty}^{\infty} = \frac{\kappa}{8\pi} \delta A. \tag{8.173}$$

This proves the result in Eq. (8.169).

There is a simple, intuitive way of understanding this result which is as follows. Let us suppose that the mass M, angular momentum J and area of the event horizon A are related by a functional form $M = M(A, J)$. Since A and J have dimensions of M^2 (in units with $G = c = 1$), it follows that the function $M(A, J)$ is homogeneous with degree (1/2). Therefore, we have

$$A\frac{\partial M}{\partial A} + J\frac{\partial M}{\partial J} = \frac{1}{2}M = \frac{\kappa}{8\pi}A + \Omega_H J. \tag{8.174}$$

The first equality follows from Euler's theorem while the second arises from Smarr's formula in Eq. (8.167). This gives

$$A\left(\frac{\partial M}{\partial A} - \frac{\kappa}{8\pi} \right) + J\left(\frac{\partial M}{\partial J} - \Omega_H \right) = 0. \tag{8.175}$$

But since A and J are free parameters, we necessarily should have

$$\frac{\partial M}{\partial A} = \frac{\kappa}{8\pi}, \qquad \frac{\partial M}{\partial J} = \Omega_H. \tag{8.176}$$

Taking the differentials of the function $M(A, J)$ and using these results we immediately get the result in Eq. (8.169).

Exercise 8.13

Zeroth law of black hole mechanics Consider a Killing vector field $\boldsymbol{\xi}$ with the corresponding surface gravity κ on a Killing horizon $\mathcal{H}$. The purpose of this exercise is to prove that the surface gravity is constant on the Killing horizon if the matter energy-momentum tensor satisfies the dominant energy condition.

(a) Let $\Psi = A^{mnr}(\xi_m \nabla_n \xi_r)$ be a scalar made from a completely antisymmetric tensor A^{mnr}. Show that Ψ vanishes on $\mathcal{H}$. Hence deduce that

$$\left(\xi_{[r} \nabla_{s]} \xi_n\right) \left(\nabla^n \xi^m\right) = \kappa \xi_{[r} \nabla_{s]} \xi^m \qquad \text{(on } \mathcal{H}). \tag{8.177}$$

(b) Since Ψ vanishes on $\mathcal{H}$, its derivative on $\mathcal{H}$ is in the direction normal to $\mathcal{H}$. Hence show that $\xi_{[m} \partial_{n]} \Psi = 0$ on $\mathcal{H}$. Thus deduce that

$$\left(\xi_n R^l{}_{sr[b} \xi_{a]} + \xi_r R^l{}_{ns[b} \xi_{a]} + \xi_s R^l{}_{rn[b} \xi_{a]}\right) \xi_l = 0. \tag{8.178}$$

Contract on r and a to obtain

$$\xi^n \xi_{[r} R^l{}_{s]nm} \xi_l = -\xi_m \xi_{[r} R^l{}_{s]} \xi_l \qquad \text{(on } \mathcal{H}). \tag{8.179}$$

(c) For any vector $\boldsymbol{v}$ we know that the scalar $\Phi \equiv \boldsymbol{v} \cdot [(\boldsymbol{\xi} \cdot \boldsymbol{\nabla})\boldsymbol{\xi} - \kappa \boldsymbol{\xi}]$ vanishes on $\mathcal{H}$ so that $\xi_{[m} \partial_{n]} \Phi = 0$ on $\mathcal{H}$. Hence obtain, using the cyclic identity of the curvature tensor,

$$\xi_m \xi_{[r} \partial_{s]} \kappa = \xi^n R^l{}_{mn[s} \xi_{r]} \xi_l = \xi^n \xi_{[r} R^l{}_{s]nm} \xi_l. \tag{8.180}$$

(d) Deduce from it the result

$$\xi_{[r} \partial_{s]} \kappa = \xi_{[s} R^l{}_{r]} \xi_l = 8\pi G \xi_{[s} T^l{}_{r]} \xi_l. \tag{8.181}$$

The dominant energy condition states that $-\xi^b T^a_b$ must be (i) either timelike or null and (ii) future directed. Using this, as well as the stationarity condition, $T_{ab} \xi^a \xi^b = 0$, show that κ must be constant on each connected component of the Killing horizon.

PROJECTS

Project 8.1

Noether's theorem and the black hole entropy

As we mentioned, it will turn out (see Chapter 14) that we can attribute an entropy to the horizon that is just one quarter of the area in Planck units for the Schwarzschild black hole. This project explores the properties of a conserved current J^a called *Noether's current* and relates it to the horizon area and thus to the entropy of the Schwarzschild black hole. The power of this idea, however, is not in the case of Einstein's theory but in more general class of theories in which the gravitational Lagrangian could be an arbitrary function of the metric and the curvature tensor. We will consider some such theories in Chapter 15 and the concept of an entropy associated with the black hole solutions in such theories will require the results in this project. This is one example of a reasonably simple idea turning out to be unexpectedly powerful.

(a) Consider a theory of gravity obtained from a generally covariant action principle involving a gravitational Lagrangian $L_{\text{grav}}(R^a{}_{bcd}, g^{ab})$ which is a scalar made from a metric, a curvature tensor and possibly derivatives of the latter. The total Lagrangian is the sum of L_{grav} and the matter Lagrangian L_m. The variation of the gravitational Lagrangian density generically leads to a surface term and hence can be expressed in the form

$$\delta(L_{\text{grav}}\sqrt{-g}) = \sqrt{-g}\left\{E_{ab}\delta g^{ab} + \nabla_a(\delta v^a)\right\}, \tag{8.182}$$

where $\nabla_a(\delta v^a)$ leads to a surface term. Convince yourself that

$$\nabla_a E^{ab} = 0 \tag{8.183}$$

identically. (That is, without assuming field equations; such results are called *off-shell*.)

(b) Concentrate now on the specific form of the variation which arises due to the infinitesimal coordinate transformation $x^i \to x^i + \xi^i$. Determine the left hand side of Eq. (8.182) for such variations and show that we now have a conservation law of the form $\nabla_a J^a = 0$ where

$$J^a \equiv \left(L_{\text{grav}}\xi^a + \delta_\xi v^a + 2E^{ab}\xi_b\right) \tag{8.184}$$

and $\delta_\xi v^a$ now represents the boundary term which arises for the specific variation of the metric in the form $\delta g^{ab} = \nabla^a \xi^b + \nabla^b \xi^a$. *This conservation is off-shell.* J^a is called the Noether current and the associated conserved charge is called the Noether charge.

(c) This J^a can be computed for any specific Lagrangian and a vector field ξ^a. Argue that, in general, the surface term will have the form

$$\delta v^a = \frac{1}{2}\alpha^{a(bc)}\delta g_{bc} + \frac{1}{2}\beta^{a(bc)}{}_d\,\delta\Gamma^d_{bc}, \tag{8.185}$$

where we use the notation $Q^{(ij)} \equiv Q^{ij} + Q^{ji}$. Specialize to the case in which ξ^i is an approximate Killing vector near an event $\mathcal{P}$. (Define the notion of an approximate Killing vector at $\mathcal{P}$ by demanding that ξ^a should satisfy the standard conditions $\nabla_{(a}\xi_{b)} = 0$; $\nabla_b\nabla_c\xi_d = R^k{}_{bcd}\xi_k$ at $\mathcal{P}$.) Show that, when ξ^a satisfies these conditions in the neighbourhood of $\mathcal{P}$, one gets $\delta_\xi v^a = 0$ and the current simplifies to

$$J^a = (L_{\text{grav}}\xi^a + 2E^{ab}\xi_b). \tag{8.186}$$

Hence show that if n_a is any vector which satisfies the condition $n_a\xi^a = 0$ at $\mathcal{P}$, we get

$$n_a J^a = (2E^{ab})n_a\xi_b. \tag{8.187}$$

(d) So far, we have not assumed the validity of the equation of motion. If L_{matt} is the matter Lagrangian with the corresponding matter action $\mathcal{A}_m$, then $\delta\mathcal{A}_m/\delta g^{ab} = -(1/2)T_{ab}\sqrt{-g}$ and the field equation is $2E_{ab} = T_{ab}$. Using this in Eq. (8.187), we get the result $n_a J^a = T^{ab}n_a\xi_b$. Show that the right hand side can be interpreted as a flux of energy through a surface with normal n_a when ξ^a is a timelike Killing vector. What do you think the left hand side is? [Answer. See Chapter 15!]

(e) Obtain the explicit expressions for E^{ab} and δv^a when the Lagrangian depends on the metric and curvature tensor but not on the derivatives of R_{abcd}. [Answer. This is a hard one but most of it is done later in Chapter 15; see Eq. (15.57).] Also argue that $\nabla_a J^a = 0$ implies that J^a can be expressed in the form $J^a = \nabla_b J^{ab}$ for an antisymmetric tensor J^{ab}.

Is J^{ab} unique for a given J^a? Combining the above results, show that J^{ab} and J^a, for the general class of theories we are considering, can be expressed in the form

$$J^{ab} = 2P^{abcd}\nabla_c\xi_d - 4\xi_d\left(\nabla_c P^{abcd}\right) \tag{8.188}$$

$$-J^a = 2\nabla_b\left(P^{adbc} + P^{acbd}\right)\nabla_c\xi_d - 2P^{abcd}\nabla_b\nabla_c\xi_d + 4\xi_d\nabla_b\nabla_c P^{abcd}, \tag{8.189}$$

where $P_{abcd} \equiv (\partial L/\partial R^{abcd})$.

(f) Compute the above expression for Einstein gravity and show that it is given by

$$J^{ab} = \frac{1}{16\pi}\left(\nabla^a\xi^b - \nabla^b\xi^a\right). \tag{8.190}$$

Let ξ^a be the timelike Killing vector in the spacetime describing a Schwarzschild black hole. Compute the Noether charge Q as an integral of J^{ab} over any two surface that is a spacelike cross-section of the Killing horizon on which the norm of ξ^a vanishes. Show that $(2\pi/\kappa)Q = (A_H/4)$, where A_H is the area of the horizon and κ is its surface gravity.

Project 8.2

Wave equation in a black hole spacetime

Consider the wave equation $\Box\Phi = 0$ for a scalar field in the Schwarzschild metric.

(a) Show that the solution can be separated in the form

$$\Phi = \frac{1}{r}\psi_\ell(r)\,Y_{\ell m}(\theta,\phi)e^{-i\omega t} \tag{8.191}$$

with ψ_ℓ satisfying the equation

$$\left[\frac{d^2}{dr^{*2}} + \omega^2 - V_\ell(r)\right]\psi_\ell(r) = 0, \tag{8.192}$$

where r^* is the tortoise coordinate and

$$V_\ell(r) = \left(1 - \frac{2M}{r}\right)\left[\frac{2M}{r^3} + \frac{\ell(\ell+1)}{r^2}\right]. \tag{8.193}$$

(b) Show that this equation does not admit static solutions which are bounded both at the horizon and at spatial infinity. This is a variant of a result known as *Price's theorem* which shows that one cannot associate a static scalar field with a black hole unlike, for example, a static (Coulomb) electric field which exists in the Reissner–Nordstrom metric.

(c) Consider a particular solution to this equation with the asymptotic behaviour

$$\psi^{\text{in}} \sim \begin{cases} e^{-i\omega r^*}, & r^* \to -\infty \\ A_{\text{out}}(\omega)e^{i\omega r^*} + A_{\text{in}}(\omega)e^{-i\omega r^*}, & r^* \to +\infty. \end{cases} \tag{8.194}$$

Show that, in the high frequency limit, we can obtain the approximate expressions

$$A_{\text{out}} \approx \frac{\Gamma\left(1 - 4i\omega M\right)(4i\omega M)^{-1/2+4i\omega M}}{\sqrt{\pi}}e^{-4i\omega M}, \qquad A_{\text{in}} \approx \frac{i\Gamma\left(1 - 4i\omega M\right)(4i\omega M)^{-1/2}}{\Gamma\left(1/2 - 4i\omega M\right)}. \tag{8.195}$$

Explain why $R = A_{\text{out}}/A_{\text{in}}$ can be thought of as a reflection amplitude for this scattering potential. Show that $|R|^2 \sim e^{-8\pi\omega M}$ as $\omega M \to +\infty$. Interpret this result.

Project 8.3

Quasi-normal modes

One particularly interesting class of solutions to the wave equation are those called *quasi-normal modes* (QNM) which decay exponentially in time at every r. Consider, for example, an outgoing positive frequency mode of the form $\exp[-i\omega(t - r^*)]$. If we want these perturbations to die down as $t \to \infty$ at any r^*, we need Im $\omega < 0$. It follows that such modes are solutions to the radial equation which *grow* exponentially as $r^* \to \pm\infty$ at any finite t. It is known that the frequencies of QNM for the Schwarzschild black hole have the structure

$$k_n = i\kappa \left(n + \frac{1}{2}\right) + \frac{\ln 3}{2\pi} \kappa + \mathcal{O}[n^{-1/2}], \qquad (8.196)$$

where κ is the surface gravity of the black hole. Your aim in this project is to obtain the imaginary part of this spectrum for large n by treating the wave equation as a one-dimensional Schrodinger equation.[3]

(a) Argue that the frequencies of QNM can be identified with the poles of the scattering amplitude $S(k)$ in the momentum space. The scattering amplitude in the Born approximation is given by the Fourier transform of the potential $V(x)$ with respect to the momentum transfer $q = k_f - k_i$. In one dimension, k_i and k_f should be parallel or antiparallel; further we can take their magnitudes to be the same for scattering in a fixed potential. Then nontrivial momentum transfer occurs only for $k_f = -k_i$ so that $q = 2k_i$ in magnitude. The scattering amplitude can now be expressed as

$$S(k) = \int_{-\infty}^{\infty} dr_* V(r_*) e^{2ikr_*}, \qquad (8.197)$$

where we have omitted irrelevant constant factors. Show that, in the case of Schwarzschild spacetime, the scattering amplitude (omitting unimportant constant prefactors) becomes, with $\theta \equiv k/\kappa, a = 2M$,

$$S(k) = \left(\frac{i}{2ka}\right)^{2-i\theta} |\Gamma(1 + i\theta)|^2 \Big[l(l + 1) \, _1F_1(1 + i\theta, i\theta, -2ika)$$

$$+ \frac{1}{2}(1 - i\theta) \, _1F_1(1 + i\theta, i\theta - 1, -2ika) \Big]$$

$$+ \Gamma(-2 + i\theta) \Big[_1F_1(3, 3 - i\theta, -2ika)$$

$$- \frac{l(l + 1)}{2ka}(2i + \theta) \, _1F_1(2, 2 - i\theta, -2ika) \Big], \qquad (8.198)$$

where $_1F_1$ is the confluent hypergeometric function and Γ is the Gamma function.

(b) From the structure of the poles, show that the imaginary parts of the QNM frequencies are

$$k_n = in\kappa \qquad \text{(for } n \gg 1\text{)}. \qquad (8.199)$$

(c) Repeat the analysis for a metric with $(1 - 2M/r)$ replaced by a function $f(r)$ having a simple zero at $r = a$. Show that the leading contribution to the scattering amplitude comes from near the horizon and is given by

$$S(k) \propto \Gamma\left(1 + i\frac{k}{\kappa}\right).$$

(8.200)

So the same idea works even for a more general case.

(d) Explain why this analysis gives the correct values for the imaginary part of the QNM. Can you devise a similar procedure to obtain the real part ?

9

Gravitational waves

9.1 Introduction

One of the key new phenomena that arises in general relativity is the existence of solutions to Einstein's equations which represent disturbances in the spacetime that propagate at the speed of light. Such solutions are called gravitational waves and this chapter will explore several features of them.[1]

9.2 Propagating modes of gravity

Within the context of special relativity, it is easy to identify a wave solution. For example, a propagating, monochromatic spherical wave will be described by an amplitude that varies in space and time as $f(t, r) \propto r^{-1} \exp[-i\omega(t - r)]$. This disturbance clearly propagates from the origin with the speed of light (which is unity in our notation) with an amplitude that decreases as $(1/r)$. Since the energy flux of a wave varies as the square of the amplitude, this wave transports a constant amount of energy across every spherical surface. Such a description can be easily made Lorentz covariant in terms of an appropriate wave vector, etc., and has an unambiguous meaning.

The situation is somewhat more complicated in the case of gravity for two (closely related) reasons. First, not all the components of the metric g_{ab} enjoy equal status in the dynamics of gravity. We saw in Section 6.3 that the g_{00} and $g_{0\alpha}$ components do not propagate in general relativity. The equations governing them are constraint equations involving G_0^0 and G_α^0 and are analogous to the equation governing the gauge dependent mode in electrodynamics. Recall that, in the case of electrodynamics *without sources*, only the two transverse degrees of freedom of the vector potential represent a genuine electromagnetic wave propagating with the speed of light. The scalar potential as well as the longitudinal part of the vector potential can be eliminated by a suitable choice of gauge. The two residual degrees of freedom represent the two physical degrees of freedom of a *massless* spin-1

399

particle, as described in Chapter 3. Since the gravitational degrees of freedom, at least in the linearized limit, should correspond to the degrees of freedom associated with a massless spin-2 particle, we will again expect only two genuine propagating degrees of freedom to exist for a gravitational wave in the absence of sources. This means we need first to disentangle the gauge degrees of freedom and identify the genuine physical degrees of freedom. In the presence of a radiating source, the situation is more complicated in the case of electrodynamics as well as gravity. In electrodynamics, if the source is an accelerated charge, say, then we will have Coulomb type fields as well as radiative modes; we again have to impose suitable gauge conditions to identify the radiative part. This is true in the case of gravity as well and – in the presence of sources – we need to separate the Coulomb type degrees of freedom from the wavelike propagating modes.

The second issue, closely related to the first one, has to do with general covariance. Since we are allowed to make arbitrary coordinate transformations in general relativity, a particular functional form of the metric coefficient – by itself – cannot be interpreted as a genuine gravitational wave disturbance. As a trivial example, note that one can generate a sinusoidal component to the metric by using a coordinate system in flat spacetime that is oscillatory with respect to an inertial frame. In flat spacetime this will, of course, be easy to identify and eliminate but in a curved spacetime disentangling coordinate effects from genuine physical effects can be more complicated.

The existence of a gravitational wave is a general feature of Einstein's theory and it is possible to obtain gravitational wave solutions to the full, nonlinear, Einstein equations. However, any kind of exact solution to Einstein's equations is hard to come by and the situation regarding gravitational waves is no exception. Fortunately, however, most of the situations in which one would like to study the gravitational waves occur when the metric disturbance is small. In that case, one can describe them using linear perturbation theory around a background metric.

The general procedure for such an approximation scheme works as follows. We write the metric in the form

$$g_{ab} = g_{ab}^{(0)} + \epsilon g_{ab}^{(1)} + \epsilon^2 g_{ab}^{(2)} + \ldots, \tag{9.1}$$

where ϵ is a bookkeeping parameter that is set to unity at the end. Such an expansion, of course, is not unique and to describe any physically relevant situation we need to supply more information regarding the various terms. The usual procedure, in the study of gravitational waves, is to introduce the two length scales λ and L which characterize the distances (and timescales) over which the perturbation and the background metric varies. We will then assume that the characteristic scale at which the perturbation varies (λ) is much smaller than the scale over which the background metric varies (L). Very often, in practical contexts, such a condition needs to be imposed in terms of the timescales rather than on *spatial* length scales. For example, consider a gravitational wave with a frequency of 10^2-10^3 Hz having a wavelength of about 300–3000 km propagating near Earth. The Earth's gravitational field is *not* constant over the wavelength scale of the gravitational wave. The typical strength of the Earth's gravitational field in dimensionless units $GM/rc^2 \approx 10^{-9}$ is very large compared with the typical amplitude ($h \approx 10^{-21}$) of a gravitational wave from astrophysical sources on Earth. Therefore, even a small spatial variation of about one part in 10^{10} due to local inhomogeneities in the mass distribution on Earth will be large compared with the gravitational wave amplitude. On the other hand, the gravitational field due to nearly Newtonian sources (like the Earth) is reasonably static and certainly does not exhibit significant temporal variations at a frequency like, say, 1 kHz. Thus, in this case, one can distinguish between the perturbation h_{ab} and the background metric $g_{ab}^{(0)}$ by looking at their characteristic timescale variation rather than at the spatial variation. We shall assume the existence of two suitable length or timescales L and λ with $\lambda \ll L$, which allows one to separate out $g_{ab}^{(0)}$ from the remaining terms.

Given such a separation for the metric, one can obtain a similar expansion for the Einstein tensor. To see the structure of the emerging equations, let us first do this purely symbolically supressing the indices. The structure of the Einstein tensor is $G \sim (\partial^2 g) + (\partial g)^2$. Substituting the expansion $g = g_0 + \epsilon g_1 + \epsilon^2 g_2$ into this and collecting together the terms involving the same powers of ϵ, we get, correct up to $\mathcal{O}(\epsilon^2)$,

$$G \sim \left[(\partial^2 g_0) + (\partial g_0)^2\right] + \epsilon \left[\partial^2 g_1 + 2\partial_0 g \, \partial g_1\right] + \epsilon^2 \left(\partial^2 g_2 + 2\partial g_0 \, \partial g_2\right) + \epsilon^2 (\partial g_1)^2.$$
$$(9.2)$$

The first two terms (zeroth and linear order in ϵ) determine the evolution of background geometry and the evolution of linear perturbations in the background geometry respectively. But note that when we go to $\mathcal{O}(\epsilon^2)$, there are two different terms. The first one is identical to the $\mathcal{O}(\epsilon)$ term with g_1 replaced by ϵg_2; the second term arises from keeping up to quadratic order in g_1. When we equate the coefficient of ϵ^2 to zero, the last term, which is quadratic in (∂g_1), will act as a source for g_2.

Going back to real expansion with tensorial indices, we get a structure very similar to the one described above:

$$G_{ab} = G_{ab}^{(0)}[g_{ab}^{(0)}] + \epsilon G_{ab}^{(1)}[g_{ab}^{(1)}; g_{ab}^{(0)}] + \epsilon^2 G_{ab}^{(1)}[g_{ab}^{(2)}; g_{ab}^{(0)}] + \epsilon^2 G_{ab}^{(2)}[g_{ab}^{(1)}; g_{ab}^{(0)}] + \dots.$$
$$(9.3)$$

Here, $G_{ab}^{(0)}[g_{ab}^{(0)}]$ is just the background Einstein tensor in the absence of perturbations. The term linear in ϵ is obtained by a second order linear operator (which depends on $g_{ab}^{(0)}$) acting on $g_{ab}^{(1)}$. When we come to quadratic order there are two contributions: the first one is of the same kind as the previous one but now with the

operator acting on $g_{ab}^{(2)}$. The second arises due to retaining up to quadratic order in
the expansion with $g_{ab}^{(1)}$ and will have terms like $(\partial g_{ab}^{(1)})^2$. When we use $G_{ab} = 0$
(in a source-free region) we get, at the zeroth and first order in ϵ the equations:

$$G_{ab}^{(0)}[g_{ab}^{(0)}] = 0; \qquad G_{ab}^{(1)}[g_{ab}^{(1)}; g_{ab}^{(0)}] = 0. \tag{9.4}$$

The first equation determines the background metric and the second equation
describes the evolution of perturbations in this background metric. Up to this order,
the description is fairly straightforward. At the next order we get an equation

$$G_{ab}^{(1)}[g_{ab}^{(2)}; g_{ab}^{(0)}] = -G_{ab}^{(2)}[g_{ab}^{(1)}; g_{ab}^{(0)}] \equiv 8\pi t_{ab}^{eff}, \tag{9.5}$$

which shows that the second order deviation in the metric, $g_{ab}^{(2)}$, is sourced by a term
that depends quadratically on $g_{ab}^{(1)}$ and could possibly be thought of as the effective energy-momentum tensor of the perturbations $g_{ab}^{(1)}$. In an approximate sense,
with some further caveats which we will discuss in Section 9.5, one can indeed
interpret this term as the energy-momentum tensor of the gravitational wave. The
situation becomes more complicated when sources are present since one needs to
carefully separate the nearly stationary parts of the source (which will contribute
to the background) from the more rapidly varying parts (which will contribute to
the generation of gravitational waves). But the key idea remains the same.

The above description suggests the following procedure for studying the gravitational waves, which we shall adopt. First, we will set $g_{ab}^{(0)} = \eta_{ab}$ and work out the
physics of gravitational waves propagating in the flat background to linear order of
accuracy in ϵ. Then we will generalize the ideas to wave propagation in a curved
background with $g_{ab}^{(0)} \neq \eta_{ab}$. This will also allow us to use the coupling of the
gravitational waves to matter and study the generation of gravitational waves. We
shall then discuss the issue of energy-momentum tensor for gravitational waves by
computing the right hand side of Eq. (9.5) and study the question of back reaction.

9.3 Gravitational waves in a flat spacetime background

In the case of gravitational waves propagating in a flat spacetime, we can express
the metric in the form $g_{ab} = \eta_{ab} + h_{ab}$, with h_{ab} being a small perturbation as
described in the beginning of Section 6.4, and most of the formalism developed
there can be used. We shall begin our discussion with this case in order to develop
the necessary physical intuition. The description of small disturbances propagating
in a *curved* background will be taken up in Section 9.4.

Let us briefly recall the results of Section 6.4. If $g_{ab} = \eta_{ab} + h_{ab}$, then, to
linear order of accuracy in h_{ab}, perturbations satisfy the differential equation (see
Eq. (6.117))

$$\Box \bar{h}_{mn} + \eta_{mn} \partial_r \partial_s \bar{h}^{rs} - \partial_n \partial_r \bar{h}_m^r - \partial_m \partial_r \bar{h}_n^r = -16\pi T_{mn}, \tag{9.6}$$

where $\bar{h}_{mn} \equiv h_{mn} - (1/2)\eta_{mn}h$. We also saw in Section 6.4 that the linear theory remains invariant under the gauge transformations produced by a vector ξ^i:

$$\bar{h}^{mr} \to \bar{h}^{mr} - \partial^m \xi^r - \partial^r \xi^m + \eta^{mr}\partial_s \xi^s. \tag{9.7}$$

Using this fact, we can *always* impose the gauge conditions (see Eq. (6.122)):

$$\partial_r \bar{h}'^{mr} = 0 \tag{9.8}$$

thereby simplifying the equation governing the perturbations to (see Eq. (6.123))

$$\Box \bar{h}^{mn} = -16\pi \, T^{mn}. \tag{9.9}$$

It is obvious from this equation that h^{mn} will possess wavelike solutions both in the presence and absence of sources. However, before we discuss those solutions, we will provide an alternative description of the linear perturbation theory which has the advantage of being completely gauge invariant.

To do this, we will use the formalism developed in Section 6.3 which showed that, even in the fully nonlinear theory, $\Box R_{bcmn}$ can be equated to a source made of derivatives of the energy-momentum tensor and nonlinear coupling terms (see Eq. (6.100)). Ignoring the nonlinear terms will lead to an equation for the curvature (see Eq. (6.101)) sourced by the second derivatives of the energy-momentum tensor:

$$\Box R_{bcmn} = 8\pi \left[\partial_b \left(\partial_m \bar{T}_{nc} - \partial_n \bar{T}_{mc}\right) - \partial_c \left(\partial_m \bar{T}_{nb} - \partial_n \bar{T}_{mb}\right)\right] \equiv 8\pi \bar{T}_{bcmn}, \tag{9.10}$$

where $\bar{T}_{ij} = T_{ij} - (1/2)g_{ij}T$ and the second equality defines $\bar{T}_{bcmn}$. This is equivalent to converting Eq. (2.124) in the electromagnetism to a wave equation for the electromagnetic field in the form

$$\Box F^{ki} = 4\pi(\partial^i J^k - \partial^k J^i). \tag{9.11}$$

Equation (9.10) is gauge invariant because the curvature tensor is *invariant* under the gauge transformations in Eq. (9.7) which can be thought of as *infinitesimal* coordinate transformations. (In contrast, the curvature tensor is *covariant* under *finite* general coordinate transformations.) Thus Eq. (9.10) provides a gauge independent description of propagating, wavelike, disturbances in a curvature.

Let us first consider propagating wave solutions to this equation in the absence of sources, when the curvature tensor satisfies the free wave equation $\Box R_{bcmn} = 0$. The basic plane wave solution to this equation can be expressed in the form $R_{abmn} = C_{abmn} \exp(ik_a x^a)$, where $k^2 = 0$ and C_{abmn} has all the symmetries of the curvature tensor. (We assume, as usual, that one has to take the real part of the expressions in the end.) In the linear theory, one can build an arbitrary solution by a linear combination of these plane waves. Since the differentiation is equivalent to multiplication by the wave vector, the Bianchi identity, in the linear limit, reduces to the algebraic equation

$$C_{bcmn}\, k_a + C_{camn}\, k_b + C_{abmn}\, k_c = 0. \tag{9.12}$$

(Note that, to linear order of accuracy, $\nabla_a R_{ijkl} \approx \partial_a R_{ijkl}$.) This equation restricts the number of independent components in C_{ijkl} and we shall now show that the only relevant components of the curvature tensor are of the form $R_{\alpha 0 \beta 0}$. To do this, let us orient the spatial coordinates such that the wave is propagating along the z-direction, giving the wave vector the components $k_s = (-\omega, 0, 0, \omega)$. Then, setting the index $c = 0$ in Eq. (9.12) gives the result

$$C_{abmn} = C_{b0mn}\, \frac{k_a}{\omega} - C_{a0mn}\, \frac{k_b}{\omega}. \tag{9.13}$$

Setting $n = 0$, we get $C_{abm0} = C_{b0m0}(k_a/\omega) - C_{a0m0}(k_b/\omega)$. Relabelling the indices as $a \to m, b \to n, m \to b, n \to 0$ and using the symmetries of the curvature tensor, we can write

$$C_{b0mn} = C_{mnb0} = C_{n0b0}\, \frac{k_m}{\omega} - C_{m0b0}\, \frac{k_n}{\omega}. \tag{9.14}$$

Substituting this back into Eq. (9.13), we can express C_{abmn} in terms of components of the form C_{n0b0}, etc. Further, in C_{n0b0}, etc., the terms with $n = 0$ or $b = 0$ vanish. This shows that all components of the curvature tensor can be expressed as sums of terms involving only components of the form $R_{\alpha 0 \beta 0}$. Imposing Einstein's equations, which require $R_{bn} = R^a{}_{ban} = 0$, leads to the condition

$$C^{\alpha}{}_{0\alpha 0} k_n k_b - C_{\alpha 0 n 0} k^{\alpha} k_b - C_{\alpha 0 b 0} k^{\alpha} k_n = 0. \tag{9.15}$$

Evaluating this expression for $n = b = 0$ gives $C^{\alpha}{}_{0\alpha 0} = 0$ which – when substituted back into Eq. (9.15) and simplified – gives $C_{z0n0} = 0 = C_{n0z0}$. Using this result in Eq. (9.14) we get $C_{z0mn} = 0 = C_{mnz0}$. Hence, it follows that the only possible nonzero components of $R_{\alpha 0 \beta 0}$ are of the form R_{x0x0} or R_{y0y0} or R_{x0y0}. Using vacuum Einstein equations $R_{00} = 0$ it is easy to show that $R_{x0x0} = -R_{y0y0}$. Therefore the only two independent degrees of freedom for the gravitational wave in empty space can be taken to be R_{x0x0} and R_{x0y0} (with $R_{y0y0} = -R_{x0x0}$). All other components, not related to these by symmetries, vanish in this case. In the case of a plane wave propagating along the z-axis, all non-vanishing components of the curvature tensor are functions of $(t - z)$. This analysis clearly demonstrates the gauge invariant propagation of curvature disturbances at the speed of light, with two independent degrees of freedom.

We will now connect up this result with a description in terms of the metric perturbations $\bar{h}_{ab}$. In the absence of sources, we can describe the general solution to Eq. (9.9) by

$$\bar{h}_{ab}(\boldsymbol{x}, t) = \int d^3k \, A_{ab}(\boldsymbol{k}) \, e^{i(\boldsymbol{k}\cdot\boldsymbol{x}-\omega t)} \tag{9.16}$$

with the usual understanding that one needs to take the real part of the complex quantities. We will now show that it is possible to impose the gauge conditions such that

$$h_{00} = h_{0\alpha} = 0; \qquad h = h^\alpha_\alpha = 0; \qquad \partial_\alpha h^{\alpha\beta} = 0. \tag{9.17}$$

(Such a gauge is called a *transverse-traceless* (*TT*) gauge.) To see this, we begin by noting that, in obtaining Eq. (9.9), we had already imposed the condition in Eq. (9.8). We can, however, still make a gauge transformation of the kind in Eq. (9.7) with $\Box\xi^a = 0$, maintaining the gauge condition in Eq. (9.8). (See the discussion following Eq. (6.122) in Section 6.4.) The general solution to the equation $\Box\xi^a = 0$ is given by

$$\xi^a = \int d^3k \, C^a(\boldsymbol{k}) \, e^{i(\boldsymbol{k}\cdot\boldsymbol{x}-\omega t)} \tag{9.18}$$

with $\omega = |\boldsymbol{k}|$. Under the gauge transformation in Eq. (9.7), the A_{ab} in Eq. (9.16) changes to

$$A_{ab} \rightarrow A'_{ab} = A_{ab} - 2ik_{(a}C_{b)} + i\eta_{ab}k^d C_d. \tag{9.19}$$

To achieve the TT gauge conditions in Eq. (9.17), we need to find a set of four functions $C^a(\boldsymbol{k})$ which satisfies the equations

$$0 = \eta^{ab} A'_{ab} = \eta^{ab} A_{ab} + 2ik^a C_a \tag{9.20}$$

$$0 = A'_{0b} = A_{0b} - iC_b k_0 - iC_0 k_b - i\delta^0_b(k^a C_a). \tag{9.21}$$

An explicit solution to these equations is given by

$$C_a = \frac{A_{bc}l^b l^c}{8i\omega^4} k_a + \frac{\eta^{bc} A_{bc}}{4i\omega^2} l_a - \frac{1}{2i\omega^2} A_{ab}l^b, \tag{9.22}$$

where $k^a = (\omega, \boldsymbol{k})$ and $l^a = (\omega, -\boldsymbol{k})$. This shows that one can *always* make a transformation to the transverse-traceless gauge as long as there is no source. The gauge conditions in Eq. (9.8) reduce the number of degrees of freedom in h_{ab} from 10 to $10 - 4 = 6$. The further gauge transformation involving the four arbitrary functions C_a reduces this further to two degrees of freedom.

We stress that while the gauge condition in Eq. (9.8) can be imposed in the context of linear perturbation theory *even in the presence of sources*, the transition to TT gauge is possible *only* in the vacuum (see Exercise 9.3 for a simple example). Even in the presence of the source, after we have imposed Eq. (9.8), we still have the freedom to perform a gauge transformation with a vector field ξ_a that satisfies $\Box \xi_a = 0$. But, in the presence of the source, we cannot use this gauge freedom to set to zero any more components of $\bar{h}_{ab}$ when $\Box \bar{h}_{ab} \neq 0$. The situation is similar to that of gauge transformations in electromagnetism. The Maxwell equations

$$\partial_m F^{mn} = \partial_m (\partial^m A^n - \partial^n A^m) = J^n \tag{9.23}$$

can *always* be reduced to the form $\Box A^n = J^n$ by imposing the Lorenz gauge condition $\partial_m A^m = 0$. But the Lorenz gauge still leaves us with the freedom to implement the change $A_m \rightarrow A_m - \partial_m \lambda$ with λ satisfying the condition $\Box \lambda = 0$. In the absence of the source we have $\Box A^n = 0$ and one can always choose a λ satisfying $\Box \lambda = 0$ and still make one of the components, say, A^0, vanish. Then, the Lorenz gauge reduces to $\nabla \cdot \mathbf{A} = 0$ which is just the transversality condition. But when $\Box A^n \neq 0$ we cannot find a λ which will satisfy $\Box \lambda = 0$ that will make one of the components of A^n vanish. The situation in the case of a gravitational wave is similar.

Using the relation between the curvature and the metric perturbation given in Eq. (6.112), one can easily show that the independent components of the curvature are given by

$$R_{\alpha 0 \beta 0} = -\frac{1}{2} \partial_0^2 h_{\alpha \beta}^{TT}, \tag{9.24}$$

thereby maintaining a one-to-one correspondence in the degrees of freedom between the metric and curvature. In the case of a plane gravitational wave propagating along the z-direction, the nonzero metric components in the TT gauge are given by

$$h_{xx}^{TT} = -h_{yy}^{TT} \equiv h_+(t - z); \qquad h_{xy}^{TT} = -h_{yx}^{TT} \equiv h_\times (t - z) \tag{9.25}$$

so that

$$R_{x0x0} = -\frac{1}{2} \partial_0^2 h_{xx}^{TT}; \qquad R_{x0y0} = -\frac{1}{2} \partial_0^2 h_{xy}^{TT}. \tag{9.26}$$

We thus see that the two independent degrees of freedom in a propagating gravitational wave can be described in terms of the two functions usually denoted by h_+ and $h_\times$. In the TT gauge they are simply related to the xx, yy and xy components of the metric perturbations. This result can be written somewhat more formally by introducing two polarization tensors $e_{\alpha\beta}^{(1)}$, $e_{\alpha\beta}^{(2)}$ for the gravitational wave. The tensor $e_{\alpha\beta}^{(1)}$ is obtained from $h_{\alpha\beta}$ by setting $h_+ = 1$, $h_\times = 0$ while $e_{\alpha\beta}^{(2)}$ is obtained by setting $h_+ = 0$, $h_\times = 1$. With this notation, we can write $h_{\alpha\beta}^{TT} = h_+ e_{\alpha\beta}^{(1)} + h_\times e_{\alpha\beta}^{(2)}$.

These two modes constitute the two linearly polarized states of the gravitational wave. One could have considered an arbitrary superpositions of these two modes to form a more general gravitational wave. For example, the right and left handed circularly polarized modes are given by

$$h_R = \frac{1}{\sqrt{2}}\left(h_+ + ih_\times\right); \qquad h_L = \frac{1}{\sqrt{2}}\left(h_+ - ih_\times\right). \tag{9.27}$$

Before we investigate the physical properties of these degrees of freedom, we will briefly comment on the issue of extracting the transverse-traceless part $h_{ab}^{TT}(t, \boldsymbol{x})$ from the linearized metric perturbations $h_{ab}(t, \boldsymbol{x})$ for a gravitational wave in the Lorenz gauge. The simplest procedure is first to calculate the curvature tensor from h_{ab} (which is a gauge invariant quantity and hence can be computed in any gauge) and then solve the equations

$$\partial_0^2 h_{\beta\gamma}^{TT} = -2R_{\beta 0\gamma 0}. \tag{9.28}$$

This can be done trivially in Fourier space (in the time coordinate) for each of the Fourier components leading to $h_{\beta\gamma}^{TT}(\omega, \boldsymbol{x}) = 2\omega^{-2} R_{\beta 0\gamma 0}(\omega, \boldsymbol{x})$. An inverse Fourier transform will now give the metric perturbations in the TT gauge.

In the case of a *plane* wave, one can also project out the TT components using the projection operator

$$P_{\beta\gamma} = \delta_{\beta\gamma} - n_\beta\, n_\gamma; \qquad n_\alpha = \frac{k_\alpha}{|\boldsymbol{k}|} \tag{9.29}$$

twice. (It is easy to verify that $P_{\beta\gamma}$ is a projection operator orthogonal to the direction of propagation and has a trace equal to 2.) In terms of this operator, the TT components of the metric are given by

$$h_{\beta\gamma}^{TT} = \left(P_{\beta\delta}\, P_{\mu\gamma} - \frac{1}{2}P_{\beta\gamma}P_{\mu\delta}\right)h^{\delta\mu} \equiv \Lambda_{\beta\gamma\delta\mu}h^{\delta\mu}. \tag{9.30}$$

The $\Lambda_{\beta\gamma\delta\mu}$ is also a projector in the sense that $\Lambda_{\alpha\beta\mu\nu}\Lambda^{\mu\nu}{}_{\sigma\rho} = \Lambda_{\alpha\beta\sigma\rho}$. Further, it is transverse in all the indices; that is the contraction of n^α with any of the indices will vanish. It is symmetric under simultaneous exchange $(\alpha\beta) \leftrightarrow (\sigma\rho)$ and traceless on the first and second pairs of indices – so that one could have used either $\bar{h}_{ab}$ or h_{ab} in the right hand side of Eq. (9.30). Note that the original perturbation should be in the Lorenz gauge (with $\Box \bar{h}_{ab} = 0$) to ensure that the TT part obeys the wave equation. The properties of $\Lambda_{\alpha\beta\mu\nu}$ makes it clear that $h_{\beta\gamma}^{TT}$ is indeed transverse and traceless. Comparing $h_{\beta\gamma}$ with $h_{\beta\gamma}^{TT}$ we see that the projection removes from the metric one 'trace part' given by

$$h_{\beta\gamma}^{T} = \frac{1}{2}P_{\beta\gamma}\left(P_{\delta\mu}\, h^{\delta\mu}\right), \tag{9.31}$$

and one 'longitudinal part' given by

$$h_{\beta\gamma}^{L} = h_{\beta\gamma} - P_{\beta\delta}\, P_{\mu\gamma}\, h^{\delta\mu} = n^\delta\, n_\gamma\, h_{\beta\delta} + n_\beta\, n^\delta\, h_{\delta\gamma} - n_\beta\, n_\gamma\left(n_\delta\, n_\mu h^{\delta\mu}\right). \tag{9.32}$$

The operation of extracting the TT component is a purely algebraic operation and hence can be carried out for any other symmetric tensor.

In the general case, the procedure is again essentially the same but the projection operator will now be an integral operator which can be written formally as

$$P_{\beta\gamma} = \delta_{\beta\gamma} - \frac{1}{\nabla^2} \partial_\beta \partial_\gamma. \tag{9.33}$$

This is equivalent to the identification $n_\delta n_\mu \rightarrow (1/\nabla^2) \partial_\delta \partial_\mu$ with $(1/\nabla^2)$ interpreted as an integral operator; that is, $A = \nabla^{-2} B$ implies that $\nabla^2 A = B$. Using these results, it is easy to verify that the change in the metric perturbation under a gauge transformation given by $\delta h_{mn} = -(\partial_n \xi_m + \partial_m \xi_n)$, has no transverse part:

$$P^{\beta\delta} P^{\gamma\mu} (\delta h_{\delta\mu}) = -P^{\beta\delta} P^{\gamma\mu} (\partial_\delta \xi_\mu + \partial_\mu \xi_\delta) = 0. \tag{9.34}$$

This shows that the transverse-traceless part is actually gauge invariant.

Exercise 9.1

Gravity wave in the Fourier space Consider a plane gravitational wave metric in the form $g_{ab} = \eta_{ab} + h_{ab}$, with $h_{ab} = A_{ab} \, e^{ikx}$. Show that the linearized curvature tensor and the Ricci tensor are given by

$$R_{smnr} = \frac{1}{2} \left(k_n k_s h_{mr} + k_r k_m h_{sn} - k_n k_m h_{sr} - k_r k_s h_{mn} \right);$$

$$R_{mn} = \frac{1}{2} \left(k_n w_m + k_m w_n - k^2 h_{mn} \right), \tag{9.35}$$

where $k^2 = k_a k^a$ and $w_m = k^r \bar{h}_{mr}$. Hence show that linearized (source-free) Einstein's equations demand $k^2 h_{mn} = k_n w_m + k_m w_n$.

(a) Use the above results to argue that, if $k^2 \neq 0$, then the curvature vanishes. Therefore, the h_{ab} in this case should correspond to a description of the flat spacetime in an oscillating coordinate system. Determine the explicit form of the coordinate transformation from the inertial coordinates to the oscillating coordinates.

(b) Show that, if $k^2 = 0$, we must have $k^r \bar{h}_{mr} = 0$ and $R_{smnr} k^r = 0$. Explain why we have a genuine gravitational wave in this case.

(c) Show that, under the gauge transformations with $\xi^m = \epsilon^m e^{ikx}$, the amplitude A^{mn} transforms as

$$A'^{mn} = A^{mn} - i\epsilon^m k^n - i\epsilon^n k^m + i\eta^{mn} \epsilon^r k_r. \tag{9.36}$$

Exercise 9.2

Effect of rotation on a TT gravitational wave Consider a gravitational wave in the TT gauge propagating along the z-axis with the amplitudes for the two polarizations denoted by h_+ and $h_\times$. Rotate the spatial coordinate system about the z-axis by an angle ψ.

(a) Show that, in the new coordinate system, the amplitudes for the two polarizations are given by

$$h'_+ = h_+ \cos 2\psi - h_\times \sin 2\psi; \qquad h'_\times = h_+ \sin 2\psi + h_\times \cos 2\psi. \tag{9.37}$$

How do the circularly polarized states h_L and h_R behave under spatial rotation?

(b) Compare this result with the corresponding situation for a transverse electromagnetic wave propagating along the z-direction.

(c) Show that when one performs a Lorentz boost along the z-direction: (i) the transverse components of the electromagnetic vector potentials (A_x^T, A_y^T) do not change but (ii) the corresponding electric field components are scaled up by the factor $[(1 - \beta)/(1 + \beta)]^{1/2}$ and (iii) the energy density and flux T^{00}, T^{0z} are scaled up by the factor $[(1 - \beta)/(1 + \beta)]$. It is usual to say that the electric field has a boost weight 1 while the energy-momentum tensor has a boost weight 2.

(d) Determine the boost weight of the metric perturbation of a TT gravitational wave.

Exercise 9.3

Not every perturbation can be TT While discussing the weak field limit of gravity in Chapter 6 we determined the external geometry of a rotating, spherical star in linear perturbation theory (see Eq. (6.127)). Show that one cannot put this metric into a TT gauge by any coordinate transformation even though it is in the Lorenz gauge. Explain why.

9.3.1 Effect of the gravitational wave on a system of particles

We will next consider the physical effects of a propagating gravitational wave on material particles. When we obtained the Schwarzschild metric as a solution to Einstein's equations, we could determine the physical effects of the curved geometry on the material particles by solving the geodesic equation

$$\frac{du^a}{d\tau} + \Gamma^a{}_{bc} u^b u^c = 0; \quad u^a \equiv \frac{dx^a}{d\tau} \tag{9.38}$$

in that metric. By similar reasoning, one might have thought that, if we solve Eq. (9.38) in the metric corresponding to a TT gravitational wave, we will be able to determine its physical effects. Curiously enough, this approach does not work, as can be seen from the following analysis.

Consider a particle that is at rest in the coordinate system in which the metric is $g_{ab} = \eta_{ab} + h_{ab}^{TT}$ (which we will call the TT coordinate system) at a given instant $\tau = \tau_0$, say. Then $u^a = (1, 0, 0, 0)$ for the particle at $\tau = \tau_0$. Further, we find from Eq. (9.38) that

$$\frac{du^\mu}{d\tau} + \Gamma^\mu_{00} = 0 \quad \text{(at } \tau = \tau_0). \tag{9.39}$$

But in the linearized theory, in the TT gauge,

$$\Gamma^\mu_{00} = \Gamma_{\mu 00} = \frac{1}{2}\left(2\partial_0 h_{\mu 0}^{TT} - \partial_\mu h_{00}^{TT}\right) = 0 \tag{9.40}$$

showing that the acceleration $(du^\mu/d\tau)$ vanishes at $\tau = \tau_0$. So if the particle was initially at rest, it will continue to remain at rest with $x^\mu = $ constant in the linear approximation after the passage of the gravitational wave. The result shows that the TT gauge condition is equivalent to choosing a set of coordinates which move *with the particles* at the lowest order of approximation. This peculiar feature arises because general coordinate transformations and the gauge transformations in the linearized theory are closely linked to each other. The choice of gauge has implications for the choice of coordinates.

It is, however, easy to see that gravitational waves *do* have measurable effects on the particles. For example, the *proper* separation between two freely falling particles will oscillate even if their *coordinate* separation remains constant. Consider two freely falling particles located at $z = 0$ but separated along the x-axis by a coordinate distance l. In the presence of a gravitational wave in the TT gauge which propagates along the z-axis, the corresponding proper distance L between these two particles is given by

$$
L = \int_0^l dx \sqrt{g_{xx}} = \int_0^l dx \left(1 + h_{xx}^{TT}(t, z = 0) \right)^{1/2}
$$
$$
\simeq \int_0^l dx \left[1 + \frac{1}{2} h_{xx}^{TT}(t, z = 0) \right] = l \left[1 + \frac{1}{2} h_{xx}^{TT}(t, z = 0) \right]. \quad (9.41)
$$

The integrals in the above expressions are from 0 to l because the coordinate separation l does not change in the presence of the wave. It is obvious from this result that the proper separation between the two particles will oscillate with a fractional change given by

$$
\frac{\delta L}{L} \simeq \frac{1}{2} h_{xx}^{TT}(t, z = 0). \quad (9.42)
$$

To bring out this effect more clearly, we will study the geodesic deviation of two nearby particles in *two different coordinate frames*. The result $x^\mu = $ constant – which we obtained earlier – holds for particles at any spatial location in the TT coordinate system. So any two nearby particles A and B, which were originally at rest with respect to each other, will continue to remain at rest with a constant separation between them; so the geodesic deviation will remain constant (see Exercise 9.4 for an explicit demonstration) in the TT coordinate system.

The situation is different if we work in the proper coordinate system for one of the particles, say A, and study the variation of the geodesic deviation in this coordinate system. We have seen in Chapter 5 that, in the proper coordinate system, $g_{ab} = \eta_{ab}$ up to quadratic order in the coordinates and, obviously, the proper coordinate system of particle A is different from the TT coordinate system. We know from Section 5.2.3, Eq. (5.28), that the separation ξ^m between two nearby geodesics satisfies the geodesic deviation equation given by

$$\frac{D^2}{D\tau^2} \xi^m = R^m{}_{nrq} u^n u^r \xi^q, \tag{9.43}$$

where $(D/D\tau) = u^a \nabla_a$. When the test particles are moving slowly, the four-velocity is approximately equal to $u^a \approx \delta_0^a$; the spatial components of the four-velocity are of the order of h_{ab}, which can be ignored since the curvature tensor is already linear in h_{ab}. Taking $u^n = (1, 0, 0, 0)$ we find that we only need to compute one component of the curvature tensor, R_{m00q}. In the TT coordinate system, this is given by

$$R_{m00q} = \frac{1}{2} \left(\partial_0 \partial_0 h_{mq}^{TT} + \partial_q \partial_m h_{00}^{TT} - \partial_q \partial_0 h_{m0}^{TT} - \partial_m \partial_0 h_{q0}^{TT} \right) = \frac{1}{2} \partial_0^2 h_{mq}^{TT} \tag{9.44}$$

since h_{m0}^{TT} vanishes. This is the curvature tensor in the TT coordinates. But since the curvature tensor is a gauge invariant concept, we would have got the same result, to linear order, in any other coordinate system including the proper coordinate system. Since the Γ_{bc}^as vanish in the proper coordinate system, the left hand side of Eq. (9.43) can be approximated as $(\partial^2 \xi^m / \partial t^2)$ to the same order of approximation. We therefore need to solve the equation

$$\frac{\partial^2}{\partial t^2} \xi^m = \frac{1}{2} \xi_q \frac{\partial^2}{\partial t^2} h_{TT}^{qm}. \tag{9.45}$$

We know that h_{ab}^{TT} can be expressed in terms of two functions h_+ and $h_\times$. To understand clearly the effects of h_+ and $h_\times$, let us first consider a situation in which $h_+ \neq 0$, $h_\times = 0$. In this case, Eq. (9.45) reduces to

$$\frac{\partial^2}{\partial t^2} \xi^x = \frac{1}{2} \xi^x \frac{\partial^2}{\partial t^2} (h_+ e^{ik_q x^q}); \quad \frac{\partial^2}{\partial t^2} \xi^y = -\frac{1}{2} \xi^y \frac{\partial^2}{\partial t^2} (h_+ e^{ik_q x^q}) \tag{9.46}$$

with the solution (to the lowest order of accuracy):

$$\xi^x = \left(1 + \frac{1}{2} h_+ e^{ik_q x^q} \right) \xi^x(0); \quad \xi^y = \left(1 - \frac{1}{2} h_+ e^{ik_q x^q} \right) \xi^y(0). \tag{9.47}$$

The particles separated in the x-direction will oscillate along the x-direction while those which are initially separated in the y-direction will oscillate along the y-direction. So if we have a ring of stationary particles in the x–y plane, they will bounce back and forth forming an ellipse with major axis along the y-direction and then along the x-direction, etc. (see Figs. 9.1–9.3). On the other hand, when we consider the case in which $h_+ = 0$, $h_\times \neq 0$, a similar analysis will give

$$\xi^x = \xi^x(0) + \frac{1}{2} h_\times e^{ik_a x^a} \xi^y(0); \quad \xi^y = \xi^y(0) + \frac{1}{2} h_\times e^{ik_a x^a} \xi^x(0). \tag{9.48}$$

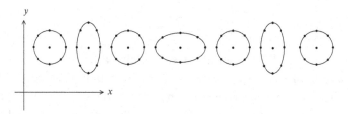

Fig. 9.1. The effect of a gravitational wave with $+$ polarization on a set of test particles originally located in a circle in the x–y plane. The deformation of circles into ellipses follows a pattern that looks like a '$+$' sign.

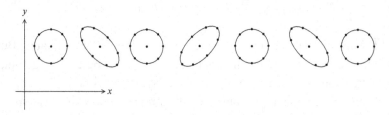

Fig. 9.2. The effect of a gravitational wave with $\times$ polarization on a set of test particles originally located in a circle in the x–y plane. The deformation of circles into ellipses follows a pattern that looks like a '$\times$' sign.

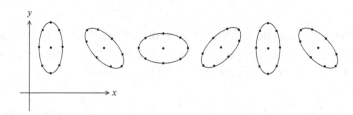

Fig. 9.3. The effect of a gravitational wave with R polarization is to distort a circle of test particles into an ellipse that rotates in a right-handed sense.

In this case, the particles bounce back and forth along 45 degree lines bisecting the x- and y-axes. (The subscripts $+$ and $\times$ in h_+ and $h_\times$ were chosen with this hindsight.) The fact that motion is confined to the x–y plane for a wave propagating along the the z-axis shows that the wave is transverse in its physical effects as well.

   To avoid misunderstandings we stress that *both* the spatial locations of particles *and* the geodesic deviation remain constant in the TT coordinates. The results in Eq. (9.47) and Eq. (9.48) hold in the proper coordinate frame of one of the particles. The fact that some part of the computation was done using $h_{\alpha\beta}^{TT}$ should not mislead one into thinking that the result is obtained in the TT coordinates. The form of R_{abcd} obtained by a computation using $h_{\alpha\beta}^{TT}$ remains valid in the proper coordinate system as well because the curvature tensor, being gauge invariant, is the same in both TT and proper coordinates to the linear order.

Exercise 9.4
Nevertheless it moves – in a gravitational wave (a) Consider a TT coordinate system in which two particles are at rest at $t = 0$. The contravariant components of the separation vector $x_A^\alpha - x_B^\alpha$ remains constant in spite of the action of the wave. Transform the components of this separation vector to the proper coordinate system and show that it matches with the result obtained in the text by using the geodesic deviation equation.

(b) Now repeat the analysis done in the text but in the TT coordinates. Let an infinitesimal spacelike vector, $l^m = (0, \mathbf{l})$, be the coordinate separation between two particles at $t = 0$. In the TT gauge, show that the equation for the geodesic deviation may be written as

$$\frac{D^2 l^m}{D\tau^2} = R^m{}_{00n} l^n = \frac{1}{2} \left(\partial_0^2 h_n^m \right) l^n. \tag{9.49}$$

Further show that, to the same order of approximation, one has the result

$$\frac{D^2 l^m}{D\tau^2} = \frac{d^2 l^m}{d\tau^2} + \frac{1}{2} \left(\partial_0^2 h_n^m \right) l^n. \tag{9.50}$$

Hence show that $l^\alpha = $ constant is a solution to the geodesic deviation equation. This demonstrates that, in the TT gauge, both the geodesic equation and the geodesic deviation equation lead to the same conclusion, as they should.

9.4 Propagation of gravitational waves in the curved spacetime

The discussion so far has been based on separating the spacetime metric as $g_{ab} = \eta_{ab} + h_{ab}$ and treating h_{ab} as a small perturbation. For most of the astrophysical applications, such an approximation is adequate if we are interested in the propagation of gravitational wave disturbances in a (nearly) flat background spacetime. There are, however, situations in which one would like to study the gravitational waves propagating in a curved spacetime. In this section we shall briefly discuss the modifications that arise to our previous results for this context.

It is certainly possible formally to separate any spacetime metric as $g_{ab} = g_{ab}^{(0)} + h_{ab}$ and treat h_{ab} as a perturbation. But since a given metric g_{ab} could be separated into a background and a perturbation in many different ways, it is not possible to treat h_{ab} as a well-defined gravitational wave in this context without further physical input. As explained earlier, in Section 9.2, this requires the existence of two well defined length scales L and λ, over which the background geometry and the perturbation vary, with $\lambda \ll L$. We shall proceed further assuming that such a separation has been introduced. Denoting all the first order perturbed quantities with a superscript 1, we find that

$$\Gamma^{i(1)}_{kl} = \frac{1}{2}\left(\nabla_l h^i_k + \nabla_k h^i_l - \nabla^i h_{lk}\right). \tag{9.51}$$

Here, as well as in what follows, the raising and lowering of indices and calculation of covariant derivatives, etc., are done using the background metric $g^{(0)}_{ab}$. Equation (9.51) is easy to obtain in a locally inertial coordinate system corresponding to $g^{(0)}_{ab}$ and then converting the ordinary derivatives into covariant derivatives; this is a valid procedure because $\Gamma^{i(1)}_{kl}$, being the difference between two connections corresponding to two metrics, $g^{(0)}_{ab}$ and g_{ab}, is a tensor (see the discussion after Eq. (6.19)). The expression for curvature can be obtained from Eq. (9.51) by noting that Γ^2 terms are ignorable in the computation to the required order. We get

$$R^{i(1)}_{klm} = \frac{1}{2}\left(\nabla_l \nabla_m h^i_k + \nabla_l \nabla_k h^i_m - \nabla_l \nabla^i h_{km} - \nabla_m \nabla_l h^i_k \right.$$
$$\left. -\nabla_m \nabla_k h^i_l + \nabla_m \nabla^i h_{kl}\right). \tag{9.52}$$

Contracting the curvature tensor, we obtain the Ricci tensor:

$$R^{(1)}_{ik} = R^{l(1)}_{ilk} = \frac{1}{2}\left(\nabla_l \nabla_k h^l_i + \nabla_l \nabla_i h^l_k - \nabla_l \nabla^l h_{ik} - \nabla_k \nabla_i h\right). \tag{9.53}$$

Similarly, the corrections to the mixed components of the Ricci tensor can be obtained using the relation

$$R^{k(0)}_{i} + R^{k(1)}_{i} = \left(R^{(0)}_{il} + R^{(1)}_{il}\right)\left(g^{kl(0)} - h^{kl}\right) \tag{9.54}$$

so that

$$R^{k(1)}_{i} = g^{kl(0)} R^{(1)}_{il} - h^{kl} R^{(0)}_{il}. \tag{9.55}$$

We will now specialize to the case in which $T_{ab} = 0$ in the region of interest; that is, we are considering the evolution of perturbations in a matter-free region, outside the sources. (The analysis can be easily generalized to a situation in which $T_{ab} \neq 0$.) Then the equation $R_{ik} = 0$ should hold order by order. Demanding $R^{(0)}_{ab} = 0$, we see that the background metric is a solution to Einstein's equations in the matter-free region. Further, $R^{(1)}_{ab} = 0$ gives the equations of motion obeyed by the perturbations

$$\nabla_l \nabla_k h^l_i + \nabla_l \nabla_i h^l_k - \nabla_l \nabla^l h_{ik} - \nabla_k \nabla_i h = 0, \tag{9.56}$$

which is the generalization of Eq. (9.6) to a curved background.

We now want to rewrite this equation in the form of a wave equation in the curved background. We saw earlier, in the case of gravitational waves propagating in the flat spacetime, that it is convenient to work with the trace-reversed metric

$\bar{h}_{ab}$. In a curved background, we shall define the corresponding quantity by $\bar{h}_{mn} = h_{mn} - (1/2)g^{(0)}_{mn}h$. Under an infinitesimal coordinate transformation, $x^a \to x^a + \xi^a$, the metric perturbations now change as

$$h_{ab} \to h_{ab} + \nabla_a \xi_b + \nabla_b \xi_a, \tag{9.57}$$

which is a direct generalization of Eq. (9.7). Using this gauge freedom we can always impose the analogue of the Lorenz gauge condition

$$\nabla_n \bar{h}^{mn} = 0. \tag{9.58}$$

To see this, note that $\nabla_n \bar{h}^{mn}$ changes under the gauge transformation as

$$\nabla_n \bar{h}^{mn} \to \nabla_n \bar{h}^{mn} + \Box \xi^m - R^m_{(0)n}\xi^n = \nabla_n \bar{h}^{mn} + \Box \xi^m, \tag{9.59}$$

which can be easily proved using Eq. (9.57). (We have used the fact that, in a vacuum, the Ricci tensor vanishes.) We can now impose the Lorenz gauge condition by choosing ξ_m to be the solution to the equation

$$\Box \xi^m = -\nabla_n \bar{h}^n_m. \tag{9.60}$$

This is a hyperbolic partial differential equation which, locally, always has solutions. Once the Lorenz condition is imposed, we can manipulate Eq. (9.56) to read

$$\Box \bar{h}_{mn} + 2R^{(0)}_{ambn}\bar{h}^{ab} = 0. \tag{9.61}$$

This equation describes the propagation of weak gravitational waves in the source-free region of a curved spacetime and is analogous to Eq. (5.135), describing the propagation of electromagnetic waves in a curved spacetime. The appearance of the curvature tensor has the same mathematical origin as the appearance of the Ricci tensor in the case of electromagnetism. Note that, if the amplitude and wavelength of h_{ab} are ϵ and λ respectively, the first term in Eq. (9.61) is $\mathcal{O}(\epsilon/\lambda^2)$ while the second term is $\mathcal{O}(\epsilon/L^2)$, where L is the scale over which the background geometry is varied. Hence, to order $\mathcal{O}(\lambda^2/L^2)$ we can ignore the second term and write the wave equation as

$$\Box \bar{h}_{mn} \simeq 0. \tag{9.62}$$

Once again, the condition in Eq. (9.60) does not completely specify the gauge and we can attempt to set $\bar{h} = 0$ by using this residual gauge freedom, as in the case of gravitational waves propagating in the flat spacetime. For this to be feasible, $\bar{h}$ has to satisfy the condition $\Box \bar{h} = 0$. However, Eq. (9.61) implies that, in the source-free region, we do have $\Box \bar{h} = 0$ and one can choose a ξ^a that is a solution to $\Box \xi^a = 0$ and impose the condition $\bar{h} = 0$. (But this is not possible in the presence of sources.) Hence Eq. (9.60) is as close to the TT gauge as one can get in the curved spacetime.

Exercise 9.5

The optics of gravitational waves Repeat the analysis of Section 5.4.4, for the gravitational waves. Assume that the perturbation is expanded in the form

$$\bar{h}_{mn} = \left(A_{mn} + \epsilon B_{mn} + \epsilon^2 C_{mn} + \cdots \right) e^{i\theta/\epsilon}, \tag{9.63}$$

where ϵ is the formal expansion parameter. Define a wave vector $k_i \equiv \partial_i \theta$, amplitude $\mathcal{A} = [(1/2) A_{mn}^* A^{mn}]^{1/2}$ and a polarization tensor $e_{mn} = A_{mn}/\mathcal{A}$. Using the wave equation and the gauge conditions, obtain the following results.

(a) The rays (defined as curves perpendicular to surfaces of constant phase) are null geodesics in the sense that $k_a k^a = 0$ and $k^b \nabla_b k^a = 0$. Further show that the polarization tensor is orthogonal to the rays ($e_{mn} k^m = 0$) and is parallel transported along the rays ($k^a \nabla_a e_{mn} = 0$).

(b) The scalar amplitude satisfies the conservation law $\nabla_i (\mathcal{A}^2 k^i) = 0$.

(c) Show that the first correction term to the amplitude obeys the equation

$$k^a B_{am} = i \nabla^a A_{am}; \qquad k^a \nabla_a B_{mn} = -\frac{1}{2} B_{mn} \nabla_a k^a + \frac{i}{2} \Box A_{mn} + i R_{ambn}^{(0)} A^{ab}. \tag{9.64}$$

(d) Show that gravitational waves of short wavelength passing through the solar system will experience the same gravitational redshift and deflection (while passing near the Sun) as the light rays.

Exercise 9.6

The $R_{ambn}^{(1)}$ is not gauge invariant, but ...   Show that under the gauge transformation in Eq. (9.57), the curvature tensor $R_{ambn}^{(1)}$ in Eq. (9.52) changes by

$$\delta R_{mnrs}^{(1)} = \xi^t \nabla_t R_{mnrs}^{(0)} + R_{tnrs}^{(0)} \nabla_m \xi^t - R_{tmrs}^{(0)} \nabla_n \xi^t + R_{mntr}^{(0)} \nabla_s \xi^t - R_{mnts}^{(0)} \nabla_r \xi^t \tag{9.65}$$

and hence is not gauge invariant. However, the change is of the order of $\mathcal{O}(\epsilon/L^2)$ while $R_{mnrs}^{(1)}$ is of the order of $\mathcal{O}(\epsilon/\lambda^2)$. Hence argue that $\delta R \simeq (\lambda/L)^2 R^{(1)}$. Does this allow us to ignore the issue of gauge invariance when $\lambda \ll L$?

Exercise 9.7

An exact gravitational wave metric Show that the metric

$$ds^2 = -2 du dv + H(u, x, y) du^2 + dx^2 + dy^2 \tag{9.66}$$

is an exact solution to Einstein's equation when H satisfies the two-dimensional Laplace equation in (x, y). Interpret this solution. Is the propagating mode in this case explicitly transverse?

9.5 Energy and momentum of the gravitational wave

In the case of a source radiating electromagnetic waves, one can unambiguously determine the energy and momentum carried away by the wave. This has to be supplied by the agency which maintains the accelerated motion of the charges in the source. Similarly, one would like to attribute some energy and momentum to the gravitational wave radiated by an isolated source. This, however, turns out to be a conceptually complicated issue. In general, propagating gravitational wave solutions are just particular kinds of solutions to Einstein's equations representing a particular class of curved spacetimes. We have already seen in Section 6.5 that one cannot associate any unique energy-momentum tensor with the gravitational field itself. Therefore, in the strict sense, one cannot consider gravitational perturbations as possessing unambiguous amounts of energy or momentum.

It is, however, possible to develop such a concept in a perturbative manner along the lines described in Section 9.2. This allows us to associate an energy-momentum tensor with the weak gravitational waves, described as perturbations propagating in a given background curved spacetime (when $g_{ab} = g_{ab}^{(0)} + h_{ab}$), as long as the wavelength of the gravitational wave is small compared with the length scale over which the background metric is varying. In such a case, the separation between the background and the perturbations is unambiguous and we can use Eq. (9.3) and Eq. (9.5) to identify the effective energy-momentum tensor of the gravitational wave as (with c and G reintroduced temporarily):

$$t_{mn} \equiv \frac{c^4}{8\pi G} \, G_{mn}^{(2)}. \tag{9.67}$$

We can now interpret Einstein's equation as having two sources, one being the conventional matter source given by T_{mn} and the other arising out of the energy-momentum tensor of the gravitational wave perturbations.

As it stands, however, there are some difficulties with such an interpretation. To begin with, $G_{ab}^{(2)}$ contains second derivatives of h_{ab} as well as terms quadratic in the first derivatives, while one would have expected the energy-momentum tensors to be just quadratic in the first derivatives. Further, while t_{mn} in Eq. (9.67) is manifestly Lorentz invariant, it is not gauge invariant. This is to be expected since the gauge invariance for the metric perturbations is intimately connected with the infinitesimal general coordinate transformations; and we already know from the discussion in Section 6.5 that one cannot determine a generally covariant energy-momentum tensor for the gravitational field.

One can get around both difficulties in the perturbative regime by the following procedure. Instead of considering the expression $G_{mn}^{(2)}$ at every spacetime event, we shall consider another quantity which is obtained by averaging each of the terms in it over a small region in the spacetime. Denoting this averaging process by the symbol $\langle \cdots \rangle$ we will redefine the energy-momentum tensor for the perturbations as

$$t_{mn} \equiv \frac{c^4}{8\pi G} \left\langle G_{mn}^{(2)} \right\rangle. \tag{9.68}$$

Fortunately, we do not have to specify the averaging procedure precisely and only need to ensure that it has certain basic properties. To begin with, because we are averaging over all the directions at each event, the average of any derivative term vanishes: that is, we demand $\langle \partial_m a \rangle = 0$. Using this result on the derivative of a product ab of two quantitites a and b, we immediately get

$$\langle (\partial_m a) b \rangle = -\langle a (\partial_m b) \rangle, \tag{9.69}$$

which shows that we can swap the derivatives inside the averaging symbol by changing its sign.

In these manipulations, the averaging process should be thought of as taking place over a region which is several wavelengths in size. Though the metric can be reduced to the Riemann normal form (see Section 5.3) around any event, we cannot do this over a larger *region* of space. The averaging over a finite region of space, therefore, retains some characteristics of metric perturbations. This argument is tentative and depends on the existence of well separated length scales; hence it cannot be considered as a rigorous proof. But one can explicitly verify that the expressions obtained by this procedure turn out to be gauge invariant. It is not possible to justify these operations in any more fundamental manner.

With the averaging process defined as above, it is straightforward to compute the energy-momentum tensor for the perturbations in Eq. (9.68). We first compute the result for propagation in a flat spacetime background $(g_{ab}^{(0)} = \eta_{ab})$ and then state the results for the more general case $(g_{ab}^{(0)} \neq \eta_{ab})$ which are quite similar. We begin by computing $R_{mn}^{(2)}$ in terms of the metric perturbations, obtaining the result

$$
\begin{aligned}
R_{mn}^{(2)} = {}&-\frac{1}{4} \left(\partial_m h^{rs} \right) \partial_n h_{rs} \\
&+ \frac{1}{2} h^{rs} \left(\partial_m \partial_s h_{nr} + \partial_n \partial_s h_{mr} - \partial_m \partial_n h_{rs} - \partial_r \partial_s h_{mn} \right) \\
&+ \frac{1}{2} \left(\partial^s h_n^r \right) \left(\partial_r h_{sm} - \partial_s h_{rm} \right) \\
&+ \frac{1}{2} \left(\partial_s h^{rs} - \frac{1}{2} \partial^r h \right) \left(\partial_m h_{nr} + \partial_n h_{mr} - \partial_r h_{mn} \right).
\end{aligned}
\tag{9.70}
$$

The left hand side is symmetric in m and n while on the right hand side some of the terms are not manifestly symmetric. Explicit calculation, however, shows that the right hand side is indeed symmetric in m and n, as it should. It is, therefore, convenient to maintain the symmetry manifest by expressing the results as a sum of this term and the one obtained by interchanging m and n and dividing by 2. We shall do this when required. We now need to average this expression which can be done in three steps. First, we use Eq. (9.69) to rewrite the product of the first derivatives in terms of the second derivatives. Next, we use Eq. (9.6) to eliminate the

$\Box h_{mn}$ terms in terms of the others. Finally, we use Eq. (9.69) again to rewrite the terms containing the second derivatives in terms of the products of first derivatives and express the results in an explicitly symmetrized form. This leads to the final result

$$\left\langle R_{mn}^{(2)} \right\rangle = \frac{1}{4} \left\langle (\partial_m h_{rs}) \, \partial_n \, h^{rs} - 2 \, (\partial_s \, h^{rs}) \, \partial_{(m} h_{n)r} + 2 \, (\partial_r h) \, \partial_{(m} h_{n)}^r - (\partial_m h) \, \partial_n h \right\rangle,$$
(9.71)

which only has terms that are quadratic in the first derivatives. Contracting this term and carrying out the above three steps once again, we can show that $\langle R^{(2)} \rangle = 0$. Finally, expressing the perturbed Einstein tensor in terms of the Ricci tensor and substituting these results into Eq. (9.68), we can obtain the expression for t_{mn}. The final result is given by (we have again set $G = c = 1$)

$$t_{mn} = \frac{1}{32\pi} \left\langle (\partial_m \bar{h}_{rs}) \, \partial_n \bar{h}^{rs} - 2 \, (\partial_s \, \bar{h}^{rs}) \, \partial_{(m} \bar{h}_{n)r} - \frac{1}{2} \, (\partial_m \bar{h}) \, \partial_n \bar{h} \right\rangle.$$
(9.72)

Much of this analysis carries through even in the case of a gravitational wave propagating in a curved background spacetime. In this case, the expression for the quadratic corrections to the curvature tensor are given by

$$R_{mn}^{(2)} = -\frac{1}{4} \, (\nabla_m h^{rs}) \, \nabla_n h_{rs}$$
$$+ \frac{1}{2} h^{rs} \, (\nabla_m \nabla_s h_{nr} + \nabla_n \nabla_s h_{mr} - \nabla_m \nabla_n h_{rs} - \nabla_r \nabla_s h_{mn})$$
$$+ \frac{1}{2} \, (\nabla^s h_n^r) \, (\nabla_r h_{sm} - \nabla_s h_{rm}) \qquad (9.73)$$
$$+ \frac{1}{2} \left(\nabla_s h^{rs} - \frac{1}{2} \nabla^r h \right) (\nabla_m h_{nr} + \nabla_n h_{mr} - \nabla_r h_{mn}).$$

This equation is the same as Eq. (9.70) with ∂_i replaced by ∇_i corresponding to the background spacetime. Proceeding further, one can compute the energy-momentum tensor exactly as before to obtain the result:

$$t_{ab} = \frac{1}{32\pi} \left\langle \nabla_a \bar{h}_{mn} \nabla_b \bar{h}^{mn} - \frac{1}{2} \nabla_a \bar{h} \nabla_b \bar{h} - (\nabla_b \bar{h}_{ma} + \nabla_a \bar{h}_{mb}) \nabla_n \bar{h}^{mn} \right\rangle.$$
(9.74)

This generalizes the result in Eq. (9.72).

The results in Eq. (9.72) and Eq. (9.74) are valid in any gauge. If we impose the Lorenz gauge as well as the condition $\bar{h} = 0$, the expression for t_{ab} reduces to

$$t_{ab} = \frac{1}{32\pi} \left\langle \nabla_a \bar{h}_{mn} \nabla_b \bar{h}^{mn} \right\rangle.$$
(9.75)

This tensor can be considered as a source for background curvature at the next order of approximation. Such a notion has to be necessarily approximate and corresponds to Eq. (9.5), which has the formal structure $R \simeq (\partial h)^2$. In terms of the typical scales, this equation corresponds to $(1/L^2) \simeq (\epsilon/\lambda)^2$. This leads to the relation $\epsilon \approx \lambda/L$ when the curvature is dominated by the gravitational waves themselves and there is no other matter contribution.

One important implication of this result is that one *cannot* develop in general relativity a systematic expansion in powers of h_{ab} *starting from flat spacetime* by writing $g_{ab} = \eta_{ab} + h_{ab}$. This is because choosing the background to be flat spacetime is equivalent to setting $(\lambda/L) = 0$ and any finite value of ϵ will then violate the condition $\epsilon \lesssim \lambda/L$. We also see that $\epsilon \ll 1$ requires $(\lambda/L) \ll 1$ and vice versa. Thus one cannot introduce the concept of a gravitational wave of arbitrarily large amplitude but varying at a length scale that is sufficiently small compared with the background scale of variation and develop a systematic perturbation theory.

In obtaining the above expressions for t_{ab}, etc., we have assumed that we are in a region of spacetime outside the sources so that $T_{ab} = 0$. If this is not the case, there will be corrections to $G^{(2)}_{ab}$, etc., owing to the presence of the source. To be consistent, we also need to separate the energy-momentum tensor of the source into parts which are varying at the scale comparable to L and parts which are varying at the scale comparable to λ. The slowly varying parts will modify the background equation while the more rapidly varying parts will act as a source for gravitational radiation, which we shall consider in the next section.

Finally, we will work out explicitly the t_{mn} for waves propagating in flat spacetime to gain some physical insight. Choosing the TT gauge we have the expression (with the c and G factors reintroduced)

$$t_{mn} = \frac{c^4}{32\pi G} \left\langle \left(\partial_m h^{TT}_{rs} \right) \partial_n h^{rs}_{TT} \right\rangle = \frac{c^4}{32\pi G} \left\langle \left(\partial_m h^{TT}_{\alpha\beta} \right) \partial_n h^{\alpha\beta}_{TT} \right\rangle, \qquad (9.76)$$

where the last result follows from the fact that $h^{m0}_{TT} = 0$. The corresponding energy flux (that is, energy passing through unit area per unit time) across a surface with normal n is given by

$$F = -ct^{0\gamma} n_\gamma. \qquad (9.77)$$

As an illustration, consider a plane gravitational wave of the form $h^{\alpha\beta}_{TT} = A^{\alpha\beta}_{TT} \cos k_a x^a$. Substituting this expression into Eq. (9.76) and averaging the oscillations over several wavelengths by using $\langle \sin^2 \theta \rangle = (1/2)$, we get the result

$$t_{mn} = \frac{c^4}{64\pi G} k_m k_n \left(A^{\alpha\beta}_{TT} A^{TT}_{\alpha\beta} \right) \qquad (9.78)$$

with the corresponding flux in the direction of $\hat{\boldsymbol{k}} = (\boldsymbol{k}/k)$ being

$$F = -ct^{0\lambda}\,\hat{k}_\lambda = -\frac{c^5}{64\pi G}\,k^0\left(k^\lambda\hat{k}_\lambda\right)\left(A^{\alpha\beta}_{TT}\,A^{TT}_{\alpha\beta}\right) = \frac{c^5}{64\pi G}\,k^0k^0\left(A^{\alpha\beta}_{TT}\,A^{TT}_{\alpha\beta}\right) = ct^{00}.$$

(9.79)

If the amplitudes of the two polarizations of the wave are h_+ and $h_\times$, then the total flux is given by

$$F = \frac{c^3}{32\pi G}\omega^2(h_+^2 + h_\times^2).$$

(9.80)

These relations are similar to the corresponding ones in the case of electromagnetic radiation because both correspond to radiation travelling at the speed of light. Using this analogy, one can introduce the notion of a graviton (similar to the notion of a photon) with a number density N and four-momentum $p^a = \hbar k^a$. In this case, the number density of gravitons in a plane wave will be proportional to the square of the amplitude just as in the case of electromagnetic radiation.

Exercise 9.8

Energy-momentum tensor of the gravitational wave from the spin-2 field In Chapter 3 we introduced a gauge invariant action for the spin-2 field in Eq. (3.28). One can obtain an energy-momentum tensor from this action, by following the general procedure outlined in Chapter 2, leading to Eq. (2.17). In this case, the resulting energy-momentum tensor will be

$$t^{mn} = -\frac{\partial L}{\partial(\partial_m h_{ab})}\,\partial^n h_{ab} + \eta^{mn}L,$$

(9.81)

where L is the Lagrangian for the spin-2 field.

(a) Obtain an explicit expression for t^{mn}. Show that it is *not* gauge invariant. (This is to be expected because we know that one cannot define a generally covariant energy-momentum tensor for the gravitational field in the full theory. In the linear limit, general coordinate transformation is closely related to the gauge transformation of the metric.)

(b) Compute t^{mn} and then impose the transverse-traceless gauge condition $\partial_m h^{mn} = 0 = h$. Average the resulting expression using the fact that, in the chosen gauge, $\Box h^{mn} = 0$ and performing the necessary integration by parts to simplify the expression. Thus show that one can reduce the expression to the form

$$t^{mn} = \frac{c^4}{32\pi G}\langle\partial^m h^{ab}\partial^n h_{ab}\rangle,$$

(9.82)

which agrees with the result obtained in the text.

Exercise 9.9

Landau–Lifshitz pseudo-tensor for the gravitational wave Compute the Landau–Lifshitz energy-momentum pseudo-tensor for the plane gravitational wave in the TT gauge and show that it matches with the expressions obtained in the text. [Hint. The computation can be simplified by taking into account the following factors. (i) Since $h = 0$, the determinant

g is -1 up to linear order and hence $\partial_l(\sqrt{-g}g^{ik}) \approx \partial_l g^{ik} \approx -\partial_l h^{ik}$. (ii) Therefore, all the nonzero terms in t^{ik} are contributed by

$$\frac{1}{2}g^{il}g^{km}g_{np}g_{qr}g^{nr}_{,l}g^{pq}_{,m} = \frac{1}{2}h^{n,i}_q h^{q,k}_n. \tag{9.83}$$

This will lead to the necessary result.]

Exercise 9.10
Gauge dependence of the energy of the gravitational waves Consider a gravitational wave of the form $h_{ab} = f''(u)yzk_ak_b$, where $u = t - x$ and $k_a = \partial_a u$. Show that, while this is a valid gravitational wave solution, its energy-momentum tensor vanishes. Argue that it is, however, possible to change the gauge and transform h_{ab} to the form $h'_{ab} = f(u)(\delta^y_a\delta^z_b + \delta^z_a\delta^y_b)$ when the energy-momentum tensor is nonzero and has the form $t_{ab} \propto f'^2 k_a k_b$.

9.6 Generation of gravitational waves

We shall now take up the study of gravitational waves in the presence of sources in the flat background spacetime, described by Eq. (9.9). We want to find the general solution to this equation for a given source T^{mn} and work out the properties of the radiation emitted by systems in different kinds of motion. Mathematically, this is identical to the problem of determining the solutions to the Maxwell equations in the presence of sources since Eq. (9.9) is identical in structure to Eq. (2.124). In fact, the problem is somewhat simpler in the case of gravitational waves since we are only interested in finding h_{ab} (which is analogous to the potential A_i).

Using our solution in Eq. (2.162), we can immediately write down the general solution to Eq. (9.9) as:

$$\bar{h}_{mn}(t, \boldsymbol{x}) = 4\int d^3\boldsymbol{y}\,\frac{1}{|\boldsymbol{x} - \boldsymbol{y}|}T_{mn}(t - |\boldsymbol{x} - \boldsymbol{y}|, \boldsymbol{y}). \tag{9.84}$$

This result is exact and quite general except for the fact that it is obtained in the Lorenz gauge. To analyse this result further, it will be convenient if we can eliminate the dependence on the retarded time by Fourier transforming on the time variable. If we denote the Fourier transform of $T_{mn}(t, \boldsymbol{y})$ in the time variable by $T_{mn}(\omega, \boldsymbol{y})$, we have the result

$$T_{mn}(t - |\boldsymbol{x} - \boldsymbol{y}|, \boldsymbol{y}) = \int_{-\infty}^{\infty}\frac{d\omega}{2\pi}T_{mn}(\omega, \boldsymbol{y})\exp[-i\omega(t - |\boldsymbol{x} - \boldsymbol{y}|)]. \tag{9.85}$$

(We will use the same symbols, like T_{mn}, in both real and Fourier space when no confusion is likely to arise.) Substituting this into Eq. (9.84) and Fourier transforming $\bar{h}_{mn}(t, \boldsymbol{x})$ to $\bar{h}_{mn}(\omega, \boldsymbol{x})$ we get the result

$$\bar{h}_{mn}(\omega, \boldsymbol{x}) = 4\int d^3\boldsymbol{y}\,\frac{e^{i\omega|\boldsymbol{x} - \boldsymbol{y}|}}{|\boldsymbol{x} - \boldsymbol{y}|}T_{mn}(\omega, \boldsymbol{y}). \tag{9.86}$$

If T_{ab} is time independent, then only the $\omega = 0$ term will contribute and this will lead to static metric perturbations which we had already discussed in Chapter 6; when $\omega = 0$, Eq. (9.86) reduces to Eq. (6.124). When T_{ab} is time dependent, we expect the metric perturbation to have a static Coulomb like part (arising from the $\omega = 0$ component) as well as a radiative part. Our interest is in obtaining the radiation field far away from the source. Formally, this could be done by the following procedure. When the source is sufficiently far away, we can expand the term $|\boldsymbol{x} - \boldsymbol{y}|^{-1}$ in a Taylor series obtaining

$$\frac{1}{|\boldsymbol{x} - \boldsymbol{y}|} = \frac{1}{r} + \frac{x^\alpha y_\alpha}{r^3} + \frac{x^\alpha x^\beta}{2r^5}(3y_\alpha y_\beta - y^2 \delta_{\alpha\beta}) + \ldots, \tag{9.87}$$

where $r = |\boldsymbol{x}|$. Retaining just the first term $(1/r)$, for example, we can approximate Eq. (9.86) by

$$\bar{h}_{mn}(\omega, \boldsymbol{x}) \simeq \frac{4}{r} \int d^3 \boldsymbol{y}\, T_{mn}(\omega, \boldsymbol{y}) e^{i\omega|\boldsymbol{x}-\boldsymbol{y}|}. \tag{9.88}$$

This result involves only the approximation that the size of the source, L, is much smaller than the distance to the field point, $r \equiv |\boldsymbol{x}|$. If we further assume that the typical speeds (v) of the source particles are small compared with c, we can also expand T_{mn} in a Taylor series as

$$T_{mn}(t - |\boldsymbol{x} - \boldsymbol{y}|, \boldsymbol{y}) = T_{mn}(t - r, \boldsymbol{y}) + \dot{T}_{mn}(t - r, \boldsymbol{y})[r - |\boldsymbol{x} - \boldsymbol{y}|]$$
$$+ \frac{1}{2}\ddot{T}_{mn}(t - r, \boldsymbol{y})[r - |\boldsymbol{x} - \boldsymbol{y}|]^2 + \cdots \tag{9.89}$$

where an overdot denotes a time derivative. If the typical timescale over which T_{mn} changes is $t_s \simeq (L/v)$, this requires the condition $(L/t_s) \ll 1$ which is the same as $v \ll 1$. Combining the expressions in Eq. (9.87) and Eq. (9.89), we can write

$$\frac{1}{|\boldsymbol{x} - \boldsymbol{y}|} T_{mn}(t - |\boldsymbol{x} - \boldsymbol{y}|, \boldsymbol{y}) = T_{mn} \left[\frac{1}{r} + \frac{x^\alpha y_\alpha}{r^3} + \frac{x^\alpha x^\beta}{2r^5}(3y_\alpha y_\beta - y^2 \delta_{\alpha\beta}) + \cdots \right]$$
$$+ \dot{T}_{mn} \left[\frac{x^\alpha y_\alpha}{r^2} + \frac{x^\alpha x^\beta}{2r^4}(3y_\alpha y_\beta - y^2 \delta_{\alpha\beta}) \right]$$
$$+ \ddot{T}_{mn} \left[\frac{x^\alpha x^\beta}{2r^3} y_\alpha y_\beta \right], \tag{9.90}$$

where the right hand side has to be evaluated at the retarded time $(t - r)$. It is now obvious that the integral in Eq. (9.84) will involve various moments of the matter distribution weighted by T_{mn}. Many of these moments will lead to static parts of the metric that we are not currently concerned with. It turns out that the lowest order, time dependent, gravitational wave part will arise essentially from the quadrupole moment, which needs to be isolated for identifying the radiation field. While this can be done in a straightforward manner using the expansion given above (see Exercise 9.11), there is a somewhat simpler route to the result which we shall follow.

To do this, we shall concentrate on the lowest order term in Eq. (9.90), retaining which we get

$$\bar{h}_{mn}(\omega, \boldsymbol{x}) = 4\frac{e^{i\omega r}}{r} \int d^3y\, T_{mn}(\omega, \boldsymbol{y}).\tag{9.91}$$

That is, we have replaced the two slowly varying factors $e^{i\omega|\boldsymbol{x}-\boldsymbol{y}|}$ and $|\boldsymbol{x} - \boldsymbol{y}|^{-1}$ in Eq. (9.86) by $e^{i\omega r}/r$ and have taken it out of the integral. We next note that, since $\partial_a T^{ab} = 0$, the integral over the source of T^{0b} is conserved and independent of time. In other words, the Fourier transform of T^{0b} with respect to the time coordinate will have only the static ($\omega = 0$) terms. The time dependent part of the metric characterizing the radiation is contained in the spatial parts $h_{\alpha\beta}$. In fact, because of the gauge condition $\partial_m \bar{h}^{mn} = 0$ (which implies $\partial_m T^{mn} = 0$ through Eq. (9.9)), we have the relation among the Fourier components given by

$$\bar{h}^{0n} = -\frac{i}{\omega}\, \partial_\alpha \bar{h}^{\alpha n}.\tag{9.92}$$

This shows that $\bar{h}^{0\beta}$ can be determined from $\bar{h}^{\alpha\beta}$. We can then use the knowledge of $\bar{h}^{0\beta} = \bar{h}^{\beta 0}$ in the above relations once again to determine $\bar{h}^{00}$. We therefore need only determine the spatial components of the metric perturbation.

We can express the integral on the right hand side of Eq. (9.91), for the spatial components, in a more conventional form, essentially using the tensor virial theorem derived in Exercise 1.17 (see Eq. (1.129)). Using this result and defining the quadrupole moment tensor of the distribution by

$$I^{\alpha\beta}(t) = \int y^\alpha y^\beta\, T^{00}(t, \boldsymbol{y})\, d^3y\tag{9.93}$$

we get the final result

$$\bar{h}_{\alpha\beta}(t, \boldsymbol{x}) = \frac{2}{r}\frac{d^2 I_{\alpha\beta}(t_r)}{dt^2},\tag{9.94}$$

where $t_r = t - r$ is the retarded time at which the right hand side is evaluated after the derivative is taken.

This result shows that the gravitational radiation from an isolated, distant ($r \gg L$), slowly moving ($v \ll l$) source is proportional to the second time derivative of the quadrupole tensor of the matter distribution. This is to be contrasted with the corresponding result in electromagnetism in which the radiative part of the field is proportional to the second derivative of the dipole moment (see Exercise 9.15 and Table 9.1 for a discussion of the electromagnetic case, exactly in analogy with the

Table 9.1. *Comparison: electromagnetic and gravitational radiation (some of the results are obtained later on in this chapter; also see Exercise 9.15)*

	Generation of electromagnetic radiation	Generation of gravitational radiation
Basic equation	$\Box A^k = -4\pi J^k$	$\Box \bar{h}^{ab} = -16\pi T^{ab}$
Condition imposed to get the basic equation	$\partial_k A^k = 0$	$\partial_a \bar{h}^{ab} = 0$
Solution (with $t_r = t - \|\boldsymbol{x} - \boldsymbol{y}\|$)	$A^k(x) = \int d^3\boldsymbol{y}\, \dfrac{J^k(t_r, \boldsymbol{y})}{\|\boldsymbol{x} - \boldsymbol{y}\|}$	$\bar{h}^{ab}(x) = 4\int d^3\boldsymbol{y}\, \dfrac{T^{ab}(t_r, \boldsymbol{y})}{\|\boldsymbol{x} - \boldsymbol{y}\|}$
Interdependence of the components due to the gauge condition	A^0 can be found from A^α	$\bar{h}^{0k}$ can be found from $\bar{h}^{\alpha\beta}$
Solution for spatial modes when $r \gg L, v \ll 1$	$A^\alpha \simeq \dfrac{1}{r}\int J^\alpha d^3\boldsymbol{y}$	$\bar{h}^{\alpha\beta} \simeq \dfrac{4}{r}\int T^{\alpha\beta} d^3\boldsymbol{y}$
Identity satisfied by the source	$\int J^\alpha d^3\boldsymbol{y} = \dfrac{d}{dt}\int J^0 y^\alpha d^3\boldsymbol{y}$	$\int T^{\alpha\beta} d^3\boldsymbol{y} = \dfrac{1}{2}\dfrac{d^2}{dt^2}\int T^{00} y^\alpha y^\beta d^3\boldsymbol{y}$
Simplified result using the identity	$A^\alpha \simeq \dfrac{1}{r}\partial_0 d^\alpha$	$h^{\alpha\beta} \simeq \dfrac{2}{r}\partial_0^2 I^{\alpha\beta}$
Radiative mode	Transverse	Transverse-traceless
Radiated energy	$\dfrac{d\mathcal{E}}{dt} = -\dfrac{2}{3}(\partial_0^2 d_\alpha)^2$	$\dfrac{d\mathcal{E}}{dt} = -\dfrac{1}{5}(\partial_0^3 Q_{\alpha\beta})^2$
Back reaction force	$f_\alpha = \dfrac{2q}{3}\partial_0^3 d_\alpha$	$f_\alpha = \dfrac{2m}{5}(\partial_0^5 Q_{\alpha\beta})\, x^\beta$

gravitational wave case). There is a simple reason why the dipole moment appears in the case of electromagnetic radiation but the quadrupole moment appears in the case of gravitational waves. The dipole moment in electromagnetism is defined using the charge distribution and, for a system made of charges with different (q/m) ratios, the dipole moment can change. In the case of gravity, the dipole moment of a system of masses

$$\boldsymbol{d} = \sum_i m_i \boldsymbol{x}_i = M\, \boldsymbol{R}_{\mathrm{CM}} \tag{9.95}$$

is proportional to the vector giving the location of the centre of mass of the system which can, at best, only move with uniform velocity for an isolated system.

Therefore, the second time derivative of the gravitational dipole moment will vanish for any isolated system of masses and there can be no dipole gravitational radiation.

Another key difference between electromagnetism and gravity is the following. In the case of electromagnetism, one can consider an arbitrary trajectory for the charged particle and obtain its electromagnetic field. In the case of gravity, there is no such thing as an arbitrarily moving mass. This is because the basic equation, Eq. (9.6), demands $\partial_a T^{ab} = 0$ which – in the case of a single particle – leads to a motion with uniform velocity. (The corresponding h_{ab} was determined in Exercise 6.14.) In electromagnetism, the field equation only requires $\partial_a J^a = 0$ which does not restrict the trajectory of the charge.

The solution in Eq. (9.94) is not in the TT gauge. The TT part of the metric can be obtained from Eq. (9.94) using the standard procedure described earlier in, say, Eq. (9.30) leading to

$$h_{\alpha\beta}^{TT}(t, \boldsymbol{x}) = \Lambda_{\alpha\beta}{}^{\gamma\lambda}(\boldsymbol{n})\frac{2}{r}\frac{d^2 I_{\gamma\lambda}(t-r)}{dt^2} \tag{9.96}$$

with $\boldsymbol{n} = \boldsymbol{r}/r$. In this equation, one can replace the quadrupole tensor $I_{\alpha\beta}$ with the reduced quadrupole tensor defined as $Q_{\alpha\beta} = I_{\alpha\beta} - (1/3)\delta_{\alpha\beta}I$. Since $\Lambda_{\alpha\beta\gamma\lambda}$ has no time dependence, it can be moved inside the time derivatives in Eq. (9.96). Now using the fact that $\Lambda_{\alpha\beta}^{\gamma\lambda}\delta_{\gamma\lambda} = 0$ we obtain the same expression for $h_{\alpha\beta}^{TT}$ with either $Q_{\alpha\beta}$ or $I_{\alpha\beta}$. Hence the TT part of the metric is completely determined by the reduced quadrupole moment $Q_{\alpha\beta}$ through the relation

$$h_{\alpha\beta}^{TT} = \Lambda_{\alpha\beta}{}^{\gamma\lambda}\frac{2}{r}\ddot{Q}_{\gamma\lambda} = \left(P_\alpha^\gamma P_\beta^\lambda - \frac{1}{2}P_{\alpha\beta}P^{\gamma\lambda}\right)\frac{2}{r}\ddot{Q}_{\gamma\lambda} \equiv \frac{2}{r}\ddot{Q}_{\gamma\lambda}^{TT}. \tag{9.97}$$

This result is important for the following reason. In general, an l-pole radiation has $(2l+1)$ separate components and hence the quadrupole radiation has $(2\times 2+1) = 5$ independent components. But the symmetric three-dimensional quadrupole tensor $I_{\alpha\beta}$ has six independent components. The above result shows that the *radiative* part contained in the TT part of the metric is determined by the *reduced* quadrupole tensor, which – being tracefree – has only five independent components.

Exercise 9.11

The TT part of the gravitational radiation from first principles The derivation in the text concentrated on the *spatial* components of the metric tensor $h_{\alpha\beta}$ and extracted its TT part by using the projection operator. But we know that, in general, a solution to Eq. (9.84) also contains a static Coulomb part and a longitudinal part. The purpose of this exercise is to solve Eq. (9.84) in a different manner and explicitly demonstrate the extraction of the TT part of the metric.

(a) In Eq. (9.84) perform a Taylor series expansion of $t - |\boldsymbol{x} - \boldsymbol{y}|$ around $t_r \equiv t - r$ and show that the solution can be approximated as

$$\bar{h}^{ab} = \frac{4}{r} \int d^3 y \left[T^{ab}(t_r, \boldsymbol{y}) + \partial_0 T^{ab}(t_r, \boldsymbol{y}) n^\alpha y_\alpha + \frac{1}{2} \partial_0^2 T^{ab}(t_r, \boldsymbol{y}) n^\alpha n^\beta y_\alpha y_\beta + \cdots \right],$$
$$(9.98)$$

where $n^\alpha = x^\alpha / r$. Define a series of moments of the matter distribution through

$$M(t_r) = \int T^{00}(t_r, y^\alpha) d^3 y, \quad M_\beta(t_r) = \int T^{00}(t_r, y^\alpha) y_\beta d^3 y,$$

$$P^\lambda(t_r) = \int T^{0\lambda}(t_r, y^\alpha) d^3 y, \quad P^\lambda_\beta(t_r) = \int T^{0\lambda}(t_r, y^\alpha) y_\beta d^3 y,$$

$$S^{\lambda\mu}(t_r) = \int T^{\lambda\mu}(t_r, y^\alpha) d^3 y \quad M_{\beta\gamma}(t_r) = \int T^{00}(t_r, y^\alpha) y_\beta y_\gamma d^3 y. \quad (9.99)$$

Use $\partial_b T^{ab} = 0$ to show that these moments satisfy the identities

$$\dot{M} = 0, \qquad \dot{M}^\gamma = P^\gamma, \qquad \dot{M}^{\beta\gamma} = P^{\beta\gamma} + P^{\gamma\beta},$$

$$\dot{P}^\beta = 0, \qquad \dot{P}^{\beta\gamma} = S^{\beta\gamma}, \qquad \frac{d^2 M^{\beta\gamma}}{dt^2} = 2 S^{\beta\gamma}. \quad (9.100)$$

Using these identities, determine the form of metric perturbations in the wave zone to be

$$\bar{h}^{00}(t, x^\alpha) = \frac{4}{r} M + \frac{4}{r} P^\beta n_\beta + \frac{4}{r} S^{\beta\gamma}(t_r) + n_\beta n_\gamma \cdots ;$$

$$\bar{h}^{0\beta}(t, x^\alpha) = \frac{4}{r} P^\beta + \frac{4}{r} S^{\beta\gamma}(t_r) n_\gamma + \cdots ;$$

$$\bar{h}^{\alpha\beta}(t, x^\alpha) = \frac{4}{r} S^{\alpha\beta}(t_r) + \cdots . \quad (9.101)$$

The last result matches with Eq. (9.94) as it should.

(b) Verify that the solution does satisfy the Lorenz gauge condition as it should. But it is not in the TT gauge. Show that a gauge transformation, with ξ^a having the components

$$\xi^0 = -\frac{1}{r} P^\gamma_{\,\gamma} - \frac{1}{r} P^{\beta\gamma} n_\beta n_\gamma, \qquad \xi^\alpha = -\frac{4}{r} M^\alpha - \frac{4}{r} P^{\alpha\beta} n_\beta + \frac{1}{r} P^\gamma_{\,\gamma} n^\alpha + \frac{1}{r} P^{\beta\gamma} n_\beta n_\gamma n^\alpha,$$
$$(9.102)$$

will explicitly transform the metric into the TT gauge as far as spatial components are concerned. That is, show that the spatial components now become $h^{TT}_{\alpha\beta}$ given in Eq. (9.96). Further show that the 00 and 0α components are now given by $\bar{h}^{TT}_{00} = 4M/r$ and $\bar{h}^{TT}_{0\alpha} = 0$. Thus all the time dependent components that represent the wave are contained in the spatial part of the metric which is explicitly transverse and traceless.

9.6.1 Quadrupole formula for the gravitational radiation

We will now combine the result obtained in Eq. (9.76) with the expression for $h^{TT}_{\alpha\beta}$ in Eq. (9.97) in order to obtain an expression for the energy radiated as a gravitational wave by a source with a given quadrupole moment. (Such a result will be analogous to Larmor's formula for radiation in the case of electromagnetism;

see Eq. (2.186).) From Eq. (9.76) we see that the flux ct_{00} of the wave can be expressed in the form (with the c and G factors temporarily reintroduced)

$$ct_{00} \equiv \frac{c^3}{32\pi G} \left\langle \dot{h}^{TT}_{\alpha\beta} \, \dot{h}^{\alpha\beta}_{TT} \right\rangle. \tag{9.103}$$

Substituting for $h^{TT}_{\alpha\beta}$ from Eq. (9.97), we get

$$ct^{00} = \frac{G}{8\pi c^5 r^2} \left\langle \dddot{Q}^{TT}_{\alpha\beta} \, \dddot{Q}^{\alpha\beta}_{TT} \right\rangle = \frac{G}{8\pi c^5 r^2} \left\langle \Lambda_{\alpha\beta\mu\nu} \Lambda^{\alpha\beta}{}_{\rho\sigma} \, \dddot{Q}^{\mu\nu} \, \dddot{Q}^{\rho\sigma} \right\rangle$$

$$= \frac{G}{8\pi c^5 r^2} \left\langle \Lambda_{\mu\nu\rho\sigma} \, \dddot{Q}^{\mu\nu} \, \dddot{Q}^{\rho\sigma} \right\rangle. \tag{9.104}$$

In arriving at the last result, we have used the fact that $\Lambda_{\alpha\beta\mu\nu}$ is a projection operator (see the discussion just after Eq. (9.30)). Working out the product using the definition of $\Lambda_{\alpha\beta\mu\nu}$ in Eq. (9.30), we get

$$\Lambda^{\beta\gamma\rho\sigma} \, \dddot{Q}_{\beta\gamma} \, \dddot{Q}_{\rho\sigma} = \left(\delta_{\beta\rho}\delta_{\gamma\sigma} - n_\beta n_\rho \delta_{\gamma\sigma} - n_\gamma n_\sigma \delta_{\beta\rho} + \frac{1}{2}n_\beta n_\gamma n_\rho n_\sigma \right) \dddot{Q}^{\beta\gamma} \, \dddot{Q}^{\rho\sigma}$$

$$= \dddot{Q}^{\beta\gamma} \, \dddot{Q}_{\beta\gamma} - 2n_\gamma \, n^\sigma \, \dddot{Q}^{\gamma\rho} \, \dddot{Q}_{\rho\sigma} + \frac{1}{2}n_\beta n_\gamma n_\rho n_\sigma \, \dddot{Q}^{\beta\gamma} \, \dddot{Q}^{\rho\sigma}. \tag{9.105}$$

The amount of radiation propagating into a solid angle $d\Omega$ can be obtained by multiplying ct_{00} by $r^2 d\Omega$. Thus we get, for the flux of radiation flowing into a solid angle $d\Omega$ in the direction indicated by the unit vector n^α to be

$$\frac{dI}{d\Omega} = \frac{G}{4\pi c^5} \left[\frac{1}{4} \left(\dddot{Q}_{\alpha\beta} \, n^\alpha n^\beta \right)^2 + \frac{1}{2} \dddot{Q}^2_{\alpha\beta} - \dddot{Q}^{\alpha\beta} \, \dddot{Q}_{\alpha\gamma} \, n_\beta n^\gamma \right] \tag{9.106}$$

with $\dddot{Q}^2_{\alpha\beta} \equiv \dddot{Q}_{\alpha\beta} \dddot{Q}^{\alpha\beta}$. Finally, the total radiation emitted in all directions is found by integrating this expression over all directions. This can be done using the formulas

$$\overline{n_\alpha n_\beta} = \frac{1}{3}\delta_{\alpha\beta}; \quad \overline{n_\alpha n_\beta n_\gamma n_\delta} = \frac{1}{15} \left(\delta_{\alpha\beta}\delta_{\gamma\delta} + \delta_{\alpha\gamma}\delta_{\beta\delta} + \delta_{\alpha\delta}\delta_{\beta\gamma} \right). \tag{9.107}$$

(These relations are easily proved by constructing the tensors with necessary symmetries in terms of the products of Kronecker deltas and determining the overall proportionality factor by contracting on all indices on both sides.) This gives the net energy loss of the system per unit time to be

$$\frac{d\mathcal{E}}{dt} = -\frac{G}{5c^5} \dddot{Q}^2_{\alpha\beta}. \tag{9.108}$$

Exercise 9.12
Flux of gravitational waves Consider an isolated, far away, source of gravitational waves. Rigorously speaking, the flux of gravitational waves should be computed from the t^{0r}

component of the energy-momentum tensor. (In the text, we computed it as ct^{00}.) Show that, for an amplitude which varies as $r^{-1}f(t-r)$, the leading order derivative with respect to r (at large r) can be related to the derivatives with respect to t. Hence show that t^{0r} will lead to exactly the same result as the one which we found.

Exercise 9.13

Original issues The quadrupole moment is defined using some specific point as the origin of the coordinate system. But, surely, the physical results should not depend on the choice of the origin. Determine how the quadrupole moment changes if the spatial origin is shifted and make sure that everything is fine.

Exercise 9.14

Absorption of gravitational waves (a) Consider a gravitational wave with $\bar{h}_{xx}^{TT} = -\bar{h}_{yy}^{TT} = A\cos\Omega(z-t)$ that is incident on a system of two masses (each of mass m) held together by a spring with a natural frequency of oscillation ω_0 and damping constant γ. Assume that the spring is oriented in the direction transverse to the z-axis. Show that, in steady state, the spring will oscillate with a displacement (around the equilibrium value L_0) given by $\xi = R\cos(\Omega t + \phi)$, where $R = (1/2)L_0\Omega^2 A/[(\omega_0 - \Omega)^2 + 4\Omega^2\gamma^2]^{1/2}$ and $\tan\phi = 2\gamma\Omega/(\omega_0^2 - \Omega^2)$.

(b) Consider now an infinite plane sheet of such harmonic oscillators located in a direction transverse to the direction of propagation of the wave. If σ is the number of oscillators per unit area, show that the work done by the gravitational wave against the damping of the oscillators will reduce the energy flux of the wave by the amount $\delta f = -(1/2)\sigma m\gamma\Omega^2 R^2$.

(c) The oscillators in the sheet will, in turn, radiate gravitational waves. Compute the total radiation from the infinite sheet of oscillators downstream and show that it is given by

$$\delta h_{xx}^{TT} = -\delta h_{yy}^{TT} = 2\pi\sigma m\Omega L_0 R\sin[\Omega(z-t) - \phi]. \tag{9.109}$$

(d) The total amplitude of the gravitational wave at some point is the sum of (i) the original amplitude and (ii) the amplitude of the radiation field produced by the sheet of oscillators. Show that this addition produces two net effects: (i) it introduces a phase shift ψ where $\tan\psi = (2\pi\sigma m\Omega L_0 R/A)\cos\phi$ and (ii) reduces the amplitude by the amount $\delta A = -(2\pi\sigma m\Omega L_0 R/A)\sin\phi$. Explain the physical origin of both effects.

(e) From the relation between energy flux of the gravitational wave and its amplitude, show that the energy lost by the gravitational wave in passing through the sheet of oscillators is precisely equal to the work done on the oscillators. Alternatively, show that if one assumes gravitational waves carry *some* energy flux F, then conservation of energy requires $\delta F/\delta A = (1/16\pi)\Omega^2 A$, implying $F = (1/32\pi)\Omega^2 A^2$. Critically examine whether this can be considered as a proof for the reality of energy flux in the gravitational waves.

9.6.2 Back reaction due to the emission of gravitational waves

The formulas obtained in the previous section for the flux of gravitational radiation and the total amount of energy radiated are valid under certain approximations which we shall recall briefly. If the source has a size L and mean internal velocity v and emits gravitational radiation with a wavelength λ, then our expressions are valid when the following conditions are satisfied. First, we must have $L/\lambda \ll 1$. This is because, when the amplitude of the motion of the source is of the order of L, the radiation is emitted at a wavelength $\lambda \approx L/v$. Hence the slow motion condition $v \ll 1$ translates into $L \ll \lambda$. Second, it is also necessary that the typical value of the Newtonian potential (in dimensionless units) is small compared to (L/λ) since one cannot resort to slow velocity, near-Newtonian, motion if this condition is violated.

Because $L \ll \lambda$, there exists a region outside the source $(r > L)$ but in the 'near zone' $(r \ll \lambda)$ in which Newtonian gravity remains valid to a sufficient level of accuracy. In this region, the Newtonian potential Φ can be expanded in terms of the multipoles as

$$\Phi = - \left(\frac{M}{r} + \frac{d_\alpha n^\alpha}{r^2} + \frac{3 Q_{\alpha\beta}(t)\, n^\alpha\, n^\beta}{2 r^3} + \cdots \right); \qquad n^\alpha = \frac{x^\alpha}{r}, \qquad (9.110)$$

where M is the total mass, d_α is the dipole moment and $Q_{\alpha\beta}$ is the reduced quadrupole moment. When we move into the radiation zone $(r \gtrsim \lambda)$, we can still describe the *static* part of the Newtonian potential in terms of mass and dipole moment – which are not involved in the emission of gravitational radiation – but the third term in the expansion of Φ, involving the traceless quadrupole moment ceases to be described in Newtonian terms. This is because retardation effects due to gravitational wave emission become significant in the radiation zone and the back reaction due to the radiation of gravitational waves will change the form of this correction. It turns out that this back reaction can be expressed by adding an effective Newtonian potential of the form

$$\Phi_{\text{react}} = \frac{1}{5} \frac{d^5 Q_{\alpha\beta}}{dt^5}\, x^\alpha\, x^\beta. \qquad (9.111)$$

The existence of such a reaction is, of course, a physical necessity since the radiation of gravitational waves is accompanied by loss of energy, angular momentum, etc., and these need to be accounted for by the radiation reaction force. We shall first provide an intuitive derivation of the reaction potential in Eq. (9.111) and then indicate the steps involved in a more formal derivation.

Let f be the 'frictional' force, acting on the particles due to the emission of gravitational radiation, defined through the relation

$$\overline{\frac{d\mathcal{E}}{dt}} = \sum f \cdot v, \qquad (9.112)$$

which equates the work done by the force f to the energy loss by the system. (Here, and in what follows, the summation is over all the particles in the system.) As in the case of the derivation of the radiation reaction force in electromagnetism (see Eq. (2.191)), the equality in Eq. (9.112) is expected to hold only as a time average for the particles executing periodic motion in a bounded region of space. The overbar in Eq. (9.112) indicates the time average over some time scale T. For any quantity that is a total time derivative, we then have

$$\overline{\frac{dQ}{dt}} \equiv \frac{1}{T} \int_0^T dt \frac{dQ}{dt} = \frac{1}{T} [Q(T) - Q(0)]. \qquad (9.113)$$

If Q takes bounded values, then this average vanishes for sufficiently large T. Using this and Eq. (9.108), one can swap time derivatives inside the average and write

$$\overline{\frac{d\mathcal{E}}{dt}} = -\frac{G}{5c^5} \overline{\dddot{Q}_{\alpha\beta} \overset{\cdots}{Q}{}^{\alpha\beta}} = -\frac{G}{5c^5} \overline{\frac{dQ^{\alpha\beta}}{dt} \frac{d^5 Q_{\alpha\beta}}{dt^5}}. \qquad (9.114)$$

Substituting the expression

$$\frac{dQ_{\alpha\beta}}{dt} = \sum m \left(x_\alpha v_\beta + x_\beta v_\alpha - \frac{2}{3} \delta_{\alpha\beta} \mathbf{r} \cdot \mathbf{v} \right) \qquad (9.115)$$

we find that

$$f_\alpha = -\frac{2Gm}{5c^5} \frac{d^5 Q_{\alpha\beta}}{dt^5} x^\beta. \qquad (9.116)$$

Clearly, one can express f_α as the gradient of the back reaction potential given in Eq. (9.111) as claimed.

While the above analysis gives the correct back reaction potential, it hides its physical origin and does not readily indicate its domain of validity. In view of the theoretical importance of this issue, we shall now provide a more formal derivation of the same by computing the metric perturbations in the near zone and determining the effective Newtonian potential that arises from it.

We begin by noting that the wave equation is invariant under time reversal $t \rightarrow -t$ and hence will possess both retarded and advanced solutions. These solutions can together be characterized by taking the metric in the form

$$\bar{h}_{\beta\gamma}(t, \boldsymbol{x}) = \frac{2}{r} \ddot{I}_{\beta\gamma} (t - \epsilon r), \qquad \epsilon = \pm 1. \qquad (9.117)$$

Further, this solution is obtained by discarding the terms that fall faster than $(1/r)$ and hence it is valid not only in the radiation zone, but also in the intermediate and near zones ($r \lesssim \lambda$ but $r > L$). The radiation reaction potential should, however, reverse in sign when we change from the retarded solutions to the advanced solutions because one is switching from energy loss to energy gain for the system. (The latter situation, represented by the advanced solution, of course, has no physical validity and is introduced here purely as a mathematical trick to easily identify the radiation reaction term.) In the near zone ($r \ll \lambda$) but outside the source, we can now expand the perturbation in powers of r obtaining

$$\bar{h}_{\beta\gamma} = 2\left[\frac{I^{(2)}_{\beta\gamma}}{r} - \epsilon I^{(3)}_{\beta\gamma} + \frac{I^{(4)}_{\beta\gamma}r}{2!} - \epsilon\frac{I^{(5)}_{\beta\gamma}r^2}{3!} + \cdots\right], \qquad I^{(n)}_{\beta\gamma} \equiv \frac{d^n I_{\beta\gamma}(t)}{dt^n}.$$

(9.118)

The metric perturbations $\bar{h}_{0\beta}$ and $\bar{h}_{00}$ can be generated from this expression by using the gauge conditions $\partial_0\bar{h}_{\beta 0} = \partial_\gamma\bar{h}^\gamma_\beta$ and $\partial_0\bar{h}_{00} = \partial_\beta\bar{h}^\beta_0$. This results in

$$\bar{h}_{0\beta} = 2\left[-\frac{I^{(1)}_{\beta\gamma}x^\gamma}{r^3} + \frac{I^{(3)}_{\beta\gamma}x^\gamma}{2!r} - \epsilon\frac{2I^{(4)}_{\beta\gamma}x^\gamma}{3!} + \frac{3I^{(5)}_{\beta\gamma}x^\gamma r}{4!} - \epsilon\frac{4I^{(6)}_{\beta\gamma}x^\gamma r^2}{5!}\right]$$

(9.119)

and

$$\bar{h}_{00} = 2\left[\frac{(3x^\beta x^\gamma - r^2\delta^{\beta\gamma})}{r^5}I_{\beta\gamma} - \frac{(x^\beta x^\gamma - r^2\delta^{\beta\gamma})}{2!r^3}I^{(2)}_{\beta\gamma} - \epsilon\frac{2}{3!}I^{(5)}_{\beta\beta}\right.$$
$$\left.+ \frac{3(x^\beta x^\gamma - r^2\delta^{\beta\gamma})}{4!r}I^{(4)}_{\beta\gamma} - \epsilon\frac{4(2x^\beta x^\gamma - r^2\delta^{\beta\gamma})}{5!}I^{(5)}_{\beta\gamma}\right].$$

(9.120)

In obtaining these, we have ignored the terms which are static or linear in time because they are not associated with radiation. The leading term, as we go close to the source, is the one in $\bar{h}_{00}$ which rises as $(1/r^3)$ and is controlled by the quadrupole tensor. This corresponds to a Newtonian potential $\Phi = -(1/2)h_{00} = -(1/4)\bar{h}_{00} = -(3/2)(Q_{\alpha\beta}\,n^\alpha\,n^\beta/r^3)$. These terms, however, are independent of ϵ and hence do not change sign when we switch from the retarded to the advanced solutions; they represent corrections to Newtonian gravity but do not contribute to the transport of energy or angular momentum.

The reaction terms due to the radiation of gravitational waves should depend linearly on ϵ and we can read off these terms from the expressions in Eq. (9.118), Eq. (9.119) and Eq. (9.120) and obtain $h_{\alpha\beta}$ from $\bar{h}_{\alpha\beta}$. We get:

$$h^{(\text{react})}_{\beta\gamma} = -2I^{(3)}_{\beta\gamma} + \frac{2}{3}I^{\lambda(3)}_\lambda\delta_{\beta\gamma} + \mathcal{O}(I^{(5)}_{\beta\gamma}r^2),$$

$$h^{(\text{react})}_{0\beta} = -\frac{2}{3}I^{(4)}_{\beta\gamma}x^\gamma + \mathcal{O}(I^{(6)}_{\beta\gamma}r^3),$$

(9.121)

$$h^{(\text{react})}_{00} = -\frac{4}{3}I^{\lambda(3)}_\lambda - \frac{1}{15}(x^\beta x^\gamma + 3r^2\delta^{\beta\gamma})I^{(5)}_{\beta\gamma}.$$

To describe these gravitational reaction terms using an effective Newtonian potential, we need to make a gauge transformation to a coordinate system in which the only dominant correction term is h_{00}. This can be achieved using the coordinate transformation $x^m_{\text{new}} = x^m_{\text{old}} + \xi^m(x)$ with

$$\xi_\beta = -I^{(3)}_{\beta\gamma}x^\gamma + \frac{1}{3}I^{\lambda(3)}_\lambda x_\beta; \qquad \xi_0 = -\frac{2}{3}I^{\lambda(2)}_\lambda + \frac{1}{6}I^{(4)}_{\beta\gamma}x^\beta x^\gamma - \frac{1}{6}I^{\lambda(4)}_\lambda r^2. \quad (9.122)$$

In the new gauge,

$$h_{\beta\gamma}^{\text{react}} = \mathcal{O}(I_{\beta\gamma}^{(5)} r^2), \qquad h_{0\beta}^{(\text{react})} = \mathcal{O}(I_{\beta\gamma}^{(6)} r^3), \qquad h_{00}^{(\text{react})} = -\frac{2}{5} Q_{\beta\gamma}^{(5)} x^\beta x^\gamma. \tag{9.123}$$

This gauge, therefore, has a purely Newtonian interpretation and we immediately obtain the earlier result $\Phi^{\text{react}} = -(1/2)h_{00}$ in Eq. (9.111). Thus the leading order radiation reaction effects, with fractional errors of the order of (λ^2/r^2) in the near zone of a Newtonian source, can be described using an effective reaction potential.

The above discussion makes it clear that the quantity $Q_{\alpha\beta}$ which appears in the emission of gravitational radiation is the same as the one which appears in the expansion of Φ in Eq. (9.110). This shows that our results for the emission of gravitational radiation are valid for any slow motion source including those which have strong internal gravity (like slowly spinning neutron stars) as long as we interpret $Q_{\alpha\beta}$ as the coefficient of the relevant term in the expansion of Φ.

When a system of bodies – in an approximately stationary motion – is emitting gravitational waves, there is also a steady loss of angular momentum from the system. In general, this rate can be written in the form

$$\overline{\frac{dM_\alpha}{dt}} = \sum \overline{(\boldsymbol{r} \times \boldsymbol{f})_\alpha} = \sum \epsilon_{\alpha\beta\gamma} \overline{x^\beta f^\gamma}, \tag{9.124}$$

where f^γ is the radiation reaction force. Using the expression in Eq. (9.116) we get

$$\overline{\frac{dM^\alpha}{dt}} = \frac{2G}{5c^5} \epsilon^{\alpha\beta}{}_\gamma \overline{Q_{\beta\delta}^{(5)} Q^{\gamma\delta}} = \frac{2G}{5c^5} \epsilon^{\alpha\beta}{}_\gamma \overline{\frac{d^3 Q_{\beta\delta}}{dt^3} \frac{d^2 Q^{\delta\gamma}}{dt^2}} \tag{9.125}$$

for the rate of emission of angular momentum by the system.

Exercise 9.15
Lessons from gravity for electromagnetism It is possible to study the simpler case of electromagnetic radiation exactly in analogy with gravitational radiation. This approach clarifies the steps in some of the derivations in the previous few sections. (Also see Table 9.1.)

(a) In the case of the electromagnetic radiation, we are interested in solving the equation $\Box A^i = -4\pi J^i$. This is already in the Lorenz gauge $\partial_i A^i = 0$. Fourier transforming the general solution

$$A^i(t, \boldsymbol{x}) = \int d^3y \frac{J^i(t - |\boldsymbol{x} - \boldsymbol{y}|, \boldsymbol{y})}{|\boldsymbol{x} - \boldsymbol{y}|} \tag{9.126}$$

in t and using the same approximations as in Eq. (9.91), obtain in Fourier space the result

$$A^\alpha(\omega, \boldsymbol{x}) \simeq \frac{e^{i\omega r}}{r} \int d^3y J^\alpha(\omega, \boldsymbol{y}), \tag{9.127}$$

which is analogous to Eq. (9.91). We need only determine the spatial components of A^i since the time component is determined by the gauge condition $\partial_i A^i = 0$ which reduces to $-i\omega A^0(\omega, \boldsymbol{y}) = \partial_\alpha A^\alpha(\omega, \boldsymbol{y})$.

(b) Use the conservation of current $\partial_i J^i = 0$ to prove the identity $J^\beta = \partial_\alpha[J^\alpha y^\beta] - i\omega J^0 y^\beta$. Use this to express the solution as

$$A^\alpha(\omega, \boldsymbol{x}) = -i\omega \frac{e^{i\omega r}}{r} d^\alpha(\omega), \tag{9.128}$$

where we have defined the dipole moment as

$$d^\alpha(t) = \int d^3y \, y^\alpha \rho(t, \boldsymbol{y}). \tag{9.129}$$

Converting back to real space, show that $\boldsymbol{A} = \dot{\boldsymbol{d}}/r$ with the right hand side evaluated at the retarded time. Also show that, within the same approximation, $A^0 = q/r$ where q is the total charge of the system. This is analogous to Eq. (9.94) for the gravitational waves.

(c) The result, however, is not in a transverse gauge. Find a gauge transformation which will eliminate the non-transverse part of $\boldsymbol{A}$. Show that the final result can be expressed in the form $\boldsymbol{A}_\perp = \dot{\boldsymbol{d}}_\perp/r$ where the transverse component is obtained using the standard projection operator.

(d) For a more challenging task, repeat the analysis of Section 9.6.2 for electromagnetism and obtain the radiation reaction force.

9.7 General relativistic effects in binary systems

We shall now apply some of the results obtained in the previous sections to study an astrophysical binary star system called[2] PSR 1913+16. This can be modelled as a two body system in which one member is a radio pulsar – from which periodic radio pulses are received and the other member is a neutron star or black hole. Since the pulsar is orbiting near a strongly gravitating compact remnant, its orbit will not be an ellipse (unlike in the case of Newtonian gravity) but will precess (see Section 7.4.1). Such a pulsar acts as an extraordinarily stable clock, located in a strong gravitational field in which the clock rate will be affected by general relativistic effects. By monitoring the orbit as well as the arrival time of pulses from such a pulsar, one can gain significant information about the general relativistic effects present in such a system. We shall first work out the effect of the emission of gravitational radiation on the orbital parameters of the system and will then describe how such a system can be used to test the predictions of general relativity.

9.7.1 Gravitational radiation from binary pulsars

Let us consider two bodies of masses m_1 and m_2 which are orbiting around the common centre of mass in a Newtonian elliptical orbit. We know from the standard Newtonian analysis, valid to the lowest order, that the orbital parameters are completely determined by the energy $E(< 0)$ and the angular momentum L. In particular, the semi-major axis a and the eccentricity e are given by

$$a = -\frac{m_1 m_2}{2E}; \qquad e^2 = 1 + \frac{2EL^2(m_1 + m_2)}{m_1^3 m_2^3}. \qquad (9.130)$$

As we shall see, such a system has a time dependent quadrupole moment and hence will lose energy and angular momentum. Differentiating a and e we get

$$\frac{da}{dt} = \frac{m_1 m_2}{2E^2} \frac{dE}{dt}; \qquad \frac{de}{dt} = \frac{m_1 + m_2}{m_1^3 m_2^3 e} \left(L^2 \frac{dE}{dt} + 2EL \frac{dL}{dt} \right). \qquad (9.131)$$

Using Eq. (9.108) and Eq. (9.125) for (dE/dt) and (dL/dt) we can determine how the orbital parameters a and e change.

We will take the orbital plane to be the x–y plane with the centre of mass at the origin and indicate the radial distances to the two masses by $r_1 = [m_2/(m_1 + m_2)]r$, $r_2 = [m_1/(m_1 + m_2)]r$ where r is the separation between the two particles. If the line joining the masses makes an angle θ with the x-axis, the orbital equation is

$$r = \frac{a(1 - e^2)}{1 + e \cos \theta}. \qquad (9.132)$$

For such a configuration, the quadrupole moment has the components

$$I_{xx} = m_1 x_1^2 + m_2 x_2^2 = \frac{m_1 m_2}{m_1 + m_2} r^2 \cos^2 \theta; \qquad I_{yy} = \frac{m_1 m_2}{m_1 + m_2} r^2 \sin^2 \theta \quad (9.133)$$

$$I_{xy} = \frac{m_1 m_2}{m_1 + m_2} r^2 \sin \theta \cos \theta; \qquad I \equiv I_{xx} + I_{yy} = \frac{m_1 m_2}{m_1 + m_2} r^2. \qquad (9.134)$$

We need to compute the third time derivative of the quadrupole moment using the fact that the orbital parameters θ and r vary as

$$\dot{\theta} = \frac{[(m_1 + m_2)a(1 - e^2)]^{1/2}}{r^2}; \qquad \dot{r} = e \sin \theta \left[\frac{m_1 + m_2}{a(1 - e^2)} \right]^{1/2}. \qquad (9.135)$$

This computation is straightforward (but tedious) and the final result is given by

$$\dddot{I}_{xx} = \frac{2m_1 m_2}{a(1 - e^2)} \left(2 \sin 2\theta + 3e \cos^2 \theta \sin \theta \right) \dot{\theta}$$

$$\dddot{I}_{yy} = -\frac{2m_1 m_2}{a(1 - e^2)} \left(2 \sin 2\theta + e \sin \theta + 3e \cos^2 \theta \sin \theta \right) \dot{\theta}$$

$$\dddot{I}_{xy} = -\frac{2m_1 m_2}{a(1 - e^2)} \left(2 \cos 2\theta - e \cos \theta + 3e \cos^3 \theta \right) \dot{\theta}. \qquad (9.136)$$

Substituting this in Eq. (9.108) (and using units with $G = c = 1$), we get

$$\frac{dE}{dt} = -\frac{1}{5}\,\dddot{Q}_{\beta\gamma}\dddot{Q}^{\beta\gamma} = -\frac{1}{5}\left(\dddot{I}_{\beta\gamma}\dddot{I}^{\beta\gamma} - \frac{1}{3}\dddot{I}^2\right) = -\frac{1}{5}\left(\dddot{I}_{xx}^2 + 2\dddot{I}_{xy}^2 + \dddot{I}_{yy}^2 - \frac{1}{3}\dddot{I}^2\right)$$

$$= -\frac{8m_1^2m_2^2}{15a^2(1-e^2)^2}\left[12(1 + e\cos\theta)^2 + e^2\sin^2\theta\right]\dot{\theta}^2. \tag{9.137}$$

This is the instantaneous energy loss due to radiation. What is more relevant is the average of this expression over an orbital period of the binary system, where the period is given by

$$T = \int_0^{2\pi} \frac{1}{\dot{\theta}}\, d\theta = \frac{2\pi a^{3/2}}{(m_1 + m_2)^{1/2}}. \tag{9.138}$$

The average value will be

$$\left\langle\frac{dE}{dt}\right\rangle = \frac{1}{T}\int_0^T \frac{dE}{dt}\, dt = \frac{1}{T}\int_0^{2\pi} \frac{dE}{dt}\frac{1}{\dot{\theta}}\, d\theta$$

$$= -\frac{32G^4}{5c^5}\frac{m_1^2m_2^2(m_1 + m_2)}{a^5(1-e^2)^{7/2}}\left(1 + \frac{73}{24}e^2 + \frac{37}{96}e^4\right). \tag{9.139}$$

In a similar manner, we can obtain the loss of angular momentum by the system to be

$$\frac{dL}{dt} = -\frac{2}{5}\epsilon^{z\alpha\beta}\ddot{Q}_\alpha^\gamma\,\dddot{Q}_{\gamma\beta} = -\frac{2}{5}\epsilon^{z\alpha\beta}\ddot{I}_\alpha^\gamma\,\dddot{I}_{\gamma\beta}$$

$$= -\frac{2}{5}\left[\ddot{I}_{xy}\left(\dddot{I}_{yy} - \dddot{I}_{xx}\right) + \ddot{I}_{xy}\left(\dddot{I}_{xx} - \dddot{I}_{yy}\right)\right]$$

$$= -\frac{8}{5}\frac{m_1^2m_2^2}{a^2(1-e^2)^2}\left[4 + 10e\cos\theta + e^2(9\cos^2\theta - 1) + e^3(3\cos^3\theta - \cos\theta)\right]\dot{\theta} \tag{9.140}$$

which, when averaged over a period, gives

$$\left\langle\frac{dL}{dt}\right\rangle = \frac{1}{T}\int_0^{2\pi}\frac{dL}{dt}\frac{1}{\dot{\theta}}\, d\theta = -\frac{32}{5}\frac{m_1^2m_2^2(m_1 + m_2)^{1/2}}{a^{7/2}(1-e^2)^2}\left(1 + \frac{7}{8}e^2\right). \tag{9.141}$$

Combining these results with Eq. (9.131) we can determine how the orbital parameters of the binary system changes. We get

$$\left\langle\frac{da}{dt}\right\rangle = \frac{2a^2}{m_1m_2}\left\langle\frac{dE}{dt}\right\rangle = -\frac{64}{5}\frac{m_1m_2(m_1 + m_2)}{a^3(1-e^2)^{7/2}}\left(1 + \frac{73}{24}e^2 + \frac{37}{96}e^4\right) \tag{9.142}$$

$$\left\langle\frac{de}{dt}\right\rangle = \frac{(m_1 + m_2)}{m_1m_2e}\left[\frac{a(1-e^2)}{m_1 + m_2}\left\langle\frac{dE}{dt}\right\rangle - \frac{(1-e^2)^{1/2}}{a^{1/2}(m_1 + m_2)^{1/2}}\left\langle\frac{dL}{dt}\right\rangle\right]$$

$$= -\frac{304}{15}\frac{m_1m_2(m_1 + m_2)e}{a^4(1-e^2)^{5/2}}\left(1 + \frac{121}{304}e^2\right). \tag{9.143}$$

The second equation shows that the eccentricity decreases due to the emission of gravitational radiation and that the orbit becomes more and more circular. Unfortunately, this effect is not directly observable in astrophysical systems. More important is the fact that the period of orbit $T \propto a^{3/2}$ decreases due to the emission of gravitational radiation by the amount $(\dot{T}/T) = (3\dot{a}/2a) = (3\dot{E}/2E)$. Using $|E| = (Gm_1m_2/2a)$, we get

$$\frac{1}{T}\frac{dT}{dt} = -\frac{96}{5}\frac{G}{c^5}\frac{M^2\mu}{a^4}f(e), \tag{9.144}$$

where $\mu = m_1m_2(m_1+m_2)^{-1}$ is the reduced mass of the system, $M = (m_1+m_2)$ is the total mass and $f(e)$ is given by

$$f(e) \equiv \left(1 + \frac{73}{24}e^2 + \frac{37}{96}e^4\right)(1 - e^2)^{-7/2}. \tag{9.145}$$

As we shall see in the next section this leads to important, testable, predictions.

Exercise 9.16
Eccentricity matters Show that, for a system with an eccentricity $e = 0.62$ (which corresponds to the binary pulsar PSR 1913+16), the correction due to eccentricity is significant and needs to be computed correctly. Do this by comparing the rate of change of the orbital period for $e = 0.62$ with that for a circular orbit with $e = 0$. Show that the results differ by nearly an order of magnitude. Further show that, if the loss of energy was computed ignoring the eccentricity but the orbit averaging was done for an elliptical orbit, then the correction factor (compared with the $e = 0$ case) will be $(1 + 15/2e^2 + 45/8e^4 + 5/16e^6)(1 - e^2)^{-7/2} \approx 25.1$.

Exercise 9.17
Getting rid of eccentric behaviour Combine Eq. (9.142) and Eq. (9.143), to obtain an average equation for da/de given by

$$\frac{da}{de} = \frac{12}{19}a\frac{1 + (73/24)e^2 + (37/96)e^4}{e(1 - e^2)[1 + (121/304)e^2]}. \tag{9.146}$$

Integrate this equation to obtain a result (with possibly the strangest of exponent):

$$a(e) \propto \frac{e^{12/19}}{1 - e^2}\left(1 + \frac{121}{304}e^2\right)^{870/2299}. \tag{9.147}$$

This function determines the rate at which the orbits become circular. Consider a binary pulsar with the initial values for semi-major axis $a_0 \approx 2 \times 10^9$ m and eccentricity $e_0 \approx 0.617$. After a long time, when the orbit has shrunk to a size $a \approx 10^3$ km $\approx 100R_S$ (where R_S is the Schwarzschild radius of a typical neutron star), estimate the eccentricity. Hence show that the orbit circularizes long before the two objects coalesce.

Exercise 9.18

Radiation from a parabolic trajectory The function $f(\epsilon)$ has an obvious singularity at $e = 1$ which is to be expected since we no longer have a bound trajectory. Work out the emission of gravitational radiation in the case of a parabolic trajectory with $e = 1$ from first principles and show that the power radiated along the trajectory is given by

$$P(r) = \frac{16G^4\mu^2 m^3}{15r^5 c^5}\left(1 + \frac{11\,l}{2r}\right), \tag{9.148}$$

where μ is the reduced mass, m is the total mass and l is the latus rectum. Integrate this expression to determine the total energy radiated in such an orbit and show that

$$E_{\text{tot}} = \frac{85\pi}{48}\frac{G\mu^2}{R}\left(\frac{v}{c}\right)^5, \tag{9.149}$$

where $v = 2(Gm/l)^{1/2}$ is the maximum velocity attained in the trajectory.

Exercise 9.19

Gravitational waves from a circular orbit A pair of identical particles of mass m are orbiting around each other in a circle of radius a centred at the origin in the x–y plane. Let the angular frequency of the orbital rotation be ω.

(a) Express T_{ab} of the particles using two delta functions and determine the metric perturbations. Show that the result can be expressed in four-dimensional notation as

$$h_{ab}(t,\boldsymbol{r}) = \frac{8m}{r}u_a u_b - \frac{8ma^2\omega^2}{r}\left[e^{+z}_{ab}\cos\left(2\omega(t-r)\right) + e^{\times z}_{ab}\sin\left(2\omega(t-r)\right)\right], \tag{9.150}$$

where e^{+z}_{ab}, etc., are the two polarization tensors for the waves propagating along the z-direction.

(b) At a sufficiently large distance, the spherical waves emitted by the system can be approximated as plane waves propagating along the radial direction. Show that, in this limit, an observer located somewhere along the z-axis will detect the wave

$$h^{TT}_{ab}(t,r\hat{z}) = -\frac{8\sqrt{2}ma^2\omega^2}{r}\,\text{Re}\left[e^{-2i\omega(t-r)}e^{Rz}_{ab}\right], \tag{9.151}$$

where e^{RZ}_{ab} is the polarization tensor for the right circularly polarized wave. In contrast, show that an observer at a large distance along the x-axis will see a linearly polarized wave

$$h^{TT}_{ab}(t,r\hat{x}) = \frac{4ma^2\omega^2}{r}\,\text{Re}\left[e^{-2i\omega(t-r)}e^{+x}_{ab}\right]. \tag{9.152}$$

Explain the physical reason for this difference.

9.7.2 Observational aspects of binary pulsars

The results described in the previous section have been applied to the study of a binary pulsar called PSR 1913+16 to verify the predictions of general relativity. This binary star system is now known extraordinarily well after 30 years of observations. In this section, we shall briefly describe some observational features related to this system and how the predictions of general relativity are verified.

Consider a pulsar of mass m_1 and a companion of mass m_2 moving about their common centre of mass. We will first relate the observed and real periods of the pulsar to each other when the general relativistic effects are important. We begin by writing

$$\frac{(\delta t)_{\text{obs}}}{(\delta t)_{\text{true}}} = \frac{(\delta t)_{\text{obs}}}{(\delta t)_{\text{stat}}} \frac{(\delta t)_{\text{stat}}}{(\delta t)_{\text{true}}}, \qquad (9.153)$$

where 'stat' refers to an observer who is stationary with respect to the centre of mass. If the observer on Earth is also stationary with respect to the centre of mass, then

$$\frac{(\delta t)_{\text{obs}}}{(\delta t)_{\text{stat}}} = \left(1 - \frac{Gm_2}{rc^2}\right)^{-1}, \qquad (9.154)$$

where r is the distance between the masses m_1 and m_2. The second factor in Eq. (9.153) is given by the standard Doppler formula

$$\frac{(\delta t)_{\text{stat}}}{(\delta t)_{\text{true}}} = \left(1 - \frac{v_1^2}{c^2}\right)^{-1/2}\left(1 + \frac{\boldsymbol{v}_1 \cdot \boldsymbol{n}}{c}\right), \qquad (9.155)$$

where $\hat{\boldsymbol{n}}$ is a unit vector pointing towards the pulsar from the Earth and $\boldsymbol{v}_1$ is the velocity of the pulsar. To the the lowest order in $\mathcal{O}(v^2/c^2)$, the net effect is given by

$$\frac{(\delta t)_{\text{obs}}}{(\delta t)_{\text{true}}} = 1 + \frac{\boldsymbol{v}_1 \cdot \boldsymbol{n}}{c} + \frac{1}{2}\frac{v_1^2}{c^2} + \frac{Gm_2}{rc^2}. \qquad (9.156)$$

Let us now consider the various terms on the right hand side. It is convenient to start with an analysis of the *Newtonian* situation in which the pulsar of mass m_1 and a companion of mass m_2 are moving in elliptical orbits about their common centre of mass, and then introduce the necessary modifications from general relativity. We choose the x–y plane as the plane of the orbit with the origin at the centre of mass. The orbital plane is taken to be inclined at an angle i with respect to the line-of-sight with the x-axis along the line of nodes (see Fig. 9.4). If ω is the angle between the periastron and the line of nodes, the position of the pulsar at any instant is given by

$$x = r_1 \cos \psi, \qquad y = r_1 \sin \psi, \qquad (9.157)$$

where

$$\psi = \omega + \phi, \qquad r_1 = \frac{a_1\left(1 - e^2\right)}{1 + e \cos \phi}. \qquad (9.158)$$

If

$$\boldsymbol{n} = \boldsymbol{e}_{z'} = \cos i\, \boldsymbol{e}_z + \sin i\, \boldsymbol{e}_y \qquad (9.159)$$

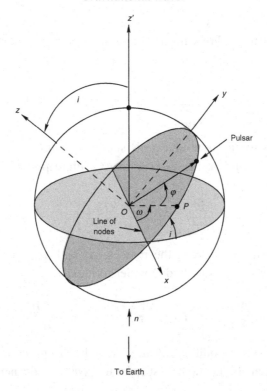

Fig. 9.4. Geometrical details of a binary pulsar orbit.

denotes the unit vector pointing from the Earth to the pulsar, then the line-of-sight component of the velocity vector of the pulsar is given by

$$v_1 \cdot n = \left(\dot{r}_1 \sin \psi + r_1 \dot{\psi} \cos \psi \right) \sin i. \tag{9.160}$$

For the elliptical orbit, $\dot{\phi}$ can be expressed as

$$\dot{\phi} = \frac{2\pi}{T (1 - e^2)^{3/2}} (1 + e \cos \phi)^2, \tag{9.161}$$

where T is the period of the orbit. Using this we obtain, after some simple algebra,

$$v_1 \cdot n = K \left[\cos (\omega + \phi) + e \cos \omega \right]; \qquad K = \frac{2\pi a_1 \sin i}{T (1 - e^2)^{1/2}}. \tag{9.162}$$

For any binary system in which the line-of-sight velocity of a star can be measured, we can use these equations to determine the following parameters: the eccentricity e and period T can be found from Eq. (9.161) which will provide the function $\phi(t)$ when integrated. Further, by studying the coefficients of $\cos \phi$ and $\sin \phi$ in

Eq. (9.162) one can obtain K and ω. Eliminating T and e from K, one determines $a_1 \sin i$, using which one can determine the *mass function*, defined as

$$f = \frac{(m_2 \sin i)^3}{(m_1 + m_2)^2} = \frac{(a_1 \sin i)^3}{G} \left(\frac{2\pi}{T}\right)^2. \tag{9.163}$$

It might seem that the Doppler shift in Eq. (9.156) can actually provide us with more information because of the additional terms of order $\mathcal{O}(v^2/c^2)$. This, however, is not true for elliptical orbits in Newtonian gravity. Using

$$v_1^2 = \dot{r}_1^2 + r_1^2 \dot{\psi}^2 = \left(\frac{2\pi}{T}\right)^2 \frac{a_1^2}{1 - e^2} \left(1 + 2e \cos \phi + e^2\right) \tag{9.164}$$

and

$$\frac{Gm_2}{r} = \frac{Gm_2^2}{(m_1 + m_2) r_1}; \qquad \left(\frac{2\pi}{T}\right)^2 = \frac{Gm_2^3}{(m_1 + m_2)^2 a_1^3} \tag{9.165}$$

we can easily show that

$$\frac{1}{2} v_1^2 + \frac{Gm_2}{r} = \beta \cos \phi + \text{ constant,} \tag{9.166}$$

where

$$\beta \equiv \frac{Gm_2^2 (m_1 + 2m_2) e}{(m_1 + m_2)^2 a_1 (1 - e^2)}. \tag{9.167}$$

Since the time dependence of this quantity is the same as the higher order term $K \cos \omega \cos \phi$ in Eq. (9.162), β will not be independently measurable. Thus, in purely Newtonian gravity, we cannot do any better.

The situation is, however, different when general relativistic effects cause the periastron to precess by the amount

$$\dot{\omega} = \frac{6\pi Gm_2}{a_1 (1 - e^2) T c^2}. \tag{9.168}$$

(This result is nontrivial. We did see, in Section 7.4.1, that an elliptical Newtonian orbit of a *test* particle in Schwarzschild geometry will exhibit general relativistic precession by the above amount. However, it is not obvious that such a result will carry through for the case of a two-body problem when the motion is studied using a reduced mass and centre of mass coordinates. Incredibly enough, this result continues to hold and is described in detail in Project 9.3. No simple reason for this feature is known.) To the lowest order, we can replace ω by $(\omega_0 + \dot{\omega}t)$ in Eq. (9.162), thereby obtaining four independent, time varying trigonometric combinations of ϕ and $\dot{\omega}t$. This allows one to separately determine from the observations the parameters K, $\omega_0, \dot{\omega}$ and β. Since $\dot{\omega}$ and β involve different combinations of the four parameters m_1, m_2, a_1 and $\sin i$, compared with the mass function, measurement of $\dot\omega$ and β allows complete determination of the orbital parameters of the binary system.

In fact, the situation is better than that. We saw earlier that two masses which are moving in an elliptic orbit with eccentricity e lose energy and angular momentum due to the radiation of gravitational waves leading to a change in the orbital period. Since all the parameters in Eq. (9.144) are now determined, one can actually *predict* the value of $\dot T$ due to gravitational radiation from the system. Observing $\dot T$ will act as a consistency check on the model. Alternatively, one can determine the masses by any combination of the observed quantities and check whether the curves in the $(m_1 - m_2)$ plane intersect on a single point.

Such an exercise was carried out for a particular pulsar PSR 1913+16 with remarkable success in verifying the predictions of general relativity. From the observed values for this pulsar, one can obtain a very precise estimate of the total mass

$$m_1 + m_2 = (2.8278 \pm 0.0007) \ M_\odot \tag{9.169}$$

in terms of solar mass. (In fact, the Newtonian gravitational constant is *not* known to a precision of four significant figures; hence the total mass can be quoted more accurately in terms of solar mass than in grams.) From other observed values one can also determine the combinations

$$\gamma = \frac{G^{2/3} m_2 (m_1 + 2m_2) e}{(m_1 + m_2)^{4/3} c^2} \left(\frac{T}{2\pi}\right)^{1/3} = (0.0007344 \text{ s}) \, m_2 (2.8278 + m_2) \tag{9.170}$$

$$\sin i = \left(\frac{2\pi}{T}\right)^{2/3} \frac{(m_1 + m_2)^{2/3} a_1 \sin i}{G^{1/3} m_2} = \frac{1.019}{m_2} \tag{9.171}$$

$$\dot T = -\frac{192\pi}{5} \frac{G^{5/3} m_1 m_2 f(e)}{c^5 (m_1 + m_2)^{1/3}} \left(\frac{2\pi}{T}\right)^{5/3} = -1.202 \times 10^{-12} m_2 (2.8278 - m_2) \tag{9.172}$$

where the masses are in solar units. The masses are about $m_1 = m_2 = (1.41 \pm 0.06)$ leading to a prediction that $\dot T = -(2.402\,58 \pm 0.000\,04) \times 10^{-12}$. This should be compared with the observed value $\dot T = -(2.425 \pm 0.010) \times 10^{-12}$. These match to within a factor of (1.002 ± 0.005), thereby providing a clear (though indirect) demonstration of the existence of gravitational waves as predicted by the general theory of relativity.

Exercise 9.20
Pulsar timing and the gravitational wave background The remarkable timing accuracy available in the case of a few binary pulsars, monitored for couple of decades, can be used

to put rigorous bounds on any possible mechanism which could affect the arrival time of the pulses. One particular example of this procedure is in the context of putting bounds on the amount of stochastic gravitational wave background that can exist in our universe. This exercise provides a simplified analysis of this effect. Consider a plane gravitational wave with a + polarization propagating along the z-axis. In TT gauge the line element will be

$$ds^2 = -dt^2 + [1 + h(u)]dx^2 + [1 - h(u)]dy^2 + dz^2, \qquad (9.173)$$

where $u = t - z$. Let a pulsar be at rest at the origin and the Earth be at rest at $r = L\hat{n}$. Radio pulses are emitted by the pulsar at its proper time $t_{\text{em}} = 0, T, 2T$, etc., where T is the time difference between the pulses. The pulse emitted at t_{em} is received at the Earth at time t_{rec}.

(a) Show that the proper time t_{rec} is related to the proper time t_{em} by

$$t_{\text{rec}} - t_{\text{em}} = \frac{1}{2}(n_x^2 - n_y^2) \int_0^L h[t_{\text{rec}} - L + (1 - n_z)r]dr \qquad (9.174)$$

to first order accuracy in h. [Hint. There are several ways of doing this but the simplest one is to use the variational principle in Eq. (4.70). Evaluate this action for a light ray having the trajectory $x^a(\rho) = [t_{\text{em}} + (1 + \delta)\rho, \hat{n}\rho]$ with $0 \leq \rho \leq L$ that differs from the true null geodesic by $\mathcal{O}(h)$. We know that, if we use a worldline which differs from the true one by $\mathcal{O}(h)$, the corresponding actions should differ by $\mathcal{O}(h^2)$. Use this fact to show that

$$\delta = \frac{n_x^2 - n_y^2}{2L} \int_0^L h[t_{\text{em}} + (1 - n_z)r]dr + \mathcal{O}(h^2). \qquad (9.175)$$

Take it from there.]

(b) Suppose that the gravitational wave had an amplitude ϵ and frequency ν. Show that the delay in the arrival time of the pulses caused by such a wave is of the order of

$$\frac{\epsilon}{4\pi\nu} \approx 3 \times 10^{-6} \text{ sec } \left(\frac{\epsilon}{10^{-13}}\right)\left(\frac{10^{-1} \text{ yr}}{\nu}\right). \qquad (9.176)$$

With the current state of art, one can achieve such a precision over a timescale of 20 years. [Hint. Take $h(u) = \epsilon \cos[2\pi\nu u]$ and perform the integral in Eq. (9.175).]

(c) The absence of any observed residuals in the timing delay can be used to put a bound on the energy density of the gravitational wave background. Show that the energy density is about $(\pi c^2/8G)\epsilon^2\nu^2$. Express the result of part (b) as a bound on the energy density.

9.7.3 Gravitational radiation from coalescing binaries

The discussion in the previous section shows that gravitational radiation is emitted by systems like PSR 1913+16. Over a period of time, this drain of energy will cause the two remnants to move towards each other and eventually coalesce together. Hence the final stages of evolution of a binary system made of two compact relativistic objects is of considerable interest both theoretically and observationally.

As the two members of the binary systems approach coalescence, large quantities of gravitational waves are expected to be emitted. Several experiments for the detection of gravitational waves rely on the emission from such coalescence of compact relativistic objects. For example, the coalescence of neutron star–neutron star (NS–NS) system at a distance of 1000 Mpc or a black hole–black hole (BH–BH) system at 3000 Mpc will produce an effective gravitational wave amplitude of about $h \simeq 10^{-22}$ in the frequency band of $(10–10^3)$ Hz; an NS–NS coalescence at 60 Mpc or a BH–BH coalescence at 200 Mpc will produce an amplitude that is about 30 times higher. The chances of such detection, in a statistical sense, depends on the rate of occurrence of such events which are difficult to estimate reliably. We shall now provide a simplified discussion of the energy spectrum from two coalescing compact objects assuming that they move towards each other *head on* rather than in a spiralling orbit.

Taking the motion to be along the x-axis with the centre of mass at the origin, we can describe the system in terms of the reduced mass $\mu = m_1 m_2/(m_1+m_2) \equiv (m_1 m_2/M)$ and the separation $x = x_1 - x_2$ with $m_1 x_1 = -m_2 x_2 = \mu x$. The emission of gravitational radiation is governed by the quadrupole moment

$$Q_{\alpha\beta} \equiv \sum_A m_A \left[x_\alpha^A x_\beta^A - \frac{1}{3}\delta_{\alpha\beta} \left(x^A\right)^2 \right]. \tag{9.177}$$

In this particular case, the components of the quadrupole tensor are given by

$$Q_{xx} = \frac{2}{3}\mu x^2, \qquad Q_{yy} = Q_{zz} = -\frac{1}{3}\mu x^2. \tag{9.178}$$

To estimate the total amount of energy which is emitted by the system, we need to compute the third time derivative of $Q_{\alpha\beta}$. From the equation of motion $\ddot{x} = -(GM/x^2)$ we get $\dot{x}^2 = (2GM/x)$ if the particles are falling freely from infinity, starting at rest. Using this, we obtain

$$\dddot{Q}_{xx} = -\frac{4}{3}G\mu M \frac{\dot{x}}{x^2} \tag{9.179}$$

and similar expressions for other components. The total amount of energy radiated per second is:

$$\frac{dE}{dt} = \frac{1}{5}\frac{G}{c^5}\left\langle \dddot{Q}_{xx}^2 + \dddot{Q}_{yy}^2 + \dddot{Q}_{zz}^2 \right\rangle = \frac{8}{15}\frac{G^3\mu^2 M^2}{c^5}\left\langle \frac{\dot{x}^2}{x^4} \right\rangle. \tag{9.180}$$

The total energy emitted when the stars start out from a large separation until they reach a separation of $x_{\min}$ will be

$$\Delta E = \int \frac{dE}{dt}\,dt = \int \frac{dE}{dt}\frac{1}{\dot{x}}\,dx = \frac{8}{15}\frac{G^3\mu^2 M^2}{c^5}(2GM)^{1/2}\int_{x_{\min}}^{\infty}\frac{dx}{x^{9/2}}, \tag{9.181}$$

where we have used $\dot{x} = (2GM)^{1/2}x^{-1/2}$. This integral diverges as $x_{\min} \to 0$; however, since the Newtonian treatment of the trajectory breaks down for $x_{\min} \lesssim$

$(2GM/c^2)$, we can possibly cut off the integral at $x_{\min} = (2GM/c^2)$, to obtain

$$\Delta E = \frac{2}{105} \frac{\mu^2 c^2}{M}. \tag{9.182}$$

In spite of the approximations used, this result agrees with more exact treatments incorporating general relativistic effects and calculated numerically. The results shows that, if $m_1 \approx m_2 \approx \mu$, about one per cent of the rest mass energy of the stars can be emitted in the form of gravitational waves.

To determine the spectrum of gravitational radiation emitted in this case, one starts by rewriting ΔE in Eq. (9.181) as

$$\Delta E = \frac{1}{5} \frac{G}{c^5} \int_{-\infty}^{\infty} d\omega \left| Q_{\alpha\beta}^{(3)}(\omega) \right|^2 = \frac{2}{5} \frac{G}{c^5} \int_0^{\infty} d\omega \left| Q_{\alpha\beta}^{(3)}(\omega) \right|^2, \tag{9.183}$$

where we have used the Parseval's theorem and limited the integration to positive values of frequency. The quantity $Q_{\alpha\beta}^{(3)}(\omega)$ denotes the Fourier transform of the third derivative of the quadrupole tensor. This gives the spectral energy distribution

$$\frac{dE}{d\omega} = \frac{2}{5} \frac{G}{c^5} \left| Q_{\alpha\beta}^{(3)}(\omega) \right|^2. \tag{9.184}$$

The Fourier transform of the quadrupole tensor can be expressed, after one integration by parts, in the form:

$$Q_{\alpha\beta}^{(3)}(\omega) = \frac{1}{(2\pi)^{1/2}} \int_{-\infty}^{\infty} \dddot{Q}_{\alpha\beta}(t) e^{i\omega t} dt$$

$$= \frac{1}{(2\pi)^{1/2}} \left[\ddot{Q}_{\alpha\beta}(t) e^{i\omega t} \Big|_{-\infty}^{\infty} - i\omega \int_{-\infty}^{\infty} \ddot{Q}_{\alpha\beta}(t) e^{i\omega t} dt \right]. \tag{9.185}$$

Since $\ddot{Q}_{xx}(t) = (4/3)(GM\mu/x)$ we need the explicit form of the trajectory to evaluate the integrals. For free fall from infinity, we can take it to be

$$x = \left(\frac{3}{2} \right)^{2/3} (2GM)^{1/3} (-t)^{2/3}, \qquad (x \to \infty \text{ as } t \to -\infty) \tag{9.186}$$

for $x > x_{\min} \equiv (2GM/c^2)$. Assuming $x = x_{\min}$ at $t = t_{\min}$, and taking the derivatives of the quadrupole tensor to be zero for $t > t_{\min}$, we see that the first term in Eq. (9.185) vanishes in both limits. So

$$Q_{xx}^{(3)}(\omega) = -\frac{i\omega}{(2\pi)^{1/2}} \int_{-\infty}^{t_{\min}} \frac{2^{7/3} (GM)^{2/3} \mu}{3^{5/3} (-t)^{2/3}} e^{i\omega t} dt$$

$$= -\frac{i\omega^{2/3} 2^{7/3} (GM)^{2/3} \mu}{(2\pi)^{1/2} 3^{5/3}} \int_{|\omega t_{\min}|}^{\infty} e^{-iy} \frac{dy}{y^{2/3}}, \tag{9.187}$$

where $y = -\omega t$. The integral can be expressed in terms of an incomplete gamma function; but for $|\omega t_{\min}| \ll 1$ we can replace the lower limit of the integration by zero, getting the value of the integral to be $\Gamma(1/3) \exp(-i\pi/6)$. Evaluating the other components in a similar fashion and combining all the results, we get:

$$\frac{dE}{d\omega} \approx \frac{2^{11/3}}{5\pi (3)^{7/3}} \Gamma^2[1/3] \frac{G}{c^5} (GM\omega)^{4/3} \mu^2, \qquad (\omega \to 0). \qquad (9.188)$$

The $\omega^{4/3}$ dependence is quite generic in the limit of $\omega \to 0$ and arises from the behaviour of $Q_{\alpha\beta}(t)$ near $t \to -\infty$, when the Newtonian approximation is quite valid.

PROJECTS

Project 9.1
Gauge and dynamical degrees of freedom

In the description of gravitational waves, we have started with the linear approximation and then imposed a sequence of gauge conditions in order to identify the physical modes. Such an approach, however, does not demonstrate explicitly that the pure gauge modes of the metric obey differential equations which do *not* allow them to propagate. This project provides a formal decomposition of the metric and Einstein's equations in the linear theory clarifying this issue. Let the metric perturbation be denoted as:

$$h_{00} = -2\Phi; \qquad h_{\alpha 0} = h_{0\alpha} = w_\alpha; \qquad h_{\alpha\beta} = 2s_{\alpha\beta} - 2\Psi\delta_{\alpha\beta}. \qquad (9.189)$$

Here Φ is a scalar with respect to purely spatial rotation, w_α behaves as a three-vector under spatial rotations and the second rank symmetric tensor $h_{\alpha\beta}$ in three dimensions is separated into a tracefree part $s_{\alpha\beta}$ and a trace.

(a) We will now write down Einstein's equations in terms of $(\Phi, \psi, w_\alpha, s_{\alpha,\beta})$. Show that the curvature tensor has the following independent components

$$R_{0\beta 0\lambda} = \partial_\beta \partial_\lambda \Phi + \partial_0 \partial_{(\beta} w_{\lambda)} - \frac{1}{2} \partial_0 \partial_0 h_{\beta\lambda}$$

$$R_{0\beta\gamma\lambda} = \partial_\beta \partial_{[\gamma} w_{\lambda]} - \partial_0 \partial_{[\gamma} h_{\lambda]\beta}$$

$$R_{\alpha\beta\gamma\lambda} = \partial_\beta \partial_{[\gamma} h_{\lambda]\alpha} - \partial_\alpha \partial_{[\gamma} h_{\lambda]\beta} \qquad (9.190)$$

with other components determined in terms of this by symmetries.

(b) To the same linear order, show that the Ricci tensor has the components

$$R_{00} = \nabla^2 \Phi + \partial_0 \partial_\gamma w^\gamma + 3\partial_0^2 \Psi$$

$$R_{0\beta} = -\frac{1}{2}\nabla^2 w_\beta + \frac{1}{2}\partial_\beta \partial_\gamma w^\gamma + 2\partial_0 \partial_\beta \Psi + \partial_0 \partial_\gamma s^\gamma_\beta$$

$$R_{\alpha\beta} = -\partial_\alpha \partial_\beta (\Phi - \Psi) - \partial_0 \partial_{(\alpha} w_{\beta)} + \Box\Psi\delta_{\alpha\beta} - \Box s_{\alpha\beta} + 2\partial_\gamma \partial_{(\alpha} s^\gamma_{\beta)}, \quad (9.191)$$

where $\nabla^2 = \delta^{\alpha\beta}\partial_\alpha \partial_\beta$, etc.

(c) Finally compute the Einstein tensor to be:

$$G_{00} = 2\nabla^2 \Psi + \partial_\gamma \partial_\lambda s^{\gamma\lambda}$$

$$G_{0\beta} = -\frac{1}{2}\nabla^2 w_\beta + \frac{1}{2}\partial_\beta \partial_\gamma w^\gamma + 2\partial_0 \partial_\beta \Psi + \partial_0 \partial_\gamma s_\beta^\gamma$$

$$G_{\alpha\beta} = (\delta_{\alpha\beta}\nabla^2 - \partial_\alpha \partial_\beta)(\Phi - \Psi) + \delta_{\alpha\beta}\partial_0 \partial_\gamma w^\gamma - \partial_0 \partial_{(\alpha} w_{\beta)}$$

$$+2\delta_{\alpha\beta}\partial_0^2 \Psi - \Box s_{\alpha\beta} + 2\partial_\gamma \partial_{(\alpha} s_{\beta)}^\gamma - \delta_{\alpha\beta}\partial_\gamma \partial_\lambda s^{\gamma\lambda}. \qquad (9.192)$$

(d) Show that the 00-component of the equation can be transformed to the form

$$\nabla^2 \Psi = 4\pi T_{00} - \frac{1}{2}\partial_\gamma \partial_\lambda s^{\gamma\lambda}. \qquad (9.193)$$

Explain why this implies that Ψ does not propagate and is determined by other quantities instantaneously. Similarly, the 0α-components of Einstein's equations can be written in the form

$$\left(\delta_{\beta\gamma}\nabla^2 - \partial_\beta \partial_\gamma\right) w^\gamma = -16\pi T_{0\beta} + 4\partial_0 \partial_\beta \Psi + 2\partial_0 \partial_\gamma s_\beta^\gamma. \qquad (9.194)$$

Again argue that w^α does not propagate.

(e) Finally, show that the spatial part of Einstein's equations can be expressed in the form

$$\left(\delta_{\alpha\beta}\nabla^2 - \partial_\alpha \partial_\beta\right) \Phi = 8\pi T_{\alpha\beta} + \left(\delta_{\alpha\beta}\nabla^2 - \partial_\alpha \partial_\beta - 2\delta_{\alpha\beta}\partial_0^2\right) \Psi - \delta_{\alpha\beta}\partial_0 \partial_\gamma w^\gamma$$

$$+ \partial_0 \partial_{(\alpha} w_{\beta)} + \Box s_{\alpha\beta} - 2\partial_\gamma \partial_{(\alpha} s_{\beta)}^\gamma - \delta_{\alpha\beta}\partial_\gamma \partial_\lambda s^{\gamma\lambda}. \qquad (9.195)$$

Argue that Φ does not propagate but this equation does propagate $s_{\alpha\beta}$. This analysis clearly demonstrates that Ψ, w^α and Φ are non-dynamical and that the only propagating degree of freedom is contained in $s_{\alpha\beta}$, which is the traceless part of the spatial component of the metric containing five degrees of freedom. Show that, under the gauge transformations in Eq. (9.7), $s_{\alpha\beta}$ transforms as $s_{\alpha\beta} \to s_{\alpha\beta} + \partial_{(\alpha}\xi_{\beta)} - (1/3)\delta_{\alpha\beta}(\partial_\mu\xi^\mu)$. Demonstrate that this gauge freedom can be used to set $\partial_\mu s^{\mu\nu} = 0$. These three conditions will make the spatial part transverse leaving $5 - 3 = 2$ degrees of freedom, which we originally identified with the two independent metric components in the TT gauge.

Project 9.2
An exact gravitational plane wave

The purpose of this project is to study a simple metric which can be interpreted as describing the propagation of a plane gravitational wave in the exact sense, without any linear approximation.

(a) Consider a metric of the form

$$ds^2 = L^2(e^{2\beta}dx^2 + e^{-2\beta}dy^2) + dz^2 - dt^2 = L^2(e^{2\beta}dx^2 + e^{-2\beta}dy^2) - du\,dv, \qquad (9.196)$$

with $u = t - z$ and $v = t + z$ and where L and β are functions of u alone. Show that the curvature tensor for this metric is nonzero in general and that Einstein's equations reduce to the single equation

$$L'' + (\beta')^2 L = 0, \qquad (9.197)$$

where the prime denotes differentiation with respect to u.

(b) Using the above result argue that, to linear order in β, the metric represents a plane gravitational wave propagating in flat spacetime with $+$ polarization along the z-axis. When the functional form of the wave amplitude, $\beta(u) = \beta(t - z)$, is arbitrary but small, interpret Einstein's equation Eq. (9.197) as determining the background curvature induced by the energy density of the gravitational wave. More precisely, show that Einstein's equations can be expressed in the form, $G_{uu}^{(bg)} = 8\pi T_{uu}^{(eff)}$, where $G_{uu}^{(bg)} = -2L''/L$ and $T_{uu}^{(eff)} = (1/4\pi)(\beta')^2$. Show that this agrees with the expression for the effective energy-momentum tensor of the gravitational wave obtained in the text.

(c) As further justification for this interpretation, consider the metric due to a plane electromagnetic wave which can be taken to be

$$ds^2 = L^2(u)\,(dx^2 + dy^2) - du\,dv. \tag{9.198}$$

Let the source for this metric be a plane electromagnetic wave with the potential $A_j = (0, \mathcal{A}(u), 0, 0)$. Show that Einstein's equation $G_{ab} = 8\pi T_{ab}$ reduces to

$$L'' + (4\pi T_{uu})L = 0; \qquad [T_{uu}]_{\text{emwave}} = \frac{(\mathcal{A}')^2}{4\pi L^2}. \tag{9.199}$$

Compare this with the case of an exact gravitational wave and interpret the result.

(d) Consider the motion of a particle in the gravitational plane wave metric. From the symmetries of the metric, determine three Killing vectors $\boldsymbol{\xi}^{(A)}$ (with $A = 1, 2, 3$) and thus obtain three conserved quantities $\boldsymbol{p} \cdot \boldsymbol{\xi}^{(A)} \equiv C^{(A)}$, where $\boldsymbol{p}$ is the four-momentum of the particle. These three constants, along with the normalization conditions $\boldsymbol{p} \cdot \boldsymbol{p} = -m^2$, reduce the problem of finding the trajectory of the particle to quadrature. Assume that the particle was originally at rest in this coordinate system and show that the particle remains at rest in the coordinate system at all times.

(e) Write down the Maxwell equations in the background of a gravitational plane wave metric and find the general solution for electromagnetic wave propagating in a spacetime curved by the gravitational wave. Pay special attention to waves propagating along the positive and negative z-axis.[3]

Project 9.3
Post-Newtonian approximation

This project obtains the Lagrangian for a two body problem in general relativity correct up to one order higher than the Newtonian approximation and uses it to determine the perihelion precession in the case of two comparable masses moving around a common centre of mass. The procedure is conceptually straightforward but algebraically challenging.[4]

(a) Convince yourself that, in order to determine the Lagrangian at the post-Newtonian order, it is sufficient if you know: the spatial components of the metric perturbations $h_{\alpha\beta}$ to the accuracy $(1/c^2)$, the mixed components $h_{0\alpha}$ to the accuracy $(1/c^3)$ and the h_{00} to accuracy $(1/c^4)$.

(b) Consider the energy-momentum tensor for a system of point particles (labelled by $A = 1, 2, \ldots$) given by

$$T^{ik} = \sum_A \frac{m_A c}{\sqrt{-g}} \frac{dx^i}{ds} \frac{dx^k}{ds} \delta(\boldsymbol{r} - \boldsymbol{r}_A). \tag{9.200}$$

Our first task is to determine its explicit form to the necessary order of accuracy when the lowest order metric is given by

$$ds^2 = -\left(1 + \frac{2}{c^2}\phi\right)c^2 dt^2 + \left(1 - \frac{2}{c^2}\phi\right)(dx^2 + dy^2 + dz^2).$$ (9.201)

Show that this gives, to the necessary order of accuracy,

$$T_{00} = \sum_A m_A c^2 \left(1 + \frac{5\phi_A}{c^2} + \frac{v_A^2}{2c^2}\right)\delta(r - r_A)$$ (9.202)

$$T_{\alpha\beta} = \sum_A m_A v_{A\alpha} v_{A\beta}\delta(r - r_A); \qquad T_{0\alpha} = -\sum_A m_A c v_{A\alpha}\delta(r - r_A),$$ (9.203)

where v is the three-dimensional velocity and $\phi_A = \phi(r_A)$. Using an expansion of the curvature tensor up to the required order, show that R_{00} can be expressed in the form

$$R_{00} = \frac{1}{2}\nabla^2 h_{00} + \frac{2}{c^4}\phi\nabla^2\phi - \frac{2}{c^4}(\nabla\phi)^2$$ (9.204)

provided we impose the gauge condition

$$\frac{\partial h_0^\alpha}{\partial x^\alpha} - \frac{1}{2c}\frac{\partial h_\alpha^\alpha}{\partial t} = 0.$$ (9.205)

This component is correct to $\mathcal{O}(1/c^4)$. Next show that

$$R_{0\alpha} = \frac{1}{2}\nabla^2 h_{0\alpha} + \frac{1}{2c^3}\frac{\partial^3\phi}{\partial t\partial x^\alpha}.$$ (9.206)

(c) Combining the above results, write down the 00-component of Einstein's equations and show that it can be manipulated to read as

$$\nabla^2\left(h_{00} - \frac{2}{c^4}\phi^2\right) = \frac{8\pi G}{c^2}\sum_A m_A\left(1 + \frac{\phi_A'}{c^2} + \frac{3v_A^2}{2c^2}\right)\delta(r - r_A).$$ (9.207)

The prime on ϕ_A indicates that, in computing the gravitational potential at the location of a given particle, we only consider the effects of all other particles and ignore the infinite self-potential. Similarly reduce the 0α-components of the equation to the form

$$\nabla^2 h_{0\alpha} = -\frac{16\pi G}{c^3}\sum_A m_A v_{a\alpha}\delta(r - r_A) - \frac{1}{c^3}\frac{\partial^2\phi}{\partial t\partial x^\alpha}.$$ (9.208)

(d) Show that the solutions to these equations (to the necessary order of accuracy) are given by

$$h_{00} = \frac{2\phi}{c^2} + \frac{2\phi^2}{c^4} - \frac{2G}{c^4}\sum_A\frac{m_A\phi_A'}{|r - r_A|} - \frac{3G}{c^4}\sum_A\frac{m_A v_A^2}{|r - r_A|}$$ (9.209)

$$h_{0\alpha} = \frac{G}{2c^3}\sum_A\frac{m_A}{|r - r_A|}[7v_{A\alpha} + (v_A \cdot n_A)n_{A\alpha}],$$ (9.210)

where n_A is a unit vector along the direction $r - r_A$. The Lagrangian for the Ath particle in the metric has the form $-m_A c(ds/dt)$. Using the expression for ds arising from the metric, expanding the square root to the required order of accuracy and dropping the constant $-m_A c^2$, obtain this Lagrangian in the form

$$L_A = \frac{m_A v_A^2}{2} + \frac{m_A v_A^4}{8c^2} - m_A c^2 \left(\frac{h_{00}}{2} + h_{0\alpha} \frac{v_A^\alpha}{c} + \frac{1}{2c^2} h_{\alpha\beta} v_A^\alpha v_A^\beta - \frac{h_{00}^2}{8} + \frac{h_{00}}{4c^2} v_A^2 \right).$$
(9.211)

(e) Varying this Lagrangian with respect to the position of a particle will give the force that is acting on a particle. The full Lagrangian for the system made of all the particles has to be constructed in such a way that when the positions are varied in the correct Lagrangian we obtain the same forces. This can be obtained for the general case by straightforward algebra. Show that, in the case of *two* bodies, such a Lagrangian is given by

$$L = \frac{m_1 v_1^2}{2} + \frac{m_2 v_2^2}{2} + \frac{G m_1 m_2}{r} + \frac{1}{8c^2} \left(m_1 v_1^4 + m_2 v_2^4 \right)$$
$$+ \frac{G m_1 m_2}{2c^2 r} \left[3(v_1^2 + v_2^2) - 7 \boldsymbol{v}_1 \cdot \boldsymbol{v}_2 - (\boldsymbol{v}_1 \cdot \boldsymbol{n})(\boldsymbol{v}_2 \cdot \boldsymbol{n}) \right] - \frac{G^2 m_1 m_2 (m_1 + m_2)}{2c^2 r^2}.$$
(9.212)

(f) Our next aim is to use this Lagrangian to study the two body problem in the post Newtonian approximation. To do this, write down the above Lagrangian as $L_0 + L_1$ in terms of the relative velocity $\boldsymbol{v} = \boldsymbol{v}_1 - \boldsymbol{v}_2$ and the relative coordinate $\boldsymbol{x} = \boldsymbol{x}_1 - \boldsymbol{x}_2$ where

$$\boldsymbol{x}_1 = \left[\frac{m_2}{m} + \frac{\mu \delta m}{2m^2} \left(v^2 - \frac{m}{r} \right) \right] \boldsymbol{x}, \qquad \boldsymbol{x}_2 = \left[-\frac{m_1}{m} + \frac{\mu \delta m}{2m^2} \left(v^2 - \frac{m}{r} \right) \right] \boldsymbol{x} \quad (9.213)$$

with $m = m_1 + m_2$, $\delta m = m_1 - m_2$ and $\mu = m_1 m_2 / m$. Show that

$$L_0 = \frac{1}{2} v^2 + \frac{Gm}{r}; \quad L_1 = \frac{1}{8} \left(1 - \frac{3\mu}{m} \right) v^4 + \frac{GM}{2r} \left(3v^2 + \frac{\mu}{m} v^2 + \frac{\mu}{m} \left(\frac{\boldsymbol{v} \cdot \boldsymbol{x}}{r} \right)^2 \right) - \frac{G^2 m^2}{2r^2}.$$
(9.214)

Write the total Hamiltonian as $H = H_0 + H_1$, where H_0 is the standard Newtonian Hamiltonian and H_1 is the post Newtonian correction. Explain why $H_1 = -L_1$. Show that, when $\boldsymbol{p}_1 = -\boldsymbol{p}_2 \equiv \boldsymbol{p}$, the Hamiltonian can be reduced to the form

$$H_{\text{rel}} = \frac{1}{2} \left(\frac{1}{m_1} + \frac{1}{m_2} \right) \boldsymbol{p}^2 - \frac{G m_1 m_2}{r} - \frac{\boldsymbol{p}^4}{8} \left(\frac{1}{m_1^3} + \frac{1}{m_2^3} \right)$$
$$- \frac{G}{2r} \left(3\boldsymbol{p}^2 \left(\frac{m_2}{m_1} + \frac{m_1}{m_2} \right) + 7\boldsymbol{p}^2 + (\boldsymbol{p} \cdot \boldsymbol{n})^2 \right) + \frac{G^2 m_1 m_2 (m_1 + m_2)}{r^2}.$$
(9.215)

Replacing $\boldsymbol{p}^2$ by $p_r^2 + L^2 / r^2$ and equating the Hamiltonian to a constant, one can obtain p_r and the radial part of the action. Show that the radial part of the action is given by

$$S_r = \int \sqrt{A + \frac{B}{r} - \frac{L^2 - 6G^2 m_1^2 m_2^2}{r^2}} \, dr, \tag{9.216}$$

where A and B are unimportant constants. From this show that the perihelion precession per orbit is given by

$$\delta\phi = \frac{6\pi G^2 m_1^2 m_2^2}{L^2} = \frac{6\pi G(m_1 + m_2)}{c^2 a(1 - e^2)}, \tag{9.217}$$

which miraculously matches with the corresponding expression for a *test* body in Schwarzschild metric. No simple reason for this coincidence is known and it is an issue worth thinking about.

10

Relativistic cosmology

10.1 Introduction

This chapter applies the general theory of relativity to the study of cosmology and the evolution of the universe. Our emphasis will be mostly on the geometrical aspects of the universe rather than on physical cosmology. However, in order to provide a complete picture and to appreciate the interplay between theory and observation, it is necessary to discuss certain aspects of the evolutionary history of the universe. We shall do this in Section 10.6 even though it falls somewhat outside the main theme of development.[1]

10.2 The Friedmann spacetime

Observations show that, at sufficiently large scales, the universe is *homogeneous* and *isotropic*; that is, the geometrical properties of the three-dimensional space: (i) are the same at all spatial locations and (ii) do not single out any special direction in space.

The geometrical properties of the space are determined by the distribution of matter through Einstein's equations. It follows, therefore, that the matter distribution should also be homogeneous and isotropic. This is certainly not true at small scales in the observed universe, where a significant degree of inhomogeneity exits in the form of galaxies, clusters, etc. We assume that these inhomogeneities can be ignored and the matter distribution may be described by a smoothed out average density in studying the large scale dynamics of the universe.

Though this is a well accepted procedure, the following subtlety must be noted: since Einstein's equations are nonlinear, there is no guarantee that the solution obtained for a distribution of a source which is averaged over some region will be the same as that obtained by first solving Einstein's equations exactly and then averaging the solution. In fact, it is possible to construct counter-examples in which the operations of averaging and solving do not commute.[2] In spite of this mathematical difficulty, we will continue to treat the geometrical features of the universe as described by a solution to Einstein's equation in which the source is averaged over a sufficiently large scale. (Also see Project 13.1.)

The assumption of homogeneity and isotropy of the 3-space singles out a preferred class of observers, viz., those observers for whom the universe appears isotropic. In other words, at cosmological scales, we *do* have a preferred Lorentz frame. (This, of course, does not contradict the notion of general covariance, which is a symmetry of the field equations. Any specific solution need not respect the full symmetry of the equations.) Another observer, who is moving with a uniform velocity with respect to this fundamental class of observers, will find the universe to be anisotropic. We will use the coordinate system appropriate to this fundamental class of observers to describe the geometry.

Our first task is to determine the form of the spacetime metric in such a coordinate system (t, x^α) in which homogeneity and isotropy are self-evident. Separating the spacetime interval gives

$$ds^2 \equiv g_{00}dt^2 + 2g_{0\alpha}dtdx^\alpha - \sigma_{\alpha\beta}dx^\alpha dx^\beta, \tag{10.1}$$

where $\sigma_{\alpha\beta}$ is a positive definite spatial metric. We first note that the isotropy of space implies that the $g_{0\alpha}$s must vanish; otherwise, they identify a particular direction in space related to the three vector v_α with components $g_{0\alpha}$. Further, in the coordinate system determined by fundamental observers, we may use the proper time of clocks carried by these observers to label the spacelike surfaces. This choice for the time coordinate t implies that $g_{00} = -1$, bringing the spacetime interval to the form

$$ds^2 = -dt^2 + \sigma_{\alpha\beta}dx^\alpha dx^\beta \equiv -dt^2 + dl^2. \tag{10.2}$$

The problem now reduces to determining the 3-metric $\sigma_{\alpha\beta}$ of a 3-space which, at any instant of time, is homogeneous and isotropic. This can be done as follows.

The assumption of isotropy implies spherical symmetry which, in turn, allows the line interval to be written in the form

$$dl^2 = a^2 \left[\lambda^2(r)dr^2 + r^2(d\theta^2 + \sin^2\theta d\phi^2) \right], \tag{10.3}$$

where $a = a(t)$ can depend only on time. The scalar curvature ^3R for this three-dimensional space is:

$$^3R = \frac{3}{2a^2r^3} \frac{d}{dr} \left[r^2 \left(1 - \frac{1}{\lambda^2} \right) \right]. \tag{10.4}$$

Homogeneity implies that all the geometrical properties are independent of r; hence ^3R must be a constant. Equating it to a constant and integrating the resulting equation we get

$$r^2 \left(1 - \frac{1}{\lambda^2}\right) = c_1 r^4 + c_2; \quad c_1, c_2 = \text{constants.} \tag{10.5}$$

To avoid a singularity at $r = 0$, we need $c_2 = 0$. Thus we get $\lambda^2 = \left(1 - c_1 r^2\right)^{-1}$. When $c_1 \neq 0$, we can rescale r and make $c_1 = 1$ or -1. This leads to the full spacetime metric:

$$ds^2 = -dt^2 + a^2(t) \left[\frac{dr^2}{1 - kr^2} + r^2(d\theta^2 + \sin^2\theta d\phi^2)\right], \tag{10.6}$$

with $k = 0, \pm 1$. The pre-factor a determines the overall scale of the spatial metric and, in general, can be a function of time: $a = a(t)$. This metric, called the *Friedmann metric*, describes a universe that is spatially homogeneous and isotropic at each instant of time. Note that the form of the metric has been determined entirely by symmetry considerations without any reference to the source T_{ik} or Einstein's equations. These geometrical considerations, however, will not allow us to determine the value of k and the form of the function $a(t)$ (called the *expansion factor*). They have to be determined using Einstein's equations once the matter distribution is specified.

The coordinates in Eq. (10.6), chosen in such a way as to make the symmetries of the spacetime self-evident, are called the 'comoving' coordinates. It is easy to show that the worldlines with $x^\alpha = $ constant are geodesics. To see this, consider a free material particle that is at rest at the origin of the comoving frame at some instant. No velocity can be induced on this particle by the gravitational field since no direction can be considered as special. Therefore the particle will continue to remain at the origin. Since spatial homogeneity allows us to choose any location as the origin, it follows that the worldlines $x^\alpha = $ constant are geodesics. (Also see Exercise 4.6(b).) Observers following these worldlines are called fundamental (or 'comoving') observers.

The spatial hypersurfaces of the Friedmann universe have positive, zero and negative curvatures for $k = +1, 0$ and -1 respectively; the magnitude of the curvature, determined from Eq. (10.4), is $(6/a^2)$ when k is nonzero. To study the geometrical properties of these spaces it is convenient to introduce a coordinate χ, defined as

$$\chi = \int \frac{dr}{\sqrt{1 - kr^2}} = \begin{cases} \sin^{-1} r & (\text{for } k = 1) \\ r & (\text{for } k = 0) \\ \sinh^{-1} r & (\text{for } k = -1). \end{cases} \tag{10.7}$$

In terms of (χ, θ, ϕ) the metric becomes

$$dl^2 = a^2 \left[d\chi^2 + S_k^2(\chi)(d\theta^2 + \sin^2\theta d\phi^2)\right], \tag{10.8}$$

where

$$S_k(\chi) = \begin{cases} \sin \chi & \text{(for } k = +1) \\ \chi & \text{(for } k = 0) \\ \sinh \chi & \text{(for } k = -1). \end{cases} \tag{10.9}$$

For $k = 0$, the space is the familiar, flat, Euclidian 3-space; the homogeneity and isotropy of this space are obvious. For $k = 1$, Eq. (10.8) represents a 3-sphere of radius a embedded in an abstract flat four-dimensional Euclidian space. Such a 3-sphere is defined by the relation

$$x_1^2 + x_2^2 + x_3^2 + x_4^2 = a^2, \tag{10.10}$$

where (x_1, x_2, x_3, x_4) are the Cartesian coordinates of an abstract four-dimensional space. We can introduce angular coordinates (χ, θ, ϕ) on the 3-sphere by the relations

$$x_1 = a \, \cos \chi \sin \theta \sin \phi; \quad x_2 = a \cos \chi \sin \theta \cos \phi;$$
$$x_3 = a \, \cos \chi \cos \theta; \quad x_4 = a \sin \chi. \tag{10.11}$$

The metric on the 3-sphere can be determined by expressing the dx_is in terms of $d\chi, d\theta$ and $d\phi$ and substituting in the line element

$$dL^2 = dx_1^2 + dx_2^2 + dx_3^2 + dx_4^2. \tag{10.12}$$

This leads to the metric

$$dL^2_{(3\text{-sphere})} = a^2[d\chi^2 + \sin^2 \chi(d\theta^2 + \sin^2 \theta d\phi^2)], \tag{10.13}$$

which is the same as Eq. (10.8) for $k = 1$. The entire 3-space of the $k = 1$ model is covered by the range of angles $[0 \le \chi \le \pi; 0 \le \theta \le \pi; 0 \le \phi < 2\pi]$ and has a finite volume:

$$V = \int_0^{2\pi} d\phi \int_0^\pi d\theta \int_0^\pi d\chi \sqrt{g} = a^3 \int_0^{2\pi} d\phi \int_0^\pi \sin \theta d\theta \int_0^\pi \sin^2 \chi d\chi = 2\pi^2 a^3. \tag{10.14}$$

The surface area of a 2-sphere, defined by χ=constant, is $S = 4\pi a^2 \sin^2 \chi$. As χ increases, S increases at first, reaches a maximum value of $4\pi a^2$ at $\chi = \pi/2$ and *decreases* thereafter. These are the properties of a 3-space which is closed but has no boundaries.

In the case of $k = -1$, Eq. (10.8) represents the geometry of a hyperboloid embedded in an *abstract* four-dimensional space with Lorentzian signature. (This space should not be confused with the physical spacetime.) Such a space is described by the line element

$$dL^2 = dx_1^2 + dx_2^2 + dx_3^2 - dx_4^2. \tag{10.15}$$

A three-dimensional hyperboloid, embedded in this space, is defined by the relation

$$x_4^2 - x_1^2 - x_2^2 - x_3^2 = a^2. \tag{10.16}$$

This 3-space can be parameterized by the coordinates (χ, θ, ϕ) with

$$x_1 = a \sinh \chi \sin \theta \sin \phi; \quad x_2 = a \sinh \chi \sin \theta \cos \phi;$$
$$x_3 = a \sinh \chi \cos \theta; \quad x_4 = a \cosh \chi. \tag{10.17}$$

Expressing the dx_is in terms of $d\chi$, $d\theta$ and $d\phi$ and substituting into Eq. (10.15), the metric on the hyperboloid can be found to be

$$dL^2_{\text{(hyperboloid)}} = a^2 \left[d\chi^2 + \sinh^2 \chi (d\theta^2 + \sin^2 \theta d\phi^2) \right], \tag{10.18}$$

which is the same as Eq. (10.8) for $k = -1$.

To cover this 3-space, we need the range of coordinates to be $[0 \leq \chi \leq \infty;$ $0 \leq \theta \leq \pi; 0 \leq \phi < 2\pi]$. This space has infinite volume, just like the ordinary flat 3-space. The surface area of a 2-sphere, defined by $\chi = $ constant, is $S = 4\pi a^2 \sinh^2 \chi$. This expression increases monotonically with χ.

Friedmann universes with $k = -1, 0$ and 1 are called 'open', 'flat' and 'closed' respectively. These terms refer to the topological nature of the 3-space. The following point, however, should be noted: our symmetry considerations (and Einstein's equations) can only determine the local geometry of the spacetime and *not* its global topology. Consider, for example, the $k = 0$ model, which has infinite volume and spatial topology of $\mathcal{R}^3$, if we allow the coordinates to take the full possible range of values: $-\infty < (x, y, z) < +\infty$. We could, however, identify the points with coordinates (x, y, z) and $(x + L, y + L, z + L)$ thereby changing the topology of this space to that of a torus. Thus, our considerations do not uniquely specify the topology of the spacetime. The choices we have made for the three cases $k = 0, \pm 1$ discussed above should only be considered as the most natural choices for these models.

The full Friedmann metric in Eq. (10.6) can be expressed in either (t, r, θ, ϕ) coordinates or in (t, χ, θ, ϕ) coordinates. Sometimes it is convenient to use a different time coordinate τ related to t by $d\tau = a^{-1}(t)dt$. In the $(\tau, \chi, \theta, \phi)$ coordinates the Friedmann metric becomes

$$ds^2 = a^2(\tau) \left[-d\tau^2 + d\chi^2 + S_k^2(\chi)(d\theta^2 + \sin^2 \theta d\phi^2) \right] \tag{10.19}$$

with $S_k^2(\chi)$ given by Eq. (10.9). In this form, all the time dependence is isolated into an overall multiplicative factor.

Exercise 10.1

Friedmann model in spherically symmetric coordinates Express the Friedmann metric in a spherically symmetric coordinate system using a set of coordinates (T, R, θ, ϕ) in which the line interval must have the form

$$ds^2 = a^2(\tau)\left[-d\tau^2 + d\chi^2 + S_k^2(\chi)\left(d\theta^2 + \sin^2\theta d\phi^2\right)\right]$$
$$= -e^\nu dT^2 + e^\lambda dR^2 + R^2\left(d\theta^2 + \sin^2\theta d\phi^2\right), \qquad (10.20)$$

where $\nu(T, R)$ and $\lambda(T, R)$ are functions of T and R. (a) Show that the relevant transformation is given by $R = ra(t), T = F(q)$ where

$$q \equiv \int^r \frac{x\,dx}{1 - kx^2} + \int^t \frac{dy}{a(y)\,\dot{a}(y)} \qquad (10.21)$$

and F is an *arbitrary* function of its argument.

(b) Determine the Newtonian approximation of Friedmann metric in which the metric is given by

$$ds^2 \cong -(1 + 2\phi_N)\,dT^2 + dR^2 + R^2(d\theta^2 + \sin^2\theta d\phi^2), \qquad (10.22)$$

where ϕ_N is the Newtonian gravitational potential. Show that

$$\phi_N(\boldsymbol{R}, t) = -\frac{1}{2}\frac{\ddot{a}}{a}R^2 \qquad (10.23)$$

in the weak gravity limit, at scales $R^2 \ll (\ddot{a}/a)^{-1}$. Provide a physical interpretation for this potential.

Exercise 10.2
Conformally flat form of the metric   The metric of the $k = 0$ Friedmann universe can be expressed in the form $g_{ik} = \Omega^2\eta_{ik}$ where η_{ik} is the flat (Lorentzian) metric and $\Omega = \Omega(T)$. Even for the $k = \pm1$ models, the metric can be reduced to the form $g_{ik} = \Omega^2\eta_{ik}$ where $\Omega = \Omega(t, \boldsymbol{x})$ now depends on spatial coordinates as well. Construct this coordinate system. [Answer. Start with the metric in conformal time coordinate $(\tau, \chi, \theta, \phi)$ as given in the first line of Eq. (10.20). Now introduce the null coordinates $\xi = (1/2)(\tau + \chi)$ and $\eta = (1/2)(\tau - \chi)$; next introduce $X = \tan\chi, Y = \tan\eta$ in the case of $k = +1$ and $X = \tanh\chi, Y = \tanh\eta$ in the case of $k = -1$. Finally, introduce the coordinates α and β with $2\alpha = X + Y$ and $2\beta = X - Y$. The metric now becomes

$$ds^2 = \frac{4a^2}{(1 \pm X^2)(1 \pm Y^2)}\left[-d\alpha^2 + d\beta^2 + \beta^2(d\theta^2 + \sin^2\theta d\phi^2)\right] \qquad (10.24)$$

where the signs $\pm$ in the conformal factor are for $k = \pm1$.

10.3 Kinematics of the Friedmann model

Several features, related to the propagation of radiation and motion of material particles in the Friedmann universe, are independent of the explicit form of $a(t)$ and we discuss these first, before studying the dynamics of the Friedmann model.

10.3.1 The redshifting of the momentum

Since $a(t)$ multiplies the spatial coordinates, any proper distance $l(t)$ between spatial locations will change with time in proportion to $a(t)$:

$$l(t) = l_0 a(t) \propto a(t). \tag{10.25}$$

In particular, the proper separation between two observers, located at constant comoving coordinates, will change with time. Let the comoving separation between two such observers be δx, so that the proper separation is $\delta l = a(t)\delta x$. Each of the two observers will attribute to the other a velocity

$$\delta v = \frac{d}{dt}\delta l = \dot{a}\delta x = \left(\frac{\dot{a}}{a}\right)\delta l. \tag{10.26}$$

Consider now a narrow pencil of (nearly monochromatic) electromagnetic radiation which crosses these two comoving observers. The time for the transit will be $\delta t = \delta l$. Let the frequency of the radiation measured by the first observer be ω. Since the first observer sees the second one to be *receding* with velocity δv, (s)he will expect the second observer to measure a Doppler shifted frequency $(\omega + \delta \omega)$ where

$$\frac{\delta \omega}{\omega} = -\delta v = -\frac{\dot{a}}{a}\delta l = -\frac{\dot{a}}{a}\delta t = -\frac{\delta a}{a}. \tag{10.27}$$

(Since the observers are separated by an infinitesimal distance of first order, δl, we can introduce a locally inertial frame encompassing both observers and apply the laws of special relativity.) This equation can be integrated to give

$$\omega(t)a(t) = \text{constant}. \tag{10.28}$$

In other words, the frequency of electromagnetic radiation changes due to expansion of the universe according to the law $\omega \propto a^{-1}$.

The above analysis did not use the wave nature of the electromagnetic radiation directly, though it is implicit in the Doppler shift formula applied in the local frame. It is, of course, possible to obtain the same result directly from the solution to Maxwell's equations in the Friedmann geometry. The dynamics of the electromagnetic field in curved spacetime is described by the action

$$\mathcal{A}_{\text{elec}} = -\frac{1}{16\pi}\int F_{ik}F^{ik}\sqrt{-g}d^4x; \quad F_{ik} = \frac{\partial A_k}{\partial x^i} - \frac{\partial A_i}{\partial x^k}. \tag{10.29}$$

(See Eq. (5.117).) Consider the conformal transformation

$$A_i \to A_i; \quad x^i \to x^i; \quad g_{ik} \to \Omega^2 g_{ik}; \quad g^{ik} \to \Omega^{-2}g^{ik}, \tag{10.30}$$

where $\Omega(x^i)$ is an arbitrary non-singular function of the spacetime coordinates. Note that $(A^i)_{\text{new}} = g_{\text{new}}^{ik}(A_k)_{\text{new}} = \Omega^{-2}A_{\text{old}}^i$. As can be directly verified, the action $\mathcal{A}_{\text{elec}}$ is invariant under conformal transformations (see Exercise 5.15) which, in turn, implies that Maxwell's equations and their solutions will be conformally invariant. The Friedmann universe in the $(\tau, \chi, \theta, \phi)$ coordinate system is conformally flat with $g_{ik} = a^2(\tau) g_{ik}^S$, where g_{ik}^S is a static metric with components independent of τ. Since the electromagnetic field is conformally invariant, solutions to the wave equation for A_i in the metric g_{ik} will be the same as the solution in the static spacetime metric g_{ik}^S and will have the time dependence

$$A_i \propto \exp(-ik\tau) = \exp\left[-ik \int \frac{dt}{a(t)}\right]. \tag{10.31}$$

Because the time derivative of the phase of the wave defines the (instantaneous) frequency, we conclude that $\omega(t)a(t) = \text{constant}$.

The above result is of considerable practical importance. It shows that if the radiation is emitted by some source at time t_e and is observed at time $t = t_o$, then its wavelength will increase if $a(t_o) > a(t_e)$. In such an expanding phase of the universe, where $a(t)$ is an increasing function of time, we can associate a redshift z with any time t in the past by the relation

$$1 + z(t) \equiv \frac{a(t_0)}{a(t)} \equiv \frac{a_0}{a(t)}, \tag{10.32}$$

where t_0 is the value of time parameter at the current epoch and a_0 is the value of expansion factor at present. Thus one can use the variables t, a or z interchangeably in a universe in which $a(t)$ is a monotonically increasing function of time.

The expansion of the universe also affects the motion of material particles. Consider again two comoving observers separated by proper distance δl. Let a material particle pass the first observer with velocity v. When it has crossed the proper distance δl (in a time interval δt), it passes the second observer whose velocity (relative to the first one) is

$$\delta u = \frac{\dot{a}}{a}\delta l = \frac{\dot{a}}{a}v dt = v\frac{\delta a}{a}. \tag{10.33}$$

The second observer will attribute to the particle the velocity

$$v' = \frac{v - \delta u}{1 - v\delta u} = v - (1 - v^2)\delta u + \mathcal{O}[(\delta u)^2] = v - (1 - v^2)v\frac{\delta a}{a}. \tag{10.34}$$

This follows from the special relativistic formula for addition of velocities which is valid in an infinitesimal region around the first observer. Rewriting this equation as

$$\delta v = -v(1 - v^2)\frac{\delta a}{a} \tag{10.35}$$

and integrating, we get

$$p = \frac{v}{\sqrt{1 - v^2}} = \frac{\text{constant}}{a}. \tag{10.36}$$

In other words, the magnitude of the three-momentum decreases as a^{-1} due to the expansion. If the particle is non-relativistic, then $v \propto p$ and the velocity itself decays as a^{-1}. This result can also, of course, be derived from the study of the geodesics in the Friedmann universe (see Exercise 10.4).

Exercise 10.3
Particle velocity in the Friedmann universe   A material particle is released with some initial velocity in a Friedmann model. (a) Show that, in the $k = -1$ model, as $t \to \infty$, the velocity of the particle approaches that of some fundamental observer but the position of the particle will be at a constant proper distance from this observer. (b) What happens in the $k = 0$ and $k = +1$ models?

Exercise 10.4
Geodesic equation in the Friedmann universe   Derive the result $p(t) \propto a(t)^{-1}$ by studying the geodesic equation in the Friedmann universe. (a) Consider a particle travelling along $\theta = $ constant, $\phi = $ constant. Show that the zeroth component of the geodesic equation reads as

$$\frac{d^2t}{ds^2} + \frac{a\dot{a}}{1 - kr^2} \left(\frac{dr}{ds} \right)^2 = 0. \tag{10.37}$$

(b) Eliminate (dr/ds) between the above equation and the first integral

$$\left(\frac{dt}{ds} \right)^2 - \frac{a^2}{1 - kr^2} \left(\frac{dr}{ds} \right)^2 = 1 \tag{10.38}$$

to obtain

$$\frac{d^2t}{ds^2} + \frac{\dot{a}}{a} \left[\left(\frac{dt}{ds} \right)^2 - 1 \right] = 0. \tag{10.39}$$

(c) Integrate this equation to obtain $a[(dt/ds)^2 - 1] = $ constant. If $p^a = (dx^a/ds)$ is the four-velocity of the particle, then the condition $p^a p_a = -1$ reduces to $[(dt/ds)^2 - \sigma_{\alpha\beta} p^\alpha p^\beta] = 1$. Show that $\sigma_{\alpha\beta} p^\alpha p^\beta \equiv |\mathbf{p}|^2 = (\text{constant}/a^2)$.

(d) The geodesic for a particle, moving in a spacetime with metric g_{ab}, can be obtained most efficiently from the Hamilton–Jacobi equation

$$g^{ab} \frac{\partial \mathcal{A}}{\partial x^a} \frac{\partial \mathcal{A}}{\partial x^b} = -m^2, \tag{10.40}$$

where $\mathcal{A}$ is the action. Write down this equation in the Friedmann metric and reduce the problem of determining the radial geodesics to quadrature.

Exercise 10.5
Generalized formula for photon redshift   This exercise generalizes the formula for the redshift of photons to a spacetime which is slightly different from that of a Friedmann

model and is described by the line interval

$$ds^2 = -dt^2 + a^2(t)\left[\delta_{\alpha\beta} - h_{\alpha\beta}(t,x)\right]dx^\alpha dx^\beta. \tag{10.41}$$

A photon emitted by a distant source propagates through this spacetime and is received by an observer at the present epoch. Repeating the analysis in the text, show that in this case

$$\frac{\delta\omega}{\omega} = -\frac{\delta a}{a} + \frac{1}{2}\dot{h}_{\alpha\beta}n^\alpha n^\beta \delta t. \tag{10.42}$$

Integrate this to obtain

$$\ln\left(\frac{\omega_2 a_2}{\omega_1 a_1}\right) = \frac{1}{2}\int_{t_1}^{t_2} \dot{h}_{\alpha\beta}n^\alpha n^\beta dt, \tag{10.43}$$

where the time integration is along the path of the ray. This result is useful for determining the effect of small inhomogeneities on the redshift of photons.

Exercise 10.6
Electromagnetism in the closed Friedmann universe Show that the closed Friedmann universe cannot contain net electric charge. Can it contain source-free electromagnetic waves of any frequency?

10.3.2 Distribution functions for particles and photons

One conclusion that can be reached immediately from Eq. (10.36) is the following. Consider a stream of particles propagating freely in the spacetime. At some time t, a comoving observer finds dN particles in a proper volume dV, all having momentum in the range $(p, p + d^3p)$. The phase space distribution function $f(x, p, t)$ for the particles is defined by the relation $dN = f\, dVd^3p$. At a later instant $(t + \delta t)$ the proper volume occupied by these particles would have increased by a factor $[a(t + \delta t)/a(t)]^3$ while the volume in the momentum space will be redshifted by $[a(t)/a(t + \delta t)]^3$, showing that the phase volume occupied by the particles does not change during the free propagation. Since the number of particles dN is also conserved, it follows that f is conserved along the streamline.

If we treat the electromagnetic radiation as consisting of photons with zero rest mass, for which $E = \hbar\omega = p$. The decay law $p \propto a^{-1}$ implies the redshift for radiation derived earlier. A few other properties of radiation can also be derived very easily by using the photon concept. Let $f_\gamma(t, x, p)$ be the phase space density of photons that is conserved during the propagation. This conserved number of photons per unit phase space volume, f_γ, can be expressed in the form

$$f_\gamma(t, x, p) = \frac{dN_\gamma}{d^3x d^3p} = \frac{dN_\gamma}{[dt_e dA_e][\omega_e^2 d\omega_e d\Omega_e]} = \frac{dN_\gamma}{[dt_r dA_r][\omega_r^2 d\omega_r d\Omega_r]}. \tag{10.44}$$

We have written the momentum space volume as $d^3p \propto p^2 dp d\Omega \propto \omega^2 d\omega d\Omega$, where $d\Omega$ is a solid angle around the direction of propagation, and $d^3x \propto dt dA$, where dA is the area normal to the direction of propagation. The subscripts e and r represent the processes of emission and reception of photons. This shows that

$$\frac{dN}{dt dA d\omega d\Omega} \frac{1}{\omega^2} = \text{invariant.} \tag{10.45}$$

Since the energy is related to the number of photons by $dE = \hbar \omega dN$ and the intensity is defined by $I \equiv (dE/dt dA d\omega d\Omega)$, it immediately follows that

$$I_{\text{rec}} = \left(\frac{\omega_{\text{rec}}}{\omega_{\text{em}}}\right)^3 I_{\text{em}}. \tag{10.46}$$

In other words, (I/ω^3) is invariant. This quantity $I(\omega)$ will have units (for example) erg s^{-1} cm^{-2} Hz^{-1}sr^{-1}. The energy density $U(\omega)$ is defined as $U = (4\pi/c)I$ and will have the dimensions erg cm^{-3} Hz^{-1}. It is clear from the invariance of $[I/\omega^3]$ that I and U vary as a^{-3} in the expanding universe. More precisely, taking the redshift of the frequency into account,

$$I[\omega(1+z); z] = I[\omega; 0](1+z)^3. \tag{10.47}$$

The total flux of radiation, obtained by integrating over all frequencies, varies as $(1+z)^4$:

$$F_{\text{total}} = \int_0^\infty I \, d\omega \propto (1+z)^4. \tag{10.48}$$

Radiation that has an intensity distribution of the form $I(\omega) = \omega^3 G(\omega/T)$ will retain the spectral shape during the expansion, with the parameter T varying with expansion as $T \propto a^{-1}$. The Planck spectrum has this form in which T corresponds to the temperature; it follows that the temperature of the radiation, which has a Planck spectrum, decreases with expansion as $T(t) \propto a(t)^{-1}$ even in the absence of any scattering or thermalization process.

10.3.3 Measures of distance

Cosmological observations are mostly based on electromagnetic radiation that is received from far away sources. Let an observer, located at $r = 0$, receive at time $t = t_0$ radiation from a source located at $r = r_e$. This radiation must have been emitted at some earlier time t_e such that the events (t_e, r_e) and $(t_0, 0)$ are connected by a null geodesic. Taking the propagation of the ray to be along $\theta = $ constant, $\phi = $ constant, we can write the equation for the null geodesic to be

$$0 = ds^2 = dt^2 - a^2(t) \frac{dr^2}{1 - kr^2}. \tag{10.49}$$

Integrating this, we can find the relation between r_e and t_e:

$$\int_{t_e}^{t_0} \frac{dt}{a(t)} = \int_0^{r_e} \frac{dr}{(1 - kr^2)^{1/2}}. \tag{10.50}$$

The left hand side is a definite function of time in a given cosmological model. Since the redshift z is also a unique function of time, we can express the left hand side as a function of z and hence r_e can be expressed as a function of z. This function $r_e(z)$ is of considerable use in observational astronomy, since it relates the radial distance of an object (from which the light is received) to the redshift at which the light is emitted. We shall see later that Einstein's equations allow us to determine $(\dot{a}/a)$ more easily than $a(t)$ itself. It is, therefore, convenient to define a quantity (called the *Hubble radius*) by

$$d_H(t) = d_H(z) \equiv \left(\frac{\dot{a}}{a}\right)^{-1}, \tag{10.51}$$

where the first equality shows that d_H can also be thought of as a function of the redshift z. This quantity allows us to convert integrals over dt to integrals over dz by using

$$dt = \left(\frac{dt}{da}\right)\left(\frac{da}{dz}\right)dz = -d_H(z)\left(\frac{dz}{1+z}\right). \tag{10.52}$$

It is now possible to write Eq. (10.50) as

$$\frac{1}{a_0}\int_0^z d_H(z)dz = S_k^{-1}(r_{\rm em}), \tag{10.53}$$

where $S_k^{-1}(x) = (\sinh^{-1}(x), x, \sin^{-1}(x))$ for $k = -1, 0, +1$ respectively (see Eq. (10.9)). Therefore $r_{\rm em}(z)$ can be written as

$$r_{\rm em}(z) = S_k(\alpha); \qquad \alpha \equiv \frac{1}{a_0}\int_0^z d_H(z)dz. \tag{10.54}$$

The $r_{\rm em}(z)$ plays an important role (i) in relating the luminosity of distant objects with the observed flux and (ii) in measuring angular sizes of distant objects. We shall now discuss these relations which are of considerable practical importance. Let $\mathcal{F}$ be the flux received from a source of luminosity $\mathcal{L}$ when the photons from the source reach us with a redshift z. The flux can be expressed as

$$\mathcal{F} = \frac{1}{(\text{Area})}\left(\frac{dE_{\rm rec}}{dt_{\rm rec}}\right). \tag{10.55}$$

Using Eq. (10.46) and the fact that the proper area at $a = a_0$, of a sphere of comoving radius $r_{\rm em}(z)$, is $4\pi a_0^2 r_{\rm em}^2$, we get

$$\mathcal{F} = \frac{1}{4\pi a_0^2 r_{\rm em}^2}\frac{1}{(1+z)^2}\left(\frac{dE_{\rm em}}{dt_{\rm em}}\right) = \frac{1}{4\pi a_0^2 r_{\rm em}^2}\frac{1}{(1+z)^2}\mathcal{L}, \tag{10.56}$$

where $\mathcal{L} = (dE_{em}/dt_{em})$ is the luminosity of the source. Since the distance to an object of luminosity $\mathcal{L}$ and flux $\mathcal{F}$ can be expressed as $(\mathcal{L}/4\pi\mathcal{F})^{1/2}$ in flat Euclidian geometry, it is convenient to define distance $d_L(z)$ called *luminosity distance* by

$$d_L(z) \equiv \left(\frac{\mathcal{L}}{4\pi\mathcal{F}}\right)^{1/2} = a_0 r_{em}(z)(1+z) = a_0(1+z)S_k(\alpha). \tag{10.57}$$

Note that, from Eq. (10.54) and Eq. (10.57) we have

$$\alpha = \frac{1}{a_0}\int_0^z d_H(z)\,dz = S_k^{-1}\left[\frac{d_L(z)}{a_0(1+z)}\right]. \tag{10.58}$$

Differentiating with respect to z gives

$$d_H(z) = \left[1 - \frac{k d_L^2(z)}{a_0^2(1+z)^2}\right]^{-1/2}\frac{d}{dz}\left[\frac{d_L(z)}{1+z}\right]. \tag{10.59}$$

Thus if $d_L(z)$ can be determined from the observations, one can determine $d_H(z)$ and thus $(\dot{a}/a)$.

Another observable parameter for distant sources is the angular diameter. If D is the physical size of the object which subtends an angle δ to the observer, then, for small δ, we have $D = r_{em}a(t_e)\delta$. The 'angular diameter distance' $d_A(z)$ for the source is defined via the relation $\delta = (D/d_A)$; so we find that

$$d_A(z) = r_{em}a(t_e) = a_0 r_{em}(t_e)(1+z)^{-1}. \tag{10.60}$$

Quite clearly $d_L = (1+z)^2 d_A$.

The proper volume of the universe spanned by the region between comoving radii r_{em} and $r_{em} + dr_{em}$ can also be expressed in terms of $d_H(z)$ and $r_{em}(z)$. This infinitesimal proper volume element is given by

$$dV = a^3 \frac{dr_{em}}{\sqrt{1 - k r_{em}^2}} r_{em}^2 \sin\theta d\theta d\phi. \tag{10.61}$$

If we interpret this volume as the volume of the universe spanned along the backward light cone by the photons which are received today with redshifts between z and $z + dz$, then we can relate dr_{em} to dt_{em} by Eq. (10.50) and obtain

$$dV = a^2 dt\left(r_{em}^2 \sin\theta d\theta d\phi\right) = \frac{a_0^2 r_{em}^2(z) d_H(z)}{(1+z)^3} dz \sin\theta d\theta d\phi. \tag{10.62}$$

To proceed further with any of these relations like Eq. (10.57), Eq. (10.60) or Eq. (10.62), we need to know the form of $d_H(z)$, which – in turn – depends on the functional form of $a(t)$. If, however, we are only interested in small z (i.e. in a small time interval $(t_0 - t_e)$) then we can Taylor expand $a(t)$ around t_0 and parametrize it by the coefficients of the Taylor expansion. It is conventional to write this expansion in the following form:

$$a(t) = a(t_0) \left[1 + \left(\frac{\dot{a}}{a} \right)_0 (t - t_0) + \frac{1}{2} \left(\frac{\ddot{a}}{a} \right)_0 (t - t_0)^2 + \cdots \right]$$

$$= a(t_0) \left[1 + H_0(t - t_0) - \frac{1}{2} q_0 H_0^2 (t - t_0)^2 + \cdots \right], \qquad (10.63)$$

where H_0 is called the Hubble constant and q_0 is called the deceleration parameter, and they are defined by

$$H_0 \equiv \left(\frac{\dot{a}}{a} \right)_{t=t_0} ; \qquad q_0 \equiv - \left(\frac{\ddot{a} a}{\dot{a}^2} \right)_{t=t_0}. \qquad (10.64)$$

Substituting this expansion in Eq. (10.50), expanding up to quadratic order in $(t - t_0)$ and r_{em} and integrating, we get the relation between r_{em} and t_e:

$$r_{em} = \frac{1}{a_0} \left[(t_0 - t_e) + \frac{1}{2} H_0 (t_e - t_0)^2 + \cdots \right]. \qquad (10.65)$$

Inverting Eq. (10.63), we can express $(t - t_0)$ in terms of $(1 + z) = (a_0/a)$:

$$(t - t_0) = -H_0^{-1} \left[z - (1 + \frac{q_0}{2}) z^2 + \cdots \right]. \qquad (10.66)$$

Finally, substituting Eq. (10.66) into Eq. (10.65) we can express r_{em} in terms of z:

$$a_0 r_{em} = H_0^{-1} \left[z - \frac{1}{2}(1 + q_0) z^2 + \cdots \right]. \qquad (10.67)$$

This shows that, to first order in z, the 'redshift velocity' $v \equiv cz$ is proportional to the proper distance $a_0 r_{em}$. We can now use this relation to express d_L (and d_A) in terms of z; we find, for example,

$$d_L(z) = a_0 r_{em}(1 + z) = H_0^{-1} \left[z + \frac{1}{2}(1 - q_0) z^2 + \cdots \right]. \qquad (10.68)$$

The quantity d_L can be determined by measuring the flux $\mathcal{F}$ for a class of objects for which the intrinsic luminosity $\mathcal{L}$ is known. If we can also measure the redshift z for these objects, then a plot of d_L against z will allow us to determine the parameter H_0. Observations suggest that

$$H_0 = 100h \, \text{km s}^{-1} \text{Mpc}^{-1}; \qquad h \simeq 0.7. \qquad (10.69)$$

From this Hubble constant H_0, which determines the rate at which the universe is expanding today, we can construct the timescale, $t_{univ} \equiv H_0^{-1} = 9.8 \times 10^9 h^{-1}$ yr and the length scale $l_{univ} \equiv ct_{univ} \cong 3000h^{-1}$Mpc. These are the characteristic scales over which global effects of cosmological expansion will be important today.

10.4 Dynamics of the Friedmann model

We shall now consider the evolution of the Friedmann model for different kinds of sources which could populate the universe. The Friedmann metric contains a constant k and a function $a(t)$, both of which can be determined via Einstein's equations

$$G^i_k = R^i_k - \frac{1}{2}\delta^i_k R = \kappa T^i_k; \qquad \kappa \equiv 8\pi G \qquad (10.70)$$

if the energy-momentum tensor for the source is specified. The assumption of isotropy implies that T^μ_0 must be zero and that the spatial components T^α_β must have a diagonal form with $T^1_1 = T^2_2 = T^3_3$; homogeneity requires all the components to be independent of the spatial coordinate x. It is conventional to write such an energy-momentum tensor as

$$T^i_k = \text{dia}[-\rho(t), p(t), p(t), p(t)]. \qquad (10.71)$$

The notation is suggested by the fact that if the source was an ideal fluid with energy density ρ and pressure p, then the energy-momentum tensor will have the above form (see Eq. (1.114)). Of course, there is no need for the source to be an ideal fluid and hence this notation is only suggestive. (In particular, there is no restriction that p should be positive.) The tensor G^i_k on the left hand side of Eq. (10.70) can be computed for the Friedmann metric in a straightforward manner. The nontrivial components are:

$$G^0_0 = -\frac{3}{a^2}(\dot{a}^2 + k), \quad G^\mu_\nu = -\frac{1}{a^2}(2a\ddot{a} + \dot{a}^2 + k)\delta^\mu_\nu. \qquad (10.72)$$

Thus Eq. (10.70) gives two independent equations

$$\frac{\dot{a}^2 + k}{a^2} = \frac{8\pi G}{3}\rho, \qquad (10.73)$$

$$\frac{2\ddot{a}}{a} + \frac{\dot{a}^2 + k}{a^2} = -8\pi Gp. \qquad (10.74)$$

These two equations, combined with the equation of state $p = p(\rho)$, completely determine the three functions $a(t)$, $\rho(t)$ and $p(t)$. We shall first discuss some general features of these equations.

We begin by stressing that Eq. (10.73) and Eq. (10.80) must be treated as equations resulting from general relativity. An attempt is sometimes made in literature to give a (pseudo) Newtonian interpretation of, for example, Eq. (10.73) in terms of Newtonian concepts, treating $(1/2)\dot{a}^2$ as kinetic energy and $(-4\pi G/3)\rho a^2$ as potential energy. Neither of these identifications has any physical justification and the similarity of Friedmann equations to Newtonian equations is purely an accident devoid of fundamental significance. Cosmology cannot be formulated in terms of Newtonian concepts (and non-relativistic expressions for kinetic and potential energies); such interpretations are misleading.

From Eq. (10.73) it follows that

$$\frac{k}{a^2} = \frac{8\pi G}{3}\rho - \frac{\dot{a}^2}{a^2} = \frac{\dot{a}^2}{a^2}\left[\frac{\rho}{(3H^2/8\pi G)} - 1\right]; \qquad H(t) \equiv \frac{\dot{a}}{a}, \qquad (10.75)$$

which suggests defining *critical density* $\rho_c(t)$ and a *density parameter* $\Omega(t)$ by

$$\rho_c(t) \equiv \frac{3H^2(t)}{8\pi G}; \qquad \Omega(t) \equiv \frac{\rho}{\rho_c}. \qquad (10.76)$$

These definitions are valid at any time t. Using these definitions and evaluating Eq. (10.75) at $t = t_0$, we get

$$\frac{k}{a_0^2} = H_0^2\left(\Omega_0 - 1\right). \qquad (10.77)$$

It is clear that $k = -1, 0$ or 1 depending on whether $\Omega_0 < 1, \Omega_0 = 1$ or $\Omega_0 > 1$ respectively. Thus Ω_0 determines the spatial geometry of the universe. When $k \neq 0$, we can find a_0 from Eq. (10.77) in terms of H_0 and Ω_0:

$$a_0 = H_0^{-1}\left(|\Omega_0 - 1|\right)^{-1/2}. \qquad (10.78)$$

The value of $\dot{a}_0$ can be fixed by

$$\dot{a}_0 = H_0 a_0 = \left(|\Omega_0 - 1|\right)^{-1/2}. \qquad (10.79)$$

Hence (H_0, Ω_0) are valid initial conditions for integrating equations Eq. (10.73), Eq. (10.74). If $k = 0$, Einstein's equations allow the scaling $a \rightarrow \mu a$, making the normalization of a arbitrary. It is conventional to take $a_0 = 1$ if $k = 0$; then the value of $(\dot{a}_0/a_0) = H_0 = \dot{a}_0$ determines $\dot{a}_0$.

Substituting for $\left(\dot{a}^2 + k\right)/a^2$ in Eq. (10.74) from Eq. (10.73) and rearranging we get:

$$\frac{\ddot{a}}{a} = -\frac{4\pi G}{3}(\rho + 3p). \qquad (10.80)$$

(This is essentially the same as Eq. (6.105) specialized to the Friedmann metric.) For normal matter, $(\rho + 3p) > 0$ implying that $\ddot{a} < 0$; i.e. the universe will have a decelerating expansion. Further, the $a(t)$ curve (which has positive $\dot{a}$ at the present epoch t_0) must be convex; in other words, a will be smaller in the past and will become zero at sometime in the past, say, at $t = t_{\text{sing}}$. It is also clear that $(t_0 - t_{\text{sing}})$ must be less than the value of the intercept $(\dot{a}/a)_0^{-1} = H_0^{-1}$. For convenience, we will choose the time coordinate such that $t_{\text{sing}} = 0$, i.e. we take $a = 0$ at $t = 0$. In that case, the present 'age' of the universe t_0 satisfies the inequality $t_0 < t_{\text{univ}}$ where

$$t_{\text{univ}} \equiv H_0^{-1} = 3.1 \times 10^{17} h^{-1}\text{s} = 9.8 \times 10^9 h^{-1}\text{yr}. \qquad (10.81)$$

As a becomes smaller the components of the curvature tensor $R^i{}_{klm}$ become larger and when $a = 0$ these components diverge. Such a singularity in the spacetime is an artefact indicating the breakdown of our theory. When the radius of curvature of the spacetime becomes comparable to the fundamental length $(G\hbar/c^3)^{1/2} \simeq 10^{-33}$ cm constructed out of G, $\hbar$ and c, quantum effects of gravity will become important, rendering the classical Einstein's equations invalid. (Of course, the theory might break down even earlier for unknown reasons.) So, in reality, t_0 is the time that has elapsed from the moment at which Einstein's equations became valid.

The quantities ρ and p are *defined* in Eq. (10.71) as the T_0^0 and T_1^1 (say) components of the energy-momentum tensor. The interpretation of p as 'pressure' depends on treating the source as an ideal fluid. The source for a Friedmann model should *always* have the form in Eq. (10.71); but, as we said before, if the source is not an ideal fluid then it is not possible to interpret the spatial components of T_k^i as pressure. It is, therefore, quite possible that the equation of state for matter at high energies do not obey the condition $(\rho + 3p) > 0$. The violation of this condition may occur at a later epoch (i.e. at larger value of a) than the epoch at which the quantum gravitational effects become important. If this happens, then the 'age of the universe' refers to the time interval since the breakdown of the condition $(\rho + 3p) > 0$.

We shall now consider the question of explicitly solving Einstein's equations. From Eq. (10.73), we see that $\rho a^3 = (3/8\pi G)a(\dot{a}^2 + k)$; differentiating this expression and using Eq. (10.74) we get

$$\frac{d}{da}(\rho a^3) = -3a^2 p. \tag{10.82}$$

Given the equation of state $p = p(\rho)$, we can integrate Eq. (10.82) to obtain $\rho = \rho(a)$. Substituting this relation into Eq. (10.73) we can determine $a(t)$.

For an equation of state of the form $p = w\rho$ with a constant w, Eq. (10.82) gives $\rho \propto a^{-3(1+w)}$; in particular, for non-relativistic matter $(w = 0)$ and radiation $(w = 1/3)$ we find $\rho_{NR} \propto a^{-3}$ and $\rho_R \propto a^{-4}$. If $w = -1$, then we find that $\rho = $ constant as the universe expands. In this case the pressure $p = -\rho$ is negative (since we must have $\rho > 0$ to maintain $(\dot{a}^2/a^2 > 0)$) and the negative pressure allows for the energy inside a volume to increase even when the volume expands. If $w = 1$, then $p = \rho$ and $\rho \propto a^{-6}$. This is called a 'stiff' equation of state. In such a medium, the 'speed of sound' $(\partial p/\partial \rho) = 1$ is the same as the speed of light.

The corresponding time evolution of $a(t)$ is easy to determine for $k = 0$. For $\rho \propto a^{-3(1+w)}$, the Friedmann equation becomes $(\dot{a}^2/a^2) \propto a^{-3(1+w)}$ or

$$\dot{a} \propto a^{-\frac{1}{2}(1+3w)}. \tag{10.83}$$

Integrating, we find:

$$a(t) \propto t^{\frac{2}{3(1+w)}} \quad \text{(for } w \neq -1\text{)}$$

$$\propto \exp(\lambda t) \quad \text{(for } w = -1\text{)}, \tag{10.84}$$

where λ is a constant. For $w = 0$, $a \propto t^{2/3}$; for $w = 1/3$, $a \propto t^{1/2}$; and for $w = 1$, $a \propto t^{1/3}$.

Some particular values of w mentioned above are of special importance. We already know that non-relativistic matter has $w = 0$, while $w = (1/3)$ corresponds to all relativistic species including radiation. The cases with $w = \pm 1$ arise naturally in the case of scalar fields which we shall briefly mention. A scalar field ϕ with potential $V(\phi)$ can be described by a Lagrangian of the form (see Eq. (2.22))

$$L = -\frac{1}{2} \partial_i \phi \, \partial^i \phi - V(\phi). \tag{10.85}$$

The energy-momentum tensor for this scalar field is (see Eq. (2.28))

$$T_{ik} = \partial_i \phi \, \partial_k \phi + g_{ik} L. \tag{10.86}$$

In a homogeneous universe, $\phi(t, x) = \phi(t)$ and only the diagonal components of $T^i_{\ k}$ remain nonzero:

$$-T^0_{\ 0} = \frac{1}{2} \dot{\phi}^2 + V(\phi); \quad T^1_1 = T^2_2 = T^3_3 = \frac{1}{2} \dot{\phi}^2 - V \tag{10.87}$$

giving

$$\rho = -T^0_{\ 0} = \frac{1}{2} \dot{\phi}^2 + V(\phi); \quad p = \frac{1}{2} \dot{\phi}^2 - V(\phi). \tag{10.88}$$

When the kinetic energy $(\dot{\phi}^2/2)$ of the field dominates over the potential energy $V(\phi)$, we get the equation of state $p = \rho$. If the potential $V(\phi)$ dominates over the kinetic energy $(\dot{\phi}^2/2)$ then $p = -\rho$. Thus the scalar field can exhibit both these equations of state in appropriate ranges.

The equation of state $p = -\rho$ also arises if the Einstein's equations are modified by adding a term $-\Lambda \delta^i_k$ on the left hand side (see Eq. (6.70)). For historical reasons the Λ introduced by such a modification is called a *cosmological constant* (see Chapter 6). Since this is completely equivalent to adding a term $\Lambda \delta^i_k$ in the right hand side as a source, we shall take the point of view that any cosmological constant can be treated as a special kind of source. Taking all these possibilities into account, the contents of the universe can be taken to be made of: (i) non-relativistic matter, (ii) relativistic matter and (iii) a cosmological constant which is also sometimes called the *vacuum energy density*.

We can use Eq. (10.76) to define the density parameter for each component of energy density in the universe, with the variables $\Omega_i \equiv \rho_i/\rho_c$ giving the fractional contribution of different components of the universe (i denoting different components) to the critical density. Observations then lead to the following results.[3]

(1) Our universe has $0.98 \lesssim \Omega_{\text{tot}} \lesssim 1.02$. The value of Ω_{tot} can be determined from the angular anisotropy spectrum of the cosmic microwave background radiation (CMBR) and these observations (combined with the reasonable assumption that $h > 0.5$) show that we live in a universe with critical density, so that $k = 0$.

(2) Observations of primordial deuterium produced in big bang nucleosynthesis (which took place when the universe was about a few minutes in age) as well as the CMBR observations show that the *total* amount of baryons in the universe contributes about $\Omega_{\text{B}} = (0.024 \pm 0.0012)h^{-2}$. Given the independent observations which fix $h = 0.72 \pm 0.07$, we conclude that $\Omega_{\text{B}} \cong 0.04$–$0.06$. These observations take into account all baryons that exist in the universe today irrespective of whether they are luminous or not. *Combined with previous item we conclude that most of the universe is non-baryonic.*

(3) A host of observations related to large scale structure and dynamics (rotation curves of galaxies, estimate of cluster masses, gravitational lensing, galaxy surveys ...) all suggest that the universe is populated by a non-luminous component of matter (dark matter; DM hereafter) made of weakly interacting massive particles that *does* cluster at galactic scales. This component contributes about $\Omega_{\text{DM}} \cong 0.20$–$0.35$ and has the simple equation of state $p_{\text{DM}} \approx 0$. Therefore, $\rho_{\text{DM}} \propto a^{-3}$ as the universe expands, which arises from the evolution of number density of particles: $\rho = nmc^2 \propto n \propto a^{-3}$.

(4) Combining the last observation with the first we conclude that there must be (at least) one more component to the energy density of the universe contributing about 70% of critical density. Early analysis of several observations indicated that this component is unclustered and has negative pressure. This is confirmed dramatically by later supernova observations. The observations suggest that the missing component has $w = p/\rho \lesssim -0.78$ and contributes $\Omega_{\text{DE}} \cong 0.60$–$0.75$. The simplest choice for such *dark energy* (DE) with negative pressure is the cosmological constant which acts like a fluid with an equation of state $p_{\text{DE}} = -\rho_{\text{DE}}$.

(5) The universe also contains radiation contributing an energy density $\Omega_{\text{R}}h^2 = 2.56 \times 10^{-5}$ today, most of which is due to photons in the CMBR. The equation of state is $p_{\text{R}} = (1/3)\rho_{\text{R}}$; so that $\rho_{\text{R}} \propto a^{-4}$. Combining it with the result $\rho_{\text{R}} \propto T^4$ for thermal radiation, it follows that $T \propto a^{-1}$. Radiation is dynamically irrelevant today but since $(\rho_{\text{R}}/\rho_{\text{DM}}) \propto a^{-1}$ it would have been the dominant component when the universe was smaller by a factor that is larger than $\Omega_{\text{DM}}/\Omega_{\text{R}} \simeq 4 \times 10^4 \Omega_{\text{DM}}h^2$.

Taking all the above observations together, we conclude that our universe has (approximately) $\Omega_{\text{DE}} \simeq 0.7, \Omega_{\text{DM}} \simeq 0.26, \Omega_{\text{B}} \simeq 0.04, \Omega_{\text{R}} \simeq 5 \times 10^{-5}$. All known observations are consistent with such an – admittedly weird – composition for the universe.

Based on the equation of state we can group these components as: (a) non-relativistic matter with $\Omega_{\rm NR} = \Omega_{\rm DM} + \Omega_{\rm B}$ and $p_{\rm NR} = 0$; (b) relativistic matter with $p_{\rm R} = (1/3)\rho_{\rm R}$; and (c) dark energy modelled by the cosmological constant with $\Omega_{\rm DE} = \Omega_\Lambda$ and $p_{\rm DE} = -\rho_{\rm DE}$. Given that the energy densities of the three components vary as $\rho_{\rm NR} \propto a^{-3}$, $\rho_{\rm R} \propto a^{-4}$ and $\rho_{\rm DE} = \rho_V = $ constant, as the universe evolves, the total energy density in the universe can be expressed as

$$\rho_{\rm total}\,(a) = \rho_{\rm R}\,(a) + \rho_{\rm NR}\,(a) + \rho_{\rm DE}\,(a)$$

$$= \rho_c \left[\Omega_{\rm R} \left(\frac{a_0}{a}\right)^4 + (\Omega_{\rm B} + \Omega_{\rm DM}) \left(\frac{a_0}{a}\right)^3 + \Omega_\Lambda \right], \qquad (10.89)$$

where ρ_c and various Ωs refer to their values at $a = a_0$. In arriving at this result we had assumed that each component of the energy density present in the universe individually satisfies the condition $\nabla_a T^a_b = 0$, which reduces to $d(\rho a^3) = -p d(a^3)$ for each of the components. This is equivalent to assuming that the components do not interact with each other. It should be noted that, without this assumption, we cannot integrate the Friedmann equation for a multi-component universe. If there are N components of matter present, with N equations of state giving $p_i = p_i(\rho_i)$, we need to determine N functions $\rho_i(t)$ as well as $a(t)$ using two Friedmann equations. This is possible only for $N + 1 = 2$; that is, for a single component universe. But when each component obeys $d(\rho a^3) = -p d(a^3)$, we obtain N additional equations. These, along with one of the Friedmann equations are enough to determine the functions $\rho_i(t)$ and $a(t)$. The second Friedmann equation is then automatically satisfied. Substituting Eq. (10.89) into Einstein's equation we get

$$\frac{\dot{a}^2}{a^2} + \frac{k}{a^2} = H_0^2 \left[\Omega_{\rm R} \left(\frac{a_0}{a}\right)^4 + \Omega_{\rm NR} \left(\frac{a_0}{a}\right)^3 + \Omega_\Lambda \right] \qquad (10.90)$$

with $\Omega_{\rm NR} = \Omega_{\rm B} + \Omega_{\rm DM}$. This equation can be cast in a more suggestive form. We write (k/a^2) as $(\Omega_{\rm tot}-1)H_0^2(a_0/a)^2$ and move it to the right hand side. Introducing a dimensionless time coordinate $\tau = H_0 t$ and writing $a = a_0 q(\tau)$ this equation becomes

$$\frac{1}{2} \left(\frac{dq}{d\tau}\right)^2 + V(q) = E, \qquad (10.91)$$

where

$$V(q) = -\frac{1}{2} \left[\frac{\Omega_{\rm R}}{q^2} + \frac{\Omega_{\rm NR}}{q} + \Omega_\Lambda q^2 \right]; \quad E = \frac{1}{2}\left(1 - \Omega_{\rm tot}\right). \qquad (10.92)$$

This equation has the structure of the first integral for motion of a particle with energy E in a potential $V(q)$. For models with $\Omega_{\rm tot} = \Omega_{\rm NR} + \Omega_\Lambda = 1$, we get

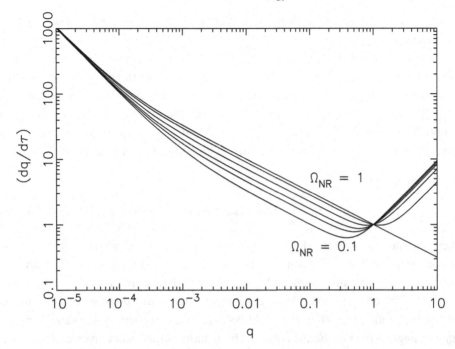

Fig. 10.1. The 'velocity' of the universe $(dq/d\tau)$ plotted against the 'position' $q = (1+z)^{-1}$ for different cosmological models with $\Omega_R = 2.56 \times 10^{-5}h^{-2}, h = 0.5, \Omega_{NR} + \Omega_\Lambda = 1$. Curves are parametrized by the values of $\Omega_{NR} = 0.1, 0.2, 0.3, 0.5, 0.8, 1.0$ going from the bottom to the top as indicated.

$E = 0$ so that $(dq/d\tau) = \sqrt{V(q)}$. Figure 10.1 shows the velocity $(dq/d\tau)$ as a function of the position $q = (1+z)^{-1}$ for such models. Several features are clear from this figure. At high redshift (small q) the universe is radiation dominated and $\dot{q}$ is independent of the other cosmological parameters; hence all the curves asymptotically approach one another at the left end of the figure. At low redshifts, the presence of the cosmological constant makes a difference and – in fact – the velocity $\dot{q}$ changes from being a decreasing function to an increasing function. In other words, the presence of a cosmological constant leads to an accelerating universe at low redshifts.

Given the definition of the Hubble parameter $H(t)$, critical density $\rho_c(t)$ and total density parameter $\Omega(t)$ at any given time t (see Eq. (10.76)), it is easy to relate $\Omega(t)$ to $\Omega(t_0) \equiv \Omega$. The relation is given by

$$\Omega(t) - 1 = \frac{\Omega - 1}{1 - \Omega + \Omega_\Lambda a^2 + \Omega_{NR} a^{-1} + \Omega_R a^{-2}}. \tag{10.93}$$

Two important conclusions follow from this equation. First, for small a the right hand side decreases rapidly and $\Omega(t) \approx 1$ in any model (other than the trivial case in which the only source is Ω_Λ). Hence in the study of the early phases of the universe we can approximate all models as those with $\Omega_{\text{tot}} = 1$. Second, a given deviation of $\Omega(t_i)$ from unity at some very early epoch $t = t_i$ will get magnified during the evolution so that $\Omega(t)$ will differ wildly from unity at late epochs. That is, we need to keep $\Omega(t)$ very close to unity at an early epoch if we find that $\Omega_{\text{tot}} = \mathcal{O}(1)$ at the present epoch. The only exception to this conclusion is in models in which Ω_{tot} is strictly equal to unity at all times – which, of course, satisfies Eq. (10.93) trivially.

Let us now consider the solutions to Eq. (10.90) or – equivalently – to Eq. (10.91). To begin with, it is clear that in the early evolution of the universe, the radiation term (Ω_R/q^2) in Eq. (10.92) will dominate the dynamics. In this limit, the solution is given by

$$\left(\frac{a}{a_0}\right) \cong \sqrt{2}\,\Omega_R^{1/4}\,(H_0 t)^{1/2}\,. \tag{10.94}$$

Thus for small a, we have $a \propto t^{1/2}$.

As the universe evolves the matter density will catch up with the radiation density and both a^{-4} and a^{-3} terms will be important. The equality of matter and radiation energies occurs at some time $t = t_{\text{eq}}$ in the past corresponding to a value $a = a_{\text{eq}}$ and redshift $z = z_{\text{eq}}$. To determine the numerical values we need to know Ω_R. Assuming that most of the energy density is in the microwave background radiation at temperature T today, we get $\rho_R = (\pi^2/15)\,(k_B^4 T^4/c^3\hbar^3)$. Dividing this by $\rho_c \simeq 1.88 \times 10^{-29} h^2\,\text{g cm}^{-3}$ we can find Ω_R. Taking (from the CMBR observations) $T = 2.73$ K gives

$$\Omega_R h^2 = 2.56 \times 10^{-5}. \tag{10.95}$$

It follows that

$$(1 + z_{\text{eq}}) = \frac{a_0}{a_{\text{eq}}} = \frac{\Omega_{\text{NR}}}{\Omega_R} \simeq 3.9 \times 10^4 (\Omega_{\text{NR}} h^2). \tag{10.96}$$

Since the temperature of the radiation grows as a^{-1}, the temperature of the universe at this epoch will be $T_{\text{eq}} = T_{\text{now}}(1 + z_{\text{eq}}) = 9.24(\Omega h^2)$ eV. For $t \ll t_{\text{eq}}$ the energy density in the universe is dominated by radiation (with $p = (1/3)\rho$) while for $t \gg t_{\text{eq}}$ the energy density is dominated by matter (with $p \simeq 0$). When both radiation and matter terms are taken into consideration, and other terms are ignored, Eq. (10.90) has the analytic solution

$$H_{\text{eq}} t = \frac{2\sqrt{2}}{3}\left[\left(\frac{a}{a_{\text{eq}}} - 2\right)\left(\frac{a}{a_{\text{eq}}} + 1\right)^{1/2} + 2\right]. \tag{10.97}$$

This equation gives $a(t)$ in terms of the two (known) parameters:

$$H_{eq}^2 \equiv \frac{16\pi G}{3}\rho_{eq} \equiv 2H_0^2\Omega_R(1+z_{eq})^4 = 2H_0^2\Omega_{NR}(1+z_{eq})^3, \qquad (10.98)$$

and

$$a_{eq} \equiv a_0(1+z_{eq})^{-1} = H_0^{-1}|(\Omega - 1)|^{-1/2}(\Omega_R/\Omega_{NR}). \qquad (10.99)$$

From Eq. (10.97) we can find the value of t_{eq}; setting $a = a_{eq}$ gives $H_{eq}t_{eq} \simeq 0.552$, or

$$t_{eq} = \frac{2\sqrt{2}}{3}H_{eq}^{-1}(2 - \sqrt{2}) \simeq 1.57 \times 10^{10}(\Omega h^2)^{-2} \text{ s}. \qquad (10.100)$$

From Eq. (10.97), we can also find two limiting forms for $a(t)$ valid for $t \gg t_{eq}$ and $t \ll t_{eq}$:

$$\left(\frac{a}{a_{eq}}\right) = \begin{cases} (3/2\sqrt{2})^{2/3}(H_{eq}t)^{2/3} \\ (3/\sqrt{2})^{1/2}(H_{eq}t)^{1/2}. \end{cases} \qquad (10.101)$$

Thus $a \propto t^{2/3}$ in the matter-dominated phase (when other contributions are negligible) and $a \propto t^{1/2}$ in the radiation-dominated phase.

At $z \ll z_{eq}$ we can ignore the radiation completely. The evolution now depends on the values of Ω_Λ and k. If $\Omega_\Lambda > \Omega_{NR}$ then the Ω_Λ term will dominate over other terms at $z \ll z_V$, where $(1 + z_V) = (\Omega_\Lambda/\Omega_{NR})^{1/3}$. Keeping only the Ω_{NR} and Ω_Λ terms we can integrate Eq. (10.90) (with $k = 0$) to determine $a(t)$. We get

$$\left(\frac{a}{a_0}\right) = \left(\frac{\Omega_{NR}}{\Omega_\Lambda}\right)^{1/3} \sinh^{2/3}\left[\frac{3}{2}\sqrt{\Omega_\Lambda}H_0t\right]; \qquad \Omega_{NR} + \Omega_\Lambda = 1. \quad (10.102)$$

When $\sqrt{\Omega_\Lambda}H_0t \ll 1$, this reduces to the matter dominated evolution with $a^3 \propto t^2$; when $\sqrt{\Omega_\Lambda}H_0t \gtrsim 1$, the growth is exponential with $a \propto \exp(\sqrt{\Omega_\Lambda}H_0t)$.

We conclude this section with a comment on another important length scale in cosmology, viz. the *horizon size*. Suppose for a moment that the universe is described by the expansion factor $a(t) = a_0t^n$, with $n < 1$ for *all* $t \geq 0$. Then, during the time interval $(0, t)$, a photon can travel a maximum coordinate distance of

$$\xi(t) = \int_0^t \frac{dx}{a(x)} = \frac{1}{a_0}\frac{t^{1-n}}{(1-n)}, \qquad (10.103)$$

which corresponds to the proper distance:

$$h(t) = a(t)\xi(t) = (1-n)^{-1}t. \qquad (10.104)$$

Numerically this quantity differs from the Hubble radius $(\dot{a}/a)^{-1} = n^{-1}t$ only by a constant factor of order unity if $a \propto t^n$ with n order of unity. Conceptually, however, they are very different entities. To avoid any possible confusion between these two quantities, we emphasize the following fact. Notice that $d_H(t)$ is a local quantity and its value at t is essentially decided by the behaviour of $a(t)$ near t; in contrast, the value of $h(t)$ depends on the entire past history of the universe. In fact, $h(t)$ depends very sensitively on the behaviour of $a(t)$ near $t = 0$ – something about which we know nothing. If, for example, $a(t) \propto t^m$ with $m \geq 1$ near $t = 0$, then $h(t)$ is infinite for all $t \geq 0$. Thus there can be several physical situations in which $h(t)$ and $d_H(t)$ differ widely; in such cases, one should examine each case carefully and decide which quantity is physically relevant.

Exercise 10.7
Nice features of the conformal time (a) Integrate the Friedmann equations for a $k = 0$ universe with matter and radiation using the conformal time τ (defined through the relation $dt = a d\tau$). (b) Integrate the Friedmann equation for a matter dominated universe with $k \neq 0$ using conformal time. Express $\Omega(t)$ and Ht in terms of τ. [Answer. (a) The Friedmann equation can now be reduced to the form

$$\left(\frac{da}{dq}\right)^2 = \Omega_{NR} a + \Omega_R; \qquad q = H_0 \tau, \tag{10.105}$$

which can be integrated to give

$$a = \sqrt{\Omega_R}(H_0 \tau) + \frac{1}{4}\Omega_{NR}(H_0 \tau)^2. \tag{10.106}$$

(b) The Friedmann equation in this case reduces to

$$d\tau = \frac{d \ln a}{(A/a - k)^{1/2}}; \qquad A \equiv \frac{8\pi G \rho a^3}{3}. \tag{10.107}$$

Integrating, we can express a in terms of τ as

$$a = \frac{A}{k} \sin^2(k^{1/2}\tau/2). \tag{10.108}$$

Note that this is valid for $k = 0, 1$. (Similar results with hyperbolic functions exist for $k = -1$.) Since $dt = a d\tau$, integration gives

$$t = \frac{A}{2k^{3/2}} \left[k^{1/2}\tau - \sin(k^{1/2}\tau)\right]. \tag{10.109}$$

The Hubble parameter and the density parameter are given by

$$H(\tau) = \frac{1}{a}\left(\frac{A}{a} - k\right)^{1/2} = \frac{k^{3/2}}{A}\frac{\cos(k^{1/2}\tau/2)}{\sin^3(k^{1/2}\tau/2)};$$

$$\Omega(\tau) = \frac{8\pi G \rho}{3H^2} = \frac{1}{1 - ka/A} = \frac{1}{\cos^2(k^{1/2}\tau/2)}. \tag{10.110}$$

The combination Ht is given by

$$Ht = \frac{\cos(k^{1/2}\tau/2)\left[k^{1/2}\tau - \sin(k^{1/2}\tau)\right]}{2\sin^3(k^{1/2}\tau/2)}, \tag{10.111}$$

which can be re-expressed in different forms.]

Exercise 10.8

Tracker solutions for scalar fields (a) Show that the field equation arising from the Lagrangian in Eq. (10.85), when the scalar field is homogeneous (i.e. $\phi(t, x) = \phi(t)$) is given by

$$\ddot{\phi} + 3H\dot{\phi} + V'(\phi) = 0. \tag{10.112}$$

Consider this equation for a potential of the form $V(\phi) = M^{4+\alpha}\phi^{-\alpha}$, where M, α are positive constants. Use natural units with $\hbar = c = 1$.

(b) If the energy density of radiation dominates over all others in the early phase so that $a(t) \propto t^{1/2}$, show that Eq. (10.112) has the solution

$$\phi = \left(\frac{\alpha(2+\alpha)^2 M^{4+\alpha}t^2}{6+\alpha}\right)^{1/(2+\alpha)}. \tag{10.113}$$

Verify that, at sufficiently early times, the energy density of the scalar field is indeed subdominant to the radiation energy density so that this is a consistent solution.

(c) Prove that this solution is an 'attractor' in the sense that any solution which comes close to it will approach it as t increases. [Hint. Introduce a small perturbation $\delta\phi$ and show that $\delta\phi$ decays as $t^{-1/4}$ for increasing t.]

(d) When the radiation energy density drops below that of matter energy density, show that the scalar field will continue to grow as $t^{2/(2+\alpha)}$. Argue that, with this behaviour, ρ_{NR} and ρ_R will eventually fall below the energy density of the scalar field. Prove that though the time at which $\rho_\phi = \rho_{NR}$ depends on M the value of the scalar field at this time is independent of M and is given by $\phi_{eq} \approx G^{-1/2}$.

(e) At very late times, show that the scalar field evolves as

$$\phi = M\left(\frac{\alpha(2+\alpha/2)t}{\sqrt{24\pi G}}\right)^{1/(2+\alpha/2)}. \tag{10.114}$$

Exercise 10.9

Horizon size The maximum proper distance a photon can travel in the interval $(0, t)$ is given by the horizon size

$$h(t) = a(t)\int_0^t \frac{dx}{a(x)}. \tag{10.115}$$

Show that, for a matter dominated universe with $\Omega_{tot} = \Omega_{NR}$, we get:

$$h(z) = H_0^{-1}(1+z)^{-1}(\Omega_{NR} - 1)^{-\frac{1}{2}}\cos^{-1}\left[1 - \frac{2(\Omega_{NR} - 1)}{\Omega_{NR}(1+z)}\right] \quad \text{(for } \Omega_{NR} > 1)$$

$$= 2H_0^{-1}(1+z)^{-\frac{3}{2}} \quad \text{(for } \Omega_{NR} = 1)$$

$$= H_0^{-1}(1+z)^{-1}(1 - \Omega_{NR})^{-\frac{1}{2}}\cosh^{-1}\left[1 + \frac{2(1 - \Omega_{NR})}{\Omega_{NR}(1+z)}\right] \quad \text{(for } \Omega_{NR} < 1)).$$

$$\tag{10.116}$$

Show also that $d_H \simeq 3H_0^{-1}\Omega_{NR}^{-\frac{1}{2}}(1+z)^{-\frac{3}{2}}$ for $(1+z) \gg \Omega^{-1}$.

Exercise 10.10

Loitering and other universes Consider a Friedmann model with (i) only Ω_{NR} and Ω_Λ being nonzero; (ii) $\Omega_{NR} \geq 0$; and (iii) Ω_Λ being positive or negative. (a) Prove that if $\Omega_\Lambda < 0$ the universe always recollapses while if $\Omega_\Lambda > 0$ and $\Omega_{NR} < 1$ the universe expands for ever. Also show that if $\Omega_{NR} > 1$ and

$$\Omega_\Lambda > 4\Omega_{NR} \left[\cos \left(\frac{1}{3} \cos^{-1} \left(\Omega_{NR}^{-1} - 1 \right) + \frac{4}{3}\pi \right) \right]^3 \tag{10.117}$$

the universe does not recollapse.

(b) Show that it is possible to adjust the values of Ω_Λ and Ω_{NR} such that the universe stays ('loiters') for a long time at a nearly constant scale factor $a = a_c$, say, at some $z > 0$. [Hint. In the model under consideration, the Hubble parameter can be expressed in the form

$$\frac{H^2(a)}{H_0^2} = \Omega_\Lambda \left(1 - a^{-2} \right) + \Omega_{NR} \left(a^{-3} - a^{-2} \right) + a^{-2}. \tag{10.118}$$

What happens when the right hand side vanishes?]

Exercise 10.11

Point particle in a Friedmann universe Consider the metric (called the *McVittie metric*) of the form

$$ds^2 = -\frac{(1-f)}{(1+f)} dt^2 + (1+f)^4 e^{g(t)} \left[1 + \frac{r^2}{4R^2} \right]^{-2} [dr^2 + r^2(d\theta^2 + \sin^2\theta d\phi^2)], \tag{10.119}$$

where

$$2f = \frac{m e^{-g(t)/2}}{r} \left[1 + \frac{r^2}{4R^2} \right]^{1/2}, \qquad R = \text{const.} \tag{10.120}$$

Write down the Einstein equation for this metric. Argue that when $m = 0$ this reduces to the Friedmann metric while, for $g = 0$ and $r^2 \ll R^2$, the solution approaches the Schwarzschild metric. This suggests that this metric could be interpreted as being due to a point particle located in a Friedmann universe. See how far you can push this interpretation.

Exercise 10.12

Collapsing dust ball revisited We saw in Chapter 7 that the exterior metric of a collapsing homogeneous dust sphere can be matched to a Schwarzschild metric. Since the interior solution is essentially a Friedmann universe, we will reanalyse this problem from a different perspective here.

(a) Consider a collapsing fluid system with an equation of state $p = w\rho$ described by an interior Friedmann metric

$$ds^2 = -dt^2 + a(t)^2 \left(dR^2 + R^2 d\Omega^2 \right), \tag{10.121}$$

where

$$a(t) \propto t^n, \qquad n = \frac{2}{3(w+1)}. \tag{10.122}$$

Show that, on introducing a new radial coordinate $r = a(t)R$, the metric becomes

$$ds^2 = -\left[1 - \left(\frac{nr}{t}\right)^2\right] dt^2 + 2\left(\frac{nr}{t}\right) dr\,dt + dr^2 + r^2 d\Omega^2. \qquad (10.123)$$

On the other hand, the standard Schwarzschild metric in the exterior region written in the form

$$ds^2 = -\left(1 - V^2\right) dt_S^2 + \frac{dr^2}{1 - V^2} + r^2 d\Omega^2, \qquad V = \pm\sqrt{\frac{2M}{r}} \qquad (10.124)$$

can be transformed to the form

$$ds^2 = -(1 - V^2)dt^2 \pm 2V\,dr\,dt + dr^2 + r^2 d\Omega^2 \qquad (10.125)$$

by choosing a new time coordinate t given by (see Project 7.2)

$$t_S = t + g(r), \qquad g' = \pm\frac{V}{1 - V^2}. \qquad (10.126)$$

Show that in the case of Schwarzschild metric g is explicitly given by

$$g = \mp 2M\left(2\sqrt{\frac{r}{2M}} + \ln\frac{\sqrt{r} - \sqrt{2M}}{\sqrt{r} + \sqrt{2M}}\right). \qquad (10.127)$$

(b) We have now expressed both the interior and exterior metrics in a similar form with

$$V = \begin{cases} \dfrac{nr}{t}, & \text{(fluid interior)}, \\[2mm] -\sqrt{\dfrac{2M}{r}}, & \text{(Schwarzschild exterior)}. \end{cases} \qquad (10.128)$$

Work out the components of Einstein's tensor for this metric and show that the energy density is given by $\rho = 3n^2/8\pi Gt^2$ in the fluid interior and by zero in the exterior.

(c) To match the two solutions we must demand that the metric function V is continuous across the surface. Show that this leads to the condition

$$r^{3/2} + \frac{\sqrt{2M}}{n}t = 0, \qquad (10.129)$$

while the geodesic equation for a zero energy particle in the Schwarzschild exterior is given by

$$r^{3/2} + \frac{3\sqrt{2M}}{2}(t - t_0) = 0. \qquad (10.130)$$

Hence show that the particle at the surface can follow a pressure free, geodesic trajectory only if $n = (2/3)$. This corresponds to a pressure free dust ball with $p = 0$. Interpret this result.

10.5 The de Sitter spacetime

A cosmological model sourced entirely by an energy-momentum tensor of the form $T^a_b = -\rho \delta^a_b$ (arising e.g. from the cosmological constant) has several special features and hence deserves a separate discussion. If the universe contains other usual forms of source (with $\rho > 0, p > 0$) as well, then at late times the evolution will be dominated essentially by the cosmological constant. We shall, however, idealize the situation by assuming that the *only* source present in the universe is a cosmological constant. Further, in view of its potential importance, we shall work in a d-dimensional spacetime with $d \geq 3$.

An energy-momentum tensor of the form $T^a_b = -\rho\,\delta^a_b$ possesses a high degree of symmetry. In addition to the homogeneity and isotropy, the spacetime is also invariant under time translation. Einstein's equation now reduces to $G^a_b = -\kappa\rho\delta^a_b$ and we are interested in the solution to this equation and its properties. In the context of cosmology, the relevant metric can be taken to be

$$ds^2 = -dt^2 + a^2(t)\left[\frac{dr^2}{1 - k(r/l)^2} + r^2 d\Omega^2_{d-2}\right], \qquad (10.131)$$

where we have introduced a length scale l for dimensional considerations and $k = 0, \pm 1$. The $a(t)$ is now dimensionless. The Friedmann equations reduces to

$$\left(\frac{\dot{a}}{a}\right)^2 = \frac{4\pi G}{d-2}\left[\frac{d}{d-1}\rho - (d-4)p\right] - \frac{kl^{-2}}{a^2} = \frac{d-2}{2(d-1)}\Lambda - \frac{kl^{-2}}{a^2} \quad (10.132)$$

and

$$\frac{\ddot{a}}{a} = -4\pi G\left(\frac{\rho}{d-1} + p\right) = \frac{d-2}{2(d-1)}\Lambda, \qquad (10.133)$$

where $T^a_b = -\rho\delta^a_b$ with $8\pi G\rho \equiv \Lambda$. These equations possess solutions for all three values of $k = 0, \pm 1$. It can be easily seen that the solutions are given by

$$a(t) = \begin{cases} \sinh(t/l), & (\text{for } k = -1) \\ \exp(\pm t/l), & (\text{for } k = 0) \\ \cosh(t/l), & (\text{for } k = +1) \end{cases} \qquad (10.134)$$

where l is related to Λ by $\Lambda = (1/2)(d-1)(d-2)l^{-2}$ and we will assume that $d \geq 3$. Expressed in this form, the geometrical features of this solution are not apparent and it is somewhat puzzling that the same source can lead to a Friedmann universe with a different value of k. This feature arises because the solution actually represents a d-dimensional hyperboloid embedded in a flat $(d+1)$-dimensional spacetime, and different cross-sections of this hyperboloid (which correspond to different spatial sections of the Friedmann universe) can have positive, negative or

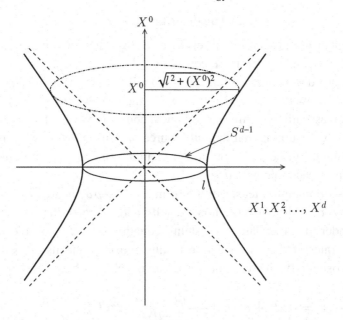

Fig. 10.2. Geometrical structure of de Sitter space in $(d+1)$ dimensions with the signature $(-, +, +, \ldots, +)$.

zero spatial curvature (see Fig. 10.2). We shall demonstrate this by explicitly constructing the hyperboloid and introducing in it the coordinate system corresponding to $k = +1$ and $k = 0$. The $k = -1$ case can be discussed along similar lines.

With this motivation, let us consider a $(d + 1)$-dimensional flat spacetime with the line element

$$ds^2 = \eta_{AB} dX^A dX^B, \qquad \eta_{AB} = \mathrm{diag}\,(-1, 1, 1, \ldots 1). \tag{10.135}$$

A d-dimensional hyperboloid in this spacetime is given by the equation

$$\eta_{AB} X^A X^B = l^2. \tag{10.136}$$

We shall now introduce a d-dimensional coordinate system on this hyperboloid using one time coordinate τ and $(d - 1)$ angular coordinates $\theta_1, \theta_2, \ldots, \theta_{d-1}$. The coordinates X^A on the hyperboloid are given in terms of τ and $\theta_1, \theta_2, \ldots, \theta_{d-1}$ by

$$X^0 = l \sinh \frac{\tau}{l}; \qquad X^\alpha = l\omega^\alpha \cosh \frac{\tau}{l}, \qquad (\alpha = 1, \ldots, d), \tag{10.137}$$

where ω^α satisfies the constraint

$$\sum_{\alpha=1}^{d} (\omega^\alpha)^2 = 1 \tag{10.138}$$

and can be expressed in terms of the angular coordinates θ_i on the sphere. It is now easy to verify that the metric in Eq. (10.135) evaluated on the hyperboloid is given by

$$ds^2 = -d\tau^2 + l^2 \cosh^2 \left(\frac{\tau}{l} \right) d\Omega_{d-1}^2. \tag{10.139}$$

This represents a Friedmann universe with $k = +1$ and $a(\tau) = l \cosh(\tau/l)$. These coordinates cover the entire hyperbola.

We can introduce a different set of coordinates so that the spatial sections are flat. In this case, we take

$$\frac{X^0}{l} = -\sinh \frac{t}{l} + \frac{(x^i/l)^2}{2} e^{t/l}$$

$$\frac{X^i}{l} = \frac{x^i}{l} e^{t/l} \qquad (i = 1, 2, \ldots, d-1)$$

$$\frac{X^d}{l} = -\cosh \frac{t}{l} - \frac{(x^i/l)^2}{2} e^{t/l}, \tag{10.140}$$

where $(x^i)^2$ stands for $(x^1)^2 + (x^2)^2 + \cdots$. Each of the coordinates x^i as well as t vary over the range $(-\infty, +\infty)$. Since $(-X^0 + X^d) = -l \exp(-t/l) \leq 0$, these coordinates cover only the upper half of the manifold. Once again, the metric in Eq. (10.135), when confined to the hyperboloid, takes the form

$$ds^2 = -dt^2 + e^{2t/l} \left[(dx^1)^2 + (dx^2)^2 + \cdots (dx^{d-1})^2 \right]. \tag{10.141}$$

This is precisely the solution we obtain by solving the Friedmann equation with $k = 0$. In this form, the time translation invariants of the spacetime geometry are obvious: on changing $t \to t + \epsilon$, the spatial coordinates are rescaled by a constant factor $\exp(\epsilon/l)$.

It is also possible to provide a completely different coordinate system for describing a spacetime sourced only by a cosmological constant. Since the source is independent of time, one must be able to describe the spacetime in static coordinates. Further, since isotropy necessarily implies spherical symmetry, there must exist a solution which is spherically symmetric and static with the cosmological constant acting as the source. Such a metric has already been discussed in Chapter 7 (see Eq. (7.45)) as well as in Chapter 8 and, in the case of d-dimensional spacetime, acquires the form

$$ds^2 = -\left[1 - \left(\frac{r}{l} \right)^2 \right] dt^2 + \frac{dr^2}{[1 - (r/l)^2]} + r^2 d\Omega_{d-2}^2. \tag{10.142}$$

This form of the metric can also be obtained from the hyperboloid in $(d+1)$-dimensional space by introducing the coordinates

$$\frac{X^0}{l} = -\sqrt{1 - \left(\frac{r}{l}\right)^2}\, \sinh\frac{t}{l}; \qquad \frac{X^d}{l} = -\sqrt{1 - \left(\frac{r}{l}\right)^2}\, \cosh\frac{t}{l}$$

$$\frac{X^i}{l} = \frac{r}{l}\,\omega^i, \qquad (i = 1, 2, \ldots, d-1), \tag{10.143}$$

where ω^i are the direction cosines. Since $(-X^0 + X^d) \leq 0$ and $(X^0 + X^d) \leq 0$ for these coordinate choices, the region with $r \leq l$ covers only a quarter of the whole manifold. As described in Chapter 8, the spacetime in these coordinates has a horizon at $r = l$.

Exercise 10.13

The anti-de Sitter spacetime Throughout the discussion in the text, we have assumed that the cosmological constant was positive; that is, the source is $T_b^a = -\rho_0 \delta_b^a$ with $\rho_0 > 0$. Repeat the analysis given in this section for the case of a spacetime with $\rho_0 < 0$. (This spacetime is called anti-de Sitter (AdS) spacetime, which has completely different geometrical properties.) In particular, show that AdS spacetimes can be embedded in the flat space of one higher dimension with mixed signature. Specifically, show that the metric of AdS spacetime in (1+2) dimension is the induced metric on the hyperboloid with the equation $-T^2 - U^2 + X^2 + Y^2 = $ constant embedded in a four-dimensional space with line element $ds^2 = -dT^2 - dU^2 + dX^2 + dY^2$.

Exercise 10.14

Geodesics in de Sitter spacetime Consider a (1+1)-dimensional de Sitter spacetime with the metric expressed in the form

$$ds^2 = \cosh^2 u\, d\phi^2 - du^2. \tag{10.144}$$

(a) Show that the geodesic equation in this spacetime can be reduced to the first integrals of the form

$$\left(\cosh^2 u\right)\dot{\phi} = K, \qquad \left(\cosh^2 u\right)\dot{u} = K^2 + L\cosh^2 u, \tag{10.145}$$

where K, L are constants and the overdot denotes a derivative with respect to an affine parameter.

(b) Show that, if $K = 0$, the geodesics are given by $\phi = $ constant. When $K \neq 0$, reduce the equation to the form $(dv/d\phi)^2 = M^2 - v^2$, where M is another constant. Integrate this equation to obtain the geodesics:

$$\tanh u = M\sin(\phi - \phi_0), \tag{10.146}$$

where ϕ_0 is a constant.

(c) Show that $M^2 > 1, M^2 = 1$ and $M^2 < 1$ correspond to timelike, null and spacelike geodesics. Plot a set of geodesics which start from the origin. Using the plot show that no two timelike geodesics will intersect again and that there are pairs of points which cannot be joined by a geodesic.

10.6 Brief thermal history of the universe

We shall next discuss briefly some key events in the evolutionary history of our universe.[4] The most well understood phase of the universe occurs when the temperature of matter (and radiation) is less than about 10^{12} K. Above this temperature, thermal production of baryons and their strong interaction are significant and somewhat difficult to model. We can ignore such complications at lower temperatures and several interesting physical phenomena take place during the (later) epochs with $T \lesssim 10^{12}$.

When the temperature of the universe is about $T \approx 10^{12}$ K, it will contain copious amount of photons and all the species of neutrinos and antineutrinos. In addition, neutrons and protons must exist at this time since there is no way of producing them later on. (This implies that physical phenomena which took place at higher temperatures should have left a small excess of baryons over antibaryons; we do not completely understand how this happened and will have to take it as an initial condition.) Since the rest mass of electrons corresponds to a much lower temperature (about 0.5×10^{10} K), there will be a large number of electrons *and positrons* at this temperature. But in order to maintain charge neutrality, we need to have a slight excess of electrons over positrons (by about 1 part in 10^9) with the net negative charge compensating the positive charge contributed by the protons.

A calculation using the known particle interaction rates shows that all these particles would have been in thermal equilibrium at this epoch. Hence standard rules of statistical mechanics allow us to determine (see Exercise 1.15) the number density (n), energy density (ρ) and the pressure (p) in terms of the distribution function f. Using the resulting energy density in the Friedmann equation and recalling that one can set $k = 0$ while studying the early phases of universe, it is easy to obtain an expression relating the temperature and time. It turns out that

$$t \approx 1 \text{ sec } \left(\frac{T}{1 \text{ MeV}} \right)^{-2} g_{\text{tot}}^{-1/2}, \tag{10.147}$$

where g_{tot} is the total number of effective spin degrees of freedom. During the subsequent evolution, two key events take place very early on.

(a) Around $T \lesssim 1.6 \times 10^{10}$ K, the reaction rate of neutrinos falls below the expansion rate of the universe and the neutrinos decouple from matter. At a slightly lower temperature, the electrons and positrons annihilate, increasing the number density of the photons. Neutrinos do not get any share of this energy since they have already decoupled from the rest of the matter. As a result, the photon temperature goes up with respect to the neutrino temperature once the e^+e^- annihilation is complete. Calculations show that this increase is approximately by a factor 1.4. Unfortunately, the current technology is not good enough to detect the primordial neutrino background in the universe. When it is detected in the future, this prediction can be a tested, thereby directly probing the conditions in the universe when it was about a few seconds old.

(b) When the temperature of the universe falls below about 0.1 MeV, primordial nucleosynthesis takes place, forming about 25 per cent by weight of helium and a tiny fraction (about 10^{-5}) of deuterium. The abundance of primordial helium and deuterium is yet another test of the model of the early universe.

These events take place within the first few minutes in the history of the universe. The next important event occurs much later (in about a few hundred thousand years) when radiation decouples from matter. Considering its importance, we shall discuss this process in some detail.

10.6.1 Decoupling of matter and radiation

In the early hot phase, the radiation will be in thermal equilibrium with matter; as the universe cools below $k_B T \simeq (\epsilon_a/10)$, where ϵ_a is the binding energy of atoms, the electrons and ions will combine to form neutral atoms and radiation will decouple from matter. This occurs at $T_{\text{dec}} \simeq 3 \times 10^3$ K. As the universe expands further, these photons will continue to exist without any further interaction with matter. It will retain the thermal spectrum since the redshift of the frequency $\nu \propto a^{-1}$ is equivalent to changing the temperature in the spectrum by the scaling $T \propto (1/a)$. It turns out that the major component of the extra-galactic background light which exists today is in the microwave band and can be fitted very accurately by a thermal spectrum at a temperature of about 2.73 K. It seems reasonable to interpret this radiation as a relic arising from the early, hot, phase of the evolving universe. This relic radiation, called *cosmic microwave background radiation* (CMBR), turns out to be a gold mine of cosmological information and has been extensively investigated in recent times. We shall discuss in some detail the formation of neutral atoms and the decoupling of photons since this is required for understanding the temperature anisotropies of CMBR discussed in Chapter 13.

The relevant reaction is $e + p \rightleftharpoons H + \gamma$ and if the rate of this reaction is faster than the expansion rate, then one can calculate the neutral fraction using Saha's equation. Introducing the fractional ionization, X_i, for each of the particle species and using the facts $n_p = n_e$ and $n_p + n_H = n_B$, it follows that $X_p = X_e$ and $X_H = (n_H/n_B) = 1 - X_e$. Saha's equation now gives

$$\frac{1 - X_e}{X_e^2} \cong 3.84\eta(T/m_e)^{3/2} \exp(B/T), \tag{10.148}$$

where $\eta = 2.68 \times 10^{-8}(\Omega_B h^2)$ is the baryon-to-photon ratio and B is the binding energy of the hydrogen atom in units with $k_B = \hbar = C = 1$. We may define T_{atom}

as the temperature at which 90 per cent of the electrons, say, have combined with protons: i.e. when $X_e = 0.1$. This leads to the condition:

$$(\Omega_B h^2)^{-1} \tau^{-\frac{3}{2}} \exp\left[-13.6\tau^{-1}\right] = 3.13 \times 10^{-18}, \qquad (10.149)$$

where $\tau = (T/1\text{eV})$. For a given value of $(\Omega_B h^2)$, this equation can be easily solved by iteration. Taking logarithms and iterating once we find $\tau^{-1} \cong 3.084 - 0.0735 \ln(\Omega_B h^2)$ with the corresponding redshift $(1 + z) = (T/T_0)$ given by

$$(1 + z) = 1367[1 - 0.024 \ln(\Omega_B h^2)]^{-1}. \qquad (10.150)$$

For $\Omega_B h^2 = 1, 0.1, 0.01$ we get $T_{\text{atom}} \cong 0.324$ eV, 0.307 eV, 0.292 eV respectively. These values correspond to the redshifts of 1367, 1296 and 1232.

Because the preceding analysis was based on equilibrium densities, it is important to check that the rate of the reactions $p + e \leftrightarrow H + \gamma$ is fast enough to maintain equilibrium. For $\Omega_B h^2 \approx 0.02$, the equilibrium condition is only marginally satisfied, making this analysis suspect. More importantly, the direct recombination to the ground state of the hydrogen atom – which was used in deriving the Saha's equation – is not very effective in producing neutral hydrogen in the early universe. The problem is that each such recombination releases a photon of energy 13.6 eV which will end up ionizing another neutral hydrogen atom that has been formed earlier. As a result, the direct recombination to the ground state does not change the neutral hydrogen fraction at the lowest order. Recombination through the excited states of hydrogen is more effective since such a recombination ends up emitting more than one photon, each of which has an energy less than 13.6 eV. Given these facts, it is necessary to analyse this as a non-equilibrium phenomenon in the expanding universe. We shall first develop the formalism for doing this and will then apply it to the current problem.

The general procedure for studying non-equilibrium abundances in an expanding universe is based on *rate equations*. Consider a reaction in which two particles 1 and 2 interact to form two other particles 3 and 4 due to a reaction of the form $1 + 2 \rightleftharpoons 3 + 4$. In general, we are interested in how the number density n_1 of particle species 1, say, changes.

We first note that, even if there is no reaction, the number density will change as $n_1 \propto a^{-3}$ due to the expansion of the universe; so what we really need to determine is the change in $n_1 a^3$. Further, the forward reaction will be proportional to the product of the number densities $n_1 n_2$ while the reverse reaction will be proportional to $n_3 n_4$. Hence we can write an equation for the rate of change of particle species n_1 as

$$\frac{1}{a^3} \frac{d(n_1 a^3)}{dt} = \mu(A n_3 n_4 - n_1 n_2). \qquad (10.151)$$

The left hand side is the relevant rate of change over and above that due to the expansion of the universe; on the right hand side, the two proportionality constants have been written as μ and $(A\mu)$, both of which, of course, will be functions of time. (The quantity μ has the units of $cm^3 \ s^{-1}$, so that $n\mu$ has the dimensions of s^{-1}; usually $\mu \simeq \sigma v$, where σ is the cross-section for the relevant process and v is the relative velocity of the reacting particles.) The left hand side has to vanish when the system is in thermal equilibrium with $n_i = n_i^{eq}$, where the superscript 'eq' denotes the equilibrium densities for the different species labelled by $i = 1$–4. This condition allows us to rewrite A as $A = n_1^{eq} n_2^{eq}/(n_3^{eq} n_4^{eq})$. Hence the rate equation becomes

$$\frac{1}{a^3} \frac{d(n_1 a^3)}{dt} = \mu n_1^{eq} n_2^{eq} \left(\frac{n_3 n_4}{n_3^{eq} n_4^{eq}} - \frac{n_1 n_2}{n_1^{eq} n_2^{eq}} \right). \tag{10.152}$$

On the left hand side, one can write $(d/dt) = Ha(d/da)$, which shows that the relevant timescale governing the process is H^{-1}. Clearly, when $H/n\mu \gg 1$ the right hand side becomes ineffective because of the (μ/H) factor and the number of particles of species 1 does not change. That is, when the expansion rate of the universe is large compared with the reaction rate, the given reaction is ineffective in changing the number of particles. This certainly does *not* mean that the reactions have reached thermal equilibrium and $n_i = n_i^{eq}$; in fact, it means exactly the opposite. The reactions are not fast enough to drive the number densities towards equilibrium densities and the number densities 'freeze out' at non-equilibrium values. (Of course, the right hand side will also vanish when $n_i = n_i^{eq}$, which is the other extreme limit of thermal equilibrium.)

To study recombination, we use Eq. (10.152) with $n_1 = n_e, n_2 = n_p, n_3 = n_H$ and $n_4 = n_\gamma$. Defining $X_e = n_e/(n_e + n_H) = n_p/n_H$, one can easily derive the rate equation for this case:

$$\frac{dX_e}{dt} = [\beta(1 - X_e) - \alpha n_b X_e^2] = \alpha \left(\frac{\beta}{\alpha}(1 - X_e) - n_b X_e^2 \right) \tag{10.153}$$

with two parameters α and β. The first term gives the photoionization rate, which produces the free electrons, and the second term is the recombination rate, which converts free electrons into hydrogen atoms; we have also used the fact $n_e = n_b X_e$, etc. Since we know that direct recombination to the ground state is not effective, the recombination rate α is the rate for capture of an electron by a proton, forming an excited state of hydrogen. To a good approximation, this rate is given by

$$\alpha = 9.78 r_0^2 c \left(\frac{B}{T} \right)^{1/2} \ln \left(\frac{B}{T} \right), \tag{10.154}$$

where $r_0 = e^2/m_e c^2$ is the classical electron radius. To integrate Eq. (10.153) we also need to know β/α. This is easy because in thermal equilibrium the right hand side of Eq. (10.153) should vanish and Saha's equation tells us the value of X_e in thermal equilibrium. On using Eq. (10.148), this gives

$$\frac{\beta}{\alpha} = \left(\frac{m_e T}{2\pi}\right)^{3/2} \exp[-(B/T)].$$

(10.155)

We can now integrate Eq. (10.153) using the variable B/T. The result shows that the actual recombination proceeds more slowly compared with that predicted by the Saha's equation. The actual fractional ionization is higher than the value predicted by Saha's equation at lower temperatures. For example, at $z = 1300$, these values differ by a factor 3; at $z \simeq 900$, they differ by a factor of 200. The value of T_{atom}, however, does not change significantly. A more rigorous analysis shows that, in the redshift range of $800 < z < 1200$, the fractional ionization varies rapidly and is given (approximately) by the formula,

$$X_e = 2.4 \times 10^{-3} \frac{(\Omega_{\text{NR}} h^2)^{1/2}}{(\Omega_B h^2)} \left(\frac{z}{1000}\right)^{12.75}.$$

(10.156)

This is obtained by fitting a curve to the solution obtained by numerical integration of the rate equation.

The formation of neutral atoms makes the photons decouple from the matter. The redshift for decoupling can be determined as the epoch at which the optical depth for photons is unity. Using Eq. (10.156), we can compute the optical depth for the photons to be

$$\tau = \int_0^t n(t) X_e(t) \sigma_T dt = \int_o^z n(z) X_e(z) \sigma_T \left(\frac{dt}{dz}\right) dz \simeq 0.37 \left(\frac{z}{1000}\right)^{14.25},$$

(10.157)

where we have used the relation $H_0 dt \cong -\Omega_{\text{NR}}^{-1/2} z^{-5/2} dz$, which is valid for $z \gg 1$. This optical depth is unity at $z_{\text{dec}} = 1072$. From the optical depth, we can also compute the probability that the photon was last scattered in the interval $(z, z + dz)$. This is given by $(\exp -\tau)\, (d\tau/dz)$, which can be expressed as

$$P(z) = e^{-\tau} \frac{d\tau}{dz} = 5.26 \times 10^{-3} \left(\frac{z}{1000}\right)^{13.25} \exp\left[-0.37 \left(\frac{z}{1000}\right)^{14.25}\right].$$

(10.158)

This $P(z)$ has a sharp maximum at $z \simeq 1067$ and a width of about $\Delta z \cong 80$. It is therefore reasonable to assume that decoupling occurred at $z \simeq 1070$ in an interval of about $\Delta z \simeq 80$.

10.7 Gravitational lensing

The deflection of light by a gravitational field discussed in Section 7.4.2 suggests that images of distant cosmic sources will be affected by the intervening gravitational field. In particular, the deflection can lead to the formation of multiple images of the source which is of astrophysical significance. We shall now develop the basic theory of this phenomenon called *gravitational lensing*.[5]

Consider a ray of light that travels from a source S to an observer O. The gravitational field along its trajectory will continuously deflect it. To simplify the calculation of this effect, we will assume that all the deflection takes place when the light crosses the 'deflector plane' L placed at some appropriate location between O and S.

It is then convenient to project all relevant quantities on to this two-dimensional plane which is perpendicular to the line connecting the source and the observer. Let s and i denote the two-dimensional vectors giving the source and image positions projected on to this two-dimensional plane; and let $d(i)$ be the (vectorial) deflection produced by the lens. The geometry of gravitational lensing is shown in Fig. 10.3. The rays are assumed to propagate in straight lines to and from the lens plane and are deflected instantaneously in the lens plane. From the geometry of the diagram it follows that $\alpha D_{LS} + \theta_s D_{OS} = \theta_i D_{OS}$. The projection of the lengths $\theta_s D_{OS}$, αD_{LS} and $\theta_i D_{OS}$ on to the lens plane gives s, d and i. So,

$$D_{OS}\, s + D_{LS}\, d = D_{OS}\, i \tag{10.159}$$

or

$$s = i - \frac{D_{LS}}{D_{OS}} d(i). \tag{10.160}$$

To estimate the deflection d, consider a bounded density distribution $\rho(x)$ producing a gravitational potential $\phi(x)$. If a light ray is moving along the z-axis, it will experience a transverse deflection by the amount

$$\frac{d}{D_{OL}} = 2 \int_{-\infty}^{\infty} dz \left[\frac{^{(2)}\nabla\phi}{c^2} \right], \tag{10.161}$$

where $^{(2)}\nabla \equiv (\partial/\partial x, \partial/\partial y)$ is the gradient in the x–y plane and the factor 2 is the result of the correction to Newtonian deflection arising from general relativity obtained, e.g. in Chapter 7. Consider now the two-dimensional divergence $^{(2)}\nabla \cdot d$. Using

$$^{(2)}\nabla^2\phi = \nabla^2\phi - \frac{\partial^2\phi}{\partial z^2} = 4\pi G\rho(x) - \frac{\partial^2\phi}{\partial z^2} \tag{10.162}$$

we get

$$^{(2)}\nabla \cdot d = \frac{2D_{OL}}{c^2} \int_{-\infty}^{\infty} dz \left[4\pi G\rho(x,y,z) - \frac{\partial^2\phi}{\partial z^2} \right] = \frac{8\pi G D_{OL}}{c^2} \Sigma(x,y), \tag{10.163}$$

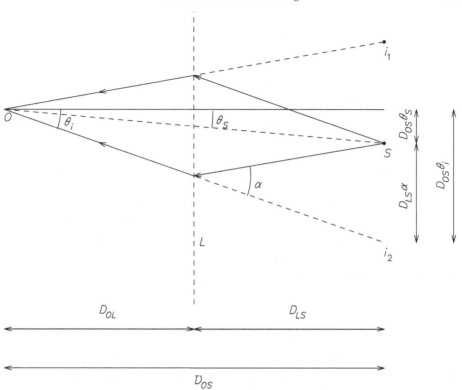

Fig. 10.3. The geometry of gravitational lensing. The source S is lensed by matter located at the plane L and produces two images i_1 and i_2 with respect to the observer O. We assume that all the deflection takes place at the plane of the lens.

where $\Sigma(x, y)$ is the surface mass density corresponding to $\rho(\boldsymbol{x})$. (The $\partial^2 \phi / \partial z^2$ term vanishes on integrations.) This two-dimensional Poisson equation has the solution:

$$\boldsymbol{d}(\boldsymbol{i}) = \frac{4 G D_{OL}}{c^2} \int d^2 x \Sigma(\boldsymbol{x}) \frac{(\boldsymbol{i} - \boldsymbol{x})}{|\boldsymbol{i} - \boldsymbol{x}|^2}, \qquad (10.164)$$

which gives the deflection $\boldsymbol{d}$ in terms of the surface density $\Sigma(\boldsymbol{x})$. This equation, together with Eq. (10.160), gives the relation between source and image position.

A more straightforward way of obtaining this result is to note that the propagation of light ray obeys the Fermat principle. It is easy to verify that the total time delay along the path, defined by an image position $\boldsymbol{i}$ and a source position $\boldsymbol{s}$, is given by

$$t(\boldsymbol{i}) = t_{\text{geom}} + t_{\text{grav}} = \frac{1 + z_L}{c} \frac{D_{OL} D_{OS}}{D_{LS}} \left[\frac{1}{2} (\boldsymbol{i} - \boldsymbol{s})^2 - \psi(\boldsymbol{i}) \right], \qquad (10.165)$$

where

$$\psi(i) = \frac{4GD_{OL}D_{LS}}{c^2 D_{OS}} \int d^2x \Sigma(x) \ln(|i - x|).$$ (10.166)

The first term in Eq. (10.165), quadratic in $(i - s)^2$ is purely geometrical and is due to the extra path length of the deflected light ray relative to the unperturbed trajectory. The second term is the time dilation introduced by the gravitational potential

$$\delta t = \int \frac{2}{c^3} |\phi| \, dl$$ (10.167)

corrected by the extra factor $(1 + z_L)$ to take into account cosmological expansion on the timescales. Fermat's principle requires that the trajectory of the light ray must be given by the extremum of the function $t(i)$ for fixed s. Ignoring overall multiplicative constants, we see that the location of the images can be obtained by calculating the extremum of the function

$$P(i) = \frac{1}{2} (i - s)^2 - \psi(i)$$ (10.168)

(treated as a function of i with fixed s). The gradient of $\psi(i)$ with respect to i is

$$\nabla \psi = \frac{4GD_{OL}D_{LS}}{c^2 D_{OS}} \int d^2x \Sigma(x) \nabla \ln(|i-x|) = \frac{4GD_{OL}D_{LS}}{c^2 D_{OS}} \int d^2x \Sigma(x) \frac{(i - x)}{|i - x|^2}.$$ (10.169)

It, therefore, follows that the equation $\nabla P = 0$ reduces to

$$i = s + \frac{4GD_{OL}D_{LS}}{c^2 D_{OS}} \int d^2x \Sigma(x) \frac{(i - x)}{|i - x|^2},$$ (10.170)

which is exactly Eq. (10.160) combined with Eq. (10.164). Hence the extrema of the function $P(i)$ gives the image positions. It is also obvious that $\psi(x)$ is proportional to the integral of the gravitational potential $\phi(x, z)$ along the z-axis and $\nabla^2_{(2)} \psi \propto \Sigma$. Equation (10.160) can now be written in component form as

$$s_a = i_a - \frac{\partial \psi}{\partial i_a} \qquad (a = 1, 2).$$ (10.171)

The time delay function $t(i)$ itself is of practical use in certain classes of observations. The time difference between two stationary points of the function $t(i)$ will give the relative time delay in light propagation between the corresponding images. If the source shows detectable variability, it will be seen in both images but with a time delay determined by the height difference between the two stationary points of the time delay surface $t(i)$ corresponding to the images.

For a smooth, spherically symmetric distribution of density (like that from a galaxy) with a central concentration, the deflection will decrease with r far away

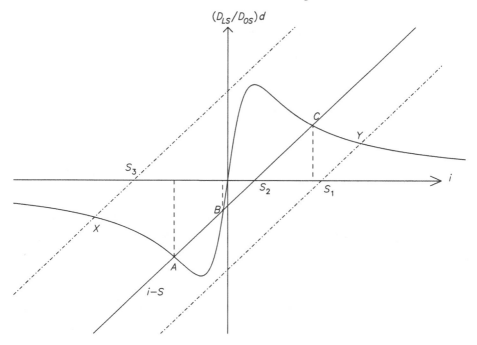

Fig. 10.4. Multiple images in the case of a spherical lens. When the source is near the centre of the lensing mass (at S_2, say), one obtains three images at the locations marked A, B and C. If the source is further away at S_1 or S_3, only one image (marked Y or X) is obtained.

from the lens. Near the origin of the lens, $\rho \approx$ constant and $\phi \propto r^2$; so the deflection will be linear in r. Hence the quantity $(D_{LS}/D_{OS})|d|$ will vary with $|i|$ roughly as shown in Fig. 10.4. The position of the images can be determined by finding the intersection of this curve with the line $y = (|i| - |s|)$. It is clear that, when the source is far away from the centre of the lens, these curves intersect only at one point giving rise to a single image. This is shown by the line through S_3 (or S_1) and the image is at X (or Y) respectively. When the source is closer to the centre, these curves intersect at three different points giving rise to three images. This is shown by the line through S_2 and the images are at A, B and C.

This lensing action can also lead to magnification of the source. This can be calculated by considering the change in the image position δi for a small change in the source position δs. The amplification will be the determinant of the matrix of the transformation between the two:

$$A = \det \left| \frac{\partial i_a}{\partial s_b} \right| = \left[\det \left| \frac{\partial s_b}{\partial i_a} \right| \right]^{-1}. \tag{10.172}$$

From Eq. (10.171) we get

$$\frac{\partial s_b}{\partial i_a} = \delta_{ab} - \frac{\partial^2 \psi}{\partial i_a \partial i_b} \equiv \delta_{ab} - \psi_{ab}. \tag{10.173}$$

The amplification as well as other properties of the lens can be conveniently studied in terms of the eigenvalues of the matrix ψ_{ab}. Since it follows from our definition (see equations Eq. (10.160) and Eq. (10.171)) that $\nabla \psi = (D_{LS}/D_{OS})d$, the two-dimensional Laplacian of ψ will be

$$^{(2)}\nabla^2 \psi = \frac{8\pi G}{c^2} \frac{D_{OL} D_{LS}}{D_{LS}} \Sigma \equiv 2\kappa. \tag{10.174}$$

The last equality defines a quantity κ (called *convergence*) in terms of the surface density Σ and the distances involved in the lensing. We thus see that $\mathrm{Tr}\,(\psi_{ab}) = \nabla^2 \psi = 2\kappa$. To characterize the matrix ψ_{ab} completely, we need only specify two more components which we take to be

$$-\gamma_1(\boldsymbol{i}) = \frac{1}{2}\left(\psi_{11} - \psi_{22}\right) \equiv \gamma(\boldsymbol{i})\cos\left[2\phi(\boldsymbol{i})\right] \tag{10.175}$$

$$-\gamma_2(\boldsymbol{i}) = \psi_{12} = \psi_{21} \equiv \gamma(\boldsymbol{i})\sin\left[2\phi(\boldsymbol{i})\right]. \tag{10.176}$$

With these definitions, we can write

$$\begin{aligned}
\psi_{ab} &= \begin{bmatrix} 1 - \kappa - \gamma_1 & -\gamma_2 \\ -\gamma_2 & 1 - \kappa + \gamma_1 \end{bmatrix} \\
&= (1 - \kappa)\begin{bmatrix} 1 & 0 \\ 0 & 1 \end{bmatrix} - \gamma \begin{bmatrix} \cos 2\phi & \sin 2\phi \\ \sin 2\phi & -\cos 2\phi \end{bmatrix}.
\end{aligned} \tag{10.177}$$

The meanings of the different terms are now clear from the above decomposition. Convergence alone will cause isotropic focusing of light rays, thereby leading to isotropic magnification of a source. The shear term, represented by γ_1 and γ_2, introduces anisotropic distortion in the image; the magnitude $\gamma = (\gamma_1^2 + \gamma_2^2)^{1/2}$ describes the total amount of shear and the angle ϕ describes the relative orientation. A circular source of unit radius will be distorted to an elliptical image with major and minor axis $(1 - \kappa - \gamma)^{-1}, (1 - \kappa + \gamma)^{-1}$. The magnification is now given by

$$A = \frac{1}{\det \psi} = \frac{1}{\left[(1 - \kappa)^2 - \gamma^2\right]}. \tag{10.178}$$

There is a simple relationship between the amplification matrix ψ_{ab} and the geometrical structure of the time delay surface $t(\boldsymbol{i})$ introduced in Eq. (10.165). In general, there are three kinds of stationary points on a two-dimensional surface defined by $t(\boldsymbol{i}) = $ constant. The nature of these stationary points will be related to the second derivatives of $t(i)$, and is proportional to the amplification matrix. If both the eigenvalues of the matrix

$$\mathcal{T} \equiv \frac{\partial^2 t(i)}{\partial i_a \partial i_b} \tag{10.179}$$

which describes the local curvature of the time delay surface are positive, the stationary point is a minimum and the images will have $\det \psi > 0$ and $\text{Tr} \, \psi > 0$. If the eigenvalues of $\mathcal{T}$ have opposite signs then we have a saddle point with $\det \psi < 0$; finally, if both the eigenvalues of $\mathcal{T}$ are negative, then $\det \psi > 0$ and $\text{Tr} \, \psi < 0$. It is obvious that when we have a minimum or a maximum, the images have positive magnification while the saddle point leads to negative magnification, which is interpreted as a reversal in the parity of the image.

Since the curvature of the surface defined by $t(i)$ measures the reciprocal of magnification, it follows that the image is strongly magnified in any direction along which the curvature is small and vice versa. The lines in the image plane on which the curvature vanishes are called *critical lines* and they correspond to infinite magnification (in the geometric optics limit, which, of course, breaks down around critical lines). The corresponding curves in the source plane are called *caustics*. We saw earlier that when the separation between the lens and the source is small, three images are formed, while when the source is away from the lens, only one image is formed. Consider now the changes in the image configuration as we slowly move the source from a position close to the lens towards larger separations. As the source moves, two of the images have to approach each other, merge and vanish. Since each of the images corresponds to a stationary point, it is clear that the curvature of the time delay surface should vanish as the images approach each other. Thus the brightest image configurations are obtained when the pair of images are close together on different sides of critical lines, just prior to merging. Critical lines and caustics provide a qualitative understanding of different image configurations and magnifications for any lensing geometry specified by $\psi(i)$. Some specific examples of gravitational lensing are examined in Project 10.1.

10.8 Killing vectors and the symmetries of the space

In obtaining the metric for the Friedmann model, we were guided by the observational input that the three-dimensional space is homogeneous and isotropic at sufficiently large scales. It turns out that such a three-dimensional space is a special case of a class of spaces known as *maximally symmetric spaces*. Further, if we relax the notion of isotropy, we obtain a larger class of spaces known as homogeneous spaces. In this section, we shall briefly discuss the notion of maximally symmetric spaces and homogeneous spaces more formally, using the concept of Killing vectors.

10.8.1 Maximally symmetric spaces

We saw in Chapter 5 (see Eq. (5.131)) that the second derivative of any Killing vector can be expressed in the form

$$\nabla_a \nabla_b \xi_n = R^m{}_{abn} \xi_m. \tag{10.180}$$

This shows that, if the Killing vector ξ_n and its first derivative $\nabla_b \xi_n$ are given, then all higher order derivatives can be determined in terms of these two. Hence a Killing vector field can be completely specified by specifying ξ_n and $\nabla_b \xi_n$ at a given point. Since $\nabla_b \xi_n = -\nabla_n \xi_b$ there are ${}^N C_2 = (1/2) N (N - 1)$ independent components in these derivatives (in an N-dimensional space) and another N components in ξ_n giving a total of $(1/2) N (N + 1)$ independently specifiable quantities. It follows that one can have only a *maximum* of $(1/2) N (N+1)$ linearly independent Killing vector fields in an N-dimensional spacetime.

A space (or a spacetime) that allows the existence of $(1/2) N (N+1)$ independent Killing vector fields is called a maximally symmetric spacetime. We shall now show that a maximally symmetric spacetime must have a constant scalar curvature with the curvature tensor being expressible in terms of scalar curvature by

$$R_{mabp} = \frac{R(g_{ap} g_{mb} - g_{ab} g_{mp})}{N(N - 1)}. \tag{10.181}$$

To prove this, we begin with the easily proved relation

$$(\nabla_i \nabla_a - \nabla_a \nabla_i) \nabla_b \xi_n = R^m{}_{nai} \nabla_b \xi_m + R^m{}_{bai} \nabla_m \xi_n, \tag{10.182}$$

which should hold for any tensor $\nabla_b \xi_a$ (see e.g. Eq. (5.21)). Using the Killing equation $\nabla_a \xi_b = -\nabla_b \xi_a$ and Eq. (10.180), we can obtain from this the relation

$$(\nabla_i R^m{}_{abn} - \nabla_a R^m{}_{ibn}) \xi_m$$
$$+ \left(R^m{}_{abn} \delta^k_i - R^m{}_{ibn} \delta^k_a + R^m{}_{bai} \delta^k_n - R^m{}_{nai} \delta^k_b \right) \nabla_k \xi_m = 0. \tag{10.183}$$

This result should hold for any Killing vector field; and in a generic space, this puts constraints on the possible values for ξ_n and $\nabla_m \xi_n$. But if the space is maximally symmetric, this relation should hold *without* any constraints on ξ_m and $\nabla_k \xi_m$. This, in turn, demands that the expressions within the two brackets in Eq. (10.183) must vanish identically. So we must have

$$\nabla_i R^m{}_{abp} = \nabla_a R^m{}_{ibp} \tag{10.184}$$

and

$$R^m{}_{abp}\delta^k_i - R^k{}_{abp}\delta^m_i - R^m{}_{ibp}\delta^k_a + R^k{}_{ibp}\delta^m_a + R^m{}_{bai}\delta^k_p - R^k{}_{bai}\delta^m_p$$
$$-R^m{}_{pai}\delta^k_b + R^k{}_{pai}\delta^m_b = 0. \tag{10.185}$$

Manipulating these results by suitable contractions, we can easily show that

$$(N-1)R^m{}_{abp} = R_{ap}\delta^m_b - R_{ab}\delta^m_p, \qquad NR^m_p = R\delta^m_p, \qquad R = \text{constant}, \tag{10.186}$$

which is equivalent to Eq. (10.181).

In the case of a three-dimensional maximally symmetric space, there will be $(1/2) \times 3 \times 4 = 6$ independent Killing vectors corresponding to three rotational and three translational degrees of freedom of a homogeneous and isotropic 3-space. Thus, the $t = $ constant spatial section of the Friedmann universe is a maximally symmetric three-dimensional space.

Exercise 10.15

Poincaré half-plane In two dimensions there are two obvious maximally symmetric spaces: viz. the plane (with zero curvature) and the surface of the sphere (with positive curvature). The third maximally symmetric space is the *Poincaré half-plane* given by the metric

$$ds^2 = \frac{a^2}{y^2}\left(dx^2 + dy^2\right), \tag{10.187}$$

where a is a constant.

(a) Show that this space has constant negative curvature $R = -(2/a^2)$.

(b) Prove that the proper length along the y-axis to $y = 0$ is infinite showing that this 'boundary' is infinitely far away.

(c) Obtain the geodesics in this spacetime and prove that they are circles with centres located along the x-axis. Obtain an explicit coordinate transformation which will express the line element in Eq. (10.187) in the standard form of a Friedmann metric.

Exercise 10.16

The Godel universe Consider a metric expressed in either of the two forms as:

$$ds^2 = a^2\left[dx^2 + \frac{1}{2}e^{2x}dy^2 + dz^2 - (e^x dy + dt)^2\right], \qquad a = \text{const.}, \tag{10.188}$$

$$ds^2 = 4a^2\left[dr^2 + dz^2 + \left(\sinh^2 r - \sinh^4 r\right)d\phi^2 - 2\sqrt{2}\,\sinh^2 r\,d\phi\,dt - dt^2\right]. \tag{10.189}$$

Show that this can be interpreted as a metric with the source made of dust as well as a negative cosmological constant term. Using the first form of the metric, show that the metric possess five different Killing vectors given by

$$\xi_1^a = (0,1,0,0), \qquad \xi_2^a = (0,0,1,0), \qquad \xi_3^a = (0,0,0,1),$$

$$\xi_4^a = (1,-y,0,0), \qquad \xi_5^a = \left(y, e^{-2x} - \frac{1}{2}y^2, 0, -2e^{-x}\right). \tag{10.190}$$

Using the second form of the metric Eq. (10.189), argue that – if ϕ is an angular coordinate – closed timelike curves will exist in this universe. This proves that one can have exact solutions to Einstein's equations with a physically reasonable source but containing closed timelike curves.

10.8.2 Homogeneous spaces

The Friedmann model of the universe was obtained from the assumption that the spatial slices of the universe can be chosen to be homogeneous and isotropic. A much wider class of spacetimes can be obtained if one considers spaces which are homogeneous but are not necessarily isotropic. The resulting cosmological models are not of direct practical utility since observations show that the spatial section of our universe is indeed highly isotropic. However, they are of interest for two indirect reasons. (i) In order to put observational constraints on the degree of anisotropy, it is necessary to work out the physical consequences in a universe that is not isotropic and compare them with observations; hence we need to study these models. (ii) They represent an interesting geometrical problem; viz. the classification of three-dimensional spaces satisfying certain symmetry conditions. We shall now describe the structure of three-dimensional homogeneous spaces and their classifications.[6]

We have seen (see Section 4.8) that translation along any vector field can be thought of as mapping the space into itself, at least in a local region where such a mapping is unique. If the vector field is a Killing vector field, then this mapping represents a symmetry of the space. For example, if the metric of a three-dimensional space can be expressed in a coordinate system such that $g_{\alpha\beta}$ is independent of a particular coordinate, say x^1, then the translation along the x^1 direction will not change the properties of the space. This is formally equivalent to the existence of a Killing vector $\xi^a = (0,1,0,0)$. If we can find *three* linearly independent Killing vectors, then we will have three different directions in which such translations can take place and we can consider such a 3-space as homogeneous. The simplest example of such a space is one with a metric of the form

$$dl^2 = a^2(t)dx^2 + b^2(t)dy^2 + c^2(t)dz^2 \tag{10.191}$$

in which the metric is independent of all the three spatial coordinates; translation along any of these directions leaves the geometrical properties of the space the same. (The space, of course, is not isotropic except when $a = b = c$.) We now have three independent Killing vector fields labelled by $B = 1, 2, 3$ and having the components $\xi_B^a = \delta_B^a$.

In general, one may not be able to determine the three Killing vectors of a homogeneous space just by inspection and we need to proceed in a different manner. Let us assume that we have a set of Killing vectors ξ_A with $A = 1, 2, 3, ...k$ in an N-dimensional space. The infinitesimal translation along a Killing vector ξ_A (or along ξ_B), changing the coordinates from x^a to x_A^a (or x_B^a), can be expressed in the form

$$x_A^a = x^a + \xi_A^a(x^i) \, d\lambda_A + \mathcal{O}(d\lambda_A^2) + \cdots \tag{10.192}$$

$$x_B^a = x^a + \xi_B^a(x^i) \, d\lambda_B + \mathcal{O}(d\lambda_B^2) + \cdots , \tag{10.193}$$

where $d\lambda_A$ and $d\lambda_B$ are two infinitesimal parameters. The first transformation maps the point P to a point P_1 while the second transformation maps the point P to a point P_2. If we perform the two transformations one after the other, the result will depend on the order in which the transformations are performed. If we perform the transformations using ξ_A first and the one using ξ_B second, the point P will be mapped to P_1 and then to a point P_3, say. Similarly, if we do the transformations in reverse order, the point P will be mapped to P_2 and then to P_4, say. It is easy to see that the difference between these two transformations can be expressed in the form

$$\Delta x_{AB}^a = (\xi_A^n \, \partial_n \xi_B^a - \xi_B^n \, \partial_n \xi_A^a) \, d\lambda_A \, d\lambda_B + \cdots = [\xi_A, \xi_B]^a \, d\lambda_A \, d\lambda_B + \cdots , \tag{10.194}$$

where we have introduced the Lie bracket (see Eq. (4.126); the procedure adopted here is identical to the one discussed there). We thus see that infinitesimal transformations commute only to the first order; at the second order, they lead to the difference obtained above.

The infinitesimal quantity Δx_{AB} is the difference of the coordinates of P_3 and P_4. We know, however, that in a homogeneous space there will exist another Killing vector (which is a linear combination of the set of Killing vectors ξ_A) that will map point P_3 to point P_4. Therefore, we must have a relation of the form

$$(\xi_A^n \, \partial_n \xi_B^a - \xi_B^n \, \partial_n \xi_A^a) = C_{AB}^D \, \xi_D^a, \qquad A, B, D = 1, \ldots, k, \tag{10.195}$$

where C_{AB}^D are a set of numerical constants called the *structure constants*. If we associate with the vectors ξ_A the operators $X_A = \xi_A^n \partial_n$ then it is easy to see that the commutators of these operators are related to the structure constants by $[X_A, X_B] = C_{AB}^D X_D$. This relation, as well as the Jacobi identity for the vector fields (see Eq. (4.139)), imply that the structure constants must satisfy the conditions

Table 10.1. *The Bianchi classification of three-dimensional homogeneous spaces*

Type	a	$n^{(1)}$	$n^{(2)}$	$n^{(3)}$
I	0	0	0	0
II	0	1	0	0
VII$_0$	0	1	1	0
VI$_0$	0	1	-1	0
IX	0	1	1	1
VIII	0	1	1	-1
V	1	0	0	0
IV	1	0	0	1
VII$_0$	a	0	1	1
III $(a = 1)$ VI$_a$ $(a \neq 1)$	a	0	1	-1

$$C_{AB}^D = -C_{BA}^D; \qquad C_{AD}^E C_{BC}^D + C_{BD}^E C_{CA}^D + C_{CD}^E C_{AB}^D = 0. \qquad (10.196)$$

The problem of classifying the homogeneous spaces therefore reduces to determining the general form of the structure constants that satisfy the above two conditions.

In the case of three-dimensional space, this is fairly easy to do. We know that $C_{\alpha\beta}^\mu$ (where we have switched to Greek subscripts, etc., from Latin capital letters since we are working in three-dimensional space) has only (3×3) independent components which can be mapped to the elements of a 3×3 matrix $N^{\mu\nu}$ by the construction $\epsilon^{\alpha\beta\nu} C_{\alpha\beta}^\mu = 2N^{\mu\nu}$. This matrix $N^{\mu\nu}$ can, in turn, be separated into a symmetric part $n^{\mu\nu}$ and an antisymmetric part which, in turn, can be expressed in terms of a vector a_α; so we obtain the decomposition:

$$N^{\mu\nu} = n^{\mu\nu} + \epsilon^{\mu\nu\alpha} a_\alpha. \qquad (10.197)$$

Expressing the structure constant in terms of $n^{\mu\nu}$ and a_α as

$$C_{\alpha\beta}^\mu = \epsilon_{\nu\alpha\beta} N^{\mu\nu} = \epsilon_{\nu\alpha\beta} n^{\mu\nu} + \delta_\beta^\mu a_\alpha - \delta_\alpha^\mu a_\beta \qquad (10.198)$$

and substituting into the Jacobi identity, we obtain the condition $n^{\alpha\beta} a_\alpha = 0$. We can always choose the coordinates such that $n^{\alpha\beta}$ is brought to diagonal form. It is clear from the above constraint that the vector a lies along one of the principal direction of $n^{\alpha\beta}$ corresponding to the eigenvalue 0. Hence we must have $an_1 = 0$ forcing either a or n_1 or both to vanish. The commutation relations now reduce to the form

$$[X_1, X_2] = -aX_2 + n_3 X_3, \qquad [X_2, X_3] = n_1 X_1, \qquad [X_3, X_1] = n_2 X_2 + aX_3.$$
$$(10.199)$$

We still have the freedom of changing the signs of X_α and rescaling them by multiplication by constants. Using this freedom we can make all the structure constants equal to ± 1 if at least one of the three quantities (a, n_2, n_3) vanishes. If none of them vanishes, we can still perform a scale transformations which leaves the ratio $a^2/n_2 n_3$ invariant.

It is now an elementary matter to list all possible three-dimensional homogeneous spaces in a manner usually called the *Bianchi classification*. These are given in Table 10.1. Of these different spaces, three are worth mentioning. The first one is Bianchi type I which was represented by a spatial metric of the form in Eq. (10.191). When $a(t) = b(t) = c(t)$ it represents the Euclidean space of zero curvature. The space of constant positive curvature (which corresponds to $k = +1$ Friedmann model) is a special case of type IX. Similarly, the space of constant negative curvature ($k = -1$ Friedmann model) is a special case of type V.

In the case of a three-dimensional homogeneous space, the Killing vector components ξ_A^n define the infinitesimal displacements of the form $\xi_A^n dx^A$ which we will write simply as $\xi_\alpha^\beta dx^\alpha$ where the Greek indices as usual run over 1, 2, 3. The infinitesimal spatial line element can now be expressed in the form

$$dl^2 = q_{\beta\mu}(t) \left(\xi_\alpha^\beta dx^\alpha \right) \left(\xi_\nu^\mu dx^\nu \right), \tag{10.200}$$

where $q_{\beta\mu}(t)$ are arbitrary functions with the spacetime metric given by

$$ds^2 = -dt^2 + dl^2 = -dt^2 + h_{\alpha\nu} dx^\alpha dx^\nu; \qquad h_{\alpha\nu} \equiv q_{\beta\mu}(t)\xi_\alpha^\beta \xi_\nu^\mu. \tag{10.201}$$

Using the general form of structure constants as classified above in Eq. (10.195), we can determine ξ_β^α and hence the metric. We shall not pursue this further since the resulting models do not seem to have direct physical relevance.

Exercise 10.17
Kasner model of the universe Consider an anisotropic spacetime with the metric $g_{ab} = $ dia $[-1, a(t), b(t), c(t)]$.
(a) Show that the vacuum Einstein equations for this spacetime can be solved by the functions $a(t) = t^{2p_1}, b(t) = t^{2p_2}, c(t) = t^{2p_3}$, where p_1, p_2, p_3 satisfy the constraints

$$p_1 + p_2 + p_3 = 1; \qquad p_1^2 + p_2^2 + p_3^2 = 1. \tag{10.202}$$

(b) Show that a parametric solutions to these equations can be expressed in the form

$$p_1(u) = \frac{-u}{1 + u + u^2}; \qquad p_2(u) = \frac{1 + u}{1 + u + u^2}; \qquad p_3(u) = \frac{u(1 + u)}{1 + u + u^2}. \tag{10.203}$$

Exercise 10.18
CMBR in a Bianchi Type I model Consider the evolution of black body radiation in a
Bianchi type I model with the spatial metric given by Eq. (10.191). Assume that, at some
very early epoch, $a = b = c$ and the universe was filled with thermalized electromagnetic radiation with some uniform temperature T_0. Determine the specific intensity $I(n, \nu)$
of the radiation measured by observers at rest in the coordinate system of the metric in
Eq. (10.191) at some arbitrary, late, time. Show that the spectrum is thermal in any given
direction but the temperature depends on the direction. [Hint. Use the three Killing vectors of the spacetime to determine the distribution function of photons. The final result is
that the specific intensity corresponds to a thermal spectrum with temperature varying as
$T(n) = T_0[(an^x)^2 + (bn^y)^2 + (cn^z)^2]^{-1/2}.$]

PROJECT

Project 10.1
Examples of gravitational lensing

This project applies the theory developed in Section 10.7 to study several specific examples
of gravitational lenses which have proved to be useful in astrophysical contexts.
(a) The simplest case corresponds to a 'sheet' of matter with a constant surface density
acting as a lens. Show that, for $\Sigma(x, y) = \Sigma_0$, Eq. (10.160) gives

$$|s| = |i| \left(1 - \frac{\Sigma_0}{\Sigma_{\text{crit}}}\right); \qquad \Sigma_{\text{crit}} \equiv \frac{c^2 D_{OS}}{4\pi G D_{OL} D_{LS}}, \qquad (10.204)$$

and that the corresponding magnification is

$$A = \left(1 - \frac{\Sigma_0}{\Sigma_{\text{crit}}}\right)^{-2}. \qquad (10.205)$$

The subscripts O, L, S stand for observer, lens and source, with D_{OS} being the distance
between the observer and the source, etc. Estimate the numerical value of Σ_{crit} in a typical
cosmological context. How does it compare with the projected surface density due to a
galaxy?
(b) Consider next a point mass M acting as a lens. Show that, in this case, we get

$$s = i\left[1 - \frac{L^2}{i^2}\right]; \qquad L^2 = \frac{4GM D_{LS} D_{OL}}{c^2 D_{OS}}. \qquad (10.206)$$

This result shows that s and i are in the same direction and their magnitudes are related
by $s = i - (L^2/i)$. Solving the quadratic equation, find the two solutions for the image
position i as

$$i_{\pm} = \frac{1}{2}\left(s \pm \sqrt{s^2 + 4L^2}\right). \qquad (10.207)$$

The length scale L is called the *Einstein ring radius*. Describe the geometry of the image configuration and the nature of the magnification as the source moves away from the lens. Estimate the angular size of Einstein ring in a typical cosmological context.

(c) The mass distribution in several astrophysical objects, like the halos around large galaxies or clusters of galaxies, is well described by the relation $M(r) \propto r$. This relation is usually written as $M(r) \cong (2\sigma^2/G)r$ for such a source which is usually called an *isothermal sphere*. Determine the Newtonian gravitational potential for this mass distribution and compute the deflection in Eq. (10.161). Show that the angular deflection is now a constant in magnitude and is given by

$$\alpha = 4\pi \left(\frac{\sigma}{c}\right)^2 \approx 2.6 \text{ arc sec } \left(\frac{\sigma}{300 \text{ km s}^{-1}}\right)^2. \tag{10.208}$$

What kind of image configuration results in this case?

(d) In the cosmological context, one is often interested in the probability of different kinds of lensing phenomena to occur when the sources are objects like, say, quasars located at large distances and the lenses are mass distributions like galaxies located somewhere along the line of sight. Making reasonable simplifying assumptions, show that the optical depth for gravitational lensing by sources of a particular kind within a redshift z_s can be expressed as

$$\tau = 16\pi^3 \left(\frac{\sigma}{c}\right)^4 n_0 \int_0^{z_s} \left(\frac{D_{OL}D_{LS}}{D_{OS}}\right)^2 d_H(z_L)\, dz_L. \tag{10.209}$$

Here, it is assumed that the comoving number density of sources is a constant (n_0) and each source is assumed to be an isothermal sphere with velocity dispersion σ. Estimate this optical depth numerically for the standard cosmological model discussed in the text.

(e) Show that the probability of a source at redshift z_s being lensed in a $\Omega_\Lambda + \Omega_{NR} = 1$ universe (relative to the corresponding probability in a $\Omega_{NR} = 1, \Omega_\Lambda = 0$ model) is given by

$$P_{\text{lens}} = \frac{15}{4} \left[1 - \frac{1}{\sqrt{1+z_s}}\right]^{-3} \int_1^{a_s} \frac{H_0}{H(a)} \left[\frac{d_A(0,a)d_A(a,a_s)}{d_A(0,a_s)}\right] da, \tag{10.210}$$

where $a_s = (1+z_s)^{-1}$ and $d_A(a_1, a_2)$ is the angular diameter distance between two events at epochs $a = a_1$ and $a = a_2$. Plot this probability as a function of Ω_Λ for $z_s = 2$ and explain how this result can be used to put a bound on Ω_Λ.

11

Differential forms and exterior calculus

11.1 Introduction

This chapter introduces the language of differential forms and exterior calculus. It will translate some of the results in the previous chapters into the language of forms and will also describe briefly the structure of gauge theories to illustrate the generality of the formalism. The emphasis will be oriented towards developing the notation and connecting it up with the more familiar developments in the earlier chapters, rather than providing a rigorous mathematical formalism starting from first principles.[1]

11.2 Vectors and 1-forms

We will begin by recalling some of the concepts like the tangent vector space, etc., introduced briefly in Chapters 1 and 4. Our aim will be to develop these ideas further in a more formal manner.

Consider a spacetime manifold in which we have introduced a coordinate system around a given event $\mathcal{P}$. A curve $\mathcal{C}$ in this region can be described by giving the four functions $x^i(\lambda)$ with the parameter λ increasing monotonically along the curve. Let $f(x)$ be a function defined on the spacetime in this region. Then the variation of this function along the curve $\mathcal{C}$ is given by

$$\frac{df}{d\lambda} = \frac{dx^i(\lambda)}{d\lambda} \frac{\partial f}{\partial x^i} \equiv V^i \, \partial_i f. \tag{11.1}$$

The four quantities $V^i = dx^i/d\lambda$ give the tangent vector to the curve at any given location. More importantly, the above relation allows us to define this vector in terms of a directional derivative operator $\boldsymbol{V} \equiv d/d\lambda$ along the path. This operator is a map from the set of smooth functions to the real line, given by $f \rightarrow \boldsymbol{V}f = df/d\lambda$. For any given function, this map produces a real number at a given event on the spacetime once the curve is specified. Further, this map obeys a linearity property

$\boldsymbol{V}(f + g) = \boldsymbol{V}f + \boldsymbol{V}g$ as well as the Leibnitz rule $\boldsymbol{V}(fg) = (\boldsymbol{V}f)g + f(\boldsymbol{V}g)$. Such a map can be used to define a tangent vector at the event $\mathcal{P}$ where the evaluations are made.

At first, it might seem a little strange that a tangent vector – which we intuitively think of as a 4-component object described by the four functions $V^i(x)$ at every event in the spacetime – is now being defined as a linear map with certain properties on the space of functions. However, the one-to-one correspondence between these two concepts becomes obvious when we note that we can always associate with any vector field $V^i(x)$, defined in the usual manner in terms of components, an operator $\boldsymbol{V}$ such that

$$\boldsymbol{V} = V^i \partial_i. \qquad (11.2)$$

In fact, the operators $\boldsymbol{e}_i = \partial_i$ play the role of the coordinate basis vectors used to define the components of $\boldsymbol{V}$ through $\boldsymbol{V} = V^i \boldsymbol{e}_i$.

There is, however, an important conceptual advantage in thinking of vectors as directional derivative operators. From the transformation rule for the components of the vector and for ∂_i, it is obvious that, under a general coordinate transformation, $V'^i \partial'_i = V^i \partial_i$ so that the operator $\boldsymbol{V}$ remains invariant under the coordinate transformations. That is, we can think of the directional derivative operator as fully analogous to the geometrical vector $\boldsymbol{V}$. In contrast, what we usually call 'the vector V^i' should be, more precisely, called 'the vector whose components are V^i'. For most physical purposes, this distinction is unnecessarily pedantic and the context will make clear whether we are referring to a vector as a geometrical object or the components of a vector in a given coordinate basis. But, for the development of the formal machinery, it is useful to maintain this distinction.

Suppose we have two different curves with parameters λ and μ that intersect at an event $\mathcal{P}$. Then we can define two directional derivatives $\boldsymbol{V} = d/d\lambda$ and $\boldsymbol{U} = d/d\mu$. We can now define the addition of two vectors at $\mathcal{P}$ through the obvious rule $(\boldsymbol{V} + \boldsymbol{U})f = \boldsymbol{V}f + \boldsymbol{U}f$. We can also multiply the tangent vectors by real numbers in a natural fashion. These properties show that the space of all tangent vectors at an event $\mathcal{P}$ in a spacetime manifold M is a well-defined vector space. This vector space is called the tangent vector space at $\mathcal{P}$ and is denoted by $T(\mathcal{P})$.

As mentioned above, the operators ∂_i form a natural basis for the tangent vector space $T(\mathcal{P})$. This basis – which is the one we have used in most of the previous chapters – is called a *coordinate basis*. The basis vectors introduced in Chapters 1 and 4 can be thought of as the tangent vectors to the coordinate lines, treated as curves on the manifold. There is, however, no compulsion that we have to use a coordinate basis and, in some applications, it may indeed be more convenient to use a non-coordinate basis. Such a non-coordinate basis $\boldsymbol{e}_a$ can be defined through the relation

$$\boldsymbol{e}_a = e_a{}^i(x)\,\partial_i, \tag{11.3}$$

where $e_a{}^i(x)$ are a set of functions which define the basis. (We will assume that det $(e_a{}^i) \neq 0$ so that this basis will properly span the vector space.) The $e_a{}^i$ can vary from point to point on the manifold, since we can choose the basis for the tangent space at each event in a completely independent manner. A simple example of non-coordinate basis arises when we describe the velocity of a particle in three-dimensional space in spherical polar coordinates (r, θ, ϕ). In a coordinate basis, the velocity vector should be

$$\boldsymbol{v} = v^i \partial_i = \dot{r}\partial_r + \dot{\theta}\partial_\theta + \dot{\phi}\partial_\phi. \tag{11.4}$$

It is, however, more conventional to write this as

$$\boldsymbol{v} = \dot{r}\partial_r + (r\dot{\theta})\left(\frac{1}{r}\partial_\theta\right) + (r\sin\theta)\dot{\phi}\left(\frac{1}{r\sin\theta}\partial_\phi\right), \tag{11.5}$$

which corresponds to using the non-coordinate basis $(\partial_r, r^{-1}\partial_\theta, (r\sin\theta)^{-1}\partial_\phi)$ so that the components become $(\dot{r}, r\dot{\theta}, \dot{\phi}r\sin\theta)$ in standard spherical polar coordinates.

Treating the vectors as differential operators allows us to define the Lie bracket of two vectors $\boldsymbol{X}$ and $\boldsymbol{Y}$ through its action on any function:

$$[\boldsymbol{X}, \boldsymbol{Y}]f = (\boldsymbol{X}\boldsymbol{Y} - \boldsymbol{Y}\boldsymbol{X})f = \boldsymbol{X}(\boldsymbol{Y}f) - \boldsymbol{Y}(\boldsymbol{X}f). \tag{11.6}$$

This gives the components of the Lie bracket in a coordinate basis to be

$$[\boldsymbol{X}, \boldsymbol{Y}]^j = X^k \partial_k Y^j - Y^k \partial_k X^j. \tag{11.7}$$

It is easy to show from the definition that the Lie bracket, or Lie derivative as it is often called, satisfies the antisymmetry property

$$\pounds_{\boldsymbol{X}}\boldsymbol{Y} \equiv [\boldsymbol{X}, \boldsymbol{Y}] = -[\boldsymbol{Y}, \boldsymbol{X}] = -\pounds_{\boldsymbol{Y}}\boldsymbol{X} \tag{11.8}$$

as well as the Jacobi identity

$$[[\boldsymbol{X}, \boldsymbol{Y}], \boldsymbol{Z}] + [[\boldsymbol{Y}, \boldsymbol{Z}], \boldsymbol{X}] + [[\boldsymbol{Z}, \boldsymbol{X}], \boldsymbol{Y}] = 0, \tag{11.9}$$

which is the same as the derivative property:

$$\pounds_{\boldsymbol{X}}[\boldsymbol{Y}, \boldsymbol{Z}] = [\pounds_{\boldsymbol{X}}\boldsymbol{Y}, \boldsymbol{Z}] + [\boldsymbol{Y}, \pounds_{\boldsymbol{X}}\boldsymbol{Z}]. \tag{11.10}$$

Also note that $[\pounds_{\boldsymbol{X}}, \pounds_{\boldsymbol{Y}}] = \pounds_{[\boldsymbol{X},\boldsymbol{Y}]}$ for actions on any vector field. The Lie brackets of the coordinate basis vectors vanish since the partial derivatives commute. More generally, the commutator $[\boldsymbol{e}_i, \boldsymbol{e}_j]$ is another vector and hence can be expanded in the same basis as $[\boldsymbol{e}_i, \boldsymbol{e}_j] = c_{ij}{}^k \boldsymbol{e}_k$, where the $c_{ij}{}^k$ are called the *structure constants* of the basis.

Our next task is to introduce the *dual space* to the tangent vector space. For any vector space T, there exists a well defined notion of a dual space T^*, which is the space of linear maps from T to the real line. In practical terms, this is realized as follows. Let Ω be an element of T^*. Then, we will specify a rule for constructing a real number from every vector V and every element Ω and denote this rule by the notation $\langle \Omega | V \rangle$. This rule should satisfy the linearity conditions

$$\langle \Omega | U + V \rangle = \langle \Omega | U \rangle + \langle \Omega | V \rangle, \qquad \langle \Omega | \lambda V \rangle = \lambda \langle \Omega | V \rangle, \tag{11.11}$$

where U and V are any two vectors and λ is a real number. We can now introduce a set of basis vectors ω^a in the dual space $T^*(\mathcal{P})$ of the tangent vector space and expand its elements as $\Omega = \Omega_a \omega^a$. By definition, the basis vectors in $T^*(\mathcal{P})$ and $T(\mathcal{P})$ satisfy the rule $\langle \omega^a | e_b \rangle = \delta_b^a$. It is now obvious that the rule for any two elements can be expressed in terms of the components as $\langle \Omega | V \rangle = \Omega_a V^a$. The dual space $T^*(\mathcal{P})$ is also sometimes called the cotangent space.

Just as the directional derivative of a function allowed us to define the elements of $T(\mathcal{P})$, the *differential* of a function provides a natural realization of the elements of $T^*(\mathcal{P})$. If $f(x)$ is a function defined on the spacetime manifold, then its differential df is defined through the rule

$$\langle df | V \rangle = V f. \tag{11.12}$$

Just as the partial derivative operators ∂_i form a natural basis for $T(\mathcal{P})$, one can consider the coordinate differentials dx^i to form the natural basis for $T^*(\mathcal{P})$. This is clear from Eq. (11.12) with the choice $f = x^i$, $V = \partial_j$ leading to

$$\langle dx^i | \partial_j \rangle = \delta_j^i \tag{11.13}$$

as expected from the dual basis. Using this, we can define an arbitrary element of $T^*(\mathcal{P})$ as $\Omega = \Omega_i dx^i$ which is completely analogous to the relation $V = V^i \partial_i$ in $T(\mathcal{P})$. Just like V, the element Ω is a geometric entity that remains invariant under the general coordinate transformation. It is usual to call any element of $T^*(\mathcal{P})$ a *1-form*.

Given the action of Lie derivatives on vectors as well as on scalars, one can define the Lie derivative on a 1-form through the relation

$$\pounds_X \langle \Omega | Y \rangle = \langle \pounds_X \Omega | Y \rangle + \langle \Omega | \pounds_X Y \rangle, \tag{11.14}$$

where Y is an arbitrary vector. This shows that the components of the Lie derivative of the 1-form in a coordinate basis should be

$$(\pounds_X \Omega)_j = X^k \partial_k \Omega_j + \Omega_k \partial_j X^k. \tag{11.15}$$

From the tangent space $T(\mathcal{P})$ and the cotangent space $T^*(\mathcal{P})$ at any given event $\mathcal{P}$, we can form direct product vector spaces by multiplying them suitably. For example, an element S of $T \otimes T$ can be expanded using the basis $e_i \otimes e_j$ as

$$S = S^{ij} e_i \otimes e_j, \tag{11.16}$$

etc. One can similarly define the elements of $T^* \otimes T^*$ or the elements of $T \otimes T^*$ all defined at an event $\mathcal{P}$, though we will not explicitly write $T(\mathcal{P}) \otimes T^*(\mathcal{P})$, etc., when the meaning is clear. The generalization to product spaces with more than two components is straightforward. The elements of these vector spaces are tensors with different number of covariant and contravariant indices. One can also think of tensors as linear functions from the appropriate product vector space to the real line. For example, a second rank tensor S allows us to construct a real number $S^{ij} V_i U_j$ from an element of $T^* \otimes T^*$, the components of which are $V_i U_j$. That is, one can think of the tensor S as a 'machine with two slots' which will output the real number $S^{ij} V_i U_j$, when two 1-forms U and V are 'inserted into its slots'. Symbolically, one indicates this by the notation $S(U, V) = S^{ij} V_i U_j$. The rule for generating the output satisfies the usual bi-linearity conditions which is obvious in terms of components. The components of the tensor themselves are obtained by using the basis forms as the input; i.e. $S(\omega^i, \omega^j) = S^{ij}$.

We can define the Lie derivative of a tensor field by assuming that the Lie derivative satisfies the Leibnitz rule while acting on tensor produces; e.g.

$$\pounds_X(S \otimes T) = \pounds_X S \otimes T + S \otimes \pounds_X T \tag{11.17}$$

and defining its action on scalar functions as $\pounds_X f = X f = \langle df | X \rangle$.

The tangent space and the cotangent space are purely local constructs which allow the manipulation of vectors at a single event. We shall next introduce the concept of covariant differentiation, by adding a further structure on the manifold. This can be done by introducing a covariant derivative operator, ∇, along a vector field with the following properties. (i) It is symmetric in the sense that $\nabla_u v - \nabla_v u = [u, v]$ for any two vector fields u and v. (ii) It is additive in two ways $\nabla_u(v + w) = \nabla_u v + \nabla_u w$ and $\nabla_{au+bn} v = a\nabla_u v + b\nabla_n v$. (iii) It satisfies the standard chain rule: $\nabla_u(fv) = f\nabla_u v + v\partial_u f$. Of these, the first rule is equivalent to the condition that the space has no torsion. We shall confine our attention to only such spaces although a somewhat more general structure can be obtained by relaxing this condition. One of the linearity conditions in (ii) and the chain rule in (iii) are obvious and the only nontrivial condition is $\nabla_{au+bn} v = a\nabla_u v + b\nabla_n v$. This condition essentially translates what we know in component form as $\nabla_X V \to X^a \nabla_a V^b$; written in the component notation, the linearity in 'differentiating vector' X^a is obvious.

More importantly, we can evaluate the covariant derivatives of any tensorial object once we know the action of ∇ on the basis vectors and basis forms. The action on the basis vectors is defined through the relation

$$\nabla_b \boldsymbol{e}_a = \boldsymbol{e}_m \Gamma^m{}_{ab}, \tag{11.18}$$

where $\Gamma^m{}_{ab}$ are the connection coefficients – but introduced without any reference to the metric yet. Using the properties listed above, one can now obtain the action of ∇ on the basis 1-forms and we get

$$\nabla_b \boldsymbol{\omega}^n = -\Gamma^n{}_{ab} \boldsymbol{\omega}^a. \tag{11.19}$$

The covariant derivative of any arbitrary tensor can now be obtained by the standard chain rule. These relations also show that the connection can be expressed in an equivalent form as $\langle \boldsymbol{\omega}^m | \nabla_b \boldsymbol{e}_a \rangle = \Gamma^m{}_{ab}$ or $\langle \nabla_b \boldsymbol{\omega}^n | \boldsymbol{e}_a \rangle = -\Gamma^n{}_{ab}$. The standard transformation law for the connection coefficients now follows from its definition.

In Chapter 4, we have been working with a coordinate basis where $\boldsymbol{e}_i = \partial_i$, etc. In a non-coordinate basis, it is clear that

$$\Gamma^m{}_{[ab]} = -\frac{1}{2} \langle \boldsymbol{\omega}^m | [\boldsymbol{e}_a, \boldsymbol{e}_b] \rangle \equiv -\frac{1}{2} c_{ab}{}^m. \tag{11.20}$$

That is, the antisymmetric part of the connection is completely determined by the structure constants. This also follows from the symmetry property of $\nabla_{\boldsymbol{u}} \boldsymbol{v} - \nabla_{\boldsymbol{u}} \boldsymbol{u} = [\boldsymbol{u}, \boldsymbol{v}]$ applied to a couple of basis vectors: $\boldsymbol{u} = \boldsymbol{e}_j; \boldsymbol{v} = \boldsymbol{e}_i$.

The connection allows the parallel transport of the vectors around closed curves and hence we can now introduce the concept of curvature by the procedure we followed in Chapter 5. To do this formally, we introduce the curvature operator

$$\mathcal{R}(\boldsymbol{A}, \boldsymbol{B}) \equiv [\nabla_{\boldsymbol{A}}, \nabla_{\boldsymbol{B}}] - \nabla_{[\boldsymbol{A}, \boldsymbol{B}]} \tag{11.21}$$

and define the components of the curvature tensor through the relation

$$R^a{}_{bcd} \equiv \langle \boldsymbol{\omega}^a | \mathcal{R}(\boldsymbol{e}_c, \boldsymbol{e}_d) \boldsymbol{e}_b \rangle. \tag{11.22}$$

It is easy to verify that, in the coordinate basis, this reduces to the definition in Eq. (5.7). In a non-coordinate basis we will have an additional term:

$$R^a{}_{bcd} = \Gamma^a{}_{bd,c} - \Gamma^a{}_{bc,d} + \Gamma^a{}_{mc} \Gamma^m{}_{bd} - \Gamma^a{}_{md} \Gamma^m{}_{bc} - \Gamma^a{}_{bm} c_{cd}{}^m, \tag{11.23}$$

where $\Gamma^a{}_{bd,c} \equiv \boldsymbol{e}_c[\Gamma^a{}_{bd}]$, etc. We see that the ideas of affine connection and curvature are logically independent of the notion of the metric.

The next level of structure that one could introduce on the manifold (making it a Riemannian manifold from an affine manifold) is a metric. Given the metric g_{ab} and two vectors $\boldsymbol{U}$ and $\boldsymbol{V}$, we can obtain a real number $g_{ab} U^a V^b = \boldsymbol{U} \cdot \boldsymbol{V}$. More formally, we can think of the metric tensor $\boldsymbol{g}$ as a machine with two slots which produces a real number from two vectors in a symmetric and bi-linear fashion. This allows us to put the components of a 1-form in one-to-one correspondence with the covariant components of the vector using the metric. Given a vector $V \in T(\mathcal{P})$, we can always choose an element Ω of $T^*(\mathcal{P})$ such that $\langle \Omega | V \rangle \equiv \Omega_b V^b$ is numerically equal to $U \cdot V$. This clearly gives a one-to-one correspondence between the elements of $T^*(\mathcal{P})$ and elements of $T(\mathcal{P})$ via the rule $\Omega_b = g_{ba} U^a$. Such an identification arises naturally when we use the notion of a metric to raise and lower the indices.

Though metric and affine connection can be introduced on a manifold as independent structures, for practical applications in general relativity, one relates these two by a consistency condition which can be stated as

$$\nabla_u (v \cdot w) = (\nabla_u v) \cdot w + v \cdot (\nabla_u w). \tag{11.24}$$

Introducing the definitions $c_{abc} \equiv c_{ab}{}^m g_{mc} \equiv \langle \omega^m | [e_a, e_b] \rangle g_{mc}$ and $\Gamma^a{}_{bc} = g^{am} \Gamma_{mbc}$ we can relate the metric and the connection by

$$\Gamma_{mbc} = \frac{1}{2} \left(g_{mb,c} + g_{mc,b} - g_{bc,m} + c_{mbc} + c_{mcb} - c_{bcm} \right), \tag{11.25}$$

where $g_{mb,c} = e_c[g_{mb}]$, etc. This reduces to the standard definition in Eq. (4.47) of Chapter 4 when we work in a coordinate basis.

In a spacetime with a metric there exists a special class of non-coordinate basis called the *orthonormal frame*, which we shall now introduce. (In fact, the two most frequently used basis vectors in a manifold correspond to the coordinate basis and the orthonormal basis.) An orthonormal basis $T^*(\mathcal{P})$ is a set of 1-forms $\omega^a = \omega^a_i dx^i$ defined in such a way that the metric tensor can be expressed in terms of ω^a_i as

$$g_{ij} = \eta_{ab} \omega^a_i \omega^b_j, \tag{11.26}$$

where $\eta_{ab} = \mathrm{dia}(-1, 1, ..., 1)$ is the standard Lorentz metric if we are working in a $(D+1)$ spacetime manifold and $\eta_{ab} = \delta_{ab}$ if we are working in a Euclidean manifold. It is obvious from Eq. (11.26) that ω^a_i can be thought of as the 'square root of the metric'. In the orthonormal frame, we have the relation

$$ds^2 = g_{ij} dx^i \otimes dx^j = \eta_{ab} \omega^a_i \omega^b_j dx^i \otimes dx^j = \eta_{ab} \omega^a \otimes \omega^b \tag{11.27}$$

so that in the orthonormal basis, the metric tensor has the components $g_{ab} = \eta_{ab}$ and all the information about the geometry is contained in ω^a_i. From Eq. (11.25), it follows that the connection is now determined by the structure functions.

The quantities ω^a_i are usually called 'vielbeins'. The choice of vielbeins for a given metric is again not unique and one can perform a rotation of the vielbein to a new set ω'^a by $\omega'^a = \Lambda^a{}_b \omega^b$, where $\Lambda^a{}_b$ is a matrix satisfying the condition $\eta_{ab} \Lambda^a{}_c \Lambda^b{}_d = \eta_{cd}$. In general, the matrix $\Lambda^a{}_b$ can be coordinate dependent, which corresponds to our freedom in performing local rotations or Lorentz transformations in the tangent space. The corresponding tangent space basis vectors are given by $\boldsymbol{e}_a = E_a^i \partial_i$, where $E_a^i w_j^a = \delta_j^i$, $E_a^i w_i^b = \delta_b^a$. If we treat the components of the vielbein w_i^a as a non-degenerate $n \times n$ matrix then E_a^i is the inverse of this matrix.

The existence of the vielbeins w_i^a with one Lorentz space index and one spacetime index allows us to define components of any tensorial quantity with either class of indices in a coordinate basis or in an orthonormal (veilbein) basis. For example, we can now expand any vector $\boldsymbol{V}$ in this (non-coordinate) basis as

$$\boldsymbol{V} = V^a \boldsymbol{e}_a = V^a E_a^i \partial_i, \tag{11.28}$$

thereby relating the components V^a in the orthonormal basis with the components in the coordinate basis by $V^i = V^a E_a^i$. The inverse relation is given by $V^a = V^i w_i^a$. In this case, we can usefully think of w_i^a as the covariant components of four different four-vectors in spacetime with a indicating which four-vector. Then $V^i w_i^a$ for each value of a can be thought of as a scalar product of the four-vectors with V^i and w_i^a, and the equation $V^a = V^i w_i^a$ can be interpreted as giving the information contained in the four components of the vector $\boldsymbol{V}$ in terms of the four scalars V^a. Under general coordinate transformations, these scalars remain invariant.

To elaborate on this point, note that, with the introduction of an orthonormal basis $\boldsymbol{e}_a = E_a^i \partial_i$, we can formally distinguish between two kinds of transformations. First is the standard general coordinate transformations $x^i \rightarrow x'^i$ under which the partial derivatives transform as $\partial_j' = (\partial x^k / \partial x'^j) \partial_k$. If we think of ∂_k and ∂_j' as two sets of *coordinate basis* vectors, then this equation gives a linear transformation of one set of coordinate basis vectors to another set of coordinate basis vectors at a given location. It is essentially this transformation that we had worked with in all the earlier chapters. (The combination $(\partial x^k / \partial x'^j) \partial_k$ can also be thought of as a non-coordinate basis in the *original coordinate system* which is sometimes convenient mathematically.) In addition, we can also consider the rotation of the orthonormal basis vectors at a given point to a new set by the rule $\boldsymbol{e}_a \rightarrow \boldsymbol{e}_a' = \Lambda_a^b(x) \boldsymbol{e}_b$ in a non-coordinate basis. The transformation matrix $\Lambda_a^b(x)$ can vary from event to event since we can think of these as local Lorentz transformations of the basis at every event. This will correspond to the change in the components given by $V^a \rightarrow V'^a = \Lambda_b^a(x) V^b$, where $\Lambda_b^a \Lambda_c^b = \delta_c^a$. If we associate a matrix Λ with Λ_b^a (so that the rows are labelled by a and the columns by b), then Λ_a^b corresponds to the elements of Λ^{-1}. If we further treat $\boldsymbol{e}_a$ as a row vector, then the transformations can be written in matrix notation as

$$e' = e \Lambda^{-1}, \qquad V' = \Lambda V. \tag{11.29}$$

When we locally rotate the basis vectors, the dual basis has to change by $w^a \rightarrow w'^a = \Lambda_b^a w^b$ in order to maintain the duality between the basis vectors. Under the general coodinate transformation, we can treat the quantities $\boldsymbol{e}_a = E_a^i \partial_i$ as scalars but they rotate amongst themselves under the local (x^i dependent) Lorentz transformations.

Having made these connections between index free notation and the standard terminology introduced in earlier chapters we shall now develop some further, useful, geometrical concepts.

11.3 Differential forms

It turns out that totally antisymmetric tensors with covariant indices play a special role in physics and mathematics and have interesting geometrical properties. These objects arise as natural generalizations of the 1-forms introduced in the previous section. We shall now define such a generalization, called a p-form, and explore its properties.

We have seen in Chapter 4 that a completely antisymmetric tensor in four dimensions cannot have more than four indices. To keep the discussion general, we shall now work in an n-dimensional space. We begin by introducing an element of $[T^*]^p \equiv T^* \otimes T^* \ldots$ (p times) by the usual definition $\Omega = \Omega_{i_1 \ldots i_p} \, \boldsymbol{dx}^{i_1} \otimes \cdots \otimes \boldsymbol{dx}^{i_p}$. Our interest is in the situation in which $\Omega_{i_1 \ldots i_p}$ is completely antisymmetric in all its indices. In that case, we can also antisymmetrize the direct product of the basis vectors. For example, in the case of $p = 2$, we define a *wedge product* by

$$\boldsymbol{dx}^i \wedge \boldsymbol{dx}^j \equiv \boldsymbol{dx}^i \otimes \boldsymbol{dx}^j - \boldsymbol{dx}^j \otimes \boldsymbol{dx}^i. \tag{11.30}$$

Similarly, for $p = 3$, we have

$$\boldsymbol{dx}^i \wedge \boldsymbol{dx}^j \wedge \boldsymbol{dx}^k \equiv \boldsymbol{dx}^i \otimes \boldsymbol{dx}^j \otimes \boldsymbol{dx}^k + \boldsymbol{dx}^j \otimes \boldsymbol{dx}^k \otimes \boldsymbol{dx}^i + \boldsymbol{dx}^k \otimes \boldsymbol{dx}^i \otimes \boldsymbol{dx}^j$$
$$- \boldsymbol{dx}^i \otimes \boldsymbol{dx}^k \otimes \boldsymbol{dx}^j - \boldsymbol{dx}^j \otimes \boldsymbol{dx}^i \otimes \boldsymbol{dx}^k - \boldsymbol{dx}^k \otimes \boldsymbol{dx}^j \otimes \boldsymbol{dx}^i, \tag{11.31}$$

etc. In general, the wedge product is made up of the direct product plus all the terms with even permutations minus the terms with odd permutations. It is obvious from the definition that the wedge product is completely antisymmetric and, in particular, $\boldsymbol{dx}^i \wedge \boldsymbol{dx}^j = -\boldsymbol{dx}^j \wedge \boldsymbol{dx}^i$. A p-form $\boldsymbol{A}$ is now defined to be an element of $[T^*]^p$ with a component expansion given by

$$\boldsymbol{A} = \frac{1}{p!} A_{i_1 \ldots i_p} \, \boldsymbol{dx}^{i_1} \wedge \cdots \wedge \boldsymbol{dx}^{i_p}. \tag{11.32}$$

The components $A_{i_1 \ldots i_p}$ are completely antisymmetric. It is clear from the above relation that if $\boldsymbol{A}$ is a p-form and $\boldsymbol{B}$ is a q-form, then we must have

$$\boldsymbol{A} \wedge \boldsymbol{B} = (-1)^{pq} \, \boldsymbol{B} \wedge \boldsymbol{A}. \tag{11.33}$$

Further, it is conventional to think of a scalar as a 0-form and an element of T^* as a 1-form.

Having defined the p-form, we shall introduce the concept of an *exterior derivative* which acts on a p-form and produces a $(p + 1)$-form. We already know that the ordinary differential operator can act on the 0-form function $f(x)$ and produce a 1-form $\mathbf{d}f$. We will generalize this notion and define the action of $\mathbf{d}$ on a p-form through the relation

$$\mathbf{d}A = \frac{1}{p!}\left(\mathbf{d}A_{i_1\ldots i_p}\right) \wedge \mathbf{d}x^{i_1} \wedge \cdots \wedge \mathbf{d}x^{i_p} \tag{11.34}$$

and treating $A_{i_1\ldots i_p}$ as an ordinary scalar function. From the standard rules of calculus, we can re-express it as

$$\mathbf{d}A = \frac{1}{p!}\left(\partial_j A_{i_1\ldots i_p}\right) \mathbf{d}x^j \wedge \mathbf{d}x^{i_1} \wedge \cdots \wedge \mathbf{d}x^{i_p}, \tag{11.35}$$

which is clearly a $(p+1)$-form. If we expand $\mathbf{d}A$ as a $(p+1)$-form by introducing its components as $(dA)_{j_1\ldots j_{p+1}}$ we can determine the components of $\mathbf{d}A$ by comparing it with Eq. (11.35). This gives

$$(dA)_{j_1\ldots j_{p+1}} = (p+1)\,\partial_{[j_1} A_{j_2\ldots j_{p+1}]}, \tag{11.36}$$

where the square brackets denote complete antisymmetrization.

It is important to note that the definition of exterior differentiation introduced above leads to generally covariant expressions in spite of the fact that Eq. (11.36) involves ordinary partial derivatives, because $\partial_{[j_1} A_{j_2\ldots j_{p+1}]} = \nabla_{[j_1} A_{j_2\ldots j_{p+1}]}$. All the extra terms involving the Christoffel symbols vanish when the indices are completely antisymmetrized in Eq. (11.36) if the Christoffel symbols are symmetric in their lower indices. We have already seen a special case of this general result in Eq. (4.105).

The utility and power of exterior differentiation arise from two of its important properties. First, it follows an analogue of the Leibnitz rule

$$\mathbf{d}(A \wedge B) = \mathbf{d}A \wedge B + (-1)^p\, A \wedge \mathbf{d}B \tag{11.37}$$

allowing us to decompose the exterior derivative of wedge products. Second, and probably the more important property, is that when $\mathbf{d}$ acts twice *on the forms* it leads to a vanishing expression which can be summarized by the rule

$$\mathbf{d}^2 \equiv 0. \tag{11.38}$$

For example, if A is a 1-form, then applying $\mathbf{d}$ twice leads to

$$\mathbf{d}^2 A = \mathbf{d}(\partial_j A_i \mathbf{d}x^j \wedge \mathbf{d}x^i) = \partial_k\partial_j A_i \mathbf{d}x^k \wedge \mathbf{d}x^j \wedge \mathbf{d}x^i = \partial_{[k}\partial_j A_{i]} \mathbf{d}x^k \wedge \mathbf{d}x^j \wedge \mathbf{d}x^i. \tag{11.39}$$

This expression vanishes because the partial derivatives commute and the complete antisymmetrization cancels out the terms pair-wise. This result is essentially an abstract generalization of the familiar statement in three-dimensional vector analysis; viz. $\nabla \times (\nabla f) = 0$ which, in component notation, is equivalent to the statement $\partial_{[i}\partial_{j]} f = 0$. The notion of exterior differentiation makes use of this fact in the form language.

Any differential form Ω that satisfies $\boldsymbol{d\Omega} = 0$ is called *closed*. Any differential form $\boldsymbol{B}$ that can be written as $\boldsymbol{B} = \boldsymbol{dA}$ is called *exact*. Our rule $\boldsymbol{d}^2 = 0$ implies that any exact form is closed. The converse is also true *but only locally*. That is, if $\boldsymbol{d\Omega} = 0$ we can always find a $\boldsymbol{\nu}$ such that $\Omega = \boldsymbol{d\nu}$ locally in the neighbourhood of any event $\mathcal{P}$. It is, however, possible that $\boldsymbol{\nu}$ becomes ill-defined somewhere else on the manifold even though Ω is everywhere non-singular.

One important physical context in which these ideas find expression is in the description of electromagnetism and – more generally – in gauge theories (see Section 11.6.2). In electromagnetism the starting point could be a 1-form $\boldsymbol{A} = A_i \boldsymbol{dx}^i$, where A_i are the standard components of the vector potential. We define the exterior derivative of this 1-form by $\boldsymbol{F} = \boldsymbol{dA}$. Since the exterior derivative automatically gives antisymmetrized derivatives, it follows that the components of $\boldsymbol{F}$, given by

$$\boldsymbol{F} = \frac{1}{2} F_{mn} \boldsymbol{dx}^m \wedge \boldsymbol{dx}^n, \tag{11.40}$$

are just the electromagnetic field tensor $F_{mn} = \partial_m A_n - \partial_n A_m$. (We recall that in the definition of F_{mn} one could use either covariant derivatives or ordinary derivatives because of the identity Eq. (4.105).) The relation $\boldsymbol{F} = \boldsymbol{dA}$ also reveals immediately that one can change $\boldsymbol{A}$ to $\boldsymbol{A}' = \boldsymbol{A} + \boldsymbol{d}f$ (where f is a scalar function) without affecting $\boldsymbol{F}$ because $\boldsymbol{d}^2 = 0$. Hence the gauge invariance of the theory is immediately manifest in our notation. Explicitly, $\boldsymbol{F}$ will have the pattern

$$\boldsymbol{F} = E_x \boldsymbol{dx} \wedge \boldsymbol{dt} + \cdots + B_x \boldsymbol{dy} \wedge \boldsymbol{dz} + \cdots. \tag{11.41}$$

From the definition $\boldsymbol{F} = \boldsymbol{dA}$ it follows that $\boldsymbol{dF} = \boldsymbol{d}^2\boldsymbol{A} = 0$. Working this out, one can easily show that

$$\boldsymbol{dF} = (\nabla \cdot B)\boldsymbol{dx} \wedge \boldsymbol{dy} \wedge \boldsymbol{dz} + (\dot{B} + \nabla \times E)_x \boldsymbol{dt} \wedge \boldsymbol{dy} \wedge \boldsymbol{dz} + \cdots = 0. \tag{11.42}$$

This is identical to two of the Maxwell equations $[\nabla \cdot B = 0, (\dot{B} + \nabla \times E) = 0]$. The power of the form language is quite obvious in this construct.

Exercise 11.1

Frobenius theorem in the language of forms Let $\boldsymbol{v}$ be a vector field on a manifold which is everywhere hypersurface orthogonal. We proved in Chapter 4 that such a vector field obeys the Frobenius theorem; see Eq. (4.115). Show that this condition can be expressed equivalently as $\boldsymbol{\Omega} \wedge \boldsymbol{d\Omega} = 0$, where $\boldsymbol{\Omega}$ is the 1-form corresponding to the vector field $\boldsymbol{v}$.

11.4 Integration of forms

The forms occur with antisymmetrized infinitesimal coordinate differentials incorporated in them and hence they can be integrated over the manifold in a natural fashion. What is more, the process of integration leads to coordinate independent results with the Jacobians arising from the coordinate transformations taken care of automatically. To see how this occurs in a simple context, consider the two-dimensional integral of a 2-form $f\boldsymbol{dx} \wedge \boldsymbol{dy}$. If the coordinates are changed from $(x, y) \rightarrow (x', y')$, then the integral transforms as

$$\int f\,\boldsymbol{dx} \wedge \boldsymbol{dy} = \int f\left(\frac{\partial x}{\partial x'}\boldsymbol{dx}' + \frac{\partial x}{\partial y'}\boldsymbol{dy}'\right) \wedge \left(\frac{\partial y}{\partial x'}\boldsymbol{dx}' + \frac{\partial y}{\partial y'}\boldsymbol{dy}'\right)$$

$$= \int f\left(\frac{\partial x}{\partial x'}\frac{\partial y}{\partial y'} - \frac{\partial x}{\partial y'}\frac{\partial y}{\partial x'}\right)\boldsymbol{dx}' \wedge \boldsymbol{dy}'. \tag{11.43}$$

We see that the antisymmetry of the wedge product automatically brings in the Jacobian of the transformation making the result covariant. More generally, an n-form

$$\boldsymbol{\Omega} = f\,\boldsymbol{dx}^1 \wedge \boldsymbol{dx}^2 \wedge \cdots \wedge \boldsymbol{dx}^n \tag{11.44}$$

will transform under change of coordinates to

$$\boldsymbol{\Omega} = f\,\frac{\partial x^1}{\partial x'^{i_1}}\frac{\partial x^2}{\partial x'^{i_2}}\cdots\frac{\partial x^n}{\partial x'^{i_n}}\,\boldsymbol{dx}'^{i_1} \wedge \boldsymbol{dx}'^{i_2} \wedge \cdots \wedge \boldsymbol{dx}'^{i_n}$$

$$= f\,\frac{\partial x^1}{\partial x'^{i_1}}\frac{\partial x^2}{\partial x'^{i_2}}\cdots\frac{\partial x^n}{\partial x'^{i_n}}\,[i_1 i_2...i_n]\boldsymbol{dx}'^1 \wedge \boldsymbol{dx}'^2 \wedge \cdots \wedge \boldsymbol{dx}'^n, \tag{11.45}$$

where $[i_1 i_2...i_n]$ is the completely antisymmetric symbol with values $0, \pm 1$ (see Eq. (4.26)). Using the result that

$$\frac{\partial x^1}{\partial x'^{i_1}}\frac{\partial x^2}{\partial x'^{i_2}}\cdots\frac{\partial x^n}{\partial x'^{i_n}}\,[i_1 i_2...i_n] = \left|\frac{\partial x}{\partial x'}\right| \tag{11.46}$$

we see that one reproduces the correct Jacobian of the transformation.

This also shows that $\boldsymbol{dx} \wedge \boldsymbol{dy}$ in the two-dimensional Cartesian space can be interpreted as an element of an infinitesimal area with a proper orientation. (The orientation arises from the fact that $\boldsymbol{dx} \wedge \boldsymbol{dy} = -\boldsymbol{dy} \wedge \boldsymbol{dx}$.) One can similarly think of multiple wedge products of coordinate differentials as giving oriented volume element in the appropriate space.

The most important result in the integration of forms is the generalization of the Stokes theorem, which can be stated as

$$\int_M d\Omega = \int_{\partial M} \Omega. \tag{11.47}$$

This, of course, incorporates the familiar Gauss theorem and Stokes theorem of three-dimensional vector calculus. In addition, it leads to two important results. First, if M is an n-manifold without a boundary (for example, a 2-manifold which is the surface of a sphere), then it follows that the integral over the manifold of any exact n-form σ vanishes. That is,

$$\int_M \sigma = \int_M d\alpha = \int_{\partial M} \alpha = 0, \tag{11.48}$$

where the last step follows from the fact that ∂M does not exist. Second, just as the exterior derivative satisfies $d^2 = 0$, the manifolds have a property $\partial^2 M = 0$, which is usually stated as 'the boundary of a boundary is zero'. If Ω is an arbitrary $(n-2)$-form, then starting with the result $d^2\Omega = 0$ and integrating twice using the Stokes theorem, we get

$$0 = \int_M d^2\Omega = \int_{\partial M} d\Omega = \int_{\partial^2 M} \Omega. \tag{11.49}$$

Since this result has to hold for any $(n-2)$-form, it follows that $\partial^2 M$ must vanish for any manifold.

While differential forms allow integration over manifolds in a natural manner, it is necessary to exercise some caution while applying these results if the coordinates are not well-defined all over the manifold. An interesting example illustrating such a pitfall is provided by the 2-form $\Omega = \sin\theta d\theta \wedge d\phi$ defined on the surface of the unit 2-sphere S^2 in terms of the polar coordinates θ and ϕ. In any local region, we can write $\Omega = d\nu$ with $\nu = -\cos\theta d\phi$. A naive application of the result $\Omega = d\nu$ will lead to an absurd conclusion

$$4\pi = \int_{S^2} \sin\theta d\theta \wedge d\phi = \int_{S^2} \Omega = \int_{S^2} d\nu = \int_{\partial S^2} \nu = 0, \tag{11.50}$$

where the last step uses the fact that S^2 has no boundary. This wrong result is obtained because the 1-form ν is not globally well defined. It is clear that ν becomes singular at $\theta = 0$ and $\theta = \pi$ (the north and south poles) where the 1-form $d\phi$ is ill defined. One can take care of this difficulty by writing $\Omega = d\nu_\pm$ where

$$\nu_\pm = (-\cos\theta \pm 1) d\phi, \tag{11.51}$$

and use ν_+ in a coordinate patch which excludes the south pole ($\theta = \pi$) and use ν_- in a coordinate patch excluding the north pole ($\theta = 0$). (The ν_+ is well defined at $\theta = 0$ because the coefficient of $d\phi$ vanishes there, etc.) These two coordinate patches completely cover S^2 in a non-singular manner. If we denote by $H_\pm$ the northern and southern hemispheres on S^2, then we have

$$\int_{S^2} \Omega = \int_{H^+} \Omega + \int_{H^-} \Omega = \int_{H^+} d\nu_+ + \int_{H^-} d\nu_-. \tag{11.52}$$

Applying Stokes' theorem individually we now get

$$\int_{S^2} \Omega = \int_{S^1} \nu_+ + \int_{(-S^1)} \nu_- = 2\pi + 2\pi = 4\pi, \tag{11.53}$$

which is the correct result. On doing the integral over the boundaries of the two hemispheres (which is, of course, the equator), one needs to be careful about the orientation of the circles to obtain the correct sign.

An interesting application of this result arises in the case of a magnetic monopole of charge g. On a sphere surrounding a magnetic monopole located at the origin, the electromagnetic field is described by the 2-form $\boldsymbol{F} = (g/4\pi)\boldsymbol{d}\cos\theta \wedge \boldsymbol{d}\phi$. The integral of this 2-form over the sphere gives g as expected. (In the usual notation, this is equivalent to having a radially directed magnetic field of strength $(g/4\pi r^2)$ everywhere on the sphere of radius r.) Let us try to determine a vector potential from which this field can be derived. Naively, one might take $\boldsymbol{A} = (g/4\pi)\cos\theta \boldsymbol{d}\phi$. However, this expression suffers from the same problem as the one encountered above; viz. that $d\phi$ is not defined at the north and south poles. To circumvent this difficulty, we should again define two different vector potentials:

$$\boldsymbol{A}_+ = \left(\frac{g}{4\pi}\right)(\cos\theta - 1)\,\boldsymbol{d}\phi; \qquad \boldsymbol{A}_- = \left(\frac{g}{4\pi}\right)(\cos\theta + 1)\,\boldsymbol{d}\phi \tag{11.54}$$

with $\boldsymbol{A}_+$ valid in a coordinate patch covering the northern hemisphere but excluding the south pole, and $\boldsymbol{A}_-$ valid in a coordinate patch covering the southern hemisphere but excluding the north pole. In the region where the two coordinate patches overlap, we find that $\boldsymbol{A}_- - \boldsymbol{A}_+ = (g/2\pi)\boldsymbol{d}\phi$ so that the two vector potentials differ by a gauge transformation. This difference leads to an interesting conclusion in quantum theory. Under a gauge transformation $A_k \to A_k + (1/q)\partial_k f$ the wave functions in quantum theory change as $\psi(x) \to e^{if}\psi$ (see Exercise 2.2). In our case, $\boldsymbol{A}_-$ and $\boldsymbol{A}_+$ are related by a gauge transformation with $f = (qg\phi/2\pi)$ and the wave functions will change under the gauge transformation by the phase factor $\exp[i(qg/2\pi)\phi]$. But we know that $\phi = 0$ and $\phi = 2\pi$ describe the same physical point. If the wave function has to be single valued, then we must necessarily have $\exp[i(qg/2\pi)2\pi] = 1$. This is possible only if $qg = 2\pi n$ for $n = 0, 1, \ldots$. This result shows that, *if* a magnetic monopole exists, *then* the electric charge must be quantized in units of $(2\pi/g)$ to maintain the single valuedness of the wave function.

Exercise 11.2
The Dirac string Express the two potentials $\mathbf{A}_\pm$ introduced above in terms of Cartesian coordinates and study their behaviour as one approaches the z-axis. Note that these potentials become singular on the z-axis for $z > 0$ and $z < 0$ in the two cases. Explain why. [Note. This is related to a structure usually called a *Dirac string* in the study of magnetic monopoles.]

Exercise 11.3
Simple example of a non-exact, closed form Consider a 1-form Ω defined in Euclidean 2-space with standard Cartesian coordinates x and y by the relation

$$\Omega = \frac{x\,dy}{x^2 + y^2} - \frac{y\,dx}{x^2 + y^2}. \tag{11.55}$$

Show that this 1-form is closed but not exact. Explain why. [Answer. Direct exterior differentiation will give $\mathbf{d}\Omega = 0$. More simply, note that $\Omega = \mathbf{d}\theta$, where $\theta = \tan^{-1}(y/x)$, then the fact $\mathbf{d}^2 = 0$ shows that Ω is closed. Integrating Ω around a circle of unit radius, we get 2π and hence it cannot be exact. This is because θ is not a single valued function on the circle and hence one cannot write $\Omega = \mathbf{d}\theta$ globally.]

11.5 The Hodge duality

We have mentioned in Chapters 1 and 4 that a completely antisymmetric covariant tensor of rank p and a completely antisymmetric covariant tensor of rank $(n - p)$ have the same number of components in an n-dimensional space. (The number of components in either case is ^nC_p.) It is this fact which allowed us to use the ϵ tensor to take the dual of a tensor of rank p and obtain a tensor of rank $(n - p)$. Since the completely antisymmetric covariant tensors lead to the notion of differential forms in a natural fashion, the duality operation also has a natural expression in terms of differential forms. To investigate this, we first define a dual basis (also called the *Hodge dual basis*) by the relation

$$*(\mathbf{d}x^{i_1} \wedge \cdots \wedge \mathbf{d}x^{i_p}) = \frac{1}{q!}\epsilon_{j_1 \ldots j_q}{}^{i_1 \ldots i_p}\,\mathbf{d}x^{j_1} \wedge \cdots \mathbf{d}x^{j_q}, \tag{11.56}$$

where $q = n - p$. (Note that the ϵ symbol has a $\sqrt{|g|}$ incorporated in its definition; see Eq. (4.26).) For any p-form Ω given by

$$\Omega = \frac{1}{p!}\Omega_{i_1 \ldots i_p}\,\mathbf{d}x^{i_1} \wedge \cdots \wedge \mathbf{d}x^{i_p} \tag{11.57}$$

we can now introduce the Hodge dual through the relation

$$*\Omega = \frac{1}{p!q!}\epsilon_{j_1 \ldots j_q}{}^{i_1 \ldots i_p}\,\Omega_{i_1 \ldots i_p}\,\mathbf{d}x^{j_1} \wedge \cdots \mathbf{d}x^{j_q}; \qquad q = n - p. \tag{11.58}$$

Obviously, $*\Omega$ is an $(n - p)$-form if Ω is a p-form. In component language, $*\Omega$ has the components

$$*(\Omega)_{j_1 \ldots j_q} = \frac{1}{p!} \epsilon_{j_1 \ldots j_q}{}^{i_1 \ldots i_p} \Omega_{i_1 \ldots i_p}, \tag{11.59}$$

which is essentially the duality operation introduced in Chapters 1 and 4. We can invert the duality operation by taking the dual again and using the properties of ϵ tensors under contraction. It is easy to show that in terms of components, we get

$$* * (\Omega)_{i_1 \ldots i_p} = (-1)^{pq+t} \Omega_{i_1 \ldots i_p}, \tag{11.60}$$

where t is the number of 'time directions' in the manifold; that is, it gives the number of negative eigenvalues of the metric tensor. (For the usual spacetime, $t = 1$.) In terms of the forms themselves, this equation translates into

$$* * \Omega = (-1)^{pq+t} \Omega. \tag{11.61}$$

One important application of the Hodge dual is in providing a way for taking the inner product of two p-forms, $\boldsymbol{A}$ and $\boldsymbol{B}$, say. We know that $*\boldsymbol{A}$ is an $(n - p)$-form and hence $*\boldsymbol{A} \wedge \boldsymbol{B}$ is an n-form. But since any n-form in an n-dimensional space should be proportional to the completely antisymmetric ϵ tensor, it follows that $*\boldsymbol{A} \wedge \boldsymbol{B}$ must be essentially made of the inner product of the components of $\boldsymbol{A}$ and $\boldsymbol{B}$. This result can be directly verified. We begin by noting that

$$*\boldsymbol{A} \wedge \boldsymbol{B}$$
$$= \frac{1}{(p!)^2 q!} \epsilon_{i_1 \ldots i_q}{}^{j_1 \ldots j_p} A_{j_1 \ldots j_p} B_{k_1 \ldots k_p} \boldsymbol{dx}^{i_1} \wedge \cdots \wedge \boldsymbol{dx}^{i_q} \wedge \boldsymbol{dx}^{k_1} \wedge \cdots \wedge \boldsymbol{dx}^{k_p}$$
$$= \frac{(-1)^t}{(p!)^2 q!} \epsilon_{i_1 \ldots i_q}{}^{j_1 \ldots j_p} A_{j_1 \ldots j_p} B_{k_1 \ldots k_p} \epsilon^{i_1 \ldots i_q k_1 \ldots k_p} \sqrt{|g|} \, \boldsymbol{dx}^1 \wedge \boldsymbol{dx}^2 \wedge \cdots \wedge \boldsymbol{dx}^n, \tag{11.62}$$

where we have used the p-dimensional version of Eq. (4.30). The contraction of the ϵs will now lead to the standard determinant tensor, using which we can obtain

$$*\boldsymbol{A} \wedge \boldsymbol{B} = \frac{1}{p!} A^{i_1 \ldots i_p} B_{k_1 \ldots k_p} \delta^{k_1 \ldots k_p}_{i_1 \ldots i_p} \sqrt{|g|} \, \boldsymbol{dx}^1 \wedge \boldsymbol{dx}^2 \wedge \cdots \wedge \boldsymbol{dx}^n$$
$$= \frac{1}{p!} A_{i_1 \ldots i_p} B^{i_1 \ldots i_p} \sqrt{|g|} \, \boldsymbol{dx}^1 \wedge \boldsymbol{dx}^2 \wedge \cdots \wedge \boldsymbol{dx}^n. \tag{11.63}$$

It is usual to write this result formally as

$$*\boldsymbol{A} \wedge \boldsymbol{B} = \frac{1}{p!} A_{i_1 \ldots i_p} B^{i_1 \ldots i_p} (*\boldsymbol{1}), \tag{11.64}$$

where the Hodge dual of the constant $\boldsymbol{1}$ is the volume n-form given by

$$\boldsymbol{\epsilon} \equiv *\boldsymbol{1} = \frac{1}{n!} \epsilon_{i_1 \ldots i_n} \boldsymbol{dx}^{i_1} \wedge \cdots \wedge \boldsymbol{dx}^{i_n} = \sqrt{|g|} \, \boldsymbol{dx}^1 \wedge \boldsymbol{dx}^2 \wedge \cdots \wedge \boldsymbol{dx}^n. \tag{11.65}$$

For example, in the Euclidean 2-space in polar coordinates with the metric $ds^2 = dr^2 + r^2 d\theta^2$, we will have $*1 = r dr \wedge d\theta$. In particular, note that the dot product of two vectors $\boldsymbol{A}$ and $\boldsymbol{B}$ can equivalently be expressed in the form

$$*(*\boldsymbol{A} \wedge \boldsymbol{B}) = A_i B^i \tag{11.66}$$

by using the forms which correspond to the two vectors.

Given the Hodge dual operator and the exterior derivative operator $\boldsymbol{d}$, it is possible to express the covariant derivatives of antisymmetric tensors in terms of them. To do this, consider the operation $*\boldsymbol{d}*$ on a p-form Ω. We know that the first Hodge dual operation leads to an $(n-p)$-form which is converted to an $(n-p+1)$-form by the exterior derivative. The second Hodge dual operation now leads to a $(p-1)$-form. Since we have operated once with the exterior derivative, the components of $*\boldsymbol{d}*\Omega$ must be essentially obtained by the first derivatives of Ω. Since $*\boldsymbol{d}*\Omega$ is a genuine tensorial object, it is obvious that it should be made of the covariant derivatives built from the components of Ω. Direct computation shows that this result is indeed true. We get

$$*\boldsymbol{d}*\Omega = \frac{(-1)^{pq+p+t+1}}{(p-1)!} \left(\nabla^k \Omega_{ki_1\ldots i_{p-1}}\right) \boldsymbol{dx}^{i_1} \wedge \cdots \wedge \boldsymbol{dx}^{i_{p-1}}, \tag{11.67}$$

where

$$\nabla^k \Omega_{k\ell_1\ldots\ell_{p-1}} = \left(\nabla_k \Omega^{km_1\ldots m_{p-1}}\right) g_{\ell_1 m_1} \cdots g_{\ell_{p-1} m_{p-1}}$$

$$= \frac{1}{\sqrt{|g|}} \partial_k \left(\sqrt{|g|}\, \Omega^{km_1\ldots m_{p-1}}\right) g_{\ell_1 m_1} \cdots g_{\ell_{p-1} m_{p-1}}. \tag{11.68}$$

In terms of components, this is equivalent to the relation

$$(*\boldsymbol{d}*\Omega)_{i_1\ldots i_{p-1}} = (-1)^{pq+p+t+1} \nabla^k \Omega_{ki_1\ldots i_{p-1}}. \tag{11.69}$$

One application of the Hodge dual is in writing the Maxwell equations in an index free notation as follows. Given the 2-form $\boldsymbol{F}$, we can take its Hodge dual to obtain $*\boldsymbol{F}$. The exterior derivative of this dual can again be computed in component notation and can be easily shown to lead to the second pair of Maxwell's equations:

$$\boldsymbol{d}*\boldsymbol{F} = \boldsymbol{d}(-B_x \boldsymbol{dx} \wedge \boldsymbol{dt} - \cdots + E_x \boldsymbol{dy} \wedge \boldsymbol{dz} + \cdots)$$

$$= \left(\frac{\partial E_x}{\partial x} + \frac{\partial E_y}{\partial y} + \frac{\partial E_z}{\partial z}\right) \boldsymbol{dx} \wedge \boldsymbol{dy} \wedge \boldsymbol{dz}$$

$$+ \left(\frac{\partial E_x}{\partial t} - \frac{\partial B_z}{\partial y} + \frac{\partial B_y}{\partial z}\right) \boldsymbol{dt} \wedge \boldsymbol{dy} \wedge \boldsymbol{dz} + \cdots$$

$$= 4\pi(\rho \boldsymbol{dx} \wedge \boldsymbol{dy} \wedge \boldsymbol{dz} - J_x \boldsymbol{dt} \wedge \boldsymbol{dy} \wedge \boldsymbol{dz} - \cdots) = 4\pi(*\boldsymbol{J}). \tag{11.70}$$

That is, we have:

$$\boldsymbol{d}(*\boldsymbol{F}) = 4\pi(*\boldsymbol{J}). \tag{11.71}$$

If we take one more exterior derivative of this equation, we obtain $4\pi \boldsymbol{d}(*\boldsymbol{J}) = \boldsymbol{d}^2(*\boldsymbol{F}) = 0$, which is equivalent to the standard result $\nabla_i J^i = 0$ describing the conservation of charge. Alternatively, we can take the dual of Eq. (11.71) and obtain $(*\boldsymbol{d}*)\,\boldsymbol{F} = (*\boldsymbol{d}*\boldsymbol{d})\,\boldsymbol{A} = 4\pi\boldsymbol{J}$. This is essentially the wave equation for the electromagnetic four potential in the form language.

11.6 Spin connection and the curvature 2-forms

Earlier, we introduced the concepts of connection and curvature through the covariant derivative operator ∇. We shall now redo the same using the language of exterior calculus, which – though equivalent to the formalism developed earlier – has some advantages and computational simplicity.

The starting point is a basis of 1-forms $\boldsymbol{\omega}^a = \omega_i^a\,\boldsymbol{d}x^i$ defined in the usual way, in a manifold on which we have introduced the coordinates x^i. As we have mentioned earlier, the choice of the basis vectors is not unique and we are free to perform a local rotation of the basis to a new set $\boldsymbol{\omega}'^a$ by $\boldsymbol{\omega}'^a = \Lambda^a{}_b\,\boldsymbol{\omega}^b$, where $\Lambda^a{}_b(x)$ is a matrix that can be coordinate dependent. This corresponds to our freedom in performing local rotations in the tangent space.

It is obvious from our construction that $\boldsymbol{\omega}^a$ is actually a vector valued 1-form; that is, it is a 1-form carrying an additional Lorentz vector index. Under local Lorentz transformations it transforms covariantly just as any four-vector should. This notion can be generalized to a tensor valued p-form in a straightforward manner. For example, we can define an object $\boldsymbol{V}^{a_1\ldots a_r}{}_{b_1\ldots b_s}$ which is a form that carries both upper and lower indices. Under the local Lorentz transformations, it transforms as

$$\boldsymbol{V}^{a_1\ldots a_r}{}_{b_1\ldots b_s} \rightarrow \boldsymbol{V}'^{a_1\ldots a_r}{}_{b_1\ldots b_s} = \Lambda^{a_1}{}_{c_1}\cdots\Lambda^{a_r}{}_{c_r}\Lambda_{b_1}{}^{d_1}\cdots\Lambda_{b_s}{}^{d_s}\boldsymbol{V}^{c_1\ldots c_r}{}_{d_1\ldots d_s}. \tag{11.72}$$

We are interested in defining a suitable generalization of the exterior derivative for such tensor valued p-forms. The straightforward procedure will not work because the result will not be covariant when $\Lambda^a{}_b$ depends on the position. This is easy to see from the example of a vector valued p-form $\boldsymbol{V}^a$. Since it transforms as $\boldsymbol{V}'^a = \Lambda^a{}_b\boldsymbol{V}^b$ we find that

$$\boldsymbol{d}\boldsymbol{V}'^a = \boldsymbol{d}(\Lambda^a{}_b\boldsymbol{V}^b) = \Lambda^a{}_b\,\boldsymbol{d}\boldsymbol{V}^b + \boldsymbol{d}\Lambda^a{}_b \wedge \boldsymbol{V}^b. \tag{11.73}$$

The existence of the second term will prevent $\boldsymbol{d}\boldsymbol{V}^b$ transforming in a covariant manner. We know, however, from our knowledge of the covariant derivative, how this problem can be circumvented. What we need to do is to introduce the concept of a covariant exterior derivative for which we need to introduce a set of quantities called *spin connections*, which are analogous to the Christoffel symbols, that will allow us to compute the action of an exterior derivative on a general 1-form basis ω^a. Once we do that, we can use the usual chain rule, etc., to determine its action on the basis vectors and thus on any arbitrary tensorial quantity. The spin connections are defined as the set of 1-forms $\mathbf{\Omega}^a{}_b = \Omega^a{}_{bi}\mathbf{d}x^i$, which are related to ω^a by the identity

$$\mathbf{d}\omega^a + \mathbf{\Omega}^a{}_b \wedge \omega^b = 0. \tag{11.74}$$

For this equation to hold covariantly in all frames, when the ω^as are rotated, the $\mathbf{\Omega}^a{}_b$ must transform according to the rule

$$\mathbf{\Omega}'^a_b = \Lambda^a{}_c \mathbf{\Omega}^c{}_d \Lambda_b{}^d - \mathbf{d}\Lambda^a{}_c \Lambda_b{}^c. \tag{11.75}$$

In matrix notation, this can be written in two equivalent forms as

$$\Omega' = \Lambda\Omega\Lambda^{-1} - d\Lambda\,\Lambda^{-1}; \qquad \Omega' = \Lambda\Omega\Lambda^{-1} + \Lambda\,d\Lambda^{-1}. \tag{11.76}$$

The second relation corresponds, in index notation, to

$$\mathbf{\Omega}'^a_b = \Lambda^a{}_c \mathbf{\Omega}^c{}_d \Lambda_b{}^d + \Lambda^a{}_c \mathbf{d}\Lambda_b{}^c. \tag{11.77}$$

One can construct a more general structure (with nonzero torsion) by *not* demanding that the expression in Eq. (11.74) vanishes but merely insisting that it transforms covariantly. We shall, however, confine our attention to connection 1-forms which satisfy Eq. (11.74).

Given the connection 1-forms, we can now introduce a Lorentz covariant exterior derivative operation D by the rule

$$\begin{aligned}
\mathbf{D}\mathbf{V}^{a_1\ldots a_r}{}_{b_1\ldots b_s} \equiv {} & \mathbf{d}\mathbf{V}^{a_1\ldots a_r}{}_{b_1\ldots b_s} \\
& + \mathbf{\Omega}^{a_1}{}_c \wedge \mathbf{V}^{ca_2\ldots a_r}{}_{b_1\ldots b_s} + \cdots + \mathbf{\Omega}^{a_r}{}_c \wedge \mathbf{V}^{a_1\ldots a_{r-1}c}{}_{b_1\ldots b_s} \\
& - \mathbf{\Omega}^c{}_{b_1} \wedge \mathbf{V}^{a_1\ldots a_r}{}_{cb_2\ldots b_s} - \cdots - \mathbf{\Omega}^c{}_{b_s} \wedge \mathbf{V}^{a_1\ldots a_r}{}_{b_1\ldots b_{s-1}c}.
\end{aligned} \tag{11.78}$$

It is a straightforward exercise to verify that this quantity does transform covariantly. The pattern here is very similar to the one encountered in the definition of covariant derivatives of tensors and the proof proceeds exactly as before. For example, in the case of an object with two Lorentz indices we have

$$\mathbf{D}\mathbf{V}^a{}_b = \mathbf{d}\mathbf{V}^a{}_b + \mathbf{\Omega}^a{}_c \wedge \mathbf{V}^c{}_b - \mathbf{\Omega}^c{}_b \wedge \mathbf{V}^a{}_c \tag{11.79}$$

$$= \mathbf{d}\mathbf{V}^a{}_b + \mathbf{\Omega}^a{}_c \wedge \mathbf{V}^c{}_b - (-1)^p \mathbf{V}^a{}_c \wedge \mathbf{\Omega}^c{}_b. \tag{11.80}$$

This can be re-expressed as a matrix equation as

$$DV = dV + \Omega \wedge V - (-1)^p V \wedge \Omega. \tag{11.81}$$

Under local rotations, we have $V \to V' = \Lambda V \Lambda^{-1}$, again in matrix notation. It is now easy to verify that

$$D'V' \equiv dV' + \Omega' \wedge V' - (-1)^p V' \wedge \Omega' = \Lambda(DV)\Lambda^{-1}, \tag{11.82}$$

which establishes the result for this specific case. In particular, demanding $\boldsymbol{D}\eta_{ab} = 0$ in an orthonormal frame (and noting $\boldsymbol{d}\eta_{ab} = 0$) we get the result $\Omega_{ab} = -\Omega_{ba}$, where $\Omega_{ab} = \eta_{ac}\Omega^c{}_b$. Hence, in an orthonormal frame of non-coordinate basis such that $g_{mn} = \eta_{mn}$, we get two conditions on $\Omega^a{}_b$, viz.

$$\boldsymbol{d}\omega^a = -\Omega^a{}_b \wedge \omega^b, \qquad \Omega_{ab} = -\Omega_{ba}. \qquad (11.83)$$

These are sufficient to determine the connection coefficients uniquely. To see this, note that the exterior derivatives of the basis vectors can be expressed in the form

$$\boldsymbol{d}\omega^a = -\frac{1}{2}\mu_{bc}{}^a\,\omega^b \wedge \omega^c, \qquad (11.84)$$

where the set of functions $\mu_{bc}{}^a$, defined by this relation, are antisymmetric in b and c. It is now straightforward to verify that the two conditions in Eq. (11.83) are satisfied by the choice

$$\Omega_{ab} = \frac{1}{2}\left(\mu_{abc} + \mu_{acb} - \mu_{bca}\right)\omega^c, \qquad (11.85)$$

where $\mu_{abc} = \eta_{cd}\mu_{ab}{}^d$. Thus, given a set of basis vectors, we can take their exterior derivative and obtain μ_{bca} using Eq. (11.84). Then Eq. (11.85) allows us to compute the spin connections.

More generally, it is useful to define the $\boldsymbol{D}$ operator such that it acts on forms like $\boldsymbol{d}$ and on vectors like ∇. It is possible to approach this identification from a more formal perspective which we shall now describe. We begin by noting that, given any function f, one can obtain a 1-form – which, for the sake of easy generalization, we will call a 'scalar valued 1-form' – through the relation

$$\langle\boldsymbol{d}f|\boldsymbol{u}\rangle = \partial_{\boldsymbol{u}}f. \qquad (11.86)$$

Here $\boldsymbol{u}$ is any arbitrary vector and this relation defines $\boldsymbol{d}f$. In a similar manner, we will define a vector valued 1-form by defining the action of the exterior derivative on a vector field $\boldsymbol{v}$ by the rule

$$\langle\boldsymbol{D}\boldsymbol{v}|\boldsymbol{u}\rangle \equiv \nabla_{\boldsymbol{u}}\boldsymbol{v}. \qquad (11.87)$$

Since this result should hold for all vectors $\boldsymbol{u}$, we can make a formal identification of a vector valued 1-form as $\boldsymbol{D}\boldsymbol{v} = \nabla\boldsymbol{v}$. The generalization to tensor valued forms is obvious through the Leibnitz rule.

We have now introduced a set of spin connections as well as the standard metric connections on the manifold. We have also seen that the same geometrical quantity (say, a vector $\boldsymbol{V}$) can be expressed either in the coordinate basis ($\boldsymbol{V} = V^i\partial_i$) or in the orthonormal basis $\boldsymbol{V} = V^a\boldsymbol{e}_a$) with $V^a = \omega^a_i V^i$. For the action of D to be consistent with the process of taking the components, we need to demand that

$Dw_j^a = 0$, where ω_j^a is treated as having a local Lorentz space index as well as a spacetime index. This condition, on expanding out as

$$\partial_i \omega_j^a + \Omega^a{}_{bi} \omega_j^b - \Gamma^k_{ij} \omega_k^a = 0 \tag{11.88}$$

gives a relation between the spin connections and the Christoffel symbols. More explicitly, we get

$$\Gamma^k{}_{ij} = E_a^k \partial_i \omega_j^a + E_a^k \Omega^a{}_{bi} \omega_j^b; \qquad \Omega^a{}_{bi} = -E_b^j \nabla_i \omega_j^a, \tag{11.89}$$

which gives the relationship between the Christoffel symbols and the spin connections.

This result also shows that, in a coordinate basis (in contrast to the orthonormal basis), we can actually identify the two structures – spin connections with metric connections. In a coordinate basis, $\omega^a = dx^a$ and the first term in Eq. (11.74) vanishes, leading to the condition

$$\Omega^a{}_{bi} dx^i \wedge dx^b = \frac{1}{2} \left(\Omega^a{}_{bi} - \Omega^a{}_{ib} \right) dx^i \wedge dx^b = 0. \tag{11.90}$$

Hence, in a coordinate basis, $\Omega^a{}_{bi}$ is symmetric in the lower indices. Further, Eq. (11.89) with $\omega_k^a = \delta_k^a$ now gives $\Omega^a{}_{bc} = \Gamma^a{}_{bc}$. So, *in a coordinate basis* we can identify the spin connections with the Christoffel symbols through

$$\Omega^j{}_l = \Gamma^j{}_{lm} dx^m. \tag{11.91}$$

If we expand the vector v in a basis as $v = v^m e_m$ then, on using Eq. (11.74), we have

$$Dv = e_m \otimes \left(dv^m + \Omega^m{}_n v^n \right). \tag{11.92}$$

It is clear that the expression in the parenthesis will be related to the covariant derivative in the coordinate basis.

Using this definition, one can obtain a compact expression for the curvature. To do that, we differentiate Eq. (11.92) once again (noting $d^2 v^m = 0$) to obtain

$$\begin{aligned} D^2 v &= De_a \wedge \left(dv^a + \Omega^a{}_n v^n \right) + e_m \otimes \left(D\Omega^m{}_n v^n - \Omega^m{}_n \wedge dv^n \right) \\ &= e_m \otimes \left(\Omega^m{}_a \wedge dv^a + \Omega^m{}_a \wedge \Omega^a{}_n v^n + d\Omega^m{}_n v^n - \Omega^m{}_a \wedge dv^a \right). \end{aligned} \tag{11.93}$$

Writing this expression as $D^2 v = e_m \otimes \Theta^m{}_n v^n$ we can identify a set of quantities called *curvature 2-form*, given by

$$\Theta^a{}_b = d\Omega^a{}_b + \Omega^a{}_c \wedge \Omega^c{}_b. \tag{11.94}$$

In matrix notation, this reads as

$$\Theta = d\Omega + \Omega \wedge \Omega. \tag{11.95}$$

It can again be directly verified that Θ transforms covariantly (that is, $\Theta' = \Lambda\Theta\Lambda^{-1}$) under local Lorentz rotations. The definition in Eq. (11.94) is analogous to the usual definition of curvature tensor in terms of Christoffel symbols. In fact, the components of Θ in a coordinate basis give the Riemann tensor via the relation

$$\Theta^a{}_b = \frac{1}{2} R^a{}_{bcd}\, \boldsymbol{d}x^c \wedge \boldsymbol{d}x^d. \tag{11.96}$$

Exercise 11.4
Dirac equation in curved spacetime The vielbein w_i^a and the spin connections are needed for expressing the covariant derivatives of spinors in curved spacetime. This is necessary to write, for example, the Dirac equation in a generally covariant manner. The purpose of this exercise is to introduce this idea for those who are familiar with the Dirac equation in the context of flat spacetime. Let γ_L^a be the standard flat spacetime Dirac matrices which satisfy the anti-commutation relation $\{\gamma_L^a, \gamma_L^b\} = 2\eta^{ab}$. The subscript L denotes that these are the matrices based on standard Lorentz transformations.

(a) Show that the curved spacetime generalization of the Dirac matrices could be taken to be $\gamma^a = w_m^a \gamma_L^m$ with $\{\gamma^a, \gamma^b\} = 2g^{ab}$.

(b) In flat spacetime, we use $\partial_m \psi$ for the derivatives of the Dirac spinor ψ since it is Lorentz covariant. In curved spacetime, we need to introduce a covariant derivative of the spinor by the relation

$$\nabla_a \psi \equiv w_a^m [\partial_m + \Omega_m]\psi, \tag{11.97}$$

where

$$\Omega_m \equiv \frac{i}{2}\Omega_{amb}\Sigma^{ab} = \frac{i}{2}\left(w_n^a \nabla_m w^{bn}\right)\Sigma_{ab}, \tag{11.98}$$

where $\Sigma_{ab} \equiv (i/4)[\gamma_a, \gamma_b]$ is the spinor representation of the generators of the Lorentz transformation. Prove this by explicitly demonstrating that $\nabla_a \psi$ transforms covariantly under coordinate transformations.

(c) Write down the generally covariant action for the Dirac field in curved spacetime.

11.6.1 Einstein–Hilbert action and curvature 2-forms

As an example of the formalism written in term of vielbeins and curvature 2-forms, we will reformulate the gravitational action principle and related topics in differential form language.[2] We saw in Chapter 6 that the Lagrangian for gravity can be taken to be proportional to the Ricci scalar R. We begin by proving that the scalar curvature in D dimensions can be expressed as follows:

$$\Theta_{ab} \wedge *(\boldsymbol{w}^a \wedge \boldsymbol{w}^b) = R\sqrt{|g|}\, d^D x. \tag{11.99}$$

Straightforward computation gives this result:

$$
\Theta^{ab} \wedge *(\omega_a \wedge \omega_b)
$$
$$
= \frac{1}{2(D-2)!} R^{ab}_{mn} \omega_a^i \omega_b^j \epsilon_{ijn_1 \cdots n_{D-2}} dx^m \wedge dx^n \wedge dx^{n_1} \wedge \cdots dx^{n_{D-2}}
$$
$$
= R^{ji}_{mn} \delta^{mn}_{ij} \sqrt{|g|} d^D x = \frac{1}{2} \left(R^{nm}_{mn} - R^{mn}_{mn} \right) \sqrt{|g|} d^D x
$$
$$
= R \sqrt{|g|}\, d^D x. \tag{11.100}
$$

At this stage, it is convenient to introduce some useful notation as follows:

$$
\epsilon^{ij} \equiv * \left(\omega^i \wedge \omega^j \right); \qquad \epsilon^{ijk} \equiv * \left(\omega^i \wedge \omega^j \wedge \omega^k \right). \tag{11.101}
$$

It is obvious from the definitions that $\epsilon_{ij} = (1/2)\epsilon_{ijkl}\, \omega^k \wedge \omega^l$ and $\epsilon_{ijk} = \epsilon_{ijkl}\, \omega^l$. Using the expansion of the curvature 2-form in Eq. (11.94) in the relation $R\epsilon = \Theta^i{}_j \wedge \epsilon_i{}^j$, we get

$$
R\epsilon = \mathbf{d} \left(\Omega^a{}_b \wedge \epsilon_a{}^b \right) + \Omega^a{}_b \wedge \mathbf{d}\epsilon_a{}^b + \Omega^a{}_s \wedge \Omega^s{}_b \wedge \epsilon_a{}^b. \tag{11.102}
$$

This expression can be simplified by using the fact that $D\epsilon_a{}^b = 0$ which, in expanded form, reads as

$$
\mathbf{D}\epsilon_a{}^b = 0 = \mathbf{d}\epsilon_a{}^b - \Omega^s{}_a \wedge \epsilon_s{}^b + \Omega^b{}_s \wedge \epsilon_a{}^s. \tag{11.103}
$$

This allows us to express the Einstein–Hilbert Lagrangian as

$$
R\epsilon = \Omega^a{}_b \wedge \Omega^s{}_a \wedge \epsilon_s{}^b + \mathbf{d} \left(\Omega^a{}_b \wedge \epsilon_a{}^b \right). \tag{11.104}
$$

This decomposition is identical to the one obtained in Eq. (6.9) as can be easily verified (see Exercise 11.6). Our next task is to vary this action. In the variation

$$
\delta(R\epsilon) = (\delta\epsilon_{ab}) \wedge \Theta^{ab} + \epsilon_{ab} \wedge \delta\Theta^{ab} \tag{11.105}
$$

we use the results

$$
\delta\epsilon_{ab} = \frac{1}{2}\epsilon_{abcd}\delta \left(\omega^c \wedge \omega^d \right) = (\delta\omega^c) \wedge \epsilon_{abc} \tag{11.106}
$$

and

$$
\delta\Theta^{ab} = \mathbf{d}\,\delta\Omega^{ab} + (\delta\Omega^a{}_l) \wedge \Omega^{lb} + \Omega^a{}_l \wedge \delta\Omega^{lb} \tag{11.107}
$$

to obtain

$$
\delta(R\epsilon) = \delta\omega^l \wedge \left(\epsilon_{abl} \wedge \Theta^{ab} \right) + \mathbf{d} \left(\epsilon_{ab} \wedge \delta\Omega^{ab} \right) + \mathbf{Q}, \tag{11.108}
$$

where

$$
\mathbf{Q} = -\mathbf{d}\epsilon_{ab} \wedge \delta\Omega^{ab} + \epsilon_{ab} \wedge \left(\delta\Omega^a{}_l \wedge \Omega^{lb} + \Omega^a{}_l \wedge \delta\Omega^{lb} \right) = 0, \tag{11.109}
$$

where the last equality follows from the fact that $\boldsymbol{D}\epsilon_{ab} = 0$. Therefore, we get the final result

$$\delta(R\epsilon) = \delta\omega^l \wedge \left(\epsilon_{abl} \wedge \Theta^{ab}\right) + \boldsymbol{d}\left(\epsilon_{ab} \wedge \delta\Omega^{ab}\right). \tag{11.110}$$

This shows that the integral over $\delta R\epsilon$ can be expressed as

$$\int \delta R\epsilon = \int \delta\omega^l \wedge \left(\epsilon_{abl} \wedge \Theta^{ab}\right) \tag{11.111}$$

except for a surface term. The expression in the integrand can be shown to be essentially the Einstein tensor (see Exercise 11.6). This proves the equality of the formalism based on curvature forms and the one discussed in Chapter 6.

Exercise 11.5
Bianchi identity in the form language Show that the Bianchi identity can be stated as $D\Theta^a{}_b = 0$. [Answer. We have, in matrix notation, $D\Theta = d\Theta + \Omega \wedge \Theta - \Theta \wedge \Omega = d\Theta + \Omega \wedge d\Omega - d\Omega \wedge \Omega$, while $d\Theta = d(d\Omega + \Omega \wedge \Omega) = d\Omega \wedge \Omega - \Omega \wedge d\Omega$ and hence $D\Theta = 0$.]

Exercise 11.6
Variation of Einstein–Hilbert action in the form language (a) By expressing the curvature forms in terms of the Christoffel symbol or otherwise, show that Eq. (11.110) is equivalent to Eq. (6.9).
   (b) Show that

$$-\frac{1}{2}\epsilon_{abl} \wedge \Theta^{bl} = G_{ab}\left(*\omega^b\right), \tag{11.112}$$

thereby completing the proof of the variation of Einstein–Hilbert action.
   (c) Show that, in two dimensions, $R\epsilon = 2\boldsymbol{d}\omega^0{}_1$; that is, it is a locally exact form. Hence its integral over a compact two-dimensional surface is a topological invariant.

Exercise 11.7
The Gauss–Bonnet term One can generalize the structure used to obtain R by taking a product of several curvature tensors. When one uses two curvature tensors, the product leads to an expression called the *Gauss–Bonnet term*.
   (a) Obtain this term by showing that

$$\Theta^{a_1 a_2} \wedge \Theta^{a_3 a_4} \wedge *(\omega_{a_1} \wedge \cdots \omega_{a_4}) = \sqrt{|g|}d^D x(R^2 - 4R^n_m R^m_n + R^{ij}_{mn}R^{mn}_{ij}). \tag{11.113}$$

[Answer.

$$\Theta^{a_1 a_2} \wedge \Theta^{a_3 a_4} \wedge *(\omega_{a_1} \wedge \cdots \omega_{a_4}) = \frac{1}{4(D-4)!}R^{a_1 a_2}_{m_1 m_2}R^{a_3 a_4}_{m_3 m_4}$$

$$\times \omega^{n_1}_{a_1} \cdots \omega^{n_4}_{a_4} \epsilon_{n_1 \cdots n_4 n_5 \cdots n_D} dx^{m_1} \wedge \cdots dx^{m_4} \wedge \cdots dx^{n_D}$$

$$= \frac{1}{4}R^{n_2 n_1}_{m_1 m_2}R^{n_4 n_3}_{m_3 m_4}\delta^{m_1 \cdots m_4}_{n_1 \cdots n_4}\sqrt{|g|}\,d^D x. \tag{11.114}$$

Simplifying the expression involving the determinant tensor, we get

$$\frac{1}{4} R^{n_2 n_1}_{m_1 m_2} R^{n_4 n_3}_{m_3 m_4} \delta^{m_1 \cdots m_4}_{n_1 \cdots n_4} = \frac{1}{4} \Big\{ R^{nm}_{mn} R^{ji}_{ij} + R^{nj}_{mn} R^{im}_{ij} - R^{nm}_{mn} R^{ij}_{ij}$$

$$- R^{nj}_{mn} R^{mi}_{ij} - R^{ni}_{mn} R^{jm}_{ij} - R^{mj}_{mn} R^{in}_{ij} - \cdots \Big\}$$

$$= (R^2 - 4 R^n_m R^m_n + R^{ij}_{mn} R^{mn}_{ij}). \tag{11.115)]}$$

(b) Consider the possibility of using the expresion in Eq. (11.113) as a Lagrangian for gravity. Show that, in $D = 4$, the variation of this Lagrangian leads only to a boundary term and hence is trivial.

Exercise 11.8
Landau–Lifshitz pseudo-tensor in the form language The purpose of this exercise is to re-write the Einstein's equation in the form language, obtained above, as a conservation law and identify the analogue of the Landau–Lifshitz energy-momentum pseudo-tensor.

(a) Introduce an 1-form corresponding to the energy-momentum tensor of matter by $\boldsymbol{T}_a \equiv T_{ab} \boldsymbol{\omega}^b$. Show that, in the presence of matter, the Einstein equations derived above will become

$$-\frac{1}{2} \epsilon_{abc} \wedge \boldsymbol{\Theta}^{bc} = 8\pi (*\boldsymbol{T}_a). \tag{11.116}$$

(b) Manipulating this expression, show that it can be re-expressed in the form

$$-\boldsymbol{d} \left(\sqrt{-g} \Omega^{bc} \wedge \epsilon^a{}_{bc} \right) = 16\pi \sqrt{-g} \left(*\boldsymbol{T}^a + *\boldsymbol{t}^a_{(\mathrm{LL})} \right), \tag{11.117}$$

where

$$*\boldsymbol{t}^a_{(\mathrm{LL})} = -\frac{1}{16\pi} \epsilon^{abcd} \left(\Omega_{sb} \wedge \Omega^s{}_c \wedge \boldsymbol{\omega}_d - \Omega_{bc} \wedge \Omega_{sd} \wedge \boldsymbol{\omega}^s \right) \tag{11.118}$$

is the analogue of the Landau–Lifshitz energy-momentum pseudo-tensor in the language of forms. It follows from this relation that

$$\boldsymbol{d} \left(\sqrt{-g} \left(*\boldsymbol{T}^a + *\boldsymbol{t}^a_{(\mathrm{LL})} \right) \right) = 0, \tag{11.119}$$

allowing an interpretation in terms of conservation laws.

(c) Also show that

$$-\frac{1}{16\pi} \boldsymbol{d} \left(\sqrt{-g} \Omega^{bc} \wedge \epsilon^m{}_{bc} \right) = \frac{1}{\sqrt{-g}} \partial_a \partial_b \lambda^{mnab} (*\boldsymbol{\omega}_n), \tag{11.120}$$

where λ^{mnab} is the pseudo potential defined in Eq. (6.180) thereby connecting up with the discussion in Section 6.5.

11.6.2 Gauge theories in the language of forms

In the case of spacetime manifold, the procedure based on curvature 2-forms is completely equivalent to the more conventional approach described in the earlier chapters. But this approach is of particular value in other contexts, like in the study of gauge theories in which one has to distinguish between the usual tangent space of the spacetime and the internal symmetry space. We shall now briefly describe the corresponding ideas in the case of non-abelian gauge theories to illustrate this aspect.

The central quantity in a gauge theory is a gauge potential A_m^α which carries a spacetime index m (varying over $0, 1, 2, 3$) and an internal space index $\alpha = 1, 2, ...N$. (In this discussion, we use Latin letters for the spacetime indices and Greek letters to denote the internal symmetry group indices, temporarily. The internal indices are summed over when they appear in pairs without any distinction over the 'upper' or 'lower' position.) The theory is usually based on a gauge group, the generators of which satisfy a Lie algebra of the form

$$[t^\alpha, t^\beta] = i f^{\alpha\beta}{}_\gamma t^\gamma, \tag{11.121}$$

where $f^{\alpha\beta}{}_\gamma$ are called the structure constants of the group. We can now construct a matrix representation for the generators t^α and also express the gauge potentials in the form of a matrix using the contraction $t^\alpha A_m^\alpha$ with a sum over the index α. Next, multiplying this structure by dx^m we introduce a *matrix valued* 1-form $A = A_m dx^m$, where A_m is an $N \times N$ matrix. The elements of this matrix will be $(A_m)_Q^P$, where P and Q vary over $1, 2, ...N$ while m is a spacetime index which varies only over $0, 1, 2, 3$. (This is one structural difference between gauge theories and gravity. The quantity analogous to $(A_m)_Q^P$ in gravity will be Christoffel symbols Γ_{mq}^p with all the indices varying over $0, 1, 2, 3$. In gauge theory we have to work with an internal symmetry space which is distinct from the tangent space of the spacetime manifold.) Since A is matrix valued, we shall confine our attention to the space of traceless and Hermitian matrices which are relevant for standard SU(N) gauge theories. The notion of gauge transformation is now generalized to the matrix transformation given by

$$A \to UAU^\dagger + UdU^\dagger, \tag{11.122}$$

where U is a unitary matrix. (This expression is identical to the second relation in Eq. (11.76) since $U^{-1} = U^\dagger$ for Hermitian matrices.) The notation used here is prevalent in mathematical literature. In contrast, physicists often use iA_m in place of A_m with an extra i-factor (see Eq. (4.163)). Thus, the gauge potential acts as the connection in gauge theories. It follows that

$$A^2 = A_m A_n dx^m \wedge dx^n = \frac{1}{2} [A_m, A_n] \, dx^m \wedge dx^n \tag{11.123}$$

is, in general, nonzero.

Our next task is to obtain the analogue of curvature which will give the generalization of the field tensor in electromagnetism. This will, in general, be a 2-form

$F = (1/2)F_{mn}dx^m \wedge dx^n$ built from A. Since there are only two possible 2-forms (dA and A^2) that can be obtained from A, we expect F to be a linear combination of these two. Taking a cue from the definition of curvature in Eq. (11.94), we shall define the field strength of the gauge field to be

$$F = dA + A^2. \tag{11.124}$$

(Again physics literature uses iF instead of F.) From Eq. (11.122), we see that the quantity dA transforms under the gauge transformations as

$$dA \to U\,dA\,U^\dagger + dU\,A\,U^\dagger - U\,A\,dU^\dagger + dU\,dU^\dagger, \tag{11.125}$$

while A^2 transforms as

$$A^2 \to U\,A^2\,U^\dagger + U\,A\,dU^\dagger - dU\,A\,U^\dagger - dU\,dU^\dagger, \tag{11.126}$$

where we have used the facts $UU^\dagger = 1$ and $U\,dU^\dagger = -dU\,U^\dagger$. Adding Eq. (11.125) and Eq. (11.126) we get

$$dA + A^2 \to U(dA + A^2)U^\dagger, \tag{11.127}$$

showing the field strength transforms covariantly as it should. In component notation, the field strength is now given by

$$F = (\partial_m A_n + A_m A_n)dx^m \wedge dx^n = \frac{1}{2}(\partial_m A_n - \partial_n A_m + [A_m, A_n])dx^m \wedge dx^n \tag{11.128}$$

so that we have

$$F_{mn} = \partial_m A_n - \partial_n A_m + [A_m, A_n] \tag{11.129}$$

(see Eq. (4.164).) Each term in this expression needs to be interpreted as a matrix with two more indices. In such an explicit form, this equation has the same structure as the one that defines Riemann curvature tensor in terms of the Christoffel symbols.

Exercise 11.9
Bianchi identity for gauge fields Prove the Bianchi identity for gauge field, given by $dF + [A, F] = 0$. Write it down explicitly in component notation. [Answer. In matrix notation, omitting the wedge symbols, we have $dF = d(dA + A^2) = dA\,A - A\,dA$ and $[A, F] = A\,dA - dA\,A$ and so $dF + [A, F] = 0$. In terms of the indices, this reads as $\epsilon^{abcd}(\partial_b F_{cd} + [A_b, F_{cd}]) = 0$.]

Exercise 11.10
Action and topological invariants in the gauge theory Consider a gauge theory described above in a four-dimensional *Euclidean* space.

(a) Show that a suitable action for this theory can be taken to be

$$\mathcal{A} = \frac{1}{2} \int \mathrm{Tr}\,(F \wedge *F). \tag{11.130}$$

Vary this action with respect to the gauge potentials A and show that the resulting equations can be expressed as

$$d * F + A \wedge *F - *F \wedge A = 0. \tag{11.131}$$

(b) A particular class of solutions to these equations are those which are self-dual in the sense that $F = *F$. For such a solution, the action is an integral over $\mathrm{Tr}\,(F \wedge F)$. Show that $\mathrm{Tr}\,(F \wedge F) = d\,\mathrm{Tr}\,(A \wedge dA) = (2/3)A \wedge A \wedge A$. Hence argue that the action for such self-dual solutions becomes a pure surface term. [Answer. Again in matrix notation, omitting the wedge symbols, we have $d\,\mathrm{Tr}\,AdA = \mathrm{Tr}\,dAdA$ and $d\,\mathrm{Tr}\,(2/3)A^3 = (2/3)$ $\mathrm{Tr}\,(dAA^2 - AdAA + A^2 dA) = 2\,\mathrm{Tr}\,dAA^2$, while $\mathrm{Tr}\,F^2 = \mathrm{Tr}\,(dA + A^2)(dA + A^2) = \mathrm{Tr}\,(dAdA + 2dAA^2)$ since $\mathrm{Tr}\,A^4 = \mathrm{Tr}\,A^3 A = -\,\mathrm{Tr}\,AA^3 = -\,\mathrm{Tr}\,A^4 = 0.$]

(c) Convert the above results into component notation by proving that $\mathrm{Tr}\,(\epsilon_{abcd}F^{ab}F^{cd}) = \partial_m J^m$ where

$$J^m = 2\epsilon^{mabc}\,\mathrm{Tr}\,\left(A_a \partial_b A_c + \frac{2}{3} A_a A_b A_c \right). \tag{11.132}$$

This is a generalization of the result in Eq. (2.107) in Chapter 2 for the case of electromagnetism.

12

Hamiltonian structure of general relativity

12.1 Introduction

We obtained Einstein's equations in Chapter 6 from an action principle in which we varied the four-dimensional metric tensor g_{ab}. The resulting equation, $G_{ab} = \kappa T_{ab}$, is generally covariant in the spacetime. We also described in Section 6.3 several peculiar features of Einstein's equations. In particular, we noticed the following. (i) The time derivatives of g_{00} and $g_{0\alpha}$ do not occur in any of the equations. (ii) No *second* time derivatives of $g_{\alpha\beta}$ occur in the time–time or space–time components of Einstein's equations. These equations contain only the first time derivatives of $g_{\alpha\beta}$. (iii) Only the space–space part of Einstein's equations involves the second time derivatives of $g_{\alpha\beta}$.

These peculiarities introduce several complications when we attempt to study Einstein's theory as describing the evolution of some well defined dynamical variables. It is clear from the above properties that one cannot treat all the components of the metric tensor on an equal footing; the real dynamics is essentially contained in the evolution of $g_{\alpha\beta}$. At the same time, the generally covariant description treats the metric as a single entity which allows for a nice geometrical interpretation of Einstein's theory. The question arises as to whether one can maintain the geometrical structure of the theory and yet perform a split of Einstein's equations, along with the dynamical variables, into space and time. We shall now describe how this can be achieved.[1] This question is of importance in clarifying the structure of the theory as well as in several other contexts, like in the study of four-dimensional spacetime embedded in a higher dimensional manifold and in certain aspects of numerical relativity.

12.2 Einstein's equations in (1+3)-form

The starting point is a scalar field $t(x^a)$ defined in the spacetime such that $t =$ constant describes a family of non-intersecting spacelike hypersurfaces with a normal $n_a \propto \partial_a t$. On each hypersurface $\Sigma(t)$ parametrized by the value of t, we introduce the spatial coordinates y^α. To connect the spatial coordinates on two different hypersurfaces, we introduce a congruence of curves that intersect these surfaces, with t being the parameter along the curve. (The curves are not necessarily geodesics; neither do they have to be orthogonal to the hypersurfaces.) This defines a mapping of events from one hypersurface to another with the same set of values y^α being given to the events intersected by the same curve. This procedure introduces a valid four-dimensional coordinate system $x^i = (t, y^\alpha)$ in the spacetime with the tangent vector $t^a = (\partial x^a/\partial t)$ to the curves satisfying the constraint $t^a \partial_a t = 1$. Further, $e_\alpha^a = (\partial x^a/\partial y^\alpha)$ acts as the projection tetrad to $\Sigma(t)$. We next introduce the unit normal to the hypersurfaces by $n_a = -N\partial_a t$ (with $n_a e_\alpha^a = 0$), where we have introduced the scalar function N (called *lapse*) to ensure proper normalization. We can decompose t^a in the form $t^a = Nn^a + N^\alpha e_\alpha^a$, where the three functions N^α form a spatial vector (called the *shift*). From the coordinate transformation $x^a = x^a(t, y^\alpha)$, it follows that

$$dx^a = t^a dt + e_\alpha^a dy^\alpha = (Ndt)n^a + (dy^\alpha + N^\alpha dt)e_\alpha^a. \tag{12.1}$$

Hence the spacetime metric, given by $ds^2 = dx_a dx^a$, simplifies to

$$ds^2 = g_{mn}dx^m dx^n = -N^2 dt^2 + h_{\alpha\beta}(dx^\alpha + N^\alpha dt)(dx^\beta + N^\beta dt), \tag{12.2}$$

where the induced 3-metric on Σ is given by

$$h_{\alpha\beta} = g_{mn}e_\alpha^m e_\beta^n = g_{\alpha\beta}. \tag{12.3}$$

This result in Eqs. (12.2) and (12.3) expresses the time–time, time–space and space–space components of the metric in terms of N, N^α and $h_{\alpha\beta}$ as $g_{00} = N^\gamma N_\gamma - N^2$; $g_{0\alpha} = N_\alpha$; $g_{\alpha\beta} = h_{\alpha\beta}$. Given these, one can compute the contravariant components of the metric in a straightforward manner and obtain

$$g^{00} = -N^{-2}, \qquad g^{0\alpha} = N^{-2}N^\alpha, \qquad g^{\alpha\gamma} = h^{\alpha\gamma} - N^{-2}N^\alpha N^\gamma. \tag{12.4}$$

It is also easy to verify that $\sqrt{-g} = N\sqrt{h}$. Thus the splitting of the spacetime into space and time is achieved mathematically by foliating the spacetime by a series of spacelike hypersurfaces $\Sigma(t)$ labelled by a coordinate t through a function $t(x^i)$ in the spacetime. As we have discussed in Chapter 4 (see the discussion around Eq. (4.31)), the induced metric on the spacelike surface $\Sigma(t)$ can also be expressed in four-dimensional notation by

$$h_{mn} = g_{mn} + n_m n_n; \qquad n_0 = -N, \qquad n_\alpha = 0, \tag{12.5}$$

where n^i is the normal to $\Sigma(t)$. This leads to the components

$$h_{00} = N^\gamma N_\gamma; \qquad h_{0\alpha} = N_\alpha \tag{12.6}$$

and, of course, the spatial components which are just $h_{\alpha\beta} = g_{\alpha\beta}$. We stress that, even though h_{mn} is a metric on the 3-space, h_{00} and $h_{0\alpha}$ are nonzero because $g_{0\alpha} \neq 0$. These nonzero $h_{0\alpha}$ components are required precisely to cancel out the effects of nonzero $g_{0\alpha}$ in various physical processes. Also note that h_b^a acts as a projection tensor onto $\Sigma(t)$.

There are three facts related to the normal vector n_a that we will keep using in the ensuing discussions:

$$h_b^a n^b = 0; \qquad n^s \nabla_m n_s = 0; \qquad n_{[m} \nabla_n n_{r]} = 0. \tag{12.7}$$

The first one follows from the definition of the induced metric; the second arises from differentiating the normalization condition $n^s n_s = -1$. Finally, since n^s is the normal to a set of hypersurfaces which foliate the spacetime, the Frobenius theorem (see Eq. (4.115)) must hold, leading to the third condition. In the coordinate system we are using, $n_i = -N\delta_i^0$ or, in the notation of differential forms, $\boldsymbol{n} = -N\boldsymbol{dt}$. The minus sign in the definition $n_0 = -N$ (with $N > 0$) is chosen so that $n^0 = +N^{-1}$ is an 'outward' pointing vector in our choice of signature. Clearly, there will be no corresponding minus sign in the case of normals to timelike surfaces.

Our next task is to define a natural covariant derivative D_m that acts on the three-dimensional vectors which are tangent to $\Sigma(t)$. The natural definition of the spatial covariant derivative of any vector X_m that satisfies the condition $X_m n^m = 0$ (which ensures that X_m is a vector tangential to Σ) is given by

$$D_m X_n = h_m^a h_n^b \nabla_a X_b. \tag{12.8}$$

The right hand side is a projection of the four-dimensional covariant derivative $\nabla_a X_b$ onto $\Sigma(t)$ using the natural projection tensor $h_m^a = \delta_m^a + n^a n_m$. In a similar manner, we will assume that D acts on scalar functions via the rule $D_m f = h_m^a \nabla_a f$. Given these two properties and the standard chain rule, we can easily determine the action of D on the contravariant vectors. We begin with the result that $D_m(X_s V^s) = V^s D_m X_s + X_s D_m V^s$. Using our definition for the action of D on scalars and covariant vectors we can obtain – by expanding out this relation and simplifying – the result

$$V^s h_m^a \nabla_a X_s + X_s h_m^a \nabla_a V^s = V^b h_m^a \nabla_a X_b + X_s D_m V^s, \tag{12.9}$$

where we have used the fact that $V^s h_s^b = V^b$ for purely spatial vectors. This relation is the same as

$$X_s h_m^a \nabla_a V^s = X_s D_m V^s. \tag{12.10}$$

We cannot simply remove X_s from both sides of this equation because this relation does not hold for all vectors X_s but only for those which satisfy the constraint $X_s n^s = 0$. But any vector that satisfies this tangentiality condition can be expressed in the form $X_s = h^r_s Y_r$, where Y_r is completely arbitrary. This allows us to obtain the result $h^r_s D_m V^s = h^r_s h^a_m \nabla_a V^s$. But by the very construction, the operator D maps spatial tensors to spatial tensors and hence the projection operator on the left hand side is redundant. This allows us to obtain the three-dimensional covariant derivative of the contravariant spatial vectors to be

$$D_m V^r = h^r_s h^a_m \nabla_a V^s. \tag{12.11}$$

Given the action of D on the contravariant and the covariant vectors as well as on the scalar functions, we can determine its action on all higher rank tensors. For example, an analysis similar to the one done above shows that its action on second rank covariant tensors is given by

$$D_m T_{rk} = h^a_m h^b_r h^s_k \nabla_a T_{bs}. \tag{12.12}$$

It is now obvious that the rule for spatial covariant derivatives is fairly straightforward; it involves taking the ordinary covariant derivative in four-dimensional space and projecting all the relevant indices onto $\Sigma(t)$ using h^a_b.

For the operation of D to be consistent, it is necessary that $D_a h_{mn} = 0$. This result can be directly verified:

$$\begin{aligned} D_a h_{mn} &= h^b_a h^r_m h^s_n \nabla_b h_{rs} = h^b_a h^r_m h^s_n \nabla_b (g_{rs} + n_r n_s) \\ &= h^b_a h^r_m h^s_n \nabla_b (n_r n_s) = h^b_a h^r_m h^s_n n_r \nabla_b n_s + h^b_a h^r_m h^s_n n_s \nabla_b n_r = 0. \end{aligned} \tag{12.13}$$

We have used the first relation in Eq. (12.7).

We now have an induced metric and a covariant derivative operator intrinsic to the 3-manifold $\Sigma(t)$ that are capable of describing the local, intrinsic properties of the spatial slices. To obtain the full information about the spacetime structure, we also need to know how the spacelike hypersurfaces $\Sigma(t)$ are embedded in the four-dimensional geometry. Intuitively, one would expect this information to be contained in the manner in which the normal to $\Sigma(t)$ varies from event to event. To quantify this variation, we will define a quantity called the *extrinsic curvature* of $\Sigma(t)$ by

$$K_{mn} = -h^a_m h^b_n \nabla_a n_b = -h^a_m \nabla_a n_n. \tag{12.14}$$

The second equality follows from writing $h^b_n = \delta^b_n + n^b n_n$ and using the second relation in Eq. (12.7). It is clear from the first equality that K_{mn} carries the information about $\nabla_a n_n$ projected onto $\Sigma(t)$. It is also obvious from this relation that $K_{ab} n^a = K_{ab} n^b = 0$. Since $n_\alpha = 0$, it follows that $K^{0j} = 0$ showing that the *contravariant* components of K^{ij} are purely spatial.

What is *not* obvious from the definition in Eq. (12.14) is that K_{mn} is symmetric in its indices. To prove the symmetry of K_{mn}, we begin with the relation

$$-K_{mn} = h_m^a \nabla_a n_n = \nabla_m n_n + n^a n_m \nabla_a n_n \qquad (12.15)$$

and expand out the last term using the Frobenius theorem, viz. the last relation in Eq. (12.7). This gives,

$$-K_{mn} = \nabla_m n_n + n^a(-n_a \nabla_n n_m - n_n \nabla_m n_a + n_m \nabla_n n_a + n_n \nabla_a n_m + n_a \nabla_m n_n), \qquad (12.16)$$

which simplifies to

$$-K_{mn} = \nabla_n n_m + n^a n_n \nabla_a n_m = h_n^a \nabla_a n_m = -K_{nm} \qquad (12.17)$$

showing that the extrinsic curvature is symmetric. Also note that we have the relation

$$-\nabla_m n_n = K_{mn} + n_m(n^a \nabla_a n_n) \equiv K_{mn} + n_m a_n. \qquad (12.18)$$

In arriving at the second relation, we have defined the acceleration corresponding to the normal vector (which, being timelike and normalized to -1, can be thought of as a valid four-velocity) by $a_n = n^a \nabla_a n_n$. Equation (12.18) shows that the covariant derivative of the normal can be decomposed into two components, one of which is tangential to $\Sigma(t)$ (and is given by K_{mn}) while the other one is normal to $\Sigma(t)$ (and is given by $n_m a_n$).

From Eq. (12.18) and Eq. (12.5) we see that the covariant spatial components are given by $K_{\alpha\beta} = -\nabla_\beta n_\alpha = -N\Gamma^0_{\alpha\beta}$. Expanding this out we get

$$K_{\alpha\beta} = \frac{1}{2N}(D_\beta N_\alpha + D_\alpha N_\beta - \partial_0 h_{\alpha\beta}), \qquad (12.19)$$

where $N_\alpha = h_{\alpha\mu} N^\mu$. In particular, if we work in a coordinate system with $N^\mu = 0$, then the extrinsic curvature has the components:

$$K_{\mu\nu} = -\frac{1}{2N}\frac{\partial h_{\mu\nu}}{\partial t}; \quad K_{0j} = K^{0j} = 0. \qquad (12.20)$$

In such a coordinate system, K_{ab} is manifestly spatial and the nonzero components measure the time derivative of $h_{\alpha\beta}$.

The geometrical significance of the extrinsic curvature is further highlighted by the fact that it can be related to the Lie derivative of the spatial metric along the normal by the relation $K_{mn} = -(1/2)\pounds_n h_{mn}$. This can again be proved by straightforward index manipulation

$$\begin{aligned}\pounds_n h_{mn} &= n^a \nabla_a h_{mn} + h_{an}\nabla_m n^a + h_{ma}\nabla_n n^a \\ &= n^a \nabla_a(n_m n_n) + g_{an}\nabla_m n^a + g_{ma}\nabla_n n^a \\ &= n^a n_m \nabla_a n_n + n^a n_n \nabla_a n_m + \nabla_m n_n + \nabla_n n_m \\ &= h_m^a \nabla_a n_n + h_n^a \nabla_a n_m = -2K_{mn}.\end{aligned} \qquad (12.21)$$

The first line is the definition of a Lie derivative; the second equality is obtained using $\nabla_a \boldsymbol{g} = 0$ and $n^s \nabla_m n_s = 0$. The fourth equality can be verified directly and the last one follows from the definition of extrinsic curvature. The quantity $\pounds_{\boldsymbol{n}} h_{mn}$ is a covariant – but foliation dependent – generalization of the time derivative of the metric.

Exercise 12.1
Extrinsic curvature and covariant derivative Let the equation to the surface $\Sigma(t)$ be given by $x^i = x^i(y^\alpha)$, which introduces the parametrization of the surface in terms of three coordinates (y^1, y^2, y^3). These functions give the projection tensors $e_\alpha^a \equiv (\partial x^a / \partial y^\alpha)$. The components of any four-dimensional tensorial quantity can be projected using e_α^a, for example, $A_i e_\alpha^i = A_\alpha$, etc. In particular, the induced metric satisfies the relation $h^{ab} = h^{\alpha\beta} e_\alpha^a e_\beta^b = g^{ab} - \epsilon n^a n^b$, where $\epsilon = \mp 1$ depending on whether $\Sigma(t)$ is spacelike or timelike.

(a) Show that the projection of the covariant derivative can be expressed as

$$e_\alpha^a e_\beta^b \nabla_b A_a = \partial_\beta A_\alpha - {}^{(3)}\Gamma_{\gamma\alpha\beta} A^\gamma, \qquad (12.22)$$

where ${}^{(3)}\Gamma_{\gamma\alpha\beta} = e_\gamma^c e_\beta^b \nabla_b e_{\alpha c}$ with $e_{\alpha c} = g_{ac} e_\alpha^a$. Also show that ${}^{(3)}\Gamma_{\gamma\alpha\beta}$ is related to $h_{\alpha\beta}$ in the usual manner.

(b) Show that the extrinsic derivative allows us to express the projection of the covariant derivative of the form

$$e_\beta^b \nabla_b V^a = e_\alpha^a D_\beta V^\alpha - \epsilon V^\alpha n^a K_{\alpha\beta}. \qquad (12.23)$$

Interpret the two terms on the right hand side.

12.3 Gauss–Codazzi equations

Given a particular foliation of the spacetime, h_{ab} and K_{ab} contain the necessary information about the intrinsic and extrinsic properties of $\Sigma(t)$. In particular, we should be able to express the full Riemann curvature tensor of the four-dimensional spacetime in terms of K_{ab} and the curvature tensor of the three-dimensional subspace. We shall now obtain these relations.

Given the covariant derivative operator D on $\Sigma(t)$, we can define the three-dimensional curvature tensor through the standard relation

$$D_m D_n X_b - D_n D_m X_b = - {}^{(3)}R^a{}_{bmn} X_a. \qquad (12.24)$$

We shall show that the three-dimensional and four-dimensional curvature tensors are related by the expression (called the *Gauss–Codazzi equation*)

$${}^{(3)}R_{abcd} = h_a^m h_b^n h_c^s h_d^t R_{mnst} + \epsilon(K_{ac}K_{bd} - K_{ad}K_{bc}), \qquad (12.25)$$

where $\epsilon = n_i n^i = -1$ for a spacelike surface. (In this form, the result is valid for a timelike hypersurface as well, though we will provide the proofs for $\epsilon = -1$.) This result, again, has a simple geometrical interpretation. The curvature of the three-dimensional surfaces has one intrinsic component which is obtained by projecting the full four-dimensional curvature tensor onto the 3-surface – given by the first term on the right hand side – and two extra terms which arise from the extrinsic properties of the embedding of the 3-surfaces in the four-dimensional space. Obtaining this result is fairly straightforward. Directly from the definition, we have:

$$
\begin{aligned}
- {}^{(3)}R_{abmn}X^a &= h_m^t h_n^l h_b^e \nabla_t \left(h_l^r h_e^s \nabla_r X_s \right) - (m \leftrightarrow n) \\
&= h_m^t h_n^l h_b^e h_l^r h_e^s \nabla_t \nabla_r X_s + h_m^t h_n^l h_b^e \left(\nabla_t h_l^r h_e^s \right) \left(\nabla_r X_s \right) - (m \leftrightarrow n) \\
&= h_m^t h_n^r h_b^s \nabla_t \nabla_r X_s \\
&\quad + h_m^t h_n^l h_b^e \left(h_l^r \nabla_t h_e^s + h_e^s \nabla_t h_l^r \right) \left(\nabla_r X_s \right) - (m \leftrightarrow n) \\
&= h_m^t h_n^r h_b^s \nabla_t \nabla_r X_s + h_m^t h_n^r h_b^e n^s \left(\nabla_t n_e \right) \left(\nabla_r X_s \right) \\
&\quad + h_m^t h_n^l h_b^s n^r \left(\nabla_t n_l \right) \left(\nabla_r X_s \right) - (m \leftrightarrow n) \\
&= h_m^t h_n^r h_b^s \nabla_t \nabla_r X_s - h_n^r n^s K_{mb} \left(\nabla_r X_s \right) \\
&\quad - h_m^t h_b^s n^r K_{tn} \left(\nabla_r X_s \right) - (m \leftrightarrow n) \\
&= h_m^t h_n^r h_b^s \nabla_t \nabla_r X_s - h_n^r n^s K_{mb} \left(\nabla_r X_s \right) - (m \leftrightarrow n). \quad (12.26)
\end{aligned}
$$

We now use the fact that, for any spatial tensor that satisfies the condition $n^s X_s = 0$ we have the relation $n^s \nabla_r X_s = -X^s \nabla_r n_s$. Using this, we obtain

$$
\begin{aligned}
- {}^{(3)}R_{abmn}X^a &= h_m^t h_n^r h_b^s \nabla_t \nabla_r X_s + h_n^r X^s K_{mb}(\nabla_r n_s) - (m \leftrightarrow n) \\
&= h_m^t h_n^r h_b^s \nabla_t \nabla_r X_s - X^s K_{mb} K_{ns} - (m \leftrightarrow n) \\
&= -h_m^t h_n^r h_b^s R_{astr} X^a - X^a K_{mb} K_{na} + X^a K_{nb} K_{ma}. \quad (12.27)
\end{aligned}
$$

We again note that this result should hold for all vectors of the form $X^a = h_b^a Y^b$ where Y^b is arbitrary. Using this and eliminating Y^b, we get the result in Eq. (12.25).

This result gives the curvature tensor, with all four indices projected onto Σ, in terms of ${}^{(3)}R^a{}_{bcd}$ and K_{ij}. To get the complete picture, we also need to find similar results for the curvature tensor with: (i) one normal component and three indices projected to Σ, and (ii) two normal components and two indices projected to Σ. We will tackle these issues now. In particular, we will prove that

$$
h_c^m h_d^n h_e^b n^a R_{mnba} = -D_c K_{de} + D_d K_{ce}, \quad (12.28)
$$

which expresses the curvature tensor with one normal component and three tangential components in terms of the derivatives of the extrinsic curvature tensor. (This expression is valid for both spacelike and timelike hypersurfaces.) Similarly, we can obtain an expression for the curvature tensor with two normal components and two spatial components as:

$$
R_{abst} h_m^a n^b h_n^s n^t = \pounds_n K_{mn} + K_{bn} K_m^b + D_m a_n + a_m a_n; \quad a^i = n^j \nabla_j n^i. \quad (12.29)
$$

Equations (12.25), (12.28) and (12.29) together express $R^a{}_{bcd}$ in terms of ${}^{(3)}R^a{}_{bcd}$ and K_{ab}. (Contraction on three indices of the curvature tensor with n^i, of course, will give zero.)

To prove Eq. (12.28), we start with the four-dimensional identity for the curvature tensor expressed in the form

$$R_{mnba}n^a = \nabla_m \nabla_n n_b - \nabla_n \nabla_m n_b = \nabla_m(-K_{nb} - n_n n^s \nabla_s n_b) - (m \leftrightarrow n).$$
(12.30)

Straightforward simplification of this expression gives

$$
\begin{aligned}
h^m_c h^n_d h^b_e n^a R_{mnba} =& - h^m_c h^n_d h^b_e \nabla_m K_{nb} - h^m_c h^n_d h^b_e (\nabla_s n_b)(\nabla_m n_n n^s) \\
& - h^m_c h^n_d h^b_e n_n n^s \nabla_m(\nabla_s n_b) - (c \leftrightarrow d) \\
=& - D_c K_{de} - h^m_c h^n_d h^b_e n^s (\nabla_s n_b)(\nabla_m n_n) - (c \leftrightarrow d) \\
=& - D_c K_{de} + h^b_e n^s (\nabla_s n_b) K_{cd} - (c \leftrightarrow d) \\
=& - D_c K_{de} + D_d K_{ce},
\end{aligned}
$$
(12.31)

which proves the result in Eq. (12.28).

To prove Eq. (12.29), we again start with the four-dimensional identity written in the form

$$R_{abst}n^t = \nabla_a \nabla_b n_s - \nabla_b \nabla_a n_s = \nabla_a(-K_{bs} - n_b a_s) - \nabla_b(-K_{as} - n_a a_s).$$ (12.32)

Contracting again with one more normal vector and simplifying, we have

$$R_{abst}n^b n^t = -n^b \nabla_a K_{bs} + h^r_a \nabla_r a_s + n^b \nabla_b K_{as} + a_s a_a.$$ (12.33)

We project the free indices onto $\Sigma(t)$ and simplify as before to obtain

$$
\begin{aligned}
R_{abst}h^a_m n^b h^s_n n^t =& -n^b h^a_m h^s_n \nabla_a K_{bs} + D_m a_n + n^b h^a_m h^s_n \nabla_b K_{as} + a_m a_n \\
=& K_{bn} h^a_m \nabla_a n^b + D_m a_n + n^b h^a_m h^s_n \nabla_b K_{as} + a_m a_n \\
=& -K_{bn} K^b_m + D_m a_n + n^b h^a_m h^s_n \nabla_b K_{as} + a_m a_n \\
=& -K_{bn} K^b_m + D_m a_n \\
& + h^a_m h^s_n(\pounds_{\mathbf{n}} K_{as} - K_{al} \nabla_s n^l - K_{ls} \nabla_a n^l) + a_m a_n \\
=& -3 K_{bn} K^b_m + D_m a_n + h^a_m h^s_n(\pounds_{\mathbf{n}} K_{as}) + a_m a_n.
\end{aligned}
$$ (12.34)

We now use the fact that K_{mn} is purely spatial so that we can write for its Lie derivative

$$
\begin{aligned}
\pounds_{\mathbf{n}} K_{mn} =& \pounds_{\mathbf{n}}(h^a_m h^s_n K_{as}) \\
=& h^a_m h^s_n(\pounds_{\mathbf{n}} K_{as}) + h^a_m K_{as}(\pounds_{\mathbf{n}} h^s_n) + h^s_n K_{as}(\pounds_{\mathbf{n}} h^a_m) \\
=& h^a_m h^s_n(\pounds_{\mathbf{n}} K_{as}) - 2 K_{ms} K^s_n - 2 K_{an} K^a_m \\
=& h^a_m h^s_n(\pounds_{\mathbf{n}} K_{as}) - 4 K_{ms} K^s_n,
\end{aligned}
$$
(12.35)

or

$$h_m^a h_n^s (\pounds_{\boldsymbol{n}} K_{as}) = \pounds_{\boldsymbol{n}} K_{mn} + 4 K_{ms} K_n^s. \tag{12.36}$$

Substituting this into Eq. (12.34), we get the result in Eq. (12.29). This equation can be simplified somewhat by noticing that

$$\begin{aligned}
D_m a_n = D_m(N^{-1} D_n N) &= -N^{-2}(D_m N)(D_n N) + N^{-1} D_m D_n N \\
&= N^{-1} D_m D_n N - a_m a_n.
\end{aligned} \tag{12.37}$$

We then get

$$R_{abst} h_m^a n^b h_n^s n^t = \pounds_{\boldsymbol{n}} K_{mn} + K_{bn} K_m^b + N^{-1} D_m D_n N. \tag{12.38}$$

We note that, while the components of curvature tensor with no indices or one index contracted with the normal (corresponding to $R_{\alpha\beta\gamma\delta}$ and $R_{0\beta\gamma\delta}$ in non-covariant notation) can be expressed in term of $^{(3)}R_{\alpha\beta\gamma\delta}$ and $K_{\alpha\beta}$, the components with two indices contracted with the normal (corresponding to $R_{0\beta0\delta}$ in non-covariant notation) will involve time derivatives of extrinsic curvature.

The results we have obtained also allow us to express the Ricci tensor and scalar curvature in a simplified form. In particular, we know from Eq. (5.64) that the complete contraction of the spatial projection of curvature tensor will give the normal components of the Einstein tensor via

$$h^{mn} h^{ab} R_{manb} = 2 n^m n^n G_{mn}. \tag{12.39}$$

Contracting our result in Eq. (12.25) appropriately, we get

$$2 n^m n^n G_{mn} = (-\epsilon)^{(3)} R + K^2 - K_{mn} K^{mn} \tag{12.40}$$

for the normal components of Einstein tensor. (In this expression we have introduced the factor ϵ such that the result is valid for both $\epsilon = -1$ and $\epsilon = +1$.) Similarly, for the component of the Einstein tensor with one tangential and one normal index, we have the result

$$n^m h_s^n G_{mn} = n^m h_s^n R_{mn} - \frac{1}{2} n_n h_s^n R = n^m h_s^n R_{mn}. \tag{12.41}$$

Using Eq. (12.28) and contracting a few indices suitably, one can obtain the result

$$h_b^n n^t G_{nt} = h_b^n n^t R_{nt} = D_b K - D_a K_b^a. \tag{12.42}$$

In a similar manner, starting with Eq. (12.25) and contracting on b and d, we get

$$h_a^m h_c^s h^{tn} R_{mnst} = {}^{(3)}R_{ac} + K_{ac} K - K_a^b K_{bc}, \tag{12.43}$$

which can be written as

$$h_a^m h_c^s R_{ms} + h_a^m n^n h_c^s n^t R_{mnst} = {}^{(3)}R_{ac} + K_{ac} K - K_a^b K_{bc}. \tag{12.44}$$

Now using Eq. (12.38), we get the result for the spatially projected components of the Ricci tensor

$$h_m^a h_n^b R_{ab} = -\pounds_{\boldsymbol{n}} K_{mn} + {}^{(3)}R_{mn} + K K_{mn} - 2K_m^s K_{sn} - N^{-1} D_m D_n N. \quad (12.45)$$

With this background, we are in a position to write down Einstein's equations in the (1+3)-form. The analogues of the time–time component and the time–space component can be obtained directly from Eq. (12.40) and Eq. (12.42). We get

$$^{(3)}R + K^2 - K_{mn} K^{mn} = 2\kappa T_{ab} n^a n^b \equiv 2\kappa\rho \qquad (12.46)$$

and

$$D_b K - D_a K_b^a = \kappa h_b^n n^t T_{nt} \equiv \kappa j_b. \qquad (12.47)$$

As described in Chapter 6, these equations are constraint equations on the metric coefficients. This is obvious from the fact that K_{ab} contains only the first time derivatives of $h_{\alpha\beta}$ (and no time derivatives of N or N^μ). Hence these equations contain only first time derivatives, thereby constraining the initial data.

The dynamics is contained in the space–space part of Einstein's equation which requires a little bit more work to express as an evolution equation. We will now show that the space–space part can be reduced to the following two equations:

$$\partial_0 h_{\alpha\beta} - \pounds_{\boldsymbol{N}} h_{\alpha\beta} = -2N K_{\alpha\beta}, \qquad (12.48)$$

where $\boldsymbol{N} = (0, N^\alpha)$, and

$$\partial_0 K_{\alpha\beta} - \pounds_{\boldsymbol{N}} K_{\alpha\beta} = N^{(3)} R_{\alpha\beta} + N \left(K K_{\alpha\beta} - 2K_\alpha^\delta K_{\delta\beta} \right)$$
$$- D_\alpha D_\beta N - 2\kappa h_\alpha^a h_\beta^b \left(T_{ab} - \frac{1}{2} g_{ab} T \right) N. \qquad (12.49)$$

Of these, Eq. (12.48) can be thought of as analogous to the definition ($\dot{q} = \partial H / \partial p$) of the canonical momentum in terms of the dynamical variables in mechanics. The second equation, Eq. (12.49), is analogous to the 'acceleration equation' ($\dot{p} = -\partial H / \partial q$).

To obtain Eq. (12.48) and Eq. (12.49) we shall begin by proving an identity which is valid for any purely spatial, second rank, tensor Q_{mn}:

$$\pounds_{\boldsymbol{n}} Q_{mn} = N^{-1} \left(\partial_0 Q_{mn} - \pounds_{\boldsymbol{N}} Q_{mn} \right), \qquad (12.50)$$

where $\boldsymbol{n} = N^{-1}(1, -N^\alpha)$ and $\boldsymbol{N} = (0, N^\alpha)$. (This result is nontrivial because the Lie derivative – unlike the covariant derivative – is not functionally linear in the subscript vector.) To prove it, we will expand the Lie derivative $\pounds_{\boldsymbol{n}}$ noting that, since it is independent of the affine connection, we can work entirely with coordinate derivatives. We have

$$\pounds_{\boldsymbol{n}} Q_{mn} = n^r \partial_r Q_{mn} + Q_{rn} \partial_m n^r + Q_{mr} \partial_n n^r = n^r \partial_r Q_{mn} - n^r \partial_m Q_{rn} - n^r \partial_n Q_{mr},$$
(12.51)

where we have used the condition $n^r Q_{rm} = 0$ to switch the derivatives from n^r to Q_{ij}. We now substitute $n^r = N^{-1}(\delta_0^r - N^r)$ and use the fact that, for a spatial vector, $Q_{0n} = N^r Q_{rn}$ to obtain

$$
\begin{aligned}
N \pounds_{\boldsymbol{n}} Q_{mn} &= \partial_0 Q_{mn} - N^r \partial_r Q_{mn} - \partial_m(N^r Q_{rn}) \\
&\quad + N^r \partial_m Q_{rn} - \partial_n(N^r Q_{mr}) + N^r \partial_n Q_{mr} \\
&= \partial_0 Q_{mn} - N^r \partial_r Q_{mn} - Q_{rn} \partial_m N^r - Q_{mr} \partial_n N^r. \quad (12.52)
\end{aligned}
$$

Using the definition of the Lie derivative again, we get

$$\pounds_{\boldsymbol{n}} Q_{mn} = N^{-1}\left(\partial_0 Q_{mn} - \pounds_{\boldsymbol{N}} Q_{mn}\right).$$
(12.53)

We now apply Eq. (12.53) to $K_{ab} = -(1/2)\pounds_{\boldsymbol{n}} h_{ab}$ to obtain the result in Eq. (12.48). To obtain Eq. (12.49), we note that we can apply our result to $\pounds_{\boldsymbol{n}} K_{ab}$ as well in Eq. (12.45). This leads to Eq. (12.49).

The four equations, Eqs. (12.46), (12.47), (12.48) and (12.49), are equivalent to the standard Einstein equations $G_{ab} = \kappa T_{ab}$ written in (1+3)-form. They suggest the following formal procedure for solving these equations. We first introduce a foliation of the spacetime that is completely arbitrary and contains the four functions $N(x)$ and $N^\alpha(x)$. The initial values for $h_{\alpha\beta}$ (which are the dynamical variables) and $K_{\alpha\beta}$ (which are analogous to the canonical momenta) are to be chosen satisfying the constraint equations Eq. (12.46) and Eq. (12.47). The dynamical equations Eq. (12.48) and Eq. (12.49) can now be used to evolve the system further. The constraint equations along with the dynamical equation completely determine all the metric coefficients.

Exercise 12.2

Gauss–Codazzi equations for a cone and a sphere The Gauss–Codazzi equation can also be applied in the case of a two-dimensional surface embedded in three-dimensional space. Work out the relevant equations for this case paying special attention to the signs and coefficients. Apply these equations to: (i) a two-dimensional sphere embedded in flat three-dimensional space; (ii) the two-dimensional surface of a cone with semi-vertical angle α embedded in flat three-dimensional space.

12.4 Gravitational action in (1+3)-form

In the previous sections, we have recast Einstein's equations in the (1+3)-form in a generally covariant (but foliation dependent) manner. More precisely, the (1+3) form of the equations now implicitly depends on the vector field $n^i(x)$ that is normal to $\Sigma(t)$ which determines the foliation of the spacetime. In Chapter 6, we obtained the same Einstein's equations by varying the four-dimensional metric g_{ab} in an action principle. Using the formalism developed in the previous sections, it should be possible to recast this action principle in a (1+3)-form and vary N, N^α and $h_{\alpha\beta}$ in the action to obtain the same equations directly in the (1+3)-form. We shall indicate in this section how this can be done. The discussion will also throw light on the alternate form of action principle discussed in Section 6.2.3.

The conventional action principle for general relativity is the Einstein–Hilbert action given by

$$A_g \equiv \frac{1}{16\pi} \int R\sqrt{-g}d^4x \tag{12.54}$$

(see Eq. (6.2) and Eq. (6.4)). In this section, we shall set $G_N = 1$, so that $\kappa = 8\pi$. We have also shown in Chapter 6 that this action can be written as the sum of a quadratic part based on a Lagrangian $L_{\rm quad}$ and a surface term arising from a total divergence; see Eq. (6.10) and Eq. (6.14). We will now redo this separation of R in a different manner using our foliation of the spacetime in terms of a series of spacelike hypersurfaces Σ with normals n^i. This will allow us to re-express the action in a different form which will also provide an interesting geometrical interpretation for the surface term.

We start from the identity $R_{abcd}n^d = (\nabla_a\nabla_b - \nabla_b\nabla_a)n_c$, which is valid for any vector field n_c, and obtain the result

$$\begin{aligned}
R_{bd}n^b n^d &= g^{ac}R_{abcd}n^b n^d = n^b\nabla_a\nabla_b n^a - n^b\nabla_b\nabla_a n^a \\
&= \nabla_a(n^b\nabla_b n^a) - (\nabla_a n^b)(\nabla_b n^a) - \nabla_b(n^b\nabla_a n^a) + (\nabla_b n^b)^2 \\
&= \nabla_i(Kn^i + a^i) - K_{ab}K^{ab} + K^a_a K^b_b,
\end{aligned} \tag{12.55}$$

where $K_{ij} = K_{ji} = -\nabla_i n_j - n_i a_j$ is the extrinsic curvature with $K \equiv K^i_i = -\nabla_i n^i$ and $K_{ij}K^{ij} = (\nabla_i n^j)(\nabla_j n^i)$. Further using

$$R = -R\,g_{ab}n^a n^b = 2(G_{ab} - R_{ab})n^a n^b\,, \tag{12.56}$$

and the Gauss–Codazzi equation (Eq. (12.40)),

$$2\,G_{ab}n^a n^b = K^a_a K^b_b - K_{ab}K^{ab} + {}^{(3)}R\,, \tag{12.57}$$

where ${}^{(3)}R$ is the scalar curvature of the three-dimensional space, we can write the scalar curvature as

$$R = {}^{(3)}R + K_{ab}K^{ab} - K^a_a K^b_b - 2\nabla_i(Kn^i + a^i) \equiv L_{\rm ADM} - 2\nabla_i(Kn^i + a^i)\,, \tag{12.58}$$

where the last equality defines – what is usually called – the *Arnowit–Deser–Misner* Lagrangian, $L_{\rm ADM} \equiv {}^{(3)}R + K_{ab}K^{ab} - K^a_a K^b_b$. It is clear from this definition and Eq. (12.19) that $L_{\rm ADM}$ is quadratic in the first time derivatives of $h_{\alpha\beta}$ and hence will provide proper dynamical equations on variation with respect to $h_{\alpha\beta}$. The action now becomes a sum of two terms, one that is quadratic in the time derivatives of $h_{\alpha\beta}$ and another that is a surface term:

$$A_g = \frac{1}{16\pi} \int_{\mathcal{V}} dt d^3x N \sqrt{h} L_{\rm ADM} - \frac{1}{8\pi} \int_{\mathcal{V}} d^4x \sqrt{-g} \nabla_i (Kn^i + a^i)$$
$$\equiv A_{\rm ADM} + A_{\rm sur}. \tag{12.59}$$

Our next task is to analyse both terms separately. First we will ignore the surface term and study the properties of $A_{\rm ADM}$, which will allow us to identify the canonical momenta conjugate to $h_{\alpha\beta}$ and rewrite the action in the Hamiltonian form. The variation of this action will again lead to Einstein's equations in the (1+3)-form. We shall be brief about this case and will quote only the relevant results since it was done in an equivalent form earlier on. Having done that, we will study the surface term in Section 12.4.2, and connect up with the analysis in Chapter 6.

12.4.1 The Hamiltonian for general relativity

We shall now concentrate on the action functional given by

$$16\pi A_{\rm ADM} = \int d^4x \sqrt{-g} L_{\rm ADM} = \int dt \int d^3x N \sqrt{h} \left[{}^{(3)}R + K_{ab}K^{ab} - K^a_a K^b_b \right] \tag{12.60}$$

without the surface term which – as we have seen – will not contribute to the equations of motion. Our aim is to understand the structure of this action functional.

It is clear that different components of the metric g_{ab} (determined by N, N^α and $h_{\alpha\beta}$) appear in this action in very different manner. The N appears in the combination $N dt$ in the integration measure and through the $(1/N)$ factor in the expansion of $K_{\alpha\beta}$. Similarly, N_α appear through $K_{\alpha\beta}$. (See Eq. (12.19).) As has been emphasized several times, the time derivatives of N and N^α do not appear in the action while the time derivatives of $h_{\alpha\beta}$ appear through $K_{\alpha\beta}$.

To obtain the variation of this action under arbitrary variation of the metric, we need to vary N, N^α and $h_{\alpha\beta}$ independently. Because of this structure, the canonical momenta conjugate to N and N^α vanishes identically and the equations $\delta A_{\rm ADM}/\delta N = 0$ and $\delta A_{\rm ADM}/\delta N^\alpha = 0$ – which will involve only the first time derivatives – will act as constraint equations. The canonical momentum corresponding to $h_{\alpha\beta}$ is nonzero and our first task is to determine the form of this momentum

$$p^{\alpha\beta} = \frac{\partial}{\partial \dot{h}_{\alpha\beta}} \left(\sqrt{-g} \, L_{\text{ADM}} \right). \tag{12.61}$$

Since $\dot{h}_{\alpha\beta}$ appears in L_{ADM} only through $K_{\alpha\beta}$ (see Eq. (12.19)), we have

$$(16\pi)p^{\alpha\beta} = \frac{\partial K_{\mu\nu}}{\partial \dot{h}_{\alpha\beta}} \frac{\partial}{\partial K_{\mu\nu}} \left(16\pi \sqrt{-g} \, L_{\text{ADM}} \right). \tag{12.62}$$

Writing the ADM Lagrangian in the form

$$16\pi \sqrt{-g} \, L_{\text{ADM}} = \left[{}^{(3)}R + \left(h^{\alpha\gamma} h^{\beta\delta} - h^{\alpha\beta} h^{\gamma\delta} \right) K_{\alpha\beta} K_{\gamma\delta} \right] N\sqrt{h} \tag{12.63}$$

we get

$$(16\pi)p^{\alpha\beta} = -\sqrt{h} \left(K^{\alpha\beta} - K \, h^{\alpha\beta} \right) \equiv \Pi^{\alpha\beta}. \tag{12.64}$$

We see that the canonical momenta conjugate to $h_{\alpha\beta}$ is essentially given by the extrinsic curvature. Given the canonical momenta, we can determine the Hamiltonian (density) through the standard relation

$$H_{\text{ADM}} = p^{\alpha\beta} \dot{h}_{\alpha\beta} - \sqrt{-g} \, L_{\text{ADM}}. \tag{12.65}$$

Working this out explicitly, we get,

$$\begin{aligned}
(16\pi)H_{\text{ADM}} &= \sqrt{h} \left(K^{\alpha\beta} - K \, h^{\alpha\beta} \right) (2NK_{\alpha\beta} - D_\beta N_\alpha - D_\alpha N_\beta) \\
&\quad - \left({}^{(3)}R + K^{\alpha\beta} K_{\alpha\beta} - K^2 \right) N\sqrt{h} \\
&= \left(K^{\alpha\beta} K_{\alpha\beta} - K^2 - {}^{(3)}R \right) N\sqrt{h} - 2 \left(K^{\alpha\beta} - K \, h^{\alpha\beta} \right) D_\beta N_\alpha \sqrt{h} \\
&= \left(K^{\alpha\beta} K_{\alpha\beta} - K^2 - {}^{(3)}R \right) N\sqrt{h} - 2D_\beta \left[\left(K^{\alpha\beta} - K \, h^{\alpha\beta} \right) N_\alpha \right] \sqrt{h} \\
&\quad + 2D_\beta \left(K^{\alpha\beta} - K \, h^{\alpha\beta} \right) N_\alpha \sqrt{h}.
\end{aligned} \tag{12.66}$$

Ignoring the total divergence term, the Hamiltonian becomes

$$(16\pi)H_{\text{ADM}} = \int_{\Sigma_\tau} \left[N \left(K^{\alpha\beta} K_{\alpha\beta} - K^2 - {}^{(3)}R \right) \right. \\
\left. + 2N_\alpha D_\beta \left(K^{\alpha\beta} - K \, h^{\alpha\beta} \right) \right] \sqrt{h} \, d^3 y. \tag{12.67}$$

In this expression, it is understood that the extrinsic curvature is expressed as a function of the canonical momenta by inverting the relation in Eq. (12.64) which will give:

$$\sqrt{h} K^{\alpha\beta} = -16\pi \left(p^{\alpha\beta} - \frac{1}{2} p \, h^{\alpha\beta} \right). \tag{12.68}$$

Finally, for the sake of completeness, we shall quote the results obtained by varying the action in the Hamiltonian approach, even though they have already been obtained in an equivalent form earlier in Section 12.3. In the Hamiltonian approach, the action functional is expressed in the form

$$\mathcal{A}_H \equiv \int_{t_1}^{t_2} dt \left[\int_{\Sigma_\tau} p^{\alpha\beta} \dot{h}_{\alpha\beta} d^3 y - H_{\text{ADM}} \right], \tag{12.69}$$

and the dynamical variables $(h_{\alpha\beta}, N^\alpha, N)$ as well as the canonical momenta $p^{\alpha\beta}$ are varied independently. To obtain the equations of motion we need the variation of the Hamiltonian with respect to these dynamical variables. The variations with respect to N^α and N are trivial to obtain while the variation with respect to $h_{\alpha\beta}$ requires more involved – though straightforward – algebra. The final result is:

$$\delta H_{\text{ADM}} = \int_{\Sigma_\tau} \left(P^{\alpha\beta} \delta h_{\alpha\beta} + Q_{\alpha\beta} \delta p^{\alpha\beta} - C \delta N - 2C_\alpha \delta N^\alpha \right) d^3 y \tag{12.70}$$

where the two 'constraint functions' are

$$C \equiv {}^{(3)}R + K^2 - K^{\alpha\beta} K_{\alpha\beta}, \qquad C_\alpha \equiv -D_\beta \left(K_\alpha^\beta - K \delta_\alpha^\beta \right), \tag{12.71}$$

and the other two functional derivatives are:

$$Q_{\alpha\beta} = 32\pi N h^{-1/2} \left(p_{\alpha\beta} - \frac{1}{2} p\, h_{\alpha\beta} \right) + 2D_{(\beta} N_{\alpha)} \tag{12.72}$$

and

$$
\begin{aligned}
P^{\alpha\beta} &= N h^{1/2} G^{\alpha\beta} - \frac{1}{2} N h^{-1/2} \left(p^{\gamma\delta} p_{\gamma\delta} - \frac{1}{2} p^2 \right) h^{\alpha\beta} \\
&\quad + 2N h^{-1/2} \left(p_\gamma^\alpha p^{\beta\gamma} - \frac{1}{2} p p^{\alpha\beta} \right) - h^{1/2} \left(D^\alpha D^\beta N - h^{\alpha\beta} D^\gamma N_\gamma \right) \\
&\quad - h^{1/2} D_\gamma \left(h^{-1/2} p^{\alpha\beta} N^\gamma \right) + 2 p^{\gamma(\alpha} D_\gamma N^{\beta)}.
\end{aligned}
\tag{12.73}
$$

Setting the variation of the action to zero (in the absence of matter), we first obtain the two constraint equations which are the time–time and time–space parts of Einstein's equations:

$$C \equiv {}^{(3)}R + K^2 - K^{\alpha\beta} K_{\alpha\beta} = 0, \qquad C_\alpha \equiv -D_\beta \left(K_\alpha^\beta - K \delta_\alpha^\beta \right) = 0. \tag{12.74}$$

The dynamics is contained in the space–space part of Einstein's equations, which – in the Hamiltonian approach – reduces to

$$\dot{h}_{\alpha\beta} = Q_{\alpha\beta}, \qquad \dot{p}^{\alpha\beta} = -P^{\alpha\beta}. \tag{12.75}$$

These equations have the same physical content as Eq. (12.48) and Eq. (12.49). Proving the equivalence is a straightforward exercise in algebra; but since we have obtained them from equivalent action principles, it is obvious that these equations must be the same.

12.4.2 The surface term and the extrinsic curvature

Having discussed the role of the ADM action, we shall go back to the expression in Eq. (12.58) and discuss the surface term that arises from integrating divergence term $\nabla_i(Kn^i + a^i)$. As expressed in Eq. (12.58), this has one term that depends on K and another which depends on the acceleration a^i of the normal vector field n^i. We will now show that the second term can also be reinterpreted entirely in terms of the extrinsic curvature of the boundary surface when the coordinates are chosen appropriately, making the right hand side of Eq. (6.41) vanish.

To do this, let us integrate Eq. (12.58) over a four-volume $\mathcal{V}$ bounded by two spacelike hypersurfaces Σ_1 and Σ_2 and a timelike hypersurface $\mathcal{S}$ and a surface at spatial infinity (see Fig. 12.1). The spacelike hypersurfaces are constant time slices with normals n^i, and the timelike hypersurface has a normal r^i orthogonal to n^i. The induced metric on the spacelike hypersurface Σ is $h_{ab} = g_{ab} + n_a n_b$, while the induced metric on the timelike hypersurface $\mathcal{S}$ is $h_{ab} = g_{ab} - r_a r_b$. The Σ and $\mathcal{S}$ intersect along a two-dimensional surface $\mathcal{Q}$, with the induced metric $\sigma_{ab} = h_{ab} - r_a r_b = g_{ab} + n_a n_b - r_a r_b$. Let the hypersurfaces $\Sigma, \mathcal{S}$ as well as their intersection 2-surface $\mathcal{Q}$ have the corresponding extrinsic curvatures K_{ab}, Θ_{ab} and q_{ab}. Doing the integrals, we get

$$\mathcal{A}_g = \frac{1}{16\pi}\int_{\mathcal{V}} d^4x \sqrt{-g}\, R = \frac{1}{16\pi}\int_{\mathcal{V}} d^4x \sqrt{-g}\, L_{\mathrm{ADM}} - \frac{1}{8\pi}\int_{\Sigma_1}^{\Sigma_2} d^3x \sqrt{h}\, K$$
$$- \frac{1}{8\pi}\int_{\mathcal{S}} dt\, d^2x\, N \sqrt{\sigma}(r_i a^i). \tag{12.76}$$

We now use the foliation condition $r_i n^i = 0$ between the surfaces (which, in general, requires the shift function to vanish), and note that

$$r_i a^i = r_i n^j \nabla_j n^i = -n^j n^i \nabla_j r_i = (g^{ij} - h^{ij})\nabla_j r_i = q - \Theta, \tag{12.77}$$

where $\Theta \equiv \Theta_a^a$ and $q \equiv q_a^a$ are the traces of the extrinsic curvature of the surfaces, when treated as embedded in the four-dimensional or three-dimensional enveloping manifolds respectively. Using Eq. (12.77) to replace $(r_i a^i)$ in the last term of Eq. (12.76), we get the result

$$\mathcal{A}_g + \frac{1}{8\pi}\int_{\Sigma_1}^{\Sigma_2} d^3x\sqrt{h}K - \frac{1}{8\pi}\int_{\mathcal{S}} dt d^2x N \sqrt{\sigma}\Theta \equiv \mathcal{A}_g - \frac{1}{8\pi}\int_{\partial\mathcal{V}} d^3x\, \epsilon\, \sqrt{h}K$$
$$= \frac{1}{16\pi}\int_{\mathcal{V}} d^4x\sqrt{-g}L_{\mathrm{ADM}} - \frac{1}{8\pi}\int_{\mathcal{S}} dt d^2x N \sqrt{\sigma}q. \tag{12.78}$$

The left hand side is now in the form of the action obtained in Eq. (6.46) which is the sum of the gravitational action $\mathcal{A}_g$ and a surface term involving the integral of the trace of the extrinsic curvature over $\partial\mathcal{V}$. (The factor ϵ takes care of the different signs for the spacelike and timelike surfaces in $\partial\mathcal{V}$.) We saw in Chapter 6 that such

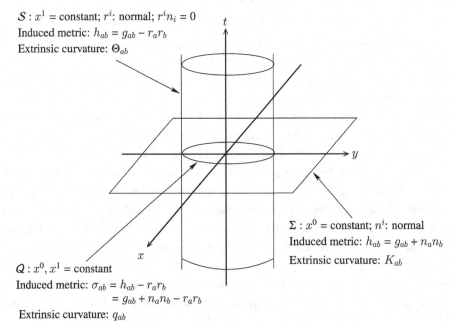

$S : x^1 = $ constant; r^i: normal; $r^i n_i = 0$

Induced metric: $h_{ab} = g_{ab} - r_a r_b$

Extrinsic curvature: Θ_{ab}

$\Sigma : x^0 = $ constant; n^i: normal

Induced metric: $h_{ab} = g_{ab} + n_a n_b$

Extrinsic curvature: K_{ab}

$Q : x^0, x^1 = $ constant

Induced metric: $\sigma_{ab} = h_{ab} - r_a r_b$

$\qquad\qquad\qquad = g_{ab} + n_a n_b - r_a r_b$

Extrinsic curvature: q_{ab}

Fig. 12.1. The surfaces bounding the region $\mathcal{V}$ that are used to define the action functional. Out of the two spacelike surfaces (Σ_1, Σ_2) and two timelike surfaces (S and another one at spatial infinity), only one of each is shown. These two surfaces intersect on a two-dimensional surface $\mathcal{Q}$ which can be thought of as embedded either in four-dimensional space or in three-dimensional space.

a sum is purely quadratic in terms of the metric tensor. In the right hand side, as it is expressed, the first term, L_{ADM}, is *not* purely quadratic in the first derivatives of the metric tensor, since it contains $^{(3)}R$, which in turn contains second derivatives of the metric tensor. We can now use a formula, analogous to Eq. (6.12), to separate the second derivatives from $^{(3)}R$. The relation is

$$^{(3)}R\sqrt{h} = {}^{(3)}L_{\text{quad}}\sqrt{h} + \partial_\mu Q^\mu, \tag{12.79}$$

where h is the determinant of the spatial metric, $^{(3)}L_{\text{quad}}$ is made from the spatial metric and its spatial derivatives, and Q^μ is the same as P^i but built from the spatial metric. The sign reflects the fact that g is negative definite while h is positive definite. What we need in Eq. (12.78) is $\sqrt{-g}\,^{(3)}R = N\sqrt{h}\,^{(3)}R$ which becomes:

$$\sqrt{-g}\,^{(3)}R = {}^{(3)}L_{\text{quad}}\sqrt{-g} + N\partial_\mu Q^\mu$$

$$= {}^{(3)}L_{\text{quad}}\sqrt{-g} - \sqrt{-g}\left(\frac{\partial_\mu N}{N}\right)\frac{\partial_\nu(hh^{\mu\nu})}{h} + \partial_\mu(NQ^\mu). \tag{12.80}$$

On integration, the last term becomes a surface integral, which can be expressed in terms of the extrinsic curvature by using a result analogous to Eq. (6.41), with vanishing shift function. This gives:

$$\int dt d^3x \partial_\mu (NQ^\mu) = \int dt d^2x NQ^\mu r_\mu = \int dt d^2x N \sqrt{\sigma} q. \tag{12.81}$$

When we substitute Eq. (12.80) into the L_{ADM} in Eq. (12.78), the terms with q cancel and we get the final result:

$$\mathcal{A}_g - \frac{1}{8\pi} \int_{\partial \mathcal{V}} d^3x \, \epsilon \sqrt{h} \, K = \frac{1}{16\pi} \int_{\mathcal{V}} d^4x \, \sqrt{-g} [(K_{ab} K^{ab} - K_a^a K_b^b)$$

$$+ {}^{(3)} L_{\text{quad}} + \frac{\partial_\mu N}{Nh} \partial_\nu (h h^{\mu\nu})], \tag{12.82}$$

which is precisely $\mathcal{A}_{\text{quad}}$. The terms with K_{ab} are quadratic in time derivatives of the spatial metric, the ${}^{(3)} L_{\text{quad}}$ has quadratic terms of spatial derivatives of the spatial metric, and the last term gives a (quadratic) cross term between spatial derivatives of the spatial metric and $g_{00,\mu}$. This is the result obtained in Section 6.2.3 by an alternative procedure taking into account only the spacelike hypersurfaces.

12.4.3 Variation of the action and canonical momenta

Finally, to complete the analysis, we shall obtain the variation of the action

$$\mathcal{A}_{\text{new}} = \mathcal{A}_g - \frac{1}{8\pi} \int_{\partial \mathcal{V}} d^3x \, \epsilon \sqrt{h} \, K \tag{12.83}$$

with respect to the metric g_{ab} and show that it correctly leads to Einstein's equations. This has already been done in an indirect manner in Section 6.2.3 (especially see Exercise 6.3) but we shall now re-do the same in a generally covariant (but foliation dependent) manner. Unlike in Exercise 6.3, now we shall not assume that $\delta g_{ab} = 0$ on the boundary and thus we will be able to identify the momentum that is canonically conjugate to h_{ab} directly from the action (instead of from the derivatives of the Lagrangian) and interpret it in terms of the extrinsic curvature. *The analysis will also allow us to interpret the Einstein–Hilbert action as a momentum space action principle.*

Consider a spacetime region $\mathcal{V}$ with a boundary $\partial \mathcal{V}$ which could be made of both timelike and spacelike pieces. The normal to the boundary is n_a with $n_a n^a = \epsilon = \pm 1$ where $\epsilon = +1$ for a timelike surface and $\epsilon = -1$ for a spacelike surface. Taking into account the ϵ factor, the induced metric on the boundary, $\partial \mathcal{V}$ is given by $h_{ab} = g_{ab} - \epsilon n_a n_b$ and the extrinsic curvature of the boundary is given by $K_{ij} = -\nabla_i n_j + \epsilon n_i a_j$. Our aim is to obtain the variation of the action in Eq. (12.83) for an arbitrary variation of g_{ab}. Since the action depends on the extrinsic curvature of the boundary, we need to know how the boundary terms change when the metric is varied. For this, we need some preliminary results which we shall first obtain.

To begin with, consider the variation δn_a of the normal when the metric is varied. We know that, if the surface is given by $\phi(x^i) = $ constant then the normal is $n_a = k\partial_a\phi$, where $k^2 = (\epsilon g^{ab}\partial_a\phi\partial_b\phi)^{-1}$. Since neither ϕ nor x^a changes under the metric variation, we have $\delta n_a = \delta k\partial_a\phi = (\delta k/k)n_a$. Using $(\delta k/k) = -(\epsilon/2)n_in_j\delta g^{ij}$, we get the result

$$\delta n_b = -\frac{1}{2}\epsilon n_b(n_in_j\delta g^{ij}).\tag{12.84}$$

Note that δn_b is in the direction of n_b. Using this fact and varying the identity $n_ih^{ij} = 0$ we get

$$n_a\delta h^{ab} = -h^{ab}\delta n_a = 0.\tag{12.85}$$

For the contravariant component of the normal, we get

$$\delta n^b = \delta(n_cg^{bc}) = n_c\delta g^{bc} - \frac{1}{2}\epsilon n^b n_in_j\delta g^{ij}.\tag{12.86}$$

From this we have the result

$$\delta n^b - \frac{1}{2}\epsilon n^b n_in_j\delta g^{ij} = \delta g^{ij}\left(\delta_i^b n_j - \epsilon n^b n_in_j\right) = n_jh_i^b\delta g^{ij},\tag{12.87}$$

which gives the variation in the direction orthogonal to n^i. Further,

$$\begin{aligned}g_{ab}\delta g^{ab} &= (h_{ab} + \epsilon n_an_b)(\delta h^{ab} + 2\epsilon n^a\delta n^b) = h_{ab}\delta h^{ab} + 2\epsilon n_b\delta n^b\\&= h_{ab}\delta h^{ab} + \epsilon n_in_j\delta g^{ij},\end{aligned}\tag{12.88}$$

where we have used the facts $n_ih^{ij} = 0$ and $n_a\delta h^{ab} = 0$. We will also need the result

$$\nabla_l V^l = D_l V^l + \epsilon n^i\nabla_i(n_l V^l) - \epsilon V^l a_l,\tag{12.89}$$

which relates the four-divergence to the three-divergence based on the covariant derivative operator D_l made from the 3-metric h_{ij}. This is easy to prove by direct manipulation:

$$\begin{aligned}D_l V^l &\equiv h_l^a h^{bl}\nabla_a V_b = (\delta_l^a - \epsilon n^a n_l)(g^{bl} - \epsilon n^b n^l)\nabla_a V_b = \nabla_l V^l - \epsilon n^a n^b\nabla_a V_b\\&= \nabla_l V^l - \epsilon n^a\nabla_a(n^b V_b) + \epsilon V_b a^b.\end{aligned}\tag{12.90}$$

Finally, we obtain a relation for $\delta K = -\delta(\nabla_i n^i)$ which will be useful. We have

$$-\delta K = \delta[\frac{1}{\sqrt{-g}}\partial_i(\sqrt{-g}n^i)] = \delta\left[\frac{1}{\sqrt{-g}}\right]\partial_i(\sqrt{-g}n^i) + \frac{1}{\sqrt{-g}}\partial_i[\delta(\sqrt{-g}n^i)].\tag{12.91}$$

Using $(\delta\sqrt{-g}/\sqrt{-g}) = -(1/2)g_{ab}\delta g^{ab}$ we can reduce this expression to the form:

$$-\delta K = -\frac{1}{2}K g_{ab}\delta g^{ab} + \nabla_i\left(\delta n^i - \frac{1}{2}n^i g_{ab}\delta g^{ab}\right).\tag{12.92}$$

The term in the bracket is given by

$$\delta n^i - \frac{1}{2}n^i g_{ab}\delta g^{ab} = \delta n^i - \frac{1}{2}n^i (h_{ab} + \epsilon n_a n_b)\,\delta g^{ab} = n_j h^i_k \delta g^{jk} - \frac{1}{2}n^i h_{ab}\delta h^{ab},$$
$$(12.93)$$

where we have used Eq. (12.87). Using now Eq. (12.89) to convert $\nabla_i(n_j h^i_k \delta g^{ik})$ and using $n_i h^i_k = 0, a_i h^i_k = a_k$, we get

$$-\delta K = -\frac{1}{2}K g_{ab}\delta g^{ab} - \frac{1}{2}\nabla_i(n^i h_{ab}\delta h^{ab}) + D_i(n_j h^i_k \delta g^{jk}) - \epsilon a_k n_j \delta g^{kj}. \quad (12.94)$$

Therefore, defining $Q^i \equiv 2n_j h^i_k \delta g^{jk}$, we get:

$$\nabla_i(n^i h_{ab}\delta h^{ab}) = 2\delta K - K g_{ab}\delta g^{ab} - 2\epsilon a_k n_j \delta g^{kj} + D_i Q^i. \quad (12.95)$$

(We will see later that terms like $D_i Q^i$ are not relevant to our discussion.) Having obtained these background results, we are now ready to turn to the main task.

We begin with the standard expression for the surface term, which arises in the variation of Hilbert action, given by

$$g^{ik}\delta R_{ik} = \nabla_a(g^{ik}\delta\Gamma^a_{ik} - g^{ia}\delta\Gamma^k_{ik}) = \nabla_a\nabla_b(-\delta g^{ab} + g^{ab}g_{ik}\delta g^{ik}). \quad (12.96)$$

The first equality is from Eq. (6.22). The second equality is the same as Eq. (6.215) and can be obtained from the first by using $\delta\Gamma^a_{bc} = \frac{1}{2}g^{ad}[\nabla_c(\delta g_{bd}) + \nabla_b(\delta g_{cd}) - \nabla_d(\delta g_{bc})]$ which, in turn, can be proved working in a local Lorentz frame and using the fact that $\delta\Gamma$ are tensors, and repeatedly using the relation $g_{ab}\delta g^{bc} = -g^{bc}\delta g_{ab}$. So we have

$$\int_{\mathcal{V}} d^4x\sqrt{-g}\,g^{ik}\delta R_{ik} = \int_{\partial\mathcal{V}} d^3x\epsilon\sqrt{h}\,n_a\nabla_b(-\delta g^{ab}+g^{ab}g_{ik}\delta g^{ik}) \equiv \int_{\partial\mathcal{V}} d^3x\epsilon\sqrt{h}\delta I,$$
$$(12.97)$$

where the factor ϵ takes care of the fact that parts of $\partial\mathcal{V}$ can be spacelike and parts can be timelike. (Since the normal n^i should be 'outward pointing', we need the $\epsilon = -1$ factor on the spacelike surfaces, as explained on page 532.) Integrating by parts, the integrand δI can be expressed in the form

$$\delta I = \nabla_b(-n_a\delta g^{ab} + n^b g_{ij}\delta g^{ij}) + (\nabla_b n_a)\delta g^{ab} + K g_{ij}\delta g^{ij}, \quad (12.98)$$

where we have used the definition $K = -\nabla_i n^i$. In the first term, $\nabla_b V^b$ with $V^b \equiv (-n_a\delta g^{ab} + n^b g_{ij}\delta g^{ij})$, we want to convert the four-dimensional divergence into a suitable three-dimensional divergence such that only the first and the third terms on the right hand side of Eq. (12.89) contribute; that is, we want to separate out the component of V^b that is orthogonal to n_b. (The motivation for this specific decomposition is the following. We will see later that terms like $D_i J^i$ do not contribute on integration over the spatial surface if J^i is a vector in the 3-surface with no component orthogonal to it; i.e. if $n_i J^i = 0$.) This can be achieved by adding and subtracting the term $n^b h_{ij} \delta h^{ij}$ to V^b. Doing this, we can express the vector V^b as

$$V^b = \left(-n_a \delta g^{ab} + n^b \left[g_{ij} \delta g^{ij} - h_{ij} \delta h^{ij} \right] + n^b h_{ij} \delta h^{ij} \right)$$
$$= \left(-n_a \delta g^{ab} + \epsilon n^b n_i n_j \delta g^{ij} \right) + \left(n^b h_{ij} \delta h^{ij} \right) = - \left(n_i h_j^b \delta g^{ij} \right) + \left(n^b h_{ij} \delta h^{ij} \right).$$

$$(12.99)$$

(In arriving at the first equality, we have used Eq. (12.88).) While evaluating $\nabla_b V^b$, we leave the second term as it is and convert the first term – which is orthogonal to n^b and hence lies purely on the boundary – by using Eq. (12.89). This gives

$$\nabla_b V^b = -D_b \left(n_i h_j^b \delta g^{ij} \right) + \epsilon a_j n_i \delta g^{ij} + \nabla_b \left(n^b h_{ij} \delta h^{ij} \right). \qquad (12.100)$$

In the second term in Eq. (12.98), we will use $\nabla_i n_j = -K_{ij} + \epsilon n_i a_j$. Carrying out these substitutions in Eq. (12.98) and simplifying, we can write δI as:

$$\delta I = -K_{ij} \delta g^{ij} + K g_{ij} \delta g^{ij} + \nabla_b \left(n^b h_{ij} \delta h^{ij} \right) + 2\epsilon n_i a_j \delta g^{ij} + D_b S^b, \quad (12.101)$$

where $S^b \equiv -n^b h_{ij} \delta h^{ij}$. We now use Eq. (12.95) to combine the second, third and fourth terms and use the fact that $K_{ij} \delta g^{ij} = K_{ij} \delta h^{ij}$, leading to

$$\delta I = -K_{ij} \delta h^{ij} + 2\delta K + D_i U^i, \qquad (12.102)$$

where $U^i \equiv S^i + Q^i$. Writing

$$2\delta K = \frac{1}{\sqrt{h}} \delta(2K\sqrt{h}) + K h_{ab} \delta h^{ab} \qquad (12.103)$$

we get the final result:

$$\delta I = \frac{1}{\sqrt{h}} \delta(2K\sqrt{h}) - (K_{ab} - K h_{ab}) \delta h^{ab} + D_i U^i, \qquad (12.104)$$

in which U^i lies entirely on the surface with $n_i U^i = 0$. On substituting back into Eq. (12.97), the integral over the 3-space of $D_i U^i$ vanishes if $\partial \mathcal{V}$ is compact. This works even when $\partial \mathcal{V}$ is made of piecewise continuous boundaries like one $t =$ constant surface connecting with another $r =$ constant surface. In such cases, the integral over the $t =$ constant surface will lead to the boundary of the r region and the integral over $r =$ constant will precisely cancel with this as long as the normals are orthogonal. Hence we get

$$\int_{\mathcal{V}} d^4 x \sqrt{-g} g^{ik} \delta R_{ik} = \delta \int_{\partial \mathcal{V}} (2K\epsilon\sqrt{h}) d^3 x - \int_{\partial \mathcal{V}} d^3 x \epsilon \sqrt{h} (K_{ab} - K h_{ab}) \delta h^{ab}.$$

$$(12.105)$$

Expressing the second term on the right hand side in terms of the variation of the covariant components, δh_{ab}, introduces an extra minus sign and we get

$$\int_{\mathcal{V}} d^4x \sqrt{-g}\, g^{ik} \delta R_{ik} = \delta \int_{\partial \mathcal{V}} (2K\epsilon \sqrt{h})d^3x + \int_{\partial \mathcal{V}} d^3x \epsilon \sqrt{h}(K^{ab} - Kh^{ab})\delta h_{ab}.$$
(12.106)

Further, we know that

$$\delta(16\pi \mathcal{A}_g) = \int_{\mathcal{V}} d^4x\, \delta(\sqrt{-g}\, R) = \int_{\mathcal{V}} d^4x \sqrt{-g}\,(G_{ab}\delta g^{ab} + g^{ab}\delta R_{ab}). \quad (12.107)$$

Using Eq. (12.106) we get:

$$\delta(16\pi \mathcal{A}_g) = \int_{\mathcal{V}} d^4x \sqrt{-g}G_{ab}\delta g^{ab} + \delta \int_{\partial \mathcal{V}} (2K\epsilon\sqrt{h})d^3x$$
$$+ \int_{\partial \mathcal{V}} d^3x \epsilon \sqrt{h}(K^{ab} - h^{ab}K)\delta h_{ab}. \quad (12.108)$$

It follows that, in the variation of $\delta \mathcal{A}_{\text{new}}$ (as defined by Eq. (12.83)), the term involving $\delta(2K\epsilon\sqrt{h})$ cancels out giving the result

$$\delta(16\pi \mathcal{A}_{\text{new}}) = \int_{\mathcal{V}} d^4x \sqrt{-g}\, G_{ab}\delta g^{ab} + \int_{\partial \mathcal{V}} d^3x \epsilon \sqrt{h}(K^{ab} - Kh^{ab})\delta h_{ab}.$$
(12.109)

The second term will be zero if $\delta h_{ab} = 0$ on $\partial \mathcal{V}$. This leads to Einstein's equations showing that $\mathcal{A}_{\text{new}}$ provides a valid variational principle if h_{ab} is kept fixed on $\partial \mathcal{V}$.

The canonical momentum p^{ab} corresponding to h_{ab} is given by the coefficient of δh_{ab} on a spacelike surface for which $\epsilon = -1$. So $p^{ab} = (16\pi)^{-1}\Pi^{ab}$, where $\Pi^{ab} \equiv -\sqrt{h}(K^{ab} - h^{ab}K)$ which, of course, agrees with the result obtained in Eq. (12.64). This fact allows us to provide an interesting interpretation to the variation of the *original* Einstein–Hilbert action. Note that, since $-h_{ab}\Pi^{ab} = \sqrt{h}(K - 3K) = -2K\sqrt{h}$, we have:

$$\delta(2K\sqrt{h}) = \Pi^{ab}\, \delta h_{ab} + h_{ab}\, \delta \Pi^{ab}. \quad (12.110)$$

Using this in Eq. (12.108) we get

$$\delta(16\pi \mathcal{A}_g) = \int_{\mathcal{V}} d^4x \sqrt{-g}G_{ab}\delta g^{ab} + \int_{\partial \mathcal{V}} d^3x\, \epsilon h_{ab}\, \delta \Pi^{ab}. \quad (12.111)$$

Thus the Einstein–Hilbert action $\mathcal{A}_g$ describes gravity *in the momentum space* and leads to the field equations when the momenta Π^{ab} are fixed at the boundaries. (See Project 6.3 for an alternative derivation of this result.) In contrast, the quadratic Lagrangian $\mathcal{A}_{\text{new}}$ describes gravity in the coordinate space with the metric h_{ab} fixed on the boundary.

12.5 Junction conditions

One of the applications of the (1+3)-formalism is in the study of *junction conditions* when the metrics on two sides of a hypersurface are driven by different matter distributions. One simple example will be a spherical configuration of matter which is collapsing. The metric inside the matter distribution is governed by the properties of the collapsing matter while the metric on the outside should satisfy a vacuum Einstein equation with $T_{ab} = 0$. The question arises as to what conditions the metric should satisfy at the boundary. We shall now see how the formalism developed in the previous sections can be exploited to describe this situation.

We begin by setting up some useful notations. Let Σ be a hypersurface with the metric being $g_{ab}^{\pm}$ on either side. (For the sake of generality we will let Σ be either timelike or spacelike.) We introduce a set of coordinates y^{α} on both sides of Σ and embed it in the four-dimensional spacetime with coordinates x^i. Let there be a congruence of geodesics through Σ intersecting it orthogonally. Any event $\mathcal{P}$ having coordinates x^i near Σ will be linked to Σ by a geodesic. We will adjust the parametrization of the geodesics by a proper time (or length, if Σ is timelike) l such that $l = 0$ corresponds to the hypersurface in question. Since every event $\mathcal{P}$ can now be attributed a proper distance $l(x^i)$ from Σ, we can think of $l(x^i)$ as a scalar field in the spacetime. The normal is now given by $n_i = \epsilon \partial_i l$ with $n^i n_i = \epsilon = \mp 1$ depending on whether the surface is spacelike or timelike. We will also use the notation of square brackets to indicate the discontinuity in a physical quantity as we cross the hypersurface; that is, $[A]$ will indicate the difference $A^+ - A^-$. With this notation, we have $[n^i] = 0$ and $[e_{\alpha}^a] = 0$, where $e_{\alpha}^a = \partial x^a / \partial y^{\alpha}$. The first condition follows from the fact that $dx^i = n^i dl$ with both l and x^i being continuous across Σ; the second condition is essentially a restatement that we use the same coordinate system on both sides of Σ.

Having set up the notation, we can now study the discontinuities in various geometrical quantities as we cross the surface. The metric can be expressed in the form

$$g_{ab} = \theta(l)g_{ab}^+ + \theta(-l)g_{ab}^-, \tag{12.112}$$

where $\theta(l)$ is the Heaviside theta function. Differentiating this relation, we will obtain a contribution from the derivative of the θ-function, which will be a Dirac delta function:

$$\partial_c g_{ab} = \theta(l)\partial_c g_{ab}^+ + \theta(-l)\partial_c g_{ab}^- + \epsilon\delta(l) [g_{ab}] n_c. \tag{12.113}$$

The last term will make the Christoffel symbols ill defined with terms proportional to $\theta(l)\delta(l)$. To avoid this ambiguity, we must demand that the discontinuity in the metric on Σ should vanish; this imposes the condition

$$[h_{\alpha\beta}] = 0. \tag{12.114}$$

Physically this requires the induced metric to be the same on both sides of Σ. We can now repeat the same analysis with the Christoffel symbols defined with the θ-function on both sides. As before, the derivative of Christoffel symbols will lead to a contribution proportional to $\delta(l)$. Hence the Riemann curvature tensor will have the form

$$R^a{}_{bcd} = \theta(l)R^{a+}{}_{bcd} + \theta(-l)R^{a-}{}_{bcd} + \delta(l)A^a{}_{bcd}, \tag{12.115}$$

where the first two terms are regular and the third term is singular with

$$A^a{}_{bcd} = \epsilon\left([\Gamma^a_{bd}]n_c - [\Gamma^a_{bc}]n_d\right). \tag{12.116}$$

We will now obtain a more explicit form for this quantity in terms of the discontinuity in the derivatives of the metric. Since the metric is well defined everywhere on Σ, the discontinuity in its derivative must be in the direction along n^a. We can, therefore, always find a tensor κ_{ab} such that $[\partial_c g_{ab}] = \kappa_{ab}n_c$ with the inverse relation given by $\kappa_{ab} = \epsilon[\partial_c g_{ab}]n^c$. Using this we can express the discontinuity in the Christoffel symbols in terms of κ_{ab} and obtain

$$A^a{}_{bcd} = \frac{\epsilon}{2}\left(\kappa^a_d n_b n_c - \kappa^a_c n_b n_d - \kappa_{bd}n^a n_c + \kappa_{bc}n^a n_d\right). \tag{12.117}$$

The corresponding discontinuous part in the Ricci tensor is determined by the quantities

$$A_{ab} \equiv A^m{}_{amb} = \frac{\epsilon}{2}\left(\kappa_{ma}n^m n_b + \kappa_{mb}n^m n_a - \kappa n_a n_b - \epsilon\kappa_{ab}\right), \tag{12.118}$$

while that in the Ricci scalar is

$$A \equiv A^a_a = \epsilon\left(\kappa_{mn}n^m n^n - \epsilon\kappa\right). \tag{12.119}$$

In Einstein's equations G_{ab} will pick up a discontinuous piece proportional to $A_{ab} - (1/2)g_{ab}A$. Therefore, we can express the energy-momentum tensor, using Einstein's equations, in the form

$$T_{ab} = \theta(l)T^+_{ab} + \theta(-l)T^-_{ab} + \delta(l)S_{ab}, \tag{12.120}$$

where $S_{ab} = (8\pi)^{-1}[A_{ab} - (1/2)g_{ab}A]$. This equation has a simple physical interpretation. The metric on the two sides of the surface are sourced by $T^\pm_{ab}$ in the usual manner. But if there is a discontinuity in the Einstein tensor, then we need an *extra* distribution of energy-momentum tensor on the surface Σ for the Einstein equations to hold everywhere.

It is possible to express the tensor S_{ab} in terms of the discontinuity in the extrinsic curvature tensor. To do this we note, from the explicit form of this tensor

$$16\pi\epsilon S_{ab} = \kappa_{ma}n^m n_b + \kappa_{mb}n^m n_a - \kappa n_a n_b - \epsilon\kappa_{ab} - (\kappa_{mn}n^m n^n - \epsilon\kappa) g_{ab}$$
(12.121)

that $S_{ab}n^b = 0$ thereby making it purely spatial. Since e^a_α act as projection tensors onto the hypersurface, S_{ab}, can be written in the equivalent form $S^{ab} = S^{\alpha\beta} e^a_\alpha e^b_\beta$. Evaluating $S_{\alpha\beta}$ explicitly, we find

$$\begin{aligned}
16\pi\epsilon S_{\alpha\beta} &= -\kappa_{ab}e^a_\alpha e^b_\beta - \epsilon\left(\kappa_{mn}n^m n^n - \epsilon\kappa\right) h_{\alpha\beta} \\
&= -\kappa_{ab}e^a_\alpha e^b_\beta - \kappa_{mn}\left(g^{mn} - h^{\mu\nu}e^m_\mu e^n_\nu\right) h_{\alpha\beta} + \kappa h_{\alpha\beta} \\
&= -\kappa_{ab}e^a_\alpha e^b_\beta + h^{\mu\nu}\kappa_{mn}e^m_\mu e^n_\nu h_{\alpha\beta}.
\end{aligned}$$
(12.122)

On the other hand, since $\nabla_b n_a = -n_c \Gamma^c_{ab}$, the discontinuity in the Christoffel symbols can be used to determine

$$[\nabla_b n_a] = \frac{1}{2}\left(\epsilon\kappa_{ab} - \kappa_{ca}n_b n^c - \kappa_{cb}n_a n^c\right).$$
(12.123)

Therefore,

$$-[K_{\alpha\beta}] = [\nabla_b n_a]e^a_\alpha e^b_\beta = \frac{\epsilon}{2}\kappa_{ab}e^a_\alpha e^b_\beta.$$
(12.124)

Using this in Eq. (12.122), we get

$$S_{\alpha\beta} = \frac{\epsilon}{8\pi}\left([K_{\alpha\beta}] - [K]h_{\alpha\beta}\right).$$
(12.125)

This condition relates the discontinuity in the extrinsic curvature to the surface energy-momentum tensor. Explicitly, we have

$$T^{ab}_\Sigma = \delta(l)S^{\alpha\beta}e^a_\alpha e^b_\beta.$$
(12.126)

As a corollary, we find that, in the absence of a surface energy-momentum tensor, K_{ij} must be continuous across the surface. One can directly verify that this condition eliminates the singularity not only in the Ricci tensor but also in the curvature tensor.

12.5.1 Collapse of a dust sphere and thin-shell

As a first application of this formalism we shall consider the collapse of a spherical region filled with pressure-free dust already discussed in Chapter 7 and Chapter 10. The inside region (which we will denote with a minus sign on the superscript) is homogeneous and isotropic and is filled with pressure-free dust. Hence the metric in this region could be taken to be that of a $k = 1$ Friedmann model with

$$ds_-^2 = -d\tau^2 + a^2(\tau)\left(d\chi^2 + \sin^2\chi d\Omega^2\right),$$
(12.127)

where τ denotes the propertime of the comoving dust particles with $x^\mu=$ constant. The Einstein equations in this region are the same as that for a Friedmann model with dust and are given by:

$$\dot{a}^2 + 1 = \frac{8\pi}{3}\rho a^2; \qquad \rho a^3 = \text{constant} \equiv \frac{3}{8\pi}a_{\max}, \qquad (12.128)$$

where we have defined the total mass content of the sphere in terms of the parameter $a_{\max}$. The solution to Eq. (12.128) can be expressed in parametric form as

$$a(\eta) = \frac{1}{2}a_{\max}(1 + \cos\eta), \quad \tau(\eta) = \frac{1}{2}a_{\max}(\eta + \sin\eta). \qquad (12.129)$$

In these coordinates, the surface of the dust sphere is located at some value $\chi = \chi_0$. We can take the collapse to begin when $\eta = \tau = 0$ and end at $\eta = \pi, \tau = (\pi/2)a_{\max}$.

The outside region – being spherically symmetric and empty – must be described by the Schwarzschild metric but in a different coordinate system. The spherical symmetry in both inside and outside regions ensures that we can take the angular coordinates on the two sphere to be the same in both the inside and outside coordinate systems. Hence, we take the outside metric to be

$$ds_+^2 = -f dt^2 + f^{-1}dr^2 + r^2 d\Omega^2, \quad f = 1 - \frac{2M}{r}, \qquad (12.130)$$

where M is yet to be determined. In the outside coordinate system, the surface Σ is described by the equations $r = R(\tau), t = T(\tau)$, where τ is the time coordinate of the inside metric. We will take the induced coordinates on Σ to be $y^\alpha = (\tau, \theta, \phi)$ in terms of which $e_0^a = u^a$ is the four-velocity of the comoving dust particle on the surface. The induced metric on Σ calculated from the inside and outside are given by

$$ds_\Sigma^2 = -d\tau^2 + a^2(\tau)\sin^2\chi_0 \, d\Omega^2 \qquad (12.131)$$

and

$$ds_\Sigma^2 = -\left(F\dot{T}^2 - F^{-1}\dot{R}^2\right)d\tau^2 + R^2(\tau)d\Omega^2, \qquad (12.132)$$

where $F \equiv (1 - 2M/R)$. Our first matching condition – viz. that the induced metric should be the same on both sides – gives the relations

$$R(\tau) = a(\tau)\sin\chi_0, \qquad F\dot{T}^2 - F^{-1}\dot{R}^2 = 1. \qquad (12.133)$$

The first equation determines the function $R(\tau)$ in terms of the known function $a(\tau)$. Using this in the second equation we can determine $T(\tau)$:

$$F\dot{T} = \sqrt{\dot{R}^2 + F} \equiv \beta(R, \dot{R}). \qquad (12.134)$$

The first equality allows us to integrate the equation and determine the function $T(\tau)$ explicitly. The second equality defines the quantity $\beta(R, \dot{R})$ for future convenience. But we know that the dust particle at the surface of the sphere has to follow a radial geodesic in the Schwarzschild metric. From our discussion in Chapter 8, it is clear (see Eq. (8.25)) that this requires $\dot{R}^2 + F(R) = E^2 = \text{constant}$, where E is the conserved energy per unit mass. Therefore we need $\beta(R, \dot{R}) = E = \text{constant}$.

We next compute the extrinsic curvature of the metric and ensure that it is the same on both sides. The outward normal to Σ, treated as a differential form, is given by $\boldsymbol{n}^- = a\boldsymbol{d}\chi$ on the inside and $\boldsymbol{n}^+ = -\dot{R}\boldsymbol{dt} + \dot{T}\boldsymbol{dr}$ on the outside. (One can easily verify the $n_a n^a = 1, n_a u^a = 0$.). From this, one can directly compute the components of the extrinsic curvature tensor both on the inside and outside:

$$K^\tau_{-\tau} = 0, \quad K^\theta_{-\theta} = K^\phi_{-\phi} = -a^{-1} \cot \chi_0 \tag{12.135}$$

and

$$K^\tau_{+\tau} = \frac{\dot{\beta}}{\dot{R}} = 0, \quad K^\theta_{+\theta} = K^\phi_{+\phi} = -\frac{\beta}{R} = -\frac{E}{R}. \tag{12.136}$$

The equality of the extrinsic curvature is therefore equivalent to the condition $E = \cos \chi_0$. It is possible to manipulate these equations to show that (see Exercise 12.3)

$$M = \frac{4\pi}{3} \rho R^3. \tag{12.137}$$

This relates the parameter in the outside Schwarzschild metric to the density inside and completes the solution.

These results can be used to obtain another interesting solution that describes the collapse of an infinitesimally thin shell of pressureless matter under its own gravity. In this case both the inside and outside geometries satisfy vacuum Einstein equations. Spherical symmetry ensures that inside is a flat spacetime (that Schwarzschild metric with $M = 0$) while the outside is a regular Schwarzschild metric with some value for M. The matching condition now requires a distribution of energy-momentum tensor $S^{ab} = \sigma u^a u^b$ on the shell surface. We will now determine the matching conditions and the equation of motion for the shell.

Using the same type of coordinates for the Schwarzschild metric on both sides (with parameters M and zero) we find, from the analysis done above, that the extrinsic curvature is given by

$$K^\tau_{\pm\tau} = -\frac{\dot{\beta}_\pm}{\dot{R}}, \quad K^\theta_{\pm\theta} = K^\phi_{\pm\phi} = -\frac{\beta_\pm}{R}, \tag{12.138}$$

where

$$\beta_+ = \sqrt{\dot{R}^2 + 1 - \frac{2M}{R}}, \quad \beta_- = \sqrt{\dot{R}^2 + 1}. \tag{12.139}$$

The difference between these components gives the surface energy-momentum tensor via Eq. (12.125). We get:

$$-\sigma = S^\tau_\tau = \frac{\beta_+ - \beta_-}{4\pi R}, \quad 0 = S^\theta_\theta = \frac{\beta_+ - \beta_-}{8\pi R} + \frac{\dot{\beta}_+ - \dot{\beta}_-}{4\pi \dot{R}}. \tag{12.140}$$

The second equation gives, on integration, $R(\beta_+ - \beta_-) = $ constant which, on substitution into the first, leads to $4\pi R^2 \sigma \equiv m$, where m is a constant. Using $(\beta_- - \beta_+) = m/R$ with the explicit expressions for $\beta_\pm$ we get the equation of motion for the shell to be

$$M = m\sqrt{1 + \dot{R}^2} - \frac{m^2}{2R}. \tag{12.141}$$

The two terms on the right hand side have the semi-Newtonian interpretation like the relativistic energy due to motion and the gravitational potential energy. With this interpretation, the mass parameter in the outside Schwarzschild metric is numerically equal to the total energy of the shell.

Exercise 12.3
Matching conditions   Verify the results quoted in the text regarding the matching conditions.

Exercise 12.4
Vacuole in a dust universe   One application of the junction condition is to describe the spacetime around a massive body located in an expanding cosmological model. Using Schwarzschild like coordinates, we can assume that a spherical massive body of radius r_0 has some interior metric for $r < r_0$ (which we are not concerned with). In the region outside the body, for $r_0 < r < r_1$, the space is empty and is described by the Schwarzschild metric. For $r > r_1$, the space is described by a Friedmann universe filled with dust. Apply the boundary conditions at $r = r_1$ and show that one can obtain a consistent solution. Obtain explicitly the matching conditions for this case.

PROJECT

Project 12.1
Superspace and the Wheeler–DeWitt equation

The Hamiltonian H_{ADM} can be given a nice interpretation along the following lines that has been very useful, for example, in the study of quantum cosmological models. For the sake of simplicity, choose the gauge in such a way that $N = 1$ and $N_\alpha = 0$, so that the spacetime metric becomes $ds^2 = -dt^2 + h_{\alpha\beta}dx^\alpha dx^\beta$. In the expression for H_{ADM} in

Eq. (12.67), only the first term contributes, giving

$$(16\pi)H_{\text{ADM}} = \int_{\Sigma_\tau} \left(K^{\alpha\beta}K_{\alpha\beta} - K^2 - {}^{(3)}R \right) \sqrt{h}\, d^3y \equiv \int d^3y\, \mathcal{H}_{\text{ADM}}. \quad (12.142)$$

(a) Show that the Hamiltonian density $\mathcal{H}_{\text{ADM}}$ (which should be treated as a function of the canonical momenta $p^{\alpha\beta}$) can be expressed in the equivalents forms as:

$$\begin{aligned}
(16\pi)^{-1}\mathcal{H}_{\text{ADM}} &= \frac{1}{\sqrt{h}} \left(p^{\alpha\beta}p_{\alpha\beta} - \frac{1}{2}p^2 \right) - \sqrt{h}\,{}^{(3)}R \\
&= \frac{1}{2}G_{\alpha\beta\gamma\delta}p^{\alpha\beta}p^{\gamma\delta} + V(h_{\alpha\beta}) \\
&= \frac{1}{2}G^{\alpha\beta\gamma\delta}p_{\alpha\beta}p_{\gamma\delta} + V(h_{\alpha\beta}),
\end{aligned} \quad (12.143)$$

where $V \equiv -\sqrt{h}\,{}^{(3)}R$ and the (so called) *Wheeler–DeWitt superspace* metric is defined as

$$G_{\alpha\beta\gamma\delta} = \frac{1}{\sqrt{h}}\left(h_{\alpha\gamma}h_{\beta\delta} + h_{\alpha\delta}h_{\beta\gamma} - 2h_{\alpha\beta}h_{\gamma\delta} \right) \quad (12.144)$$

with the inverse being

$$G^{\alpha\beta\gamma\delta} = \frac{1}{4}\sqrt{h}\left(h^{\alpha\gamma}h^{\beta\delta} + h^{\alpha\delta}h^{\beta\gamma} - 2h^{\alpha\beta}h^{\gamma\delta} \right), \quad (12.145)$$

where

$$G_{\alpha\beta\gamma\delta}\, G^{\mu\nu\gamma\delta} = \delta^\mu_{(\alpha}\delta^\nu_{\beta)}. \quad (12.146)$$

The Wheeler–deWitt metric can be thought of as a metric in the space of 3-geometries $h_{\alpha\beta}$.

(b) Consider a special case in which the metric $h_{\alpha\beta}$ is diagonal with elements $g_A = (h_{11}, h_{22}, h_{33})$, where $A = 1, 2, 3$. Show that the Wheeler–DeWitt metric can be thought of as a matrix G_{AB} and one can introduce a line interval in the superspace by $dL^2 = G_{AB}dg^A dg^B$, where the summation also includes an integration over coordinates. Show that the 3-geometries which obey Einstein's equations satisfy an equation in superspace which is a 'geodesic equation' with an extra 'force term' on the right hand side which depends on ${}^{(3)}R$.

(c) Most of the work involving superspace uses the concept of a *mini-superspace* which is obtained by assuming that the metric $h_{\alpha\beta}$ is only a function of time. All homogeneous cosmological models, for example, fall into this category. As a specific example, consider a metric $h_{\alpha\beta}(t)$ which is diagonal and is expressed in terms of three other functions of time $(\Omega(t), \beta_+(t), \beta_-(t))$ by the relations

$$h_{11} = \exp\left\{ 2\left(-\Omega + \beta_+ + \sqrt{3}\,\beta_- \right) \right\}; \quad h_{22} = \exp\left\{ 2\left(-\Omega + \beta_+ - \sqrt{3}\,\beta_- \right) \right\};$$
$$h_{33} = \exp\left\{ 2\left(-\Omega - 2\beta_+ \right) \right\}. \quad (12.147)$$

Write down the superspace metric in terms of the 'coordinates' $(\Omega(t), \beta_+(t), \beta_-(t))$. How does the geodesic equation look?

(d) Most quantum cosmological models, based on superspace, are constructed by treating $h_{\alpha\beta}(t)$ as the finite number of dynamical degrees of freedom and constructing a quantum *mechanical* description for the same. Classically, we know from Eq. (12.74) that the constraint equations require the vanishing of the gravitational Hamiltonian $\mathcal{H}_{\text{ADM}}$. In quantum theory, this can be incorporated by demanding that the wave function $\Psi(h_{\alpha\beta})$

satisfies the equation $\mathcal{H}_{\text{ADM}}\Psi = 0$, which is called the Wheeler–DeWitt equation. Write down the Wheeler–DeWitt equation for the wave function $\Psi(\Omega, \beta_+, \beta_-)$ in the example considered in part (c) above. How do you think one should study the evolution of the wave function in the absence of a time coordinate? How can one interpret the wave function of the universe obtained as a solution to the above equation?

13

Evolution of cosmological perturbations

13.1 Introduction

The evolution of the homogeneous universe was described in Chapter 10. Following up on that, we shall now turn to the study of the formation of structures in the universe. The key idea is that if there were small fluctuations in the energy density in the early universe, then the gravitational instability could amplify them leading – eventually – to structures like galaxies, clusters, etc. The most popular model for generating the initial fluctuations is based on the paradigm that, if the very early universe went through a phase of accelerated expansion (called the *inflationary phase*), then the quantum fluctuations of the field driving the inflation could lead to fluctuations in the energy density. (We will discuss this idea in detail in Chapter 14.) When the perturbations are small, one can use the linear perturbation theory to study its growth. The observations of cosmic microwave background radiation (CMBR) at $z \simeq 10^3$ show that the fractional perturbations in the energy density were quite small (about 10^{-4}–10^{-5}) at $z \approx 1000$ when the matter and radiation decoupled. Hence, linear perturbation theory can be used to make clear predictions about the state of the universe at $z \approx 1000$ and to study the anisotropies in the CMBR. This will be one key application of the formalism developed in this chapter.[1]

13.2 Structure formation and linear perturbation theory

The basic idea behind linear perturbation theory in cosmology is well defined and simple. We perturb the background Friedmann metric by $g_{ik}^{FRW} \rightarrow g_{ik}^{FRW} + h_{ik}$ and also perturb the source energy-momentum tensor by $T_{ik}^{FRW} \rightarrow T_{ik}^{FRW} + \delta T_{ik}$. Linearizing the Einstein equations, one can relate the perturbed quantities by an equation of the form $\mathcal{L}(g_{ik}^{FRW})h_{ik} = \delta T_{ik}$, where $\mathcal{L}$ is a second order linear differential operator depending on the background metric g_{ik}^{FRW}. Since the background is maximally symmetric, one can separate out time and space dependencies of the perturbed quantities; for example, in the case of the spatially flat Friedmann model, simple Fourier transform with respect to x can be used for this purpose. We can then write down the equation for any given mode, labelled by a wave vector $\boldsymbol{k}$ in the Fourier space in the form

$$\mathcal{L}(a(t), \boldsymbol{k})h_{ab}(t, \boldsymbol{k}) = \delta T_{ab}(t, \boldsymbol{k}) \tag{13.1}$$

and study its evolution. We can also label each mode by a (time dependent) wavelength normalized to the present value $\lambda(t) = (2\pi/k)(1+z)^{-1}$, and associate with it the mass of non-relativistic particles contained in a sphere of radius $\lambda(t)/2$, defined by

$$M(\lambda) \equiv \frac{4\pi\rho(t)}{3}\left[\frac{\lambda(t)}{2}\right]^3 = \frac{4\pi\rho_0}{3}\left(\frac{\lambda_0}{2}\right)^3 = 1.5 \times 10^{11} M_\odot (\Omega_{\mathrm{NR}}h^2)\left(\frac{\lambda_0}{1\,\mathrm{Mpc}}\right)^3. \tag{13.2}$$

This quantity remains constant for a given mode as the universe expands.

The behaviour of the amplitude of the perturbation at any given mode depends on the relative value of its proper wavelength $\lambda(t)$ compared with the Hubble radius $d_H(t) \equiv (\dot{a}/a)^{-1}$. Since the Hubble radius grows as $d_H(t) \propto t$ while the wavelength of the mode grows as $\lambda(t) \propto a(t) \propto (t^{1/2}, t^{2/3})$ in the radiation dominated (RD) and matter dominated (MD) phases, it follows that $\lambda(t) > d_H(t)$ at sufficiently early times. When $\lambda(t) = d_H(t)$, we say that the mode is *entering the Hubble radius*. Since the comoving Hubble radius at $z = z_{\mathrm{eq}}$ (when the energy densities of matter and radiation are equal) is

$$\lambda_{\mathrm{eq}} \simeq \left(\frac{H_0^{-1}}{\sqrt{2}}\right)\left(\frac{\Omega_R^{1/2}}{\Omega_{NR}}\right) \simeq 14\mathrm{Mpc}(\Omega_{\mathrm{NR}}h^2)^{-1} \tag{13.3}$$

it follows that the modes with $\lambda_0 > \lambda_{\mathrm{eq}}$ enter the Hubble radius in the MD phase while the more relevant modes with $\lambda < \lambda_{\mathrm{eq}}$ enter in the RD phase.

Thus, for a given mode, we can identify three distinct phases. First, very early on, when $\lambda > d_H, z > z_{\mathrm{eq}}$ the wavelength is bigger than the Hubble radius and the dynamics is described by general relativity. At this stage, the universe is radiation dominated, gravity is the only relevant force and the perturbations are linear. Next, when $\lambda < d_H$ and $z > z_{\mathrm{eq}}$ one can describe the dynamics (partly) by Newtonian considerations. The perturbations are still linear and the universe is radiation dominated. Finally, when $\lambda < d_H, z < z_{\mathrm{eq}}$ we have a matter dominated universe in which we can use the Newtonian formalism; but at this stage – when most astrophysical structures form – we need to grapple with nonlinear astrophysical processes. Our aim will be to obtain the equations of linear perturbation theory and use them to evolve the perturbations from a very early epoch through these three phases till $z \simeq 10^3$. (We will not discuss the nonlinear evolution of perturbations.) We will begin by obtaining the relevant perturbation equations.

13.3 Perturbation equations and gauge transformations

When the metric is perturbed to the form: $g_{ab} \rightarrow g_{ab} + h_{ab}$ the 10 degrees of freedom in h_{ab} can be expressed in terms of $(h_{00}, h_{0\alpha} \equiv w_\alpha, h_{\alpha\beta})$ with $10 = 1 + 3 + 6$ in obvious correspondence. In terms of these variables, the line interval can be written as

$$ds^2 = a^2(\eta) \left[-(1 + 2\Phi) \, d\eta^2 + 2w_\alpha d\eta dx^\alpha + (\delta_{\alpha\beta} + 2h_{\alpha\beta}) dx^\alpha dx^\beta \right], \quad (13.4)$$

where we are using the conformal time coordinate η defined via $dt = a d\eta$ and Φ, ψ, w_α are functions of η and $\boldsymbol{x}$. (The conformal time was denoted by τ in Chapter 10. The factor 2 in $h_{\alpha\beta}$ is introduced for later convenience. We shall work, throughout this chapter, with the spatially flat background Friedmann model.) Our next task is to decompose the spatial vector w^α and the spatial tensor $h_{\alpha\beta}$ into their irreducible components with respect to spatial rotations. We know that any three-vector field $\boldsymbol{w}(\boldsymbol{x})$ can be split as $\boldsymbol{w} = \boldsymbol{w}^\perp + \boldsymbol{w}^\parallel$ in which $\boldsymbol{w}^\parallel \equiv \nabla\Phi^\parallel$ is curl-free (and carries one degree of freedom) while $\boldsymbol{w}^\perp$ is divergence-free (and has two degrees of freedom). This result is obvious in the Fourier space since we can write the Fourier transform $\boldsymbol{w}(\boldsymbol{k})$ of any vector field $\boldsymbol{w}(\boldsymbol{x})$ as a sum of two terms, one along $\boldsymbol{k}$ and one transverse to $\boldsymbol{k}$ as:

$$\boldsymbol{w}(\boldsymbol{k}) = \boldsymbol{w}^\parallel(\boldsymbol{k}) + \boldsymbol{w}^\perp(\boldsymbol{k}) = \boldsymbol{k} \left(\frac{\boldsymbol{w}(\boldsymbol{k}) \cdot \boldsymbol{k}}{k^2} \right) + \left[\boldsymbol{w}(\boldsymbol{k}) - \boldsymbol{k} \left(\frac{\boldsymbol{k} \cdot \boldsymbol{w}(\boldsymbol{k})}{k^2} \right) \right]; \quad (13.5)$$

with $\boldsymbol{k} \times \boldsymbol{w}^\parallel = 0$; $\boldsymbol{k} \cdot \boldsymbol{w}^\perp = 0$. Fourier transforming back, we can obtain a separation of $\boldsymbol{w}(\boldsymbol{x})$ into curl-free and divergence-free parts and can write it as

$$w_\alpha = w_\alpha^\perp + \partial_\alpha \Phi^\parallel; \qquad \partial^\alpha w_\alpha^\perp = 0. \quad (13.6)$$

Similar decomposition works for $h_{\alpha\beta}$ and in this case we can write:

$$h_{\alpha\beta} = -\psi\delta_{\alpha\beta} + \left(\nabla_\alpha U_\beta^\perp + \nabla_\beta U_\alpha^\perp \right) + \left(\nabla_\alpha \nabla_\beta - \frac{1}{3}\delta_{\alpha\beta}\nabla^2 \right) \Phi_1 + h_{\alpha\beta}^{\perp\perp}. \quad (13.7)$$

The $U_\alpha^\perp$ is divergence-free (with two degrees of freedom) and $h_{\alpha\beta}^{\perp\perp}$ is traceless and divergence-free (with two degrees of freedom); the variable ψ carries the information contained in the trace h_α^α. Thus the most general perturbation h_{ab} (with 10 degrees of freedom) can be decomposed into the form:

$$h_{ab} = (h_{00}, w_\alpha, h_{\alpha\beta}) = [\Phi, (\Phi^\parallel, \boldsymbol{w}^\perp), (\psi, \Phi_1, U_\alpha^\perp, h_{\alpha\beta}^{\perp\perp})], \quad (13.8)$$

with the number of degrees of freedom given by $[1, (1, 2), (1, 1, 2, 2)]$.

The perturbations of the metric introduced above are identical in structure to the gravitational wave perturbations propagating in a curved background that we discussed in Chapter 9. There we noted that, in order to give physical meaning to the perturbations, it is necessary to eliminate the gauge ambiguities – which was done by choosing the transverse-traceless gauge. A similar problem arises in the case of cosmological perturbations. Since infinitesimal coordinate transformations can change the functional form of the metric perturbations, one cannot attribute direct physical meaning to all the components of h_{ab} without a further choice of gauge. To decide about the gauge choices, we first need to determine how the components of h_{ab} mix with each other under an infinitesimal coordinate transformation. We know that, under $x^i \rightarrow \bar{x}^i = x^i + \xi^i(x^i)$, the transformation of the metric is described by Eq. (4.135). Once again, it is convenient to separate the *spatial* part ξ^α of the four-vector $\boldsymbol{\xi} = (\xi^0, \xi^\alpha)$ into curl-free and divergence-free vectors by writing

$$\xi^\alpha = \partial^\alpha \xi + \xi^{\perp \alpha}; \qquad \partial^\alpha \xi_\alpha^\perp = 0. \tag{13.9}$$

Then, it is easy to show that, under the infinitesimal coordinate transformations, the perturbations transform to a new set, indicated by an overbar with

$$\bar{\Phi} = \Phi - \frac{\dot{a}}{a}\xi^0 - \dot{\xi}^0, \quad \bar{\Phi}^{\|} = \Phi^{\|} + \xi^0 - \dot{\xi}, \quad \bar{\psi} = \psi + \frac{1}{3}\partial_\alpha \xi_\perp^\alpha + \frac{\dot{a}}{a}\xi^0, \quad \bar{\Phi}_1 = \Phi_1 - \xi, \tag{13.10}$$

$$\bar{w}_\alpha^\perp = w_\alpha^\perp - \dot{\xi}_\alpha^\perp, \quad \bar{U}_\alpha^\perp = U_\alpha^\perp - \frac{1}{2}\xi_\alpha^\perp, \quad \bar{h}_{\alpha\beta}^{\perp\perp} = h_{\alpha\beta}^{\perp\perp}. \tag{13.11}$$

(We will use the over-dot to denote $(d/d\eta)$ so that the standard Hubble parameter is $H = (1/a)(da/dt) = \dot{a}/a^2$.) So clearly, the functional form of the perturbations cannot be given an invariant meaning since it can change under the coordinate transformations except for $h_{\alpha\beta}^{\perp\perp}$. The fact that $h_{\alpha\beta}^{\perp\perp}$ is gauge invariant is, of course, obvious from our discussion of gravitational waves in Chapter 9.

There are two ways of getting around this ambiguity. The first procedure is to construct another set of variables that actually remain invariant under the coordinate transformations. Obviously, there is an infinite set of such gauge invariant variables and we shall choose the ones given by:

$$\Phi_A = \Phi + \frac{1}{a}\frac{\partial[a(\Phi^{\|} - \dot{\Phi}_1)]}{\partial\eta}, \quad \Phi_H = \frac{\dot{a}}{a}(\Phi^{\|} - \dot{\Phi}_1) - \psi - \frac{1}{3}\partial^\alpha \partial_\alpha \Phi_1,$$

$$\Psi_\alpha = w_\alpha^\perp - 2\dot{U}_\alpha^\perp \tag{13.12}$$

and $h_{\alpha\beta}^{\perp\perp}$ – which is anyway gauge invariant. It can be directly verified – using Eq. (13.10) and Eq. (13.11) – that these variables are indeed gauge invariant. One can then work out the equations satisfied by these gauge invariant variables and study their evolution.

An alternative, and simpler, procedure is to use the freedom in the choice of coordinates *to fix the gauge* and then work with the variables that have direct physical meaning in that gauge. Once again, several such gauge choices are possible and we shall choose one which maintains a direct connection with the gauge invariant variables. This gauge, called the *Poisson gauge*, is obtained by imposing the four conditions $\Phi^{\parallel} = \Phi_1 = 0$ and $U_\alpha^\perp = 0$, thereby leaving six degrees of freedom in $(\Phi, \psi, \boldsymbol{w}^\perp, h_{\alpha\beta}^{\perp\perp})$ as nonzero. From Eq. (13.12), it follows that, in the Poisson gauge, these variables actually match with their gauge invariant counterparts; i.e. $\Phi_A = \Phi$; $\Phi_H = \psi$; $\Psi_\alpha = w_\alpha^\perp$, which is a nice feature of the Poisson gauge. An explicit transformation which will convert perturbations in an arbitrary gauge to the Poisson gauge can be effected by the four-vector ξ^a with components:

$$\xi^0 = \dot{\Phi}_1 - \Phi^{\parallel}, \quad \xi = \Phi_1, \quad \xi_\alpha^\perp = 2U_\alpha^\perp. \tag{13.13}$$

The perturbed line element in the Poisson gauge takes the form:

$$ds^2 = a^2(\eta)\left[-\{1+2\Phi\}\,d\eta^2 + 2w_\alpha^\perp\,d\eta\,dx^\alpha + \{(1-2\psi)\,\delta_{\alpha\beta} + 2h_{\alpha\beta}^{\perp\perp}\}\,dx^\alpha\,dx^\beta\right]. \tag{13.14}$$

It is now straightforward to compute G_b^a for this metric. (One possibility is to note that the metric is now of the form $g_{ab} = a^2(\eta)[\eta_{ab} + h_{ab}]$. The curvature components for the metric in the square brackets have been computed in Project 9.1 and the effect of conformal transformations is described in Exercise 5.7.) In the perturbed components, $\delta G_{00}, \delta G_{0\alpha}, \delta G_{\alpha\beta}$, it is natural to decompose $\delta G_{0\alpha}, \delta G_{\alpha\beta}$ into parts that arise from the scalar, vector and tensor degrees of freedom of the metric. We will also use the abbreviations

$$\mathcal{H} = \dot{a}/a, \qquad \Upsilon = \mathcal{H}^2\left(1 - 2\ddot{a}a/\dot{a}^2\right) = -\mathcal{H}^2\left(1 + 2\dot{\mathcal{H}}/\mathcal{H}^2\right) \tag{13.15}$$

to express the perturbations in G_b^a, separated into different components. First, the 00 component is contributed only by the scalar degrees of freedom of the metric:

$$\frac{a^2}{2}\,\delta G_0^0 = -\nabla^2\psi + 3\mathcal{H}\left(\dot{\psi} + \mathcal{H}\Phi\right). \tag{13.16}$$

As for the 0α components, we separate them into a longitudinal (parallel to $\boldsymbol{k}$ in Fourier space) part contributed by the scalars ψ and Φ and transverse (perpendicular to $\boldsymbol{k}$ in Fourier space) part contributed by the vector $w_\alpha^\perp$:

$$\frac{a^2}{2}\,[\delta G_\alpha^0]_{\parallel} = -\partial_\alpha\left(\dot{\psi} + \mathcal{H}\Phi\right); \qquad \frac{a^2}{2}\,[\delta G_\alpha^0]_\perp = \frac{1}{4}\nabla^2 w_\alpha^\perp. \tag{13.17}$$

(The full δG_α^0 is the sum of these two terms.) Similarly, the $\alpha\beta$ component can be separated into a transverse-traceless part contributed by the tensor perturbations $(h_{\alpha\beta}^{\perp\perp})$ of the metric, a transverse part that comes from the vector perturbations $(w_\alpha^\perp)$ and a part which comes from the scalars (ψ, Φ). We find:

$$\frac{a^2}{2}\,[\delta G_\beta^\alpha]_S = \frac{1}{2}\left[\left\{-2\Upsilon\Phi + 2\mathcal{H}\left(\dot{\Phi} + 2\dot{\psi}\right) + 2\ddot{\psi} - \nabla^2\left(\psi - \Phi\right)\right\}\delta_\beta^\alpha \right. $$
$$\left. + \left(\partial^\alpha\partial_\beta\left(\psi - \Phi\right)\right)\right] \tag{13.18}$$

$$\frac{a^2}{2} [\delta G_{\alpha\beta}]_V = -\frac{1}{2} \left[(\partial_0 + 2\mathcal{H}) \, \partial_{(\alpha} w_{\beta)}^\perp \right] \tag{13.19}$$

$$\frac{a^2}{2} [\delta G_{\alpha\beta}]_T = -\frac{1}{2} \left[(\Box - 2\mathcal{H}\partial_0) \, h_{\alpha\beta}^{\perp\perp} \right], \tag{13.20}$$

where the subscripts S, V and T stand for the contributions from the scalar, vector and tensor parts of the metric perturbations and $\Box$ is constructed from the flat metric.

We can perform a similar perturbation analysis of the source, T_b^a, and the results, in general, will depend on the nature of the source and the nature of the perturbations. For our purpose, we shall assume that the source is an ideal fluid with $T_b^a = (\rho + p)u^a u_b + p\delta_b^a$ and that the perturbations are in the variables u^a, ρ and p without any new anisotropic terms being introduced. We will denote the fractional perturbations in ρ and p by δ, δ_p respectively so that the perturbed energy density and pressure are

$$\rho + \delta\rho \equiv \rho(1 + \delta), \qquad p + \delta p \equiv p(1 + \delta_p). \tag{13.21}$$

In the unperturbed Friedmann universe, $u^i = (u^0, 0, 0, 0) = (a^{-1}, 0, 0, 0)$; the perturbations will modify u^0 and introduce a spatial component to the fluid velocity $v = u/u^0 = (dx/d\eta)$ which is a first order perturbation. Writing $u^\alpha = u^0 v^\alpha \approx a^{-1} v^\alpha$ and using $u^i u_i = -1$ we can determine u^0 and thus all the components of u^a to the desired order of accuracy. We get

$$u^0 = \frac{1}{a}(1 - \Phi); \quad u_0 = -a(1 + \Phi); \quad u^\alpha = \frac{1}{a}v^\alpha; \quad u_\alpha = a(v_\alpha + w_\alpha), \tag{13.22}$$

where $v_\alpha = \delta_{\alpha\beta}v^\beta$. It will again be convenient to separate v into transverse and longitudinal parts by writing $v_\alpha \equiv \partial_\alpha v + v_\alpha^\perp$. Using these expressions, we can now determine the first order perturbations to the energy-momentum tensor. We get:

$$\delta T_0^0 = -\rho\delta; \qquad \delta T_\beta^\alpha = (p\delta_p) \, \delta_\beta^\alpha$$
$$[\delta T_\alpha^0]_\parallel = (\rho + p)\partial_\alpha v; \quad [\delta T_\alpha^0]_\perp = (\rho + p)(v_\alpha^\perp + w_\alpha^\perp). \tag{13.23}$$

If the source is made of different fluids (like dark matter and radiation) the δT_b^a is obtained by adding up the different components, like $\delta\rho = \rho_1\delta_1 + \rho_2\delta_2 + \cdots$, etc.

We can now study the evolution of the perturbations using the Einstein equations $\delta G_b^a = \kappa \delta T_b^a$. From the structure of δT_b^a in Eq. (13.23) we note that the perturbation δT_β^α is isotropic (proportional to δ_β^α) and arises from scalar δp. So in the space–space part of the equations $\delta G_b^a = \kappa \delta T_b^a$ we can equate all the parts of δG_β^α that are *not* isotropic to zero. This allows us to draw several conclusions.

First, we can set the expression in Eq. (13.20) to zero, leading to a perturbation equation for $h_{\alpha\beta}^{\perp\perp}$ which decouples from the rest:

$$\left(\partial_0^2 + 2\mathcal{H}\partial_0 - \nabla^2\right) h_{\alpha\beta}^{\perp\perp} = 0. \tag{13.24}$$

This represents the propagation of gravitational wave perturbations in the Friedmann universe. Writing

$$h_{\alpha\beta}^{\perp\perp} = \frac{v(\eta)}{a} e_{\alpha\beta}^{\perp\perp} \exp(i\boldsymbol{k}\cdot\boldsymbol{x}), \qquad (13.25)$$

where we have substituted the Fourier mode (plane wave) expansion for the spatial part and introduced the constant polarization tensor $e_{\alpha\beta}^{\perp\perp}$, we find that $v(\eta)$ satisfies the simpler equation

$$\ddot{v} + \left(k^2 - \frac{\ddot{a}}{a}\right)v = 0. \qquad (13.26)$$

We will have occasion to study this equation in the context of inflationary models in the next chapter. (Also see Exercise 13.2.)

Second, equating the expression in Eq. (13.19) to zero allows us to integrate and determine the time dependence of $w_\alpha^\perp$ to be $w_\alpha^\perp \propto a^{-2}$ showing that the vector perturbations decay as the universe expands. We now use this result in the second equation in Eq. (13.17) to determine the time dependence of δG_α^0 to be $\delta G_\alpha^0 \propto a^{-4}$. Using the form of δT_α^0 in Eq. (13.23), we find that the time dependence of $(v_\alpha^\perp + w_\alpha^\perp)$ is given by:

$$(v_\alpha^\perp + w_\alpha^\perp) \propto \frac{1}{a^4(\rho+p)}. \qquad (13.27)$$

In the radiation dominated (RD) phase $(\rho+p) \propto a^{-4}$, and in the matter dominated (MD) phase $(\rho+p) \propto a^{-3}$. So this expression remains constant in the RD phase and decays as $1/a$ in the MD phase. In all these phases $w_\alpha^\perp \propto a^{-2}$ decays; so we conclude that the *transverse* part of the velocity field remains constant in the RD phase and decays as $1/a$ in the MD phase. (Note that in a small region of proper size $l \propto a$, the transverse velocity $v^\perp$ will introduce an angular momentum $J \propto (\rho l^3)\, l v_\perp \propto \rho a^4 v_\perp$. Conservation of J then leads to the correct scaling law for $v_\perp$ in RD and MD phases.)

So, we are left with scalar perturbations and the longitudinal perturbations sourced by $v_\alpha \propto \partial_\alpha v$. Again, since δT_β^α is proportional to δ_β^α, we can equate the part in Eq. (13.18) which is *not isotropic* to zero leading to $\partial_\alpha \partial^\beta (\psi - \Phi) = 0$. We can therefore take $\psi = \Phi$ without loss of generality.

Using these facts, and ignoring for the moment the vector and tensor perturbations, we can simplify the equations significantly. At this stage, it is convenient to switch to Fourier space. The corresponding equations in the Fourier space are obtained by the substitution $\partial_\alpha \rightarrow ik_\alpha$. Using the same symbols for, say, $\Phi(t, \boldsymbol{x})$ or its spatial Fourier transform $\Phi(t, \boldsymbol{k})$, the 00 and 0α components of Einstein's equations for our perturbed metric (in Fourier space, for a multi-component source) now reads as:

$$k^2\Phi + 3\frac{\dot{a}}{a}\left(\dot{\Phi} + \frac{\dot{a}}{a}\Phi\right) = -4\pi Ga^2\sum_A \rho_A \delta_A \equiv -4\pi Ga^2 \rho_{bg}\delta_{total} \quad (13.28)$$

$$\dot{\Phi} + \frac{\dot{a}}{a}\Phi = -4\pi Ga^2\sum_A(\rho + p)_A v_A, \quad (13.29)$$

where A denotes different components like dark matter, radiation, etc. The trace of Einstein's equations gives

$$3\frac{\dot{a}}{a}\dot{\Phi} + 2\frac{\ddot{a}}{a}\Phi - \frac{\dot{a}^2}{a^2}\Phi + \ddot{\Phi} = 4\pi Ga^2\sum_A \delta p_A. \quad (13.30)$$

Using Eq. (13.29) in Eq. (13.28) we can get a modified Poisson equation which is purely algebraic:

$$-k^2\Phi = 4\pi Ga^2\sum_A\left(\rho_A\delta_A - 3\left(\frac{\dot{a}}{a}\right)(\rho_A + p_A)v_A\right), \quad (13.31)$$

which once again emphasizes the fact that in the relativistic theory, both pressure and density act as a source of gravity. As far as the growth of matter perturbations are concerned, we can ignore the vector and tensor perturbations and work with a simpler metric of the form:

$$ds^2 = a^2(\eta)[-(1 + 2\Phi)d\eta^2 + (1 - 2\Phi)\delta_{\alpha\beta}dx^\alpha dx^\beta] \quad (13.32)$$

with just one perturbed scalar degree of freedom in Φ.

Exercise 13.1

Synchronous gauge Another standard gauge, widely used in the literature, corresponds to a gauge choice with $\Phi = w = w_\alpha^\perp = 0$ so that $g_{00}(= -a^2)$ and $g_{0\alpha}(= 0)$ remain unperturbed.

(a) Prove that the infinitesimal four-vector $\boldsymbol{\xi} = (\xi^0, \partial_\alpha\xi + \xi_\alpha^\perp)$ which will lead to this gauge from an arbitrary gauge is determined by

$$\xi^0 = \frac{1}{a}\int d\eta\, a\, \Phi, \quad \xi = \int d\eta\, \xi^0 + \int d\eta\, w, \quad \xi_\alpha^\perp = \int d\eta\, w_\alpha^\perp. \quad (13.33)$$

(b) This choice, however, does not fix the gauge completely. Show that a further transformation by a vector with the form

$$\xi^0 = \frac{1}{a}\xi^{(1)}(x^\alpha), \quad \xi = \xi^{(1)}(x^\alpha)\int\frac{d\eta}{a}, \quad \xi_\alpha^\perp = \xi_\alpha^\perp(x^\alpha) \quad (13.34)$$

will maintain the gauge condition.

(c) Suppose you have the solution to the perturbation equation in the Poisson gauge in terms of the (gauge invariant) variables Φ_A, Φ_H. Show that the corresponding solution in the synchronous gauge is given by

$$\Phi_l = \int \frac{1}{a} \left(\int^{\eta} a \, \Phi_A \, d\eta_1 \right) d\eta, \qquad \psi_S = \Phi_H + \frac{a'}{a^2} \int a \, \Phi_A \, d\eta. \qquad (13.35)$$

Exercise 13.2

Gravitational waves in a Friedmann universe The Eq. (13.26) can be solved in terms of elementary functions for several interesting cases like matter dominated, radiation dominated and cosmological constant dominated universes.

(a) Show that, in the radiation dominated case, the solution which is well behaved at $\eta = 0$ is given by $v \propto \eta^{-1} \sin(k\eta)$ while in the universe dominated by a cosmological constant with $\dot{a}/a^2 = H = $ constant, the two independent solutions can be taken to be the real and imaginary parts of

$$v = a \left(1 + \frac{ik}{aH} \right) \exp \left(\frac{-ik}{Ha} \right). \qquad (13.36)$$

What is the solution for the matter dominated case?

(b) More generally, prove that the *asymptotic* solutions to Eq. (13.26) are given by

$$v \simeq C_1 a + C_2 a \int \frac{d\eta}{a^2} \qquad \text{(for } k\eta \ll 1)$$

$$= \exp(\pm ik\eta) \qquad \text{(for } k\eta \gg 1). \qquad (13.37)$$

Exercise 13.3

Perturbed Einstein tensor in an arbitrary gauge (a) Show that, for a metric in Eq. (13.4), the Einstein tensor has the component

$$-(a^2/2)G_0^0 = (3/2)\mathcal{H}^2 + \left[\nabla^2 \Psi - 3\mathcal{H} \left(\dot{\Psi} + \mathcal{H}\Phi \right) - \mathcal{H}(\nabla \cdot w) + (1/2) \, \partial_\alpha \partial_\beta h_{\alpha\beta} \right] \qquad (13.38)$$

$$-(a^2/2)G_\alpha^0 = \left[\partial_\alpha \left(\dot{\Psi} + \mathcal{H}\Phi \right) + (1/4) \left[\nabla \times (\nabla \times w) \right]_\alpha + (1/2)\partial_0 \partial_\beta h_{\alpha\beta} \right] \qquad (13.39)$$

$$(a^2/2)G_0^\alpha = \left[\partial_\alpha \left(\dot{\Psi} + \mathcal{H}\Phi \right) + (1/4) \left[\nabla \times (\nabla \times w) \right]_\alpha + (1/2)\partial_0 \partial_\beta h_{\alpha\beta} \right.$$
$$\left. + \left(\mathcal{H}^2 - \dot{\mathcal{H}} \right) w_\alpha \right] \qquad (13.40)$$

$$(a^2/2)G_\beta^\alpha = (\Upsilon/2)\delta_\beta^\alpha$$
$$+ (1/2) \left[-2\Upsilon\Phi + 2\mathcal{H} \left(\dot{\Phi} + 2\dot{\Psi} \right) + 2\ddot{\Psi} - \partial_\rho \partial_\sigma h_{\rho\sigma} \right] \delta_\beta^\alpha$$
$$+ (1/2) \left[\left(\partial_\alpha \partial_\beta - \delta_{\alpha\beta} \nabla^2 \right) (\Psi - \Phi) - (\partial_0 + 2\mathcal{H}) \left(\partial_{(\alpha} w_{\beta)} - \delta_{\alpha\beta} \nabla \cdot w \right) \right]$$
$$- (1/2) \left[(\Box - 2\mathcal{H}\partial_0) \, h_{\alpha\beta} + 2\partial_\rho \partial_{(\beta} h_{\alpha)\rho} \right], \qquad (13.41)$$

where

$$\nabla \cdot w \equiv \partial_\alpha w_\alpha, \quad (\nabla \times w)_\alpha \equiv \epsilon_{\alpha\beta\sigma} \partial_\beta w_\sigma, \quad \Box \equiv -\partial_0^2 + \nabla^2 \qquad (13.42)$$

and

$$\mathcal{H} = \dot{a}/a, \qquad \Upsilon = \mathcal{H}^2 \left(1 - 2\ddot{a}a/\dot{a}^2\right) = -\mathcal{H}^2 \left(1 + 2\dot{\mathcal{H}}/\mathcal{H}^2\right). \qquad (13.43)$$

The spatial indices are raised and lowered using Kronecker delta and hence their positions are irrelevant.

(b) Use these expressions to compute the components of the Einstein tensor in the synchronous gauge and write down the corresponding Einstein equations.

13.3.1 Evolution equations for the source

The evolution of Φ as well as δT_b^a can be now determined by directly solving Einstein's equations $\delta G_b^a = \kappa \delta T_b^a$ given above. But it turns out to be more convenient – and transparent – to use the equations of motion for the matter variables, since we are eventually interested in the matter perturbations. We shall now write down the continuity and Euler equations for the source, obtained by expanding out $\nabla_a T_b^a = 0$ in the perturbed metric. The continuity equation $\nabla_a T_0^a = 0$ becomes:

$$\dot{\rho}_{\text{tot}} + 3 \left(aH - \dot{\Phi}\right) (\rho_{tot} + p_{tot}) = -\nabla_\alpha \left[(\rho_{tot} + p_{tot})v^\alpha\right], \qquad (13.44)$$

where ρ_{tot} and p_{tot} denote the total density and pressure; $\rho_{\text{tot}} = \rho(1 + \delta)$, etc. Since the momentum flux in the relativistic case is $(\rho_{tot} + p_{tot})v^\alpha$, all the terms in the above equation are intuitively obvious, except probably the $\dot{\Phi}$ term. To see the physical origin of this term, note that the perturbation in Eq. (13.32) changes the factor in front of the spatial metric from a^2 to $a^2(1 - 2\Phi)$ so that $\ln a \to \ln a - \Phi$; hence the *effective* Hubble parameter from $(\dot{a}/a)$ to $(\dot{a}/a) - \dot{\Phi}$, which explains the extra $\dot{\Phi}$ term. (Note that $aH = (da/dt) = (\dot{a}/a) = \mathcal{H}$ in our notation.) This, of course, is an exact equation for matter variables in the perturbed metric given by Eq. (13.32) up to $\mathcal{O}(v^2)$; but we only need the terms which are of linear order in $\delta\rho, \delta p$, etc. Writing the curl-free velocity part as $v^\alpha = -\nabla^\alpha v$, the *linearized* equations, for perturbations in dark matter (with $p = 0$) and radiation (with $p = (1/3)\rho$) are explicitly given by:

$$\dot{\delta}_m = \frac{d}{d\eta}\left(\frac{\delta n_m}{n_m}\right) = \nabla^2 v_m + 3\dot{\Phi}; \quad \frac{3}{4}\dot{\delta}_R = \frac{d}{d\eta}\left(\frac{\delta n_R}{n_R}\right) = \nabla^2 v_R + 3\dot{\Phi}, \qquad (13.45)$$

where n_m and n_R are the number densities of dark matter particles and photons. In Fourier space, these equations become:

$$\dot{\delta}_m = \frac{d}{d\eta}\left(\frac{\delta n_m}{n_m}\right) = -k^2 v_m + 3\dot{\Phi}; \quad \frac{3}{4}\dot{\delta}_R = \frac{d}{d\eta}\left(\frac{\delta n_R}{n_R}\right) = -k^2 v_R + 3\dot{\Phi}.$$

$$(13.46)$$

Note that these equations imply

$$\frac{d}{d\eta}\left[\frac{\delta n_R}{n_R} - \frac{\delta n_m}{n_m}\right] = \frac{d}{d\eta}\left[\delta \ln\left(\frac{n_R}{n_m}\right)\right] = \frac{d}{d\eta}\left[\delta\left(\ln\left(\frac{s}{n_m}\right)\right)\right] = -k^2(v_R - v_m),$$
$$(13.47)$$

where $s \propto n_R$ is the entropy of the photon gas. For long wavelength perturbations (in the limit of $k \to 0$), this will lead to $\delta(s/n_m) \simeq 0$; that is, the evolution of the perturbation conserves the entropy per particle in the long wavelength limit.

Let us next consider the Euler equation $\nabla_a T^a_\beta = 0$ which has the general form:

$$\partial_\eta[(\rho_{tot} + p_{tot})v^\alpha] = -(\rho_{tot} + p_{tot})\nabla^\alpha\Phi - \nabla^\alpha p - 4aH(\rho_{tot} + p_{tot})v^\alpha, \quad (13.48)$$

where $\partial_\eta = d/d\eta$. Once again each of the terms is simple to interpret. The $(\rho_{tot} + p_{tot})$ arises because the pressure also contributes to inertia in a relativistic theory and the factor 4 in the last term on the right hand side arises because the term $v^\alpha\partial_\eta(\rho_{tot} + p_{tot})$ on the left hand side needs to be compensated. Taking the linear limit of this equation, for dark matter and radiation, we get:

$$\dot{v}_m = \Phi - aHv_m; \quad \dot{v}_R = \Phi + \frac{1}{4}\delta_R. \quad (13.49)$$

Thus we now have four equations in Eqs. (13.46) and (13.49) for the five variables $(\delta_m, \delta_R, v_m, v_R, \Phi)$. All we need to do is to pick one more from Einstein's equations to complete the set. We shall now describe the evolution of perturbations determined by these equations.

To get a feel for the solutions to these perturbation equations, let us consider a spatially flat universe dominated by a single component of matter with the equation of state $p = w\rho$ and $\delta p = w\delta\rho$. (A purely radiation dominated universe, for example, will have $w = 1/3$.) In this case the Friedmann equation gives $\rho \propto a^{-3(1+w)}$ and

$$\frac{\dot{a}}{a} = \frac{2}{(1+3w)\eta}; \quad \frac{\ddot{a}}{a} = \frac{2(1-3w)}{(1+3w)^2\eta^2}. \quad (13.50)$$

We now need to solve Eq. (13.28) and Eq. (13.30) with $\delta p = w\delta\rho$. Eliminating the term $4\pi Ga^2\delta\rho$ between these two equations and simplifying the rest, we can get an equation entirely in terms of the potential Φ which has the form:

$$\ddot{\Phi} + \frac{6(1+w)}{1+3w}\frac{\dot{\Phi}}{\eta} + k^2w\Phi = 0. \quad (13.51)$$

The second term is the damping due to the expansion of the universe while the last term is the pressure support that will lead to oscillations in Φ. Clearly, the factor $k\eta$ determines which of these two terms dominates. When the pressure term dominates $(k\eta \gg 1)$, we expect oscillatory behaviour while when the background expansion dominates ($k\eta \ll 1$), we expect the growth to be suppressed. This is precisely what happens. The exact solution to Eq. (13.51) is given in terms of the Bessel functions

$$\Phi(\eta) = \frac{C_1(\boldsymbol{k}) J_{\nu/2}(\sqrt{w}k\eta) + C_2(\boldsymbol{k}) Y_{\nu/2}(\sqrt{w}k\eta)}{\eta^{\nu/2}}; \quad \nu = \frac{5 + 3w}{1 + 3w}. \tag{13.52}$$

From the theory of Bessel functions, we know that:

$$\lim_{x \to 0} J_{\nu/2}(x) \simeq \frac{x^{\nu/2}}{2^{\nu/2}\Gamma(\nu/2 + 1)}; \quad \lim_{x \to 0} Y_{\nu/2}(x) \propto -\frac{1}{x^{\nu/2}}. \tag{13.53}$$

This shows that, to obtain a finite value for Φ as $\eta \to 0$, we need to set $C_2 = 0$. This gives the gravitational potential to be

$$\Phi(\eta) = \frac{C_1(\boldsymbol{k}) J_{\nu/2}(\sqrt{w}k\eta)}{\eta^{\nu/2}}; \quad \nu = \frac{5 + 3w}{1 + 3w}. \tag{13.54}$$

The corresponding density perturbation will be:

$$\delta = -2\Phi - \frac{(1 + 3w)^2 k^2 \eta^2}{6} \frac{C_1(\boldsymbol{k}) J_{\nu/2}(\sqrt{w}k\eta)}{\eta^{\nu/2}}$$
$$+ (1 + 3w)\sqrt{w}k\eta \frac{C_1(\boldsymbol{k}) J_{(\nu/2)+1}(\sqrt{w}k\eta)}{\eta^{\nu/2}}. \tag{13.55}$$

To understand the nature of this solution, we note that the Hubble radius $d_H = (\dot{a}/a)^{-1} \propto \eta$ and $kd_H \propto d_H/\lambda \propto k\eta$. So the argument of the Bessel function is just the ratio (d_H/λ). From the theory of Bessel functions, we know that for small values of the argument $J_\nu(x) \propto x^\nu$ is a power law while for large values of the argument it oscillates with a decaying amplitude:

$$\lim_{x \to \infty} J_{\nu/2}(x) \sim \frac{\cos[x - (\nu - 1)\pi/4]}{\sqrt{x}}. \tag{13.56}$$

Hence, for modes which are still outside the Hubble radius ($k \ll \eta^{-1}$), we have a constant amplitude for the potential and density contrast:

$$\Phi \approx \Phi_i(\boldsymbol{k}); \quad \delta \approx -2\Phi_i(\boldsymbol{k}), \tag{13.57}$$

where the subscript 'i' denotes the initial values. That is, the perturbation is frozen (except for a decaying mode) at a constant value. On the other hand, for modes which are inside the Hubble radius ($k \gg \eta^{-1}$), the perturbation is rapidly oscillatory (if $w \neq 0$). The pressure is effective at small scales and leads to acoustic oscillations in the medium.

A special case of the above result occurs for the flat, matter dominated, universe with $w = 0$. In this case, we need to take the $w \to 0$ limit and the exact solution is indeed a constant $\Phi = \Phi_i(\boldsymbol{k})$ (plus a decaying mode $\Phi_{decay} \propto \eta^{-5}$ which diverges as $\eta \to 0$ and hence is not physically admissible). The corresponding density perturbations is:

$$\delta = -(2 + \frac{k^2\eta^2}{6})\Phi_i(\boldsymbol{k}). \tag{13.58}$$

This result shows that the density perturbation is again frozen at large scales but grows at small scales:

$$\delta = \begin{cases} -2\Phi_i(\boldsymbol{k}) = \text{constant} & (k\eta \ll 1) \\ -\frac{1}{6}k^2\eta^2\Phi_i(\boldsymbol{k}) \propto \eta^2 \propto a & (k\eta \gg 1). \end{cases} \tag{13.59}$$

We will use these results in our later discussion.

13.4 Perturbations in dark matter and radiation

We shall now move on to the more realistic case of a multi-component universe consisting of radiation and collisionless dark matter. (We are ignoring the baryons, which are gravitationally subdominant, in our discussion.) It is convenient to use $y = a/a_{eq}$ as an independent variable rather than the conformal time coordinate η. The background expansion of the universe, described by the function $a(t)$, can be equivalently expressed as (see Exercise 10.6):

$$y \equiv \frac{\rho_M}{\rho_R} = \frac{a}{a_{eq}} = x^2 + 2x, \qquad x \equiv \left(\frac{\Omega_{NR}}{4a_{eq}}\right)^{1/2} H_0\eta. \tag{13.60}$$

It is also useful to define a critical wave number k_c by:

$$k_c^2 = \frac{H_0^2\Omega_{NR}}{a_{eq}} = 4(\sqrt{2}-1)^2\eta_{eq}^{-2} = 4(\sqrt{2}-1)^2 k_{eq}^2; \qquad k_c^{-1} = 19(\Omega_{NR}h^2)^{-1}\text{Mpc}, \tag{13.61}$$

which essentially sets the comoving scale (k_{eq}) corresponding to matter–radiation equality except for a numerical factor. Note that $2x = k_c\eta$ and $y \approx k_c\eta$ in the RD phase while $y = (1/4)(k_c\eta)^2$ in the MD phase.

We now manipulate Eqs. (13.46), (13.49), (13.28) and (13.29) governing the growth of perturbations by essentially eliminating the velocity. This leads to three equations:

$$y\Phi' + \Phi + \frac{1}{3}\frac{k^2}{k_c^2}\frac{y^2}{1+y}\Phi = -\frac{1}{2}\frac{y}{1+y}\left(\delta_m + \frac{1}{y}\delta_R\right) \tag{13.62}$$

$$(1+y)\delta_m'' + \frac{2+3y}{2y}\delta_m' = 3(1+y)\Phi'' + \frac{3(2+3y)}{2y}\Phi' - \frac{k^2}{k_c^2}\Phi \tag{13.63}$$

$$(1+y)\delta_R'' + \frac{1}{2}\delta_R' + \frac{1}{3}\frac{k^2}{k_c^2}\delta_R = 4(1+y)\Phi'' + 2\Phi' - \frac{4}{3}\frac{k^2}{k_c^2}\Phi \tag{13.64}$$

for the three unknowns Φ, δ_m, δ_R with the prime denoting the derivative with respect to $y = a/a_{eq}$. To solve these equations and determine the growth of perturbations we need suitable initial conditions. The initial conditions need to be imposed very early on, at $y = y_i$, when the modes are much bigger than the Hubble radius, which corresponds to the $y \ll 1, k \to 0$ limit. In this limit, the above equations become:

$$y\Phi' + \Phi \approx -\frac{1}{2}\delta_R; \quad \delta_m'' + \frac{1}{y}\delta_m' \approx 3\Phi'' + \frac{3}{y}\Phi'; \quad \delta_R'' + \frac{1}{2}\delta_R' \approx 4\Phi'' + 2\Phi'. \quad (13.65)$$

We will take $\Phi(y_i, k) = \Phi_i(k)$ as a specified function that needs to be determined by the processes that generate the initial perturbations. The first equation in Eq. (13.65) then shows that we can take $\delta_R = -2\Phi_i$ for $y_i \to 0$. Further, Eq. (13.47) shows that adiabaticity is respected at these scales (i.e. $\delta(s/n_m) = 0$) and we can take $\delta_m = (3/4)\delta_R = -(3/2)\Phi_i$. Then Eq. (13.62) determines Φ' if $(\Phi, \delta_m, \delta_R)$ are given. Finally we use the last two equations in Eq. (13.65) to set $\delta'_m = 3\Phi', \delta'_R = 4\Phi'$, Thus we can take the initial conditions at some $y = y_i \ll 1$ to be

$$\Phi(y_i, k) = \Phi_i(k); \quad \delta_R(y_i, k) = -2\Phi_i(k); \quad \delta_m(y_i, k) = -(3/2)\Phi_i(k), \quad (13.66)$$

along with $\delta'_m(y_i, k) = 3\Phi'(y_i, k); \ \delta'_R(y_i, k) = 4\Phi'(y_i, k)$.

Given these initial conditions, it is fairly easy to integrate the equations forward in time and the numerical results are shown in Figs. 13.1, 13.2 and 13.3. (In the figures k_{eq} is defined to be $a_{eq}H_{eq}$ for simplicity.) To understand the nature of the evolution, it is, however, useful to obtain some analytic approximations to Eqs. (13.62)–(13.64), which is what we will do next.

13.4.1 Evolution of modes with $\lambda \gg d_H$

Let us begin by considering very large wavelength modes corresponding to the $k\eta \to 0$ limit. In this case adiabaticity is respected and we can set $\delta_R \approx (4/3)\delta_m$. Then Eqs. (13.62) and (13.63) become

$$y\Phi' + \Phi \approx -\frac{3y + 4}{8(1 + y)}\delta_R; \quad \delta'_R \approx 4\Phi'. \quad (13.67)$$

Differentiating the first equation and using the second to eliminate δ_R, we get a second order equation for Φ. Fortunately, this equation has an exact solution thereby leading to

$$\Phi = \Phi_i \frac{1}{10y^3}\left[16\sqrt{(1+y)} + 9y^3 + 2y^2 - 8y - 16\right]; \quad \delta_R \approx 4\Phi - 6\Phi_i. \quad (13.68)$$

(We will describe a simple way to obtain this solution in Section 13.4.4.) The initial condition on δ_R is chosen such that it goes to $-2\Phi_i$ initially when $\Phi \to \Phi_i$. The solution shows that, as long as the mode is bigger than the Hubble radius, the potential changes very little; it is constant initially as well as in the final matter dominated phase. At late times ($y \gg 1$) we see that $\Phi \approx (9/10)\Phi_i$ so that Φ decreases only by a factor (9/10) during the entire evolution if $k \to 0$ is a valid approximation.

13.4.2 Evolution of modes with $\lambda \ll d_H$ in the radiation dominated phase

When the mode enters the Hubble radius in the radiation dominated phase, we can no longer ignore the pressure term. The pressure makes the radiation density contrast oscillate and the gravitational potential, driven by this, also oscillates with a decaying amplitude. An approximate procedure to describe this phase is to solve the coupled $\delta_R - \Phi$ system, ignoring δ_m (which is subdominant) and *then* determine δ_m using the form of Φ.

When δ_m is ignored, the problem reduces to the one solved earlier for the single component case in Eqs. (13.54) and (13.55) with $w = 1/3$ giving $\nu = 3$. Since $J_{3/2}$ can be expressed in terms of trigonometric functions, the solution given by Eq. (13.54) with $\nu = 3$, simplifies to

$$\Phi = \Phi_i \frac{3}{l^3 y^3} \left[\sin(ly) - ly \cos(ly) \right]; \quad l^2 = \frac{2k^2}{3k_c^2}. \tag{13.69}$$

(Note that as $y \to 0$, we have $\Phi = \Phi_i$, $\Phi' = 0$.) This solution shows that once the mode enters the Hubble radius, the potential decays in an oscillatory manner. For $ly \gg 1$, the potential becomes $\Phi \approx -3\Phi_i(ly)^{-2} \cos(ly)$. In the same limit, we get

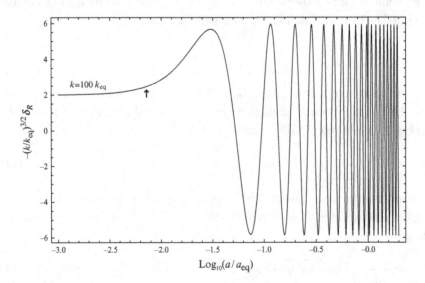

Fig. 13.1. Evolution of δ_R for a mode with $k = 100k_{\rm eq}$. The mode remains frozen outside the Hubble radius with $(k/k_{\rm eq})^{3/2}(-\delta_R) \approx (k/k_{\rm eq})^{3/2}2\Phi = 2$ (the normalization is arbitrary and chosen for convenience so that $\Phi_i = -1$ in Eq. (13.57)) and oscillates when it enters the Hubble radius. The epoch at which the mode enters the Hubble radius is indicated by an arrow. The oscillations are well described by Eq. (13.70) with an amplitude of 6.

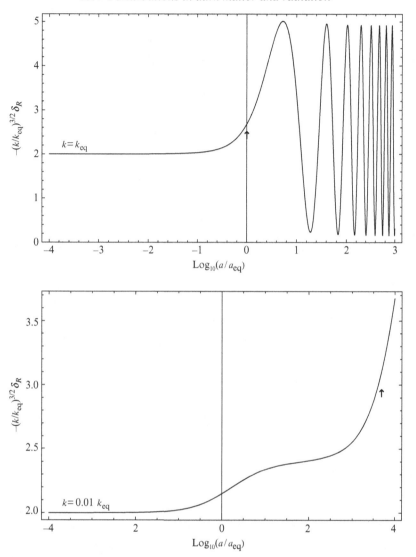

Fig. 13.2. Evolution of δ_R for two modes $k = k_{eq}$ and $k = 0.01\, k_{eq}$. The modes remain frozen outside the Hubble radius at $(-\delta_R) \approx 2$ and oscillates when they enter the Hubble radius. The mode in the bottom panel stays outside the Hubble radius for most of its evolution and hence changes very little. The arrow denotes the epoch at which the mode enters the Hubble radius.

from Eq. (13.55) that

$$\delta_R \approx -\frac{2}{3}k^2\eta^2\Phi \approx -2l^2y^2\Phi \approx 6\Phi_i\cos(ly). \qquad (13.70)$$

(This is analogous to the result in Eq. (13.58) for the radiation dominated case.) This oscillation is seen clearly in Fig. 13.1 and in the top panel of Fig. 13.2. The amplitude of oscillations is accurately captured by Eq. (13.70) for the $k = 100k_{eq}$ mode but not for $k = k_{eq}$; this is to be expected since the mode is not entering the Hubble radius in the radiation dominated phase. (Of course, the perturbation equations require $\Phi \ll 1$, etc. But once the solution is found, the overall normalization is irrelevant.)

Let us next consider the matter perturbations during this phase. They grow, driven by the gravitational potential determined above. When $y \ll 1$, Eq. (13.63) becomes:

$$\delta_m'' + \frac{1}{y}\delta_m' = 3\Phi'' + \frac{3}{y}\Phi' - \frac{k^2}{k_c^2}\Phi \equiv J(y). \tag{13.71}$$

The Φ is essentially determined by radiation and satisfies Eq. (13.51); using this, we can rewrite Eq. (13.71) as

$$\frac{d}{dy}(y\delta_m') = -9(\Phi' + \frac{1}{3}l^2 y\Phi) = yJ(y). \tag{13.72}$$

The general solution to the homogeneous part of Eq. (13.72) (obtained by ignoring the right hand side) is $(c_1 + c_2 \ln y)$; hence the general solution to this equation is

$$\delta_m = (c_1 + c_2 \ln y) + \int_0^y \frac{d\bar{y}}{\bar{y}} \int_0^{\bar{y}} dy_1\, y_1\, J(y_1). \tag{13.73}$$

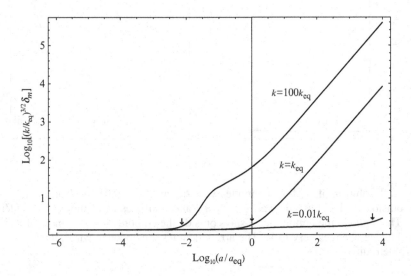

Fig. 13.3. Evolution of $|\delta_m|$ for three different modes. The modes are labelled by their wave numbers and the epochs at which they enter the Hubble radius are shown by small arrows. All the modes remain frozen when they are outside the Hubble radius and grow linearly in the matter dominated phase once they are well inside the Hubble radius. The mode that enters the Hubble radius in the radiation dominated phase grows logarithmically until $y = y_{eq}$. These features are well approximated by Eqs. (13.73) and (13.78).

At very early times, the contribution from the integral is small and our initial conditions require $c_2 = 0$ and $c_1 = -(3/2)\Phi_i$. Further, one can show through an integration by parts that the expression reduces to

$$\delta_m = -\frac{3}{2}\Phi_i - \int_0^y dx\, x\, J(x) \ln\left(\frac{x}{y}\right). \tag{13.74}$$

The dominant contribution to the integral comes from the region around $ly \approx 1$. So we get, approximately:

$$\delta_m \cong -\frac{3}{2}\Phi_i - \int_0^\infty dx\, x J(x) \ln x - (\ln y) \int_0^\infty dx\, x\, J(x) \equiv A + B \ln y. \tag{13.75}$$

Hence, after the mode enters the Hubble radius, we expect it to have the form

$$\delta_m(y) \approx k_1 \Phi_i \ln(k_2 y) \tag{13.76}$$

with $A/B = \ln k_2$ and $k_1 \Phi_i = B$, where k_1 and k_2 are constants which can be determined by doing the integrals in Eq. (13.73) but are irrelevant for our purpose. The key conclusion is that for $y \ll 1$ the growing mode varies as $\ln y$ and dominates over the rest; hence we conclude that matter perturbations, driven by Φ, grow logarithmically during the radiation dominated phase for modes which are inside the Hubble radius.

13.4.3 Evolution in the matter dominated phase

Finally let us consider the matter dominated phase, in which we can ignore the radiation and concentrate on Eq. (13.62) and Eq. (13.63). When $y \gg 1$ these equations become:

$$y\Phi' + \Phi \approx -\frac{1}{2}\delta_m - \frac{k^2 y}{3k_c^2}\Phi; \qquad y\delta_m'' + \frac{3}{2}\delta_m' = -\frac{k^2}{k_c^2}\Phi. \tag{13.77}$$

These two equations have a simple solution which we found earlier (see Eq. (13.59)):

$$\Phi = \Phi_\infty = \text{const.}; \qquad \delta_m = -2\Phi_\infty - \frac{2k^2}{3k_c^2}\Phi_\infty y \simeq y. \tag{13.78}$$

In this limit, the matter perturbations grow linearly with expansion: $\delta_m \propto y \propto a$. In fact, this is the most dominant growth mode in the linear perturbation theory.

Exercise 13.4
Meszaros solution It is possible to obtain an interesting solution for modes which have entered the Hubble radius in the radiation dominated phase that connects smoothly the logarithmic growth (in RD) to the linear growth (in MD) for matter perturbation. This exercise develops this solution.

(a) For modes inside the Hubble radius, perturbations in the radiation oscillate due to pressure support and can be ignored. In this limit, show that Eq. (13.62), Eq. (13.63) and Eq. (13.64) can be combined to give

$$\delta'' + \frac{2 + 3y}{2y(y + 1)} \delta' - \frac{3}{2y(y + 1)} \delta = 0, \tag{13.79}$$

where $y = a/a_{eq}$ and primes indicate derivatives with respect to y. [Hint. Keep the dominant term for large k/k_c and ignore δ_R.]

(b) One solution to Eq. (13.79), given by $\delta_1(y) = y + (2/3)$, can be written down by inspection. Use the Wronskian condition to obtain the second solution to be

$$\delta_2(y) = \delta_1(y) \ln \left[\frac{\sqrt{1 + y} + 1}{\sqrt{1 + y} - 1} \right] - 2\sqrt{1 + y}. \tag{13.80}$$

(c) The general solution is a linear combination of the two in the form

$$\delta(k, y) = C_1 \delta_1(y) + C_2 \delta_2(y). \tag{13.81}$$

Write down the asymptotic limits of this solution for $y \ll 1$ and $y \gg 1$ and compare with the discussion in the text.

(d) Consider now the solution for the modes that enter the Hubble radius $y = y_H \ll 1$. In this case, we can match the solution obtained above to the one obtained in Eq. (13.76) by equating both δ and δ' at $y = y_H$. Hence determine C_1 and C_2 in terms of k_1 and k_2 and write down the form of $\delta_m(y)$ at late times with the constants determined in terms of the integrals over $J(x)$ in Eq. (13.73).

13.4.4 An alternative description of the matter–radiation system

Before proceeding further, we will describe an alternative procedure for studying the perturbations in dark matter and radiation that has certain advantages. In the previous discussion, we used the perturbations in the energy density of radiation (δ_R) and matter (δ_m) as the dependent variables. Instead, we will now use perturbations in the *total* energy density, δ, and the perturbations in the entropy per particle, σ, as the new dependent variables. In terms of δ_R, δ_m, these new variables are defined as:

$$\delta \equiv \frac{\delta\rho_{tot}}{\rho_{tot}} = \frac{\rho_R \delta_R + \rho_m \delta_m}{\rho_R + \rho_m} = \frac{\delta_R + y\delta_m}{1 + y}; \quad y = \frac{\rho_m}{\rho_R} = \frac{a}{a_{eq}} \tag{13.82}$$

$$\sigma \equiv \left(\frac{\delta s}{s} \right) = \frac{3\delta T_R}{T_R} - \frac{\delta\rho_m}{\rho_m} = \frac{3}{4}\delta_R - \delta_m = \frac{\delta n_R}{n_R} - \frac{\delta n_m}{n_m}. \tag{13.83}$$

Given the equations for δ_R, δ_m, one can obtain the corresponding equations for the new variables (δ, σ) by straightforward algebra. It is convenient to express them as

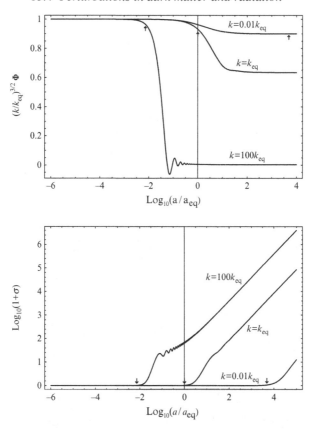

Fig. 13.4. Top panel: the evolution of the gravitational potential Φ for three different modes. The wave number is indicated by the label and the epoch at which the mode enters the Hubble radius is indicated by a small arrow. The top-most curve is for a mode which stays outside the Hubble radius for most of its evolution and is well described by Eq. (13.68). The other two modes show the decay of Φ after the mode has entered the Hubble radius in the radiation dominated epoch as described by Eq. (13.69). Bottom panel: evolution of entropy perturbation (see Eq. (13.83) for the definition). The entropy perturbation is essentially zero till the mode enters the Hubble radius and grows afterwards, tracking the dominant energy density perturbation.

two coupled equations for Φ and σ (instead of for δ and σ). After some long but direct algebra we get:

$$y\Phi'' + \frac{y\Phi'}{2(1+y)} + 3(1+c_s^2)\Phi' + \frac{3c_s^2\Phi}{4(1+y)} + c_s^2\frac{k^2}{k_c^2}\frac{y}{1+y}\Phi = \frac{3c_s^2\sigma}{2(1+y)} \quad (13.84)$$

$$y\sigma'' + \frac{y\sigma'}{2(1+y)} + 3c_s^2\sigma' + \frac{3c_s^2y^2}{4(1+y)}\frac{k^2}{k_c^2}\sigma = \frac{c_s^2y^3}{2(1+y)}\left(\frac{k}{k_c}\right)^4\Phi, \quad (13.85)$$

where we have defined

$$c_s^2 \equiv \frac{(4/3)\rho_R}{4\rho_R + 3\rho_m} = \frac{1}{3}\left(1 + \frac{3}{4}\frac{\rho_m}{\rho_R}\right)^{-1} = \frac{1}{3}\left(1 + \frac{3}{4}y\right)^{-1}. \qquad (13.86)$$

These equations show that the entropy perturbations σ and the gravitational poten-tial Φ (which is directly related to the total energy density perturbations) act as sources for each other. The coupling between the two arises through the right hand sides of Eq. (13.84) and Eq. (13.85). We also see that if we set $\sigma = 0$ as an ini-tial condition, this is preserved to $\mathcal{O}(k^4)$ and – for long wavelength modes – Φ evolves independently of σ. The solutions to these coupled equations obtained by numerical integration are shown in Fig.13.4 (bottom panel). The entropy pertur-bation σ remains close to zero till the mode enters the Hubble radius and grows afterwards tracking either δ_R or δ_m depending on which is the dominant energy density perturbation.

To illustrate the behaviour of Φ, let us consider the adiabatic perturbations at large scales with $\sigma \approx 0, k \to 0$; then the gravitational potential satisfies the equation:

$$y\Phi'' + \frac{y\Phi'}{2(1+y)} + 3(1+c_s^2)\Phi' + \frac{3c_s^2\Phi}{4(1+y)} = \frac{3c_s^2\sigma}{2(1+y)} \approx 0 \qquad (13.87)$$

which has the two independent solutions:

$$f_1(y) = 1 + \frac{2}{9y} - \frac{8}{9y^2} - \frac{16}{9y^3}, \qquad f_2(y) = \frac{\sqrt{1+y}}{y^2}, \qquad (13.88)$$

both of which diverge as $y \to 0$. We need to combine these two solutions to find the general solution, keeping in mind that the general solution should be non-singular and become a constant (say, unity) as $y \to 0$. This fixes the linear combination uniquely:

$$f(y) = \frac{9}{10}f_1 + \frac{8}{5}f_2 = \frac{1}{10y^3}\left[16\sqrt{(1+y)} + 9y^3 + 2y^2 - 8y - 16\right]. \quad (13.89)$$

Multiplying by Φ_i we get the solution that was found earlier (see Eq. (13.68)).

Given the form of Φ and $\sigma \simeq 0$ we can determine all other quantities. In particular, we get:

$$\delta_R = \frac{-2(1+y)d(y\Phi)/dy + y\sigma}{1 + (3/4)y} \simeq -\frac{2(1+y)}{1 + (3/4)y}\frac{d}{dy}(y\Phi). \qquad (13.90)$$

The corresponding velocity field, which we quote for future reference, is given by

$$v_\alpha = -\frac{3c_s^2}{2(\dot{a}/a)}(1+y)\partial_\alpha\left(\frac{d(y\Phi)}{dy}\right). \qquad (13.91)$$

We conclude this section by mentioning another useful result related to Eq. (13.84) which is quite useful. When $\sigma \approx 0$, the equation for Φ can be re-expressed as

$$a\frac{d\mathcal{R}}{da} = -\frac{2c_s^2}{3}\frac{k^2}{a^2 H^2}\frac{\rho}{\rho + p}\Phi \tag{13.92}$$

where we have defined:

$$\mathcal{R} \equiv \frac{2}{3}\frac{\rho}{\rho + p}\frac{a}{\dot{a}}\left(\dot{\Phi} + \frac{\dot{a}}{a}\Phi\right) + \Phi = \frac{H}{\rho + p}\frac{ik^\alpha}{k^2}\delta T_\alpha^0 + \Phi. \tag{13.93}$$

(The i in the last expression factor arises because of converting a spatial gradient to the k space; of course, when everything is done correctly, all physical quantities will be real.) Other equivalent, alternative forms for $\mathcal{R}$, which are useful are:

$$\mathcal{R} = \frac{2}{3[1 + w(a)]}\frac{d}{da}(a\Phi) + \Phi = \frac{H^2}{a(\rho + p)}\frac{d}{dt}\left(\frac{a\Phi}{H}\right). \tag{13.94}$$

For modes that are much bigger than the Hubble radius, $(k/aH) \to 0$, and the right hand side of Eq. (13.92) vanishes showing that $\mathcal{R}$ is conserved. When $\mathcal{R} =$ constant, we can integrate Eq. (13.94) easily to obtain:

$$\Phi = c_1\frac{H}{a} + c_2\left[1 - \frac{H}{a}\int_0^a \frac{da'}{H(a')}\right]. \tag{13.95}$$

This is the easiest way to obtain the solution quoted earlier in Eq. (13.68).

The conservation law for $\mathcal{R}$ also allows us to understand in a simple manner our previous result that Φ deceases only by a factor $(9/10)$ as we evolve the equations from $y \ll 1$ to $y \gg 1$ if the mode remains bigger than the Hubble radius. Let us compare the values of $\mathcal{R}$ early in the radiation dominated phase and late in the matter dominated phase. From the first equation in Eq. (13.94) (using $\Phi' \approx 0$) we find that, in the radiation dominated phase, $\mathcal{R} \approx (1/2)\Phi_i + \Phi_i = (3/2)\Phi_i$; late in the matter dominated phase, $\mathcal{R} \approx (2/3)\Phi_f + \Phi_f = (5/3)\Phi_f$. Hence the conservation of $\mathcal{R}$ gives $\Phi_f = (3/5)(3/2)\Phi_i = (9/10)\Phi_i$, which was the result obtained earlier. The expression in Eq. (13.95) also gives the form of Φ at late times in the Λ dominated or curvature dominated universe.

One key feature that arises from the study of linear perturbation theory is that the perturbations Φ, δ_R and δ_m grow by very different amounts. The Φ either changes very little or decays; the δ_R grows in amplitude only by a factor of a few. The physical reason, of course, is that the amplitude is frozen at super-Hubble scales and the pressure prevents the growth at sub-Hubble scales. In contrast, δ_m, which is pressureless, grows logarithmically in the radiation dominated era and linearly during the matter dominated era. Since the later phase lasts for a factor of 10^4 in expansion, we get a fairly large amount of growth in δ_m.

Exercise 13.5

Growth factor in an open universe We know that, in a matter dominated universe with $\Omega_{\text{NR}} = 1$, the density perturbations grow as $\delta_m \propto a$ for $a \gg a_{\text{eq}}$. The growth factor in the case of an open universe with $\Omega_{\text{NR}} < 1$, is different. One way to compute it is to use Eq. (13.95) with the appropriate $H(a)$ for the open universe and then determine δ from Φ.

(a) Show that this will give a growing mode

$$\delta_1(a, \Omega_{\text{NR}}) = \frac{5\Omega_{\text{NR}}}{2(1 - \Omega_{\text{NR}})} \left[3 \frac{\sqrt{1+x}}{x^{3/2}} \ln\left(\sqrt{1+x} - \sqrt{x}\right) + 1 + \frac{3}{x} \right], \qquad (13.96)$$

with $x = (1 - \Omega_{\text{NR}})a/\Omega_{\text{NR}}$ and arbitrary normalization. Compute and plot the growth factor $f(\Omega_{\text{NR}}) \equiv \delta(1, \Omega_{\text{NR}})/\delta(a_{\text{eq}}, \Omega_{\text{NR}})$ as a function of Ω_{NR}.

(b) Compare the above result with the growth factor for a universe dominated by cosmological constant at late times with $\Omega_{\text{NR}} + \Omega_{\Lambda} = 1$.

13.5 Transfer function for the matter perturbations

We now have all the ingredients to evolve the matter perturbation from an initial value $\delta = \delta_i$ at $y = y_i \ll 1$ to the current epoch $y = y_0 = a_{\text{eq}}^{-1}$ in the matter dominated phase at $y \gg 1$. Initially, the wavelength of the perturbation will be bigger than the Hubble radius and the perturbation will essentially remain frozen. When it enters the Hubble radius in the radiation dominated phase, it begins to grow but only logarithmically (see Section 13.4.2) until the universe becomes matter dominated. In the final matter dominated phase, the perturbation grows linearly with the expansion factor. The relation between the final and initial perturbation can be obtained by combining these results.

Usually, one is interested more in the power spectrum $P_k(t)$ and in the power per logarithmic band in k-space $\Delta_k(t)$. These quantities are defined in terms of $\delta_k(t)$ through the equations

$$P_k(t) \equiv |\delta_k(t)|^2; \quad \Delta_k^2(t) \equiv \frac{k^3 P_k(t)}{2\pi^2}. \qquad (13.97)$$

It is, therefore, convenient to study the evolution of $k^{3/2}|\delta_k|$ since its square will immediately give the power per logarithmic band Δ_k^2 in k-space.

Let us first consider a mode which enters the Hubble radius in the radiation dominated phase at the epoch $a = a_{\text{enter}}(k)$. The condition $d_H(a_{\text{enter}}) = (2\pi/k)a_{\text{enter}}$ leads to the scaling relation, $a_{\text{enter}}/k \propto t_{\text{enter}} \propto a_{\text{enter}}^2$ from which we find that $a_{\text{enter}} \propto k^{-1}$ giving $y_{\text{enter}} = (k_{\text{eq}}/k)$. Hence

$$\frac{\Delta(k, a = 1)}{\Delta_{\text{ent}}(k)} = \frac{\delta(k, a = 1)}{\delta_{\text{ent}}(k)} = \frac{1}{a_{\text{eq}}} \ln\left(\frac{a_{\text{eq}}}{a_{\text{enter}}}\right) \propto \ln\left(\frac{k}{k_{\text{eq}}}\right), \qquad (13.98)$$

where the two factors $-\ln(a_{\text{eq}}/a_{\text{enter}})$ and $1/a_{\text{eq}}$ – give the growth in the radiation and matter dominated phases respectively. Let us next consider the modes that enter the Hubble radius in the matter dominated phase. In this case, the scaling relation is $a_{\text{enter}}/k \propto t_{\text{enter}} \propto a_{\text{enter}}^{3/2}$ so that $y_{\text{enter}} = (k_{\text{eq}}/k)^2$. Hence

$$\frac{\Delta(k, a = 1)}{\Delta_{\text{ent}}(k)} = \frac{\delta(k, a = 1)}{\delta_{\text{ent}}(k)} = \frac{1}{a_{\text{enter}}} \propto k^2, \tag{13.99}$$

where the factor $(1/a_{\text{enter}})$ takes care of the growth from $a = a_{\text{enter}}$ to $a = 1$.

To proceed further and determine the k-dependence of $k^{3/2}\delta$ at $a = 1$, we need to know the k-dependence of the perturbation when it enters the Hubble radius which, of course, is related to the mechanism that generates the initial power spectrum. The most natural choice will be that all the modes enter the Hubble radius with a constant value for $\Delta_{\text{ent}}(k)$ at the time of entry. This would imply that the physical perturbations are scale invariant at the time of entering the Hubble radius, a possibility that was suggested by Zel'dovich and Harrison (decades before inflation was invented!). We will see in Chapter 14 that this is also true for perturbations generated by inflation and hence is a reasonable assumption, at least in such models. Therefore we shall assume

$$\Delta_{\text{ent}}^2(k) \propto k^3 |\delta_{\text{ent}}(k)|^2 = k^3 P_{\text{ent}}(k) = C = \text{constant}. \tag{13.100}$$

Using this result in Eq. (13.98) and Eq. (13.99), we find that the power per logarithmic band at the present epoch is given by

$$\Delta^2(k, a = 1) \propto k^3 \left| \delta_m(k, a = 1) \right|^2 \propto \begin{cases} k^4 & (\text{for } k \ll k_{\text{eq}}) \\ (\ln k)^2 & (\text{for } k \gg k_{\text{eq}}) \end{cases}. \tag{13.101}$$

The corresponding power spectrum $P(k) \propto k^{-3}\Delta_k^2$ is

$$P(k, a = 1) \propto \left| \delta_m(k, a = 1) \right|^2 \propto \begin{cases} k & (\text{for } k \ll k_{\text{eq}}) \\ k^{-3}(\ln k)^2 & (\text{for } k \gg k_{\text{eq}}) \end{cases}. \tag{13.102}$$

The form for $P(k)$ shows that the evolution imprints the scale k_{eq} on the power spectrum even though the initial power spectrum is scale invariant. For $k < k_{\text{eq}}$ (for large spatial scales), the primordial form of the spectrum is preserved and the evolution only increases the amplitude while preserving the shape. For $k > k_{\text{eq}}$ (for small spatial scales), the shape is distorted and in general the power is suppressed in comparison with larger spatial scales. This result arises because modes with small wavelengths enter the Hubble radius in the RD phase and grow very little between a_{enter} and a_{eq} and have to wait till the universe becomes matter dominated in order to grow. This is in contrast to modes with large wavelengths which enter the Hubble radius in the MD phase and grow linearly from $a = a_{\text{enter}}$ itself. It is this effect which suppresses the power at small wavelengths (for $k > k_{\text{eq}}$) relative to power at larger wavelengths.

13.6 Application: temperature anisotropies of CMBR

We shall now apply the formalism we have developed to understand the temperature anisotropies in the cosmic microwave background radiation (CMBR) – which is probably the most useful application of *linear* perturbation theory. We shall begin by first developing the general formulation and the terminology used to describe the temperature anisotropies in CMBR.

Towards every direction in the sky, $n = (\theta, \psi)$ we can define a fractional temperature fluctuation $Q(n) \equiv (\Delta T/T)(\theta, \psi)$ in the CMBR. Expanding this quantity in spherical harmonics on the sky plane as well as in terms of the spatial Fourier modes, we get the two relations:

$$Q(n) \equiv \frac{\Delta T}{T}(\theta, \psi) = \sum_{l,m}^{\infty} a_{lm} Y_{lm}(\theta, \psi) = \int \frac{d^3 k}{(2\pi)} Q(k) e^{ik \cdot nL}, \quad (13.103)$$

where $L = \eta_0 - \eta_{\mathrm{LSS}}$ is the distance to the last scattering surface (LSS) from which we are receiving the radiation. The last equality in Eq. (13.103) allows us to define the expansion coefficients a_{lm} in terms of the temperature fluctuation in the Fourier space $Q(k)$. The standard expansion of plane waves in terms of the spherical harmonics

$$e^{ik \cdot nL} = 4\pi \sum_{l,m} i^l j_l(kL) Y_{lm}^*(k) Y_{lm}(k) \quad (13.104)$$

now gives

$$a_{lm} = \int \frac{d^3 k}{(2\pi)^3} (4\pi) i^l Q(k) j_l(kL) Y_{lm}(\hat{k}), \quad (13.105)$$

where $\hat{k}$ is the unit vector. Next, let us consider the angular correlation function of the temperature anisotropy, which is given by

$$\mathcal{C}(\alpha) = \langle Q(n) Q(m) \rangle = \sum \sum \langle a_{lm} a_{l'm'}^* \rangle Y_{lm}(n) Y_{l'm'}^*(m), \quad (13.106)$$

where the wedges $\langle \cdots \rangle$ denote an ensemble average and α is the angle between the directions n and m. For a Gaussian random field of initial fluctuations we can express the ensemble average as $\langle a_{lm} a_{l'm'}^* \rangle = C_l \delta_{ll'} \delta_{mm'}$. Using Eq. (13.105), we get a relation between C_l and $Q(k)$ given by:

$$C_l = \frac{2}{\pi} \int_0^\infty k^2 dk \, |Q(k)|^2 \, j_l^2(kL). \quad (13.107)$$

Further, Eq. (13.106) now becomes:

$$\mathcal{C}(\alpha) = \sum_l \frac{(2l+1)}{4\pi} C_l P_l(\cos \alpha). \quad (13.108)$$

The result in Eq. (13.108) can be interpreted as giving the contribution of different multipoles to the temperature anisotropy. The lowest order term is the dipole with $l = 1$. However, this term is not directly observable because the motion of our local group through the CMBR leads to a large $l = 1$ dipole contribution (about $\Delta T/T \approx 10^{-3}$) in the temperature anisotropy (see Exercise 2.22). Hence, in the analysis of CMBR anisotropies, the $l = 1$ term is subtracted out and the leading term is the quadrupole with $l = 2$. The quadrupole contribution is given by

$$\left(\frac{\Delta T}{T}\right)^2_Q = \frac{5}{4\pi}C_2. \tag{13.109}$$

We also see that the root-mean-square value of temperature fluctuations is given by

$$\left(\frac{\Delta T}{T}\right)^2_{\text{rms}} = C(0) = \frac{1}{4\pi}\sum_{l=2}^{\infty}(2l+1)\,C_l. \tag{13.110}$$

Both these expressions (in Eq. (13.109) and Eq. (13.110)) can be explicitly computed if we know $Q(k)$ from the perturbation theory.

For a given value of l, the measured value of C_l is the average over all $m = -l, \ldots - 1, 0, 1, \ldots l$. For a Gaussian random field, one can also compute the variance around this mean value. It can be shown that this variance in C_l is $2C_l^2/(2l+1)$ (see Exercise 13.6). In other words, there is an intrinsic root-mean-square fluctuation in the observed, mean, value of the C_ls which is of the order of $\Delta C_l/C_l \approx (2l+1)^{-1/2}$. It is not possible for any CMBR observations that measure the C_ls to reduce the uncertainty below this intrinsic variance – usually called the *cosmic variance*. For large values of l, the cosmic variance is usually subdominant to other observational errors but for low l this is the dominant source of uncertainty in the measurement of C_ls. In fact, the current CMBR observations are only limited by cosmic variance at low l.

Exercise 13.6
Cosmic variance The C_ls defined above are the averages over $m = -l, -(l-1), \ldots, (l-1), l$ for any given value of l. For a Gaussian random field, compute the variance around this mean value and show that $(\Delta C_l)^2 = 2C_l^2/(2l+1)$. [Hint. First prove that $\langle |a_{lm}|^4 \rangle = 3C_l^2$. Then use the fact that the C_ls which are measured are actually the averages defined as

$$\hat{C}_l = \frac{1}{2l+1}\sum_{m=-l}^{l}|a_{lm}|^2. \tag{13.111}$$

Now compute the quantity $\langle (\hat{C}_l - C_l)^2 \rangle$ to obtain the result.]

13.6.1 The Sachs–Wolfe effect

As an illustration of the formalism developed above, let us compute the C_ls for low l that will be contributed essentially by fluctuations at large spatial scales. Since these fluctuations will be dominated by gravitational effects, at scales bigger than the Hubble radius, we can ignore the complications arising from baryonic physics and compute these using the results obtained earlier.

We begin by noting that the redshift law for the photons in the unperturbed Friedmann universe, $\nu_0 = \nu(a)/a$, gets modified to the form $\nu_0 = \nu(a)/[a(1 + \Phi)]$ in a perturbed Friedmann universe owing to the contribution of Φ to the gravitational time dilation. Further, since the energy density of the radiation is $\rho_R \propto T^4$, we have $\delta_R \equiv (\delta\rho_R/\rho_R) = 4(\delta T/T)$, giving $T = \langle T \rangle(1 + \delta_R/4)$ where $\langle T \rangle$ is the mean temperature. So the argument of the Planck spectrum will now be:

$$\frac{\nu_0}{T_0} = \frac{\nu(a)}{aT_0(1 + \Phi)} = \frac{\nu(a)}{a\langle T_0\rangle[1 + (\delta_R/4)](1 + \Phi)} \cong \frac{\nu(a)}{a\langle T_0\rangle[1 + \Phi + (\delta_R/4)]}.$$
(13.112)

This is equivalent to a net temperature fluctuation of

$$\left(\frac{\Delta T}{T}\right)_{\text{obs}} = \frac{1}{4}\delta_R + \Phi$$
(13.113)

at large scales. (Note that the observed $\Delta T/T$ is not just $(\delta_R/4)$ as one might have naively imagined.) To proceed further, we recall our solution for the gravitational potential (in Eq. (13.68)) valid at large scales:

$$\Phi = \Phi_i \frac{1}{10y^3}\left[16\sqrt{(1 + y)} + 9y^3 + 2y^2 - 8y - 16\right]; \quad \delta_R = 4\Phi - 6\Phi_i.$$
(13.114)

At $y = y_{\text{dec}}$ we can take Φ to be given by the the asymptotic solution $\Phi_{dec} \approx (9/10)\Phi_i$ with sufficient accuracy. Hence we get

$$\left(\frac{\Delta T}{T}\right)_{\text{obs}} = \left[\frac{1}{4}\delta_R + \Phi\right]_{\text{dec}} = 2\Phi_{\text{dec}} - \frac{3}{2}\Phi_i \approx 2\Phi_{\text{dec}} - \frac{3}{2}\frac{10}{9}\Phi_{\text{dec}} = \frac{1}{3}\Phi_{\text{dec}}.$$
(13.115)

We thus obtain the simple result that the observed temperature fluctuations at very large scales are simply related to the fluctuations of the gravitational potential at these scales. (This contribution is called the *Sachs–Wolfe effect*.)

Note that the regions in the last scattering surface which has $\Phi < 0$ are the overdense regions containing hotter radiation in the sense that $\delta_R/4 = -(2/3)\Phi > 0$ when $\Phi < 0$. But these photons lose an energy Φ while climbing out of these deeper potential wells, leading to the observed temperature fluctuations $\Delta T/T =$

$(1/3)\Phi < 0$ when $\Phi < 0$. So the large scale *hot spots* seen in the sky correspond to *underdense* regions at the time recombination.

Fourier transforming Eq. (13.115) we get $Q(\boldsymbol{k}) = (1/3)\Phi(\boldsymbol{k}, \eta_{\mathrm{LSS}})$ where η_{LSS} is the conformal time at the last scattering surface. It follows from Eq. (13.107) that the contribution to C_l from the gravitational potential is

$$C_l = \frac{2}{\pi} \int_0^\infty k^2 dk |Q(k)|^2 j_l^2(kL) = \frac{2}{\pi} \int_0^\infty \frac{dk}{k} \frac{k^3 |\Phi_k|^2}{9} j_l^2(kL) \qquad (13.116)$$

with

$$L = \eta_0 - \eta_{\mathrm{LSS}} \approx \eta_0 \approx 2(\Omega_{\mathrm{NR}} H_0^2)^{-1/2} \approx 6000 \, \Omega_{\mathrm{NR}}^{-1/2} \, h^{-1} \, \mathrm{Mpc}. \qquad (13.117)$$

For a scale invariant spectrum, $k^3 |\delta_k|^2$ is a constant independent of k. (We will see later in Chapter 14 that inflation can generate such a scale invariant spectrum.) Since $\delta \approx -2\Phi$ at the large scales (see Eq. (13.78); the extra correction term in Eq. (13.78) being about 3×10^{-4} for $k \approx L^{-1}, y = y_{\mathrm{dec}}$) it follows that $k^3 |\Phi|^2$ is a constant at large scales. In this case, it is conventional to introduce a constant amplitude A and write:

$$\Delta_\Phi^2 \equiv \frac{k^3 |\Phi_k|^2}{2\pi^2} = A^2 = \mathrm{constant}. \qquad (13.118)$$

Substituting this form into Eq. (13.116) and evaluating the integral over j_l^2 using

$$\int_0^\infty \frac{dz}{z} j_l^2(z) = \frac{1}{2l(l+1)} \qquad (13.119)$$

we get the final result:

$$\frac{l(l+1)C_l}{2\pi} = \left(\frac{A}{3}\right)^2. \qquad (13.120)$$

As an application of this result, let us consider the observations of the COBE satellite which measured the temperature fluctuations for the first time in 1992. This satellite obtained the root-mean-square fluctuations and the quadrupole after smoothing the temperature fluctuations with a Gaussian window over an angular scale of about $\theta_c \approx 10°$. Hence the observed values are slightly different from those in Eq. (13.109) and Eq. (13.110). We have, instead,

$$\left(\frac{\Delta T}{T}\right)_{\mathrm{rms}}^2 = \frac{1}{4\pi} \sum_{l=2}^\infty (2l+1) C_l \exp\left(-\frac{l^2 \theta_c^2}{2}\right); \qquad \left(\frac{\Delta T}{T}\right)_Q^2 = \frac{5}{4\pi} C_2 e^{-2\theta_c^2},$$

$$(13.121)$$

due to the smoothing at the scale θ_c. Using Eqs. (13.118) and (13.120) we find that

$$\left(\frac{\Delta T}{T}\right)_Q \cong 0.22A; \qquad \left(\frac{\Delta T}{T}\right)_{\mathrm{rms}} \cong 0.51A. \qquad (13.122)$$

Given these two measurements, one can verify that the fluctuations are consistent with the scale invariant spectrum by checking their ratio. Further, the numerical value of the observed $(\Delta T/T)$ can be used to determine the amplitude A. One finds that $A \approx 3.1 \times 10^{-5}$ which sets the scale of fluctuations in the gravitational potential at the time when the perturbation enters the Hubble radius. The dimensionless measure of density fluctuations now becomes $k^3 P(k)/2\pi^2 \simeq (k/0.074\,h$ $\text{Mpc}^{-1})^4$.

Incidentally, note that the solution $\delta_R = 4\Phi - 6\Phi_i$ corresponds to $\delta_m = (3/4)\delta_R = 3\Phi - (9/2)\Phi_i$. At $y = y_{\text{dec}}$, taking $\Phi_{\text{dec}} = (9/10)\Phi_i$, we get $\delta_m = 3\Phi_{\text{dec}} - (9/2)(10/9)\Phi_{\text{dec}} = -2\Phi_{\text{dec}}$. Since $(\Delta T/T)_{\text{obs}} = (1/3)\Phi_{\text{dec}}$ we get $\delta_m = -6(\Delta T/T)_{\text{obs}}$. This shows that the amplitude of matter perturbations is a factor six larger than the amplitude of the temperature anisotropy for our (adiabatic) initial conditions. In several other models, one gets $\delta_m = \mathcal{O}(1)(\Delta T/T)_{\text{obs}}$. So, to reach a given level of nonlinearity in the matter distribution at the present epoch, these models will require higher values of $(\Delta T/T)_{\text{obs}}$ at decoupling. This fact allows such models to be observationally ruled out.

There is another useful result which we can obtain from Eq. (13.107) along the same lines as we derived the Sachs–Wolfe effect. Whenever $k^3|Q(k)|^2$ is a slowly varying function of k, we can pull this factor out of the integral in Eq. (13.107) and use Eq. (13.119) to the integral. This will give the result, for any slowly varying $k^3|Q(k)|^2$, to be

$$\frac{l(l+1)C_l}{2\pi} \approx \left(\frac{k^3|Q(k)|^2}{2\pi^2}\right)_{kL \approx l}. \qquad (13.123)$$

This is applicable even when different processes contribute to temperature anisotropies as long as they add in quadrature. While not very accurate, it allows one to estimate the physical effects which operate at different scales rapidly.

PROJECT

Project 13.1
Nonlinear perturbations and cosmological averaging

We know that the energy-momentum tensor of the real universe, $T_{ab}(t, \boldsymbol{x})$, is inhomogeneous and anisotropic and will lead to a very complicated form for the metric g_{ab} if only we could solve the exact Einstein's equations $G_{ab}[g] = \kappa T_{ab}$. The metric describing the large scale structure of the universe should be obtained by averaging this exact solution over a large enough scale, to get $\langle g_{ab}\rangle$. But what we actually did in Chapter 10, for example, is to average the energy-momentum tensor *first* to get $\langle T_{ab}\rangle$ and *then* solve Einstein's equations. But since $G_{ab}[g]$ is a nonlinear function of the metric, $\langle G_{ab}[g]\rangle \neq G_{ab}[\langle g\rangle]$ and there is a discrepancy. This is most easily seen by writing

$$G_{ab}[\langle g\rangle] = \kappa[\langle T_{ab}\rangle + \kappa^{-1}(G_{ab}[\langle g\rangle] - \langle G_{ab}[g]\rangle)] \equiv \kappa[\langle T_{ab}\rangle + T_{ab}^{corr}]. \qquad (13.124)$$

If – based on observations – we take $\langle g_{ab} \rangle$ to be the standard Friedman metric, this equation shows that it has, as its source, *two* terms. The first is the standard average energy-momentum tensor and the second is a purely geometrical correction term $T_{ab}^{corr} = \kappa^{-1}(G_{ab}[\langle g \rangle] - \langle G_{ab}[g] \rangle)$, which arises because of nonlinearities in the Einstein's theory that leads to $\langle G_{ab}[g] \rangle \neq G_{ab}[\langle g \rangle]$. If this correction term is large, we are missing out something significant in the standard cosmological models. There are occasional suggestions in the literature that this is indeed the case and, in fact, the correction term could possibly account for what is conventionally interpreted as dark energy. This project (which is somewhat open ended and requires you to do additional reading) invites you to study this issue and conclude that the effect is not important.

(a) Consider the perturbed Friedmann metric

$$ds^2 = -(1 + 2\phi)dt^2 + a(t)^2(1 - 2\phi)|d\boldsymbol{x}|^2 \,, \tag{13.125}$$

where $|d\boldsymbol{x}|^2$ is the flat 3-space metric in Cartesian coordinates. Further assume that we are studying a matter dominated era. Argue that the correction terms must be of the form $a^{-2}\langle \nabla_x \phi \cdot \nabla_x \phi \rangle$. Show that, if linear perturbation theory is valid at all scales and the models are compatible with the results described in Section 13.6.1, then the effect is indeed small and ignorable.

(b) The really interesting situation arises, of course, when the density contrast has become nonlinear at small scales and some structure has already formed. The question is whether the nonlinearity at small scales can produce a significant effect at large scales. The result will depend on the kind of power spectrum for matter perturbations that is assumed. In usual cosmological models this is well approximated by a spectrum called the BBKS spectrum[2] which has the asymptotic form that was obtained in Eq. (13.102). Show that, for this spectrum, the effect is still small.

(c) A simple way to model this effect is to study the collapse of a spherically symmetric overdense region in a matter dominated universe. This formalism has already been developed in Section 7.6. The metric in Eq. (7.181) can be expressed in the form

$$ds^2 = -dt^2 + \frac{R'^2 dr^2}{1 - k(r)r^2} + R^2 d\Omega^2 \,, \tag{13.126}$$

with a prime denoting a derivative with respect to r. Assume that initial conditions (at an epoch when the scale factor was a_i) are a perturbation around a $\Omega_{\rm tot} = 1$ solution, so that $|k(r)r^2| \ll 1$. We will take $R_i(r) = r$ and $\dot{R}_i(r) = H_i r$, where $H_i = H_0 a_i^{-3/2}$ and a dot denotes a derivative with respect to t. Explain why this initial condition is realistic. Show that at *any later time*, one can choose two coordinates $(t_{(1)}, r_{(1)})$ in place ot (t, r) so as to transform this metric to the form

$$ds^2 = -(1 + 2\phi)dt_{(1)}^2 + a(t_{(1)})^2(1 - 2\psi)\left(dr_{(1)}^2 + r_{(1)}^2 d\Omega^2\right) \,, \tag{13.127}$$

where $a(x) \equiv (3H_0 x/2)^{2/3}$, using the coordinate transformation

$$r_{(1)} = \frac{R(t, r)}{a(t)}(1 + \xi(t, r)) \;;\; t_{(1)} = t + \xi^0(t, r) \,, \tag{13.128}$$

where $|\xi|, |\xi^0 H| \ll 1$, $(H = 2/3t)$, *provided* the comoving peculiar velocity $v \equiv \partial_t(R/a)$ remains non-relativistic. In the process derive the explicit expressions for ϕ, ψ, ξ^0 and ξ given by

$$\xi^{0\prime} = avR' \; ; \; \xi' = \frac{1}{2}\left(k(r)r^2 + (av)^2\right)\frac{R'}{R}$$

$$\psi = \xi^0 H + \xi \; ; \; \phi = -\dot{\xi}^0 + \frac{1}{2}(av)^2. \tag{13.129}$$

Although the expressions for ϕ and ψ appear to be different, the Einstein equations in this Newtonian gauge show that in fact these functions are equal at the leading order. For any specific model this can also be explicitly verified.

(c) Interpret the above solution as showing that a perturbation series in the matter density contrast becomes a late time perturbation series in the matter velocities, even when the density contrast becomes nonlinear.

[Note. Averaging such a model using a well defined averaging scheme, one can show that the order-of-magnitude estimates made earlier are quite accurate and that nonlinear effects from small scales do not contribute significantly to the backreaction at late times in a matter dominated era.[3]]

14

Quantum field theory in curved spacetime

14.1 Introduction

This chapter describes some interesting results that arise when one studies standard quantum field theory in a background spacetime with a nontrivial metric.[1] It turns out that the quantum field theory in a curved spacetime (or a non-inertial coordinate system) with a horizon exhibits some peculiar *and universal* properties. In particular, the study of the quantum field suggests that the horizon is endowed with a temperature $T = \kappa/2\pi$ (in natural units with $\hbar = c = k_B = 1$), where κ is the surface gravity of the horizon. This result can be viewed from very different perspectives, not all of which can be proved to be rigorously equivalent to one another. In view of the importance of this result, most of this chapter will concentrate on obtaining this result using different techniques and interpreting it physically. The latter part of the chapter will develop quantum field theory in an external Friedmann universe and will apply that formalism to study the generation of perturbations during the inflationary phase of the universe.

14.2 Review of some key results in quantum field theory

Fortunately, most of the important results we are interested in can be obtained with a minimum amount of background knowledge in quantum field theory. In order to set the stage, we shall rapidly review the necessary concepts in this section.

Quantum field theory attempts to describe particles as excitations of an underlying field. Consider a physical state with $n(\mathbf{k})$ particles – each of mass m, momentum $\mathbf{k}$ and energy $\omega_k = (k^2 + m^2)^{1/2}$. Such a state can be specified by providing a set of integers $n_1, n_2, \ldots$ for different values of momenta $\mathbf{k}_1, \mathbf{k}_2, \ldots$. We denote by the function $n(\mathbf{k})$ such a distribution so that, for every relevant value of $\mathbf{k}$, the function $n(\mathbf{k})$ is an integer giving the number of particles with momentum $\mathbf{k}$. (In the continuum limit, we will interpret $n(\mathbf{k})d^3\mathbf{k}$ as the number of particles with momentum between $\mathbf{k}$ and $\mathbf{k} + d\mathbf{k}$. We will not bother to define such a continuum limit rigorously since it is unnecessary for our purpose.) In quantum theory, such a state in the Hilbert space can be denoted by the symbol

$$|\psi\rangle = |n_1, n_2, \ldots\rangle = |\{n(\boldsymbol{k})\}\rangle \tag{14.1}$$

and is completely described by a function $n(\boldsymbol{k})$. The energy E and momentum $\boldsymbol{P}$ of this quantum state should be those contributed by all the particles together and we expect

$$E = \sum n(\boldsymbol{k})\omega_k; \qquad \boldsymbol{P} = \sum n(\boldsymbol{k})\boldsymbol{k}, \tag{14.2}$$

where the sum is over all $\boldsymbol{k}$. We know that physical processes can lead to emission and absorption of particles (for example, photons) and thus our formalism should be capable of describing transitions between states having different numbers of particles.

The results in Eq. (14.1) and Eq. (14.2) emphasize the (obvious) fact that the quantum state describing a collection of particles can be specified by a set of integers. When the number of particles corresponding to a given momentum $\boldsymbol{k}$ changes by unity, the energy of the state must change by ω_k. This property is exhibited by states of a simple harmonic oscillator with frequency ω_k and unit mass. If we consider a bunch of such harmonic oscillators described by the Hamiltonian

$$H = \sum \frac{1}{2} \left[p_{\boldsymbol{k}}^2 + \omega_k^2 q_{\boldsymbol{k}}^2 \right] \tag{14.3}$$

then the general eigenstate of the Hamiltonian will have exactly the same energy spectrum as the one given by the first equation in Eq. (14.2). This suggests that we should take a closer look at a collection of harmonic oscillators, each labelled by a vector $\boldsymbol{k}$, in order to describe states containing a certain number of particles.

Classically, such a system of oscillators will be described by an action of the form

$$\mathcal{A} = \sum \int dt \frac{1}{2} \left[\dot{q}_{\boldsymbol{k}}^2 - (k^2 + m^2) q_{\boldsymbol{k}}^2 \right] \rightarrow \int \frac{d^3 \boldsymbol{k}}{(2\pi)^3} \int dt \frac{1}{2} \left[\dot{q}_{\boldsymbol{k}}^2 - \omega_k^2 q_{\boldsymbol{k}}^2 \right]. \tag{14.4}$$

In arriving at the second expression, we have replaced the sum over the oscillators by an integral over the vector $\boldsymbol{k}$ and rescaled $q_{\boldsymbol{k}}$ so as to absorb any extra measure of integration. Remarkably enough, this action in Eq. (14.4) can be obtained from the *Lorentz invariant* action for a scalar field

$$\mathcal{A} = -\frac{1}{2} \int d^4 x \left[\partial_i \phi \partial^i \phi + m^2 \phi^2 \right]. \tag{14.5}$$

To establish this connection, we do a spatial Fourier transform of the field $\phi(t, \boldsymbol{x})$ and write

$$\phi(t, \boldsymbol{x}) = \int \frac{d^3 \boldsymbol{k}}{(2\pi)^3} Q_{\boldsymbol{k}}(t) \exp[i\boldsymbol{k} \cdot \boldsymbol{x}]. \tag{14.6}$$

As Q_k is complex, we have two degrees of freedom corresponding to the real and imaginary parts of Q_k for each k. If we write $Q_k = (A_k + iB_k)$, then, since Φ is a real scalar field, we can relate the variables for k to that for $-k$ as $A_k^* = A_{-k}$ and $B_k^* = -B_{-k}$. Evidently, only half the modes constitute independent degrees of freedom. Therefore, we can work with a new set of real modes q_k for all values of k with a suitable redefinition, say, by taking $q_k = A_k$ for one half of k and $q_{-k} = B_k$ for the other half. This will lead to the action in Eq. (14.4) with real values for q_k. We will usually assume that this is taken care of and work with an action in Eq. (14.4).

The action in Eq. (14.5) *turns out to be* relativistically invariant because the dispersion relation $\omega_k = (k^2 + m^2)^{1/2}$ is quadratic and relativistically invariant. Mathematically this allows us to combine the quadratic term in the time derivative $\dot{q}_k^2$ (which is present in *any* Lagrangian) with the $k^2 q_k^2$ term (which is special to harmonic oscillator potential) to obtain the relativistically invariant $\partial_i \phi \partial^i \phi$ term in Eq. (14.5). Note that the classical dynamics of a collection of oscillators knows nothing about Lorentz invariance, except through this dispersion relation.

It follows that the dynamics of a scalar field $\phi(t, x)$ described by the action in Eq. (14.5) will be identical to that of an infinite set of harmonic oscillators each labelled by a wave vector k. So the quantum theory of the field ϕ can be now obtained by constructing the quantum theory of each of the harmonic oscillators. For example, the ground state of the quantum field can be obtained by multiplying together the ground state wave functions of all the oscillators:

$$\Psi\left[\phi(x)\right] = \prod_k \left(\frac{\omega_k}{\pi}\right)^{1/4} \exp\left(-\frac{1}{2}\omega_k |q_k|^2\right) = \bar{N} \exp\left[-\frac{1}{2}\int \frac{d^3 k}{(2\pi)^3} \omega_k |q_k|^2\right],$$

$$(14.7)$$

where $\bar{N}$ is a normalization factor independent of q_k which we will ignore for the moment. The argument of the exponent can be expressed in a more useful form in real space in terms of $\nabla \phi$. This expression is particularly simple for a massless scalar field with $m = 0$ and $\omega_k = |k|$. Writing $\omega_k |q_k|^2 = k^2 |q_k|^2 / |k|$ and using the fact that ikq_k is essentially the Fourier spatial transform of $\nabla \phi$, we get

$$\int \frac{d^3 k}{(2\pi)^3} \omega_k |q_k|^2 = \int \frac{d^3 k}{(2\pi)^3} \frac{|k|^2 |q_k|^2}{|k|} = \frac{1}{2\pi^2} \int d^3 x \int d^3 y \left\{\frac{\nabla_x \phi \cdot \nabla_y \phi}{|x - y|^2}\right\}.$$

$$(14.8)$$

Substituting this into Eq. (14.7) and taking the modulus, we get the probability distribution in the ground state to be

$$\mathcal{P}[\phi(x)] = |\Psi[\phi(x)]|^2 = N \exp\left\{-\frac{1}{2\pi^2} \int\int d^3 x\, d^3 y \frac{\nabla_x \phi \cdot \nabla_y \phi}{|x - y|^2}\right\} \quad (14.9)$$

with $N = |\bar{N}|^2$. This expression suggests that the vacuum state of the field can host zero point fluctuations of the field variable ϕ. The probability that one detects a particular field configuration $\phi(\boldsymbol{x})$ when the field is in the vacuum state can be obtained by evaluating the value of $\mathcal{P}$ for this particular functional form $\phi(\boldsymbol{x})$. (The result is independent of time because of the stationarity of the vacuum state.) Given the ambiguity in the overall normalization factor N, which is formally divergent, this probability should be interpreted as a relative probability. That is, the ratio $\mathcal{P}_1/\mathcal{P}_2$ will give the relative probability between two field configurations characterized by the functions $\phi_1(\boldsymbol{x})$ and $\phi_2(\boldsymbol{x})$. With this interpretation, it is clear that the vacuum state in quantum field theory can exhibit nontrivial phenomena due to field fluctuations. The same comments apply for any other quantum state.

Another aspect of the vacuum state, related to the existence of the vacuum fluctuations, is the fact that the vacuum state has nonzero energy density. We know from elementary quantum mechanics that the ground state of a harmonic oscillator has an energy $(1/2)\omega_k$. Hence the ground state of the quantum field has the energy E_0 which is the sum of the ground state energies of all the oscillators

$$E_0 = \frac{1}{2} \int \frac{d^3\boldsymbol{k}}{(2\pi)^3} \, \omega_k = \frac{1}{2} \int \frac{d^3\boldsymbol{k}}{(2\pi)^3} \, \sqrt{|\boldsymbol{k}|^2 + m^2}, \tag{14.10}$$

which diverges as k^4 for large k. There is no satisfactory solution known to this divergence problem. One ad-hoc way of tackling it is to redefine the Hamiltonian of each of the oscillators by subtracting out $(1/2)\omega_k$. While this procedure works to a certain extent (and can be implemented in a more sophisticated manner), it is intrinsically unsatisfactory and it is not clear why gravity ignores this contribution. The probability distribution for different field configurations in the vacuum clearly demonstrates the existence of vacuum fluctuations with an associated nonzero energy $(1/2)\omega_k$ per mode. Since one expects vacuum field fluctuations to exist in the quantum theory of fields, it is somewhat artificial to set the energy corresponding to vacuum fluctuations to zero by hand. We shall see later on that the dynamics of the vacuum state is highly nontrivial; the infinite energy of the vacuum is the first of the series of counter-intuitive results we come across in quantum theory.

The above discussion focused on the Schrödinger equation for the study of the harmonic oscillator. In the Heisenberg picture, the Hamiltonian H for an oscillator can be expressed in the form

$$\frac{H}{\omega} = \frac{1}{2}(P^2 + Q^2) = \frac{1}{\sqrt{2}}(Q - iP)\frac{1}{\sqrt{2}}(Q + iP) - \frac{i}{2}[Q, P] \equiv a^\dagger a + \frac{1}{2} \equiv n + \frac{1}{2}, \tag{14.11}$$

where we have rescaled the coordinate and momentum by defining $P = p/\sqrt{\omega}$; $Q = \sqrt{\omega}q$ and introduced the creation (a) and annihilation ($a^\dagger$) operators by the definition $a = (1/\sqrt{2})(Q + iP)$. From $[q, p] = i$ it follows that $[a, a^\dagger] = 1$; the standard analysis now shows that the eigenstates of $a^\dagger a$ can be labelled by integers, giving the energy spectrum to be $E_n = \omega(n + 1/2)$.

In the Heisenberg picture the operators evolve in time and the equation of motion reads $i\dot{a} = [a, H] = \omega[a, a^\dagger a] = \omega a$ with the solutions

$$a(t) = a(0)e^{-i\omega t} \equiv ae^{-i\omega t}; \quad a^\dagger(t) = a^\dagger(0)e^{+i\omega t} \equiv a^\dagger e^{i\omega t}, \tag{14.12}$$

where we have simplified the notation by writing $a(0) = a$, etc. The coordinate q evolves as

$$q(t) = \frac{1}{\sqrt{2\omega}}(a^\dagger(t) + a(t)) = \frac{1}{\sqrt{2\omega}}\left[ae^{-i\omega t} + a^\dagger e^{i\omega t}\right]. \tag{14.13}$$

It is conventional to call $e^{-i\omega t}$ the positive frequency (energy) modes and to call $e^{+i\omega t}$ the negative frequency modes. In the expansion of $q(t)$ the coefficient of positive frequency mode gives the annihilation operator and the coefficient of the negative frequency mode gives the creation operator.

All these can, again, be generalized to a quantum field by treating each harmonic oscillator in the Heisenberg picture and introducing a creation ($a_{\boldsymbol{k}}$) and annihilation ($a_{\boldsymbol{k}}^\dagger$) operator for each oscillator with $[a_{\boldsymbol{k}}, a_{\boldsymbol{l}}^\dagger] = \delta(\boldsymbol{k} - \boldsymbol{l})$, etc. The time evolution of the field operator is then given by

$$\phi(t, \boldsymbol{x}) = \int \frac{d^3 \boldsymbol{k}}{(2\pi)^{3/2}\sqrt{2\omega_{\boldsymbol{k}}}}\left[a_{\boldsymbol{k}}\, e^{ikx} + a_{\boldsymbol{k}}^\dagger\, e^{-ikx}\right] \tag{14.14}$$

with $kx = k_a x^a$. Again, in the expansion of $\phi(t, \boldsymbol{x})$ the coefficient of positive frequency mode (e^{ikx}) gives the annihilation operator and the coefficient of the negative frequency mode (e^{-ikx}) gives the creation operator.

The canonical momentum for the field is defined as $\pi(t, \boldsymbol{x}) \equiv \dot{\phi}(t, \boldsymbol{x})$ (see Chapter 2, Eq. (2.25)). Using the commutation relation for creation and annihilation operators, it is now easy to verify that the field variables obey the commutation rules:

$$[\phi(\boldsymbol{x}, t), \pi(\boldsymbol{y}, t)] = i\delta(\boldsymbol{x} - \boldsymbol{y}); \quad [\phi(\boldsymbol{x}, t), \phi(\boldsymbol{y}, t)] = [\pi(\boldsymbol{x}, t), \pi(\boldsymbol{y}, t)] = 0. \tag{14.15}$$

In fact, the usual approach to quantum field theory begins with a quantum field obeying the field equation derived from the action in Eq. (14.5) and imposes these commutation rules on the field operator and the canonical momentum. One then shows that the field can be expanded as in Eq. (14.14) with the creation and annihilation operators obeying the standard commutation relations and obtains the particle interpretation through the integer eigenvalues of the $a_{\boldsymbol{k}}^\dagger a_{\boldsymbol{k}}$ operator.

14.2.1 *Bogolyubov transformations and the particle concept*

With future applications in mind, we shall now introduce an important concept called the *Bogolyubov transformations*. The mode expansion in Eq. (14.14) can be equivalently written in the form

$$\phi(t, \boldsymbol{x}) = \int \frac{d^3 \boldsymbol{k}}{(2\pi)^{3/2}} \left[a_{\boldsymbol{k}} v_{\boldsymbol{k}}(t) e^{i\boldsymbol{k} \cdot \boldsymbol{x}} + a_{\boldsymbol{k}}^\dagger v_{\boldsymbol{k}}^*(t) e^{-i\boldsymbol{k} \cdot \boldsymbol{x}} \right], \qquad (14.16)$$

where $v_{\boldsymbol{k}}(t)$ is the solution to the differential equation $\ddot{v}_{\boldsymbol{k}} + \omega_{\boldsymbol{k}}^2 v_{\boldsymbol{k}} = 0$ with the normalization condition

$$i(\dot{v}_{\boldsymbol{k}} v_{\boldsymbol{k}}^* - v_{\boldsymbol{k}} \dot{v}_{\boldsymbol{k}}^{*\prime}) = 1. \qquad (14.17)$$

Consider now a new set of functions $u_{\boldsymbol{k}}(t)$ defined through the relations

$$u_{\boldsymbol{k}}(t) = \alpha_{\boldsymbol{k}} \, v_{\boldsymbol{k}}(t) + \beta_{\boldsymbol{k}} \, v_{\boldsymbol{k}}^*(t), \qquad (14.18)$$

where $\alpha_{\boldsymbol{k}}$ and $\beta_{\boldsymbol{k}}$ are constant complex coefficients obeying the condition

$$|\alpha_{\boldsymbol{k}}|^2 - |\beta_{\boldsymbol{k}}|^2 = 1. \qquad (14.19)$$

It is now easy to verify that $u_{\boldsymbol{k}}$ satisfies the same normalization condition and the harmonic oscillator equation as $v_{\boldsymbol{k}}$. We can, therefore, expand the scalar field in terms of $u_{\boldsymbol{k}}$s as well with a new set of creation and annihilation operators. That is, we can write

$$\phi(t, \boldsymbol{x}) = \int \frac{d^3 \boldsymbol{k}}{(2\pi)^{3/2}} \left[b_{\boldsymbol{k}} u_{\boldsymbol{k}}(t) e^{i\boldsymbol{k} \cdot \boldsymbol{x}} + b_{\boldsymbol{k}}^\dagger u_{\boldsymbol{k}}^*(t) e^{-i\boldsymbol{k} \cdot \boldsymbol{x}} \right], \qquad (14.20)$$

where the new creation $(b_{\boldsymbol{k}})$ and annihilation $(b_{\boldsymbol{k}}^\dagger)$ operators are related to the old ones $a_{\boldsymbol{k}}$ and $a_{\boldsymbol{k}}^\dagger$ by

$$a_{\boldsymbol{k}} = \alpha_{\boldsymbol{k}} b_{\boldsymbol{k}} + \beta_{\boldsymbol{k}}^* b_{\boldsymbol{k}}^\dagger; \qquad b_{\boldsymbol{k}} = \alpha_{\boldsymbol{k}}^* a_{\boldsymbol{k}} - \beta_{\boldsymbol{k}}^* a_{\boldsymbol{k}}^\dagger. \qquad (14.21)$$

It is easy to verify that $b_{\boldsymbol{k}}$ and $b_{\boldsymbol{k}}^\dagger$ obey the same commutation rules as $a_{\boldsymbol{k}}$ and $a_{\boldsymbol{k}}^\dagger$ when Eq. (14.19) is satisfied. Hence, the field will still obey the commutation rules in Eq. (14.15).

We now have two different ground states for this system. The first one $|0\rangle_v$ is the ground state annihilated by all the $a_{\boldsymbol{k}}$s, while the second one $|0\rangle_u$ is the ground state annihilated by all the $b_{\boldsymbol{k}}$s:

$$a_{\boldsymbol{k}}|0\rangle_v = 0 \qquad b_{\boldsymbol{k}}|0\rangle_u = 0. \qquad (14.22)$$

The crucial point is that the ground state $|0\rangle_v$ is *not* a vacuum state with respect to the new creation and annihilation operators $b_{\boldsymbol{k}}$ and $b_{\boldsymbol{k}}^{\dagger}$. The new number operator $b_{\boldsymbol{k}}^{\dagger}b_{\boldsymbol{k}}$ has a nonzero expectation value in the original ground state $|0\rangle_v$ given by:

$$_v\langle 0|b_{\boldsymbol{k}}^{\dagger}b_{\boldsymbol{k}}|0\rangle_v = |\beta_{\boldsymbol{k}}|^2. \tag{14.23}$$

Both $|0\rangle_v$ and $|0\rangle_u$ are equally valid ground states but correspond to different sets of harmonic oscillators. The first one corresponds to the solutions to the harmonic oscillator equation of the form

$$v_{\boldsymbol{k}} = \frac{1}{\sqrt{2\omega_{\boldsymbol{k}}}} e^{-i\omega_{\boldsymbol{k}}t}, \tag{14.24}$$

which is what we started with. The second set of harmonic oscillators, which leads to the new ground state $|0\rangle_u$, corresponds to the mode functions (as they are called) which behave as

$$u_{\boldsymbol{k}} = \frac{1}{\sqrt{2\omega_{\boldsymbol{k}}}} \left(\alpha_{\boldsymbol{k}} e^{-i\omega_{\boldsymbol{k}}t} + \beta_{\boldsymbol{k}} e^{i\omega_{\boldsymbol{k}}t} \right). \tag{14.25}$$

Suppose one starts with the field and its canonical momentum as the fundamental quantities and imposes the quantization condition in Eq. (14.15). We can now expand the field in terms of infinitely many classes of mode functions like the $u_{\boldsymbol{k}}$s and $v_{\boldsymbol{k}}$s leading to infinitely many choices for creation and annihilation operators and ground states. The Bogolyubov transformation given above is just one simple case of many possibilities. (For example, we could have mixed up the creation and annihilation operators belonging to different $\boldsymbol{k}$ vectors.)

This raises an interesting question of principle. What is the physical vacuum state, the excitation of which will correspond to the physical particle? From our analysis given above, it is clear that one cannot answer this question without additional inputs. To arrive at one such physical input, we note that the modes $v_{\boldsymbol{k}}$ are made of purely positive frequency functions $e^{-i\omega t}$ while the modes $u_{\boldsymbol{k}}$ involving a nonzero $\beta_{\boldsymbol{k}}$ mixes up the positive and negative frequency modes. In the standard quantum field theory based on the Lorentz group, one makes the *additional assumption* that the physical particles correspond to modes which are purely positive frequency in the time coordinate. It can easily be shown that this criterion is Lorentz invariant and all inertial observers will agree on the definition of a vacuum state and that of a particle.

One can immediately see that this assumption will create difficulties when one studies quantum field theory in an external field or when one attempts to maintain general covariance in formulating the notion of a vacuum state. We shall see several examples of these difficulties in the coming sections but it is worthwhile to discuss it briefly to bring out the key issues which are involved.

To begin with, let us consider quantum field theory in an external potential which is varying with time. This could be done, for example, by assuming that the parameter m that appears in Eq. (14.5) is not a constant but some specified function of time. This will make the frequency of the harmonic oscillators $\omega_k(t)$ time dependent. The time evolution will now, in general, mix up positive and negative frequency modes. That is, if we start with a solution that behaves as a purely positive frequency mode (like v_k) at some given instant of time, then it will evolve into a mixture of positive and negative frequency modes (like u_k) at any later instant in time. In particular, let us assume that the frequency of the harmonic oscillator was asymptotically constant with a value ω_1 as $t \to -\infty$ and with a value ω_2 as $t \to +\infty$. From the general theory of second order differential equations, we know that a mode which behaves as v_k with $\omega = \omega_1$ at early times will end up as u_k with $\omega = \omega_2$ at late times. This implies that the vacuum state at early times $|0\rangle_v$ will appear to contain particles at late times because of Eq. (14.23). In this particular case – which we will encounter in the context of quantum field theory in the Friedmann universe – one would interpret the phenomena as being due to genuine particle production by the external potential.

A more subtle effect – which occurs even in the case of non-inertial coordinates in flat spacetime – arises when there exist two timelike Killing vector fields $(\partial/\partial t)$ and $(\partial/\partial T)$ corresponding to two natural time coordinates, say T and t, that are available in a given region of spacetime and we allow for general coordinate transformations. One can then construct positive frequency modes $e^{-i\omega T}$ with respect to T and the corresponding vacuum state or, alternatively, one can use the modes $e^{-i\omega t}$ (which are positive frequency modes with respect to t) and construct a (different) vacuum state. In general, the transformation between the coordinates will be a nonlinear function and the expansion of $e^{-i\omega T}$ in terms of the second set of modes will involve a superposition of positive and negative frequency modes. It follows that the vacuum state defined using the T coordinate will not be a vacuum state defined with respect to t coordinates. One concludes that the notion of the vacuum state and that of a particle are *not* generally covariant concepts though they are Lorentz invariant concepts. This is, possibly, one of the key lessons from the study of quantum field theory in a nontrivial background metric.

14.2.2 Path integrals and Euclidean time

The time evolution in quantum field theory for the free field is quite trivial and follows directly from the time evolution of the individual harmonic oscillators. However, bearing future applications in mind, we shall describe the time evolution from a different perspective by using a *path integral kernel*. It is known in quantum mechanics that the net probability amplitude $K(2; 1)$ for the particle to go from the event $\mathcal{P}_1$ to the event $\mathcal{P}_2$ is obtained by adding up the amplitudes for all the paths connecting the events

$$K(\mathcal{P}_2; \mathcal{P}_1) \equiv K(t_2, q_2; t_1, q_1) = \sum_{\text{paths}} \exp[i\mathcal{A}(\text{path})], \qquad (14.26)$$

where $\mathcal{A}$(path) is the action evaluated for a given path connecting the end points $\mathcal{P}_1$ and $\mathcal{P}_2$. The addition of the *amplitudes* allows for the quantum mechanical interference between the paths. The quantity $K(t_2, q_2; t_1, q_1)$ contains the full dynamical information about the quantum mechanical system. Given $K(t_2, q_2; t_1, q_1)$ and the initial amplitude $\psi(t_1, q_1)$ for the particle to be found at q_1, we can compute the wave function $\psi(t, q)$ at any later time by the usual rules for combining the amplitudes:

$$\psi(t, q) = \int dq_1 K(t, q; t_1, q_1) \psi(t_1, q_1). \qquad (14.27)$$

We shall now obtain a relation between the ground state wave function of the system and the path integral kernel which will prove to be useful.

In the conventional approach to quantum mechanics, using the Heisenberg picture, we will describe the system in terms of the position and momentum operators $\hat{q}$ and $\hat{p}$. Let $|q, t\rangle$ be the eigenstate of the operator $\hat{q}(t)$ with eigenvalue q. The kernel – which represents the probability amplitude for a particle to propagate from (t_1, q_1) to (t_2, q_2) – can be expressed, in more conventional notation, as the matrix element

$$K(t_2, q_2; t_1, q_1) = \langle q_2, t_2 | t_1, q_1 \rangle = \langle q_2, 0 | \exp[-iH(t_2 - t_1)] | 0, q_1 \rangle, \quad (14.28)$$

where H is the time independent Hamiltonian describing the system. This relation allows one to represent the kernel in terms of the energy eigenstates of the system. We have

$$\begin{aligned}
K(T, q_2; 0, q_1) &= \langle q_2, 0 | \exp -iHT | 0, q_1 \rangle \\
&= \sum_{n,m} \langle q_2 | E_n \rangle \langle E_n | \exp -iHT | E_m \rangle \langle E_m | q_1 \rangle \\
&= \sum_n \psi_n(q_2) \psi_n^*(q_1) \exp(-iE_n T), \qquad (14.29)
\end{aligned}$$

where $\psi_n(q) = \langle q | E_n \rangle$ is the nth energy eigenfunction of the system under consideration. Equation (14.29) allows one to express the kernel in terms of the eigenfunctions of the Hamiltonian. For any Hamiltonian which is bounded from below it is convenient to add a constant to the Hamiltonian so that the ground state – corresponding to the $n = 0$ term in the above expression – has zero energy. We shall assume that this is done. Next, we will analytically continue the expression in Eq. (14.29) to imaginary values of T by writing $iT = T_E$. (The subscript

E stands for Euclidean; the special relativistic line element now changes from $-dT^2 + d\mathbf{X}^2$ to $dT_E^2 + d\mathbf{X}^2$, converting the Riemannian space to Euclidean space.) The Euclidean kernel obtained from Eq. (14.29) has the form

$$K_E(T_E, q_2; 0, q_1) = \sum_n \psi_n(q_2)\psi_n^*(q_1)\exp(-E_n T_E). \qquad (14.30)$$

Suppose we now set $q_1 = q$, $q_2 = 0$ in the above expression and take the limit $T_E \to \infty$. In the large time limit, the exponential will suppress all the terms in the sum except the one with $E_n = 0$ which is the ground state for which the wave function is real. We therefore obtain the result

$$\lim_{T \to \infty} K(T, 0; 0, q) \approx \psi_0(0)\psi_0(q_1) \propto \psi_0(q). \qquad (14.31)$$

Hence the ground state wave function can be obtained by analytically continuing the kernel into imaginary time and taking a suitable limit. The proportionality constant is irrelevant since it can always be obtained by normalizing the wave function $\psi_0(q)$. This result holds for any closed system with bounded Hamiltonian.

For actions which are quadratic in the dynamical variables (like in the case of the harmonic oscillator) we can make further progress. In this case, we can express the kernel completely in terms of the classical trajectory as

$$K(t_2, q_2; t_1, q_1) = N(t_1, t_2)\exp\frac{i}{\hbar}\mathcal{A}_c(t_2, q_2; t_1, q_1), \qquad (14.32)$$

where $\mathcal{A}_c$ is the action evaluated for a classical trajectory and $N(t_2, t_1)$ is a normalization factor which can be expressed in terms of $\mathcal{A}_c$ (but we will not require its explicit form). We see that the q-dependence of the kernel is only through the classical action $\mathcal{A}_c$. (The above result is *exact* for systems with quadratic actions with even time dependent coefficients but we will use it only for time independent systems.) We now use this fact to express the kernel in Eq. (14.31) in terms of the classical action for a system with a quadratic Lagrangian. If we analytically continue the action to imaginary values of time, then the ground state wave function of such a system can be expressed in terms of the Euclidean action evaluated on a classical trajectory. Explicitly, combining Eq. (14.31) and Eq. (14.32), we have the result

$$\psi_0(q) \propto \exp\left[-\mathcal{A}_E\left(T_E = \infty, 0; T_E = 0, q\right)\right] \propto \exp\left[-\mathcal{A}_E\left(\infty, 0; 0, q\right)\right]. \qquad (14.33)$$

This result shows that, if we find the classical Euclidean solution which describes a trajectory that starts at q at $T_E = 0$ and goes to $q = 0$ at very late times, then the ground state wave function of the quantum system can be expressed in terms of the classical Euclidean action for this trajectory.

Since the ground state of a free quantum field theory is obtained by the product of the ground states of the oscillators, the above result carries through to field theory as well. We now have

$$\Psi_0(\phi(\boldsymbol{x})) \propto \exp\left[-\mathcal{A}_E\left(T_E = \infty, 0; T_E = 0, \phi(\boldsymbol{x})\right)\right]. \tag{14.34}$$

Very often, it is convenient to use the formula for the individual oscillators and obtain the final result by integrating over all the oscillators in the argument of the exponential.

The analytic continuation to imaginary values of time has close mathematical connections with the description of systems in a thermal bath. To see this, consider the mean value of some observable $\mathcal{O}(q)$ of a quantum mechanical system. If the system is in an energy eigenstate described by the wave function $\psi_n(q)$, then the expectation value of $\mathcal{O}(q)$ can be obtained by integrating $\mathcal{O}(q)|\psi_n(q)|^2$ over q. If the system is in a thermal bath at temperature β^{-1}, described by a canonical ensemble, then the mean value has to be computed by averaging over all the energy eigenstates *as well* with a weightage $\exp(-\beta E_n)$. In this case, the mean value can be expressed as

$$\langle\mathcal{O}\rangle = \frac{1}{Z}\sum_n \int dq\,\psi_n(q)\mathcal{O}(q)\psi_n^*(q)\,e^{-\beta E_n} \equiv \frac{1}{Z}\int dq\,\rho(q,q)\mathcal{O}(q), \tag{14.35}$$

where Z is the partition function and we have defined a *density matrix* $\rho(q,q')$ by

$$\rho(q,q') \equiv \sum_n \psi_n(q)\psi_n^*(q')\,e^{-\beta E_n} \tag{14.36}$$

in terms of which we can rewrite Eq. (14.35) as

$$\langle\mathcal{O}\rangle = \frac{\mathrm{Tr}\,(\rho\mathcal{O})}{\mathrm{Tr}\,(\rho)}, \tag{14.37}$$

where the trace operation involves setting $q = q'$ and integrating over q. This standard result shows how $\rho(q,q')$ contains information about both thermal and quantum mechanical averaging. Comparing Eq. (14.36) with Eq. (14.29) we find that the density matrix can be immediately obtained from the Euclidean kernel by:

$$\rho(q,q') = K_E(\beta, q; 0, q') \tag{14.38}$$

with the Euclidean time acting as inverse temperature.

As an example that will be useful later, we quote the explicit form of this result in the case of a harmonic oscillator. The action for the harmonic oscillator can be obtained easily from the classical solution (see Exercise 14.1). Analytically continuing the expression to imaginary time, we find that the thermal density matrix of the system can be expressed in the form

$$\rho(q, q') = N \exp - \left\{ \frac{\omega}{2} \left[\frac{\cosh(\beta\omega)}{\sinh(\beta\omega)} (q^2 + q'^2) - \frac{2qq'}{\sinh(\beta\omega)} \right] \right\}$$

$$= N \exp \left[-\frac{\omega}{4} \left((q - q')^2 \coth \left(\frac{\beta\omega}{2} \right) + (q + q')^2 \tanh \left(\frac{\beta\omega}{2} \right) \right) \right],$$

(14.39)

with $N^2 = (\omega/2\pi) \operatorname{cosech} \beta\omega$. The equivalence of the two expressions is easy to verify by using standard trigonometric identities.

Exercise 14.1

Path integral kernel for the harmonic oscillator Consider a harmonic oscillator with the Lagrangian $L = (1/2)[\dot{q}^2 - \omega^2 q^2]$.

(a) Obtain a classical solution to the equations of motion such that $q(0) = q_i$ and $q(T) = q_f$. Evaluate the classical action for this trajectory and show that it is given by

$$\mathcal{A}_c = \frac{\omega}{2 \sin \omega T} \left[(q_i^2 + q_f^2) \cos \omega T - 2q_i q_f \right].$$

(14.40)

(b) This result will allow you to express the path integral kernel for the harmonic oscillator in the form $K(q_f, T; q_i, 0) = F(T) \exp[(i/\hbar)\mathcal{A}_c]$. Explain why the kernel must satisfy the constraint

$$K(q_3, t_3; q_1, t_1) = \int_{-\infty}^{\infty} dq_2 \, K(q_3, t_3; q_2, t_2) \, K(q_2, t_2; q_1, t_1).$$

(14.41)

(Note that the integral is only over q_2 and yet the result in the left hand side is independent of t_2.) Determine $F(T)$ using this fact and show that

$$F(T) = \left(\frac{\omega}{2\pi i \hbar \sin \omega T} \right)^{1/2}.$$

(14.42)

Hence show that the harmonic oscillator kernel has the form

$$K(q_f, T, q_i, 0) = \left(\frac{\omega}{2\pi i \hbar \sin \omega T} \right)^{1/2} \exp \left[\frac{i}{\hbar} \frac{\omega}{2 \sin \omega T} \left\{ (q_i^2 + q_f^2) \cos \omega T - 2q_i q_f \right\} \right].$$

(14.43)

Use this to obtain the result in Eq. (14.39).

14.3 Exponential redshift and the thermal spectrum

Having developed the preliminary results that are required for our study, we shall now start exploring several aspects of quantum field theory in curved spacetime. As explained in Section 14.1, one of the key results of quantum field theory in curved spacetime is the emergence of thermodynamic features in spacetimes with horizons. This, in turn, arises because of the exponential redshift suffered by the wave modes in spacetimes with horizons. We shall therefore begin our discussion establishing the relation between the two.

Consider an inertial reference frame S and an observer who is moving at a speed v along the x-axis in this frame. If her trajectory is $X = vT$ the clock she is carrying will show the proper time $\tau = T/\gamma$, where $\gamma = (1 - v^2)^{-1/2}$. Combining these results we can write the trajectory in parametrized form as $t(\tau) = \gamma\tau$; $X(\tau) = \gamma v\tau$. Let us suppose that a monochromatic plane wave $\phi(T, X) = \exp -i\Omega(T - X)$ exists at all points in the inertial frame. At any given X, it oscillates with time as $e^{-i\Omega T}$ so that Ω is the frequency as measured in S. The moving observer, of course, will measure how the ϕ changes with respect to her proper time. This is easily obtained by substituting the trajectory $T(\tau) = \gamma\tau$; $X(\tau) = \gamma v\tau$ into the function $\phi(T, X)$ obtaining $\phi[T(\tau), X(\tau)] = \phi[\tau]$. Simple calculation gives

$$\phi[T(\tau), X(\tau)] = \phi[\tau] = \exp -i\tau\Omega\gamma(1 - v) = \exp -i\tau\Omega\sqrt{\frac{1 - v}{1 + v}}. \tag{14.44}$$

Clearly, the moving observer sees the wave changing in time with a frequency

$$\Omega' \equiv \Omega\sqrt{\frac{1 - v}{1 + v}}. \tag{14.45}$$

A monochromatic wave will therefore appear to be a monochromatic wave but with a Doppler shifted frequency – which is a standard result in special relativity derived in a slightly different manner.

We will now use the same procedure for an observer moving along the x-axis in a uniformly *accelerated* trajectory given by (see Eq. (3.65))

$$X(\tau) = \frac{1}{\kappa}\cosh(\kappa\tau); \quad T(\tau) = \frac{1}{\kappa}\sinh(\kappa\tau). \tag{14.46}$$

Proceeding exactly in analogy with Eq. (14.44) we can determine how the *accelerated* observer will view the monochromatic wave. We get:

$$\phi[T(\tau), X(\tau)] = \phi[\tau] = \exp \frac{i}{\kappa}[\Omega\exp -\kappa\tau] \equiv \exp i\theta(\tau). \tag{14.47}$$

Unlike in the case of uniform velocity, we now find that the phase $\theta(\tau)$ of the wave itself is decreasing exponentially with time. Defining the instantaneous frequency by $\omega(\tau) = -d\theta/d\tau$, we find that an accelerated observer will see the wave with an instantaneous frequency that is getting exponentially redshifted:

$$\omega(\tau) = \Omega\exp(-\kappa\tau). \tag{14.48}$$

The power spectrum of this wave in Eq. (14.47) is given by $P(\nu) = |f(\nu)|^2$ where $f(\nu)$ is the Fourier transform of $\phi(\tau)$ with respect to τ:

$$\phi(\tau) = \int_{-\infty}^{\infty} \frac{d\nu}{2\pi} f(\nu)e^{-i\nu\tau}; \quad f(\nu) = \int_{-\infty}^{\infty} d\tau\phi(\tau)e^{i\nu\tau}. \tag{14.49}$$

Because of the exponential redshift, this power spectrum will *not* vanish for $\nu < 0$. Evaluating this Fourier transform (see Exercise 14.2) one gets

$$f(\nu) = \left(\frac{1}{\kappa}\right)\left(\frac{\Omega}{\kappa}\right)^{i\nu/\kappa}\Gamma(-i\nu/\kappa)e^{\pi\nu/2\kappa}. \qquad (14.50)$$

Obtaining the modulus $|f(\nu)|^2$ by using the identity $\Gamma(z)\Gamma(-z) = -\pi/z\sin(\pi z)$, we get the remarkable result that the power, per logarithmic band in frequency, at negative frequencies is a Planckian at temperature $T = (\kappa/2\pi)$:

$$\nu|f(-\nu)|^2 = \frac{\beta}{e^{\beta\nu} - 1}; \quad \beta = \frac{2\pi}{\kappa}. \qquad (14.51)$$

Although $f(\nu)$ in Eq. (14.50) depends on Ω, the power spectrum $|f(\nu)|^2$ is independent of Ω; monochromatic plane waves of any frequency (as measured by the inertial observers at $X = $ constant) will appear to have a Planckian power spectrum in terms of the (negative) frequency ν, defined with respect to the proper time of the accelerated observer.

We saw earlier (see Exercise 8.4) that waves propagating from a region near the horizon $\mathcal{H}$ will undergo exponential redshift. An observer detecting this exponentially redshifted radiation at late times $(T \to \infty)$, originating from a region close to $\mathcal{H}$ will attribute to this radiation a Planckian power spectrum given by Eq. (14.51). This result lies at the foundation of associating temperature with horizons.

The Planck spectrum in Eq. (14.51) is in terms of the frequency and $\beta = (2\pi c/\kappa)$ in normal units has the (correct) dimension of time; no $\hbar$ appears in the result. If we now switch the variable from frequency to energy, invoking the basic tenets of quantum mechanics, and write $\beta\nu = (\beta/\hbar)(\hbar\nu) = (\beta/\hbar)E$, then one can identify a temperature $k_BT = (\kappa\hbar/2\pi c)$ which scales linearly with $\hbar$. An astronomer measuring frequency rather than photon energy will see the spectrum in Eq. (14.51) as Planckian without any quantum mechanical input.

Exercise 14.2
Power spectrum of a wave with exponential redshift The thermal nature of the power spectrum for a wave with exponential redshift was obtained by evaluating the second integral in Eq. (14.49) with $\phi(\tau)$ given by Eq. (14.47). This kind of integral appears in several contexts in the study of quantum field theory in curved spacetime. This exercise obtains this result.
(a) Show that

$$\int_0^{+\infty} x^{s-1}e^{-bx}dx = \exp(-s\ln b)\,\Gamma(s), \qquad (14.52)$$

when Re $b > 0$ and Re $s > 0$.

(b) Consider next the integral which we need to evaluate and express it in the form

$$F(\omega, \Omega) \equiv \int_{-\infty}^{+\infty} \frac{du}{2\pi} \exp\left(i\Omega u + i\frac{\omega}{\kappa} e^{-\kappa u}\right) = \frac{1}{2\pi\kappa} \int_0^{+\infty} dx\, x^{-(i\Omega/\kappa)-1}\, e^{(i\omega/\kappa)x},$$

$$(14.53)$$

where $x \equiv e^{-\kappa u}$. This integral has the same structure as the one in Eq. (14.52) with $b = -(i\omega/\kappa)$, $s = -(i\Omega/\kappa)$. To ensure convergence, consider b and s to be the limit of $b = -(i\omega/\kappa) + \epsilon$, $s = -(i\Omega/\kappa) + \epsilon$, with $\epsilon \to 0^+$. Using

$$\ln b = \lim_{\epsilon \to +0} \ln\left(-\frac{i\omega}{\kappa} + \epsilon\right) = \ln\left|\frac{\omega}{\kappa}\right| - i\frac{\pi}{2}\,\text{sign}\left(\frac{\omega}{\kappa}\right) \qquad (14.54)$$

show that

$$F(\omega, \Omega) = \frac{1}{2\pi\kappa} \exp\left[\frac{i\Omega}{\kappa} \ln\left|\frac{\omega}{\kappa}\right| + \frac{\pi\Omega}{2\kappa}\,\text{sign}\left(\frac{\omega}{\kappa}\right)\right] \Gamma\left(-\frac{i\Omega}{\kappa}\right); \qquad (14.55)$$

and

$$F(\omega, \Omega) = F(-\omega, \Omega)\exp\left(\frac{\pi\Omega}{\kappa}\right), \qquad (14.56)$$

where $\omega > 0$, $\Omega > 0$. Use these to obtain the results in Eq. (14.50).

14.4 Vacuum state in the presence of horizons

We shall now discuss the formulation of quantum field theory in a static spacetime with a bifurcation horizon. The key new feature which arises from this study is the following. The vacuum state of the quantum field, defined in the global (Kruskal like) coordinate system behaves like a thermal state as far as observers confined to one side of the horizon. The temperature of the bath turns out to be $T = \kappa/2\pi$, where κ is the surface gravity of the horizon. Given the fundamental importance of this result, we shall describe it from different perspectives starting from the simplest context which arises in flat spacetime itself.

As we saw in the previous section, the vacuum state of a quantum field exhibits this thermal phenomenon when viewed in the Rindler coordinate system appropriate for an accelerated observer. To exhibit this result without unnecessary complications we shall start our study in a (1+1) spacetime. Consider the two-dimensional Minkowski spacetime with the line element:

$$ds^2 = -dT^2 + dX^2. \qquad (14.57)$$

The complete manifold is covered by the range $(-\infty < T < +\infty, -\infty < X < +\infty)$ for these coordinates. We now introduce two sets of coordinate patches, (x, t) and (x', t'), on the regions $X > |T|$ ("Right", $\mathcal{R}$) and $-X > |T|$ ("Left", $\mathcal{L}$) by the transformations:

$$X = \kappa^{-1} e^{\kappa x} \cosh \kappa t; \quad T = \kappa^{-1} e^{\kappa x} \sinh \kappa t \qquad \text{(in } \mathcal{R}) \qquad (14.58)$$

$$X = -\kappa^{-1} e^{\kappa x'} \cosh \kappa t'; \quad T = -\kappa^{-1} e^{\kappa x'} \sinh \kappa t' \qquad \text{(in } \mathcal{L}). \qquad (14.59)$$

All the coordinates (x, t), (x', t') vary from $(-\infty, +\infty)$. The metric Eq. (14.57) in terms of (x, t) (or (x', t')) has the form

$$ds^2 = e^{2\kappa x}(-dt^2 + dx^2). \qquad (14.60)$$

We have already encountered this coordinate system in Section 3.5.1, where we showed that it represents the proper reference frame for the uniformly accelerated observer. Further, in Section 8.7, we have seen that the Rindler and inertial coordinate systems bear the same kinematic relation with each other as the Schwarzschild and Kruskal–Szekeres coordinate system. In fact, we will see that the discussion in this section carries over in a straightforward manner to all other static spacetimes with a bifurcation horizon.

Consider a scalar field $\phi(X, T)$, described by the generally covariant action

$$A = -\frac{1}{2} \int d^2 x \sqrt{-g} g^{ik} \partial_i \phi \partial_k \phi = \frac{1}{2} \int_{-\infty}^{+\infty} dT \int_{-\infty}^{+\infty} dX \left[\left(\frac{\partial \phi}{\partial T} \right)^2 - \left(\frac{\partial \phi}{\partial X} \right)^2 \right] \qquad (14.61)$$

in terms of the global Minkowski coordinates. In the region $\mathcal{R}$, the action can be written in terms of x and t in the *same form*. This is because, in two dimensions, $\sqrt{-g} g^{ik} = \eta^{ik}$ for the line element in Eq. (14.60). So:

$$A_R = \frac{1}{2} \int_{-\infty}^{+\infty} dt \int_{-\infty}^{+\infty} dx \left[\left(\frac{\partial \phi}{\partial t} \right)^2 - \left(\frac{\partial \phi}{\partial x} \right)^2 \right]. \qquad (14.62)$$

We shall construct the quantum theory for this field in the Schrödinger picture by decomposing the field into harmonic oscillator modes. Consider a field configuration $\phi(X)$ on the $T = 0$ hypersurface, expanded in Fourier space as

$$\phi(X) = \int_{-\infty}^{+\infty} \frac{dK}{2\pi} q_K e^{iKX}. \qquad (14.63)$$

The vacuum state in the Minkowski coordinates is obtained by multiplying the vacuum state for each of the harmonic oscillators with the coordinate variables q_K and thus is described by the functional (see Eq. (14.7))

$$\Psi[\phi(X)] = \Psi[(q_K)] = \prod_K \left(\frac{\omega_K}{\pi} \right)^{1/4} \exp \left(-\frac{1}{2} \omega_K |q_K|^2 \right)$$

$$= N \exp \left[-\frac{1}{2} \int_0^\infty \frac{dK}{2\pi} |K| |q_K|^2 \right]. \qquad (14.64)$$

Let us next consider the description of the same quantum state in terms of the Rindler coordinates defined on the $T = 0$ hypersurface. In particular, we want to know how the ground state described by Eq. (14.64) appears when described in the Rindler frame. The hypersurface $T = 0$ is covered completely by $(t = 0, x)$ for $X > 0$ and $(t' = 0, x')$ for $X < 0$. So, in the regions $\mathcal{R}$ and $\mathcal{L}$, the field can be decomposed into Fourier modes as

$$\phi(X > 0) = \phi(x) = \int_{-\infty}^{+\infty} \frac{dk}{2\pi} a_k e^{ikx} \qquad (\text{in } \mathcal{R}) \qquad (14.65)$$

$$\phi(X < 0) = \phi(x') = \int_{-\infty}^{+\infty} \frac{dk}{2\pi} b_k e^{ikx'} \qquad (\text{in } \mathcal{L}). \qquad (14.66)$$

Using the relations Eq. (14.63), Eq. (14.65) and Eq. (14.66), we can express q_K in terms of a_k and b_k. After some simple algebra, we get

$$q_K = \int_{-\infty}^{+\infty} \phi(X) e^{-iKX} dX = \int_{-\infty}^{+\infty} \frac{dk}{2\pi} \left[a_k f(k, -K) + b_k f(k, +K) \right], \qquad (14.67)$$

where the function $f(k, K)$ is defined by the integral

$$f(k, K) = \int_{-\infty}^{+\infty} dx \exp \left[\kappa x + i \left(kx + \left(\frac{K}{\kappa} \right) e^{\kappa x} \right) \right]. \qquad (14.68)$$

This function can be expressed in terms of the gamma function of the imaginary argument (see Exercise 14.2); but we will not need its explicit form.

Using Eq. (14.67) we can express the wave function in Eq. (14.64) in terms of a_ks and b_ks. A detailed but straightforward calculation gives

$$\Psi[a_k, b_k] = N \exp \left[-\frac{1}{2} \int_0^\infty \frac{dk}{2\pi} (P(k)(|a_k|^2 + |b_k|^2) - Q(k)(a_k^* b_k + a_k b_k^*)) \right] \qquad (14.69)$$

with

$$P(k) = \coth \left(\frac{\pi \omega_k}{\kappa} \right), \quad Q(k) = \operatorname{cosech} \left(\frac{\pi \omega_k}{\kappa} \right); \quad \omega_k = |k|. \qquad (14.70)$$

This wave functional contains exactly the same amount of information as the vacuum state wave functional in Eq. (14.64). In inertial coordinates, the ground state wave functional is described in terms of the harmonic oscillators q_K while, in the Rindler coordinates, it is described in terms of two sets of harmonic oscillators a_k, b_k.

An interesting situation arises if we now try to calculate the expectation value of any observable quantity $\mathcal{O}$ which depends *only* on the a_ks. Note that the field configuration at any event in $\mathcal{R}$ can be obtained by evolving the field configuration in the semi-infinite line $T = 0, X > 0$. Since the field configuration on the semi-infinite line is completely determined by a_ks, it follows that one can compute any local observable made from the quantum field in $\mathcal{R}$ entirely in terms of a_ks. An observer confined to the region $\mathcal{R}$ will, therefore, have her local observables made out of a_ks alone. The expectation value of any such observable is given by

$$\langle \mathcal{O} \rangle = \int \prod_k da_k \int \prod_k db_k \Psi^*(a_k, b_k) \mathcal{O}(a_k) \Psi(a_k, b_k)$$

$$\equiv \int \prod_k da_k \rho(a_k, a_k) \mathcal{O}(a_k) = \mathrm{Tr}(\rho\mathcal{O}), \tag{14.71}$$

where we have defined

$$\rho(a_k', a_k) \equiv \int \prod_k db_k \Psi^*(a_k', b_k) \Psi(a_k, b_k). \tag{14.72}$$

This density matrix ρ is obtained by integrating out degrees of freedom contained in the wave function that are irrelevant for a given observer. (Such a procedure is usually resorted to in statistical mechanics while integrating over unobserved microstates of the system, thereby converting a description based on a wave function into a description in terms of a density matrix.) Evaluating the integrals in Eq. (14.72) using Eq. (14.69), we find that the density matrix is now given by

$$\rho(a_k', a_k) \equiv N \exp\left[-\frac{1}{2} \int_{-\infty}^{+\infty} \frac{dk}{2\pi} \frac{\omega_k}{4} \left((a_k - a_k')^2 \coth\left(\frac{\omega_k}{2T}\right)\right.\right.$$
$$\left.\left. + (a_k + a_k')^2 \tanh\left(\frac{\omega_k}{2T}\right)\right)\right] \tag{14.73}$$

with $T = \kappa/2\pi$. This is precisely the density matrix for a system of harmonic oscillators in a thermal bath with temperature $T = (\kappa/2\pi)$ (see Eq. (14.39)). It follows that, as far as observation of variables confined to the region $\mathcal{R}$ is concerned, the expectation values of observables are given by $\mathrm{Tr}\,(\rho\mathcal{O})$, where ρ is a thermal density matrix. So, clearly, the vacuum state appears as a thermal state for observers confined to the region $\mathcal{R}$. We stress that in this approach the thermal description arises because we have integrated out the variables b_k which are hidden by a horizon from an observer confined to $\mathcal{R}$.

Exercise 14.3
Casimir effect We have seen that the thermal effects in the Rindler frame arise because the vacuum functional has a different form when expressed in terms of the Rindler modes. A somewhat similar result, called the *Casimir effect*, arises in a different context. It is experimentally observed that two plane, parallel, conducting plates located with a separation L attract each other with a force given by $F = -(\pi^2/240)(\hbar c/L^4)$. This result arises essentially because the vacuum functional of the electromagnetic field, in the presence of two plates, is different from the vacuum functional without the plates. This result also illustrates the effect of the ground state energy density of the harmonic oscillator modes.

(a) Show that the vacuum functional for a scalar field in (1+1) spacetimes in both cases (i.e. with and without the plates) can be expressed in the form

$$\Psi\left[\phi(X), 0\right] = N \exp\left[-\int dX dY \nabla\phi(X)\nabla\phi(Y)\mathcal{G}(X, Y)\right] \qquad (14.74)$$

where, in the absence of the plates,

$$\mathcal{G}(X, Y) = \frac{1}{2}\int \frac{dK}{(2\pi)} \frac{\exp iK(X - Y)}{|K|} \qquad (14.75)$$

and in the presence of plates at $X = 0$ and $X = L$,

$$\mathcal{G}(X, Y) = \frac{1}{2}\sum_{n=1}^{\infty}\left(\frac{L}{n\pi}\right)\exp\left[\left(\frac{in\pi}{L}\right)(X - Y)\right]. \qquad (14.76)$$

In this case, we assume that $\phi(X)$ vanishes at $X = 0$ and $X = L$. The above result shows that the vacuum functional in the two cases are quite different.

(b) Show that the difference in the ground state energy in the two cases can be expressed in the form

$$\Delta E = \frac{1}{2}\sum_{n=1}^{\infty}\left(\frac{n\pi}{L}\right) - \frac{1}{2}\int_0^{\infty} dn\left(\frac{n\pi}{L}\right). \qquad (14.77)$$

Both the expressions are divergent but their difference can be computed if we use the prescription

$$\int_0^{\infty} n\, dn - \sum_{n=0}^{\infty} n = \lim_{\lambda \to 0}\left(\int_0^{\infty} ne^{-\lambda n}dn - \sum_{n=0}^{\infty} ne^{-\lambda n}\right) = \frac{1}{12}. \qquad (14.78)$$

Prove this result and hence obtain $\Delta E = -(\pi/24L)$.

(c) Generalize the above results for an electromagnetic field in a (1+3)-dimensional, flat spacetime.

14.5 Vacuum functional from a path integral

We shall now obtain the same result – viz. that the global vacuum state appears as a thermal state for observers confined to $\mathcal{R}$ – in a more elegant manner which will serve two important purposes. (i) It will demonstrate that the density matrix in the region $\mathcal{R}$ in the Rindler frame has the explicit form $\rho = \exp(-\beta H_R)$, where $\beta = 2\pi/\kappa$ and H_R is the Hamiltonian in the Rindler frame. (ii) It will show the connection between the thermal nature of the vacuum state and the Euclidean description of quantum field theory.[2]

On the $T = t = 0$ hypersurface one can define a vacuum state $|\text{vac}\rangle$ of the theory by giving the field configuration for the whole range of $-\infty < X < +\infty$. This field configuration, however, separates into two disjoint sectors when one uses the Rindler coordinate system. Therefore, we now need to specify both the field configuration $\phi_R(X)$ for $X > 0$ and $\phi_L(X)$ for $X < 0$ to match the initial data

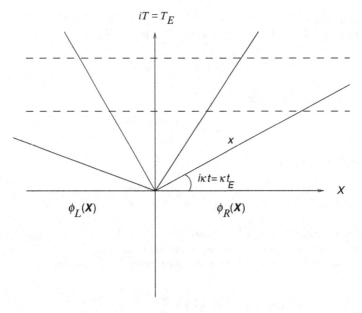

Fig. 14.1. Analytic extension to the imaginary time in two different time coordinates in the presence of a horizon. When one uses the path integral to determine the ground state wave functional on the $T_E = 0$ surface, one needs to integrate over the field configurations in the upper half ($T_E > 0$) with a boundary condition on the field configuration on $T_E = 0$ (see Eq. (14.34)). This can be done either by using a series of hypersurfaces parallel to the horizontal axis (shown by broken lines) or by using a series of hypersurfaces corresponding to the radial lines. Comparing the two results, one can show that the ground state in one coordinate system appears as a thermal state in the other.

in the global coordinates; given these data, the vacuum state is specified by the functional $\langle \text{vac} | \phi_L, \phi_R \rangle$.

Let us next consider the *Euclidean* sector corresponding to the (T_E, X) plane where $T_E = iT$. The quantum field theory in this plane can be defined along standard lines. The analytic continuation in t, however, is a different matter. Writing the metric in Eq. (14.60) as

$$ds^2 = e^{2\kappa x}(-dt^2 + dx^2) = -\kappa^2 l^2 dt^2 + dl^2, \tag{14.79}$$

where $l = \kappa^{-1} e^{\kappa x}$, we see that this metric becomes the one in polar coordinates on Euclidean continuation to ($\kappa t_E = i\kappa t, x$). That is,

$$ds^2 = \kappa^2 l^2 dt_E^2 + dl^2 \tag{14.80}$$

with t_E becoming the angular coordinate having a periodicity of $(2\pi/\kappa)$ as shown in Fig. 14.1. While the evolution in T_E will take the field configuration from $T_E = 0$ to $T_E \to \infty$, the same time evolution gets mapped in terms of t_E into evolving the 'angular' coordinate t_E from 0 to $2\pi/\kappa$. (This is clear from Fig. 14.1.)

It is obvious that the entire upper half-plane $T > 0$ is covered in two completely different ways in terms of the evolution in T_E compared with evolution in t_E. In (T_E, X) coordinates, we vary X in the range $(-\infty, \infty)$ for each T_E and vary T_E in the range $(0, \infty)$. In (t_E, x) coordinates, we vary x in the range $(0, \infty)$ for each t_E and vary t_E in the range $(0, \pi/\kappa)$. This fact allows us to prove that

$$\langle \text{vac} | \phi_L, \phi_R \rangle \propto \langle \phi_L | e^{-\pi H/\kappa} | \phi_R \rangle \tag{14.81}$$

as we shall see below. If we denote the proportionality constant by C, then the normalization condition

$$1 = \int \mathcal{D}\phi_L \, \mathcal{D}\phi_R \, |\langle \text{vac} | \phi_L, \phi_R \rangle|^2 = \int \mathcal{D}\phi_L \, \mathcal{D}\phi_R \, \langle \text{vac} | \phi_L, \phi_R \rangle \langle \phi_L, \phi_R | \text{vac} \rangle$$

$$= C^2 \int \mathcal{D}\phi_L \, \mathcal{D}\phi_R \, \langle \phi_L | e^{-\pi H_R/\kappa} | \phi_R \rangle \, \langle \phi_R | e^{-\pi H_R/\kappa} | \phi_L \rangle = C^2 \, \text{Tr} \left(e^{-2\pi H_R/\kappa} \right) \tag{14.82}$$

fixes the proportionality constant C, allowing us to write Eq. (14.81) in the form

$$\langle \text{vac} | \phi_L, \phi_R \rangle = \frac{\langle \phi_L | e^{-\pi H/\kappa} | \phi_R \rangle}{\left[\text{Tr}(e^{-2\pi H/\kappa}) \right]^{1/2}}. \tag{14.83}$$

To provide a simple proof of Eq. (14.81), let us consider the ground state wave functional $\langle \text{vac} | \phi_L, \phi_R \rangle$ in the extended spacetime expressed as a path integral. From the results of Section 14.2.2 we know that the ground state wave functional can be represented as a Euclidean path integral of the form

$$\langle \text{vac} | \phi_L, \phi_R \rangle \propto \int_{T_E = 0; \phi = (\phi_L, \phi_R)}^{T_E = \infty; \phi = (0,0)} \mathcal{D}\phi \, e^{-A}, \tag{14.84}$$

where $T_E = iT$ is the Euclidean time coordinate. From Fig. 14.1 it is obvious that this path integral could also be evaluated in the polar coordinates by varying the angle $\theta = \kappa t_E$ from 0 to π. When $\theta = 0$ the field configuration corresponds to $\phi = \phi_R$ and when $\theta = \pi$ the field configuration corresponds to $\phi = \phi_L$. Therefore Eq. (14.84) can also be expressed as

$$\langle \text{vac} | \phi_L, \phi_R \rangle \propto \int_{\kappa t_E = 0; \phi = \phi_R}^{\kappa t_E = \pi; \phi = \phi_L} \mathcal{D}\phi \, e^{-A}. \tag{14.85}$$

But in the Heisenberg picture, 'rotating' from $\kappa t_E = 0$ to $\kappa t_E = \pi$ is a time evolution governed by the Rindler Hamiltonian H_R. So the path integral Eq. (14.85) can be represented as a matrix element of the Rindler Hamiltonian H_R giving us the result

$$\langle \text{vac} | \phi_L, \phi_R \rangle \propto \int_{\kappa t_E = 0; \phi = \phi_R}^{\kappa t_E = \pi; \phi = \phi_L} \mathcal{D}\phi \, e^{-A} = \langle \phi_L | e^{-(\pi/\kappa) H_R} | \phi_R \rangle, \tag{14.86}$$

proving Eq. (14.81). Normalizing the result properly gives Eq. (14.83).

From this result, we can show that for operators $\mathcal{O}$ made out of variables having support in $\mathcal{R}$, the vacuum expectation values $\langle \text{vac}|\,\mathcal{O}(\phi_R)|\text{vac}\rangle$ become thermal expectation values. This arises from the straightforward algebra of inserting a complete set of states appropriately:

$$\langle \text{vac}|\,\mathcal{O}(\phi_R)|\text{vac}\rangle = \sum_{\phi_L} \sum_{\phi_R^{(1)},\phi_R^{(2)}} \langle \text{vac}|\phi_L,\phi_R^{(1)}\rangle\langle\phi_R^{(1)}|\mathcal{O}(\phi_R)|\phi_R^{(2)}\rangle\langle\phi_R^{(2)},\phi_L|\text{vac}\rangle$$

$$= \sum_{\phi_L} \sum_{\phi_R^{(1)},\phi_R^{(2)}} \frac{\langle\phi_L|e^{-(\pi/\kappa)H_R}|\phi_R^{(1)}\rangle\langle\phi_R^{(1)}|\mathcal{O}|\phi_R^{(2)}\rangle\langle\phi_R^{(2)}|e^{-(\pi/\kappa)H_R}|\phi_L\rangle}{Tr(e^{-2\pi H_R/\kappa})}$$

$$= \frac{Tr(e^{-2\pi H_R/\kappa}\mathcal{O})}{Tr(e^{-2\pi H_R/\kappa})}. \tag{14.87}$$

Thus, tracing over the field configuration ϕ_L in the region behind the horizon leads to a thermal density matrix $\rho \propto \exp[-(2\pi/\kappa)H]$ for the observables in $\mathcal{R}$. This is identical to the result obtained earlier in Eq. (14.71) but clearly identifies the relevant Hamiltonian to be H_R.

The main ingredients that have gone into this result are the following. (i) The singular behaviour of the (t, x) coordinate system near $x = 0$ divides the $T = 0$ hypersurface into two separate regions. (ii) In terms of *real* (t, x) coordinates, it is not possible to distinguish between the points (T, X) and $(-T, -X)$ but the *complex* transformation $t \to t \pm i\pi$ maps the point (T, X) to the point $(-T, -X)$. That is, a rotation in the complex plane (Re t, Im t) encodes the information contained in the full $T = 0$ plane.

The results in Eq. (14.83) and Eq. (14.87) are completely general and we have *not* assumed any specific Lagrangian for the field. In particular, we have not assumed that we are working with a free, non-interacting, quantum field – unlike in the previous section. This shows the generality of the thermal behaviour produced by the existence of the horizon.

It is, however, possible to give a more explicit demonstration of this result for any *free* field theory in static spacetimes with a bifurcation horizon. (In particular, we will discuss $(1 + 3)$-dimensional Rindler coordinates in flat spacetime.) To do this, we begin by noting that in any spacetime, with a metric which is independent of the time coordinate and $g_{0\alpha} = 0$, the wave equation for a massive scalar field $(\Box - m^2)\phi = 0$ can be separated in the form $\phi(t, \boldsymbol{x}) = \psi_\omega(\boldsymbol{x})e^{-i\omega t}$ with the modes $\psi_\omega(\boldsymbol{x})$ satisfying the equation

$$\frac{|g_{00}|}{\sqrt{-g}}\partial_\alpha(\sqrt{-g}\,g^{\alpha\beta}\partial_\beta\psi_\omega) - m^2|g_{00}|\psi_\omega = -\omega^2\psi_\omega. \tag{14.88}$$

The normalization of the modes ψ_ω may be chosen using the conserved scalar product:

$$(\psi_\omega, \psi_\nu) \equiv \int d^3x \sqrt{-g} |g^{00}| \psi_\omega \psi_\nu^* = \delta(\omega - \nu). \tag{14.89}$$

Using this relation in the field equation, it can easily be deduced that

$$\int d^3x \sqrt{-g} \partial_\alpha \psi_\omega^* \partial^\alpha \psi_\nu = \omega^2 \delta_{\omega\nu}. \tag{14.90}$$

Expanding the field as $\phi(t, x) = \sum_\omega q_\omega(t) \psi_\omega(x)$ and substituting into the free field action, we find that the action again reduces to that of a sum of harmonic oscillators:

$$A = -\frac{1}{2} \int \sqrt{-g}\, dt\, d^3x\, (\partial_a \phi \partial^a \phi + m^2 \phi^2) = \frac{1}{2} \sum_\omega \int dt \left[|\dot{q}_\omega|^2 - (\omega^2 + m^2)|q_\omega|^2 \right]. \tag{14.91}$$

Since this result is applicable in any static metric, we can decompose the field into oscillators either in terms of the (T, X) coordinates or in terms of the two sets of (t, x) coordinates on $\mathcal{R}$ and $\mathcal{L}$.

So, on the $T = 0$ surface, we expand the field in terms of a set of mode functions $F_\Omega(X, X_\perp)$ with coefficients Q_Ω; that is, $\phi = \sum_\Omega Q_\Omega F_\Omega(X, X_\perp)$. Similarly, the field can be expanded in terms of a set of modes in $\mathcal{R}$ and $\mathcal{L}$:

$$\phi(X > 0, X_\perp) = \sum_\omega a_\omega f_\omega(X, X_\perp); \quad \phi(X < 0, X_\perp) = \sum_\omega b_\omega g_\omega(X, X_\perp). \tag{14.92}$$

The functional integral in Eq. (14.84) now reduces to the product over a set of independent harmonic oscillators and thus the ground state wave functional can again be expressed in the form

$$\Psi[Q] = \langle \text{vac} | \phi(X) \rangle = \prod_\Omega \langle \text{vac} | Q_\Omega \rangle \propto \exp\left[-\sum_\Omega A_E(T_E = \infty, 0; T_E = 0, Q_\Omega) \right], \tag{14.93}$$

where A_E is the Euclidean action with the boundary conditions as indicated. On the other hand, we have shown that this ground state functional is the same as $\langle \phi_R, \kappa t_E = \pi | \phi_L, \kappa t_E = 0 \rangle$. Hence

$$\Psi[a, b] = \langle \text{vac} | \phi(X) \rangle \propto \exp\left[-\sum_\omega A_E(\kappa t_E = \pi, a_\omega; \kappa t_E = 0, b_\omega) \right]. \tag{14.94}$$

The Euclidean action for a harmonic oscillator q with boundary conditions $q = q_1$ at $t_E = 0$ and $q = q_2$ at $t_E = \beta$ is given by (see Exercise 14.1)

$$A_E(q_1, 0; q_2, \beta) = \frac{\omega}{2} \left[\frac{\cosh \omega\beta}{\sinh \omega\beta}(q_1^2 + q_2^2) - \frac{2q_1 q_2}{\sinh \omega\beta} \right]. \tag{14.95}$$

Equation (14.93) corresponds to $\beta = \infty, q_2 = 0, q_1 = Q_\Omega$ giving $[A_E(T_E = \infty, 0; T_E = 0, Q_\Omega)] = (\Omega/2)Q_\Omega^2$ leading to the standard ground state wave functional. The more interesting one is, of course, the one in Eq. (14.94) corresponding to $\beta = (\pi/\kappa), q_1 = a_\omega, q_2 = b_\omega$. This gives

$$A_E(a_\omega, 0; b_\omega, (\pi/\kappa)) = \frac{\omega}{2} \left[\frac{\cosh(\pi\omega/\kappa)}{\sinh(\pi\omega/\kappa)}(a_\omega^2 + b_\omega^2) - \frac{2a_\omega b_\omega}{\sinh(\pi\omega/\kappa)} \right]. \quad (14.96)$$

An observer confined to $\mathcal{R}$ will again have observables made out of $a_\omega s$ alone. Let $\mathcal{O}(a_\omega)$ be any such observable. The expectation value of $\mathcal{O}$ in the state Ψ is given by

$$\langle\mathcal{O}\rangle = \int \prod_\omega da_\omega \int \prod_\omega db_\omega \Psi^*(a_\omega, b_\omega)\mathcal{O}\Psi(a_\omega, b_\omega)$$

$$\equiv \int \prod_\omega da_\omega \rho(a_\omega, a_\omega)\mathcal{O}(a_\omega) = \text{Tr}(\rho\mathcal{O}), \quad (14.97)$$

where

$$\rho(a'_\omega, a_\omega) \equiv \int \prod_\omega db_\omega \Psi^*(a'_\omega, b_\omega)\Psi(a_\omega, b_\omega)$$

$$= C \exp - \sum_\omega \left\{ \frac{\omega}{2} \left[\frac{\cosh(2\pi\omega/\kappa)}{\sinh(2\pi\omega/\kappa)}(a_\omega^2 + a_\omega'^2) - \frac{2a_\omega a'_\omega}{\sinh(2\pi\omega/\kappa)} \right] \right\}$$

$$(14.98)$$

is a thermal density matrix corresponding to the temperature $T = (\kappa/2\pi)$ (see Eq. (14.39)).

The fact that the exponential in the density matrix in Eq. (14.98) is very similar to that in Eq. (14.96), with π replaced by 2π, is noteworthy and this result can be obtained more directly from an alternative argument. The matrix element of ρ can be expressed as the integral

$$\langle\phi'_R|\rho|\phi''_R\rangle = \int \mathcal{D}\phi_L\langle\phi_L\phi'_R|\text{vac}\rangle\langle\text{vac}|\phi_L\phi''_R\rangle. \quad (14.99)$$

Each of the two terms in the integrand can be expressed in terms of A_E using Eq. (14.85). In one of them, we shall take $\kappa t_E = \epsilon$ (with ϵ being infinitesimal and positive) at the lower limit of the integral and in the other, we will take $\kappa t_E = -\epsilon$ at the lower limit of the integral. Hence the product which occurs in the integrand of Eq. (14.99) can be thought of as evolving the field from a configuration ϕ''_R at $\kappa t_E = +\epsilon$ to a configuration ϕ'_R at $\kappa t_E = -\epsilon$ rotating in κt_E in the anti-clockwise direction from ϵ to $(2\pi - \epsilon)$. In the limit of $\epsilon \to 0$, this is same as evolving the system by the angle $\kappa t_E = 2\pi$. So we can set $\beta = (2\pi/\kappa), q_1 = a_\omega, q_2 = a'_\omega$ in Eq. (14.95) leading to Eq. (14.98). In arriving at Eq. (14.96) we have evolved the same system from $\kappa t_E = 0$ to $\kappa t_E = \pi$ in order to go from $x > 0$ to $x < 0$. This explains the correspondence between Eq. (14.98) and Eq. (14.96).

The essence of the results obtained in this section is the following. Suppose a metric, when analytically continued to imaginary values of time by $t \to it = \tau$, exhibits periodicity in τ with period β. Then we can associate a temperature $T = \beta^{-1}$ with this spacetime. We will need this feature in Chapters 15 and 16.

Exercise 14.4
Bogolyubov coefficients for (1+1) Rindler coordinates In the case of (1+1) dimension, the metric in inertial and Rindler coordinates can be expressed in the form

$$ds^2 = -dU\,dV = -e^{\kappa(v-u)}\,du\,dv, \tag{14.100}$$

where $U = T - X$, $V = T + X$, $u = t - x$, $v = t + x$ and $\kappa U = -e^{-\kappa u}$, $\kappa V = e^{\kappa v}$. Consider a massless scalar field in these two coordinates and write down the normalized mode functions exploiting the fact that they are plane waves in 1+1 spacetime. Expand the positive frequency mode in the inertial coordinate in terms of the positive and negative frequency modes of the Rindler coordinate by writing

$$\frac{1}{\sqrt{\omega}} e^{-i\omega U} = \int_0^\infty \frac{d\Omega}{\sqrt{\Omega}} \left(\alpha_{\Omega\omega} e^{-i\Omega u} - \beta^*_{\Omega\omega} e^{i\Omega u} \right). \tag{14.101}$$

Using

$$\int_{-\infty}^\infty e^{i(\Omega-\Omega')u}\,du = 2\pi\delta(\Omega-\Omega') \tag{14.102}$$

show that

$$\begin{Bmatrix} \alpha_{\Omega\omega} \\ \beta_{\Omega\omega} \end{Bmatrix} = \int_{-\infty}^\infty e^{\mp i\omega U + i\Omega u}\,du = \pm\frac{1}{2\pi}\sqrt{\frac{\Omega}{\omega}} \int_{-\infty}^0 (-aU)^{-\frac{i\Omega}{\kappa}-1} e^{\mp i\omega U}\,dU \tag{14.103}$$

$$= \pm\frac{1}{2\pi\kappa}\sqrt{\frac{\Omega}{\omega}} e^{\pm\frac{\pi\Omega}{2\kappa}} \exp\left(\frac{i\Omega}{\kappa} \ln\frac{\omega}{\kappa} \right) \Gamma\left(-\frac{i\Omega}{\kappa} \right). \tag{14.104}$$

Also show that

$$|\alpha_{\Omega\omega}|^2 = e^{2\pi\Omega/\kappa}|\beta_{\Omega\omega}|^2. \tag{14.105}$$

Using this, compute the mean number of Rindler particles in the inertial vacuum.

Exercise 14.5
Bogolyubov coefficients for (1+3) Rindler coordinates Consider a positive frequency mode in the inertial coordinates

$$\Phi(T, \mathbf{Z}) \propto \exp[-i\Omega T + iKX + i\mathbf{K}_\perp \cdot \mathbf{X}_\perp]; \quad (\Omega > 0) \tag{14.106}$$

with frequency Ω and wave vector $\mathbf{K}$, where $\Omega^2 = K^2 + K_\perp^2 + m^2$. Here m is the mass of the scalar field, K is the component of the wave vector along X and $\mathbf{K}_\perp$ is the wave vector in the transverse Y–Z plane. Define two variables μ, θ through the relations $\Omega = \mu\cosh\theta$; $K = \mu\sinh\theta$ and express this mode in the Rindler coordinates in the form

$\Phi = F(t, x) \exp[i\boldsymbol{K}_\perp \cdot \boldsymbol{X}_\perp]$. Let the Fourier transform of $F(t, x)$ with respect to t be defined through

$$F(t, x) = \int_{-\infty}^{\infty} \frac{d\omega}{2\pi} e^{-i\omega t} Q(\omega, \mu\, e^x); \qquad Q(\omega, \alpha) \equiv \int_0^{\infty} \frac{ds}{s} s^{i\omega} e^{-i(\alpha/2)[s-(1/s)]},$$

(14.107)

where we have set $\kappa = 1$ for simplicity.

(a) Show that

$$Q(-\omega, \alpha) = \begin{cases} e^{-\pi\omega} Q^*(\omega, \alpha) & \text{(for } \alpha > 0) \\ e^{\pi\omega} Q^*(\omega, \alpha) & \text{(for } \alpha < 0) \end{cases}$$

(14.108)

allowing one to write

$$F(t, x) = \int_0^{\infty} \frac{d\omega}{2\pi} [Q(\omega, \mu x) e^{-i\omega t} + e^{-\pi\omega} Q^*(\omega, \mu x) e^{i\omega t}].$$

(14.109)

(b) Write the field mode in the inertial coordinates, containing creation and annihilation operators $(A, A^\dagger)$, in terms of the creation and annihilation operators $(a, a^\dagger)$ appropriate to the (t, x) coordinates as

$$AF + A^\dagger F^* = \int_0^{\infty} \frac{d\omega}{2\pi} \left[(A + A^\dagger e^{-\pi\omega}) Q e^{-i\omega t} + \text{h.c.} \right] = \int_0^{\infty} \frac{d\omega}{2\pi} N \left[aQe^{-i\omega t} + \text{h.c.} \right],$$

(14.110)

where N is a normalization constant. Using the condition $[a, a^\dagger] = 1$, show that $N = [1 - \exp(-2\pi\omega)]^{-1/2}$. Hence show that the number operator corresponding to Rindler mode has a thermal expectation value in the inertial vacuum state:

$$\langle \text{vac} | a^\dagger a | \text{vac} \rangle = N^2 e^{-2\pi\omega} = (e^{2\pi\omega} - 1)^{-1}.$$

(14.111)

(c) Expand the mode functions $\phi(T, \boldsymbol{X})$ in R and L in terms of mode functions corresponding to the Rindler coordinate with creation and annihilation operators $a_1, a_1^\dagger$ and $a_2, a_2^\dagger$ obtaining

$$\phi(\boldsymbol{Z}, T) = \sum_\omega \left(\phi_\omega(\boldsymbol{x}_\perp, x) a_1(\omega) e^{-i\omega\tau} + \text{h.c.} \right) \quad (x > 0)$$

$$\phi(\boldsymbol{Z}, T) = \sum_\omega \left(\phi_\omega(\boldsymbol{x}_\perp, -x) a_2(\omega) e^{+i\omega\tau} + \text{h.c.} \right) \quad (x < 0).$$

(14.112)

Let $|n, n\rangle$ be a state with n Rindler particles in the left wedge and n Rindler particles in the right wedge. Expand the vacuum state as the sum $|\text{vac}\rangle = \sum_n C_n |n, n\rangle$ and show that the coefficient C_n must satisfy the recursion relation $C_{n+1} = e^{-\omega\pi} C_n$. Hence show that

$$|\text{vac}\rangle = C \int ds\, |s\rangle_L |s\rangle_R e^{-\pi n\omega_s},$$

(14.113)

where $|s\rangle_R$ is the eigenstate of the Hamiltonian in R, $|s\rangle_L$ is the eigenstate of the Hamiltonian in L and C is a normalization constant. Interpret this result.

Exercise 14.6

Rindler vacuum and the analyticity of modes In the previous problem, one notices that a particular combination $A + e^{-\pi\omega} A^\dagger$ emerges quite naturally in Eq. (14.110). There is a simple and elegant way of understanding this result in terms of analytic properties of mode function. To do this, consider (for simplicity) the $(1+1)$ spacetime with inertial coordinates (T, X) and the Rindler coordinates (t, x) and (t', x') in $\mathcal{R}$ and $\mathcal{L}$ related through Eq. (14.58) and Eq. (14.59). For simplifying the notation, we shall just use (t, x) as coordinates in both $\mathcal{R}$ and $\mathcal{L}$ with the understanding that they are related to inertial coordinates differently depending on the region.

(a) Convince yourself that $g_k^{(1)} = (4\pi\omega)^{-1/2} e^{-ik(t-x)}$ is the positive frequency mode in $\mathcal{R}$ and $g_k^{(2)} = (4\pi\omega)^{-1/2} e^{+ik(t+x)}$ is the positive frequency mode in $\mathcal{L}$.

(b) Express $g_k^{(1)}$ in terms of the inertial coordinates T, X obtaining

$$\sqrt{4\pi\omega}\, g_k^{(1)} = \kappa^{i\omega/\kappa}(-T + X)^{i\omega/\kappa}. \tag{14.114}$$

Show that the analytic continuation of the right hand side is straightforward but it does not match with $g_k^{(2)}$ in $\mathcal{L}$. On the other hand, show that the combination

$$h_k^{(1)} = \frac{1}{\sqrt{2\sinh(\pi\omega/a)}}\left(e^{\pi\omega/2a} g_k^{(1)} + e^{-\pi\omega/2a} g_{-k}^{(2)*}\right) \tag{14.115}$$

is well-defined along the whole $T = 0$ surface.

(c) We know that the positive frequency modes in the inertial frame $\exp[-i\omega(T - X)]$ is analytic and bounded for complex $(T - X)$ as long as $\text{Im}\,(T - X) \leq 0$. Show that this also holds for $h_k^{(1)}$ with a proper branch cut to define the imaginary power. Argue from this result that $h_k^{(1)}$ can be expressed as a linear combination of purely positive frequency inertial modes and hence they will lead to the standard inertial vacuum.

(d) Complete the analysis by using the modes $h_k^{(1)}$ and the resulting Bogolyubov transformation to obtain the number density of Rindler particles in the inertial vacuum.

Exercise 14.7
Response of an accelerated detector This exercise explores a closely related, but conceptually distinct, effect, viz. that a detector travelling along a uniformly accelerated trajectory in the Minkowski vacuum will behave as though it is at rest inside a thermal bath at temperature $\kappa/2\pi$. This result is fairly straightforward to obtain along the following lines. A system made of a scalar field and a detector can be described by a Hamiltonian $H = H_{\text{field}}(\Phi) + H_{\text{det}} + H_{\text{int}}$, where H_{field} is the Hamiltonian for a free massless scalar field, H_{det} is the Hamiltonian describing the internal structure of the detector and H_{int} is the interaction Hamiltonian which couples the field and the detector. We will assume that H_{det} leads to definite internal energy levels for the detector and we will concentrate on two of those levels, viz. a ground state with energy E_0 and an excited state with energy E_1. The coupling term is assumed to be given by

$$H_{\text{int}} = \mu(\tau)\,\Phi\,[x(\tau)], \tag{14.116}$$

where $\mu(\tau)$ is an observable corresponding to the detector which evolves with respect to the proper time τ of the detector and couples to the scalar field Φ locally. The $x^i(\tau)$ is the worldline of the detector. (One can obtain an intuitive physical picture by thinking of the scalar field as analogous to the electromagnetic radiation field, H_{det} as the Hamiltonian of an atom with discrete energy levels and the interaction term as analogous to the coupling between the current J^i and the vector potential A_i. This allows the radiation field to cause excitation of the electron in the atom from the ground state to an excited state, which could be thought of as the detection of the photon by an atom acting as the detector.)

(a) Let the initial state of the field be the inertial vacuum $|0\rangle$ and the initial state of the detector be $|E_0\rangle$. Because of the coupling, let us assume that the field makes a transition to some state $|\psi\rangle$ while the detector makes the transition to the excited state $|E_1\rangle$. Show that the probability for this event, to the lowest order in the perturbation theory, is given by

$$P(E) = |\mathcal{M}|^2 \int\limits_{-\infty}^{\infty} d\tau \int\limits_{-\infty}^{\infty} d\tau' \, e^{-iE(\tau-\tau')} \, G\left[x(\tau), x(\tau')\right], \tag{14.117}$$

where $|\mathcal{M}|^2 \equiv |\langle E_1|\mu(0)|E_0\rangle|^2$, $E \equiv E_1 - E_0$ and

$$G\left[x(\tau), x(\tau')\right] = \langle 0|\Phi\left[x(\tau)\right] \Phi\left[x(\tau')\right]|0\rangle. \tag{14.118}$$

This probability can be evaluated for any given trajectory $x^i(\tau)$ of the detector.

(b) Show that, in the case of a uniformly accelerated trajectory with an acceleration κ, $G[x(\tau), x(\tau')]$ is only a function of $\Delta\tau \equiv \tau - \tau'$ and is given by

$$G(\Delta\tau) = -\frac{1}{4\pi^2} \frac{(\kappa/2)^2}{\sinh^2[\kappa(\Delta\tau - i\epsilon)/2]}. \tag{14.119}$$

(c) In this case, we can remove one of the integrals in Eq. (14.117) leading to the rate of transitions so that the transition probability *rate* for the detector becomes

$$\mathcal{R}(E) = \int\limits_{-\infty}^{\infty} d\Delta\tau \, e^{-iE\Delta\tau} \, G(\Delta\tau). \tag{14.120}$$

Show that, in the limit of $\kappa \to 0$, we get $\mathcal{R} \propto \theta(-E)$ so that the detector does not get excited when $E > 0$. On the other hand, show that when $\kappa \neq 0$ we get a rate of excitation which is proportional to

$$\mathcal{R}(E) \propto \left(\frac{E}{e^{2\pi E\kappa^{-1}} - 1}\right). \tag{14.121}$$

This is precisely what one would have found in the case of a detector immersed in a thermal bath with temperature $\kappa/2\pi$.

14.6 Hawking radiation from black holes

The description in the previous sections shows that the vacuum state defined in a coordinate system – which covers the full manifold – appears as a thermal state to an observer who is confined to part of the manifold partitioned by a horizon. This result (and the analysis) will hold for any static spacetime with a bifurcation horizon, like the Schwarzschild spacetime, de Sitter spacetime, etc. All these cases describe a situation in thermal *equilibrium* at a temperature $T = \kappa/2\pi$ (where κ is the surface gravity of the horizon) as far as an observer confined to the region R is concerned.

A completely different phenomenon arises in the case of a dynamical situation like the one in which a spherically symmetric massive body, say, collapses to form a black hole. In this case, time reversal invariance is explicitly broken. The study of a quantum field theory in such a context shows that, at late times, there will be a flux of radiation flowing towards the future null infinity with a Planckian spectrum corresponding to a temperature $\kappa/2\pi$. This process is called *black hole evaporation*.

We stress that this result is conceptually different from associating a temperature with the horizon. In the case of a Rindler spacetime, for example, there is no steady flux of radiation propagating towards future null infinity even though an observer confined to the region $\mathcal{R}$ will interpret the expectation values of operators as thermal averages corresponding to a temperature $\kappa/2\pi$. This corresponds to a situation which *is* time reversal invariant characterized by thermal equilibrium. The black hole evaporation, in contrast, is an irreversible process.

To understand how this arises in the case of a time dependent metric, we shall first consider the quantum field theory in the coordinate system suitable for an observer travelling with a *variable* acceleration (see Project 3.4). The transformation from the flat inertial coordinates $(T, \boldsymbol{X})$ to the proper coordinates $(t, \boldsymbol{x})$ of an observer with variable acceleration is effected by $Y = y, Z = z$ and

$$X = \int' \sinh \chi(t) dt + x \cosh \chi(t); \quad T = \int' \cosh \chi(t) dt + x \sinh \chi(t),$$
(14.122)

where the function $\chi(t)$ is related to the time dependent acceleration $g(t)$ by $g(t) = (d\chi/dt)$. The form of the metric in the accelerated frame is remarkably simple:

$$ds^2 = -(1 + g(t)x)^2 dt^2 + dx^2 + dy^2 + dz^2.$$
(14.123)

We will treat $g(t)$ as an arbitrary function except for the limiting behaviour $g(t) \to 0$ for $t \to -\infty$ and $g(t) \to g_0 = \text{constant}$ for $t \to +\infty$. Hence, at early times, the line elements in Eq. (14.123) represent the standard inertial coordinates and the positive frequency modes $\exp(-i\omega t)$ define the standard Minkowski vacuum. At late times, the metric goes over to the Rindler coordinates and we are interested in knowing how the initial vacuum state will be interpreted at late times. The wave equation $(\Box - m^2)\phi = 0$ for a massive scalar field can be separated in the transverse coordinates as $\phi(t, x, y, z) = f(t, x)e^{ik_y y}e^{ik_z z}$, where f satisfies the equation

$$-\frac{1}{(1 + g(t)x)}\frac{\partial}{\partial t}\left(\frac{1}{(1 + g(t)x)}\frac{\partial f}{\partial x}\right) = M^2 f$$
(14.124)

with $M^2 \equiv m^2 + k_y^2 + k_z^2$. It is possible to solve this partial differential equation with the ansatz

$$f(x, t) = \exp i \left(\int \alpha(t) dt + \beta(t)x\right),$$
(14.125)

where α and β satisfy the equations $\alpha^2(t) - \beta^2(t) = M^2$; $\dot{\beta} = g(t)\alpha$; $\dot{\alpha} = g(t)\beta$; these are solved uniquely in terms of $\chi(t)$ to give $\alpha(t) = M \cosh[\chi(t) - \eta]$; $\beta(t) = M \sinh[\chi(t) - \eta]$, where η is another constant. The final solution for the mode labelled by $(\boldsymbol{k}_\perp, \eta)$ is now given by

$$f_{k_y k_z \eta}(x, t) = N \exp -iM \left[\int \cosh(\chi - \eta) dt + x \sinh(\chi - \eta) \right], \quad (14.126)$$

where N is suitable normalization factor. For the limiting behaviour we have assumed for $g(t)$, we see that $\chi(t)$ vanishes at early times and varies as $\chi(t) \approx (g_0 t + \text{constant})$ at late times. Correspondingly, the mode f will behave as

$$f(x, t) \rightarrow N \exp -iM \left[t \cosh \eta - x \sinh \eta \right] \quad (14.127)$$

at early times $(t \rightarrow -\infty)$ which is just the standard Minkowski positive frequency mode with $\omega = M \cosh \eta$, $k_x = M \sinh \eta$. (We also note from this limiting behaviour that $N = (2\omega)^{-1/2}$ though this is irrelevant for our purpose.) Let us now consider the late time behaviour of the mode which is given by

$$f(x, t) \rightarrow \exp -i \left[(M/2g_0)(1 + g_0 x) e^{g_0 t} \right]. \quad (14.128)$$

On the other hand, the observer moving with a uniform acceleration g_0 at late times will use as positive frequency modes those which evolve as $\exp(-i\nu t)$. It is clear that the mode in Eq. (14.128) is not a purely positive frequency mode with respect to the time coordinate t which the uniformly accelerated observer uses at late times. One can now decompose the mode in Eq. (14.128) in terms of positive and negative frequency components with two Bogolyubov coefficients (α, β). The $|\beta^2|$ will then give the particle content in this quantum state (which was a vacuum state for the observer in the asymptotic past) in the asymptotic future. The Bogolyubov transformations between the mode in Eq. (14.128) and modes which vary as $\exp(-i\nu t)$ will involve exactly the same mathematics as in Eq. (14.51). We will get a thermal spectrum at late times. (Though we went through the algebra in some detail, the corresponding result in (1+1) dimension can be obtained very simply by transforming a mode, $\exp(-i\Omega U)$, to the coordinates used by an observer moving along an arbitrary trajectory given by Eq. (3.89). Assuming that $\chi \rightarrow 0$ at very early times, and $\chi \rightarrow g_0 t$ at late times, we immediately get the required result.)

While this example illustrates the basic idea that is involved, it cannot be considered as describing genuine particle production since the spacetime is actually flat. We shall now work out the corresponding result for a scalar field in the time dependent metric generated by collapsing matter. The scalar field can be decomposed into positive and negative frequency modes in the usual manner. We choose these modes in such a way that at early times they correspond to a vacuum state. In the presence of collapsing matter, these modes evolve at late times to those with exponential redshift, thereby leading to thermal behaviour.

To work this out, we need an explicit model for the collapsing matter. Since only the exponential redshift of the modes at late times is relevant as far as the thermal spectrum is concerned, the result should be independent of the detailed nature of the collapsing matter. So we we shall choose a simple model for the formation of the black hole, based on a spherical distribution of mass M that collapses under its own weight. Further, the angular coordinates do not play a significant role in this analysis, allowing us to work in the two-dimensional (t, r) subspace.

The line element outside and inside the collapsing, spherically symmetric, distribution of matter is taken to be

$$ds^2 = \begin{cases} -C(r)\,du\,dv & \text{(exterior)} \\ -B(U, V)\,dU\,dV & \text{(interior)} \end{cases} \tag{14.129}$$

where

$$u = t - \xi + R_0^*; \quad v = t + \xi - R_0^*; \quad \xi = \int dr\, C^{-1}; \tag{14.130}$$

$$U = \tau - r + R_0, \qquad V = \tau + r - R_0. \tag{14.131}$$

The R_0 and R_0^* are constants related in the same manner as r and ξ; the outside metric, of course, is just the Schwarzschild metric with $C = (1 - 2M/r)$ and ξ being the tortoise coordinate.

Let us assume that, for $\tau \leq 0$, matter was at rest with its surface at $r = R_0$ and for $\tau > 0$, it collapses inward along the trajectory $r = R(\tau)$. The coordinates have been chosen so that at the onset of collapse ($\tau = t = 0$) we have $u = U = v = V = 0$ at the surface. Let the coordinate transformations between the interior and exterior be given by the functional forms $U = f(u)$ and $v = h(V)$. Matching the geometry along the trajectory $r = R(\tau)$, it is easy to show that

$$\frac{dU}{du} = \frac{df}{du} = (1 - \dot{R})C \left(\left[BC(1 - \dot{R}^2) + \dot{R}^2 \right]^{1/2} - \dot{R} \right)^{-1} \tag{14.132}$$

$$\frac{dv}{dV} = \frac{dh}{dV} = \frac{1}{C(1 + \dot{R})} \left(\left[BC(1 - \dot{R}^2) + \dot{R}^2 \right]^{1/2} + \dot{R} \right) \tag{14.133}$$

where $\dot{R}$ denotes $dR/d\tau$ and U, V and C are evaluated along $r = R(\tau)$. Also note that, since $\dot{R} < 0$ for the collapsing shell, we should take $(\dot{R}^2)^{1/2} = -\dot{R}$.

We now introduce a massless scalar field in this spacetime that satisfies the equation $\Box \phi = 0$. As the modes of the scalar field propagate inwards they will reach $r = 0$ and re-emerge as out-going modes. In the (t, r) plane, this requires reflection of the modes on the $r = 0$ line, which corresponds to $V = U - 2R_0$. Since the modes vanish for $r < 0$, continuity requires $\phi = 0$ at $r = 0$. The solutions to the two-dimensional wave equations $\Box \phi = 0$ which (i) vanish on the line

$V = U - 2R_0$ and (ii) reduce to the standard exponential form in the remote past, can be determined by noting that, along $r = 0$, we have

$$v = h(V) = h[U - 2R_0] = h[f(u) - 2R_0], \qquad (14.134)$$

where the square bracket denotes functional dependence. Hence the solution is

$$\Phi = \frac{i}{\sqrt{4\pi\omega}} \left(e^{-i\omega v} - e^{-i\omega h[f(u) - 2R_0]} \right). \qquad (14.135)$$

Given the trajectory $R(\tau)$, one can integrate Eq. (14.132) to obtain $f(u)$ and use Eq. (14.135) to completely solve the problem. This will describe time dependent particle production from some collapsing matter distribution and – in general – the results will depend on the details of the collapse.

The analysis, however, simplifies considerably and a universal character emerges if the collapse proceeds to form a horizon on which $C \to 0$. Near $C = 0$, equations (14.132) and (14.133) simplify to

$$\frac{dU}{du} \approx \frac{\dot{R} - 1}{2\dot{R}} C(R); \quad \frac{dv}{dV} \approx \frac{B(1 - \dot{R})}{2\dot{R}}, \qquad (14.136)$$

where we have used the fact that $(\dot{R}^2)^{1/2} = -\dot{R}$ for the collapsing solution. Further, near $C = 0$, we can expand $R(\tau)$ as $R(\tau) = R_h + \nu(\tau_h - \tau) + \mathcal{O}[(\tau_h - \tau)^2]$, where $R = R_h$ at the horizon and $\nu = -\dot{R}(\tau_h)$. Integrating Eq. (14.136) treating B as approximately constant, we get

$$\kappa u \approx -\ln |U + R_h - R_0 - \tau_h| + \text{const}, \qquad (14.137)$$

where $\kappa = (1/2)(\partial C/\partial r)_{R_h}$ is the surface gravity and

$$v \approx \text{constant} - BV(1 + \nu)/2\nu. \qquad (14.138)$$

It is now clear that: (i) the relation between v and V is linear and hence holds no surprises; it also depends on B, and (ii) the relation between U and u, which can be written as $U \propto \exp(-\kappa u)$, is universal (independent of B) and signifies the exponential redshift we have alluded to several times. The late time behaviour of out-going modes can now be determined using Eq. (14.137) and Eq. (14.138) in Eq. (14.135). We get:

$$\Phi \cong \frac{i}{\sqrt{4\pi\omega}} \left(e^{-i\omega v} - \exp \left(i\omega \left[ce^{-\kappa u} + d \right] \right) \right) \qquad (14.139)$$

where c, d are constants. This mode with exponential redshift, when expressed in terms of $\exp(-i\nu u)$ as

$$\Phi_\omega(u) = \int_0^\infty \frac{d\nu}{2\pi} \left[\alpha_{\omega\nu} e^{-i\nu u} + \beta_{\omega\nu} e^{i\nu u} \right] \qquad (14.140)$$

will lead to a thermal distribution of particles with temperature $T = \kappa/2\pi$. For the case of a black hole, if we take $\kappa = 1/4M$, then we get

$$\alpha_{\omega\nu} = \frac{-2iM\nu e^{-i\omega d}e^{i\nu t_0}}{2\pi\sqrt{\nu\omega}}\left(\frac{-e^{-(t_0+d)/4M}}{\omega c}\right)^{-4iM\nu}e^{2M\pi\nu}\Gamma(-4iM\nu) \tag{14.141}$$

and $\beta_{\omega\nu} = e^{-4M\pi\nu}\alpha^*_{\omega\nu}$. (The relevant Fourier transforms can be worked out by using the results of Exercise 14.2.) Note that these quantities *do* depend on c, d, t_0, etc; but the modulus

$$|\beta_{\omega\nu}|^2 = \frac{1}{2}\frac{4M}{[\exp(8\pi M\nu) - 1)]} \tag{14.142}$$

is independent of these factors. The mathematics is essentially the same as in Eqs. (14.50) and (14.51). This shows that the vacuum state at early times will be interpreted as containing a thermal spectrum of particles at late times. This particular result has several intriguing features that we will comment on. (We will discuss some further aspects of this in Chapter 16.)

Our result implies that when a spherically symmetric configuration of matter collapses to form a black hole, observers at large distances will receive thermal radiation of particles from the black hole at late times. (It is possible to prove this more formally by considering the expectation values of the energy-momentum tensor of the scalar field; this will demonstrate the flux of energy to large distances). In Chapter 8 we saw that – in the case of rotating black holes – certain modes of a propagating wave can acquire an enhancement of energy, which we called super-radiance. If we think of the super-radiance as stimulated emission of radiation by the black hole in some modes, owing to the presence of the incoming wave, it seems natural to expect spontaneous emission of radiation in various modes by the black hole in quantum theory. The black hole evaporation can be thought of as spontaneous emission of particles that survives even in the limit of zero angular momentum of the black hole.

It seems natural to assume that the source of this energy radiated to infinity is the mass of the collapsing structure. Strictly speaking, this is an extrapolation from our result because it involves changes in the background metric – which was parametrized by M – owing to the effect of the radiation, while our original result was based on a *test* scalar field in a *fixed* background metric. We shall, nevertheless, make this assumption and explore its consequences. Given the temperature of the black hole $T(E) = 1/8\pi M$ as a function of the energy $E = M$, we can integrate the expression $dS = dE/T(E)$ to define an 'entropy' $S(E)$ for the black hole:

$$S = \int_0^M (8\pi E)dE = 4\pi M^2 = \frac{1}{4}\left(\frac{A_{hor}}{L_P^2}\right), \tag{14.143}$$

where $A_{hor} = 4\pi(2M)^2$ is the area of the $r = 2M, t = $ constant surface and $L_P^2 = G\hbar/c^3$ is the square of the Planck length. This result shows that the entropy obtained by integrating the $T(E)$ is proportional to the area of the horizon.

This result connects up with several classical results about black holes that we discussed in Chapter 8. First, we know that the area of the horizon does not decrease during classical processes involving the black holes, which suggests an analogy between horizon area and entropy. We saw examples of this in Chapter 8 but this result can be proved in complete generality in classical theory allowing us to interpret it as the second law of thermodynamics involving black holes. Second, Eq. (8.99) obtained in Chapter 8 – which becomes, in the case of the Schwarzschild metric the relation $\delta M = (\kappa/8\pi)\delta A$ – now acquires a direct thermodynamic interpretation. The factor $(\kappa/2\pi)$ and $A/4$ can indeed be identified with physical temperature and entropy. Note that the classical analysis in Chapter 8 can identify these quantities only up to a multiplicative factor. If we redefine the temperature as $c_1(\kappa/2\pi)$ and entropy as $(1/4c_1)A$, with some positive constant c_1, the product TdS remains the same. On the other hand, the analysis of quantum fields in the Schwarzschild metric allows us to determine the temperature and entropy uniquely. We now know that the entropy is one-quarter of the area of the horizon expressed in Planck units. (There is also an ambiguity in the overall additive constant to entropy which we have settled in Eq. (14.143) by taking $S = 0$ when $M = 0$. This might appear reasonable but recall that $T \to \infty$ when $M \to 0$; flat spacetime, treated as the $M \to 0$ limit of the Schwarzschild metric, has infinite temperature rather than zero temperature. Hence, it is worth emphasizing that this is a specific assumption.). Another result, obtained in Chapter 8 – viz. that the surface gravity remains constant over the horizon – can now be interpreted as the zeroth law of black hole thermodynamics.

The association of a temperature with a horizon is completely straightforward and well founded. As we saw earlier, this result arises even in the case of a Rindler horizon in flat spacetime. Extrapolating further and attributing an entropy to the horizon (through, for example, Eq. (14.143)) seems reasonable but raises several intriguing new questions. Should we attribute an entropy density – i.e. entropy per unit area – to all horizons including the Rindler horizon? What are the degrees of freedom associated with this entropy? Is it a property of the geometry or is it a property of the quantum fields that live on the geometry? These and related questions are still somewhat open and we will discuss some of them and their wider implications in Chapter 16.

Exercise 14.8

Horizon entropy and the surface term in the action The entropy of the horizon obtained by using quantum field theory in a background geometry can be provided with an intrinsic meaning in terms of the boundary term in the action functional (which we have completely ignored in arriving at the field equations!). This exercise derives it in the simplest context and we will discuss this aspect in detail in Chapters 15 and 16. Consider a spacetime in which the metric near the horizon can be approximated in the Rindler form

$$ds^2 = -\kappa^2 \xi^2 dt^2 + d\xi^2 + dx_\perp^2. \tag{14.144}$$

We want to evaluate the surface term in the action obtained from L_{sur} in Eq. (6.14) on the horizon in the *Euclidean sector* obtained by setting $t = i\tau$. (Note that the horizon maps to the origin of the t–x plane in the Euclidean sector.) To do this, evaluate the integral over the surface $\xi = \epsilon$ and then take the limit $\epsilon \to 0$. Show that the Euclidean surface action is equal to $(1/4)A_\perp$, where $A_\perp$ is the transverse area. [Answer. We need to integrate $\sqrt{h} n_c V^c$ where $\sqrt{h} = \epsilon \sqrt{\sigma}$ and σ is the determinant of the metric in the transverse coordinates. In the integral, the range of τ is $(0, \beta = 2\pi/\kappa)$. Using $\sqrt{h} = \kappa\epsilon$ and $V^\xi = -2/\epsilon$, we find that $16\pi A_{\text{sur}} = 4\pi A_\perp$.]

14.7 Quantum field theory in a Friedmann universe

We shall next take up the study of quantum field theory in a $k = 0$ Friedmann universe with an arbitrary expansion factor $a(t)$. This analysis leads to issues which are quite different from those we have studied so far. We shall first develop the formalism in a general manner (to the extent one can) and then describe the results for a specific case which is of relevance in the study of inflationary universe and structure formation.[3]

14.7.1 General formalism

Consider a spatially flat Friedmann universe described by the line-element

$$ds^2 = -dt^2 + a^2(t)\, dx^2 = a^2(\eta)\left(-d\eta^2 + dx^2\right), \tag{14.145}$$

where t is the cosmic time, $a(t)$ is the scale factor and η denotes the conformal time with $d\eta = dt/a(t)$. We will consider a scalar field $\phi(\eta, x)$ described by the action

$$\mathcal{A}[\phi] = -\frac{1}{2}\int d^4x\, \sqrt{-g}\, \partial_i\phi\, \partial^i\phi. \tag{14.146}$$

The homogeneity and isotropy of the Friedmann metric in Eq. (14.145) allows us to decompose the scalar field ϕ in terms of the spatial Fourier modes as

$$\phi(x) = \int \frac{d^3k}{(2\pi)^3}\, Q_k(\eta)\, e^{ik\cdot x}. \tag{14.147}$$

The functions Q_k are complex with $Q_k^* = Q_{-k}$. As described after Eq. (14.6), we can, instead, work with a set of *real* variables q_k. Then, in terms of suitably defined real variables q_k, the action (14.146) can be expressed as follows:

$$\mathcal{A}[\phi] = \frac{1}{2} \int d^3k \int d\eta \, a^2 \left(\dot{q}_k^2 - k^2 q_k^2 \right),$$ (14.148)

where the overdot denotes differentiation with respect to the conformal time η and
$k = |\mathbf{k}|$. The action in Eq. (14.148) describes a collection of independent oscillators with time dependent mass a^2 and frequency k. It is obvious that the quantum
field theory for the scalar field can be constructed by working out the quantum
mechanics of each of the *time dependent* harmonic oscillators. Given the quantum
state $\psi_k[q_k, t_i(k)]$ for the mode q_k at an initial time $t_i(k)$, one can obtain the state
at a later time t by

$$\psi_k(q_k, t) = \int dQ_k \, K[q_k, t; \, Q_k, t_i(k)] \, \psi_k[Q_k, t_i(k)],$$ (14.149)

where $K(q_k, t; \, Q_k, t_i)$ is the (path integral) kernel for an oscillator with time
dependent parameters, which can be written down in terms of the classical solution
(see Eq. (14.32)). To keep our discussion general, we have allowed for the possibility that the initial quantum state of each of the oscillators is specified at different
times so that t_i can (in general) depend on $\mathbf{k}$. Of course, when the initial states of
all the modes are specified at the same time, $t_i(\mathbf{k}) = t_i$ will be independent of $\mathbf{k}$.
The *dynamics* of the system is completely specified by the kernel K.

It is obvious that the mathematics of a free scalar field in a Friedmann universe
is trivial and is no more complicated than that of an oscillator with time dependent
parameters. The real interest in this problem arises from two *conceptual* issues.
(i) To make any predictions, we need to know $\psi_k(q_k, t)$, which, in turn, requires
knowing the initial wave function at $t = t_i$. We have absolutely no idea – in the
case of the real universe – what to use for $\psi_k[Q_k, t_i(k)]$ and so different choices
(usually called 'vacuum states') can be investigated. (ii) There is an infinite number
of such oscillators, leading to the standard, unresolved, issues of regularization in
quantum field theory that we mentioned on page 594.

More explicitly, we can construct the quantum theory of each of the modes by the
following procedure. In the Schrödinger picture, the scalar field ϕ can be quantized
by quantizing each independent oscillator q_k. The Hamiltonian corresponding to
the kth oscillator is given by

$$H_k = \frac{p_k^2}{2 \, a^2} + \frac{1}{2} a^2 k^2 q_k^2,$$ (14.150)

where p_k is the momentum conjugate to the coordinate q_k. Therefore, each of the
oscillators satisfies the Schrödinger equation

$$i \frac{\partial \psi_k}{\partial \eta} = -\frac{1}{2 \, a^2} \frac{\partial^2 \psi_k}{\partial q_k^2} + \frac{1}{2} a^2 k^2 q_k^2 \, \psi_k$$ (14.151)

and the complete quantum state of the field is described by a wave function that is a product of ψ_k for all k.

It is possible to make further progress if we make a reasonable choice for the form of the quantum state. As we do not expect a large scale, spatially inhomogeneous, classical scalar field to be present in the universe, it is conventional to assume that the expectation value $\langle \psi_k | q_k | \psi_k \rangle$ vanishes in the quantum state of the field for $k \neq 0$. (Since $\langle \psi_k | q_k | \psi_k \rangle$ satisfies the classical equations of motion, this condition will be satisfied at all times if suitable initial conditions are imposed at an early epoch.) Next, we note that there exists a class of solutions to Eq. (14.151) that are *form invariant* in the sense that the q_k dependence of the wave function $\psi_k(q_k, \eta)$ is the same at all η. A simple solution that has such a property is an exponential of a quadratic function of q_k. When $\langle \psi_k | \hat{q}_k | \psi_k \rangle = 0$, the mean value of the Gaussian vanishes and the quantum state of the mode can be described by the wave function

$$\psi_k (q_k, \eta) = N_k(\eta) \, \exp - \left[R_k(\eta) \, q_k^2 \right], \tag{14.152}$$

where $N_k(\eta)$ and $R_k(\eta)$ are complex quantities. The normalization condition on the wave function then relates N_k and R_k as follows: $|N_k|^2 = \left[(R_k + R_k^*) / \pi \right]^{1/2}$. Therefore, the only nontrivial aspect of the quantum state is encoded in the time dependence of the function $R_k(\eta)$. Substituting the ansatz in Eq. (14.152) into the Schrödinger equation (Eq. (14.151)), we can obtain an equation for R_k. Introducing a function $\mu_k(\eta)$ through the relation $R_k = - \left(i \, a^2/2 \right) \left(\dot{\mu}_k / \mu_k \right)$, it is easy to show that μ_k satisfies the differential equation

$$\ddot{\mu}_k + 2 \frac{\dot{a}}{a} \dot{\mu}_k + k^2 \mu_k = 0, \tag{14.153}$$

which is the same as the classical equation of motion satisfied by the oscillator variable q_k. Hence, if we can solve the classical equation of motion for the scalar field in a background spacetime, then the quantum evolution of the wave function in Eq. (14.152) can be completely determined. In terms of cosmic time, Eq. (14.153) becomes

$$\frac{d^2 \mu_k}{dt^2} + 3H \frac{d \mu_k}{dt} + \frac{k^2}{a^2} \mu_k = 0; \qquad H \equiv \frac{\dot{a}}{a^2} = \frac{1}{a} \frac{da}{dt}. \tag{14.154}$$

Since the classical equation for μ_k is a second order differential equation, it requires two initial conditions for its solution. These initial conditions are, again, related to the freedom in the choice of the initial quantum state for the field. We will say more about this later on.

The formalism developed above completely solves the problem of quantum field theory in a Friedmann universe. Once a choice is made for the initial conditions which are required to solve Eq. (14.153), we can obtain the wave function at any other later time. If this wave function is expanded in terms of the eigenstates of the instantaneous Hamiltonian, it is possible to define the particle content of the state using such an expansion. We shall not pursue this further since such a procedure is inherently ambiguous in a time dependent background. It turns out, however, that a more useful quantity (especially in the study of an inflationary universe) is the power spectrum per logarithmic interval in k defined by the relation

$$P_\phi(k, \eta) = \frac{k^3 \langle q_k^2 \rangle}{2\pi^2} = \frac{k^3}{2\pi^2} \int_{-\infty}^{\infty} dq_k \, |\psi_k(q_k, \eta)|^2 \, q_k^2. \qquad (14.155)$$

Using Eq. (14.152) and expressing R_k in terms of μ_k, this result can be expressed in the form

$$P_\phi(k, \eta) = \frac{k^3}{2\pi^2} \left(\frac{|\mu_k|^2}{W(k)} \right), \qquad (14.156)$$

where $W(k)$ is a (time independent) constant determined by the Wronskian condition for μ_k, and defined as

$$W(k) = ia^2 \left(\mu_k \, \dot\mu_k^* - \dot\mu_k \, \mu_k^* \right). \qquad (14.157)$$

Since the differential equation (14.153) has real coefficients, if s_k is a solution, so is s_k^* and the general solution is a linear superposition of the form, say, $\mu_k = [\mathcal{A}(k) \, s_k + \mathcal{B}(k) \, s_k^*]$. The quantum state $\psi_k(q_k, \eta)$, however, depends only on R_k which is independent of the overall scaling of μ_k. This feature translates into the power spectrum as well; a global scaling of μ_k also changes the Wronskian $W(k)$ leaving $[|\mu_k|^2/W(k)]$ in the power spectrum in Eq. (14.156) invariant. Hence, we can ignore the overall scaling in μ_k. We shall therefore set $\mathcal{A}(k)$ to unity and choose the Wronskian $W(k)$ to be

$$W(k) = 1 - |\mathcal{B}(k)|^2. \qquad (14.158)$$

Then, the power spectrum in Eq. (14.156) reduces to

$$P_\phi(k, \eta) = \frac{k^3}{2\pi^2} \left(\frac{[1 + |\mathcal{B}(k)|^2] \, |s_k|^2 + 2 \, \mathrm{Re.} \, [\mathcal{B}(k) \, s_k^{*2}]}{1 - |\mathcal{B}(k)|^2} \right). \qquad (14.159)$$

14.7.2 Application: power law expansion

As an illustration of this formalism and with future applications in mind, we shall study the quantum theory in the case of a Friedmann universe in which the scale factor grows as a power law in time. We shall take

$$a(t) = a_0 \, t^p, \qquad (14.160)$$

where $p > 1$. In terms of the conformal time η, this scale factor can be written as

$$a(\eta) = (-\mathcal{H}\,\eta)^{(\gamma+1)}, \tag{14.161}$$

where γ and $\mathcal{H}$ are given by

$$\gamma = -\left(\frac{2p-1}{p-1}\right) \quad \text{and} \quad \mathcal{H} = (p-1)\,a_0^{1/p}. \tag{14.162}$$

By definition we take $-\infty < \eta < 0$ and $\gamma \le -2$, with $\gamma = -2$ corresponding to exponential growth $a(t) \propto \exp(Ht)$. Note that $H = \dot{a}/a^2 = (\gamma+1)(-\mathcal{H})^{-(\gamma+1)}\eta^{-(\gamma+2)}$ so that $H = \mathcal{H}$ when $\gamma = -2$; so $\mathcal{H}$ is indeed the Hubble constant in the limit of de Sitter expansion.

On substituting $\mu_k = (f_k/a)$ in Eq. (14.153), we find that f_k satisfies the following equation which we have encountered earlier as Eq. (13.26):

$$\ddot{f}_k + \left[k^2 - \left(\frac{\ddot{a}}{a}\right)\right] f_k = 0. \tag{14.163}$$

For $a(\eta)$ in Eq. (14.161) this reduces to

$$\ddot{f}_k + \left[k^2 - \frac{1}{\eta^2}\left(\nu^2 - \frac{1}{4}\right)\right] f_k = 0; \qquad \nu^2 = \left(\gamma + \frac{1}{2}\right)^2. \tag{14.164}$$

The general solution to this differential equation can be written in terms of Hankel functions of order ν as

$$f_k(\eta) = \frac{\sqrt{\pi\eta}}{2}\left(e^{-(i\pi\gamma/2)}\,H_\nu^{(1)}(k\eta) + \mathcal{B}(k)\,e^{(i\pi\gamma/2)}\,H_\nu^{(2)}(k\eta)\right), \quad \nu = -\left(\gamma + \frac{1}{2}\right), \tag{14.165}$$

where $H_\nu^{(1)}$ and $H_\nu^{(2)}$ are the Hankel functions of the first and the second kind, respectively. The k-dependent constant $\mathcal{B}(k)$ is to be fixed by choosing suitable initial conditions for each of the modes. The overall normalization is chosen such that the Wronskian in Eq. (14.157) is related to $\mathcal{B}(k)$ by Eq. (14.158).

To make further progress, we need to choose the form of the function $\mathcal{B}(k)$. One natural choice is obtained by assuming that very high frequency modes (with $|k\eta| \to \infty$) should behave as though they are in flat spacetime because they probe length scales which are small compared with the curvature scale and hence can be studied in a local inertial frame. This condition implies that the wave function in Eq. (14.152) should have the following asymptotic form:

$$\lim_{(k/aH)\to\infty} \psi_k\,(q_k, \eta) \to \left(\frac{k\,a^2}{\pi}\right)^{1/4} \exp - \left(\frac{k}{2}\,a^2\,q_k^2 + i\frac{k}{2}\eta\right), \tag{14.166}$$

when $a =$ constant, which, in turn, requires that, as $k \to \infty$, we need to have $R_k \to k\,a^2/2$ and $N_k \to (k\,a^2/\pi^{1/4})\,e^{-ik\eta/2}$. This condition is satisfied if we choose the boundary condition such that $f_k \to (e^{ik\eta}/\sqrt{2k})$ as $k \to \infty$. Using the asymptotic form of the Hankel function, it is easy to see that this can be achieved by setting $\mathcal{B}(k) = 0$ in Eq. (14.165), so that we have

$$f_k(\eta) = \left(\frac{\sqrt{\pi\eta}}{2} \right) e^{-(i\gamma\pi/2)} H_\nu^{(1)}(k\eta); \qquad \nu = -\left(\gamma + \frac{1}{2} \right). \tag{14.167}$$

This choice corresponds to a quantum state that is known as the *Bunch–Davies vacuum*. Note that, according to Eq. (14.158), $\mathcal{B}(k) = 0$ implies $W(k) = 1$. Therefore, on substituting the above f_k in Eq. (14.156) we find that

$$\mathcal{P}_\phi(k, \eta) = \frac{k^3}{2\pi^2} \frac{|f_k|^2}{a^2} = \frac{k^3}{8\pi\mathcal{H}} |\mathcal{H}\eta|^{2\nu} \left| H_\nu^{(1)}(k\eta) \right|^2. \tag{14.168}$$

We will see later that what is relevant in the case of structure formation is the k-dependence of $\mathcal{P}_\phi$ at the time the mode leaves the Hubble radius; that is, when $(k/a) = d_H^{-1}$ (called the moment of 'Hubble exit') which gives $k\eta = |\gamma + 1|$. At this instant, we get:

$$\mathcal{P}_\phi(k) = C_\nu \left(\frac{\mathcal{H}^2}{2\pi^2} \right) \left(\frac{k}{\mathcal{H}} \right)^{3-2\nu}, \qquad \text{(at } k = aH), \tag{14.169}$$

where C_ν is given by

$$C_\nu = \left(\frac{\pi}{4} \right) \left| H_\nu^{(1)} \left[\frac{1}{2} - \nu \right] \right|^2 \left| \left(\frac{1}{2} - \nu \right) \right|^{2\nu}. \tag{14.170}$$

A result, with the same k-dependence as in Eq. (14.169) but with a different numerical coefficient, can be obtained in a different context. Since $H_\nu(x)$ has the limiting form $H_\nu(x) \to -(i\Gamma(\nu)/\pi)(x/2)^{-\nu}$ for small x, we get from Eq. (14.168) that

$$\mathcal{P}_\phi(k, \eta) \simeq D \left(\frac{\mathcal{H}^2}{2\pi^2} \right) \left(\frac{k}{\mathcal{H}} \right)^{3-2\nu}; \qquad D = \frac{|\Gamma(\nu)|^2 2^{-2\nu}}{4\pi} \tag{14.171}$$

for $|k\eta| \ll 1$. Thus, at super-Hubble scales, $\mathcal{P}_\phi$ is independent of η and is frozen in time. It has the same k-dependence as $\mathcal{P}_\phi$ at the Hubble exit. Hence both Eq. (14.169) and Eq. (14.171) are used interchangeably in the literature.

With future applications in mind, we will obtain the $p \to \infty$ limit of the expression in Eq. (14.169). If we put $p = 1/\epsilon_1$, then as $\epsilon_1 \to 0$, we have

$$\gamma = -\frac{2 - \epsilon_1}{1 - \epsilon_1} \simeq -2 - \epsilon_1; \qquad \nu = \frac{3 - \epsilon_1}{2(1 - \epsilon_1)} \simeq \frac{3}{2} + \epsilon_1 \tag{14.172}$$

and

$$H \simeq \frac{\dot{a}}{a^2} = (-1 + \epsilon_1)(-\mathcal{H})\eta^{\epsilon_1} \simeq \mathcal{H}\eta^{\epsilon_1} \simeq \mathcal{H}k^{-\epsilon_1}, \tag{14.173}$$

where the last expression is valid at the Hubble exit corresponding to $k\eta = 1$. Hence, in this limit,

$$\mathcal{P}_\phi(k)\Big|_{k\eta=1} \simeq \left(\frac{\mathcal{H}^2}{2\pi^2}\right)\left(\frac{k}{\mathcal{H}}\right)^{-2\epsilon_1} \approx \left(\frac{H^2}{2\pi^2}\right), \qquad (14.174)$$

where we have used Eq. (14.173) and the second relation in Eq. (14.162), in the limit of $\epsilon_1 \to 0$. We have also used the fact that $C_{3/2} = 1$ from the definition in Eq. (14.170). Note that the weak k-dependence of $\mathcal{P}_\phi$ at the Hubble exit arises from the weak k-dependence of H at the Hubble exit given by Eq. (14.173).

The situation corresponding to $\epsilon_1 = 0, a(t) \propto \exp(Ht)$ deserves a separate discussion since it represents the de Sitter universe. In this case, we can take $Ht = -\ln(1 - H\eta)$, $a(\eta) = (1 - H\eta)^{-1}$, with a shift in the origin of η, so that when $H \to 0$, η goes over to t. In this case, Eq. (14.154) has the simple analytic solution

$$\mu_k = -i(2k^3)^{-1/2}He^{ik/H}(1+ix)e^{-ix}; \qquad x = \frac{k}{Ha}, \qquad (14.175)$$

which goes over to $\mu_k \to (2k)^{-1/2}\exp(ik\eta)$ when $(H/k) \to 0$ which is our boundary condition. For this choice, the Wronskian W in Eq. (14.157) is unity and we get

$$\mathcal{P}_\phi(k, a) = \frac{k^3}{2\pi^2}|\mu_k|^2 = \frac{H^2}{4\pi^2}\left(1 + \frac{k^2}{a^2H^2}\right). \qquad (14.176)$$

At the Hubble exit, $(k/aH) = 1$, this gives

$$\mathcal{P}_\phi(k)\Big|_{k=aH} = \frac{H^2}{2\pi^2}, \qquad (14.177)$$

which is independent of k. This result, of course, matches Eq. (14.174) in the limit $\epsilon \to 0$.

14.8 Generation of initial perturbations from inflation

In the description of linear perturbation theory in Chapter 13, we assumed that some small perturbations existed in the early universe that were amplified through gravitational instability. To provide a complete picture of structure formation we need a mechanism for the generation of these initial perturbations. One such mechanism is provided by the inflationary scenario, which allows for the quantum fluctuations in a scalar field driving the inflation to provide classical energy density perturbations at a late epoch. (Originally inflationary scenarios were suggested as a pseudo-solution to certain pseudo-problems; these are only of historical interest today and the only reason to take the possibility of an inflationary phase in the early universe seriously is because it provides a mechanism for generating the initial perturbations.) We shall now discuss how this can come about.

14.8.1 Background evolution

The basic assumption in the inflationary scenario is that the universe underwent a rapid – nearly exponential – expansion for a brief period of time in the very early universe. The simplest way of realizing such a phase – as we shall see – is to postulate the existence of a scalar field with a nearly flat potential. The energy-momentum tensor for such a scalar field

$$T_n^m = -\nabla_n\phi\nabla^m\phi + \left[\frac{1}{2}(\nabla\phi)^2 + V(\phi)\right]\delta_n^m \qquad (14.178)$$

can be taken to be the sum of a homogeneous part, $T_b^a(\eta)$, which is independent of x (and drives the background expansion of the universe), and a small perturbation $\delta T_b^a(\eta, x)$, which depends on space and time and acts as the seed of initial energy density fluctuations. The perturbed Einstein equations $G_b^a + \delta G_b^a = 8\pi\kappa(T_b^a + \delta T_b^a)$ are now satisfied by taking $G_b^a = 8\pi\kappa T_b^a$ and $\delta G_b^a = 8\pi\kappa\delta T_b^a$. We shall first describe the background evolution driven by the homogeneous part of the energy density given by the components:

$$^{(0)}T_0^0 = \frac{1}{2a^2}\dot{\phi}_0^2 + V(\phi_0)\,, \quad ^{(0)}T_\alpha^0 = 0\,, \quad ^{(0)}T_\beta^\alpha = \left[-\frac{1}{2a^2}\dot{\phi}_0^2 + V(\phi_0)\right]\delta_\beta^\alpha\,.$$

$$(14.179)$$

In this case, it is convenient to work with the cosmic time $dt = a(\eta)d\eta$ for describing the background evolution. The dynamics of the universe, driven by a scalar field source, is then described by:

$$\frac{1}{a^2}\left(\frac{da}{dt}\right)^2 = H^2(t) = \frac{8\pi\kappa}{3}\left[V(\phi) + \frac{1}{2}\left(\frac{d\phi}{dt}\right)^2\right]; \qquad \frac{d^2\phi}{dt^2} + 3H\frac{d\phi}{dt} = -\frac{dV}{d\phi}.$$

$$(14.180)$$

If the potential is nearly flat for a certain range of ϕ, we can introduce the 'slow roll-over' approximation, under which these equations can be approximated by the set:

$$H^2 \simeq \frac{8\pi\kappa V(\phi)}{3}; \qquad 3H\frac{d\phi}{dt} \simeq -V'(\phi), \qquad (14.181)$$

where the prime on V denotes differentiation with respect to ϕ. For this slow roll-over to last for a reasonable length of time, we need to assume that the terms ignored in the Eq. (14.180) – to arrive at Eq. (14.181) – are indeed small. This can be quantified in terms of the parameters:

$$\epsilon_1(\phi) = \frac{1}{16\pi\kappa}\left(\frac{V'}{V}\right)^2; \qquad \epsilon_2(\phi) = \frac{1}{8\pi\kappa}\frac{V''}{V}, \qquad (14.182)$$

which must remain small compared with unity. Typically the inflation ends when this assumption breaks down. If such an inflationary phase lasts up to some time t_{end} then the universe would have undergone an expansion by a factor $\exp N(t)$ during the interval (t, t_{end}) where

$$N \equiv \ln \frac{a(t_{\text{end}})}{a(t)} = \int_t^{t_{\text{end}}} H \, dt \simeq 8\pi\kappa \int_{\phi_{\text{end}}}^{\phi} \frac{V}{V'} \, d\phi. \tag{14.183}$$

One usually takes $N \simeq 65$ or so.

It is not difficult to obtain *exact* solutions for $a(t)$ with such rapid expansion by tailoring the potential for the scalar field. In fact, given any $a(t)$ and thus a $H(t) = a^{-1}(da/dt)$, one can determine a potential $V(\phi)$ for a scalar field such that the expressions in Eq. (14.180) are satisfied. One can verify that this can be done by the choice:

$$V(t) = \frac{1}{16\pi\kappa} H \left[6H + \frac{2}{H} \frac{dH}{dt} \right]; \quad \phi(t) = \int dt \left[-\frac{1}{4\pi\kappa} \frac{dH}{dt} \right]^{1/2}. \tag{14.184}$$

Given any $H(t)$, these equations give $(\phi(t), V(t))$ and thus implicitly determine the necessary $V(\phi)$. As an example, note that a power law inflation, $a(t) = a_0 t^p$ (with $p \gg 1$), is generated by

$$V(\phi) = V_0 \exp \left(-\sqrt{\frac{16\pi\kappa}{p}} \, \phi \right), \tag{14.185}$$

while an exponential of power law

$$a(t) \propto \exp(A t^f), \quad f = \frac{\beta}{4+\beta}, \quad 0 < f < 1, \quad A > 0 \tag{14.186}$$

can arise from

$$V(\phi) \propto \phi^{-\beta} \left(1 - \frac{\beta^2}{48\pi\kappa} \frac{1}{\phi^2} \right). \tag{14.187}$$

Thus generating a rapid expansion in the early universe is trivial if we are willing to postulate scalar fields with tailor made potentials, as is often done in the literature.

For future reference, we quote the following *exact* relations, which are easy to prove. First we note that ϵ_1 and ϵ_2 control the time derivatives of $H(t)$ through

$$\epsilon_1 = \frac{dH^{-1}}{dt}; \quad \epsilon_1 - \epsilon_2 = \frac{1}{2H} \frac{(d^2 H/dt^2)}{(dH/dt)}. \tag{14.188}$$

(For the power law inflation discussed in the previous section with $a \propto t^p$, we have $\epsilon_1 = 1/p$ and $\epsilon_2 = 2/p$.) Further, one can easily show (using e.g. Eq. (14.184) that

$$\frac{d\epsilon_1}{dt} = 2H\epsilon_1(2\epsilon_1 - \epsilon_2); \quad 4\pi\kappa\dot{\phi}_0^2 = (a^2 H^2)\epsilon_1. \tag{14.189}$$

We will have occasion to use these later.

14.8.2 Perturbations in the inflationary models

Let us next consider the evolution of perturbations in a universe that underwent exponential inflation. The linear perturbations in the energy-momentum tensor of the scalar field are given by

$$\delta T_0^0 = a^{-2} \left[-\dot{\phi}_0^2 \Phi + \dot{\phi}_0 \delta \dot{\phi} + V'(\phi) a^2 \delta \phi \right],$$

$$\delta T_\alpha^0 = a^{-2} \dot{\phi}_0 \delta \partial_\alpha \phi,$$

$$\delta T_\beta^\alpha = a^{-2} \left[+\dot{\phi}_0^2 \Phi - \dot{\phi}_0 \delta \dot{\phi} + V'(\phi) a^2 \delta \phi \right] \delta_\beta^\alpha, \qquad (14.190)$$

where we are using the notation of Chapter 13. We need to equate this δT_b^a to the perturbed Einstein tensor δG_b^a calculated in Section 13.3. From the structure of the δT_b^a in Eq. (14.190), it is obvious that we can ignore the vector and tensor perturbations of the metric. Further, since δT_β^α is proportional to δ_β^α, it follows that we can set $\psi = \Phi$ for the metric perturbations. Then the equations for the perturbation in the scalar field $\delta \phi$ and the gravitational potential Φ reduce to

$$\nabla^2 \Phi - 3\mathcal{H}\dot{\Phi} - (\dot{\mathcal{H}} + 2\mathcal{H}^2)\Phi = 4\pi\kappa(\dot{\phi}_0 \delta \phi + V'(\phi) a^2 \delta \phi), \qquad (14.191)$$

$$\dot{\Phi} + \mathcal{H}\Phi = 4\pi\kappa\dot{\phi}_0 \delta \phi, \qquad (14.192)$$

$$\ddot{\Phi} + 3\mathcal{H}\dot{\Phi} + (\dot{\mathcal{H}} + 2\mathcal{H}^2)\Phi = 4\pi\kappa(\dot{\phi}_0 \delta \phi - V'(\phi) a^2 \delta \phi). \qquad (14.193)$$

(It might seem that we have three equations for two variables Φ and $\delta \phi$. However, by differentiating Eq. (14.192), it is easy to see that Eq. (14.193) is satisfied when Eq. (14.192) is satisfied. Therefore, we have only two equations for the two unknowns.) By straightforward manipulation these equations can be converted into second order differential equations for $\delta \phi$ and Φ. The equation for the perturbations in the scalar field becomes

$$\ddot{\delta\phi} + 2\mathcal{H}\dot{\delta\phi} - \nabla^2 \delta \phi + V''(\phi) a^2 \delta \phi = 4\dot{\phi}_0 \dot{\Phi} - 2V'(\phi) a^2 \Phi, \qquad (14.194)$$

where the right hand side, made of the gravitational perturbations, acts as a source for $\delta \phi$. The equation for the gravitational perturbation can be cast in the form

$$\ddot{\Phi} + 2\frac{d}{d\eta}\left(\ln \frac{a}{\dot{\phi}_0} \right) \dot{\Phi} - \nabla^2 \Phi + 2\dot{\phi}_0 \left(\frac{d}{d\eta} \frac{\mathcal{H}}{\dot{\phi}} \right) \Phi = 0. \qquad (14.195)$$

During the inflationary phase the $a(t)$ grows exponentially and hence the wavelength of any perturbation will also grow with it exponentially. The Hubble radius, on the other hand, will remain constant. It follows that one can have a situation in which a given mode has a wavelength smaller than the Hubble radius at the beginning of the inflation but grows and becomes bigger than the Hubble radius as inflation proceeds. It is conventional to say that a perturbation of comoving wavelength λ_0 'leaves the Hubble radius' when $\lambda_0 a = d_H$ at some time $t = t_{\text{exit}}(\lambda_0)$. For $t > t_{\text{exit}}$ the wavelength of the perturbation is bigger than the Hubble radius. Eventually the inflation ends and the universe becomes radiation dominated. Then the wavelength will grow at a rate ($\propto t^{1/2}$) slower than the Hubble radius ($\propto t$) and will enter the Hubble radius again during $t = t_{\text{enter}}(\lambda_0)$. We saw in Chapter 13 that the final power spectrum can be related to the amplitude of the perturbations at $t = t_{\text{enter}}(\lambda)$. So our first task is to relate the amplitude of the perturbation at $t = t_{\text{exit}}(\lambda_0)$ with the perturbation at $t = t_{\text{enter}}(\lambda_0)$.

We know that for modes bigger than Hubble radius, we have the conserved quantity (see Eq. (13.93) given by

$$\mathcal{R} \equiv \Phi + \frac{2}{3} \frac{(\mathcal{H}^{-1}\dot{\Phi} + \Phi)}{1 + w}. \tag{14.196}$$

Other equivalent forms for this expression (given in Chapter 13; see Eq. (13.93)) are

$$\mathcal{R} = \frac{2}{3} \frac{\rho}{\rho + p} \frac{a}{\dot{a}} \left(\dot{\Phi} + \frac{\dot{a}}{a}\Phi\right) + \Phi = \frac{H}{\rho + p} \frac{ik^\alpha}{k^2}\delta T_\alpha^0 + \Phi. \tag{14.197}$$

When the modes re-enter the Hubble radius, the universe is radiation dominated and $\mathcal{R}_{\text{entry}} \approx (3/2)\Phi$. On the other hand, during inflation, we have seen that we can write the scalar field as the sum of a dominant homogeneous part plus a small, spatially varying fluctuation: $\phi(t, \boldsymbol{x}) = \phi_0(t) + \delta\phi(t, \boldsymbol{x})$ so that δT_α^0 in Eq. (14.197) is contributed by $\delta\phi$ (see Eq. (14.190)). But it is easy to see that Φ is negligible at $t = t_{\text{exit}}$ because

$$\Phi \sim \frac{4\pi\kappa a^2}{k^2}\delta\rho \sim \frac{4\pi\kappa}{H^2}\delta\rho \sim \frac{\delta\rho}{\rho} \sim \left(\frac{\rho + p}{\rho}\right)\left(\frac{\delta\rho}{\rho + p}\right) \ll \frac{H}{(\rho + p)}\frac{\dot{\phi}_0}{a}\delta\phi_k. \tag{14.198}$$

Therefore,

$$\mathcal{R}_{\text{exit}} \approx \frac{H}{(\dot{\phi}_0^2/a^2)}\left[-\frac{\dot{\phi}_0\delta\phi_k}{a}\right] = -aH\frac{\delta\phi_k}{\dot{\phi}_0}. \tag{14.199}$$

Using the conservation law $\mathcal{R}_{\text{exit}} = \mathcal{R}_{\text{entry}}$, we get

$$\Phi_k\Big|_{\text{entry}} = \left[\frac{2}{3}\left(\frac{aH}{\dot{\phi}_0}\right)\delta\phi_k\right]_{\text{exit}}. \tag{14.200}$$

Thus, given a perturbation of the scalar field $\delta\phi_k$ during inflation, we can compute the value of the gravitational potential perturbations Φ_k at the time of re-entry, which – in turn – can be used to compare the theory with the observations.

The manner in which the constancy of $\mathcal{R}$ determines the relation between the perturbations at the time of exit of the Hubble radius and at the time of entry into the Hubble radius is noteworthy. The sum of the two terms in Eq. (14.197), of course, remains almost constant when the mode is outside the Hubble radius. But initially,

Φ is negligible and $\mathcal{R}$ is entirely contributed by the term containing δT^0_α. After the end of inflation, this term starts decreasing and Φ starts increasing, keeping $\mathcal{R}$ constant. At late times, $\mathcal{R}$ is contributed predominantly by Φ. It is actually possible to construct a hybrid field made of $\delta\phi$ and Φ with its own dynamical equation, derived from a Lagrangian, and reproduce the above results. The reason such an approach – in which a scalar field degree of freedom and a gravitational degree of freedom can be linearly combined – works is because the gravitational degree of freedom Φ does not represent gravitational radiation and is similar to the Coulomb degree of freedom in electrodynamics. While a quantum of a scalar field *cannot* be combined with a graviton in quantum theory, there is no such restriction in combining a non-dynamical degree of freedom of gravity with $\delta\phi$.

The conservation of $\mathcal{R}$ therefore allows us to relate the power spectrum $\mathcal{P}$ at the time of re-entry into the Hubble radius to the power spectrum at the time of exit of the Hubble radius

$$\mathcal{P} = \frac{k^3}{2\pi^2}|\Phi_k|^2_{\text{entry}} = \left(\frac{k^3}{2\pi^2}\right)\frac{4}{9}\left(\frac{aH}{\dot{\phi}_0}\right)^2_{\text{exit}}|\delta\phi_k|^2_{\text{exit}}. \tag{14.201}$$

Using the second relation in Eq. (14.189) this can be expressed in the form

$$\mathcal{P} = \left(\frac{16\pi\kappa}{9}\right)\left[\frac{1}{\epsilon_1}\frac{k^3|\delta\phi_k|^2}{2\pi^2}\right]_{\text{exit}}. \tag{14.202}$$

If the slow roll-over inflation is close to de Sitter, $(k^3|\delta\phi_k|^2)_{\text{exit}} \propto H^2$ and the k dependence of $\mathcal{P}$ will be due to the weak time dependence of $[H^2(t_{\text{exit}})/\epsilon_1(t_{\text{exit}})]$ because t_{exit} will have weak dependence on k.

To determine $\delta\phi_k$ during the inflationary phase, we need to solve Eq. (14.194). We note that, during the relevant epoch, $\Phi \approx a\epsilon_1 H\delta\phi_k/\dot{\phi}_0$, with $\epsilon_1 = -\dot{H}/H^2$ being the slow roll-over parameter. Using $|4\dot{\phi}_0\dot{\Phi}| \ll |2V'a^2\Phi|$ on super-Hubble scales and using $3H\dot{\phi} \approx -aV'$, we can approximate Eq. (14.194) by

$$\ddot{\delta\phi}_k + 2\mathcal{H}\dot{\delta\phi}_k + \left[k^2 + a^2V''(\phi) + 6a^2\epsilon_1 H^2\right]\delta\phi_k = 0. \tag{14.203}$$

Introducing the variable $\delta\chi_k \equiv a\,\delta\phi_k$, this equation reduces to

$$\ddot{\delta\chi}_k + \left[k^2 - \frac{1}{\eta^2}\left(\nu^2 - \frac{1}{4}\right)\right]\delta\chi_k = 0, \qquad \nu^2 = \frac{9}{4} + 9\epsilon_1 - 3\epsilon_2, \tag{14.204}$$

which is identical to Eq. (14.164) so that the general solution is given by

$$\delta\chi_k = \sqrt{-\eta}\left[c_1(k)H_\nu^{(1)}(k\eta) + c_2(k)H_\nu^{(2)}(k\eta)\right]. \tag{14.205}$$

This is as far as one can proceed along the classical lines. The two constants c_1 and c_2 can depend on k and, until they are determined by some independent reasoning, one cannot predict the k-dependence of the power spectrum at the time of re-entry.

It is at this stage that one invokes quantum mechanical considerations in order to fix the k-dependence. The important – and conceptually difficult – question is how we can obtain a *c-number* field $\delta\phi_k$ from a quantum scalar field. There is no simple answer to this question but one possible way of doing it is as follows. Let us start with the quantum operator for a scalar field decomposed into the Fourier modes with $\hat{q}_{\boldsymbol{k}}(t)$ denoting an infinite set of operators:

$$\hat{\phi}(t, \boldsymbol{x}) = \int \frac{d^3 k}{(2\pi)^3} \hat{q}_{\boldsymbol{k}}(t) e^{i\boldsymbol{k}\cdot\boldsymbol{x}}. \tag{14.206}$$

We choose a quantum state $|\psi\rangle$ such the expectation value of $\hat{q}_{\boldsymbol{k}}(t)$ vanishes for all nonzero $\boldsymbol{k}$ so that the expectation value of $\hat{\phi}(t, \boldsymbol{x})$ gives the homogeneous mode that drives the inflation. The quantum fluctuation around this homogeneous part in a quantum state $|\psi\rangle$ is given by

$$\sigma_{\boldsymbol{k}}^2(t) = \langle\psi|\hat{q}_{\boldsymbol{k}}^2(t)|\psi\rangle - \langle\psi|\hat{q}_{\boldsymbol{k}}(t)|\psi\rangle^2 = \langle\psi|\hat{q}_{\boldsymbol{k}}^2(t)|\psi\rangle. \tag{14.207}$$

It is easy to verify that this fluctuation is just the Fourier transform of the two-point function in this quantum state:

$$\sigma_{\boldsymbol{k}}^2(t) = \int d^3 x \langle\psi|\hat{\phi}(t, \boldsymbol{x} + \boldsymbol{y})\hat{\phi}(t, \boldsymbol{y})|\psi\rangle e^{i\boldsymbol{k}\cdot\boldsymbol{x}}. \tag{14.208}$$

Since $\sigma_{\boldsymbol{k}}$ characterizes the quantum fluctuations, it seems reasonable to introduce a c-number field $\delta\phi(t, \boldsymbol{x})$ by the definition

$$\delta\phi(t, \boldsymbol{x}) \equiv \int \frac{d^3 k}{(2\pi)^3} \sigma_{\boldsymbol{k}}(t) e^{i\boldsymbol{k}\cdot\boldsymbol{x}}. \tag{14.209}$$

This c-number field will have a *c-number power spectrum*, which has the same form as the *quantum* fluctuations of the scalar field. Hence we may take this as our *definition* of an equivalent classical perturbation. (There are more sophisticated ways of getting this result but none of them is fundamentally more sound that the elementary definition given above.)

We now have all the ingredients in place and we can use the formalism developed in Section 14.7.2. Taking the quantum state to be the one in Eq. (14.152), we invoke the assumption that, when $H \to 0$, the mode functions $\delta\phi_k$ should reduce to the standard mode functions of flat spacetime field theory and should take the form $(2k)^{-1/2} \exp(ik\eta)$. Using this criterion, it is easy to see that we need to impose on the solution in Eq. (14.205) the boundary conditions

$$c_2 = 0; \qquad c_1 = \frac{\sqrt{\pi}}{2} \exp[-i\pi/2(\nu + (1/2))]. \tag{14.210}$$

This allows us to choose the fluctuations in Eq. (14.205) to be

$$\delta\chi_k = \frac{\sqrt{\pi}}{2} e^{i(\nu+\frac{1}{2})\frac{\pi}{2}} \sqrt{-\eta} H_{\nu}^{(1)}(k\eta), \tag{14.211}$$

which is identical to Eq. (14.167) except for the definition of ν, which is now given by Eq. (14.204). (For a power law, slow roll-over with $\epsilon_1 = 1/p$, $\epsilon_2 = 2\epsilon_1$, Eq. (14.204) gives $\nu^2 = (9/4) + 3\epsilon_1$ which, of course, matches with the second relation in Eq. (14.172).) The relevant factor we need is

$$\frac{k^3|\delta\phi_k|^2}{2\pi^2} = \frac{k^3}{2\pi^2 a^2}\frac{|\eta|}{4}\frac{\pi}{4}\left|H_\nu^{(1)}(k\eta)\right|^2, \tag{14.212}$$

evaluated at the Hubble exit $k = aH$. In the case of slow roll-over inflation, it is easy to show from Eq. (14.188) that $aH \simeq -(1 + \epsilon_1)\eta^{-1}$ so that $k = aH$ corresponds to $k\eta \simeq 1$. In this limit, using Eq. (14.170) with $\nu = 3/2$, we get

$$\left(\frac{k^3|\delta\phi_k|^2}{2\pi^2}\right)_{k=aH} = \left[\frac{H^2}{2\pi^2}\right]_{\text{exit}}, \tag{14.213}$$

which is the same as Eq. (14.174). Substituting into Eq. (14.202) we get the final result

$$\mathcal{P} = \left(\frac{2}{9}\right)\left(\frac{4\kappa}{\pi}\right)\left[\frac{H^2}{\epsilon_1}\right]_{\text{exit}}. \tag{14.214}$$

The k-dependence of this expression arises from the weak k-dependence of the factor (H^2/ϵ_1) because it is evaluated at $t = t_{\text{exit}}(k)$. (For the power law, slow roll-over case, ϵ_1 is a constant and all the time dependence – and k-dependence – arises from H^2 as we saw in Eq. (14.213).) If we write $(H^2/\epsilon_1) \propto k^m$ at $k = aH$ then the index m can be obtained as the logarithmic derivative

$$m = \frac{d}{d\ln k}[\ln H^2 - \ln\epsilon_1]. \tag{14.215}$$

We first note that

$$\left.\frac{d\ln H}{d\ln k}\right|_{k=aH} = \left[\frac{k}{H}\frac{dH}{d\eta}\right]\left(\frac{d\eta}{dk}\right)\bigg|_{k=aH}. \tag{14.216}$$

We now use the facts $\dot{H} = -aH^2\epsilon_1$, $(d\eta/dk)_{\text{exit}} = -[d(aH)^{-1}/dk]_{\text{exit}} = k^{-2}$ to obtain

$$\left.\frac{d\ln H}{d\ln k}\right|_{k=aH} = -\frac{k}{H}\frac{aH^2\epsilon_1}{k^2}\bigg|_{k=aH} = -\epsilon_1, \tag{14.217}$$

which matches with Eq. (14.173) as it should. The logarithmic derivative of ϵ_1 can be similarly obtained using the first equation in Eq. (14.189) to get $(d\ln\epsilon_1/d\ln k) = 2(2\epsilon_1 - \epsilon_2)$. Substituting into Eq. (14.215) we find that the k-dependence is contained in the index

$$m \equiv \frac{d\ln(\mathcal{P})}{d\ln k} \simeq -6\epsilon_1 + 2\epsilon_2. \tag{14.218}$$

As a check, note that, for power law inflation with $\epsilon_1 = 1/p$, $\epsilon_2 = 2/p$, this gives $m = -2/p = -2\epsilon_1$ which matches with the result in Eq. (14.174). For sufficiently small ϵ_1 and ϵ_2 we find that the spectrum is scale invariant in the sense that $k^3 |\Phi_k|^2_{\text{entry}}$ is independent of k. (As in the case of Eq. (14.171), here also $k^3 |\delta\phi_k|^2$ in Eq. (14.212) gets frozen in time at super-Hubble scales. Using either the expression at $k = aH$ or the expression for $k \ll aH$ leads to the same final k-dependence.)

It is sometimes claimed in the literature that a scale invariant spectrum is a prediction of inflation. *This is simply wrong.* One has to make several *other* assumptions including an all important choice for the quantum state (about which we know nothing) to obtain a scale invariant spectrum. In fact, one can prove that, given any power spectrum $P(k)$, one can find a quantum state such that this power spectrum is generated.[4] So whatever results are obtained by observations can be reconciled with inflationary generation of perturbations.

A qualitative way of understanding the k-dependence of the final result is as follows. We first note that, to the lowest order, we have

$$\mathcal{P} \sim k^3 |\Phi_k|^2 \sim \frac{H^6}{(V')^2} \sim \left(\frac{\kappa^3 V^3}{V'^2} \right). \tag{14.219}$$

Let us define the deviation from the scale invariant index by $m = (d \ln \mathcal{P}/d \ln k)$. Using

$$\frac{d}{d \ln k} = a \frac{d}{da} = \frac{\dot{\phi}}{aH} \frac{d}{d\phi} = -\frac{1}{8\pi\kappa} \frac{V'}{V} \frac{d}{d\phi} \tag{14.220}$$

one finds that

$$-m = 6\epsilon_1 - 2\epsilon_2, \tag{14.221}$$

which is the same as Eq. (14.218). Thus, as long as ϵ_1 and ϵ_2 are small we do have $m \approx 0$; what is more, given a potential one can estimate ϵ_1 and ϵ_2 and thus the deviation m.

The same process that generates quantum fluctuations of the scalar field can also generate spin-2 perturbations during the inflationary phase. If we take the normalized gravitational wave amplitude as $h_{ab} = \sqrt{16\pi\kappa} \, e_{ab}\phi$, the function ϕ behaves like a scalar field. The normalization factor $\sqrt{16\pi\kappa}$ is dictated by the fact that the action for the perturbation should reduce to that of a spin-2 field described in Chapter 3. We have already seen in Chapter 13 that the tensor mode satisfies the same equation as the perturbations in the scalar field. (Compare Eq. (13.26) with Eq. (14.163).) The analysis, therefore, proceeds exactly as before and we find that the power spectrum of the gravitational waves is

$$\mathcal{P}_{\text{grav}}(k) \cong \frac{k^3 |h_k|^2}{2\pi^2} = 32 \, \kappa \, H^2. \tag{14.222}$$

From Eq. (14.217) we see that $m = -2\epsilon_1$ for the gravitational waves. Comparing the results for scalar and tensor modes,

$$\mathcal{P}_{\text{scalar}} \propto \frac{H^6}{(V')^2} \propto \left(\frac{\kappa^3 V^3}{V'^2} \right); \quad \mathcal{P}_{\text{tensor}} \propto \kappa H^2 \propto \kappa^2 V, \qquad (14.223)$$

we get $(\mathcal{P}_{\text{tensor}}/\mathcal{P}_{\text{scalar}})^2 \approx 16\pi\epsilon \ll 1$. Further, if $-m \approx 4\epsilon$ (see Eq. (14.221) with $\epsilon_1 \approx \epsilon_2$, which is a valid assumption for most models) we have the relation $(\mathcal{P}_{\text{tensor}}/\mathcal{P}_{\text{scalar}})^2 \approx \mathcal{O}(3)|m|$ which connects three quantities, all of which are independently observable in principle. If these are actually measured in the future it could act as a consistency check of the inflationary paradigm.

In any inflationary scenario the modes with reasonable proper size today originated from sub-Planck length scales early on. A scale λ_0 today will be

$$\lambda_{end} = \lambda_0 \frac{a_{end}}{a_0} = \lambda_0 \frac{T_0}{T_{end}} \approx \lambda_0 \times 10^{-28} \qquad (14.224)$$

at the end of inflation (if inflation took place at GUT scales) and

$$\lambda_{begin} = \lambda_{end} A^{-1} \approx \lambda_0 \times 10^{-58} (A/10^{30})^{-1} \qquad (14.225)$$

at the beginning of inflation if the inflation changed the scale factor by $A \simeq 10^{30}$. Clearly, $\lambda_{begin} < L_P$ for $\lambda_0 < 3$ Mpc! Most structures in the universe today correspond to transplanckian scales at the start of the inflation. It is not clear whether we can trust standard physics at early stages of inflation or whether transplanckian effects will lead to some observable effects.

Exercise 14.9

Gauge invariance of $\mathcal{R}$ Show that a change in the time slicing introduced by $t \to t + \Delta t$ will produce a perturbation in any scalar quantity (like the scalar field ϕ) by the amount $\Delta\phi = -\dot{\phi}\Delta t$. Also show that the change in the time slicing modifies the scalar spatial perturbations in the metric by an amount $\delta\psi = H\Delta t$. Hence show that, when $\psi = \Phi$, the variable $\mathcal{R}$ is gauge invariant under the change in the time slicing.

Exercise 14.10

Coupled equations for the scalar field perturbations Show that the equations governing Φ and $\delta\phi$ can be expressed in Fourier space in the form

$$\frac{d\Phi_k}{dt} + H\Phi_k = \frac{1}{2} \left(\frac{d\phi_0}{dt} \right) \delta\phi_k \qquad (14.226)$$

$$\frac{d}{dt} \left(\frac{\delta\phi_k}{(d\phi_0/dt)} \right) = \left(1 - \frac{2k^2}{a^2(d\phi_0/dt)^2} \right) \Phi_k \qquad (14.227)$$

where we have set $8\pi\kappa = 1$. (In this form the perturbation equations are easy to integrate numerically.)

PROJECTS

Project 14.1

Detector response in stationary trajectories

In Exercise 14.7 we discussed the response of a detector in a uniformly accelerated trajectory and found that it behaves like a detector that is stationary in a thermal bath. One of the crucial inputs in this derivation was the conversion of the total probability in Eq. (14.117) to a rate in Eq. (14.120). This, in turn, was possible only because $G[x(\tau), x(\tau')]$ – which is, in general, a function of both τ and τ' – turned out to be a function of only $\tau - \tau'$. There are wide varieties of trajectories for which this result holds and in each of them one can obtain a *rate* of excitation for the detector. Different classes of trajectories lead to different kinds of spectrum showing that the thermal spectrum seen by an accelerated detector is just one amongst several possibilities.

(a) Argue that, if the trajectory $x^i(\tau)$ is an integral curve of a timelike Killing vector field in the spacetime, then $G[x(\tau), x(\tau')] = G(\tau - \tau')$.

(b) We know that there are 10 independent Killing vector fields in flat spacetime. A suitable linear combination of them with the coefficients chosen such that it is timelike in a given region of spacetime will give an explicit realization of the above idea. As an example, consider a Killing vector field with the components

$$\xi^i(x) = (1 + \kappa x, \kappa t - \lambda y, \lambda x - \rho z, \rho y), \tag{14.228}$$

where κ, λ and ρ are three constants. Consider the trajectories obtained by (i) $\lambda = \rho = 0$, (ii) $\rho = 0$ and (iii) $\lambda = \kappa, \rho = 0$. In each case, construct the corresponding trajectory (i.e., the integral curves of ξ^i) explicitly and determine the conditions under which it remains timelike. Construct the proper coordinate system for the observers following these trajectories and the metric in these non-inertial coordinates.

(c) Obtain the Green function $G(\tau - \tau')$ in each of these cases and calculate its Fourier transform with respect to $\tau - \tau'$, thereby determining the detector response.

(d) Finally, work out the quantum field theory for a massless scalar field in each of the non-inertial coordinate systems determined in (b) above and work out the Bogolyubov coefficients between the positive frequency mode functions in these coordinate systems and the inertial coordinate system. You will find that, in the case of uniformly accelerated frame, the Bogolyubov coefficients as well as the detector response leads to a thermal spectrum but not in general. Interpret this result.[5]

Project 14.2

Membrane paradigm for the black holes

This is an open-ended project which invites you to explore a particular perspective on black hole physics and assertain for yourself its power and limitations.[6] The basic idea stems from the fact that, for an outside observer, the black hole horizon acts as a one-way membrane. Hence, as far as any physical process that is taking place outside the horizon is concerned (irrespective of how close to the horizon it occurs), we can think of the horizon itself as a physical system ('membrane') with specific properties. The main task is to determine the physical properties of the membrane so that we obtain consistent physics outside the horizon. We shall mostly concentrate on electromagnetic processes.

(a) Consider an observer A located just outside the horizon of a Schwarzschild black hole with fixed coordinates ($r = 2M + \epsilon, \theta, \phi$). Introduce the proper coordinate system, naturally adapted to such an observer. Consider another observer B who is freely falling into the black hole in a radial trajectory and introduce the local inertial frame appropriate to this observer. Write down the explicit coordinate transformation between these two frames which is valid around the location of the first observer. Are these coordinate transformations well behaved at the horizon?

(b) Write down Maxwell's equations and the Lorentz force law appropriate for the two observers A and B in $(1+3)$ notation. Make sure that the equations are expressed in terms of the electric and magnetic fields (as well as other variables) as measured by the respective observers. (You may find the results of Exercise 5.16 useful.)

(c) The electric and magnetic field components, as measured by any freely falling observer, should be finite on the horizon. Derive the boundary conditions this implies for the electric and magnetic fields as measured by the oberver A. Show that the tangential components of all electromagnetic fields look similar to that of an in-going electromagnetic wave near the horizon. Provide a physical interpretation of this result.

(d) The electromagnetic field lines of a charged particle kept stationary in the frame of observer A were derived in Project 5.2. Show that this is consistent with the assumption that the horizon surface is endowed with an effective charge density. Estimate the charge density. Describe qualitatively how the field lines change if the charge moves slowly a small distance in a tangential direction. Also show that the horizon can be attributed a surface resistivity $(4\pi/c) \approx 377$ ohms, which is the impedence of free space. Explain why this is reasonable.

(e) See whether you can explain the thermodynamic properties of a black hole (temperature, entropy, etc.) in terms of the membrane paradigm. If so, what are the degrees of freedom responsible for the entropy of the black hole? How do the thermal phenomena appear to a freely falling observer, in contrast to a stationary observer?

Project 14.3
Accelerated detectors in curved spacetime

We have seen in the text that thermal effects arise in two separate contexts. First, a uniformly accelerated detector, moving with an acceleration κ in flat spacetime, behaves as though it is immersed in a thermal bath with temperature $T = \kappa/2\pi$. Second, certain spacetimes with bifurcation horizons like Schwarzschild spacetime, de Sitter spacetime, etc., exhibit thermal properties. This project investigates the effects of combining these two phenomena in some specific contexts.

(a) Prove that an inertial detector will detect thermal radiation with temperature $H/2\pi$ in a de Sitter universe. Now show that an accelerated detector in the de Sitter universe will detect a thermal bath with the temperature $T = (H^2 + \kappa^2)^{1/2}/2\pi$, where κ is the magnitude of the proper acceleration. Interpret this result.[7] [Hint. Use the embedding of the de Sitter spacetime in the five-dimensional space judiciously.]

(b) We know (see Project 7.1) that the Schwarzschild metric can also be thought of as an induced metric on a surface embedded in a six-dimensional space. Can you obtain a result similar to the one in part (a) above by using this embedding?

(c) Consider a more general case of a uniformly accelerated trajectory in an arbitrary curved spacetime. Is there a way of generalizing the above results at least approximately in this context?

15

Gravity in higher and lower dimensions

15.1 Introduction

Once we venture into spacetime dimensions other than four, the possible theoretical models increase enormously and it is impossible (and, to a certain extent, unnecessary) to do justice to all of them in a comprehensive manner. Hence, in this chapter, we will confine our attention to a *few selected topics* which deal with the description of gravitational field in dimensions other than four.

The motivation to study gravity in $D < 4$ is quite different from the motivation to study gravity in $D > 4$. The interest in two and three spacetime dimensions arises from the hope that the simplified models in these lower dimensions will provide a better understanding of gravity in four dimensions as well as help us to appreciate those features of gravity which are closely tied to the fact that the spacetime has four dimensions. While these models are interesting, it is probably fair to say that the study of lower dimensional models has not significantly enhanced our understanding of general relativity in four dimensions. Therefore, our discussion of these models will be quite brief.[1]

The motivation to study models in $D > 4$ comes from two key factors. First, there is a very natural class of gravitational theories (called *Lanczos–Lovelock* models) which exist in higher dimensions and share several properties of Einstein's theory in four dimensions. Second, it is not inconceivable that the real world does have more than four dimensions with the extra dimensions either compactified or dynamically structured in a suitable manner so as to have escaped experimental detection so far. We shall describe both these aspects in this chapter but will provide more details as regards the Lanczos–Lovelock models.

15.2 Gravity in lower dimensions

15.2.1 *Gravity and black hole solutions in* $(1+2)$ *dimensions*

We shall begin our discussion with the structure of Einstein's theory in (1+2) dimensions. The natural action for this theory could be taken to be

$$A[g] = \frac{1}{16\pi G} \int d^3x \sqrt{-g} \, (R - 2\Lambda + L_m), \qquad (15.1)$$

where R is the scalar curvature of the (1+2)-dimensional spacetime, Λ is the cosmological constant and L_m is the matter Lagrangian. In natural units with $c = \hbar = 1$, the gravitational constant G should have the dimension of length (to make the action dimensionless) in $(1+2)$ dimensions. It is convenient to choose the remaining freedom in units to set $8G = 1$. Varying the action we obtain Einstein's equations in (1+2) dimensions to be

$$R_{ab} - \frac{1}{2}g_{ab} \, (R - 2\Lambda) = \pi T_{ab}, \qquad (15.2)$$

where a, b, etc., range over $0, 1, 2$. We have already seen in Section 5.6.2 (see Eq. (5.194)) that the curvature tensor is completely determined by the Ricci tensor in three dimensions. Hence the distribution of matter, described by $T_{ab}(x)$, completely determines the curvature tensor of the spacetime purely algebraically. It follows that in regions outside of matter in the spacetime, where $T_{ab} = 0$, the curvature tensor vanishes and the spacetime is locally flat (if $\Lambda = 0$) or locally de Sitter (if $\Lambda > 0$) or locally anti-de Sitter (if $\Lambda < 0$). There are no propagating modes in three-dimensional gravity. The absence of a gravitational field outside of matter distribution also shows that there is no natural Newtonian approximation to this theory.

In spite of this, rather trivial, structure it is possible to find some interesting solutions to Eq. (15.2). One such solution that has been extensively investigated in the literature is called the *Banados–Teitelboim–Zanelli* (BTZ) black hole.[2] This is a solution to Eq. (15.2) with $T_{ab} = 0$ and $\Lambda \equiv -(1/\ell^2) < 0$. The metric is given by

$$ds^2 = -Fdt^2 + \frac{dr^2}{F} + r^2(d\phi - \Omega dt)^2, \qquad (15.3)$$

where

$$F(r) = -M + \frac{r^2}{\ell^2} + \frac{J^2}{4r^2}, \qquad \Omega(r) = -\frac{J}{2r^2}, \qquad |J| \le M\ell. \qquad (15.4)$$

The parameter ($M > 0$) is analogous to the black hole mass and J is analogous to the angular momentum. The spacetime has two horizons at $r = r_{\pm}$ where

$$r_{\pm}^2 = \frac{M\ell^2}{2} \left\{ 1 \pm \left[1 - \left(1 - \frac{J}{M\ell} \right)^2 \right]^{1/2} \right\}. \qquad (15.5)$$

Just as in the case of the Kerr metric, the metric in Eq. (15.3) has two Killing vectors $\xi_{(t)}$ and $\xi_{(\phi)}$ corresponding to translations in t and ϕ. The static limit for the spacetime occurs at $r_{\text{stat}} = M^{1/2}\ell$, which is located outside the event horizon r_+. The surface gravity of the event horizon (using which one could develop the thermodynamics of BTZ black holes exactly as in the case of four-dimensional black hole spacetimes described in Chapter 14) is given by $\kappa = (r_+^2 - r_-^2)/\ell^2 r_+$.

Since the BTZ metric is a solution with $T_{ab} = 0, \Lambda < 0$ the curvature is completely determined by Λ and the spacetime must be locally the same as an anti-de Sitter (AdS) space (see Exercise 10.13.) The key difference between the BTZ solution and the AdS space is in their global topology. We know that an AdS spacetime can be embedded in a flat space of one higher dimension with the metric

$$ds^2 = -dT^2 - dU^2 + dX^2 + dY^2. \tag{15.6}$$

The hypersurface described by the equation

$$-T^2 - U^2 + X^2 + Y^2 = -\ell^2 \tag{15.7}$$

has constant negative curvature and thus is a solution to vacuum Einstein's equation with a negative cosmological constant. Locally, the embedded surface can be expressed in the form

$$T = \sqrt{\frac{r^2}{M} - \ell^2} \sinh \frac{\sqrt{M}}{\ell} t, \qquad U = \frac{r}{\sqrt{M}} \cosh \left(\sqrt{M}\,\phi\right)$$

$$X = \sqrt{\frac{r^2}{M} - \ell^2} \cosh \frac{\sqrt{M}}{\ell} t, \qquad Y = \frac{r}{\sqrt{M}} \sinh \left(\sqrt{M}\,\phi\right). \tag{15.8}$$

In these coordinates, the induced metric becomes

$$ds^2 = -\left(-M + \frac{r^2}{\ell^2}\right) dt^2 + \frac{dr^2}{-M + (r^2/\ell^2)} + r^2 d\phi^2. \tag{15.9}$$

Comparing this expression with the line-element in Eq. (15.2), we see that this corresponds to a non-rotating BTZ black hole with $J = 0$. But for this interpretation to be globally valid, we need to make an identification of ϕ and $\phi + 2\pi$. It is this identification which makes the topology of the BTZ black hole different from that of the AdS spacetime.

An alternate way of expressing this result is to note that the mass parameter M can be removed from Eq. (15.9) by rescaling the coordinate t and r. However, this rescaling will change the period of ϕ in an M dependent manner. Hence the topology is sensitive to M even though the metric can be made independent of this parameter. It is this feature which leads to part of the interest in BTZ black holes.

15.2.2 Gravity in two dimensions

In the case of (1+1) dimensions, the Einstein tensor identically vanishes (see Section 5.6.1) and hence one cannot obtain nontrivial solutions to Einstein equations in a natural fashion. It is, however, possible to reduce some of the higher dimensional theories (like the four-dimensional gravity) to a two-dimensional system by confining one's attention to spacetimes with certain symmetries. For example, if we restrict ourselves to spherically symmetric spacetimes in four dimensions we can perform an integration over the angular coordinates in the action functional and obtain an effective, two-dimensional, action which involves integration over t and r coordinates of a reduced Lagrangian. Such an action, however, will *not* be in the form of an Einstein–Hilbert action but the resulting theory can be considered as an effective two-dimensional gravity. (Some of these features were explored in D dimensions in Exercise 7.1.) We will not discuss such models in this chapter.

15.3 Gravity in higher dimensions

The existence of more than three spatial dimensions has been often postulated in physical theories in order to exploit the extra features which such models bring in. Any such model obviously has to possess a mechanism by which the extra dimensions are rendered unobservable at moderate energy scales. Consider, for example, the description of a relativistic particle of mass m arising as the quanta of a scalar field ϕ in a spacetime with (4+1) dimensions. Denoting the ordinary three dimensions with coordinates x^α and the extra spatial coordinate by y, the free-field equation can be written as

$$\partial^2_{(5)}\phi - m^2\phi \equiv (-\partial^2_t + \partial_\alpha\partial^\alpha + \partial^2_y)\phi - m^2\phi^2 = 0 \qquad (15.10)$$

with $\alpha = 1, 2, 3$ covering the ordinary three spatial dimensions. This equation possesses plane wave solutions which can propagate in the y-direction *as well* and hence will modify the dispersion relation of particles. This will, of course, lead to observable consequences which are not seen in the real world. The simplest way of getting around this difficulty is the following. We postulate that the extra spatial dimension is compact and has a small scale length. For example, let us consider the possibility that the topology of the space is $\mathcal{R}^3 \times S^1$, where the S^1 has a radius R. In this case, we identify the coordinates y and $y+2\pi R$ and the acceptable solutions to Eq. (15.10) can be expressed in the form

$$\phi(x^a, y) = \sum_n \phi_n(x^a) \exp\left(i\frac{ny}{R}\right), \qquad (15.11)$$

where n is an integer ensuring the condition that $\phi\,(x^a, y) = \phi\,(x^a, y + 2\pi R)$. Substituting Eq. (15.11) into Eq. (15.10) we find that ϕ_n satisfies the equation

$$\Box\,\phi_n - \left(m^2 + \frac{n^2}{R^2}\right)\phi_n = 0. \tag{15.12}$$

This is just the standard wave equation for a particle with the mass term m^2 replaced by $m^2 + n^2/R^2$. Hence the dispersion relation for the waves will now be given by $\omega^2 = k^2 + m^2 + n^2/R^2$. Let us suppose that we are studying physics at energy scales below some threshold ω_{max}. It is clear that, unless $\omega_{\text{max}} > (1/R)$, we cannot ensure the validity of this dispersion relation except by choosing $n = 0$. Hence, physics at energy scales below $(1/R)$ will essentially be described by the standard dispersion relation $\omega^2 = k^2 + m^2$. That is, we cannot probe any features of the compact dimension unless we go to energy scales greater than $(1/R)$ so that we can excite at least the $n = 1$ mode which 'winds' around the compact dimension. So, by keeping R sufficiently small, one can escape any experimental bound.

In the above discussion we postulated that the extra dimension is compact and introduced an effective length scale R thereby preventing the modes from exploring the extra dimension. It turns out that one can also do this by dynamically confining all matter fields (other than gravity) within a small effective length in the extra dimensions without actually postulating that the extra dimensions are compact. We shall see some examples of this in Section 15.3.2.

The above features also influence the Newtonian approximation of gravitational theories in D dimensions. With three spatial dimensions, the Newtonian potential energy has the form $V(r) = G_{(4)}m_1m_2/r \equiv (m_1m_2/M_{(4)}^2)(1/r)$, where $M_{(4)}$ is the standard Planck mass and $G_{(4)}$ is the Newtonian constant in $(1+3)$ dimensions. (In arriving at the second equation, we have used natural units with $\hbar = c = 1$.) When we generalize this result to D dimensions, the potential will be modified to $V(r) = (m_1m_2/M_{(D)}^{D-2})(1/r^{D-3})$ at small scales ($r \lesssim R$). But we have already seen that, at low energies, we cannot probe the compact dimensions. So, for $r \gg R$, the extra dimensions cannot be probed by the gravitational potential and it should take the form $V(r) = (m_1m_2/M_{(D)}^{D-2})(1/R^{D-4})(1/r)$ in the Newtonian theory. This result shows that $M_{(D)}^2 = M_{(4)}^2/[M_{(D)}R]^{D-4}$ allowing for the intriguing possibility that $M_{(D)}$ (the Planck mass in the full D-dimensional spacetime) can be much smaller than $M_{(4)} \approx 10^{19}$ GeV if $M_{(D)}R$ is large enough. This fact has also motivated some of the investigations into D-dimensional gravitational theories.

We shall begin our study of higher dimensional gravity by describing some aspects of black hole solutions in higher dimensions.

15.3.1 *Black holes in higher dimensions*

Let us consider Einstein's theory in $D > 4$ dimensions, taking the action to be

$$A[g] = \int d^D x \sqrt{-g} \left(\frac{1}{16\pi} R + L_m \right). \tag{15.13}$$

Here we are using units with $c = G = 1$; note that, in D dimensions, G should have the dimensions of (length)$^{D-2}$ to make action dimensionless in natural units. The L_m stands for the matter Lagrangian and we have set the cosmological constant to zero for simplicity. Varying the action we get the Einstein equation in D-dimensional spacetime to be

$$R_{ab} - \frac{1}{2} g_{ab} R = 8\pi T_{ab}, \tag{15.14}$$

where a, b, etc., vary over $(0, 1, 2, \ldots, D-1)$

In four dimensions we know that there are black hole solutions with the metric components $-g_{00} = g_{rr}^{-1} = 1 - h_{00}$, where $h_{00} = (2M/r)$ is the solution in the weak field approximation. Remarkably enough this idea generalizes to D dimensions. The metric for a spherically symmetric black hole in D dimensions, which is a solution to Eq. (15.14), has the form

$$ds^2 = -B(r)dt^2 + B^{-1}(r)\, dr^2 + r^2 d\Omega_{D-2}^2, \tag{15.15}$$

where $d\Omega_{D-2}^2$ in the last term is the metric on the unit sphere in $(D-2)$ dimensions and

$$B(r) = 1 - \frac{16\pi M}{(D-2)A_{D-2}} \frac{1}{r^{D-3}}, \tag{15.16}$$

with $A_{D-2} = (2\pi^{(D-1)/2})/\Gamma\left((D-1/2)\right)$ being the area of a unit sphere in $(D-2)$ space. In $D = 4$ this reduces to the standard Schwarzschild metric. In fact, the result generalizes to *charged* black holes as well with $B(r)$ now given by

$$B(r) = 1 - \frac{16\pi M}{(D-2)A_{D-2}} \frac{1}{r^{D-3}} + \frac{2Q^2}{(D-2)(D-3)} \frac{1}{r^{2(D-3)}}. \tag{15.17}$$

The corresponding electromagnetic field is obtained from the vector potential having the form $A_b\, dx^b = (Q/(D-3)[1/r^{D-3}]\, dt$. From the asymptotic structure of the metric one can identify Q as the charge and M as the mass. (These results were obtained earlier in a different manner in Exercise 7.4, with a cosmological constant also added to the source.)

15.3.2 *Brane world models*

We shall next consider another class of solutions that arise in higher dimensional gravity and have attracted considerable attention in recent years in the cosmological context.[3] For simplicity, we shall illustrate the concepts in the case of five-dimensional spacetime. In these models (usually called *brane world models*), one describes our four-dimensional world as embedded in a five-dimensional spacetime which has one extra spacelike dimension. Unlike in the discussion presented in the beginning of Section 15.3, we do *not* assume in these models that the extra dimension is compact. Instead, the solutions to the five-dimensional Einstein equations are chosen in such a way that the *effective* size of the fifth dimension (which is *formally* infinite in extent) is small. The non-gravitational physics of matter fields is confined to the four-dimensional sector (usually called a *brane*), while the gravity is allowed to probe the five-dimensional region (usually called *bulk*).

The embedding of a subspace in a larger space has already been studied in Chapter 12 in the context of three-dimensional surfaces embedded in four-dimensional spacetime. These ideas generalize in a direct and obvious manner to higher dimensions. Consider a brane with the metric q_{ab} embedded in the five-dimensional space with the metric g_{ab}. If the normal to the brane is n^a, then the two metrics are related by $q_{ab} = g_{ab} - n_a n_b$. Our first task is to obtain the equations obeyed by the metric on the brane (which will replace the standard Einstein equations) from the Einstein equations for the five-dimensional spacetime. To do this, we start with the Gauss–Codacci equation relating the curvature components of the brane and the bulk given by

$$^{(4)}R^a{}_{bcd} = {}^{(5)}R^m{}_{nrs} q^a_m q^n_b q^r_c q^s_d + K^a_c K_{bd} - K^a_d K_{bc},$$ (15.18)

$$D_n K^n_m - D_m K = {}^{(5)}R_{rs} n^s q^r_m.$$ (15.19)

(These are analogous to Eq. (12.25) and Eq. (12.28) of Chapter 12.) The extrinsic curvature of the brane is as usual defined by $K_{mn} = q^a_m q^b_n \nabla_a n_b$ with $K \equiv K^m_m$. The D_m is the covariant derivative operator compatible with the metric q_{mn}. Contracting Eq. (15.18) on a and c we get

$$^{(4)}R_{mn} = {}^{(5)}R_{rs} q^r_m q^s_n - {}^{(5)}R^a{}_{bcd} n_a q^b_m n^c q^d_n + K K_{mn} - K^a_m K_{na}.$$ (15.20)

Contracting this further to find a Ricci scalar, we can construct the Einstein tensor for the four-dimensional subspace to be

$$^{(4)}G_{mn} = \left[{}^{(5)}R_{rs} - \frac{1}{2} g_{rs} {}^{(5)}R\right] q^r_m q^s_n + {}^{(5)}R_{rs} n^r n^s q_{mn}$$ (15.21)

$$+ K K_{mn} - K^r_m K_{nr} - \frac{1}{2} q_{mn}(K^2 - K^{ab} K_{ab}) - A_{mn},$$

where we have defined the quantity

$$A_{mn} \equiv {}^{(5)}R^a{}_{brs} n_a n^r q^b_m q^s_n.$$ (15.22)

We now substitute the five-dimensional Einstein equations, $^{(5)}R_{ab} - (1/2)g_{ab}$ $^{(5)}R = \kappa^2_{(5)} T_{ab}$ (where $\kappa^2_{(5)}$ is related to the five-dimensional gravitational constant), into Eq. (15.21). It is also convenient to decompose the five-dimensional curvature tensor into a part related to the Ricci tensor and a part related to the Weyl tensor (see Eq. (5.65)) as

$$^{(5)}R_{manb} = \frac{2}{3}(g_{m[n}{}^{(5)}R_{b]a} - g_{a[n}{}^{(5)}R_{b]m}) - \frac{1}{6}g_{m[n}g_{b]a}{}^{(5)}R + {}^{(5)}C_{manb}. \quad (15.23)$$

Putting all these together, Eq. (15.21) becomes

$$^{(4)}G_{mn} = \frac{2\kappa^2_{(5)}}{3}\left(T_{rs}q^r_m q^s_n + \left(T_{rs}n^r n^s - \frac{1}{4}T^r_r\right)q_{mn}\right) \quad (15.24)$$

$$+ KK_{mn} - K^s_m K_{ns} - \frac{1}{2}q_{mn}\left(K^2 - K^{ab}K_{ab}\right) - E_{mn},$$

where

$$E_{mn} \equiv {}^{(5)}C^a{}_{brs}n_a n^r q^b_m q^s_n. \quad (15.25)$$

From the definition it is clear that E_{mn} is traceless. Equation (15.19) can now be written in the form

$$D_n K^n_m - D_m K = \kappa^2_{(5)} T_{rs}n^s q^r_m. \quad (15.26)$$

To proceed further, it is simpler to make a choice of coordinates similar to the one with vanishing lapse and shift functions in the case of (1+3) decomposition. In such a case we can take the five-dimensional line-element to be

$$ds^2 = d\chi^2 + q_{mn}dx^m dx^n, \quad (15.27)$$

where χ denotes the coordinate normal to the brane (with $n_m dx^m = d\chi$) and we shall assume that the brane is located at $\chi = 0$. (Such a coordinate choice is always possible locally.) In this case, it is obvious that $a^m = n^n \nabla_n n^m = 0$. As regards the energy-momentum tensor in five dimensions, we shall assume that the bulk contributes only a cosmological constant while the normal matter contribution arises only from the brane. This will allow us to write the energy-momentum tensor in the form

$$T_{mn} = -\Lambda g_{mn} + S_{mn}\delta(\chi); \qquad S_{mn} = -\sigma q_{mn} + \tau_{mn}. \quad (15.28)$$

The form of S_{mn} shows that σ can be thought of as being analogous to the cosmological constant in the brane or equivalently as the tension of the brane in five dimensions. The Dirac delta function in T_{mn}, indicating the fact that normal matter lives only on the brane, shows that we have to impose junction conditions on the brane to take care of the discontinuity when we cross the $\chi = 0$ surface. These have already been derived in Section 12.5. Denoting by a square bracket the discontinuity in a physical quantity when we cross a brane, equations analogous to Eq. (12.114) and Eq. (12.125) now reduce to

$$[q_{mn}] = 0, \qquad [K_{mn}] = -\kappa_{(5)}^2 \left(S_{mn} - \frac{1}{3} q_{mn} S \right). \tag{15.29}$$

These equations, however, determine only the *difference* in the physical quantities, for example the extrinsic curvature, on the two sides of the brane. It is usual to assume further that the solutions we are interested in do not distinguish between the two sides of the brane, allowing us to impose a Z_2 symmetry with respect to the brane. This assumption will allow us uniquely to determine the extrinsic curvature to be

$$K_{mn}^+ = -K_{mn}^- = -\frac{1}{2}\kappa_{(5)}^2 \left(S_{mn} - \frac{1}{3} q_{mn} S \right). \tag{15.30}$$

Because of the Z_2 symmetry, we can now evaluate the physical quantities on either side of the brane. We now substitute Eq. (15.30) into Eq. (15.24) to obtain the equation governing gravity on the brane:

$$^{(4)}G_{mn} = -\Lambda_4 q_{mn} + 8\pi G_N \tau_{mn} + \kappa_{(5)}^4 \pi_{mn} - E_{mn}, \tag{15.31}$$

where $\Lambda_4 = (1/2)\kappa_{(5)}^2 \left(\Lambda + (1/6)\kappa_{(5)}^2 \sigma^2 \right)$, $G_N = \left(\kappa_{(5)}^4 \sigma/48\pi \right)$, and

$$\pi_{mn} = -\frac{1}{4}\tau_{ma}\tau_n{}^a + \frac{1}{12}\tau\tau_{mn} + \frac{1}{8}q_{mn}\tau_{ab}\tau^{ab} - \frac{1}{24}q_{mn}\tau^2. \tag{15.32}$$

Expressed in this form, Eq. (15.31) resembles the standard four-dimensional Einstein equations with different kinds of sources on the right hand side. On the right hand side of Eq. (15.31) the first term is the four-dimensional cosmological constant which is given in terms of the five-dimensional cosmological constant and a term involving the brane tension. The second term is the standard source term in Einstein's equation defined in terms of an effective Newtonian constant. The third term is a correction to the second term which essentially involves the square of the energy-momentum tensor. This term and the last term – which arises from the Weyl tensor in five dimensions – are the nontrivial effects induced in the four-dimensional dynamics due to the existence of the extra dimensions.

In this form, Eq. (15.31) is not closed because we still need to determine the form of E_{mn}. It is actually possible to manipulate the five-dimensional Einstein equations to find the evolution equation for E_{mn}; we shall, however, not bother to do this because, in most cases of interest, one works with the ansatz $E_{mn} = 0$, which turns out to be consistent. It should, however, be noted that E_{mn} carries information about the gravitational field of the bulk. When it does *not* vanish, it is not freely specifiable and will be constrained by the evolution of the matter on the brane. But when the bulk is taken to be, say, AdS spacetime as is usually done, one can set $E_{mn} = 0$ consistently.

As an example in the study of these equations we shall consider a simple brane cosmological model. This is constructed by assuming that the bulk is a five-dimensional anti-de Sitter spacetime with a negative cosmological constant $\Lambda = -6/\ell^2$, and the brane is described by a standard Friedmann model with a spatial curvature parameter $k = 0, \pm 1$ and an expansion factor $a(t)$. The energy-momentum tensor on the brane is taken to be $\tau_{mn} = \text{dia}\,(-\rho, p, p, p)$. In this case it is easy to show that Eq. (15.31) reduces to the equation

$$H^2 = \left(\frac{\dot{a}}{a}\right)^2 = \left(\frac{\kappa_{(5)}^4}{36}\sigma^2 - \mu^2\right) + \frac{\kappa_{(5)}^4}{18}\sigma\rho + \frac{\kappa_{(5)}^4}{36}\rho^2 - \frac{k}{a^2} + \frac{C}{a^4}, \qquad (15.33)$$

where $\mu \equiv \ell^{-1}$ is the AdS parameter in mass units and C is a constant. (The term C/a^4 arises from the Weyl tensor on the bulk.) The second Friedmann equation can be traded off as usual for the conservation law $d(\rho a^3) = -pda^3$ which can be integrated if the equation of state is given.

An interesting solution to these equations can be obtained if we fine-tune the brane tension and the bulk cosmological constant by choosing $(\kappa_{(5)}^2/6)\sigma = \mu$. Then the first term on the right hand side of Eq. (15.33) vanishes and one can identify the four-dimensional Newtonian constant as

$$\kappa_4^2 \equiv 8\pi G = \frac{\kappa_{(5)}^4}{6}\sigma = \kappa_{(5)}^2\mu. \qquad (15.34)$$

With this identification, the second term on the right hand side of Eq. (15.33) dominates at low densities and leads to the standard Friedmann equation, while the third term is a correction to this equation which becomes significant for $\rho \gtrsim \sigma$. Though the last term in Eq. (15.33) – which arises from the Weyl tensor in the bulk – can be nonzero in general, one can consistently make the choice $C = 0$ if the bulk spacetime is anti-de Sitter. If we assume that $k = 0$ (that is, the brane is a spatially flat Friedmann universe) and that $C = 0$, Eq. (15.33) reduces to

$$H^2 = \frac{8\pi G}{3}\rho\left(1 + \frac{\rho}{2\sigma}\right). \qquad (15.35)$$

This can be easily integrated for a standard equation of state of the form $p = w\rho$. The conservation law now gives $\rho = \rho_0 a^{-q}$, $q \equiv 3(w+1)$ leading to the equation

$$\frac{\dot{a}^2}{a^2} = \frac{\kappa_{(5)}^4}{18}\sigma\rho_0 a^{-q} + \frac{\kappa_{(5)}^4}{36}\rho_0^2 a^{-2q} \qquad (15.36)$$

which has the solution

$$a^q = \frac{q^2}{4}\beta t^2 + q\sqrt{\xi}t, \qquad (15.37)$$

where

$$\beta = \frac{\kappa_{(5)}^4}{18}\sigma\rho_0, \qquad \xi = \frac{\kappa_{(5)}^4}{36}\rho_0^2. \qquad (15.38)$$

It is clear from this expression that there is a transition in the expansion rate of the universe at a timescale of the order of $\mu^{-1} = \ell$. At earlier times, we have a high energy regime with a behaviour $a \propto t^{1/q}$, while at late times we have a low energy regime with a standard evolution $a \propto t^{2/q}$. For example, if we treat matter on the brane to be relativistic with $w = 1/3, q = 4$ then the high energy regime is characterized by $a \propto t^{1/4}$ while the low energy regime has the expansion rate $a \propto t^{1/2}$ as in the standard Friedmann model. This is merely one representative solution and it is obvious that Eq. (15.33) has a much richer class of solutions depending on the different assumptions one makes.

Finally we comment on the choice we have made for the bulk, making it an AdS spacetime. A simpler choice, of course, would have been to assume $\Lambda = 0, k = 0$ and possibly $C = 0$ in the original equation. In this case, it is easy to verify that the expansion will be driven by the term which is quadratic in the energy density and the usual term (which is linear in energy density) will be absent. If we take matter to be made of relativistic particles with $w = 1/3$, this will lead to the behaviour $a(t) \propto t^{1/4}$ in the radiation dominated case in contrast to the usual $a(t) \propto t^{1/2}$. Such behaviour, however, cannot be reconciled with the observations of nucleosynthesis and hence is ruled out. In fact, nucleosynthesis also puts a bound on the parameters in the solution obtained above in Eq. (15.37) and requires the mass scale associated with the brane tension $\sigma^{1/4}$ to be greater than about 1 MeV. However, this bound is weaker than the bounds obtained in the laboratory which test for deviations from $(1/r^2)$ dependence of the gravitational force. These bounds require ℓ to be less than about 1 mm which puts a stronger constraint on the string tension.

15.4 Actions with holography

So far we have been concerned with $D(> 4)$ dimensional spacetimes in which gravity is described by the Einstein–Hilbert Lagrangian. We shall now consider a more general class of Lagrangians that can exist in higher dimensional spacetime and exhibit several unique properties. These Lagrangians are polynomials in the curvature tensor; that is, they involve the sum of products of curvature tensors with the indices contracted in a specific manner. Since the curvature tensor involves second derivatives, $\partial^2 g$, of the metric tensor, a term involving the product of, say, three curvature tensors will have a term that is cubic in $\partial^2 g$. One would have normally thought that this will lead to equations of motion involving derivatives which are of higher order than two in the dynamical variables – and hence would have been unacceptable. However, remarkably enough, it is possible to construct Lagrangians which are polynomials in the curvature tensor which lead to equations of motion that only involve up to second derivatives of the dynamical variables. (That is, these equations of motion will involve only functions of curvature tensor but not terms involving, say, the first derivative of the curvature tensor which could bring in third derivatives of the metric, etc.) To understand this result, it is desirable to concentrate on some peculiar features of the Einstein–Hilbert action and generalize them in a specific manner. We shall now turn to this task.[4]

We begin by recalling some peculiar features of the Einstein–Hilbert action in D dimensions, given by

$$A_{EH} = \int_V d^D x \sqrt{-g}\, L_{EH} = \int_V d^D x \sqrt{-g}\, R \tag{15.39}$$

where we have ignored the overall normalization factor $(16\pi)^{-1}$. As we saw in Chapter 6, this Lagrangian can be written in the form

$$L_{EH} \equiv Q_a{}^{bcd} R^a{}_{bcd}; \qquad Q_a{}^{bcd} = \frac{1}{2}(\delta_a^c g^{bd} - \delta_a^d g^{bc}). \tag{15.40}$$

The tensor $Q_a{}^{bcd}$ is the only fourth rank tensor that can be constructed from the metric (alone) that has all the symmetries of the curvature tensor. In addition it has zero divergence on all indices, $\nabla_a Q^{abcd} = 0$, etc. Since the curvature tensor $R^a{}_{bcd}$ (with one upper and three lower indices) can be expressed entirely in terms of Γ^i_{kl} and $\partial_j \Gamma^i_{kl}$ *without requiring* g^{ab}, the L_{EH} expressed as in Eq. (15.40) clearly separates the dependence on the metric (through $Q_a{}^{bcd}$ alone) from the dependence on connection and its derivative (through $R^a{}_{bcd}$). With g^{ab}, Γ^i_{kl} and $R^a{}_{bcd}$ (instead of g_{ab} and its first and second derivatives) treated as independent variables, the vacuum Einstein equations take a very simple form:

$$\left(\frac{\partial \sqrt{-g} L_{EH}}{\partial g^{ij}}\right) = R^a{}_{bcd}\left(\frac{\partial \sqrt{-g} Q_a{}^{bcd}}{\partial g^{ij}}\right) = 0. \tag{15.41}$$

That is, Einstein's equations arise through ordinary partial differentiation of the Lagrangian density with respect to g^{ab}, keeping Γ^i_{kl} and $\partial_j \Gamma^i_{kl}$ as constant.

If we raise one index of the curvature tensor, the Einstein–Hilbert Lagrangian can be written as

$$L_{EH} \equiv \delta_{ab}^{cd} R^{ab}_{cd}; \qquad \delta_{ab}^{cd} = \frac{1}{2}(\delta_a^c \delta_b^d - \delta_a^d \delta_b^c), \tag{15.42}$$

where δ_{ab}^{cd} is the alternating ('determinant') tensor. The importance of this form lies in the fact that it allows the generalization to a Lagrangian containing a product of, say, m curvature tensors, which – as we shall see – will share many properties of the Einstein–Hilbert action.

Since L_{EH} is linear in the second derivatives of the metric, it is clear that $\sqrt{-g}L_{EH}$ can be written as a sum $L_{\text{bulk}} + L_{\text{sur}}$, where L_{bulk} is quadratic in the first derivatives of the metric and L_{sur} is a total derivative which leads to a surface term in the action. From our discussion in Chapter 6 we know that this separation is given by

$$\sqrt{-g}L_{\text{EH}} = 2\partial_c\left[\sqrt{-g}Q_a{}^{bcd}\Gamma^a_{bd}\right] + 2\sqrt{-g}Q_a{}^{bcd}\Gamma^a_{dk}\Gamma^k_{bc} \equiv L_{\text{sur}} + L_{\text{bulk}} \quad (15.43)$$

with

$$L_{\text{bulk}} = 2\sqrt{-g}Q_a{}^{bcd}\Gamma^a_{dk}\Gamma^k_{bc}; \qquad L_{\text{sur}} = 2\partial_c\left[\sqrt{-g}Q_a{}^{bcd}\Gamma^a_{bd}\right] \equiv \partial_c\left[\sqrt{-g}V^c\right], \quad (15.44)$$

where the last equality defines the D-component object V^c. (We introduced this separation in Eq. (6.9); here we have defined $L_{\text{bulk}} = \sqrt{-g}L_{\text{quad}}$.) The key non-trivial result regarding the Einstein–Hilbert action is a simple relation between L_{bulk} and L_{sur}, allowing L_{sur} to be determined completely by L_{bulk}. (See Project 6.3; we will call actions with this property *holographic*). Using g_{ab} and $\partial_c g_{ab}$ as the independent variables in L_{bulk} one can easily prove (in $D > 2$ dimensions) that:

$$L_{\text{sur}} = -\frac{1}{[(D/2) - 1]}\partial_i\left(g_{ab}\frac{\partial L_{\text{bulk}}}{\partial(\partial_i g_{ab})}\right). \quad (15.45)$$

It is surprising that the Einstein–Hilbert action possesses such a holographic relation between the surface and bulk terms, since we never demanded any such feature while choosing it as the action describing gravity. We will now show that there is a more general class of Lagrangians (called *Lanczos–Lovelock Lagrangians*) which share many of the properties including the holographic relation. (In the next chapter we shall describe how these results can be understood in an alternative perspective that interprets gravity as an emergent phenomenon.)

Since the relations like Eq. (15.45) are linear in the Lagrangian, it follows that if two Lagrangians individually satisfy these relations, so will their sum with arbitrary constant coefficients. This allows us to investigate the individual terms in a sum of terms separately and also allows us to ignore relative coupling constants between them. With this motivation, we shall consider Lagrangians of the form

$$\sqrt{-g}L = \sqrt{-g}Q_a{}^{bcd}R^a{}_{bcd}; \qquad \nabla_c Q_a{}^{bcd} = 0, \quad (15.46)$$

where Q^{abcd} has the algebraic symmetries of the curvature tensor. From Eq. (15.43), which did not use the *explicit form* of Q^{abcd} but only the condition $\nabla_c Q^{abcd} = 0$, we see that *all* such Lagrangians can be separated into bulk and surface terms:

$$\sqrt{-g}L = 2\partial_c\left[\sqrt{-g}Q_a{}^{bcd}\Gamma^a_{bd}\right] + 2\sqrt{-g}Q_a{}^{bcd}\Gamma^a_{dk}\Gamma^k_{bc} \equiv L_{\text{sur}} + L_{\text{bulk}}. \quad (15.47)$$

Let us next consider the possible fourth rank tensors Q^{abcd}, which (i) have the symmetries of curvature tensor; (ii) are divergence-free; and (iii) made from g^{ab} and $R^a_{\ bcd}$. If we do not use the curvature tensor, then we have just one choice made from the metric given in Eq. (15.40) and this will lead to the Einstein–Hilbert action. Next, if we allow for $Q_a^{\ bcd}$ to depend *linearly* on curvature (so that the Lagrangian is quadratic in curvature), then we have the following additional choice for the tensor with required symmetries:

$$Q^{abcd} = R^{abcd} - G^{ac}g^{bd} + G^{bc}g^{ad} + R^{ad}g^{bc} - R^{bd}g^{ac}. \tag{15.48}$$

The corresponding Lagrangian in Eq. (6.17) now becomes:

$$\begin{aligned}
L_{GB} &= \frac{1}{2}\left(g_{ia}g^{bj}g^{ck}g^{dl} - 4g_{ia}g^{bd}g^{ck}g^{jl} + \delta^c_a\delta^k_i g^{bd}g^{jl}\right)R^i_{\ jkl}R^a_{\ bcd} \\
&= \frac{1}{2}\left[R^{abcd}R_{abcd} - 4R^{ab}R_{ab} + R^2\right].
\end{aligned} \tag{15.49}$$

This is just the Gauss–Bonnet Lagrangian, the variation of which is a pure divergence in four dimensions but not in higher dimensions (see Exercise 11.7).

Both the Einstein–Hilbert Lagrangian and the Gauss–Bonnet Lagrangian can be written in a condensed notation using alternating tensors as:

$$L_{EH} = \delta^{13}_{24}R^{24}_{13}; \qquad L_{GB} = \delta^{1357}_{2468}R^{24}_{13}R^{68}_{57}, \tag{15.50}$$

where the numeral n actually stands for an index a_n, etc., and the δs are the alternating tensors of the required order. (See Exercise 1.6 for the definition and properties of alternating tensors.) The obvious generalization leads to the mth order Lanczos–Lovelock Lagrangian:

$$L_{(m)} = \delta^{1357...2k-1}_{2468...2k}R^{24}_{13}R^{68}_{57}...R^{2k-2\,2k}_{2k-3\,2k-1}; \qquad k = 2m, \tag{15.51}$$

where $k = 2m$ is an even number. (Note that $m = 1$ gives the Einstein–Hilbert Lagrangian while $m = 2$ gives the Gauss–Bonnet Lagrangian.) The $L_{(m)}$ is clearly a homogeneous function of degree m in the curvature tensor R^{ab}_{cd} so that it can also be expressed in the form

$$L_{(m)} = \frac{1}{m}\left(\frac{\partial L_{(m)}}{\partial R^a_{\ bcd}}\right)R^a_{\ bcd} \equiv \frac{1}{m}P_a^{\ bcd}R^a_{\ bcd}, \tag{15.52}$$

where we have defined $P_a^{\ bcd} \equiv (\partial L_{(m)}/\partial R^a_{\ bcd})$ so that $P^{abcd} = mQ^{abcd}$. It can be directly verified that for these Lagrangians:

$$\nabla_a P^{abcd} = 0 = \nabla_a Q^{abcd} \tag{15.53}$$

so that these Lagrangians belong to the general class described by Eq. (15.46). The proof proceeds as follows. For the mth Lanczos–Lovelock Lagrangian $L_{(m)} =$

$R^{c_1 d_1}_{a_1 b_1} \ldots R^{c_m d_m}_{a_m b_m} \delta^{a_1 b_1 \ldots a_m b_m}_{c_1 d_1 \ldots c_m d_m}$ we have the result:

$$P^{ab}_{cd} = \frac{\partial L_{(m)}}{\partial R^{cd}_{ab}} = m R^{c_2 d_2}_{a_2 b_2} \ldots R^{c_m d_m}_{a_m b_m} \delta^{ab a_2 b_2 \ldots a_m b_m}_{cd c_2 d_2 \ldots c_m d_m}. \tag{15.54}$$

Therefore,

$$\nabla_a P^{ab}_{cd} = m \nabla_a \left(R^{c_2 d_2}_{a_2 b_2} \ldots R^{c_m d_m}_{a_m b_m} \right) \delta^{ab a_2 b_2 \ldots a_m b_m}_{cd c_2 d_2 \ldots c_m d_m}. \tag{15.55}$$

We note that the derivatives of the curvature tensor appearing in the expression are rendered completely antisymmetric in all the lower indices owing to the contraction with the alternating tensor. But the Bianchi identity states that $\nabla_{[a} R^{c_2 d_2}_{a_2 b_2]} = 0$ and thus we get $\nabla_a P^{ab}_{cd} = 0$. Because of the symmetries, P^{abcd} is divergence-free in *all* indices. Therefore, Lanczos–Lovelock Lagrangians belong to the class described by Eq. (15.46) and hence they allow a separation into bulk and surface terms as given by Eq. (15.47).

Note that one can add Lagrangians for different m with arbitrary coefficients c_m to get $L = L_{(1)} + c_2 L_{(2)} + c_3 L_{(3)} + \cdots$. Because of the antisymmetry of $\delta^{13 \ldots 2k-1}_{24 \ldots 2k}$ it follows that we must have $4m \leq D$ for the alternating tensor not to vanish identically. It can be shown that, when $4m = D$, the variation of $L_{(m)}$ is a pure surface term and will not contribute to the equations of motion and hence we need $D \geq 4m + 1$ to get a nontrivial effect. (See Exercise 11.7 for an illustration of this feature for $m = 2, D = 4$.) Therefore, if we are working in a spacetime of D dimensions, there are only a finite number of Lanczos–Lovelock Lagrangians that can be added together to describe the dynamics; the highest order Lanczos–Lovelock Lagrangian which is used must satisfy $4m_{\max} \leq D$ to have any nontrivial effect. In four dimensions, this uniquely selects Einstein gravity corresponding to $m = 1$. We shall now prove a host of relations for this class of Lagrangians.

The first result is that the equations of motion for these Lagrangians take a particularly simple form similar to Eq. (15.41). To see this, let us consider a *general* action of the form

$$A = \int_V d^D x \sqrt{-g} \, L[g^{ab}, R^a{}_{bcd}] \tag{15.56}$$

in which we have ignored higher derivatives of $R^a{}_{bcd}$ for simplicity. The variation of this action can leads to the following result:

$$\delta A = \delta \int_V d^D x \sqrt{-g} \, L = \int_V d^D x \sqrt{-g} \, E_{ab} \delta g^{ab} + \int_V d^D x \sqrt{-g} \, \nabla_j \delta v^j, \tag{15.57}$$

where

$$\sqrt{-g} E_{ab} \equiv \left(\frac{\partial \sqrt{-g} L}{\partial g^{ab}} - 2\sqrt{-g} \nabla^m \nabla^n P_{amnb} \right); \qquad P_a{}^{bcd} \equiv \left(\frac{\partial L}{\partial R^a{}_{bcd}} \right) \tag{15.58}$$

and

$$\delta v^j \equiv [2P^{ibjd}(\nabla_b \delta g_{di}) - 2\delta g_{di}(\nabla_c P^{ijcd})]. \tag{15.59}$$

To prove Eq. (15.57), let us consider the variation of the quantity $L\sqrt{-g}$, where L is a generally covariant scalar made from g^{ab} and $R^a{}_{bcd}$. We can express its variation in the form

$$
\begin{aligned}
\delta\left(L\sqrt{-g}\right) &= \left(\frac{\partial L\sqrt{-g}}{\partial g^{ab}}\right) \delta g^{ab} + \left(\frac{\partial L\sqrt{-g}}{\partial R^a{}_{bcd}}\right) \delta R^a{}_{bcd} \\
&= \left(\frac{\partial L\sqrt{-g}}{\partial g^{ab}}\right) \delta g^{ab} + \sqrt{-g}P_a{}^{bcd} \delta R^a{}_{bcd}.
\end{aligned}
\tag{15.60}
$$

The term $P_a{}^{bcd} \delta R^a{}_{bcd}$ is generally covariant and hence can be evaluated in the local inertial frame using

$$
\begin{aligned}
\delta R^a{}_{bcd} &= \nabla_c\left(\delta\Gamma^a_{db}\right) - \nabla_d\left(\delta\Gamma^a_{cb}\right) = \frac{1}{2}\nabla_c\left[g^{ai}\left(-\nabla_i\delta g_{db} + \nabla_d\delta g_{bi} + \nabla_b\delta g_{di}\right)\right] \\
&\quad - \{\text{term with } c \leftrightarrow d\}.
\end{aligned}
\tag{15.61}
$$

When this expression is multiplied by $P_a{}^{bcd}$ the middle term $g^{ai}\nabla_d\delta g_{bi}$ does not contribute because of the antisymmetry of P^{ibcd} in i and b. The other two terms contribute equally and we get a similar contribution from the term with c and d interchanged. Hence we get

$$P_a{}^{bcd}\delta R^a{}_{bcd} = 2P^{ibcd}\nabla_c\nabla_b(\delta g_{di}). \tag{15.62}$$

Manipulating the covariant derivative, this can be re-expressed in the form

$$P_a{}^{bcd}\delta R^a{}_{bcd} = 2\nabla_c\left[P^{ibcd}\nabla_b\delta g_{di}\right] - 2\nabla_b\left[\delta g_{di}\nabla_c P^{ibcd}\right] + 2\delta g_{di}\nabla_b\nabla_c P^{ibcd}. \tag{15.63}$$

Combining this with the first term in Eq. (15.60) and rearranging the expression, we can easily obtain Eq. (15.57).

The result in Eq. (15.57) is completely general. We now see that the equations of motion simplify significantly for a subclass of Lagrangians which satisfy Eq. (15.53) and are given by

$$\frac{1}{\sqrt{-g}}\frac{\partial\sqrt{-g}L}{\partial g^{ab}} = \frac{1}{2}T_{ab}, \tag{15.64}$$

where the right hand side arises from the variation of matter action. In the absence of matter, just setting the *ordinary derivative* of Lagrangian density with respect to g^{ab} to zero will give the equations of motion, as in the case of the Einstein–Hilbert action. Expanding out Eq. (15.64) and using $(\partial\sqrt{-g}/\partial g^{ab}) = -(1/2)\sqrt{-g}\,g_{ab}$

(which follows from Eq. (4.34)) as well as Eq. (15.54) we get:

$$\left(\frac{\partial L}{\partial R^{kl}_{ij}}\frac{\partial R^{kl}_{ij}}{\partial g^{ab}} - \frac{1}{2}g_{ab}L\right) = \left(P^{ij}_{kb}R^k{}_{aij} - \frac{1}{2}g_{ab}L\right)$$

$$= \left(P_b{}^{kij}R_{akij} - \frac{1}{2}g_{ab}L\right) = \frac{1}{2}T_{ab}. \tag{15.65}$$

In arriving at the first equality we have used the fact that while differentiating $R^{kl}_{ij} = g^{lm}R^k{}_{mij}$ we should keep $R^k{}_{mij}$ fixed. Clearly, the field equations do not involve more than second derivatives of the metric but (for $m \geq 2$) they are nonlinear functions of the second derivatives.

Before proceeding further, we should stress that we get sensible equations of motion from the action in Eq. (15.56) only if we set $(\delta v^j)n_j = 0$ at the boundary. For a general Lagrangian the counter-terms one needs to add are quite complicated but it appears that one can have a well defined action functional (see Project 15.1).

Exercise 15.1

Field equations in the Gauss–Bonnet theory   For $m = 2$, the Lanczos–Lovelock Lagrangian reduces to the Gauss–Bonnet Lagrangian. Adding this to the standard Einstein–Hilbert Lagrangian, the full action for the theory becomes

$$\mathcal{A} = \int d^D x \sqrt{-g} \left[\frac{1}{16\pi}(R + \alpha\mathcal{L}_{GB})\right] + \mathcal{A}_{matter}, \tag{15.66}$$

where R is the D-dimensional Ricci scalar; and $\mathcal{L}_{GB}$ is the Gauss–Bonnet Lagrangian which has the form

$$\mathcal{L}_{GB} = R^2 - 4R_{ab}R^{ab} + R_{abcd}R^{abcd}. \tag{15.67}$$

Vary this action and show that the equations of motion can be expressed in the form $G_{ab} + \alpha H_{ab} = 8\pi T_{ab}$, where G_{ab} is the Einstein tensor and

$$H_{ab} \equiv 2\left[RR_{ab} - 2R_{aj}R^j{}_b - 2R^{ij}R_{aibj} + R_a{}^{ijk}R_{bijk}\right] - \frac{1}{2}g_{ab}\mathcal{L}_{GB}. \tag{15.68}$$

Exercise 15.2

Black hole solutions in the Gauss–Bonnet theory   Show that the Gauss–Bonnet field equations derived in Exercise 15.1 admit black hole solutions with the metric:

$$ds^2 = -f(r)dt^2 + f^{-1}(r)dr^2 + r^2 d\Omega^2_{D-2}, \tag{15.69}$$

with a source $T^t_t = T^r_r \equiv \epsilon(r)/8\pi$. Reduce the Gauss–Bonnet equations to the single differential equation

$$rf' - (D-3)(1-f) + \frac{\bar{\alpha}}{r^2}(1-f)\left[2rf' - (D-5)(1-f)\right] = \frac{2\epsilon(r)}{D-2}r^2, \tag{15.70}$$

where $\bar{\alpha} = (d-3)(d-4)\alpha$. Solve this equation explicitly for $D = 5$.

Exercise 15.3

Analogue of Bianchi identity in the Lanczos–Lovelock theories Let $E_{ab} \equiv P_b{}^{kij} R_{akij} - \frac{1}{2} g_{ab} L$. Since it arises from the variation of a generally covariant scalar with respect to g^{ab}, it follows that $\nabla_a E_b^a = 0$.

(a) Prove this directly from the definition of $P_b{}^{kij}$ and L. That is, show explicitly that

$$\nabla_a \left(P_b{}^{kij} R^a{}_{kij} \right) = \frac{1}{2} \partial_b L. \tag{15.71}$$

(b) Suppose you are given that

$$2P_b{}^{kij} R^a{}_{kij} n^b n_a = T_b^a n^b n_a \tag{15.72}$$

for *all* null vectors n_a. Show that this implies the equations

$$2E_b^a = \left[2P_b{}^{ijk} R^a{}_{ijk} - \delta_b^a L \right] = T_b^a + \Lambda \delta_b^a, \tag{15.73}$$

where Λ is a constant. [Answer. Equation (15.72) implies that $2P_b{}^{ijk} R^a{}_{ijk} - T_b^a = -\lambda \delta_b^a$, where $\lambda(x)$ is some scalar. Rewriting this equation as $2E_b^a + (L + \lambda) \delta_b^a = T_b^a$, taking the divergence and using $\nabla_a E_b^a = 0 = \nabla_a T_b^a$, we get $\partial_b(L + \lambda) = 0$. Therefore, $\lambda = -L +$ constant leading to the result in Eq. (15.73).]

It also follows from Eq. (15.64) that, for the mth Lanczos–Lovelock Lagrangian, $L_{(m)}$ (given by Eq. (15.51)), the trace of the equations of motion is proportional to the Lagrangian:

$$E \equiv g^{ab} E_{ab} = \frac{g^{ab}}{\sqrt{-g}} \frac{\partial \sqrt{-g} L_{(m)}}{\partial g^{ab}} = -[(D/2) - m] L_{(m)}. \tag{15.74}$$

This *off-shell* relation (that is, a relation that remains valid even when field equations are not imposed) is easy to prove from the fact that we need to introduce m factors of g^{ab} in Eq. (15.51) to proceed from $R^a{}_{bcd}$ to R^{ab}_{cd} and that $\sqrt{-g}$ is a homogeneous function of g^{ab} of degree $-D/2$. Using this in Eq. (15.65), we get $L_{(m)} = -2E/(D - 2m) = -T/(D - 2m)$, which allows us to rewrite the field equation in the form

$$2P_b{}^{ijk} R^a{}_{ijk} = T_b^a - \delta_b^a \frac{1}{(D - 2m)} T. \tag{15.75}$$

This is similar in structure to $R_b^a = 8\pi(T_b^a - \delta_b^a(1/2)T)$, to which it reduces in $D = 4$ for the $m = 1$ case.

Further, we can prove (by straightforward but rather tedious algebra) that for *any* Lagrangian which depends on the metric and curvature tensor:

$$g_{ab} \frac{\delta L \sqrt{-g}}{\delta(\partial_i g_{ab})} = -2\sqrt{-g} \left(\frac{\partial L}{\partial R_{nbid}} \right) \Gamma_{nbd} = -2\sqrt{-g} P^{nbid} \Gamma_{nbd}, \tag{15.76}$$

where the *Euler derivative* occurring on the left hand side, denoted using δ, is defined as

$$\frac{\delta K[\phi, \partial_i \phi, \ldots]}{\delta \phi} = \frac{\partial K[\phi, \partial_i \phi, \ldots]}{\partial \phi} - \partial_a \left[\frac{\partial K[\phi, \partial_i \phi, \ldots]}{\partial(\partial_a \phi)} \right] + \cdots \qquad (15.77)$$

for any function $K[\phi, \partial_i \phi, \ldots]$. In the case of Lanczos–Lovelock Lagrangians, $P^{nbid} = mQ^{nbid}$ so that we get the relation:

$$\partial_i \left[g_{ab} \frac{\delta \sqrt{-g} \, L_{(m)}}{\delta(\partial_i g_{ab})} \right] = -m L_{\text{sur}}^{(m)}. \qquad (15.78)$$

This shows that m times the surface term is indeed of the '$d(qp)$' structure described in Project 6.3, provided the momentum is defined using the *total* Lagrangian L and the Euler derivative.

We shall now take up the generalization of the relation Eq. (15.45) (between L_{bulk} and L_{sur}) for the Lanczos–Lovelock case when $Q_a{}^{bcd}$ depends on the metric as well as the curvature. We will prove that:

$$[(D/2) - m]L_{\text{sur}}^{(m)} = -\partial_i \left[g_{ab} \frac{\delta L_{\text{bulk}}^{(m)}}{\delta(\partial_i g_{ab})} + \partial_j g_{ab} \frac{\partial L_{\text{bulk}}^{(m)}}{\partial(\partial_i \partial_j g_{ab})} \right]. \qquad (15.79)$$

The proof of Eq. (15.79) is based on a simple homology argument and combinatorics, generalizing a corresponding proof for Eq. (15.45) in the Einstein–Hilbert case, given in Project 6.3. Consider any Lagrangian $L(g_{ab}, \partial_i g_{ab}, \partial_i \partial_j g_{ab})$ and let $E^{ab}[L]$ denote the Euler–Lagrange function resulting from L:

$$E^{ab}[L] \equiv \frac{\partial L}{\partial g_{ab}} - \partial_i \left[\frac{\partial L}{\partial(\partial_i g_{ab})} \right] + \partial_i \partial_j \left[\frac{\partial L}{\partial(\partial_i \partial_j g_{ab})} \right]. \qquad (15.80)$$

Forming the contraction $g_{ab}E^{ab}$ and manipulating the terms, we can rewrite this equation as:

$$g_{ab}E^{ab}[L] = g_{ab}\frac{\partial L}{\partial g_{ab}} + (\partial_i g_{ab})\frac{\partial L}{\partial(\partial_i g_{ab})} + (\partial_i \partial_j g_{ab})\frac{\partial L}{\partial(\partial_i \partial_j g_{ab})}$$
$$- \partial_i \left[g_{ab}\frac{\delta L}{\delta(\partial_i g_{ab})} + \partial_j g_{ab}\frac{\partial L}{\partial(\partial_i \partial_j g_{ab})} \right]. \qquad (15.81)$$

We will now apply this relation to the bulk Lagrangian $L_{\text{bulk}}^{(m)} = 2\sqrt{-g}Q_a{}^{bcd}\Gamma_{dj}^a\Gamma_{bc}^j$ corresponding to the mth order Lanczos–Lovelock Lagrangian. (Hereafter, we will simplify the notation by just calling it L_{bulk}; it is understood that we are dealing with the mth order Lanczos–Lovelock Lagrangian throughout.) Since both $L_{(m)} \equiv L$ and L_{bulk} lead to the same equations of motion, $E^{ab}[L] = E^{ab}[L_{\text{bulk}}]$. Hence, using Eq. (15.74), we find the left hand side of Eq. (15.81) to be $[(D/2) - m]\sqrt{-g}L$. (The sign change is because we are now varying g_{ab} while in Eq. (15.74) we varied g^{ab}.) We will next show that the first three terms on the right hand side of Eq. (15.81) add up to give $[(D/2) - m]L_{\text{bulk}}$. Bringing this term to the left hand side and using $L_{\text{sur}} = \sqrt{-g}L - L_{\text{bulk}}$ will then lead to Eq. (15.79).

To prove these facts, let us write $L_{\text{bulk}}/\sqrt{-g}$ entirely in terms of $g^{ab}, \partial_i g_{ab}$ and $\partial_j \partial_i g_{ab}$ by multiplying out completely. In any given term, let us assume there are n_0 factors of g^{ab}, n_1 factors of $\partial_i g_{ab}$ and k factors of $\partial_i \partial_j g_{ab}$. Then homogeneity implies that for this particular term (labelled by k, which is the number of $\partial_i \partial_j g_{ab}$ that occur in it), the first three terms on the right hand side of Eq. (15.81) are given by

$$g_{ab}\frac{\partial L_{\text{bulk}}^{(k)}}{\partial g_{ab}} = [(D/2) - n_0]L_{\text{bulk}}^{(k)}; \qquad (\partial_i g_{ab})\frac{\partial L_{\text{bulk}}^{(k)}}{\partial(\partial_i g_{ab})} = n_1 L_{\text{bulk}}^{(k)};$$

$$(\partial_i \partial_j g_{ab})\frac{\partial L_{\text{bulk}}^{(k)}}{\partial(\partial_i \partial_j g_{ab})} = k L_{\text{bulk}}^{(k)}. \tag{15.82}$$

(In the first relation $D/2$ comes from the $\sqrt{-g}$ factor and the sign flip on n_0 is because of switching over from g^{ab} to g_{ab}.) Since all the indices – the two upper indices from each g^{ab}, three lower indices from each $\partial_i g_{ab}$, four lower indices from each $\partial_j \partial_i g_{ab}$ – are to be contracted out, we must have $2n_0 = 3n_1 + 4k$ which fixes n_0 in terms of n_1 and k. We next note that $Q_a{}^{bcd}$ is made of $(m - 1)$ factors of curvature tensor and each curvature tensor has the structure $R \simeq [\partial^2 g + (\partial g)^2]$. If we multiply out $(m - 1)$ curvature tensors, a generic term in the product will have k factors of $\partial^2 g$ and $(m - 1 - k)$ factors of $(\partial g)^2$. In addition, the two Γs in $L_{\text{bulk}} \simeq Q\Gamma\Gamma$ will contribute two more factors of (∂g). So, for this generic term, $n_1 = 2(m - 1 - k) + 2 = 2(m - k)$. Using our relation $2n_0 = 3n_1 + 4k$, we find $n_0 = 3m - k$. Substituting into Eq. (15.82) we get

$$g_{ab}\frac{\partial L_{\text{bulk}}^{(k)}}{\partial g_{ab}} = [(D/2) - 3m + k]L_{\text{bulk}}^{(k)}; \qquad (\partial_i g_{ab})\frac{\partial L_{\text{bulk}}^{(k)}}{\partial(\partial_i g_{ab})} = 2(m - k)L_{\text{bulk}}^{(k)};$$

$$(\partial_i \partial_j g_{ab})\frac{\partial L_{\text{bulk}}^{(k)}}{\partial(\partial_i \partial_j g_{ab})} = k L_{\text{bulk}}^{(k)}. \tag{15.83}$$

Though each of these terms depends on k, the sum of the three terms is independent of k leading to the same contribution from each term. So we get:

$$g_{ab}\frac{\partial L_{\text{bulk}}}{\partial g_{ab}} + (\partial_i g_{ab})\frac{\partial L_{\text{bulk}}}{\partial(\partial_i g_{ab})} + (\partial_i \partial_j g_{ab})\frac{\partial L_{\text{bulk}}}{\partial(\partial_i \partial_j g_{ab})} = [(D/2) - m]L_{\text{bulk}}. \tag{15.84}$$

Substituting this in Eq. (15.81), transferring these terms to the left hand side and using $L\sqrt{-g} - L_{\text{bulk}} = L_{\text{sur}}$, we get the result in Eq. (15.79).

The result in Eq. (15.79) is the appropriate generalization of Eq. (15.45) in the case of Einstein–Hilbert action and has a similar (generalized) '$d(qp)$' structure.

We shall next show that the surface term is closely related to: (a) the horizon entropy and (b) the Noether charge associated with the diffeomorphism invariance of the action.

15.5 Surface term and the entropy of the horizon

Surface terms in actions sometimes assume special significance in a theory and this is particularly true for Einstein–Hilbert action. In this case, one can relate the surface term in the action to the entropy of the horizons, if the solution possesses a bifurcation horizon. More generally, if the metric near the horizon can be approximated as a Rindler metric, then one can obtain the result that the entropy per unit transverse area is $(1/4)$ in Einstein's theory. To see this, we only need to evaluate the surface contribution

$$S_{\text{sur}} = 2 \int d^D x \, \partial_c \left[\sqrt{-g} Q^{abcd} \Gamma_{abd} \right] = 2 \int d^D x \, \partial_c \left[\sqrt{-g} Q^{abcd} \partial_b g_{ad} \right] \quad (15.85)$$

for a metric in the Rindler approximation

$$ds^2 = -\kappa^2 x^2 dt^2 + dx^2 + dx_\perp^2, \quad (15.86)$$

where $x_\perp$ denotes $(D-2)$ transverse coordinates. For the static metric, the time integration in Eq. (15.85) is trivial and involves multiplication by the range of integration. Since the surface gravity of the horizon (located at $x = 0$) is κ, the natural range for time integration is $(0, \beta)$ where $\beta = 2\pi/\kappa$. This is most easily seen in the Euclidean sector in which there is a natural periodicity. If we analytically continue the time coordinate in the metric in Eq. (15.86) to $t_E \equiv it$, we have $ds_E^2 = x^2 d(\kappa t_E)^2 + dx^2 + \cdots$ near $x = 0$. This is identical to the metric in polar coordinates and to avoid a conical singularity at $x = 0$ the t_E must be in the range $(0, 2\pi/\kappa)$. (Also see Section 14.5.) If $n_i = (0, 1, 0 \ldots)$ is a unit vector in the x-direction, the contribution on the $x = \epsilon$ surface, where ϵ is infinitesimal, is

$$S_{\text{sur}} = 2\beta \int d^{D-2} x_\perp \sqrt{-g} \, n_c Q^{abcd} \partial_b g_{ad}. \quad (15.87)$$

Clearly, only the $b = x, a = d = 0$ term contributes. Writing $Q^{0x0x} = g^{00} g^{xx} Q_{x0}^{x0}$, a simple calculation shows that

$$S_{\text{sur}} = -8\pi \int_{\mathcal{H}} d^{D-2} x_\perp \sqrt{\sigma} \left(Q_{0x}^{0x} \right), \quad (15.88)$$

where σ is the determinant of the metric in the transverse space. (It is preferable to do this computation at the surface $x = \epsilon$ and take the limit of $\epsilon \to 0$ at the end of the calculation to get the contribution from the horizon $\mathcal{H}$.) What we have evaluated in Eq. (15.88) is the contribution of the integral from *one* surface, which is taken to be the location of the horizon near which the Rindler approximation is valid.

In the case of the Einstein–Hilbert action, $Q^{abcd} = (1/32\pi)\left[g^{ac}g^{bd} - g^{ad}g^{bc}\right]$, where we have added the correct normalization to ensure that $Q^{abcd}R_{abcd} = R/16\pi$. So we get:

$$Q^{0x}_{0x} = \frac{1}{32\pi}; \qquad S_{\text{sur}} = -\frac{1}{4}\mathcal{A}_\perp \tag{15.89}$$

showing that the surface contribution is one quarter of the transverse area $\mathcal{A}_\perp$ of the horizon which is the same as horizon entropy, except for the sign. (Also see Exercise 14.8.) The sign can be understood from the fact that the contribution is calculated at the lower limit of the integral with the normal vector pointing towards positive x, while the outward pointing normal of an $x = \epsilon$ boundary, to a region outside the horizon, should be in the direction of negative x.

We will now show that the above result, relating the boundary term in the action to the entropy of horizons, continues to hold for Lanczos–Lovelock Lagrangians with our definition of L_{sur}, thereby providing a thermodynamic interpretation for our holographic separation of Lanczos–Lovelock Lagrangians.

To do this we need an expression for the entropy of the horizon in a general context when the Lagrangian depends of $R^a{}_{bcd}$ in a nontrivial manner. Such a formula can be obtained by interpreting entropy as the Noether charge associated with diffeomorphism invariance as discussed in Project 8.1. We shall briefly recall this approach and then use this definition.

In any generally covariant theory, the infinitesimal coordinate transformations $x^a \rightarrow x^a + \xi^a$ lead to conservation of a Noether current which depends on ξ^a (see Eq. (8.184)). As we saw in Project 8.1, the expression for the Noether current simplifies considerably when ξ^a is an (approximate) Killing vector. For a Killing vector ξ^a, the Noether current becomes, on-shell,

$$J^a = \left(T^{aj} + g^{aj}L\right)\xi_j. \tag{15.90}$$

Therefore, for any vector k_a that satisfies $k_a\xi^a = 0$, we have the result

$$(k_a J^a) = T^{aj}k_a\xi_j. \tag{15.91}$$

The change in this quantity, when T^{aj} changes by a small amount δT^{aj}, will be $\delta(k_a J^a) = k_a\xi_j\delta T^{aj}$. It is this relation that is the key to the interpretation of the Noether charge as the entropy. When some amount of matter with an energy-momentum tensor δT^{aj} crosses the horizon, the corresponding energy flux can be thought of as being given by the integral of $k_a\xi_j\delta T^{aj}$ over the horizon, where ξ^a is the Killing vector field corresponding to the bifurcation horizon and k_a is a vector orthogonal to ξ^a which can be taken as the normal to a timelike surface, infinitesimally away from the horizon. (Such a surface is sometimes called a 'stretched horizon' and is defined by the condition $N = \epsilon$, where N is the lapse function with

$N = 0$ representing the horizon.) In the $(D-1)$-dimensional integral over this surface, one coordinate is just time; since we are dealing with a static situation, the time integral reduces to multiplication by the range of integration. Based on our discussion earlier leading to Eq. (15.87), we assume that this range is $(0, \beta)$ where $\beta = 2\pi/\kappa$ and κ is the surface gravity of the horizon. Thus, on integrating $\delta(k_a J^a)$ over the horizon we get

$$\delta \int_{\mathcal{H}} d^{D-1}x \sqrt{h}(k_a J^a) = \int_{\mathcal{H}} d^{D-1}x \sqrt{h}k_a\xi_j\delta T^{aj} = \beta \int_{\mathcal{H}} d^{D-2}x\sqrt{h}k_a\xi_j\delta T^{aj}$$

$$\tag{15.92}$$

where the integration over time has been replaced by a multiplication by $\beta = (2\pi/\kappa)$. The integral over δT^{aj} is the flux of energy δE through the horizon so that $\beta\delta E$ can be interpreted as the rate of change of the entropy. This suggests an expression for entropy given by[5]

$$S_{\text{Noether}} = \beta\mathcal{N} = \beta \int d^{D-1}\Sigma_a J^a = \beta \int d^{D-2}\Sigma_{ab}J^{ab}, \tag{15.93}$$

where we have again replaced the time integration by multiplication by β and introduced the antisymmetric tensor J^{ab} by $J^a = \nabla_b J^{ab}$; such a J^{ab} is assured to exist because of the fact $\nabla_a J^a = 0$. In the final expression the integral is over any surface with $(D-2)$ dimension which is a spacelike cross-section of the Killing horizon on which the norm of ξ^a vanishes.

We want to evaluate the Noether charge $\mathcal{N}$ corresponding to the current J^a for a static metric with a bifurcation horizon and a Killing vector field $\xi^a = (1, \mathbf{0})$; $\xi_a = g_{a0}$. The location of the horizon is given by the vanishing of the norm, $\xi^a\xi_a = g_{00}$, of this Killing vector. Using these facts as well as the relations $\nabla_c\xi^d = \Gamma^d_{c0}$, etc., we find that $J^{ab} = 2P^{abcd}\nabla_c\xi_d = -2P_d^{0ab}\Gamma^d_{c0}$. Therefore the Noether charge is given by

$$\mathcal{N} = \int_t d^{D-1}x \sqrt{-g}\, J^0 = \int_t d^{D-1}x\, \partial_b(\sqrt{-g}J^{0b})$$

$$= \int_{t,r_H} d^{D-2}x \sqrt{-g}\, J^{0r}, \tag{15.94}$$

in which we have ignored the contributions arising from b when it ranges over the transverse directions. This is justifiable when transverse directions are compact or, in the case of the Rindler approximation, when nothing changes along the transverse direction. In the radial direction, we have again confined ourselves to the contribution at $r = r_H$, which is taken to be the location of the horizon. We have:

$$J^{0r} = 2P^{0rcd}\nabla_c\xi_d = 2P^{0rcd}\partial_c\xi_d = 2P^{0rcd}\partial_c g_{d0} = 2P^{cdr0}\partial_d g_{c0}. \tag{15.95}$$

The second equality arises from $P^{0rcd} = -P^{0rdc}$ and the rest follows from the symmetries of P^{abcd}. So

$$\mathcal{N} = 2 \int_{t,r_H} d^{D-2}x \sqrt{-g}\, P^{cdr0} \partial_d g_{c0} = 2m \int_{t,r_H} d^{D-2}x \sqrt{-g}\, Q^{cdr0} \partial_d g_{c0}.$$

(15.96)

The entropy, computed from the Noether charge approach is thus given by

$$S_{\text{Noether}} = \beta \mathcal{N} = 2\beta m \int_{t,r_H} d^{D-2}x \sqrt{-g}\, Q^{cdr0} \partial_d g_{c0}.$$

(15.97)

We will now show that this is the same result one obtains by evaluating our surface term on the horizon except for a proportionality constant. In the stationary case, the contribution of the surface term on the horizon is given by

$$S_{\text{sur}} = 2 \int d^D x\, \partial_c \left[\sqrt{-g} Q^{abcd} \partial_b g_{ad} \right] = 2 \int dt \int_{r_H} d^{D-2}x \sqrt{-g}\, Q^{abrd} \partial_b g_{ad}.$$

(15.98)

Once again, taking the integration along t to be in the range $(0, \beta)$ and ignoring transverse directions, we get

$$S_{\text{sur}} = 2\beta \int_{r_H} d^{D-2}x \sqrt{-g}\, Q^{abr0} \partial_b g_{a0}.$$

(15.99)

Comparing with Eq. (15.96), we find that

$$S_{\text{Noether}} = m S_{\text{sur}}.$$

(15.100)

The overall proportionality factor m has a simple physical meaning. Equation (15.78) tells us that the quantity $m L_{\text{sur}}$, rather than L_{sur}, has the '$d(qp)$' structure, and it is this particular combination which plays the role of entropy, as to be expected. (We will comment on some of the deeper implications of these results in the next chapter.)

Finally, we obtain an *explicit* expression for the horizon entropy in the Lanczos–Lovelock models which turns out to be quite interesting. From Eq. (15.88) we see that this is determined by $Q^{0x}_{0x} = Q^{x0}_{x0}$. Let us consider this quantity Q^{x0}_{x0} for the mth order Lanczos–Lovelock action, given by

$$Q^{x0}_{x0} = \frac{1}{16\pi} \frac{1}{2^m} \delta^{x0a_3...a_{2m}}_{x0b_3...b_{2m}} \left(R^{b_3 b_4}_{a_3 a_4} ... R^{b_{2m-1} b_{2m}}_{a_{2m-1} a_{2m}} \right) \Bigg|_{x=\epsilon},$$

(15.101)

where we have added a normalization which gives Einstein's action for $m = 1$ and will *define* $Q^{x0}_{x0} = 1/16\pi$ for the $m = 0$ case. We have also indicated that we are evaluating this expression in the Rindler limit of the horizon, as in Eq. (15.88). The presence of both 0 and x in each row of the alternating tensor forces all other indices to take the values $2, 3, ..., D-1$. In fact, we have $\delta^{x0a_3...a_{2m}}_{x0b_3...b_{2m}} = \delta^{A_3 A_4...A_{2m}}_{B_3 B_4...B_{2m}}$

with $A_i, B_i = 2, 3, ..., D - 1$ (the remaining combinations of Kronecker deltas on expanding out the alternating tensor are all zero since $\delta_A^0 = 0 = \delta_A^x$ and so on). Hence Q_{x0}^{x0} reduces to

$$Q_{x0}^{x0} = \frac{1}{2} \left(\frac{1}{16\pi} \frac{1}{2^{m-1}} \right) \delta_{B_3 B_4 ... B_{2m}}^{A_3 A_4 ... A_{2m}} \left(R_{A_3 A_4}^{B_3 B_4} ... R_{A_{2m-1} A_{2m}}^{B_{2m-1} B_{2m}} \right) \Bigg|_{x=\epsilon} . \tag{15.102}$$

In the $\epsilon \to 0$ limit therefore, recalling that $R_{CD}^{AB} |_{\mathcal{H}} = {}^{(D-2)} R_{CD}^{AB} |_{\mathcal{H}}$, we find that Q_{x0}^{x0} is essentially the Lanczos–Lovelock Lagrangian of order $(m-1)$:

$$Q_{x0}^{x0} = \frac{1}{2} L_{(m-1)}, \tag{15.103}$$

and the entropy $S_{\text{Noether}} = m S_{\text{sur}}$ becomes

$$S_{\text{Noether}}^{(m)} = -4\pi m \int_{\mathcal{H}} d^{D-2} x_\perp \sqrt{\sigma} L_{(m-1)}, \tag{15.104}$$

where we have restored the superscript (m) in the last expression. This is the entropy in the Lanczos–Lovelock theory.

It is interesting to observe that the entropy of the mth order Lanczos–Lovelock theory is an integral over the *Lagrangian* of $(m-1)$th order. For $m = 1$ (Einstein gravity), the $L_{(0)}$ is a constant giving an entropy proportional to the transverse area; for $m = 2$ (Gauss–Bonnet gravity), the entropy is proportional to the integral of R over transverse direction. If we take a linear combination of actions, then the entropy is given by the sum

$$S = \sum_{m=1}^{K} c_m S^{(m)} = -\sum_{m=1}^{K} 4\pi m c_m \int_{\mathcal{H}} d^{D-2} x_\perp \sqrt{\sigma} L_{(m-1)}, \tag{15.105}$$

of each order contribution.

Exercise 15.4
Entropy as the Noether charge The formal definition of entropy of a bifurcation horizon from the Noether charge is given by

$$S = \frac{2\pi}{\kappa} \oint_\Sigma \left(\frac{\delta L}{\delta R_{abcd}} \right) \epsilon_{ab} \epsilon_{cd} d\Sigma, \tag{15.106}$$

where κ is the surface gravity of the horizon and the $(D-2)$-dimensional integral is on a spacelike bifurcation surface with ϵ_{ab} denoting the bivector normal to the bifurcation surface. The variation is performed as if R_{abcd} and the metric are independent. The whole expression is evaluated on a solution to the equation of motion. Show that this matches with the expression used in the text. Prove that, for a metric of the form

$$ds^2 = -f(r)dt^2 + f^{-1}(r)dr^2 + R^2(r)d\Omega_{D-2}^2 \tag{15.107}$$

with $f(r_H) = 0$, this expression reduces to

$$S = \frac{8\pi}{\kappa} \oint_{r_H,t} \left(\frac{\delta L}{\delta R_{rtrt}} \right) R^{D-2} d\Omega^2_{D-2}$$

(15.108)

and agrees with the result in the text.

PROJECT

Project 15.1
Boundary terms for the Lanczos–Lovelock action

In the text, we mentioned the issue of making the action for Lanczos–Lovelock theories well defined so that a proper variational principle can be constructed. As we discussed in Chapter 6, the variation of a gravitational action will usually contain the normal derivatives of the variations of the metric on the boundary. While the tangential derivatives of the metric variations automatically vanish when the metric is fixed on the boundary, we need to explicitly cancel out the normal derivatives by adding a suitable boundary term. This boundary term will not – and need not – be unique but we must find at least one suitable term to make the variation of the action well defined. This is an algebraically intensive project which invites you to do this for Lanczos–Lovelock models.

(a) As a warm-up, consider the $m = 2$ Lanczos–Lovelock model which is the Gauss–Bonnet theory with the Lagrangian density L_{GB} given by Eq. (15.49). Consider an action in which the bulk term is an integral of $(R - 2\Lambda + 2\alpha L_{GB})$, where α is a constant. Show that a suitable boundary term for such a theory, which should be added to the action, can be taken to be

$$A_{\text{bound}} = 2 \int_{\partial V} d^{D-1}x \sqrt{|h|} \left[K + 4\alpha (J - 2K_{ab}\mathcal{G}^{ab}) \right]$$

(15.109)

where J is the trace of

$$J_{ab} = \frac{1}{3} \left(2KK_{ac}K^c_b + K_{cd}K^{cd}K_{ab} - 2K_{ac}K^{cd}K_{db} - K^2 K_{ab} \right)$$

(15.110)

and $\mathcal{G}_{ab}$ is the $(D-1)$-dimensional Einstein tensor built out of the induced metric h_{ab} on the boundary. One way of doing this is to explicitly vary the boundary term and the action and prove that the normal derivatives cancel out. A somewhat more general approach is to list all the possible terms one can write on the boundary, compute their variations and fix the relative coefficients so that the required terms can be cancelled.[6]

(b) Show that an alternative expression for the integrand of the boundary term in the case of Einstein's theory and Gauss–Bonnet theory are:

$$C_1 = 2\delta^{i_1}_{j_1} K^{j_1}_{i_1} = 2K$$

(15.111)

$$C_2 = 4 \int_0^1 dt \delta^{i_1 i_2 i_3}_{j_1 j_2 j_3} K^{j_1}_{i_1} \left(\frac{1}{2} R^{j_2 j_3}_{i_2 i_3} - t^2 K^{j_2}_{i_2} K^{j_3}_{i_3} \right)$$

$$= 2\delta^{i_1 i_2 i_3}_{j_1 j_2 j_3} K^{j_1}_{i_1} \left(R^{j_2 j_3}_{i_2 i_3} - \frac{2}{3} K^{j_2}_{i_2} K^{j_3}_{i_3} \right).$$

(15.112)

(c) For the general Lanczos–Lovelock theory, the above expressions can be generalized in the form

$$C_n = 2n \int_0^1 dt \delta^{i_1 i_2 i_3 \ldots i_{2n-1}}_{j_1 j_2 j_3 \ldots j_{2n-1}} K^{j_1}_{i_1} \left(\frac{1}{2} R^{j_2 j_3}_{i_2 i_3} - t^2 K^{j_2}_{i_2} K^{j_3}_{i_3} \right)$$

$$(\ldots) \left(\frac{1}{2} R^{j_{2n-2} j_{2n-1}}_{i_{2n-2} i_{2n-1}} - t^2 K^{j_{2n-2}}_{i_{2n-2}} K^{j_{2n-1}}_{i_{2n-1}} \right). \qquad (15.113)$$

See whether you can verify this result[7] and, better still, interpret it.

16

Gravity as an emergent phenomenon

16.1 Introduction

The description of gravity based on Einstein's general theory of relativity is quite satisfactory in most respects. It has been repeatedly verified experimentally as regards those features which could be directly tested while other parts of it are conceptually very elegant and beautiful. Nevertheless, it is obvious that this theory is fundamentally flawed or – at the least – incomplete.

Such a conclusion emerges from the fact that there exist well defined situations in which the theory is incapable of predicting the future evolution of the dynamical variables owing to the development of singularities. To see this concretely, consider the example of a collapsing sphere of dust described in Chapter 8. An observer comoving along with the dust particle will find that the trajectory of the dust particle hits a singularity (at which the curvature and density diverge) *within finite proper time τ as shown by the observer's clock*. In other words, the observer can not ascertain beforehand her future evolution for arbitrarily large values of τ using Einstein's theory of gravity. As another example, consider the standard description of our universe in terms of a Friedmann model discussed in Chapter 10. For reasonable values of the parameters of the model at the present moment – which are determined observationally – the theory is incapable of describing the state of the universe, say, 20 billion years ago for any equation of state for high density matter having positive pressure and energy density. This, again, is because of the fact that Friedmann equations – when integrated backwards – predict a singularity for such a model within a cosmic time of $t < 20$ billion years. By any standard criteria for judging a theory, these should be treated as serious conceptual problems indicating a breakdown of the theory.

Since the theory seems to work when the local curvature of the spacetime is moderate, it is possible that Einstein's theory of gravity needs to be replaced by a better – and as such unknown – description when the curvature becomes

670

sufficiently large. It is possible that such a modification arises naturally when one succeeds in providing a quantum field theoretical description of gravity. At the same time, it is also conceivable that Einstein's theory requires modifications owing to other reasons well before quantum gravitational effects become important. The situation as regards either such modification is completely unclear at present even though several candidate models for such a description exist.

The purpose of this last chapter is to describe what are likely to be some conceptual ingredients of any such modifications of gravity (due to quantum gravitational effects or otherwise). Broadly speaking, any such description will go beyond general relativity and will treat it as an emergent, low energy, phenomena. We will begin by describing the nature of the emergent phenomena and then provide a description of gravity from a new perspective.

16.2 The notion of an emergent phenomenon

To understand what is involved in such a description, let us examine a more familiar situation in Newtonian physics. It is well known that many of the phenomena in fluid mechanics can be understood in terms of dynamical equations involving the density $\rho(t, x)$, velocity $v(t, x)$, etc., which are well defined when the fluid is treated as a continuum. We, however, know that the real fluid is made of discrete molecules with empty space in between. The macroscopic variables like density, velocity, etc., have no microscopic significance and certainly cannot be used to understand the microscopic physics of molecules making up the fluid. In this description, one thinks of fluid mechanics as an emergent phenomenon with its own dynamical variables and corresponding equations of motion that are valid at scales much bigger than the typical intermolecular separations.

Analogously, one could imagine the description of spacetime in terms of $g_{ab}(t, x)$ as an emergent phenomenon valid at scales large compared with some critical length scale which, possibly – but not necessarily – could be the Planck length $L_P = (G\hbar/c^3)^{1/2}$. It is then conceivable that all of the general theory of relativity is similar to the description of, say, fluid mechanics. The variables $g_{ab}(t, x)$ (being analogous to $\rho(t, x)$, $v(t, x)$, etc.) have no significance in the microscopic description of spacetime. Just as the proper description of molecules of a fluid requires the introduction of new degrees of freedom and a theoretical formalism (based on quantum mechanics), the microscopic description of spacetime will require the introduction of new degrees of freedom and theoretical formalism. These new degrees of freedom and theoretical description could be widely different from the one based on $g_{ab}(t, x)$, just as the quantum description of molecules will be quite different from that of a fluid based on $\rho(t, x)$, $v(t, x)$, etc. In that case, quantization of $g_{ab}(t, x)$ itself is not of use in unravelling the microscopic structure of spacetime any more than quantizing the density and velocity of a fluid will help us to understand molecular dynamics.

There is, however, one key link between the microscopic and emergent phenomena that we can exploit. A fluid or a gas is known to exhibit *thermal phenomena* which involve the concepts of temperature and transfer of heat energy. If the fluid is treated as a continuum and is described by $\rho(t, x)$, $v(t, x)$, etc., all the way down to microscopic scales, then it is *not* possible for it to exhibit thermal phenomena. As first stressed by Boltzmann, the heat content of a fluid arises due to random motion of discrete microscopic structures which *must* exist in the fluid. These new degrees of freedom – which we now know are related to the actual molecules – make the fluid capable of storing energy internally and exchanging it with the surroundings. Given an apparent continuum phenomenon which exhibits temperature, Boltzmann could infer the existence of underlying discrete degrees of freedom.

The Boltzmann interpretation of thermal behaviour has two other attractive features. First, while the existence of the discrete degrees of freedom is vital in such an approach, the exact nature of the degrees of freedom is largely irrelevant. For example, whether we are dealing with argon molecules or helium molecules is largely irrelevant in the formulation of gas laws and such differences can be taken care of in terms of a few well-chosen numbers (like the specific heat). This suggests that such a description will have a certain amount of robustness and independence as regards the precise nature of microscopic degrees of freedom.

Second, the entropy of the system arises because of our ignoring the microscopic degrees of freedom. Turning this around, one can expect the form of entropy functional to encode the nature of microscopic degrees of freedom. If we can arrive at the appropriate form of entropy functional, in terms of some effective degrees of freedom, then we can expect it to provide the correct description. (Incidentally, this is why thermodynamics needed no modification due to either relativity or quantum theory. An equation like $TdS = dE + PdV$, for example, will have universal applicability as long as effects of relativity or quantum theory are incorporated in the definition of $S(E, V)$ appropriately.) We have already seen that the horizons which arise in general relativity are endowed with temperatures which shows that, at least in this context, some microscopic degrees of freedom are coming into play. This suggests that one could use a thermodynamic description to link the standard description of gravity with the statistical mechanics of – as yet unknown – microscopic degrees of freedom using a suitably defined entropy functional.

We shall see in this chapter how these ideas can be developed further and how it provides the explanation for several features which must otherwise be thought of as purely algebraic accidents in the conventional description of gravity.

16.3 Some intriguing features of gravitational dynamics

Since the thermodynamics of horizons seems to hold the key link between microscopic and emergent descriptions of gravity – just as in the case of a fluid the existence of temperature leads us to infer the existence of discrete degrees of freedom – we will first take a closer look at some of these features in the conventional description. We will highlight several curious connections between horizon thermodynamics and gravitational dynamics which are unexplained in the conventional description of gravity, in order to motivate an alternative perspective.

16.3.1 Einstein's equations as a thermodynamic identity

Consider a static, spherically symmetric horizon, in a spacetime described by a metric:

$$ds^2 = -f(r)c^2dt^2 + f^{-1}(r)dr^2 + r^2d\Omega^2. \tag{16.1}$$

The location of the horizon is given by the simple zero of the function $f(r)$, say at $r = a$. The Taylor series expansion of $f(r)$ near the horizon $f(r) \approx f'(a)(r - a)$ shows that the metric reduces to the Rindler metric near the horizon in the $r-t$ plane (see Eq. (8.7)) with the surface gravity $\kappa = (c^2/2)f'(a)$. Then, an analytic continuation to imaginary time allows us to identify the temperature associated with the horizon to be (see Chapter 14)

$$k_BT = \frac{\hbar cf'(a)}{4\pi}, \tag{16.2}$$

where we have introduced the normal units. In the standard Minkowski coordinates, the Rindler horizon is located at $X^2 - c^2T^2 = 0$, which gets mapped to the origin $X^2 + c^2T_E^2 = 0$ in the Euclidean sector. So the region close to the Rindler horizon is localized around the origin in the Euclidean sector, allowing us to provide a local description in the Euclidean sector.

The association of temperature in Eq. (16.2) with the metric in Eq. (16.1) only requires the conditions $f(a) = 0$ and $f'(a) \neq 0$. The discussion so far has not assumed anything about the dynamics of gravity or Einstein's field equations. We shall now take the next step and write down the Einstein equation for the metric in Eq. (16.1), which is given by $(1 - f) - rf'(r) = -(8\pi G/c^4)Pr^2$, where $P = T_r^r$ is the radial pressure (see Eq. (7.41)). When evaluated on the horizon $r = a$ we get the result

$$\frac{c^4}{G}\left[\frac{1}{2}f'(a)a - \frac{1}{2}\right] = 4\pi Pa^2. \tag{16.3}$$

If we now consider two solutions to the Einstein equations differing infinitesimally in the parameters such that horizons occur at two different radii a and $a + da$, then multiplying Eq. (16.3) by da, we get:

$$\frac{c^4}{2G}f'(a)a\,da - \frac{c^4}{2G}da = P(4\pi a^2 da). \tag{16.4}$$

The right hand side is just PdV, where $V = (4\pi/3)a^3$ is what is called the areal volume which is the relevant quantity when we consider the action of pressure on a surface area. In the first term, we note that $f'(a)$ is proportional to the horizon temperature in Eq. (16.2). Rearranging this term slightly and introducing an $\hbar$ factor *by hand* into an otherwise classical equation to bring in the horizon temperature, we can rewrite Eq. (16.4) as

$$\underbrace{\frac{\hbar c f'(a)}{4\pi}}_{k_B T}\underbrace{\frac{c^3}{G\hbar}d\left(\frac{1}{4}4\pi a^2\right)}_{dS} - \underbrace{\frac{1}{2}\frac{c^4 da}{G}}_{-dE} = \underbrace{Pd\left(\frac{4\pi}{3}a^3\right)}_{P\,dV}. \tag{16.5}$$

The labels below the equation indicate a natural interpretation for each of the terms and the whole equation now becomes $TdS = dE + PdV$ allowing us to read off the expressions for entropy and energy:

$$S = \frac{1}{4L_P^2}(4\pi a^2) = \frac{1}{4}\frac{A_H}{L_P^2}; \quad E = \frac{c^4}{2G}a = \frac{c^4}{G}\left(\frac{A_H}{16\pi}\right)^{1/2}, \tag{16.6}$$

where A_H is the horizon area and $L_P^2 = G\hbar/c^3$. This remarkable result deserves several comments.

To begin with, the result shows that Einstein's equations can be re-interpreted as a thermodynamic identity for a virtual displacement of the horizon by an amount da. The uniqueness of the factor $P(4\pi a^2)da$, where $4\pi a^2$ is the proper area of a surface of radius a in spherically symmetric spacetimes, implies that we cannot carry out the same exercise by multiplying Eq. (16.3) by some other arbitrary factor $F(a)da$ instead of just da in a natural fashion. This, in turn, uniquely fixes both dE and the combination TdS. The product TdS is classical and is independent of $\hbar$ and hence we can determine T and S only within a multiplicative factor. (We encountered a similar situation in Chapter 14; see page 624.) Fortunately, the Euclidean extension of the metric allows us to determine the form of T independently, thereby removing this ambiguity in the overall multiplicative factor. Thus, given the structure of the metric in Eq. (16.1) and Einstein's equations, we can determine T, S and E uniquely. The fact that $T \propto \hbar$ and $S \propto 1/\hbar$ is analogous to the situation in classical thermodynamics in contrast with statistical mechanics. The TdS in thermodynamics is independent of Boltzmann's constant while statistical mechanics will lead to $S \propto k_B$ and $T \propto 1/k_B$.

Second, we obtained in Chapter 14 an expression in Eq. (16.2) for the temperature of the horizon, in terms of the surface gravity, using the periodicity in the Euclidean sector (see the discussion in Section 14.5). But recall that this temperature was obtained by studying a quantum field in the externally specified metric *without any reference to Einstein's equations.* We then gave a tentative argument based on energy conservation in black hole evaporation (see Eq. (14.143)) to obtain an expression for the entropy of black holes that turned out to be one quarter of the area of the black hole horizon in units of L_P^2. We now find that the same result arises *without any reference to an externally specified quantum field theory but on using Einstein's equations.* Further, we have only used the form of the metric and Einstein's equations close to the horizon; hence the argument is valid for a much larger class of spacetimes – for example, the de Sitter spacetime – for which the notion of 'evaporation' is non-existent or ill defined in the conventional approach.

This strongly suggests that the association of entropy and temperature with a horizon is quite fundamental and is actually connected with the dynamics (encoded in Einstein's equations) of the gravitational field. The fact that the quantum field theory in a spacetime with a horizon exhibits thermal behaviour should then be thought of as a *consequence* of a more fundamental principle. We shall return to this issue in the next section.

Finally, recall that we also obtained in Chapter 14 the laws of black hole dynamics in a form very similar to the first law of thermodynamics; see, for example, Eq. (8.99). For spherically symmetric spacetimes this law relates the change in the mass of a black hole to the change in the entropy by $TdS = dM$. For any vacuum solution to Einstein's equations, with $P = 0$, Eq. (16.5) will also reduce to this form with $E = M$. But, in general, in spite of the superficial similarity, Eq. (16.5) is *different* from the conventional first law of black hole thermodynamics due to the presence of the PdV term. The difference is easily seen, for example, in the case of a Reissner–Nordstrom black hole for which $T_r^r = P$ is nonzero due to the presence of a nonzero electromagnetic energy-momentum tensor on the right hand side of Einstein's equations. If a *chargeless* particle of mass dM is dropped into a Reissner–Nordstrom black hole, then Eq. (8.99) will give $TdS = dM$. But in Eq. (16.5), the energy term, defined as $E \equiv a/2$, changes by $dE = (da/2) = (1/2)[a/(a - M)]dM \neq dM$. (This is in contrast to the Schwarzschild metric for which $T_b^a = 0$, making the PdV term vanish so that $TdS = dE = dM$.) It is easy to see, however, that for the Reissner–Nordstrom black hole, the combination $dE + PdV$ is precisely equal to dM making sure $TdS = dM$. So we need the PdV term to get $TdS = dM$ from Eq. (16.5) when a *chargeless* particle is dropped into a Reissner–Nordstrom black hole. More generally, if da arises due to changes dM and dQ, it is easy to show that Eq. (16.5) gives $TdS = dM - (Q/a)dQ$, where the second term arises from the electrostatic contribution. This ensures that Eq. (16.5) is perfectly consistent with the standard first law of black hole dynamics in those contexts in which both are applicable but $dE \neq dM$ in general.

16.3.2 Gravitational entropy and the boundary term in the action

The above result suggests that one can attribute an entropy to the horizon by using only the Einstein equations and periodicity in imaginary time. This idea, which appears acceptable in the light of null surfaces blocking information and acting as a one-way membrane, can be made more precise as follows. We begin with the assertion that *all observers have a right to describe physics using an effective theory based only on the variables they can access.* In the study of particle physics models, this concept forms the cornerstone of what is usually called renormalization group theory. To describe particle interactions at 10 GeV in the lab, we usually do not need to know what happens at 10^{14} GeV in theories which have predictive power. In the absence of such a principle, very high energy phenomena (which are unknown from direct experiments in the lab) will affect the low energy phenomena that we are attempting to study. In the context of a theory involving a nontrivial metric of spacetime, we need a similar principle to handle the fact that different observers will have access to different regions of a general spacetime. If a class of observers perceive a horizon, they should still be able to do physics using only the variables accessible to them without having to know what happens on the other side of the horizon.

This, in turn, implies that there should exist a mechanism that will encode the information in the region $\mathcal{V}$ that is inaccessible to a particular observer at the boundary $\partial\mathcal{V}$ of that region. One possible way of ensuring this is to add a suitable boundary term to the action principle which will provide additional information content for observers who perceive a horizon. Such a procedure leads to three immediate consequences.

(a) If the theory is generally covariant, so that observers with horizons (for example, uniformly accelerated observers using a Rindler metric) need to be accommodated in the theory, such a theory *must* have an action functional that contains a surface term. We saw in Chapter 6 that the generally covariant action in Einstein's theory did contain a surface term. The present approach explains the logical necessity for such a surface term in a generally covariant theory that was not evident in the standard approach adopted in Chapter 6.

(b) If the surface term encodes information which is blocked by the horizon, then there must be a simple correspondence between the bulk term and surface term in the action. We found that this is indeed the case; there is a peculiar (unexplained) relationship between L_{bulk} and L_{sur}:

$$\sqrt{-g}L_{\text{sur}} = -\partial_a \left(g_{ij} \frac{\partial \sqrt{-g}L_{\text{bulk}}}{\partial(\partial_a g_{ij})} \right) \tag{16.7}$$

(see Project 6.3; Eq. (6.237)). This shows that the gravitational action is 'holographic' with the same information being coded in both the bulk and surface terms.

In fact, in any local region around an event, it is the surface term which contributes to the action. We have seen in Section 5.3.1 that one can introduce, in the neighbourhood of any event, the Riemann normal coordinates in which $g \simeq \eta + R x^2$, $\Gamma \simeq R x$. In the gravitational Lagrangian $\sqrt{-g}R \equiv \sqrt{-g}L_{\text{bulk}} + \partial_a P^a$ (see Eq. (6.9)) with $L_{\text{bulk}} \simeq \Gamma^2$ and $\partial P \simeq \partial\Gamma$, the L_{bulk} term vanishes in this neighbourhood while $\partial_a P_b \simeq R_{ab}$, leading to

$$\int_{\mathcal{V}} d^4x R\sqrt{-g} \approx \int_{\mathcal{V}} d^4x \partial_a P^a \approx \int_{\partial\mathcal{V}} d^3x n_a P^a, \tag{16.8}$$

showing that in a small region around the event in the Riemann normal coordinates, gravitational action can be reduced to a pure surface term.

(c) Further, if the surface term encodes information which is blocked by the horizon, then it should actually lead to the entropy of the horizon. In other words, we should be able to compute the horizon entropy by evaluating the surface term. This is indeed true and we have already demonstrated in Section 15.5 (see Eq. (15.89)) that the surface term gives the horizon entropy for any metric for which near-horizon geometry has the Rindler form.

It should be stressed that this connection between horizon entropy and the surface term in the action is a mystery within the conventional approach. In the usual approach, we *ignore* the surface term completely (see Chapter 6; or cancel it with a counter-term) and obtain the field equation from the bulk term in the action. Any solution to the field equation obtained by this procedure is logically independent of the nature of the surface term. But we find that when the *surface term* (which was ignored) is evaluated at the horizon that arises in any given solution, it does correctly give the entropy of the horizon! This is possible only because there is a relationship, given by Eq. (16.7), between the surface term and the bulk term which is again an unexplained feature in the conventional approach to gravitational dynamics. Since the surface term has the thermodynamic interpretation as the entropy of horizons, and is related holographically to the bulk term, we are again led to an indirect connection between spacetime dynamics and horizon thermodynamics.

16.3.3 Horizon thermodynamics and Lanczos–Lovelock theories

So far we have been discussing the issue of horizon thermodynamics within the context of Einstein's theory. But we saw in Chapter 15 that some of these features hold for a much wider class of theories than Einstein gravity, such as Lanczos–Lovelock models. For example, we saw in Chapter 15 that the Lanczos–Lovelock action has the same holographic structure and – again – the entropy of the horizons is related to the surface term of the action. Further, the connection between the field equations and the thermodynamic relation $TdS = dE + PdV$ is not restricted to Einstein's theory (GR) alone, but is in fact true for the case of the generalized, higher derivative, Lanczos–Lovelock gravitational theory in D dimensions as well (see Exercise 16.1). We will, in fact, see later that the field equations derived from *any* generally covariant action have a thermodynamic interpretation.

These facts suggest that a wide class of models describing gravity exhibits an intriguing thermodynamic interpretation. In the conventional approach, these models are introduced without any reference to spacetime thermodynamics which is obtained as a *consequence* of some special solutions in the theory. In such an approach, all the results described in this section need to be thought of as unexplained, curious, algebraic accidents having no deep significance. What is more, such an approach ignores the lesson we should have learnt from Boltzmann, viz. that thermal behaviour demands the existence of microscopic degrees of freedom and the macroscopic behaviour of a system exhibiting thermal behaviour must be then thought of as an emergent phenomena. We shall now see how an alternative perspective on gravity can be developed that incorporates these features in an elegant and natural manner.

Exercise 16.1

Gauss–Bonnet field equations as a thermodynamic identity The equation governing gravitational dynamics reduces to a thermodynamic identity near the horizon in all Lanczos–Lovelock theories.[1] This exercise explores the result for Gauss–Bonnet theory and invites you to try the harder case of Lanczos–Lovelock theories.

(a) Write down the Gauss–Bonnet field equations for a metric in Eq. (15.69) at $r = a$, where $f(r)$ vanishes.

(b) Show that this equation can be reduced to the form $TdS = dE + PdV$, where $T = f'(a)/4\pi$ is the horizon temperature and the entropy S and the energy E are given by

$$S = \frac{A_{D-2}}{4} a^{D-2} \left[1 + \left(\frac{D-2}{D-4} \right) \frac{2\,\bar{\alpha}}{a^2} \right], \qquad (16.9)$$

$$E = \frac{(D-2)A_{D-2}a^{D-3}}{16\pi} \left(1 + \frac{\bar{\alpha}}{a^2} \right), \qquad (16.10)$$

where $A_{D-1} = 2\pi^{D/2}/\Gamma(D/2)$ is the area of a unit $(D-1)$-sphere and $\bar{\alpha} = (d-3)(d-4)\alpha$.

(c) Show that the entropy in Eq. (16.9) is exactly the same as the one obtained earlier in Eq. (15.104). Interpret the expression for the energy.

(d) For a challenge, extend the results for a black hole solution in a general Lanczos–Lovelock theory.

16.4 An alternative perspective on gravitational dynamics

The description of gravity in conventional general relativity has two separate ingredients. The first one ('gravity tells us how matter should move') is based on the principle of equivalence and provides a procedure for calculating the effect of gravity on all other material systems. The principle of equivalence, as we pointed out in Chapter 3, suggests that any gravitational field can be described by modifying the line interval to the form $ds^2 = g_{ab}(t, \boldsymbol{x})dx^a dx^b$. The influence of gravity on other material systems can then be incorporated by the minimal coupling (discussed in Chapter 5) which involves the replacements $\eta_{ab} \to g_{ab}$, $d^4x \to \sqrt{-g}d^4x$, $\partial_a \to \nabla_a$, etc., in the action principle. The elegance and beauty usually attributed to Einstein's theory arises from this feature, viz. that all the gravitational influence on material systems can be given a geometrical foundation.

The second aspect of general relativity ('matter tells us how spacetime should curve') involves the field equation of gravity, which allows us to determine $g_{ab}(t, \boldsymbol{x})$ from the matter distribution. Unfortunately, *we do not have an equally elegant principle* (like the principle of equivalence) to determine the dynamical equations of gravity. The standard approach, described in Chapter 6, based on a Lagrangian proportional to $(R - 2\Lambda)$, is merely the simplest possible choice. Several other alternatives to this Lagrangian can be provided, most of which can be designed to agree with observational constraints. Since many of the conceptual problems in general relativity are linked to the actual dynamical equation, the absence of a guiding principle to determine the dynamics is quite unsatisfactory.

Our discussion in the previous section showed that the conventional approach also leads to several unexpected connections between horizon thermodynamics and gravitational dynamics not only in Einstein's theory but also in Lanczos–Lovelock models. This suggests that *one may be able to find a thermodynamic principle from which the gravitational dynamics of spacetime can be obtained.* Our first aim will be to unravel the key ingredients of such a dynamical principle. To do this, we shall re-examine certain aspects of standard general relativity and will identify these ingredients. In particular, we would like the dynamical principle, leading to the equations of motions of the theory, to be closely linked to the principle of equivalence itself. So we shall begin with certain consequences of the principle of equivalence that arise in any geometrical description of gravity.

The first point to note is that horizons are inevitable in any theory of gravity obeying the principle of equivalence. This principle of equivalence implies that trajectories of light will be affected by gravity. So in any theory that links gravity to spacetime dynamics, we can have nontrivial null surfaces which block information from a certain class of observers. Therefore, the existence of horizons is a feature which will exist in any geometrical theory of gravity and is fairly independent of the field equations (i.e., dynamics) of the theory.

Further, the key property of a horizon – viz. it acts as a one-way membrane as regards the flow of information – is *always* an observer (trajectory) dependent concept, even when the horizon itself can be given a purely geometrical definition. For example, the $r = 2M$ surface in Schwarzschild geometry (which can be defined as the boundary of the causal past of $\mathcal{I}^+$; see Chapter 8) acts as a horizon *only* for the class of observers who choose to stay at $r > 2M$; observers falling into the black hole will have access to more information and the $r = 2M$ surface is not, operationally speaking, a horizon for them. In general, we can consider a congruence $\mathcal{C}$ of timelike curves and define the horizon $\mathcal{H}(\mathcal{C})$ associated with $\mathcal{C}$ to be the boundary of the union of the causal past of $\mathcal{C}$. In such an approach, $\mathcal{H}(\mathcal{C})$ will act 'locally' as a horizon for a class of observers.

To make this notion more formal, we can proceed as follows. We choose any event $\mathcal{P}$ and introduce a local inertial frame (LIF) around it with Riemann normal coordinates $X^a = (T, \boldsymbol{X})$ and orient the spatial coordinates such that – near $\mathcal{P}$ – the light cones, in the $X - T$ plane, say, are described by $T = \pm X$. Next we transform from the LIF to a local Rindler frame (LRF) coordinates x^a by accelerating along the X-axis with an acceleration κ by the usual transformation. This LRF and its local horizon $\mathcal{H}$ will exist within a region of size $L \ll \mathcal{R}^{-1/2}$ as long as $\kappa^{-1} \ll \mathcal{R}^{-1/2}$, where $\mathcal{R}$ is a typical component of curvature tensor of the background spacetime. This condition can always be satisfied by taking a sufficiently large κ. This procedure introduces a class of uniformly accelerated observers who will perceive the null surface $T = \pm X$ as the local Rindler horizon $\mathcal{H}$. (The local nature of this notion is obvious in the Euclidean sector in which the horizon maps to the origin.) Such a notion of the local Rindler horizon is again completely independent of the field equations and relies only on the principle of equivalence.

The existence of such horizons that act as a one-way membrane for some local Rindler observers raises an interesting conundrum when we recall that all observers have the right to describe physics using an effective theory based only on the variables she can access. Now an observer can let some matter flow across $\mathcal{H}$ making the entropy of matter disappear, thereby reducing the total amount of entropy accessible to the outside observer who perceives $\mathcal{H}$ as a horizon.[2] Such a violation of the second law of thermodynamics can be avoided only if we *demand* that the horizon should have an entropy which should increase when energy flows across it. If energy dE flows across a hot horizon with temperature T then $dE/T = dS$ should be the change in the entropy of the horizon.

We therefore conclude that all null surfaces which could locally act as one-way membranes should have an (observer dependent) entropy associated with them. Again all these results are independent of the field equations of gravity and rely only on (i) the principle of equivalence, (b) the existence of observers who perceive a null surface as a horizon and (c) the notion that all observers can do physics with the variables accessible to them.

We next notice that, instead of dropping matter across the horizon, one can consider a virtual, infinitesimal, displacement of the $\mathcal{H}$ normal to itself engulfing some matter. We need only consider infinitesimal displacements because the entropy of the matter is not 'lost' until it crosses the horizon; when the matter is at an infinitesimal distance from the horizon, one can still retrieve the entropy of matter. All the relevant physical processes take place at a region very close to the horizon. At this stage, an infinitesimal displacement of $\mathcal{H}$ normal to itself will engulf some matter and make its entropy disappear. Thus, some entropy will be lost to the outside observers unless displacing a piece of the local Rindler horizon should cost some entropy. Interestingly enough, one can prove that this entropy balance condition holds at every event in the spacetime in *any* theory of gravity which is generally covariant. The key to this result is the expression for the Noether current in Eq. (15.90), which uses only the fact that the theory is generally covariant.

We again begin by choosing a local inertial frame (LIF) around some event $\mathcal{P}$ with coordinates X^a such that $\mathcal{P}$ has the coordinates $X^a = 0$ in the LIF. We let k^a be a future directed null vector at $\mathcal{P}$ and align the coordinates of the LIF such that it lies in the $X-T$ plane at $\mathcal{P}$. One can now construct, in the neighbourhood of $\mathcal{P}$, a local Rindler frame (LRF) along the lines described in Exercise 8.12 with acceleration κ. The metric near the origin reduces to the form in Eq. (15.86), where $(t, x, \boldsymbol{x}_\perp)$ are the coordinates of LRF. Let ξ^a be the approximate Killing vector corresponding to translation in the Rindler time such that the vanishing of $\xi^a \xi_a \equiv -N^2$ characterizes the location of the local horizon $\mathcal{H}$ in LRF. As usual, we shall do all the computation on a timelike surface infinitesimally away from $\mathcal{H}$. This surface has $N = $ constant and is usually called a 'stretched horizon'. (It can be defined more formally using the orbits of ξ^a and the plane orthogonal to the acceleration vector $a^i = \xi^b \nabla_b \xi^i$.) The local temperature on the stretched horizon will be $\kappa/2\pi N$ so that $\beta_{\mathrm{loc}} = \beta N$. Let the timelike unit normal to the stretched horizon be r_a.

An infinitesimal displacement of a local patch of the stretched horizon in the direction of r_a by an infinitesimal proper distance ϵ will change the proper volume by $dV_{\mathrm{prop}} = \epsilon \sqrt{\sigma} d^{D-2}x$, where σ_{ab} is the metric in the transverse space. The flux of energy through the surface will be proportional to $T_b^a \xi^b r_a$ and the corresponding entropy flux can be obtained by multiplying the energy flux by β_{loc}. Hence the 'loss' of matter entropy is $\delta S_m = \beta_{\mathrm{loc}} \delta E = \beta_{\mathrm{loc}} T^{aj} \xi_a r_j dV_{\mathrm{prop}}$.

To find the change in the gravitational entropy, we use the Noether current J^a corresponding to the local Killing vector ξ^a. From Eq. (15.90), for any vector v_a, we can write the relation

$$v_a J^a = \xi_j v_a T^{aj} + L(\xi_j v^j). \tag{16.11}$$

Taking $v^a = r^a$ and multiplying by $\beta_{\text{loc}} = \beta N$, we get

$$\beta_{\text{loc}} r_a J^a = \beta_{\text{loc}} \xi_b r_a T^{ab} + \beta N (r_a \xi^a) L. \tag{16.12}$$

As the stretched horizon approaches the true horizon, we know that $N r^a \rightarrow \xi^a$ (see Exercise 8.12) and $\beta \xi^a \xi_a L \rightarrow 0$, making the last term vanish. So

$$\delta S_{\text{grav}} \equiv \beta \xi_a J^a dV_{\text{prop}} = \beta T^{aj} \xi_a \xi_j dV_{\text{prop}} = \delta S_m, \tag{16.13}$$

showing the validity of local entropy balance for any β. In this limit, ξ^i also goes to $\kappa \lambda k^i$, where λ is the affine parameter associated with the null vector k^a we started with (see Exercise 8.12) and all the reference to the LRF goes away. It is clear that the properties of the LRF are relevant conceptually to define the intermediate notions, but the essential result that we needed was Eq. (15.90) which can be interpreted now as providing a local entropy balance.

Incidentally, the Noether current relation can also be used to provide an alternative interpretation of the energy balance along the following lines. Since J^0 is the Noether charge density, $\delta S = \beta_{\text{loc}} u_a J^a dV_{\text{prop}}$ can be interpreted as the entropy associated with a volume dV_{prop} as measured by an observer with four-velocity u^a, where $\beta_{\text{loc}} = \beta N$ is the local redshifted temperature and N is the lapse function. If we consider observers moving along the orbits of the Killing vector ξ^a then $u^a = \xi^a / N$ and we get

$$\delta S_{\text{grav}} = \beta N u_a J^a dV_{\text{prop}} = \beta [\xi_j \xi_a T^{aj} + L(\xi_j \xi^j)] dV_{\text{prop}}. \tag{16.14}$$

As one approaches the horizon, $\xi^a \xi_a \rightarrow 0$, making the second term vanish and we get

$$\delta S_{\text{grav}} = \beta (\xi_j \xi_a T^{aj}) dV_{\text{prop}}. \tag{16.15}$$

In the same limit ξ^j will become proportional to the original null vector k^j we started with. As we shall see below, for any null vector k_j, one can interpret the right hand side as the matter entropy present in a proper volume dV_{prop}. So this equation can again be thought of as an entropy balance equation.

The above results imply that the field equations arising from any generally covariant action can be given a thermodynamic interpretation; that is, we assumed the validity of the field equations and derived the local entropy balance. Our aim, however, is to obtain the field equations from a dynamical principle rather than *assume* the field equations. (One cannot, unfortunately, reverse the above argument because the interpretation of J^a as the entropy current assumes the validity of the field equations as discussed in Chapter 15 page 665.)

To achieve this aim, we need to assume a suitable form of entropy functional for gravity S_{grav} in terms of the normal to the null surface which is independent of the field equations. Then it seems natural to demand that the dynamics should follow from the extremum prescription $\delta[S_{grav} + S_{matter}] = 0$ for *all null surfaces in the spacetime*, where S_{matter} is the matter entropy.

The form of S_{matter} is easy to ascertain from the previous discussion. If T_{ab} is the matter energy-momentum tensor in a general $D(\geq 4)$-dimensional spacetime then an expression for matter entropy can be taken to be

$$S_{matter} = \int_{\mathcal{V}} d^D x \sqrt{-g} T_{ab} n^a n^b, \tag{16.16}$$

where n^a is a null vector field. It is easy to show that S_{matter} reduces to the standard expression for matter entropy in, for example, $D = 4$. In the LRF, an infinitesimal spacetime region will contribute in Eq. (16.16) an amount $T_{ab} n^b n^b \, N \sqrt{h} \, d^3 x \, dt = \delta E \, N dt$, which on integration over t in the range $(0, \beta)$ where $\beta^{-1} = T = (\kappa/2\pi)$ gives $\delta S_{matter} = \beta N \delta E = \beta_{loc} T_{ab} n^a n^b \sqrt{h} \, d^3 x$ when the energy flows across a surface with normal n^a. Integrating, we get an expression for S_{matter} to which Eq. (16.16) reduces in the LRF. More explicitly, if T_{ab} is due to an ideal fluid at rest in the LIF, $T_{ab} n^a n^b$ will contribute $(\rho + P)$, which – by the Gibbs–Duhem relation – is just $T_{local} s$, where s is the entropy density and $T_{local}^{-1} = \beta N$ is the properly redshifted temperature with $\beta = 2\pi/\kappa$ being the periodicity of the Euclidean time coordinate. Then

$$\int dS = \int \sqrt{h} d^3 x s = \int \sqrt{h} d^3 x \beta_{loc}(\rho + P) = \int \sqrt{h} N d^3 x \beta(\rho + P)$$
$$= \int_0^\beta dt \int d^3 x \sqrt{-g} \, T^{ab} n_a n_b, \tag{16.17}$$

which matches with Eq. (16.16) in the appropriate limit.

We next need an expression for the gravitational entropy S_{grav}. We shall postulate this to be a quadratic expression in the derivatives of the normal:

$$S_{grav} = -4 \int_{\mathcal{V}} d^D x \sqrt{-g} P_{ab}{}^{cd} \nabla_c n^a \nabla_d n^b, \tag{16.18}$$

where the explicit form of $P_{ab}{}^{cd}$ is ascertained below. The expression for the total entropy, which we need to extremize now becomes:

$$S[n^a] = - \int_{\mathcal{V}} d^D x \sqrt{-g} \left(4 P_{ab}{}^{cd} \nabla_c n^a \nabla_d n^b - T_{ab} n^a n^b \right). \tag{16.19}$$

We next address the crucial conceptual difference between the extremum principle introduced here and the conventional one. Usually, given a set of dynamical variables n_a and a functional $S[n_a]$, the extremum principle will give a set of equations for the dynamical variable n_a. Here the situation is different and we expect the variational principle to hold for *all* null vectors n^a thereby leading to a condition on the *background metric*. Obviously, the functional in Eq. (16.19) must be rather special to accomplish this and one needs to impose restrictions on $P_{ab}{}^{cd}$ (and on T_{ab}, though that condition turns out to be trivial) to achieve this. It turns out – as we shall see below – that two conditions are sufficient to ensure this. First, the tensor P_{abcd} should have the same algebraic symmetries as the Riemann tensor R_{abcd} of the D-dimensional spacetime. Second, we need

$$\nabla_a P^{abcd} = 0, \tag{16.20}$$

as well as $\nabla_a T^{ab} = 0$ which is anyway satisfied by any matter energy-momentum tensor. In a complete theory, the explicit form of P^{abcd} will be determined by the long wavelength limit of the microscopic theory just as the elastic constants can – in principle – be determined from the microscopic theory of the lattice. In the absence of such a theory, we need to determine P^{abcd} by general considerations. We shall first determine the form of P^{abcd} before extremizing the expression in Eq. (16.19).

Essentially, we need to determine the most general tensor in D-dimensions built from the metric and curvature tensor that has the symmetries of the curvature tensor and is divergence free in all the indices. We avoid derivatives of R_{abcd} in order to avoid equations of motion which are of higher degree than two. Such a P^{abcd} can be expressed as a series in the powers of derivatives of the metric as:

$$P^{abcd}(g_{ij}, R_{ijkl}) = c_1 \overset{(1)}{P}{}^{abcd}(g_{ij}) + c_2 \overset{(2)}{P}{}^{abcd}(g_{ij}, R_{ijkl}) + \cdots , \tag{16.21}$$

where $c_1, c_2, \ldots$ are coupling constants. The lowest order term depends only on the metric with no derivatives. The next term depends (in addition to the metric) linearly on curvature tensor and the next one will be quadratic in curvature, etc. It is not difficult to determine the mth order term which satisfies our constraints. It is given by

$$\overset{(m)}{P}{}^{cd}_{ab} \propto \delta^{cda_3\ldots a_{2m}}_{abb_3\ldots b_{2m}} R^{b_3 b_4}_{a_3 a_4} \cdots R^{b_{2m-1} b_{2m}}_{a_{2m-1} a_{2m}} = \frac{\partial L_{(m)}}{\partial R^{ab}_{cd}}, \tag{16.22}$$

where $\delta^{cda_3\ldots a_{2m}}_{abb_3\ldots b_{2m}}$ is the alternating tensor and the last equality shows that it can be expressed as a derivative of the mth order Lanczos–Lovelock Lagrangian, given by

$$L = \sum_{m=1}^{K} c_m L_{(m)} \ ; \ L_{(m)} = \frac{1}{16\pi} 2^{-m} \delta^{a_1 a_2 \ldots a_{2m}}_{b_1 b_2 \ldots b_{2m}} R^{b_1 b_2}_{a_1 a_2} R^{b_{2m-1} b_{2m}}_{a_{2m-1} a_{2m}} , \tag{16.23}$$

where the c_m are arbitrary constants and $L_{(m)}$ is the mth order Lanczos–Lovelock Lagrangian. The normalization factors are chosen for future convenience and we assume $D \geq 2K+1$. (See Chapter 15; Eq. (15.51).) It is conventional to take $c_1 = 1$ so that $L_{(1)}$, which gives Einstein gravity, reduces to $R/16\pi$. The normalizations $m > 1$ are somewhat ad hoc for an individual $L_{(m)}$ since the c_ms are unspecified at this stage.

Let us pause for a moment and take a closer look at the structure that is emerging. The lowest order term in Eq. (16.21) (which – as we shall see later – leads to Einstein's theory) is

$$\overset{(1)}{P}{}^{ab}_{cd} = \frac{1}{16\pi} \frac{1}{2} \delta^{ab}_{cd} = \frac{1}{32\pi} (\delta^a_c \delta^b_d - \delta^a_d \delta^b_c). \tag{16.24}$$

In fact, if we assume that P^{abcd} is to be built *only* from the metric, this choice is unique. The first order term which arises from the Gauss–Bonnet Lagrangian in Eq. (16.23) is:

$$\overset{(2)}{P}{}^{ab}_{cd} = \frac{1}{16\pi} \frac{1}{2} \delta^{aba_3a_4}_{cdb_3b_4} R^{b_3b_4}_{a_3a_4} = \frac{1}{8\pi} \left(R^{ab}_{cd} - G^a_c \delta^b_d + G^b_c \delta^a_d + R^a_d \delta^b_c - R^b_d \delta^a_c \right) \tag{16.25}$$

(see Eq. (15.48)) and similarly for all the higher orders terms.

Having determined the form of P^{abcd}, we now proceed to obtain the field equations which determine the metric by extremizing the total entropy. In our paradigm, these field equations for the metric arise from extremizing S with respect to variations of the null vector field n^a and demanding that the resulting condition holds for *all null vector fields*. Varying the normal vector field n^a after adding a Lagrange multiplier function $\lambda(x)$ for imposing the condition $n_a \delta n^a = 0$, we get

$$-\delta S = 2 \int_{\mathcal{V}} d^D x \sqrt{-g} \left(4 P_{ab}{}^{cd} \nabla_c n^a \left(\nabla_d \delta n^b \right) - T_{ab} n^a \delta n^b - \lambda(x) g_{ab} n^a \delta n^b \right), \tag{16.26}$$

where we have used the symmetries of $P_{ab}{}^{cd}$ and T_{ab}. An integration by parts and the condition $\nabla_d P_{ab}{}^{cd} = 0$ lead to

$$-\delta S = 2 \int_{\mathcal{V}} d^D x \sqrt{-g} \left[-4 P_{ab}{}^{cd} (\nabla_d \nabla_c n^a) - (T_{ab} + \lambda g_{ab}) n^a \right] \delta n^b$$
$$+ 8 \int_{\partial \mathcal{V}} d^{D-1} x \sqrt{h} \left[k_d P_{ab}{}^{cd} (\nabla_c n^a) \right] \delta n^b, \tag{16.27}$$

where k^a is the D-vector field normal to the boundary $\partial \mathcal{V}$ and h is the determinant of the intrinsic metric on $\partial \mathcal{V}$. As usual, in order for the variational principle to be well defined, we require that the variation δn^a of the vector field should vanish on the boundary. The second term in Eq. (16.27) therefore vanishes, and the condition that $S[n^a]$ be an extremum for arbitrary variations of n^a then becomes

$$2P_{ab}{}^{cd}\left(\nabla_c\nabla_d - \nabla_d\nabla_c\right)n^a - (T_{ab} + \lambda g_{ab})n^a = 0, \qquad (16.28)$$

where we used the antisymmetry of $P_{ab}{}^{cd}$ in its upper two indices to write the first term. The definition of the Riemann tensor in terms of the commutator of covariant derivatives reduces the above expression to

$$\left(2P_b{}^{ijk}R^a{}_{ijk} - T^a_b + \lambda\delta^a_b\right)n_a = 0, \qquad (16.29)$$

and we see that the equations of motion *do not contain* derivatives with respect to n^a which is, of course, the crucial point. This peculiar feature arose because of the symmetry requirements we imposed on the tensor $P_{ab}{}^{cd}$. We require that the condition in Eq. (16.29) holds for *arbitrary* vector fields n^a. This equation is identical to $2P_b{}^{kij}R^a{}_{kij}n^bn_a = T^a_b n^b n_a$, which we have already encountered in Exercise 15.3, where we saw that it is equivalent to[3] the field equations:

$$16\pi\left[P_b{}^{ijk}R^a{}_{ijk} - \frac{1}{2}\delta^a_b L_{(m)}\right] = 8\pi T^a_b + \Lambda\delta^a_b. \qquad (16.30)$$

These are identical to the field equations for Lanczos–Lovelock gravity with a cosmological constant arising as an undetermined integration constant (see Eq. (15.65)). To the lowest order, when we use Eq. (16.24) for $P_b{}^{ijk}$, Eq. (16.30) reduces to Einstein's equations. More generally, we get Einstein's equations with higher order corrections which are to be interpreted as emerging from the derivative expansion of the action functional as we probe smaller and smaller scales. Remarkably enough, we can derive not only Einstein's theory but even the Lanczos–Lovelock theory from an extremum principle involving the null normals *without varying g_{ab} in an action functional!* (This also means that we do not need to worry about the boundary terms discussed in Project 15.1.)

To gain a bit more insight into this approach, let us consider the on-shell value of the extremum functional. The term 'on-shell' refers to satisfying the relevant equations of motion, which in this case are given by Eq. (16.29). Manipulating the covariant derivatives in Eq. (16.19), we can easily show that

$$-S|_{\text{on-shell}} = 4\int_{\partial\mathcal{V}} d^{D-1}x\sqrt{h}\,k_a\left(P^{abcd}n_c\nabla_b n_d\right), \qquad (16.31)$$

where we have manipulated a few indices using the symmetries of P^{abcd}. In the case of a *stationary* horizon which can be locally approximated as Rindler space-time, one can interpret this on-shell value as the entropy of the horizon. To do this, we will use a limiting procedure and provide a physically motivated choice of n^a based on the local Rindler frame. As usual, we shall introduce the LIF and RIF and take the normal to the stretched horizon (at $N = \epsilon$) to be r_a. In the coordinates used in Eq. (15.86), we have the components:

$$n_a = (0, 1, 0, 0, \ldots) \; ; \; n^a = (0, 1, 0, 0, \ldots) \; ; \; \sqrt{h} = \kappa\epsilon\sqrt{\sigma}, \qquad (16.32)$$

where σ is the metric determinant on the $t = $ constant, $N = $ constant surfaces. This vector field n^a is a natural choice for evaluation of S on shell if we evaluate the surface integral on a surface with $N = \epsilon = $ constant, and take the limit $\epsilon \to 0$ at the end of the calculation. In the integrand of Eq. (16.31) for the entropy functional, we use $d^{D-1}x = dt d^{D-2}x_\perp$, and $\nabla_b n_d = -\Gamma^a{}_{bd}n_a = -\Gamma^x{}_{bd}$, of which only $\Gamma^x{}_{00} = \kappa^2\epsilon$ is nonzero. The integrand for the mth order term in Eq. (16.31) can be evaluated as follows:

$$\sqrt{h}\, k_a \left(4P^{abcd} n_c \nabla_b n_d \right) = \kappa\epsilon\sqrt{\sigma} \left(4P^{xbxd}\nabla_b n_d \right) = \kappa\epsilon\sqrt{\sigma} \left(-4P^{x0x0}\Gamma^x{}_{00} \right)$$

$$= \kappa^3\epsilon^2\sqrt{\sigma} \left(-4P^{x0x0} \right) = \kappa^3\epsilon^2\sqrt{\sigma} \left(-4mg^{00}g^{xx}Q^{x0}_{x0} \right)$$

$$= \kappa\sqrt{\sigma} \left(4mQ^{x0}_{x0} \right), \qquad (16.33)$$

where $Q^{abcd} = (1/m)P^{abcd}$. The rest of the calculation proceeds exactly as from Eq. (15.101) to Eq. (15.104) in Chapter 15 and we thus find that Eq. (16.31) gives the horizon entropy.

Finally, we comment on another important difference between the conventional approach of deriving the gravitational field equations and the alternative perspective described here in which the metric was *not* varied in the action to obtain the field equations. In the standard approach, one starts with an action

$$\mathcal{A}_{\text{tot}} = \int d^D x \, \sqrt{-g} \, (L_{\text{grav}} + L_m) \qquad (16.34)$$

and varies (i) the matter degrees of freedom to obtain the equations of motion for matter and (ii) the metric g^{ab} to obtain the field equations of gravity. The equations of motion for matter remain invariant if one adds a constant, say $-\rho_0$, to the matter Lagrangian, which is equivalent to adding a constant ρ_0 to the Hamiltonian density of the matter sector. Physically, this symmetry reflects the fact that the zero level of the energy is arbitrary in the matter sector and can be set to any value without leading to observable consequences. (We used this fact, for example, to set the zero point energies of the quantum oscillators to zero in Chapter 14.) However, gravity breaks this symmetry belonging to the matter sector. A shift $L_m \to L_m - \rho_0$ will change the energy-momentum tensor T^a_b which acts as the source of gravity by a term proportional to $\rho_0 \delta^a_b$ (see Eq. (6.70)). Therefore, having a nonzero baseline for the energy density of matter is equivalent to a theory with a cosmological constant which – in turn – will lead to observable consequences. If we interpret the evidence for dark energy in the universe as resulting from the cosmological constant, then

Gravity as an emergent phenomenon

its value has to be fine tuned to enormous accuracy to satisfy the observational constraints. It is not clear why a particular parameter in the low energy matter sector has to be fine tuned in such a manner.

In the alternative perspective described in this chapter, the action in Eq. (16.19) is clearly invariant under the shift $L_m \rightarrow L_m - \rho_0$ since it only introduces a term $-\rho_0 n_a n^a = 0$ for any null vector n_a. In other words, one *cannot* introduce the cosmological constant as a low energy parameter in the action in this approach. We saw, however, that the cosmological constant reappears as an integration constant when the equations are solved. The integration constants which appear in a particular solution have a completely different conceptual status compared with the parameters that appear in the action describing the theory. It is much less troublesome to choose a fine tuned value for a particular integration constant in the theory if observations require us to do so. From this point of view, the cosmological constant problem is considerably less severe when we view gravity from the alternative perspective.

Notes

Chapter 1

1 Throughout the book, the Latin indices $a, b, \ldots, i, j \ldots$, etc., run over $0, 1, 2, 3$ with the 0-index denoting the time dimension and $(1, 2, 3)$ denoting the standard space dimensions. The Greek indices, $\alpha, \beta, \ldots$, etc., will run over $1, 2, 3$. The symbol $\equiv$ is used to indicate that the equation defines a new variable or notation.

Some of the textbooks and review articles referred to in more than one chapter in this book are listed below. They will be referred to hereafter as **R(1)**, **R(2)**, ..., etc.

R(1) L. D. Landau and E. M. Lifshitz, *The Classical Theory of Fields, Course of Theoretical Physics*, Volume 2 (Pergamon Press, UK, 1975).

R(2) T. Padmanabhan, *Theoretical Astrophysics, Volume I: Astrophysical Processes*, Chapter 3 (Cambridge University Press, Cambridge, 2000).

R(3) H. P. Robertson and T. W. Noonan, *Relativity and Cosmology* (W. B. Saunders Company, 1968).

R(4) C. W. Misner, K. S. Thorne and J. A. Wheeler, *Gravitation* (Freeman and Company, New York, 1973).

R(5) A. Gupta and T. Padmanabhan (1998), *Radiation from a charged particle and radiation reaction – revisited*, Phys. Rev, **D57**, 7241.

R(6) T. Padmanabhan, *From Gravitons to Gravity: myths and reality*, Int. J. Mod. Phys., **D 17**, 367–398 (2008) [gr-qc/0409089].

R(7) J. Plebański and A. Krasiński, *An Introduction to General Relativity and Cosmology*, (Cambridge University Press, Cambridge, 2006).

R(8) M. Nakahara, *Geometry, Topology and Physics* (IOP Publishing Ltd, 1990).

R(9) T. Padmanabhan, *Gravity and thermodynamics of horizons*, Phys. Reports, **406**, 49 (2005) [gr-qc/0311036].

R(10) A. Mukhopadhyay and T. Padmanabhan, *Holography of Gravitational Action Functionals*, Phys. Rev., **D 74**, 124023 (2006) [hep-th/0608120].

R(11) T. Padmanabhan, *Theoretical Astrophysics, Volume III: Galaxies and Cosmology*, (Cambridge University Press, Cambridge, 2002).

R(12) S. Dodelson, *Modern Cosmology* (Elsevier, 2003).

R(13) R. Durrer, *The Cosmic Microwave Background* (Cambridge University Press, Cambridge, 2008).

R(14) T. Padmanabhan (2006), *Advanced topics in Cosmology: A pedagogical introduction, Graduate School in Astronomy*: X Special Courses at the National Observatory of Rio de Janeiro (X CCE) AIP Conference Proceedings, 843, 111–166 [astro-ph/0602117].

R(15) V. Mukhanov, *Physical Foundations of Cosmology* (Cambridge University Press, Cambridge, 2005).

2 Special relativity is discussed in several textbooks exhaustively. See, for example: **R(1)** and Chapter 3 of **R(2)**. A detailed description of several conceptual issues and apparent paradoxes in special relativity can be found in **R(3)**.

3 The action principle in classical mechanics is described in, for example, L. D. Landau and E. M. Lifshitz, *Mechanics, Course of Theoretical Physics*, Volume 1 (Pergamon Press, UK, 1976); also see Chapter 2 of **R(2)**.

4 The Lorentz group in discussed in several standard textbooks on Lie group theory. We will be following the approach in Chapter 41 of **R(4)**.

5 For an alternative treatment, see pages 66–69 of **R(3)**. There is a purely geometrical approach by which one can understand Thomas precession; see, for example, T. Padmanabhan, *Foucault meets Thomas*, Resonance, **13**, 706–715 (2008); http://www.ias.ac.in/resonance/August2008/p706-715.pdf

Chapter 2

1 The description of electromagnetic field in four-dimensional notation is presented in an elegant manner in **R(1)**. Also see Chapter 3 of **R(2)**.

2 Note that it is Lorenz and not Lorentz. In electromagnetism, this gauge was first used by L. V. Lorenz in 1867 when the more famous H. A. Lorentz was just 14 years old. We shall give the credit to Lorenz and use the correct spelling. The original reference is L. Lorenz, *On the Identity of the Vibrations of Light with Electrical Currents*, Philos. Mag. **34**, 287–301 (1867).

3 The derivation given here is based on Chapter 4 of **R(2)**; also see **R(5)**.

4 This formula was first obtained and interpreted by O. Heaviside, *The radiation from an electron moving in an elliptic or any other orbit*, Nature, **69** 342, (1904); also see O. Heaviside, *Electromagnetic Theory, Vol. III* (The Electrician Printing and Publishing Co., London, 1912). It is also given in R. P. Feynman, R. B. Leighton and M. Sands, *Feynman Lectures in Physics*, Volume 1 (Addison-Wesley, USA, 1963) without the derivation. A simple derivation is available in A. R. Janah, T. Padmanabhan and T. P. Singh, *On Feynman's formula for the electromagnetic field of an arbitrarily moving charge*, Am. Jour. Phys., **56**, 1036 (1988).

5 There is a large amount of literature on this subject and detailed references to earlier works can be found in E. Eriksen and O. Gron, *Electrodynamics of Hyperbolically Accelerated Charges: I. The Electromagnetic Field of a Charged Particle with Hyperbolic Motion*, Ann. Phys. **286**, 320 (2000).

Chapter 3

1 For a somewhat different but enlightening perspective leading to the same conclusions, see pages 148–173 of **R(3)**.

2 The discussion here closely follows **R(6)**.

3 The discussion here follows Chapter 7 of **R(4)**.

4 A careful discussion of this thought experiment was originally given by H. Bondi. See, for example, H. Bondi, *Relativity theory and gravitation*, Eur. J. Phy., **7**, 106 (1986).

5 See J. Schwinger, *Particles, Sources and Fields*, Volume I (Perseus Books, USA, 1998), page 83.

Chapter 4

1 All standard textbooks on general relativity describe tensor analysis to the extent required. A concise but excellent discussion is available in **R(1)**, while a much more mathematical description is given in **R(4)**.

2 For more details see, for example, R. Utiyama and T. Fukuyama, *Gravitational field as a generalized gauge field*, Prog. Theo. Phys., **45**, 612 (1971); T. W. B. Kibble, *Lorentz invariance and the gravitational field*, J. Math. Phys., **2**, 212 (1961).

Chapter 5

1 The proof can be found, for example, on page 59 of **R(7)**.
2 The discussion here follows closely the one in **R(1)**. A more detailed description of Petrov classification can be found in the Chapter 11 of **R(7)**. Also see H. Stephani, *Relativity* (Cambridge University Press, Cambridge, 2004), Chapter 32.
3 The concept of holonomy is described in any modern book on differential geometry; see, for example, **R(8)**.
4 For a detailed discussion, see **R(5)**.

Chapter 6

1 The various aspects of this exercise are discussed in detail in **R(6)**.
2 This gauge goes under different names such as the Hilbert gauge, DeDonder gauge, etc. It appears that, historically, this gauge was first suggested by Einstein in a communication to de Sitter.
3 For a detailed discussion of these issues, see **R(6)**.
4 These results are discussed nicely in some of the old, classical texts on general relativity like, for example, C. Moller, *Theory of Relativity* (Claredon Press, Oxford 1952).
5 See for example, T. Padmanabhan, *Cosmology and Astrophysics through Problems* (Cambridge University Press, Cambridge, 1996), problem 5.9.
6 See **R(9)**.
7 See **R(10)**.

Chapter 7

1 It is fairly straightforward to write a program which will trace the orbit based on the form of $V_{\text{eff}}(r)$. Such programs can be found on the web, for example, at the sites:
http:/www.fourmilab.to/gravitation/orbits;
http://www.astro.queensu.ca/~musgrave/cforce/blackhole.html.
2 See C. Frondsal (1959), *Completion and embedding of the Schwarzschild solution*, Phys. Rev. **116**, 778.
3 See Matt Visser (2005), *Heuristic approach to the Schwarzschild geometry*, Int. J. Mod. Phys., **D14**, 2051.
4 The original references in which this problem is tackled are: B. S. DeWitt and R. W. Brehme, *Radiation damping in a gravitational field*, Ann. Phys., **9**, 220, (1960); C. M. DeWitt and B. S. DeWitt, *Falling charges*, Physics, **1**, 3, (1964); J. M. Hobbs, *A Vierbein formalism of radiation damping*, Ann. Phys., **47**, 141 (1968). For a more recent discussion, see R. Y. Chiao, *The interface between quantum mechanics and general relativity*, arXiv: quant-ph/0601193 and the references therein.

Chapter 8

1 Most of the modern textbooks on general relativity describe the physics of black holes. In particular, see Chapters 31–34 of **R(4)**; see also V. P. Frolov and I. D. Novikov, *Black Hole Physics* (Kluwer Academic Publishers, 1998).
2 (a) The Kerr black hole is described in both **R(1)** and **R(4)**. Also see M. Visser (2007), *The Kerr spacetime: A brief introduction*, arXiv:0706.0622; introductory chapter in the book *The Kerr Spacetime*, edited by David Wiltshire, Matt Visser and Susan Scott (Cambridge University Press, Cambridge, 2009).
(b) A simple but adequate discussion from a modern point of view is available in M. P. Hobson, G. P. Efstathiou and A. N. Lasenby, *General Relativity: An Introduction for Physicists* (Cambridge University Press, Cambridge, 2006), Chapter 13.

3 For a review of quasi-normal modes, see for example, K. D. Kokkotas and B. Schmidt, *Quasi-Normal Modes of Stars and Black Holes*, Living Rev. Relativity, **2**, 2 (1999); http://www.livingreviews.org/lrr-1999-2. This project is based on T. Padmanabhan, *Quasi normal modes: A simple derivation of the level spacing of the frequencies*, Class. Quan. Grav., **21**, L1 (2004) [gr-qc/0310027]; and T. Roy Choudhury and T. Padmanabhan, *Quasi normal modes in Schwarzschild-de Sitter spacetime: A simple derivation of the level spacing of the frequencies*, Phys. Rev., **D 69**, 064033 (2004) [gr-qc/0311064].

Chapter 9

1 Most textbooks on general relativity discuss gravitational waves to a certain extent. An excellent, modern discussion of this subject can be found in M. Maggiore, *Gravitational Waves Volume 1: Theory and Experiments* (Oxford University Press, 2007).
2 The numbers 1913+16 refer to the coordinates of the celestial object in terms of what are called the right ascension and declination. The right ascension is measured in hours and minutes over a 24 hour cycle and declination is measured in angular coordinates. Astronomers tend to use traditional and practical coordinate systems rather than more logical ones!
3 For the solution, see T. Padmanabhan (1977), *Solutions of scalar and electromagnetic wave equations in the field of gravitational and electromagnetic waves*, Pramana, **9**, 371. The metrics for gravitational and electromagnetic plane waves are discussed in Chapter 35 of **R(4)**.
4 We follow the discussion in **R(1)**. This is also discussed quite nicely on pages 268–271 of **R(3)**.

Chapter 10

1 The current popularity enjoyed by cosmology as a research area has led to the genesis of several textbooks on the subject. The topics described here, as well as physical cosmology, are discussed in much greater detail in **R(11)**, **R(12)** and **R(13)**.
2 For a specific example, see T. Padmanabhan, *Gravitational field of the quantised electromagnetic plane wave*, Gen. Rel. Grav. **19**, 927 (1987).
3 More details can be found in the books mentioned in **R(11)**, **R(12)** and **R(13)**. Also see **R(14)** for detailed references.
4 More details can be found, for example, in the review **R(14)**.
5 For a more detailed discussion, see P. Schneider, J. Ehlers and E. E. Falco, *Gravitational Lenses* (Springer, 1999).
6 Homogeneous spaces is described concisely in **R(1)**. A nice discussion of these topics as well as their application to cosmological models can be found in M. P. Ryan and L. C. Spatilly, *Homogeneous Relativistic Cosmologies* (Princeton University Press, 1975).

Chapter 11

1 The techniques of differential geometry are described in detail in **R(4)**. A more elementary description is available in B. F. Schutz, *Geometrical Methods of Mathematical Physics* (Cambridge University Press, Cambridge, 1980). A more sophisticated treatment is available in **R(8)**.
2 These aspects are discussed, though somewhat concisely, in N. Straumann, *General Relativity With Applications to Astrophysics* (Springer, 2004).

Chapter 12

1 A very geometric description of this subject can be found in Chapter 21 of **R(4)**. Also see E. Poisson, *A Relativist's Toolkit: The Mathematics of Black-Hole Mechanics* (Cambridge University Press, Cambridge, 2004).

Chapter 13

1 More details on these topics can be found in **R(11)**, **R(12)** and **R(13)**.
2 This power spectrum is described in any of the standard textbooks in cosmology like **R(11)**, **R(12)** and **R(13)**.
3 This project was suggested by Aseem Paranjape. More details and references to other related work can be found in Aseem Paranjape and T. P. Singh, *Cosmic Inhomogeneities and the Average Cosmological Dynamics*, Phys. Rev. Lett. **101**, 181101 (2008) [arXiv:0806.3497].

Chapter 14

1 More detailed description of these topics can be found in:
   (a) N. D. Birrell and P. C. W. Davies, *Quantum Fields in Curved Space* (Cambridge University Press, 1984)
   (b) V. Mukhanov and S. Winitzki, *Introduction to Quantum Effects in Gravity* (Cambridge University Press, 2007)
   (c) S. A. Fulling, *Aspects of Quantum Field Theory in Curved Spacetime* (Cambridge University Press, 1989)
   (d) R. M. Wald, *Quantum Field Theory in Curved Spacetime and Black Hole Thermodynamics* (University of Chicago Press, 1994)
   as well as in **R(9)** and **R(15)**.
2 We follow the approach of T. D. Lee, (1986), *Are Black Holes Black Bodies?* Nucl. Phys. **B264**, 437. Also see, **R(9)**.
3 More detailed discussion is available in, for example, **R(15)**.
4 For an explicit construction, see for e.g., L. Sriramkumar and T. Padmanabhan *Initial state of matter fields and trans-Planckian physics: Can CMB observations disentangle the two?* Phys. Rev. D, **D 71**, 103512 (2005) [gr-qc/0408034].
5 See L. Sriramkumar and T. Padmanabhan , *Probes of vacuum structure of quantum fields in classical backgrounds,* Int. Jour. Mod. Phys. **D 11**, 1 (2002) [gr-qc/9903054] and references therein.
6 Most of the answers to the issues raised here (and much more!) can be found in the book: *Black Holes: The Membrane Paradigm*, Eds. K. S. Thorne, R. H. Price and D. A. Macdonald (Yale University Press, 1986).
7 See S. Deser and O. Levin, *Accelerated Detectors and Temperature in (Anti) de Sitter Spaces*, Class. Quant. Grav., **14**, L163–L168 (1997).

Chapter 15

1 More detailed discussion of lower dimensional gravity is available in the following books.
   (a) J. D. Brown, *Lower Dimensional Gravity* (World Scientific, 1988).
   (b) S. Carlip *Quantum Gravity in 2+1 Dimensions* (Cambridge University Press, 2003).
2 See, for example, S. Carlip, *The (2 + 1)-dimensional black hole*, Class. Quant. Grav. **12** (1995), 2853–2880.
3 Brane world models are described in detail, for example, in the following book: P. D. Mannheim, *Brane-Localized Gravity* (World Scientific, 2005). Also see Roy Maartens, *Brane-World Gravity*, Living Rev. Relativity **7** 7 (2004): http://www.livingreviews.org/lrr-2004-7.
4 We will follow closely the discussion in **R(10)**.
5 This is known as Wald entropy; see R. M. Wald, *Black hole entropy is Noether charge*, Phys. Rev. **D48**, 3427 (1993) [gr-qc/9307038]; V. Iyer and R. M. Wald, *Comparison of the Noether charge and Euclidean methods for computing the entropy of stationary black holes*, Phys. Rev. **D52**, 4430 (1995) [gr-qc/9503052].

6 The result is available in several papers and the form quoted here is taken from S. C. Davis, *Generalized Israel junction conditions for a Gauss–Bonnet brane world*, Phys. Rev. **D67**, 024030 (2003). A *tour de force* analysis of all available boundary terms and their variations is available in T. S. Bunch, *Surface terms in higher derivative gravity*, J. Phys. A **14**, L139 (1981).

7 This way of writing the counter-terms can be found, for example, in O. Miskovic and R. Olea, *Thermodynamics of Einstein–Born–Infeld black holes with negative cosmological constant*, JHEP 0710:028, (2007), in which one can find references to other related papers.

Chapter 16

1 The general proof of this assertion can be found in: Aseem Paranjape, Sudipta Sarkar and T. Padmanabhan, *Thermodynamic route to field equations in Lanczos–Lovelock gravity*, Phys. Rev., **D 74**, 104015 (2006) [hep-th/0607240].

2 This is related to the famous question first posed by Wheeler to Bekenstein: What happens if you mix cold and hot tea and pour it down a horizon, erasing all traces of the 'crime' in increasing the entropy of the world? This is based on what Wheeler told me in 1985, from *his* recollection of events; it is also mentioned in his book, J. A. Wheeler, *A Journey into Gravity and Spacetime*, (Scientific American Library, NY, 1990), page 221. I have heard somewhat different versions from other sources.

3 The details of the argument and calculations can be found in the following references.

(a) T. Padmanabhan, *Dark Energy and Gravity*, Gen. Rel. Grav., **40**, 529–564 (2008) [arXiv:0705.2533].

(b) T. Padmanabhan and Aseem Paranjape, *Entropy of Null Surfaces and Dynamics of Spacetime*, Phys. Rev., **D75**, 064004 (2007) [gr-qc/0701003].

Index

Printed in the United States
By Bookmasters